P9-ARG-840

MARINE ECOLOGY

A Comprehensive, Integrated Treatise on Life in Oceans and Coastal Waters

Volume I	ENVIRONMENTAL FACTORS
Volume II	PHYSIOLOGICAL MECHANISMS
Volume III	CULTIVATION
Volume IV	DYNAMICS
Volume V	OCEAN MANAGEMENT

MARINE ECOLOGY

A Comprehensive, Integrated Treatise on Life in Oceans and Coastal Waters

Editor
OTTO KINNE
Biologische Anstalt Helgoland
Hamburg, Federal Republic of Germany

VOLUME V
Ocean Management
Part 3: Pollution and Protection of the Seas—Radioactive Materials, Heavy Metals and Oil

A Wiley–Interscience Publication

1984
JOHN WILEY & SONS
Chichester · New York · Brisbane · Toronto · Singapore

Library of Congress Cataloging in Publication Data

(Revised for vol. V, pts. 3 & 4)
Kinne, Otto
Marine ecology.
Vol. V published: Chichester, New York, Wiley
Includes bibliographies.
CONTENTS: v. I. Environmental factor. 3 v.–v. II
Physiological mechanisms. 2. v—[etc.]—v. V. Ocean management. 4. v.
1. Marine ecology—Collected works. I. Title.
QH541.5.S3K5 574.5′2636 79-121779

ISBN 0 471 90216 0

British Library Cataloguing in Publication Data

Marine ecology
Vol. V: Ocean management
Pt. 3
1. Marine ecology
I. Kinne, Otto
574.5′2636 QH541.5.S3

ISBN 0 471 90216 0

Typeset by Preface Ltd, Salisbury, Wiltshire
Printed by The Pitman Press, Bath, Avon.

FOREWORD
to
VOLUME V: OCEAN MANAGEMENT

'Ocean Management', the last volume of *Marine Ecology*, describes and evaluates all the essential information available on structures and functions of interorganismic coexistence; on organic resources of the seas; on pollution of marine habitats and on the protection of life in oceans and coastal waters. The volume consists of four parts:

Part 1

Zonations and Organismic Assemblages

Chapter 1: Introduction to Part 1—Zonations and Organismic Assemblages
Chapter 2: Zonations
Chapter 3: General Features of Organismic Assemblages in the Pelagial and Benthal
Chapter 4: Structure and Dynamics of Assemblages in the Pelagial
Chapter 5: Structure and Dynamics of Assemblages in the Benthal
Chapter 6: Major Pelagic Assemblages
Chapter 7: Specific Pelagic Assemblages
Chapter 8: Major Benthic Assemblages
Chapter 9: Specific Benthic Assemblages

Part 2

Ecosystems and Organic Resources

Chapter 1: Introduction to Part 2—Ecosystems and Organic Resources
Chapter 2: Open-Ocean Ecosystems
Chapter 3: Coastal Ecosystems
3.1: Brackish Waters, Estuaries and Lagoons
3.2: Flow Patterns of Energy and Matter

Part 3

Pollution and Protection of the Seas: Radioactive Materials, Heavy Metals and Oil

Part 4

Pollution and Protection of the Seas: Pesticides, Domestic Wastes and Thermal Deformations

The culmination of *Marine Ecology*, Volume V has its roots in, and draws much of its basic substance from, the preceding volumes of the treatise:

Volume I ('Environmental Factors') is concerned with the most important environmental factors operating in oceans and coastal waters and their effects on micro-organisms, plants and animals.

Volume II ('Physiological Mechanisms') reviews the information available on the mechanisms involved in the synthesis and transversion of organic material; in thermo-regulation, ion- and osmoregulation; in evolution and population genetics; and in organismic orientation in space and time.

Volume III ('Cultivation') comprehensively assesses the art of maintaining, rearing, breeding and experimenting with marine organisms under environmental and nutritive conditions which are, to a considerable degree, controlled.

Volume IV ('Dynamics') summarizes and critically evaluates the knowledge available on the production, transformation and decomposition of organic matter in the marine environment, as well as on food webs and population dynamics.

Of necessity somewhat heterogeneous in concept and coverage, 'Ocean Management' introduces the readers to fields of applied marine ecology, i.e. to man's use of oceans and coastal waters for his own ends. In order to provide a solid basis for a sound assessment of the sea's man-supporting qualities, Parts 1 and 2 deal comprehensively with the basic multi-specific units encountered and with the resources which they constitute. After summarizing our current knowledge on the large variety of organismic groupings in the form of zonations and assemblages, and after considering the structures and functions of major marine ecosystems, the significance of these units and their components as resources utilizable for food or for raw materials are reviewed in depth. Parts 3 and 4, finally, evaluate man's potentially destructive impact. They focus on the different facets of pollution and critically assess measures—currently applied and considered practicable in the future—for protecting life in oceans and coastal waters from detrimental human influences.

Selected as early as 1965, the title 'Ocean Management' still seems too ambitious, if not somewhat misleading—even though, as anticipated, a host of new data and many new and important insights into the machinery of ecological systems have been brought to light in the meantime. Environmental management requires the concerted, judicious, responsible application of science and technology for the protection and control of those properties of ecosystems, species, resources, or areas that are regarded as absolute requirements for the continued support of civilized human societies*. Our knowledge has remained insufficient for an objective, exact definition of such properties. Hence, maintenance of a high degree of organismic and environmental diversity, and maximum possible conservation of natural conditions is deemed essential to avoid or to reduce irreversible long-term damage.

The means and aims of ocean management receive detailed attention in Part 2. Our capacity for true management is still restricted to narrowly-defined areas and to specific organisms, i.e. to a few heavily exploited or otherwise especially endangered species. Marine ecosystem management remains problematic; it is in need of much more basic ecological knowledge than is at present available.

The recent trend of using English as the international scientific language—in itself of great significance for international communication and cooperation—has often been disadvantageous to scientists with insufficient command of that language. It has frequently diminished the representativeness and distorted the emphasis of the total information actually available. While organizing and editing *Marine Ecology*, I have therefore attempted to include scientists from countries which made important contributions to the field of marine ecology, but whose scientists largely use non-English languages. In this way I wanted to underline the international status and significance of marine ecological research and the need to draw from different sources in order to provide the best possible representation of the state-of-the-art. Of course, I had to pay for this: an enormous amount of time and effort had to be invested in translation and manuscript improvement.

While I am writing this last foreword of the treatise, *Marine Ecology* is nearing completion. It is my sincere wish to thank again all those who have supported me during the many years of planning, carrying out and finalizing this *magnum opus*. From 1965 to 1981, the work on *Marine Ecology* has taken up most of my evenings, weekends and holidays. Who can blame me for feeling relieved?

As was the case with previous volumes of this treatise, I have received much help and advice while working on Volume V. With profound gratitude I acknowledge the close and fruitful cooperation with all contributors; the support, patience and confidence of the publishers; and, last but not least, the technical assistance of Monica Blake, Angela Giraldi, Alice Langley, Julia Maxim, Seetha Murthy, Sherry Stansbury and Helga Witt.

O.K.

*KINNE, O. 1980: *14th European Marine Biology Symposium 'Protection of Life in the Sea'*: Summary of symposium papers and conclusions. *Helgoländer Meeresunters.*, **33**, 732–761.

CONTENTS OF VOLUME V, PART 3

CONTRIBUTORS
TO
VOLUME V, PART 3

BRYAN, G. W. *Marine Biological Association of the United Kingdom, The Laboratory, Citadel Hill, Plymouth, PL1 PB2, England.*

JOHNSTON, R. *24 Hilton Drive, Aberdeen, AB2 2NP, Scotland.*

KINNE, O. *Biologische Anstalt Helgoland (Zentrale), Notkestraße 31, 2000 Hamburg 52, Federal Republic of Germany.*

WOODHEAD, D. S. *Ministry of Agriculture, Fisheries and Food, Directorate of Fisheries Research, Fisheries Radiobiological Laboratory, Hamilton Dock, Lowestoft, Suffolk, NR32 1DA, England.*

OCEAN MANAGEMENT

Marine Ecology Vol. V, Part 3
Edited by Otto Kinne

1. INTRODUCTION TO PART 3—POLLUTION AND PROTECTION OF THE SEAS: RADIOACTIVE MATERIALS, HEAVY METALS, AND OIL

O. KINNE

(1) General Aspects

Pollution and protection of the seas are the most essential themes of ocean management. Based on, and guided by, ecological knowledge, ocean management comprises the judicious and responsible application of scientific and technological knowledge with the aim to achieve the maximum degree of ecosystem protection commensurate with the highest sustainable quality of living for mankind. The need for ocean management stems from—often conflicting—ethical, scientific, social, and economic attitudes of man. Aspects of ocean management concerned with the planned and controlled exploitation of living marine resources have been considered in Part 2 of Volume V.

Parts 3 and 4 of Volume V, 'Ocean Management', deal with the investigation, comprehension, monitoring, and management of man's detrimental impacts, actual and potential, on the seas and their biota. A gigantic multiple resource, a buffer of air quality and overall ecological balance, and a regulator of the earth's climates, the seas are the final receptacle for much of man's wastes. Increasing concern of scientists and the general public, and—often belated—political and administrative attention, have turned pollution and protection of the seas into key issues of basic and applied marine ecology. While the knowledge and concepts necessary for restricting, monitoring, and managing pollution have their roots in the foundation elaborated by basic research (Volumes I to IV), applied pollution studies increasingly stimulate basic research and often challenge or modify what has been brought to light thus far. Such interaction and cooperation between basic and applied marine ecology are likely to remain essential factors for much-needed progress.

The principal requirements for successful environmental management in oceans and coastal waters are: (i) long-term ecological research; (ii) world-wide international cooperation (national territorial borders are irrelevant for the functioning of ecosystems, and damage caused by one nation may affect living systems in another); (iii) adequate interpretation and transposing of scientific knowledge into legislation and effective control measures.

Originally I had planned to accommodate the topic 'Pollution and Protection of the Seas in one part (Part 3) of Volume V. However, the large amount of information available forced editor and publisher to double the printing space initially envisaged.

Even now the coverage of human activities which may affect, deform or damage natural processes of life in the seas is incomplete. We have focused on aspects which have received most attention and/or pose especially severe threats. Part 3 covers radioactive materials (Chapter 2), heavy metals and their compounds (Chapter 3), and oil pollution and its management (Chapter 4); following the introduction (Chapter 5), Part 4 deals with pesticides and technical organic chemicals (Chapter 6), domestic wastes (Chapter 7), and thermal deformations (Chapter 8). Together, Parts 3 and 4 provide the most comprehensive review of marine pollution research thus far available. Based on Volumes I to IV, the 4 parts of Volume V present an exhaustive, critical state-of-the-art-account that can effectively assist in the development of concepts and mechanisms for managing the threat to marine life arising from anthropogenic interferences. The enormous extent and intensity of these interferences are a function of the world-wide, almost logarithmic increase in human population size and the rapidly growing impact potential *per capita,* especially in the so-called 'developed countries'. For general assessments of pollution impact, control, and management in the seas consult KINNE and AURICH (1968), RUIVO (1972), GOLDBERG (1976), JOHNSTON (1976), KINNE and BULNHEIM (1980), and GERLACH (1981).

(2) Terms and Concepts

The term (marine) 'pollution' has been defined repeatedly. I quote here 3 increasingly detailed versions, all of which are basically in line with the connotation of the term employed in Parts 3 and 4 of Volume V of *Marine Ecology*:

(i) 'Human activities causing negative* effects on human health, resources, amenities or ecosystems' (KINNE, 1968, p. 518, 1980a, p. 734).
(ii) 'Any kind of man-made reduction in the quality of the marine environment' (WORLD HEALTH ORGANIZATION, early version).
(iii) 'Introduction by man, directly or indirectly, of substances or energy into the marine environment (including estuaries), resulting in such deleterious effects as harm to living resources, hazards to human health, hindrance to marine activities including fishing, impairment of quality for use of sea water and reduction of amenities' (GESAMP† I/11, Paragraph 12, 1969, with slight amendments in the Updated Memorandum on GESAMP, 4 August, 1972; see also GESAMP VIII/11, Paragraphs 74 and 75, 1976).

The emphasis on man in these definitions is not only essential for pointing to the ultimate source of pollution, but also for distinguishing the term pollution as defined above from 'natural pollution'. Examples of natural pollution are volcanic activities, intensive dust storms, natural oil seeps, and thermal deformations due to extensive climatic (hydrographic) change. Usually, man judges the seriousness of pollution dam-

* In regard to ecosystems, 'negative' refers here to detrimental effects causing impairment in such ecological parameters as survival, growth, reproduction, energy procurement, stress endurance, competition, orientational and behavioural competence, diversity, abundance and system homoeostasis.

† Joint *G*roup of *E*xperts on the *S*cientific *A*spects of *M*arine *P*ollution, representing IMCO/FAO/UNESCO/WMO/WHO/IAEA/UN.

age by the degree to which it affects his health and general welfare. Related to the term 'pollution', the term 'contamination' denotes a reduction in quality of matter, including food, used by man and other organisms.

Definition (iii) neglects the fact that potentially deleterious anthropogenic effects on nature are not restricted to 'introduction'. There are 4 principal types of impact inflicted by man on natural ecosystems (KINNE, 1982: removal, change, addition, and mixing) (Fig. 1-1). Removal comprises harvesting of selected wild organisms (especially fisheries; for details see Volume V: GULLAND (1983) as well as exploitation of non-living materials such as water, coal, metals, and oil (mining, drilling). Change involves modification of natural flow patterns of energy and matter, e.g. by promoting processes and organisms useful to man and by suppressing processes and organisms (competitors, predators, disease agents) which interfere with such exploitation (agriculture, aquaculture; for details see Volume III: KINNE and ROSENTHAL, 1977). Addition refers to the introduction into natural ecosystems of materials and energy considered waste (i.e. pollution *sensu stricto;* for details consult Parts 3 and 4 of Volume V). Mixing involves the translocation (transplantation, deportation) of originally separated ecosystem components—often over considerable distances—both unintended (via transportation in general) and intended (use of imported organisms in agriculture and aquaculture; for details consult Volume III: KINNE and ROSENTHAL, 1977; and ROSENTHAL, 1978, 1980).

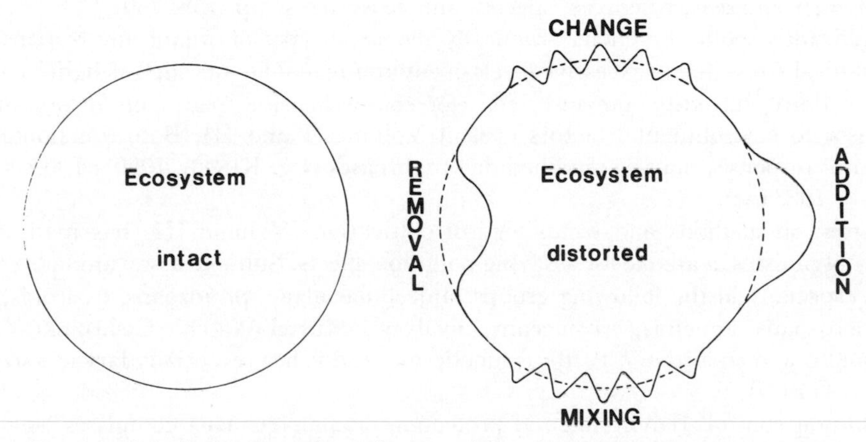

Fig. 1-1: Principal types of man's potentially detrimental impact on nature. Removal (e.g. of fishery products, coal, metals, oil), change (modification of natural flow patterns of energy and matter (e.g. agriculture, aquaculture), addition (input of pollutants and thermal energy), and mixing (translocation of ecosystem components) tend to distort natural ecosystem dynamics. (Based on KINNE, 1982)

Removal, change, and mixing entail elements of pollution. Thus, modern fisheries may cause large-scale injuries and mortalities among organisms unintentionally caught during normal fishing procedures, as well as damage to habitats by fishing gear. Such effects of fishery activities qualify as 'harm to living resources' (see also Volume V: GULLAND (1983)). Change may destabilize or otherwise impair populations or ecosystems thus reducing their capacity for impact compensation or impairing 'the quality of the marine environment'. Mixing can lead to modifications in the receiving ecosystem (e.g. population explosions in species originally foreign to the target ecosystem) and thus cause 'negative effects on resources, amenities or ecosystems'.

In situ effects of pollution depend on the pattern of pollutant release in space and time. Point release (discharge) usually causes initial intensities of impact different from non-point release. Point-discharge effects are a function of local hydrographical, geological and biological conditions; they often depend acutely on timing, e.g. discharge in summer or winter, at high tide or low tide, continuous or interruptive.

Negative ecological consequences of pollution manifest themselves in detrimental deviations from the normal state of individuals, populations or ecosystems. At the individual and population levels such deviations involve impairments, quantifiable in terms of functions (e.g. reductions in survival, growth, reproduction, energy procurement, stress endurance, disease resistance, competitive capacity, abundance or distribution) and of structures (e.g. abnormalities in morphological differentiation, meristic characters, architecture of organs, tissues, and cells; proliferative disorders). At the ecosystem level, negative deviations comprise distortions in the flow patterns of energy and matter, or in species composition, or reductions in system stability that ultimately diminish the system's man-supporting qualities (pp. 1096 and 1097).

The principal criteria for assessing biological pollution effects, as well as measures of pollution control and environmental management, have been reviewed briefly by KINNE (1980a, pp. 751–760). He distinguishes 4 categories of environmental management: the management of areas, resources, species, and ecosystems (pp. 758–760).

Significantly, pollution elicits practically the same types of organismic responses as does natural stress due to excessive levels of environmental factors such as light, temperature, salinity, turbidity, pressure, etc. (for comprehensive treatments of organismic responses to environmental factors consult Volumes I and II). Both functional and structural responses may be modified due to diseases (e.g. KINNE, 1980b; LAUCKNER, 1980a–j, 1983a–c).

Progress in methods and technology of cultivation (Volume III) has made more marine organisms available for assaying pollution effects. Suitable assay animals can be found especially in the following groups: unicellular algae, protozoans, hydroids, rotifers, bryozoans, annelids, crustaceans, bivalves ('Mussel Watch': GOLDBERG, 1975; GOLDBERG and co-authors, 1978), echinoderms, and fishes (especially larvae); see also PHILLIPS (1977).

'Pollution control' (environmental protection or management) comprises legal and other organizational arrangements aiming at the prevention, reduction or adjustment (to levels considered acceptable) of negative pollution effects. Based on scientific studies, marine pollution control presently focuses on: (i) Banning extremely dangerous substances from being released into the sea. The banned substances, the so-called 'black list', include chlorinated hydrocarbons, other halogenated organic compounds, and substances capable of forming organohalogen compounds in the marine environment (Chapter 6), unless they are degradable or biologically innocuous; mercury and cadmium compounds (Chapter 3); plastics; degradation-resistant oils and hydrocarbons derived from petroleum (Chapter 4); and certain categories of radioactive wastes (Chapter 2). (ii) Limiting the release of problem substances and making their input into the seas subject to special permission. The list of problem substances—the 'grey list'—includes organophosphorus, organosilicon and organotin compounds, pure phosphorus, degradable oils and hydrocarbons derived from petroleum, compounds of arsenic, chromium, copper, lead, nickel, and zinc, all radioactive wastes not included in (i) (for details consult Chapter 2), as well as taint substances which impair the taste of seafood. (iii) Developing technologies of manufacture that eliminate or reduce the detrimental effects of potential pollutants. (iv) Recycling, i.e. reuse of water, materials, and commodities via reprocessing and resource recovery. (v) Controlling the status of the environment by long-term studies of natural ecological processes and superimposed man-made disturbances in an attempt to assess the consequences of pollution and of the management measures adopted.

Ultimately, pollution control can only attain a high degree of effectiveness if developed and enforced internationally, preferably at the global level, and if it includes air, land, fresh water as well as the seas. The definition of safety limits must take into account local situations. Thus a harbour, or a ship canal may be allowed higher levels of pollution than the ecologically sensitive mud flats, the spawning area of a commercially valuable fish, or the open sea.

Major international conventions aiming at pollution control in oceans and coastal waters are the Oslo, London, Helsinki and Paris Conventions and the new IMCO Convention. There now exist some 40 regional, national, and international arrangements which include aspects of pollution control in oceans and coastal waters.

(3) The Most Dangerous Groups of Pollutants

As has been pointed out previously (KINNE, 1980a, p. 750) the most dangerous group of pollutants comprises pesticides and related chemical compounds (Chapter 6). These substances are largely or entirely foreign to the marine environment and are usually designed to kill or incapacitate certain organisms (e.g. insects). Since nature has no or only very limited means for metabolizing or degrading these substances, they tend to accumulate and to exert long-term effects. Heavy metals (Chapter 3) rank second as potential hazards, particularly in estuaries and near the coast. Present concentrations of radioactive materials (Chapter 2) are considered to be below the threshold values for detrimental effects. Oil, i.e. petroleum and its derivatives (Chapter 4), represents a product of nature, remobilized and thus made amenable to ecosystem degradation. Nevertheless, heavy oil spills have caused severe local damage. Refined oil, due to addition of toxic substances, tends to exert more detrimental effects than natural oil. The ecological significance of 'oil-pollution-derived' organic compounds that may interfere with chemical ecosystem integration and chemical communication among organisms remains to be investigated more fully. A number of synthetic organic compounds resist decomposition and thus cause long-term interference. While thermal deformations (Chapter 8) may represent a significant additional stress to the estuarine and coastal flora and fauna, especially near power-station outfalls, there appears to be no immediate danger yet to marine life in general. Domestic wastes (Chapter 7), finally, can cause local eutrophication, reduction of amenities, odour nuisance and related hazards, but are unlikely to inflict wider damage, unless they contain toxic substances.

Sudden exposure to the most dangerous pollutants, even if these are only short-lived, often exerts greater immediately demonstrable effects than long-term exposure to subcritical pollution stress. The latter leaves time for compensatory regulations and adaptations and thus tends to mask any detrimental consequences that may occur.

(4) The Most Endangered Areas

The most endangered sea areas are those near the primary source of pollutant release: industrial sites on the coasts, especially in harbours, estuaries, and bays. Economically attractive in terms of trade, traffic and pollutant release, estuaries, mud flats, sheltered bays and fjords tend to be particularly vulnerable to pollution. Estuaries and mud flats have been shown to trap, retain, and accumulate pollutants in their sediments. At the same time they have important ecological functions for recruiting and supporting life in adjacent sea and land areas. For details consult KINNE and BULNHEIM (1980) and the following chapters. Other highly endangered areas are the ecologically highly sensitive coral-reef and mangrove systems. Sea areas more or less separated from major oceanic water bodies, such as the Mediterranean, Baltic, and North Seas tend to suffer more than water bodies in the open oceans. Limited water exchange in these seas is often combined with supranormal pollutant inputs from major industrial nations.

(5) Pollution Management

It is, of course, legitimate to consider oceans and coastal waters as a resource for a variety of human endeavours such as fisheries, navigation, recreation, aquaculture, and waste disposal. The point is that the degree to which a given resource is exploited must be subject to a balanced compromise between different use-interests and be commensurate with nature's capacity of compensation. Critical long-term damage to marine resources must be avoided. While the information presently available is still insufficient for a sound interpretation and comprehension of ecological dynamics in the seas (Volume IV), there can be little doubt that the development of civilized human societies has depended—and will continue to depend—on intact marine, as well as limnic and terrestrial ecosystems.

As has been emphasized previously (KINNE, 1980a, p. 757), to protect, maintain or manage natural ecosystems does not necessarily mean to aim for the conservation of a specific ecosystem function or structure. A major feature of ecosystems is their ability to adapt. Dynamic variability and evolutionary change—not consistency—are the essential vectors of ecosystems. These properties facilitate survival under changing conditions. Apparently, functions and structures of ecosystems have been subject to change ever since the first forms of life organized themselves into multispecific systems, i.e. into dynamic entities, characterized by increasingly complex relations among interacting and interdependent biotic and abiotic components, linked by energy flow and material cycling, by exchange between biotic and abiotic, and among biotic constituents, as well as by homoeostatic control mechanisms (Volume IV: KINNE, 1978, p. 1). Ecosystems—the basic carrying units of life on earth—have changed and will continue to change. However, it is obvious that ecosystems must have certain qualities in order to be able to support civilized human communities as we know them today. A desert is an ecosytem, a biological sewage-processing plant is an ecosystem, the interstitial micro- and meioflora and fauna in the sand filter of a large sea-water aquarium tend to form an ecosystem—but none of these systems can support man.

The ultimate aim of monitoring, comprehending, and managing man's potentially destructive impact on natural ecosystems must be, therefore, to avoid critical hastening of change and irreversible damage that leads to a reduction or elimination of the system's man-supporting qualities. Changes should be avoided or be reduced to a level that can be expected to be reversible within a reasonable period of time, say within years or decades. As has already been said, the present status of marine ecological knowledge is inadequate for a detailed comprehension of ecosystem dynamics (Volume IV), let alone for a definition of the man-supporting system properties which require protection in order to safeguard human existence in the future. As long as our knowledge is insufficient for an exact definition of those functions and structures of ecosystems which are essential for supporting human societies, we must formulate and enforce measures that assure the continued existence of a high degree of organismic and environmental diversity, and the conservation of a large variety of gene pools and habitats, thus reducing the risk of irreversible damage.

At a time when some politicians of the major industrial nations begin to shy away from, or pay only lip service to, their paramount responsibilities in the field of environmental protection, we must state here clearly and firmly: no area of political and administrative activity or of research is more important for, and is a better investment towards, the long-term future of humanity than are ecosystem research and environmental protection. Man's unique ecological role has its price. If man wants to be around on this globe for more than the next 4 or 5 generations he must transfer a major portion of the enormous funds presently spent on 'defence' into increasing his capabilities for investigating, protecting, and managing natural multi-species systems on land as well as in limnic and marine waters.

The restriction of pollution management to man-supporting system qualities will not be enough for those who insist on protective measures for specific animal or plant populations threatened by extinction. While such protection goals are commendable, they are unattainable—except where ruthless exploitation or hunting represent the major impact factor—for reasons already pointed out above (p. 1096). In addition, such goals suffer from lack of objectivity. Of the many hundreds of thousands of species which populate oceans and coastal waters, only a very few large-sized and generally well-known animals—such as birds, seals, dolphins and whales—have been able to arouse public emotions in their support, and rightly so. But then who fights for bacteria, phytoplankters, protozoans, nematodes or copepods? All of these, and numerous others, are functionally much more important ecosystem components than are the few large predators at the top of the food pyramid. In fact, the very life of these top predators depends, first of all, on the well-being of the smaller, less-known, less-publicized and apparently less-loved organisms. Critical damage to even a few of these groups of organisms may 'automatically' end the existence of 'high-sympathy animals'. To avoid misinterpretation: I fully endorse actions for the protection of birds, seals, dolphins, and whales, but I would like to encourage a wider frame of reference.

In spite of considerable efforts, it has remained difficult—in most cases impossible—to differentiate between natural variations and superimposed man-made effects. With some degree of certainty, such differentiation can presently be made only in cases of very extreme impacts that cause advanced stages of deformation in populations and ecosystems. For proper identification and management of pollution effects, it is a prime requirement that we know nature's own long-term variabilities. Lack of financial support for routine studies and insufficient attention by marine ecologists to physical, chemical, and biological processes in the seas as a function of time have deprived us of the baseline data necessary now for assessing pollution effects with a sufficient degree of soundness. Monitoring programmes are meaningful only in combination with long-term basic research. Natural variations in North Sea phytoplankton dynamics, for example, may have wavelengths of *ca.* 5 yr (GILLBRICHT, 1981, p. 24) or much longer (GILLBRICHT, pers. comm.). We need more funds for extensive, sound, long-term programmes in order to keep our hands on the pulse of nature, to recognize her normal dynamics and to monitor early signs of disease due to man's activities. And we need more institutionalized feedback and control of monitoring data, management procedures, and environmental predictions. In a letter to me (11 Jan., 1983). DON REISH writes:

> 'Monitoring data often appear only on laboratory sheets or in annual reports and never become part of the scientific literature. There is a serious need for a critical evaluation on at least a regional basis of all such data generated to determine whether or not it is answering our questions. Monitoring programs are generally funded under the control of administrators who may have only a limited knowledge of ecosystems. They tend to fund the programs with a minimum amount of money in order to comply with the regulations. Impact studies are made prior to an environmental change, but after the project is completed, nobody looks to see if the predictions were right or wrong.'

The most important general guidelines for pollution management are: (i) pollution control is easiest and most effective at the input level: control and manage man's input and you have little to worry about having to manage the environment; (ii) pollution prevention is cheaper than the repair of pollution-caused damage;* (iii) repair of man-made environmental deformation requires much more knowledge than damage prevention. Effective, ecologically sound remedial mechanisms—other than simple pollution restriction and removal of polluted materials—are not yet available; (iv) for defining the quantities and qualities of permissible input levels, extensive continuous monitoring is necessary in order to record and evaluate the responses to pollution in populations and ecosystems.

For many pollutants the most obvious and most effective means of management is *ad hoc* treatment of objectionable materials at the production site (at-the-source pollution control). Industrial production procedures must, from the beginning, pay more attention to material selection, possibilities for reutilization (recycling), safe storage and transformation of by-products into less toxic substances. In addition, optimization of pollutant dispersion systems, as well as careful selection of discharge locations and times, can effectively reduce the initial impact of pollution.

The essential prerequisites for managing man's impact on nature may be summarized as follows: (i) Reliable, world-wide information on pollutant sources, types and input rates covering air, land, fresh and marine waters. (ii) Knowledge of the fate of released pollutants: (a) changes in physico-chemical characteristics due to dilution, evaporation, hydrolysis, radiation, sorption to particulate matter and biota, deposition, accumulation, inactivation, remobilization, metabolization, release, excretion, etc.; (b) pollutant derivatives resulting from these processes, their properties and potentials as harmful substances; (c) cumulative ecological consequences of (a) and (b) under field conditions.

* The restoration of the Thames estuary, for example, turned out to be exceedingly expensive.

(iii) Elaboration of criteria with early-warning capabilities at the population and ecosystem levels. (iv) Determination of contaminant-specific biological effects and selection of suitable assay organisms.

There is great need for laboratory studies on population responses conducted under realistic experimental conditions, i.e. at concentration levels and with physico-chemical states of pollutants actually encountered in the seas. Combined with *in situ* test-pollution experiments and field studies, such work will provide better insights into the processes which determine the total environmental impact, the ensuing organismic compensations and the final result, i.e. regulation, adaptation, changes in functions and structures, reversible or irreversible damage.

(6) Major Shortcomings

Many activities unfolding under the heading 'environmental protection' entirely miss the point. They focus on the protection of the 'human environment', not realizing that man depends in many ways on other organisms which provide the basis for his existence. Protection of the 'human environment' means nothing else than to change further the living conditions on this earth in one-sided favour of one species—*Homo sapiens*—usually at the expense of other forms of life. In other words, 'environmental protection' as understood by many politicians, industrialists and large portions of the general public is part of the detrimental process of expanding the unique dominance of man—the very process that made 'environmental protection', as understood by ecologists, necessary in the first place! Here we have a perfect example of a concept conversion—of how different interests produce different meanings and strive for different ends—under the same headline.

In most fields of impact assessment the majority of the attempts to assess pollution effects on marine organisms have focused on individuals instead of on populations, and many have used death as end-point measure, instead of more sensitive and ecologically more meaningful sublethal responses. Mortality studies are useful for a first gross evaluation of toxicity levels and for interspecific comparison of tolerance limits (which often tend to decrease with increasing complexity of the organisms tested, e.g. bacteria > protozoans > invertebrates > fishes). However, the significance of mortality studies for interpreting ecological *in situ* dynamics is rather limited. We need to know more about long-term pollution effects on fully acclimated organisms with special attention being paid to rates of growth and reproduction, to physiological performance and to behaviour, as well as to competitive interactions under natural conditions, taking into account the total resulting amount of stress encountered. How does an otherwise subcritical radiation stress, for example, affect a population simultaneously exposed to increased levels of temperature, salinity, or pollutants such as heavy metals, oil, pesticides, and domestic wastes? It is here that the ground on which to build predictions and management procedures is soft and weak, and unable to carry the often large constructions of hypotheses, theories, and models that claim ecological validity.

Much of what we know today on the responses of marine organisms to environmental stress, natural and man-induced, has been obtained from unstabilized (not fully acclimated) test organisms kept under ecologically inadequate environmental and nutritive conditions. In many cases the test organisms were apparently parasitized, infected or injured (Volume II: KINNE, 1975, p. 2; Volume III: KINNE, 1976, p. 8; KINNE, 1977).

Indeed, the amount of knowledge produced on the short-term performance of dying, diseased, shocked and malnourished marine animals is impressive.

Pollutants are not chemically pure substances. They are mixtures of different substances whose composition may vary already at the production site—usually without notice to pollution researchers. Additional changes—due to ageing, use, handling, etc., as well as to impurities—sometimes more dangerous than the main pollutant itself—may further modify the physico-chemical identity of a pollutant even before its release into the sea. Once in contact with sea water and its biota, pollutants again tend to change their forms and properties. Of course, such changes influence and modify also the pollutant's biological effects.

In nature, an organism is usually not confronted with a single pollutant, but with different pollutants acting in concert. The combined effects of several pollutants may be larger or smaller—and qualitatively different—from the sum of individually assessed component effects. In much of the current pollution-effect research, these facts have received insufficient attention. On the other hand, it is of course important—and sometimes the only key available for assessing pollution effects—to determine the most toxic component(s) in a pollutant mixture and their ecological consequences.

Pollution effects may manifest themselves in biotic and abiotic components of an ecosystem. Since both components are ecologically interrelated in a multitude of exchange and equilibrium phenomena, they must be considered together. Much of the published work at hand has neglected this point.

Two aspects, in particular, have received insufficient attention:

(i) The need to assess pollution effects against the background of long-term *in situ* variability. Nowhere in the world are data available on the dynamics of marine communities that would reach back more than 2 or 3 decades and cover a sufficiently dense net of vital data. Worse than that: all modern 'monitoring programmes' which have come to my attention continue to be grossly inadequate. In order to detect potential damage against 'natural background noise', and early enough to allow for corrections, the net of stations to be covered and the diversity and intensity of measurements to be made must be increased far above present schedules.

(ii) Nature's compensatory mechanisms (at the individual, population, and ecosystem levels) tend to conceal damage very effectively (KINNE, 1980a, pp. 753, 754, 759–760). However, the total build-up of stress—resulting from natural plus man-made factors—may not be without ultimate penalty. With most of the criteria presently used, the response levels of no harm and heavy, or even irreversible damage may be much closer together than is generally assumed. The critical point of switching to sudden collapse or radical change must become subject to intensive research.

Most administrative concepts and measures of environmental management are thus far restricted to the management of different use-interests. In these cases, the management goal is to reduce the pollution impact to levels compatible with continued resource exploitation. The fisheries industry, for example, considers pollution acceptable so long as the annual harvest-levels of certain species are sustained. To make use-interests the prime criterion for pollution control is not acceptable to the ecologist. Reliable landing statistics of the fisheries are difficult to obtain—for several reasons, including commercial competition and fishermen's dominating interest in quick profit which tends to obscure their ability to recognize long-term requirements for environmental protection (KORRINGA, 1980). The fisheries exploit the seas, as do industry, shipping, tourism, recreation services, and others. All these human endeavours have responsibilities for environmental protection and their say in resource partitioning, but the definition of ecological safety limits, the evaluation of criteria, and the elaboration of the scientific basis for programmes intended to protect life in the seas 'must be a matter for scientists free of direct commercial interests and pressures' (KINNE, 1980a, p. 757).

For the future, major shortcomings are likely to increase, even beyond the present level, in the fields of administrative and political decision making—especially when it comes to planning sufficiently ahead or to enforcing unpopular restrictions. Large sectors of governments are so much concerned with competition, power games, and self-interest that they may not be able to respond adequately to the many problems produced by modern industrialized societies—including pollution and environmental management. Even the few politicians who seem to have sufficient far-sightedness cannot act more than a step ahead of current public opinion. Are our present patterns of government and of public opinion-making still adequate for coping with the consequences of the way of life modern man has adopted? Can they guide humanity safely into the future?

(7) Comments on Chapters 2 to 4

Chapter 2: Contamination due to Radioactive Materials

While our first assessment of the effects of ionizing radiation on marine organisms (Volume I: CHIPMAN, 1972a,b,c) concentrated on responses to radiation as a natural environmental factor, the present review updates that assessment and extends the approach to focus on the ecological consequences of both the introduction of man-made radionuclides into marine environments and the concomitant increases in the radiation exposure of aquatic organisms. Chapter 2 considers basic aspects of radioactivity and radiation (nature and quantification of radioactivity; interaction of radiation with matter and living tissues; problems of dosimetry); measures of radiation protection and management of waste disposal; concentrations measured in the marine environment; fate of discharged radionuclides; extent of human exposure; and potential effects on marine organism.

Based on intensive research at 4 representative sites (Blackwater estuary, Oyster Creek, northeast Irish Sea, and northeast Atlantic deep ocean dumpsite), the most essential conclusion drawn is that the waste-management procedures thus far employed to ensure that the radiation exposure of human populations does not exceed prescribed limits also provide a sufficient measure of protection for populations of marine organisms. Even at the maximum dose rates estimated for the well-documented situation arising from the Windscale discharge into the northeast Irish Sea only very minor effects can be expected to occur. At the other 3 disposal sites the estimated dose rates appear to be too low to elicit any detectable response. With the possible exception of chromosome aberrations, such responses tend to be difficult to demonstrate and to differentiate both from natural variabilities and from interfering responses to changes in other environmental factors of either natural or anthropogenic origin. The available evidence indicates that even if there were minor effects in individual organisms, these would not become apparent as a response at higher levels of organization—populations and ecosystems.

However, the use of coastal waters and oceans as repositories for at least a part of the radioactive wastes, generated by the continued industrial exploitation of radioactivity and nuclear energy, will probably tend to increase. The detrimental effects of radiation on living tissues, the absence of natural mechanisms capable of 'degrading' radioactive materials, and the often very long half-lives of man-made radionuclides disposed of into the seas could pose long-term problems. It is, therefore, necessary to expand our knowledge of the ecological effects of radioactive contamination, to monitor carefully the impact of radionuclides on marine systems, and to enhance and complement our present capability for impact management.

In general, the degree of the potentially negative impact of radioactive wastes released into the seas is a function of the *in situ* concentrations attained. While water movement (Volume I) tends to dilute and disperse the radionuclides, geochemical and biological processes tend to reconcentrate them, thus increasing the exposure of marine organisms and, ultimately, man.

In living tissues, the absorption of radiation produces excited and ionized atoms and molecules, a proportion of which initiate biochemical change. While progress has been slow in analysing the development of these initial changes to an overt expression of damage, quantitative relationships have been firmly established between the absorbed radiation dose and such end-point phenomena as cell mortality, chromosome damage, mutations, etc. Since even small doses to humans entail a finite risk, the International Commission on Radiological Protection (ICRP) recommends that all unnecessary radiation exposure be avoided; no practice be adopted unless its introduction produces a net benefit; all exposures be kept as low as reasonably achievable; and the dose equivalent to individuals shall not exceed the limits recommended for the appropriate circumstances by ICRP.

For the management of marine disposals of radioactive wastes, 3 approaches have been employed: point of discharge control (p. 1131), specific activity approach (p. 1132), and critical pathway analysis (p. 1134). The prime objective of these 3 approaches is the limitation of human exposure. Hence radionuclide disposals must be considered in terms of the extent of the resulting exposure of man. Of the numerous sources of man-made radionuclides entering the marine environment as waste, the nuclear fuel cycle exerts by far the most pronounced effect. The industrial manufacture of consumer goods incorporating radionuclides, medical procedures and research activities represent only minor contributions. In regard to the nuclear fuel cycle, reactor operation, fuel reprocessing at coastal sites, and the disposal of radioactive wastes into the deep ocean constitute the major sources of contamination. Once released into the marine environment, the behaviour of the radionuclides is modified, especially by dispersion (p. 1155), interaction with marine organisms (p. 1166), and interaction with sediments (p. 1178).

In the marine environment, the dose rates to marine organisms due to releases of man-made radionuclides range from less than the natural background for typical nuclear power stations in routine operation to a few tens of mrem h^{-1} for the Windscale discharge into the northeast Irish Sea. In contrast to the situation for humans, where the response of the individual to radiation is the important criterion, for marine organisms it is the response at the population level which is of consequence, especially the capacity of the population to sustain itself through reproduction and competition under the total amount of stress encountered, including that due to radioactive materials. Unfortunately, there is very little information available on the responses of populations of marine micro-organisms, plants and animals to radiation in the presence of additional stresses, e.g. extreme temperature, heavy metals or detergents, and under ecologically valid conditions. Here lies fertile ground for future research.

In respect of fecundity, the most radiosensitive aquatic organisms thus far examined are the fishes. While there is considerable intraspecific variability, minor damage can be expected in fish gonads at dose rates exceeding 0·1 rad h^{-1}. This, however, lies considerably above the maximum dose rate estimated for benthic fish in contaminated environments (Table 2-30, p. 1206). Since it has been an area of disagreement, the effects of radiation on developing fish embryos is discussed in considerable detail. The radiosensitivity tends to decline during embryogenesis and attention must, therefore, be focused on the most sensitive phases immediately following fertilization. The lowest dose rate at which detrimental long-term effects have been documented appears to be 0·21 rad h^{-1}. Although responses at the cellular level—e.g. mitotic index, chromosome aberrations, and changes in immuno-competence—show developing fish embryos to be quite sensitive to radiation damage, embryo abnormality and mortality, and larval mortality and growth rate—perhaps more important criteria for the total population—do not indicate a particularly high radiosensitivity. More research is required to evaluate the significance of the various forms of radiation damage for the total population and to define safe limits more explicitly.

The limited data available indicate that marine organisms exhibit radiation-induced mutations of types and with a sensitivity similar to those of their more intensively studied terrestrial counterparts. It seems reasonable to assume a mutation rate in the range 10^{-3} to 10^{-4} gamete^{-1} rad^{-1}. Hence, at the dose rates presently prevailing in contaminated marine environments, detrimental effects would not be expected.

Responses to increased radiation exposure thus far documented in natural populations suggest the absence of detrimental genetic effects. Even if these occur, the severity of genetic damage to be expected at the population level is likely to remain well within the capacity of repair through natural selection.

Chapter 3: Pollution due to Heavy Metals and their Compounds

In spite of the danger to life of supranormal heavy-metal levels—documented in numerous experimental studies—detrimental *in situ* effects confidentially attributable to metallic contamination have become known only from heavily polluted coastal areas. And only in the case of the Minamata disease in Japan has metal-contaminated seafood caused severe, permanent illness in man.

Nevertheless, at the present state-of-the-art, such apparent scarcity or lack of demonstrable negative deviations from the normal state under increased *in situ* levels of heavy metals does not mean metal pollution is harmless. A long list of insufficiencies and uncertainties obscures the picture: problems in analytical procedures; insufficiencies in monitoring programmes; insufficient recruitment and identification of sensitive organisms and processes; difficulties in detecting deleterious effects against the background of natural variation and in disclosing cause-and-effect relation under complex, natural conditions; absence in many contaminated areas of natural baseline data, necessary for comparison and impact assessment; insufficient knowledge on the biological availability of different types and forms of metals and on the fate and effects of metals in organisms; lack of information on responses to metal pollution in the majority of marine organisms, especially those of no immediate commercial value.

There also is need for further refinement of concepts, tools, and methods. We must make our instrumentarium sensitive and reliable enough for coping with a matter as delicate as is the definition and management of ecologically significant sublethal long-term consequences of heavy-metal pollution.

In many cases, only crude and obvious effects have been recognized. We still know too little about the combined effects of several stressors acting in concert—such as extreme intensities of natural environmental factors, nutritional deficiencies, parasitation and different types of other pollutants. It is, of course, the total resulting stress encountered that determines the organism's response. Some marine bacteria, plants, and animals exhibit remarkably efficient capacities for detoxification, regulation, and adaptation. Only when such compensatory mechanisms fail can measurable changes in performance and ecological potential be expected to reveal themselves. Masking of potentially detrimental effects of supranormal metal levels must also be expected at the population level, where selection and migration may obscure response or damage at the individual level. While behaviour and chemical communication may be of considerable ecological significance in nature (see also Chapter 4), little is known yet on possible interferences due to metal stress. Our present knowledge on the toxicity for marine organisms of heavy metals is almost entirely based on laboratory experiments, conducted under unrealistically high test concentrations.

Major shortcomings of the present concept and instrumentarium for assessing *in situ* consequences of increasing levels of heavy metals and their components include the following: (i) insufficient sensitivity of monitoring programmes; (ii) lack of metal specificity of analytical methods; (iii) analytical difficulties in situations of multi-factor stress exposure; (iv) insufficient awareness of the often high efficiencies of organismic compensatory mechanisms (regulation, adaptation, selection) which tend to mask sublethal, but possibly near-critical effects—especially in complex forms of life and at supra-individual levels.

Of the metals considered in Chapter 3, Fe, Cu, Zn, Co, Mn, Mo, Se, Cr, Ni, V (possibly also As and Sn) are essential for the normal functioning of biological processes. Several of these metals constitute essential parts of enzymes, others of haemoglobin (Fe), haemocyanin (Cu) or of structural body elements. However, at significantly increased concentration levels, these 'essential metals' tend to become toxic. Hence, they are considered here along with the 'non-essential' metals Ag, Al, Be, Cd, Hg, Pb, Sb, Ti, and Tl. Since the *in situ* concentrations of trace metals are very low, the possibilities of man-made contamination are high. Marine organisms commonly accumulate trace metals by concentration factors of 10^3 to 10^5; hence, in polluted waters, tissue concentrations may attain toxic levels. For man, excessive intake of metal-contaminated seafood may result in health problems. In this context, the potentially most dangerous metals are mercury, cadmium, and lead. In several parts of the world the consumption of mercury-contaminated fish has led to increases in human Hg tissue levels approaching the assumed critical threshold for the Minamata disease. Regular consumption of contaminated seafood can readily lead to critical accumulation, both in man and marine animals. In the sea, the more abundant zinc and copper may sometimes constitute a greater hazard to marine species than mercury or cadmium.

One of the main objectives of Chapter 3 is to assess and to compare the evidence for the toxicity to marine organisms of increased metal concentrations obtained in the laboratory and at sea. The reviewer devotes considerable attention to sources, inputs, fate and bioaccumulation of metals, and to the effects of elevated metal concentrations.

Man's activities contribute to metal input into the sea via rivers and outfalls, atmospheric fallout, dumping, marine mining, and drilling from ships. The most contaminated sites are estuaries, fjords, and bays. Further from land, atmospheric input contributes more than originally assumed. Estuaries often act as traps where metal inputs are deposited in sediments, from which pollutants may be released again over long periods of time. In the remobilization of metals trapped in sediments, microbial methylation of mercury (and possibly other metals) turned out to be of particular importance.

For assessing metal effects on aquatic organisms, biological availability, detoxification, accumulation and active or passive release from the body play a basic role. The biologically most available states of metals in sea water are still uncertain but are thought to include organic forms like methyl mercury and the free ions of some metals. Concentrations of the latter depend on the total metal concentration (largely a function of salinity and pH), and on inorganic as well as organic complexation (a function of the organic ligand). The ecological significance of organic complexes varies from near-prevention of metal uptake, to no effect, to uptake enhancement. In sediment surfaces metal availability appears to be controlled by metal-binding substrates (e.g. iron oxides or organics). The prime source of metal uptake is ambient water for bacteria, phytoplankton and seaweed, but food for birds and marine mammals. In most of the invertebrates examined, water-movement routes near or within the body seem important and relationships to mechanisms of osmo-and ion regulation may be assumed. Mechanisms of detoxification include: binding to non-specific high-molecular-weight proteins or polysaccharides, and to low-molecular-weight proteins of the metallothionein type; immobilization via intracellular inclusion and incorporation in shell, skeleton, fur or feathers; demethylation of methyl mercury and conversion to less toxic organic compounds. Mechanisms of active or passive release from the body include: diffusion; excretion; and losses via alimentary tract, moulting or gamete release.

Biomagnification—the increasing accumulation of metals at higher trophic levels via the food chain—appears to play a less important role than often predicted. While there are cases where the bodies of predators contain higher heavy-metal levels than their prey, appreciable biomagnification has been documented only for methyl mercury.

First-order indicator species for biological monitoring programmes are seaweeds and molluscs. Presumably, several different types of organisms are required for assessing contamination levels due to dissolved, particulate, and sediment-bound metals.

Chapter 3 features a detailed documentation of biological effects of increased ambient metal concentrations. Experimental studies reveal that mercury, copper, and silver are not only the most rapidly absorbed metals, but also the most toxic ones. Particularly sensitive biological parameters include fecundity, embryonic and larval development, oxygen consumption and enzyme functions, as well as growth of individuals and populations. Evidence presented for some phytoplankters, crustaceans, and fishes suggests that detrimental cadium effects could occur at concentrations recorded at sea; however for most species such damage need not be expected. Potentially toxic zinc and copper concentrations have been measured in several sea areas. *In situ* concentrations of lead never seem to have approached toxic levels. In regard to the other metals there is little evidence for problems, except under exceptional conditions.

For developing and refining management procedures of heavy-metal pollution we need (i) reliable information on input data; (ii) more field studies examining metal effects under natural conditions; (iii) contaminant-specific pollution indices; (iv) studies analysing changes in the chemical identity of metal pollutants released into the sea;

forms of metals most readily metabolized; pollution consequences at the population and ecosystem levels; (v) carefully designed, long-term monitoring programmes in ecological key areas receiving minimum, medium, and maximum pollution loads—both at national and international levels. The management goal must be defined more clearly. More directivity and uniformity is needed for agreement on target components of the ecosystem concerned, exposure standards, safety margins, and standardization of measurement procedures.

The intensities and time scales of physico-chemical processes of heavy-metal metabolization in ecosystems require investigation. Safety margins and management of pollution input will ultimately depend on the rate of ecosystem processes which counteract pollutant accumulation as a function of time. Apparently, reduction in metal input may quite rapidly result in concomitant decreases in the metal load in organisms; however, sediment-bound metals tend to require much longer periods of time for abatement.

Chapter 4: Oil Pollution and its Management

In its broadest sense, the term 'oil' refers to diverse, separable liquids produced by plants and animals. Natural oils are widely distributed and, together with carbohydrates, proteins, and fats, form the nutritional basis for man and most animals. 'Crude oil' or 'petroleum'* is a complex mixture of hydrocarbons and other remnants of life, deposited beneath the earth's surface several hundred million years ago. Petroleum comprises materials in gaseous, liquid, and solid states. Vast amounts of petroleum have aggregated in the pore spaces of sedimentary rocks. As natural gas and processed fossil fuels they now provide by far the largest portion of modern man's energy budget. Very occasionally petroleum escapes naturally to the surface from crustal faults.

Petroleum originates from primordial organic materials, removed from ecosystem recycling by burial and then exposed to conditions of elevated temperatures (up to 80 °C) and pressure in the absence of air. Turned into a greasy or tarry fluid, the materials became accumulated in geological reservoirs. The large variety of complex molecular structures discovered in petroleums of different origin mirrors differences in the composition of the primordial donators—diatoms, blue-green algae, foraminiferans, other protozoans, terrestrial plants *m* and in the local processes of partial degradation, chemical maturation, and storage. While no 2 crude oils from different or even neighbouring sources are completely identical, all crude oils are derived essentially from only 2 elements: carbon (80 to 87%) and hydrogen (10 to 15%). The resulting compounds can be subdivided into 3 basic groups: paraffins or methanes (general formula: C_nH_{2n+2}) a widely variable proportion of cycloparaffins or naphthenes (C_nH_{2n}), and aromatics (C_nH_{2n-6}) together with their alkyl derivatives and other hydrocarbons.

The extremely numerous hydrocarbons in crude oil are blended in an infinite number of different mixtures which, in addition, contain varying small amounts of other substances—such as sulphur, nitrogen, oxygen, salts, and trace metals—and are associated with siliceous skeletal remains, spores, resins, coal, and lignite.

One major problem in studying oil pollution in the marine environment is that none of the crude oils and only few of the oil products have ever been comprehensively analysed, i.e. to single hydrocarbon components. Partial analysis indicates that a crude oil consists of

* 'Petroleum', from the Greek *petra* and the Latin *oleum,* literally means 'rock oil'.

from many thousands up to a million individual components. Additional problems for *in situ* pollution assessments result from the fact that crude oils seldom come into contact with marine organisms in a single physico-chemical state. Also, exposure to sea water immediately initiates a series of changes which, again, are incompletely known. In regard to oil inputs to the sea from air, land, and rivers virtually nothing is certain about the chemical identity of the oil products which arise from day-to-day activities, except for the most durable hydrocarbons. Under such conditions, how can conclusive results be obtained in regard to the biological effects of crude-oil components in the seas? Let alone the need to separate such effects from biological consequences due to concomitant stress from other pollutants, extreme intensities of natural environmental factors and the whole range of interorganismic relationships! Unless we settle for—admittedly unsatisfactory—gross simplifications the problems encountered are virtually insolvable.

From its ancient storage beds, petroleum is being pumped to the earth's surface, distributed and consumed in unimaginably large amounts. Components and derivatives of petroleum are now present everywhere and constantly contaminate air, soil, lakes, rivers, and seas. All organisms on earth are exposed to enhanced concentrations of petroleum components and their derivatives. The most spectacular effects of oil pollution manifest themselves in major spills due to tanker or drilling accidents. Pictures of oiled, dying birds and of devastated coastal areas, especially of beaches and harbours, have repeatedly shocked the world. Major oil spills have killed billions of animals and plants in the waters near the input sites, along the path of escaped oil drifting at the sea's surface, and wherever spilled oil has reached the coastline. There is practically no sea area and no coastline entirely free of some form of petroleum products.

Fortunately, oil—itself a product of nature—is not as foreign to the marine environment as are, for example, the man-made pesticides (Volume V, Part 4). Unlike pesticides, detergents, organometals and wastes with high levels of toxic metals, crude oils are virtually non-toxic when reduced to their elements under aerobic conditions. Once removed from its protective, subterraneous storage site, oil becomes amenable to degradation, especially due to the activities of micro-organisms. Reintegration of oil released into marine ecosystems appears to be the major reason for the absence of demonstrable accumulating oil-pollution effects on a global scale. A total of more than 200 species (60 genera) of bacteria, filamentous fungi, and yeasts are capable of oxidizing hydrocarbons. Many of these can live in the marine environment.

At sea, major oil spills have been shown to reduce the reproductive capacities of local phyto- and zooplankton communities and hence the nutritional support of commercially used seafood organisms, such as fish and shellfish. Stranded oil has devastated wide beach areas and killed, damaged or incapacitated innumerable coastal plants and animals. In experiments, even minute traces of oil have been shown to cause a variety of metabolic impairments: reductions in growth; disease resistance; tolerance to environmental stress; changes in behaviour; and disturbances at the enzyme level.

Nevertheless, the evidence at hand witnesses that, except for areas directly affected by heavy spills or major industrial activities, negative (i.e. detrimental) effects of oil pollution have remained marginal.

Oil dispersants, and other substances formulated to combat oil pollution—as well as drilling muds, biocides, chemical additives, lubricants, etc. employed by the oil industry—tend to cause more severe effects than the oil itself. For an assessment regarding the responses of marine organisms to these substances consult p. 1499 ff.

While there is no unequivocal evidence for irreversible long-term damage of oil pollution at the population or ecosystem levels, there is reason to warn that we may not yet be able to see the whole picture. Long-lived crude-oil hydrocarbons, their derivatives and specific macromolecular components of degrading oil may cause difficult-to-detect damage by interfering with biochemical avenues of ecosystem integration and with chemical communication among aquatic organisms. Is the present scarcity of demonstrable negative effects at sublethal levels a consequence of insufficient knowledge? Are our analytical methods inadequate? Are we unable to detect potentially detrimental effects before damage has occurred? Since constant oil contamination represents an active demand on organisms to adjust to its presence, does this constitute a widespread suboptimal condition of existence?

While scientific understanding of oil pollution cannot be complete and exact, this does not afford an excuse for less than the best methods and efforts to prevent and combat incursion of oil wherever it arises by responsible industrial, local environmental and national or international pollution management.

Oil pollution research and management still suffer from: (i) lack of a comprehensive concept of the factors contributing to oil pollution, and of internationally integrated policies regarding oil spill abatement—most authorities fail to grasp the meaning of the term 'oil pollution' and are aware only of major oil spills; (ii) deficiencies in existing controls and insufficient harmonization and integration of local, national, and international monitoring activities; (iii) lack of world-wide management policies covering pollution of air, land, fresh water, and the seas. While some oil and shipping industries strive to meet their responsibilities, many governmental administrators and experts seem content to perpetuate outdated practices relating to the regulation, reporting, and measurement of chronic inputs. There is need for improvements in the partnership between scientific experts and administrators, for concerted efforts to utilize oil more efficiently and for more effective recycling of waste oils.

One promising aspect of oil-pollution assessment is the use of indicator organisms. Thus the responses to oil pollution of the well-investigated mussel *Mytilus edulis* have yielded encouraging results ('Mussel Watch Programme', p. 1518 ff). Hydrocarbon-determinations in *M. edulis* corresponded broadly with the known history of local oil input and provided insights into the natural cleansing capacity of biotopes subsequent to an oil spill. In addition to the mussel's ability to record oil exposure, its responses to toxic polycyclic aromatic hydrocarbons open the door to sound evaluations of oil-product toxicities.

Recent progress in crude-oil analysis (p. 1485) has provided new instrumentation for identifying different sources of crude oil. The 'fingerprinting' technique, developed in the United States, is considered scientifically and legally sound, and provides a means for tracing the sources of an illegal oil spill. Nevertheless, great care is needed in the choice of methods for monitoring oil pollution 'and much of the published information needs careful scrutiny on the basis of the exact methods of sampling and analysis employed' (p. 1492).

The international efforts to control and manage marine oil pollution have a long history. The first convention was drafted in Washington in 1926, but never adopted. Since then several conventions have initiated national legislation aiming at the restriction and control of oil pollution. The latest IMCO Marine Pollution 1978 Convention still awaits signing by a number of major nations (p. 1538). Options for remedial actions are reviewed on p. 1539 ff.

Chapter 4 considers most of the information presently available on oil-pollution research and outlines possible ways and means of oil-pollution management. As is the case with other sources of pollution, prevention is easier to manage, less costly and more effective than damage correction. The reviewer focuses his attention on the sources of oil input to the sea; chemical aspects of oil pollution; ecological consequences of pollution due to oil and oil dispersants; and the control and management of oil pollution. He reminds us that oil is a finite, non-renewable resource. Within 50 yr or so petroleum will be scarce, and somewhat later the oil flow will cease altogether. Parallel to this long-term development, pollution due to petroleum release will decline and so will its effects on the sea's living and non-living components.

Acknowledgements The first draft of this chapter has been circulated among several colleagues. Their comments and criticisms, especially those of Drs R. Johnston, D. S. Woodhead, and D. J. Reish, have improved and complemented this account. I am grateful for this help. Angela Giraldi, Seetha Murthy and Helga Witt have rendered untiring technical assistance in the completion of this chapter and, in fact, of Volume V.

Literature Cited (Chapter 1)

CHIPMAN, W. A. (1972a). Ionizing radiation. Bacteria, fungi and blue-green algae. In O. Kinne (Ed.), *Marine Ecology,* Vol. I, Environmental Factors, Part 3. Wiley, London. pp. 1585–1604.

CHIPMAN, W. A. (1972b). Ionizing radiation. Plants. In O. Kinne (Ed.), *Marine Ecology,* Vol. I, Environmental Factors, Part 3. Wiley, London. pp. 1605–1620.

CHIPMAN, W. A. (1972c). Ionizing radiation. Animals. In O. Kinne (Ed.), *Marine Ecology,* Vol. I, Environmental Factors, Part 3. Wiley, London. pp. 1621–1657.

GERLACH, A. (1981). *Marine Pollution: Diagnosis and Therapy,* Springer-Verlag, Heidelberg.

GILLBRIGHT, M. (1981). Hydrographie, Nährstoffe und Phytoplankton bei Helgoland. *Jahresbericht 1980,* Biologische Anstalt Helgoland, Hamburg. pp. 23–27.

GOLDBERG, E. D. (1975). The mussel watch—a first step in global marine monitoring. *Mar. Pollut. Bull., N.S.,* **6,** 111.

GOLDBERG, E. D. (1976). *Strategies for Marine Pollution Monitoring,* Wiley, New York.

GOLDBERG, E. D., BOWEN, V. T., FARRINGTON, J. W., HARVEY, G., MARTIN, J. H., PARKER, P. L., RISEBROUGH, R. W., ROBERTSON, W., SCHNEIDER, E., and GAMBLE, E. (1978). The mussel watch. *Environ. Conserv.,* **5**(2), 101–125.

GULLAND, J. A. (1983). World resources of fisheries and their management. In O. Kinne (Ed.) *Marine Ecology*, Vol. V, Ocean Management, Part 2. Wiley, Chichester. pp. 839–1061.

JOHNSTON, R. (1976). *Marine Pollution,* Academic Press, London.

KINNE, O. (1968). International Symposium 'Biological and hydrographical problems of water pollution in the North Sea and Adjacent Waters: Closing address.' *Helgoländer wiss. Meeresunters.*, **17**, 518–522.

KINNE, O. (1975). Introduction to Volume II. In O. Kinne (Ed.), *Marine Ecology,* Vol. II, Physiological Mechanisms, Part 1. Wiley, London. pp. 1–8.

KINNE, O. (1976). Introduction to Volume III. In O. Kinne (Ed.), *Marine Ecology*, Vol. III, Cultivation, Part 1. Wiley, London. pp. 1–17.

KINNE, O. (1977). Cultivation of animals: research cultivation. In O. Kinne (Ed.), *Marine Ecology,* Vol. III, Cultivation, Part 2. Wiley, Chichester. pp. 579–1293.

KINNE, O. (1978). Introduction to Volume IV. In O. Kinne (Ed.), *Marine Ecology,* Vol. IV, Dynamics. Wiley, Chichester. pp. 1–11.

KINNE, O. (1980a). 14th European Marine Biology Symposium 'Protection of Life in the Sea': summary of symposium papers and conclusions. *Helgoländer Meeresunters.*, **33,** 732–761.

KINNE, O. (1980b). Diseases of marine animals: general aspects. In O. Kinne (Ed.), *Diseases of Marine Animals,* Vol. I. Wiley, Chichester. pp. 13–73.

KINNE, O. (1982). Aquakultur und die Ernährung von morgen. *Spektrum der Wissenschaft*, **12,** 46-57.

KINNE, O and AURICH, H. (Eds) (1968). 'Internationales Symposium 1967: Biologische und hydrographische Probleme der Wasserverunreinigung in der Nordsee und angrenzunden Gewässern. Helgolander wiss. Meeresunters. **17**, 1–530.

KINNE, O. and BULNHEIM, H.-P. (Eds.) (1980). 14th European Marine Biology Symposium 'Protection of Life in the Sea'. *Helgoländer Meeresunters.*, **33**, 1–772.

KINNE, O. and ROSENTHAL, H. (1977). Cultivation of animals: commercial cultivation (aquaculture). In O. Kinne (Ed.), *Marine Ecology,* Vol. III, Cultivation, Part 3. Wiley, Chichester. pp. 1321–1398.

KORRINGA, P. (1980). Management of marine species. *Helgoländer Meeresunters.*, **33**, 641–661.

LAUCKNER, G. (1980a). Diseases of marine Protozoa. In O. Kinne (Ed.), *Diseases of Marine Animals,* Vol. I. Wiley, Chichester. pp. 75–134.

LAUCKNER, G. (1980b). Diseases of Mesozoa. In O. Kinne (Ed.), *Diseases of Marine Animals,* Vol. I. Wiley, Chichester. pp. 135–137.

LAUCKNER, G. (1980c). Diseases of Porifera. In O. Kinne (Ed.), *Diseases of Marine Animals,* Vol. I. Wiley, Chichester. pp. 139–165.

LAUCKNER, G. (1980d). Diseases of Cnidaria. In O. Kinne (Ed.), *Diseases of Marine Animals,* Vol. I. Wiley, Chichester. pp. 167–237.

LAUCKNER, G. (1980e). Diseases of Ctenophora. In O. Kinne (Ed.), *Diseases of Marine Animals,* Vol. I. Wiley, Chichester. pp. 239–253.

LAUCKNER, G. (1980f). Diseases of Tentaculata. In O. Kinne (Ed.), *Diseases of Marine Animals,* Vol. I. Wiley, Chichester. pp. 255–259.

LAUCKNER, G. (1980g). Diseases of Sipunculida, Priapulida and Echiurida. In O. Kinne (Ed.), *Diseases of Marine Animals,* Vol. I. Wiley, Chichester. pp. 261–278.

LAUCKNER, G. (1980h). Diseases of Platyhelminthes. In O. Kinne (Ed.), *Diseases of Marine Animals*, Vol. I. Wiley, Chichester. pp. 279–302.

LAUCKNER, G. (1980i). Diseases of Nemertea. In O. Kinne (Ed.), *Diseases of Marine Animals,* Vol. I. Wiley, Chichester. pp. 303–309.

LAUCKNER, G. (1980j). Diseases of Mollusca: Gastropoda. In O. Kinne (Ed.), *Diseases of Marine Animals,* Vol. I. Wiley, Chichester. pp. 311–424.

LAUCKNER, G. (1983a). Diseases of Mollusca: Bivalvia. In O. Kinne (Ed.), *Diseases of Marine Animals,* Vol. II. Biologische Anstalt Helgoland, Hamburg. pp. 477–961.

LAUCKNER, G. (1983b). Diseases of Mollusca: Placophora. In O. Kinne (Ed.), *Diseases of Marine Animals,* Vol. II. Biologische Anstalt Helgoland, Hamburg. pp. 963–977.

LAUCKNER, G. (1983c). Diseases of Mollusca: Scaphopoda. In O. Kinne (Ed.), *Diseases of Marine Animals,* Vol. II. Biologische Anstalt Helgoland, Hamburg. pp. 979–973.

PHILLIPS, D. J. H. (1977). The use of biological indicator organisms to monitor trace metal pollution in marine and estuarine environments—a review. *Environ. Pollut.,* **13,** 281–316.

ROSENTHAL, H. (1978). Bibliography on transplantation of aquatic organisms and its consequences on aquaculture and ecosystems. *Spec. Publs. European Mariculture Soc.*, **3,** 1–146.

ROSENTHAL, H. (1980). Implications of transplantations to aquaculture and ecosystems. *Mar. Fish. Rev.*, **42,** 1-14.

RUIVO, M. (Ed.) (1972). *Marine Pollution and Sea Life*, Fishing News (Books) Ltd, Surrey.

Marine Ecology Vol. 5, Part 3
Edited by Otto Kinne

2. CONTAMINATION DUE TO RADIOACTIVE MATERIALS

D. S. WOODHEAD

(1) Introduction

The significance of ionizing radiation as a factor in the marine environment has been considered earlier in *Marine Ecology* (Volume I: CHIPMAN, 1972). CHIPMAN concluded (p. 1642):

> 'So far as is known, the levels of natural ionizing radiation present in oceans and coastal waters elicit no specific responses by marine animals; they cause no observable damages or injuries.'

This conclusion was based on a comprehensive review of the available literature which detailed the effects observed in aquatic organisms exposed to much increased radiation intensities under controlled laboratory conditions. Concerned primarily with ionizing radiation as a natural factor in the marine environment, CHIPMAN made only brief mention of the increased concentrations of radionuclides in seas and coastal waters arising from human activities. These have included the atmospheric and underwater testing of nuclear weapons and the disposal into the sea of the low-level radioactive wastes generated by the exploitation of nuclear power and the industrial and medical uses of radionuclides. The negotiation and partial implementation of an international treaty prohibiting the testing of nuclear weapons in the atmosphere has greatly reduced the input to the marine environment from this source and the corresponding degree of contamination is declining. However, the inputs due to the peaceful exploitation of radioactivity and nuclear energy continue and the projected expansion of electricity generation based on nuclear power in many countries makes it very probable that such inputs of artificial (i.e. man-made) radionuclides will be a factor in the marine environment for the foreseeable future.

The harmful effects that radiation could produce in living tissue were recognised very soon after the discovery of X-rays and radioactivity at the end of the nineteenth century. However, the potential benefits to be gained both in diagnosis and, somewhat later, in therapy through the controlled induction of damage, led to the rapid adoption of radiation as a tool by the medical profession despite the attendant risks. In order to minimize the hazard to both medical personnel and patients, studies were initiated in two interrelated areas: firstly, it was necessary to develop a means of determining the amount of energy absorbed from the radiation field by a tissue, organ or whole organism, i.e. to quantify the insult in terms of absorbed dose; secondly, a better understanding of the biological effects of radiation and their relation to absorbed dose was required. The results of these early studies, and particularly the establishment of satisfactory and reproducible methods of dosimetry, provided a basis both for improved radiotherapy practice and for recommendations for the limitation of occupational exposure. Hence, radiological safety procedures were developed which ensured compliance with exposure limits while permitting the continued utilization of radiation and radioactive materials. The rapid expansion of these activities over the past 35 yr has led to the extension of radiological safety into the field of public health. Where there is a potential for the exposure of the general public to either radioactivity or radiation, techniques have been developed to assess the degree of such exposure and thus to allow the possibility of its control within defined limits. In addition, there has also been a parallel development of concern, albeit of lesser degree, as to the effects of the increasing scale of these activities on both the living and non-living resources of the environment.

The use of the sea as a repository for wastes of any form may be legitimately regarded as a resource. However, the extent to which this resource may be exploited must be assessed within the perspective of other, perhaps conflicting, demands on the marine environment, e.g. as a source of food and raw materials, for leisure, etc. For radioactive wastes disposed of into the sea, the hazards will usually be related to the concentrations attained. In general, hydrographic processes will operate to dilute and disperse the radionuclides, whereas geochemical and biological processes will tend to reconcentrate them—thus, perhaps, increasing the potential for exposure of marine organisms and, ultimately, man. The capacity of a particular marine site to accept radioactive wastes without unacceptably harmful consequences clearly depends on many factors and a considerable understanding of the environmental behaviour of the individual radionuclides making up the waste is necessary before disposal can be authorized.

The control of marine disposals of radioactive wastes is based upon the need to ensure that the consequent exposure of humans, either as individuals or as populations, does not exceed acceptable limits. The exposure of populations of marine organisms has been of secondary concern; nevertheless, rational resource management requires that this hazard also be assessed and included in any consideration of the overall acceptability of a waste disposal practice.

(2) Radioactivity and Radiation

(a) Nature of Radioactivity

Radioactivity is the physical manifestation of the unstable state of the nuclei of a proportion of the atoms in a radioactive material. The instability arises from an excess of energy in the nucleus. In most cases this is reduced by the spontaneous emission of more or less energetic α- or β-particles which are often accompanied by γ-rays. The emission of an α-particle reduces both the mass and the electrical charge of the nucleus whereas β-decay has a very small effect on the nuclear mass but increases or decreases the nuclear charge for negative electron (β^-) or positron (β^+) emission respectively. In each case the change in nuclear charge means that the daughter nucleus is that of a different element and it may or may not be stable. The γ-rays arise from the rearrangement of the energy states of the daughter nucleus. Somewhat simplified decay schemes for radionuclides of interest in an environmental context are illustrated in Fig. 2-1 including examples of the less common decay mode via electron capture (EC).

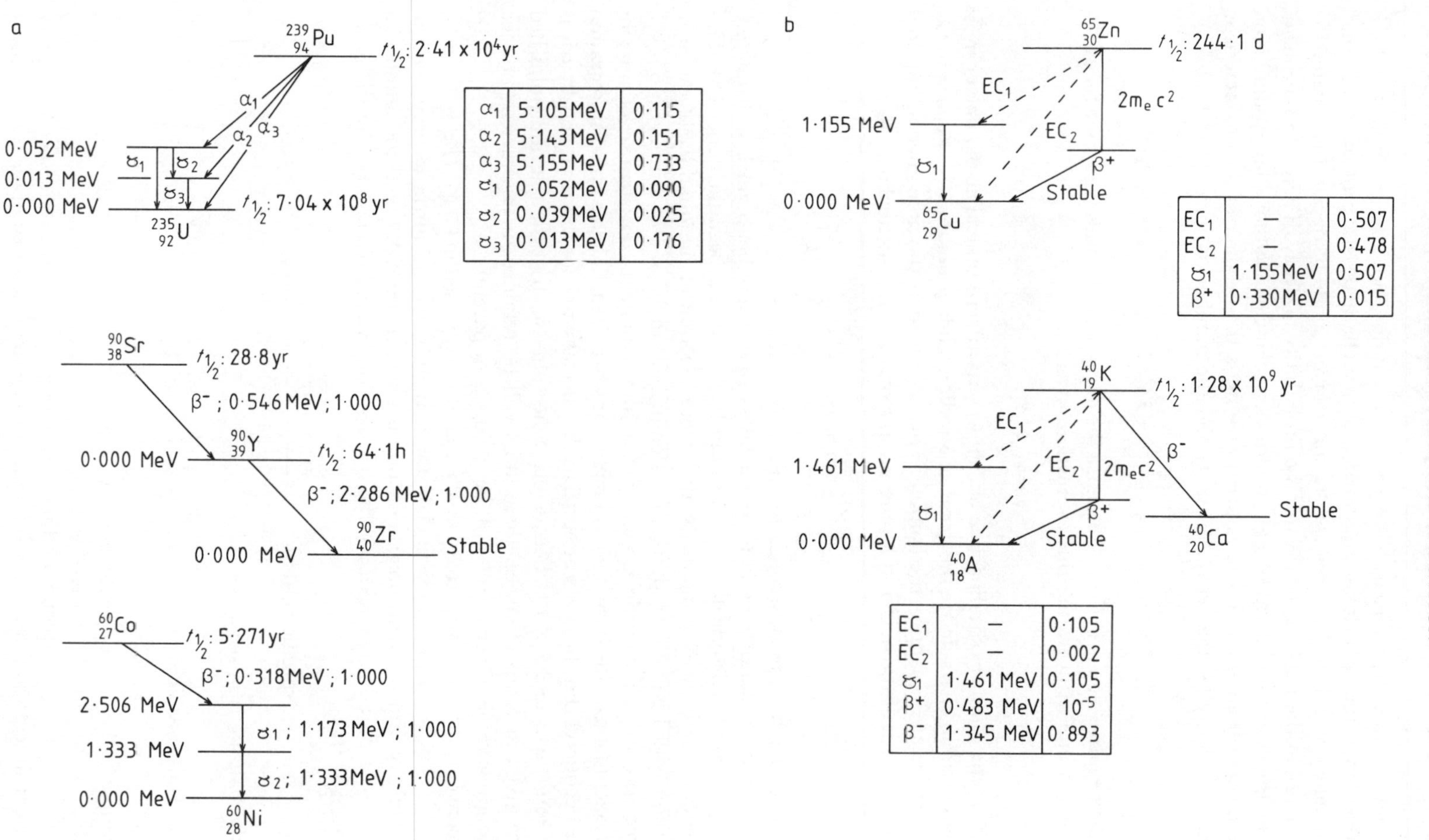

Fig. 2-1: Examples of radioactive decay schemes. (After LEDERER and SHIRLEY, 1978)

Although a radioactive nucleus has excess energy it does not lose this immediately; it is effectively in a state of unstable equilibrium with a characteristic probability of decay per unit time. As a broad generalization, for a given mode of decay, the greater the excess energy in the nucleus the greater the probability of decay. There is no means of predicting the moment of decay of a given radioactive nucleus but for a large population of these nuclei the number of decays per unit time is the product of the characteristic probability of decay (λ, the decay constant) and the number of nuclei (N):

$$\text{i.e.} \frac{dN}{dt} = -\lambda N$$

where the negative sign is necessary because the decay rate declines with time following the reduction in N. Integrating this expression gives:

$$N = N_0 e^{-\lambda t}$$

where N_0 is the number of radioactive nuclei present at zero time. The parameter most commonly used to characterize the decay of a radionuclide is not the decay constant (λ) but a closely related quantity, the half-life. This is defined as the time interval required for the decay of half the nuclei in a pure sample of a given radionuclide, i.e.

$$\frac{N_0}{2} = N_0 e^{-\lambda t_{\frac{1}{2}}} \text{ whence } t_{\frac{1}{2}} = \frac{0.693}{\lambda}$$

Values of $t_{\frac{1}{2}}$ vary from microseconds to billions of years.

Radioactivity is quantified in terms of the decay rate or number of nuclear disintegrations per unit time. Historically, the unit of radioactivity has been the curie (Ci). This was originally defined in terms of the disintegration rate of the quantity of radon in equilibrium with 1 g of pure radium, and is equivalent to $3{\cdot}7 \times 10^{10}$ disintegrations s^{-1}. More recently, as part of the development of an internationally consistent set of units for general scientific use, the basic unit of activity has been redefined as one disintegration s^{-1}. The becquerel (Bq) has been adopted as the special name for this unit, and 1 curie is exactly equivalent to $3{\cdot}7 \times 10^{10}$ becquerels. Since almost all of the relevant published data are given in terms of curies, these units will be retained in this review, but the corresponding value in becquerels will be included in parentheses.

The concentration of radioactivity in a material is expressed in Ci g^{-1} (Bq kg^{-1}) or Ci l^{-1} (Bq l^{-1}), but for a pure radionuclide the activity g^{-1} is an intrinsic property: the specific activity. Since the disintegration rate is given by $0{\cdot}693N/t_{\frac{1}{2}}$ and the number of atoms g^{-1} of a pure radionuclide by the quotient of Avogadro's number by the gram atomic weight (W_A)

$$\text{specific activity} = \frac{0.693 \times 6.025 \times 10^{23}}{t_{\frac{1}{2}} \times W_A \times 3.7 \times 10^{10}} \text{ Ci g}^{-1}$$

where the half-life is in seconds;

$$\text{thus specific activity} = \frac{1.128 \times 10^{13}}{t_{\frac{1}{2}} W_A} \text{ Ci g}^{-1} \left(\frac{4.174 \times 10^{26}}{t_{\frac{1}{2}} W_A} \text{ Bq kg}^{-1} \right)$$

For example: ^{239}Pu has a specific activity of $6{\cdot}20 \times 10^{-2}$ Ci g^{-1} ($2{\cdot}29 \times 10^{12}$ Bq kg^{-1}) while that of the much shorter-lived natural α-emitter ^{210}Po is $4{\cdot}48 \times 10^{3}$ Ci g^{-1} ($1{\cdot}66 \times 10^{17}$ Bq kg^{-1}). A second, and more common, usage of the term specific activity relates the disintegration rate of a given radionuclide to the amount of the same element (including both stable and unstable isotopes) present in a material. The units remain pCi g^{-1} (Bq kg^{-1}) but there is no uniquely defined value. For an environmental sample for example, the value depends upon the degree of contamination, and can change in relation to both the input and environmental conditions. This usage is most often found in studies in radiochemistry and radiobiology where the very small gravimetric quantities of a radionuclide which can be easily quantified by radiometric methods are used as tracers. For example, 100 pCi (3·7 Bq) of ^{65}Zn corresponds to $1{\cdot}22 \times 10^{-14}$g (or $1{\cdot}13 \times 10^{8}$ atoms).

(b) Interaction of Radiation with Matter

Before considering the means by which the radioactivity of a material may be quantified, it is necessary to examine how the radiations emitted during nuclear disintegration interact with matter. This is also fundamental to a discussion of the biological effects of radiation. Charged α- and β-particles traversing matter may interact with both the negatively charged orbital electrons and the positively charged nuclei of the constituent atoms. The energy loss from the incident particle through the action of the electrical forces can be shown to be inversely proportional to the mass of the interacting particle; therefore, the majority of the energy is transferred to the orbital electrons. The magnitude of the transfer is also dependent on the strength of the force and the time period during which it acts. The energy gained by the orbital electron may result in either excitation, or separation from the parent atom, i.e. ionization. For a given energy, an α-particle has a much lower velocity than a β-particle because of the difference in mass ($m_\alpha : m_e = 7300 : 1$); this factor, together with the greater charge (2 : 1) means that in any interaction an α-particle is more likely to lose sufficient energy to cause excitation and ionization than a β-particle. Thus the path of an α-particle is characterized by rapid energy loss, very dense ionization and, in unit density tissue, a very short path length not exceeding about 100 μm. Also, due to the greater mass of the α-particle compared with that of the orbital electrons, the path is relatively straight; only a close encounter with the electric charge on the equivalent or greater mass of an atomic nucleus can cause α-particle scattering. The path of a β-particle, on the other hand, is characterized by slow energy loss, sparse ionization and frequent wide-angle scattering; the total path length in unit density tissue may be several centimetres for the higher energy β-particles.

Although the γ-ray, being uncharged, does not continuously lose energy along its track through ionization, it does have certain probabilities of interaction with orbital electrons via 2 processes. Through the photoelectric effect, the incident γ-ray is totally absorbed in ejecting a bound electron from an atom. The difference in energy between the incident γ-ray and the bound electron appears as kinetic energy. The Compton scattering process involves an elastic collision with a free or loosely bound electron in which the electron receives a fraction of the energy of the incident γ-ray. The scattered γ-ray, of lower energy, undergoes further Compton or photoelectric interactions until totally absorbed. In both cases, the kinetic energy of the free electron produced by the interaction is dissipated through excitation and ionization. The relative importance of these 2 processes depends on the γ-ray energy and the atomic numbers of the elements making up the absorbing material. At higher γ-ray energies (>1·02 MeV) and more commonly in materials containing high atomic number elements, there is a third process leading to γ-ray absorption: pair production. The incident γ-ray interacts with the electric field of the nucleus and materializes into a negative electron and a positron. The energy difference between the γ-ray and the energy equivalent of the rest mass of the 2 electrons (1·02 MeV) appears as kinetic energy which is once again dissipated as excitation and ionization along the particle tracks. The stochastic nature of the interactions between γ-rays and the absorbing material means that γ-rays can penetrate considerable distances, of the order of tens of centimetres of unit density tissue, before absorption occurs.

The positrons produced by radioactive decay and through γ-ray absorption via pair production have only a transitory existence. Their kinetic energy is normally lost through excitation and ionization and the resulting, slow-moving positron combines with a negative electron, cancelling the electric charge, and converting the rest mass of the 2 electrons into 2 γ-rays, each with an energy of 0·511 MeV. This is the so-called annihilation radiation.

The energy of the incident radiation is not completely degraded at this stage of the absorption process. The excited and ionized atoms which have been produced gradually revert to a stable ground state by the rearrangement of the orbital electrons and the emission of lower energy electromagnetic radiation (low energy X-rays, ultra-violet and visible light). These, too, are absorbed, yielding atomic and molecular vibrations which are ultimately indistinguishable from those due to the thermal energy content of the absorber. Where the absorber consists of more than 1 element, a proportion of the incident energy may be taken up to produce an altered chemical composition.

(c) Quantification of Radioactivity

The quantification of the radioactivity in a sample requires the detection and enumeration of the radiations characteristic of each radionuclide present. Since the primary product of radiation absorption is ionization, essentially all detection systems in general use depend upon the direct or indirect measurement of ionization. Although a given sample may contain α-, β- and β–γ-active radionuclides it is not possible to evaluate the concentrations of each by a single technique. Each radiation type is best treated with a specialized analytical approach dictated by its physical properties.

The decay scheme shown in Fig. 2-1 for ^{239}Pu is a fairly typical example of the radiations produced by the majority of α-emitting radionuclides: there are several distinct components to the α-spectrum and the accompanying γ-rays are of relatively low energy. The α-particle energy spectrum provides the best basis for identifying the α-emitters present in a sample; the γ-rays generally do not permit easy discrimination between these radionuclides. The short range of the α-particles in solid materials means that only those emitted from nuclei within this distance of the surface can leave the sample and, also, that the flux at the surface of the sample contains α-particles with all energies from zero to the maximum emission energy. Both the low concentrations usually present in environmental samples (hence a low surface flux) and the degraded energy spectrum make it necessary to use radiochemical procedures to separate and concentrate the α-emitting radionuclides from the bulk of the sample. This has the additional advantage that chemical discrimination between elements can be applied. The separated elements are usually electroplated or evaporated on to a small metal disc. The former technique is preferable since it gives a thinner, more uniform source.

The ionization produced by, and hence the energy of, the α-particles emitted from these sources may be determined by 2 different methods. Firstly, the ionization is

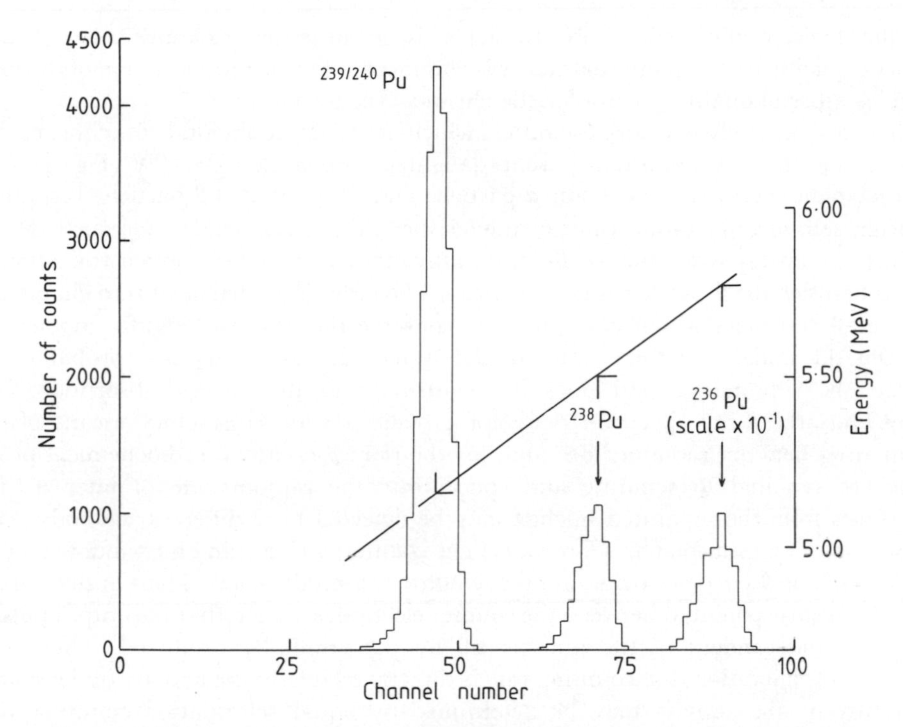

Fig. 2-2: Alpha-particle energy spectrum of plutonium extracted from mussels, *Mytilus edulis*, sampled from the Windscale pipeline. (Ministry of Agriculture, Fisheries and Food, Directorate of Fisheries Research, unpublished)

generated in the gas in an ion chamber, and the ions collected by applying a voltage between 2 electrodes. The number of ion pairs produced is proportional to the energy of the α-particle and when these are collected a corresponding electrical pulse is generated in the electrical circuit. This is then amplified, sorted as to amplitude (equivalent to α-particle energy) and stored to generate an energy spectrum. Alternatively, the α-particles may be absorbed by, and hence produce ionization in, a silicon, surface barrier, semiconductor detector. Essentially, this operates as a solid-state ion chamber producing an electrical pulse with an amplitude proportional to the α-particle energy. As before, the pulses can be processed to generate an energy spectrum. In practice, the latter is the most commonly used technique. A typical spectrum from the plutonium extracted from an edible mussel *Mytilus edulis* sample collected on the Cumbrian coast in the vicinity of the discharge from the fuel reprocessing factory at Windscale (England) is given in Fig. 2-2. Before the information contained in the spectrum can be used to derive the concentrations of the plutonium isotopes in the original sample, 2 further data are required. These are the recovery efficiency of the chemical procedure used to separate the plutonium from the sample, and the counting efficiency of the semiconductor detector. The first is determined by adding to the sample a measured quantity of a plutonium isotope known not to be present initially, i.e. a yield tracer or monitor; the second is obtained by counting, in the same geometry, a standard source prepared under identical conditions. In the example shown, the yield tracer was ^{236}Pu, the chemical yield, 45·6%

and the detector efficiency, 22·5%. In fact, it is not necessary to know either of these factors explicitly when an internal tracer is employed, but the determination of the yield provides a useful quality control for the chemical separation.

There are relatively few pure β-emitters which are of environmental consequence, but they include the very important parent–daughter combination ^{90}Sr–^{90}Y (Fig. 2-1). In contrast to the energy spectra of both α-particles and γ-rays, that of β-particles is continuous from zero to a maximum value characteristic of the radionuclide concerned. This is because the energy available for β-decay, although constant for a given transition, is shared between the β-particle and a neutrino. The latter is a particle of zero charge and very small rest mass which is required to conserve the intrinsic angular momentum between the initial and final states of the system; it has a very low probability of interaction with matter and does not contribute to the energy absorption from decay. Thus the β-ray spectrum does not provide a very satisfactory means of discriminating between radionuclides and, for the pure β-emitters, radiochemical procedures are required to separate and concentrate the radionuclide of interest. The β-particles from the separated nuclide may be detected by 2 different methods. After preparation of a thin, planar source a Geiger counter with a thin end window may be employed. The β-particles penetrate the window and produce ionization in the counter gas. The electric potential between the counter electrodes is such that the output pulse is effectively independent of the number of ion pairs initially produced. Thus, after appropriate calibration, the counting rate is directly related to the activity of the source. Alternatively, the activity may be determined by liquid scintillation counting. The separated radionuclide is converted to a form which is compatible with the organic solvent base of the scintillator, and the 2 are mixed in small plastic or glass vials. A part of the excitation and ionization energy deposited in the liquid is transferred to activator molecules present as a solute in the solvent base. These return to the ground state with the emission of light, which is detected by a photomultiplier and converted to an electrical pulse. Again, after appropriate calibration, the pulse rate is related to the source activity. In general, low-activity sources must be evaluated with gas-filled counters since lower backgrounds are obtainable. As for the α-emitters, the use of chemical separation procedures requires the addition of a yield tracer to the sample.

In the case of ^{90}Sr, the separated radionuclide, together with the yield tracer (γ-emitting ^{85}Sr) and stable strontium carrier, are obtained as the chloride in acid solution. This is γ-counted (see below) to determine the chemical yield, and is left for at least 2 wk to allow the short-lived ^{90}Y daughter to grow back into equilibrium with the ^{90}Sr parent. The greater energy of the β-particles from ^{90}Y decay allows easier detection of this radionuclide and gives a quantitative determination of the parent activity. The ^{90}Y is stripped from the solution by coprecipitation with ferric hydroxide after the addition of a small quantity of ferric chloride ($\approx$2 mg iron) followed by ammonium hydroxide. Filtration and drying produces a thin, planar source suitable for Geiger counting. The decline of the source activity, as shown by sequential counts, with the characteristic half-life of ^{90}Y (64·1 h) provides additional confirmation of the radionuclide identity and repeat determinations may be made after the regrowth of the ^{90}Y in the final solution.

The much greater penetrating power of γ-rays, compared with that of either α- or β-particles, together with the emission of line spectra characteristic of the radionuclides, reduces the need for complex sample preparation. In most cases all that is required is drying to reduce the mass (and bulk of biological samples) followed by homogenization. There are 2 detectors which are commonly used for γ-ray spectroscopy: the thallium-activated sodium iodide [NaI(Tl)] scintillation crystal and the lithium-drifted germanium [Ge(Li)] semiconductor. The high effective atomic number of sodium iodide gives good γ-ray detection efficiency, but the overall efficiency of conversion of the initial ionization energy to photoelectrons at the cathode of the photomultiplier is relatively low: perhaps 2×10^3 photoelectrons from the absorption of a 1 MeV γ-ray which produced some $2{\cdot}5 \times 10^4$ ion pairs in the crystal. Due to the random nature of the processes involved in the conversion there is quite a large variation in the number of photoelectrons obtained for a constant γ-ray energy, and hence the energy discrimination is relatively poor. The great advantage of the Ge(Li) semiconductor detector is the excellent energy resolution which is attainable. This derives from the low energy required to produce an ion pair in the semiconductor and the fact that the ionization is detected directly as a pulse of electric current. The improvement over the NaI(Tl) crystal–photomultiplier combination, in terms of the initial number of electrons constituting the current, is of the order of 10^2. The one disadvantage stems from the small volume of the available detectors and hence the relatively low γ-ray detection efficiency.

In Fig. 2-3 are shown the γ-ray energy spectra obtained with a NaI(Tl)-based spectrometer for a sample of edible mussels collected from the vicinity of the Windscale discharge and for the detector background. The difference between these, the true spectrum of the contaminants in the sample, is given in Fig. 2-3b, where it is apparent that there are 2 major and several minor peaks. Also given is the energy calibration curve. The identification and quantification of the radionuclides contributing to this spectrum are based upon a knowledge of the complete spectra, obtained under standard assay conditions identical to the sample, for each of the radionuclides which can be expected to be present. Two examples, for ^{106}Ru and ^{137}Cs, are given in Fig. 2-3c. The problems raised by the relatively poor energy resolution of the NaI(Tl)-based spectrometer can be gauged by the presence of a partial overlap between one of the peaks from each spectrum. It is for this reason that the information contained in the complete spectrum is employed in the analysis rather than just that in the peaks corresponding to the complete absorption of the characteristic γ-rays. Qualitatively, however, it is reasonably clear that both of these radionuclides are present in the mussel sample. The quantitative analysis of the spectrum is made by a computer which sequentially combines the standard spectra in the proportions necessary to generate a best fit to the sample spectrum. The minimization of the sum of squares of the differences between the computed and measured spectra in each channel is used as the criterion of goodness of fit (SALMON, 1961, 1963). In addition to ^{106}Ru and ^{137}Cs, lesser quantities of ^{60}Co, ^{95}Zr–^{95}Nb, ^{134}Cs, and ^{144}Ce were assessed to be present in the mussel sample.

The γ-ray energy spectrum obtained with a Ge(Li) semiconductor spectrometer for a similar sample of edible mussels is given in Fig. 2-4. The much superior energy resolution of this analytical system is immediately apparent; the ^{106}Ru (0·62 MeV) and ^{137}Cs (0·66 MeV) photopeaks which partially overlapped in the NaI(Tl) spectrum are here completely resolved. It is also possible to differentiate quantitatively between the parent–daughter pair, ^{95}Zr–^{95}Nb. Because of the improved energy resolution, the individual radionuclides can be quantified simply by using the total number of counts in the photopeaks after interpolating and subtracting the background continuum.

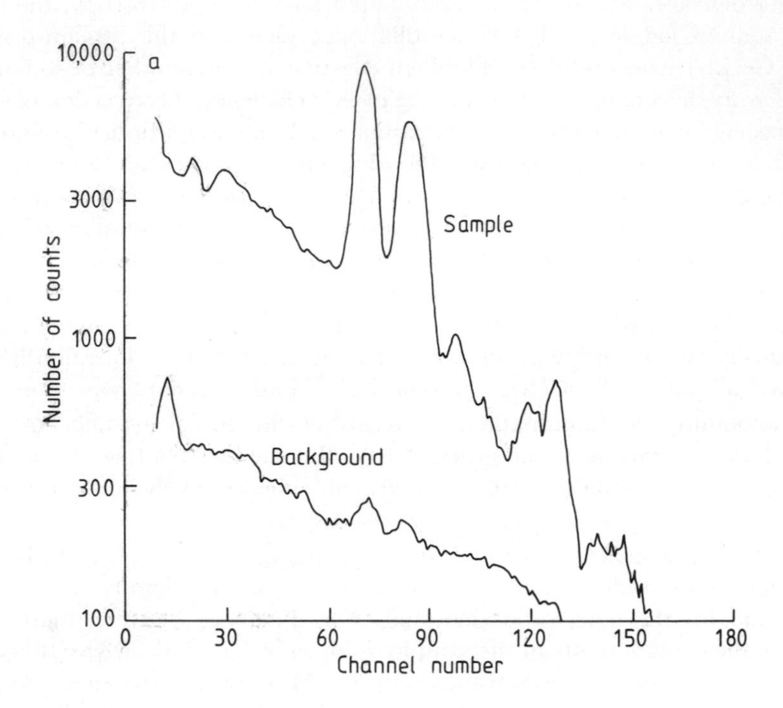
a
Number of counts
10,000
3000
1000
300
100
Sample
Background
0
30
60
90
120
150
180
Channel number

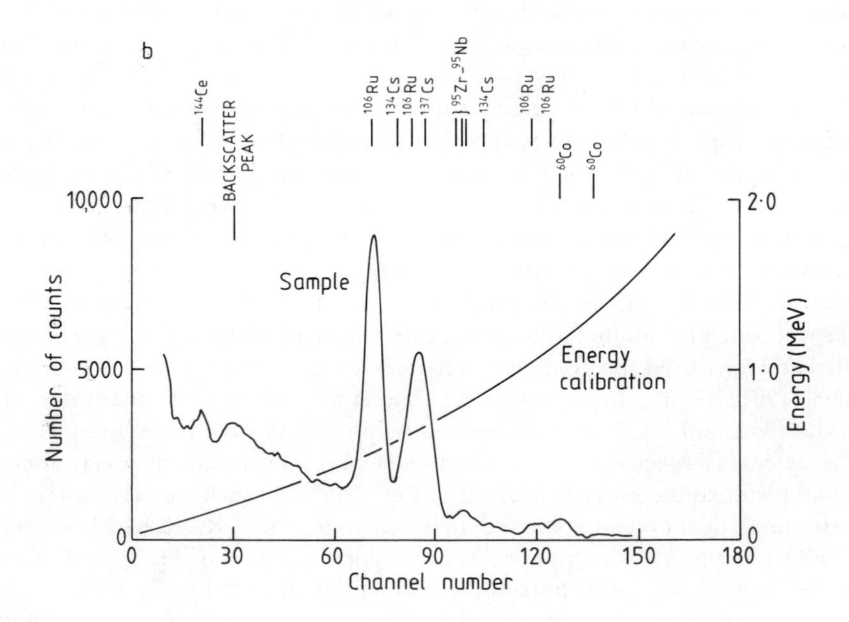
b
144Ce
BACKSCATTER PEAK
106Ru
134Cs
106Ru
137Cs
95Zr-95Nb
134Cs
106Ru
106Ru
60Co
60Co
10,000
5000
0
Number of counts
Sample
Energy calibration
2·0
1·0
0
Energy (MeV)
0
30
60
90
120
150
180
Channel number

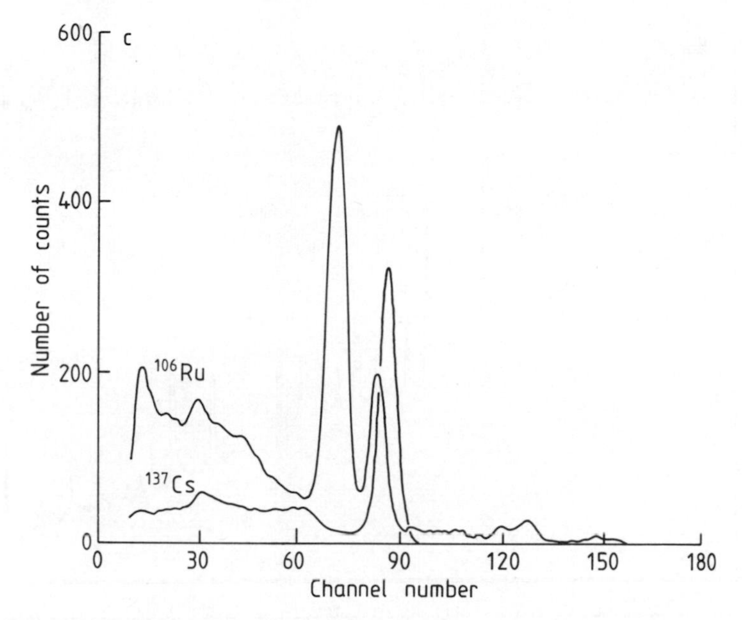

Fig. 2-3: Gamma-ray energy spectra. (a) Obtained with a NaI(Tl) spectrometer for mussels, *Mytilus edulis*, sampled from the Windscale pipeline, together with a background spectrum; (b) spectrum with background subtracted plus an energy calibration and peak identification; (c) standard γ-ray energy spectra for ^{106}Ru and ^{137}Cs. Ministry of Agriculture, Fisheries and Food, Directorate of Fisheries Research, unpublished)

The discussion of measurement techniques has necessarily been rather brief, but further information on analytical schemes, with particular emphasis on marine samples, is available (IAEA, 1970, 1975; DUTTON and co-authors, 1974).

(d) Interaction of Radiation with Living Tissue

The primary products of the absorption of radiation in living tissue are, as discussed in the preceding Section (b), excited and ionized atoms and molecules. In the complex biochemical environments which exist in tissues and, more importantly, within cells, it is reasonable to suppose that this absorbed energy will not simply be degraded to thermal vibration, but that some proportion will initiate chemical change. To a first approximation it can be assumed that all the excitation and ionization is generated in water molecules since these constitute the greater part of most living tissues. Thus:

$$H_2O + \text{energy} \rightarrow H_2O^+ + e^-$$

The ionized water molecule is unstable and breaks up to give a hydrogen ion and a hydroxyl radical, i.e.:

$$H_2O^+ \rightarrow H^+ + OH^{\cdot}$$

The free electron produced by the initial ionization is rapidly hydrated:

$$e^- + H_2O^-$$

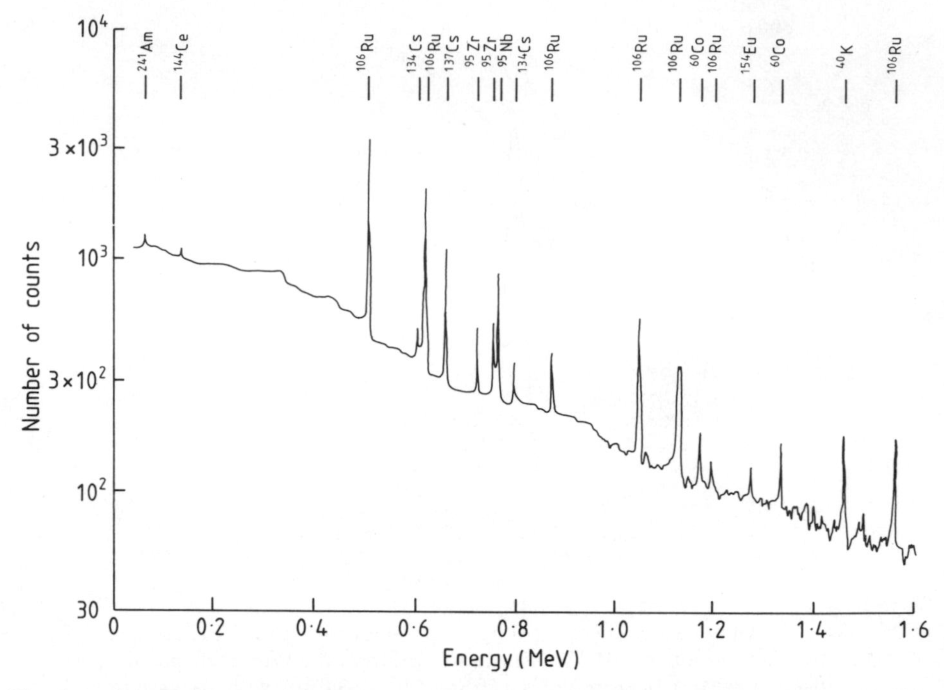

Fig. 2-4: Gamma-ray energy spectrum obtained with a Ge(Li) spectrometer for mussels, *Mytilus edulis*, sampled at Windscale, and peak identification. Ministry of Agriculture, Fisheries and Food, Directorate of Fisheries Research, unpublished)

giving a negative ion which breaks up to produce a hydrogen radical and a hydroxyl ion:

$$H_2O^- \rightarrow OH^- + H^{\cdot}$$

In oxygenated systems, which is effectively the case in most tissues, the following reaction produces a hydroperoxy radical:

$$H^{\cdot} + O_2 \rightarrow HO_2^{\cdot}$$

The hydrogen and hydroxyl ions produced make a negligible contribution to the respective concentrations in the cell. Since the ionization potential of water is 12·6 eV compared with the 30 to 35 eV (depending on the incident particle type and energy) required to produce an ion pair in water, several excited molecules must be generated for each one ionized. At the higher levels of excitation the molecule may become disrupted:

$$\begin{aligned} H_2O + \text{energy} &\rightarrow H_2O^{*} \\ &\rightarrow H^{\cdot} + OH^{\cdot} \end{aligned}$$

These free hydroxyl and hydroperoxy radicals can interact with, and oxidize, biological molecules generating an initial chemical lesion. The severity of the eventual biological effect arising from the chemical lesion depends upon the complex and competing interactions between the normal repair mechanisms and the development of the damage via the metabolic processes within the cell. A basic problem in radiobiology concerns the mechanisms whereby the relatively few initial ionization and excitation events ($\approx 10^6$), induced in a cell by a lethal radiation dose, can be amplified to cause such a catastrophic effect (BACQ and ALEXANDER, 1961; ELKIND and WHITMORE, 1967). Not surprisingly, progress in elucidating these mechanisms at the cellular level has been rather slow. However, the connection has been firmly established between the end biological effect which can be readily quantified, e.g. organism or cell mortality, chromosome damage, mutations, etc., and the quantity of energy absorbed in the biological material. Thus it is clear that radiation dosimetry, including the distribution of energy deposition in time and space, is of fundamental importance to the assessment of radiation hazards to either human populations or the natural environment.

(e) Radiation Dosimetry

A direct measurement of the radiation energy absorbed within a biological entity is impossible and all methods of radiation dosimetry are more or less indirect. In the early days of the medical uses of radiation the primary requirement was for a means of controlling exposure from penetrating X-rays and γ-rays (from ^{226}Ra and its daughter nuclides), and a system evolved in which radiation fields could be quantified by means of a measurement of the ionization produced in air. This approach was formalized and standardized at the second International Congress of Radiology held in Stockholm in 1928 with the definition of the unit of X-ray dose, the roentgen (R), as follows (SPIERS, 1956; p. 7):

> 'The roentgen is the quantity of X-radiation which, when the secondary electrons are fully utilized and the wall effect of the chamber is avoided, produces in 1 cm^3 of atmospheric air at 0 °C and 76 cm of mercury pressure such a degree of conductivity that one electro-static unit of charge is measured at saturation current.'

This definition, although subsequently extended to include γ-radiation, has remained essentially unchanged to the present. The quantity of which the roentgen is the unit is called exposure.

In addition to indicating a means of measurement, the definition also implies an energy absorption per unit mass of air (and air alone). Since the energy deposition is due to the ionization produced by the secondary electrons, and since the energy required to generate an ion pair in air is nearly independent of electron energy except at very low energies, the absorbed energy per roentgen is effectively independent of the X- or γ-ray energy and is currently accepted to be 86·9 ergs g^{-1} ($8{\cdot}69 \times 10^{-6}$ J g^{-1}) (ATTIX and ROESCH, 1968). It must be emphasized that this is a derived quantity dependent upon a knowledge of the energy required to produce 1 ion pair in air, and would, therefore, alter as the value for this parameter, presently 33·7 eV/ion pair, is refined.

It has already been noted in the preceding Section (b) that the absorption of energy from an X- or γ-ray field arises from energy-dependent interactions between the radiation and orbital electrons and that the relative probability of the interactions varies with atomic number. Therefore, it is to be expected that a radiation field yielding an exposure of 1 roentgen, i.e. 86·9 ergs g^{-1} air, would not necessarily deposit the same quantity of energy g^{-1} of another material placed at the same point in the radiation field. In soft tissue, for example, the energy deposition would be 96 erg g^{-1} at 0·3 MeV photon energy and would show some variation with photon energy, while in bone the energy deposition would be increased due to the higher effective atomic number (ATTIX and ROESCH,

1968). Thus the quantification of a radiation field in roentgens provides only an indirect indication of the energy absorbed in a biological target. The increased use of radionuclides, fissile materials, and particle accelerators increased the possibilities for exposure to particulate radiations which produce similar biological effects due to ionization but for which the roentgen unit, by definition, is inapplicable. To overcome these difficulties and in recognition of the significant correlation between absorbed energy and resultant effect a new quantity, absorbed dose, to be expressed in units of rads, was established. One rad was defined as the absorbed dose of any ionizing radiation which was accompanied by the release by ionizing particles of 100 ergs of energy g^{-1} of absorbing material at the point of interest. In the International System of Units the new unit of radiation absorbed dose is the Gray (Gy) which corresponds to an energy absorption of 1 J kg^{-1}; thus 1 Gy is the equivalent of 100 rad.

Further investigation of the biological effects of various radiation types (or qualities) demonstrated that a given effect resulted from different absorbed doses of the individual radiation types; that is, there is a difference in biological effectiveness. This led to the introduction of another unit, the rem, to quantify the biologically effective dose which, for a given radiation type, is related to the corresponding absorbed dose by a factor called the relative biological effectiveness (RBE) as follows:

$$\text{Biologically effective dose (rems)} = \text{Absorbed dose (rads)} \times \text{RBE}$$

For a given radiation type, the RBE is the inverse ratio of the absorbed dose in rads required to produce a given effect to the absorbed dose of 250 keV X-rays (the reference radiation type) required to produce an identical effect. The value obtained for the RBE depends primarily on the nature of the radiation being investigated, e.g. α-particles, β-particles, neutrons, protons, etc. but it is also modified by such secondary factors as the radiation energy, dose rate, and the nature of the biological effect under consideration.

Due to these complexities, the biologically effective dose and RBE have found their primary application in radiobiological studies. However, in radiological protection it is also necessary to quantify the radiation dose in a manner which is directly related to the potential hazard. It had been noted that the RBE value was partially correlated with the density of ionization generated by the particles depositing the energy, that is, the quantity designated as linear energy transfer (LET) and measured in keV (energy) per micron (μm) of track length. Hence the RBE was replaced by the quality factor (QF), and appropriate values for ranges of LET were arbitrarily chosen from the literature on the differential effects of radiation on small mammals (ICRP, 1977). Thus, in radiological protection, the biologically effective dose, renamed the dose equivalent, became:

$$\text{Dose equivalent (rems)} = \text{Absorbed dose (rads)} \times \text{QF}.$$

In the International System of Units the new unit of dose equivalent is the Sievert (Sv) and is equivalent to 100 rems.

At this point a dilemma arises. The biologically effective dose and RBE have been essentially restricted for use in laboratory studies of the biological effects of radiations and quality factors have, strictly, only been defined for human radiation protection. In any assessment of the effects of radiation in the natural environment however, it is necessary to take account of the differential responses to equal absorbed doses of different radiations, that is, a factor equivalent to RBE or QF is required to modify either measured or calculated absorbed doses in rads (Gy). A very limited amount of data have been obtained in laboratory experiments concerning the responses of aquatic organisms to different radiations, but the few RBE values determined suggest that the differential responses are similar to those observed in mammals (HYODO-TAGUCHI and co-authors, 1973; HYODO-TAGUCHI and MARUYAMA, 1977; EGAMI and IJIRI, 1979). Therefore, it seems reasonable to employ the QF values (derived from RBE values) recommended for human exposures for estimating the biologically effective doses in aquatic environments while recognizing that this procedure is open to argument. Similarly, the term 'biologically effective dose' will be employed here since it describes precisely the quantity in question and there is no good reason for adding further to an already large terminology.

In respect of environmental dosimetry it is an estimation of the biologically effective dose from α-particles which is of importance. The linear energy transfer varies with the α-particle energy along its track and an accurate estimate of the dose requires a knowledge of the α-particle energy spectrum incident on the target volume. This requirement implies a detailed knowledge of the spatial distribution of the α-active radionuclides relative to the target—information which is generally unavailable. In such a situation it has been conservatively recommended that a QF of 20 is applicable (ICRP, 1977), and this will be adopted as appropriate in the remainder of this review. Also it will be assumed that the QF for γ-rays and β-particles is unity.

The means used to determine the absorbed dose or dose rate depends on many factors including radiation type, target, and dose rate. For γ-radiation and, to a much lesser extent, for β-radiation from external sources, instrumental methods can be employed. Generally, these are based on a radiation detector whose energy response has been modified to correspond either to that of air if it is to measure the exposure in an X- or γ-ray field or the absorbed dose in air in a mixed photon and β-ray field, or to that of tissue if it is to indicate the absorbed dose in tissue. The radiation detector may be an ion chamber, Geiger counter or scintillation crystal together with the associated electronics. A more recent development which has proved to be very useful in determining the integrated absorbed doses to both humans and other organisms from γ- and β-radiation in contaminated environments is the thermoluminescent dosimeter. Lithium fluoride is the thermoluminescent material of choice since it has good sensitivity, the stored signal shows very little fading with time and, most significantly, the energy response to γ-radiation is very close to that of soft tissue. The lithium fluoride may be used in the form of micro-crystals, either loose or embedded in a teflon matrix, or as large single crystals. The size and geometry of the dosimeter can be adapted to each particular application. A small proportion of the electrons released from valence states in the crystals as a result of the absorption of ionizing radiation, become trapped in higher energy states from which a direct return to the ground state has a very low probability. These electrons may subsequently be freed from the traps by heating the crystals and they then return to the ground state with the emission of light. The total light output is found to be proportional to dose over a wide range of doses. The low penetrating power of α-particles means that the dose from external sources is only of significance for tissues very close (<100 μm) to the body surface and that the relative geometry of the source and target are very important. Thus the α-dose rate is usually calculated from a measurement of the α-particle flux at the source surface together with a known or assumed α-particle energy distribution (for a full discussion of dose measurement the interested reader is referred to ATTIX and ROESCH, 1966 and ATTIX and TOCHILIN, 1969).

For internal sources the estimates of the absorbed dose rate must be calculated from measurements or estimates of the tissue concentrations and distributions of the radionuclides. These methods will be discussed more fully in later sections.

(3) Radiological Protection

The existence of a hazard associated with ionizing radiation was appreciated very soon after the discovery of X-rays and radioactivity through the occurrence of radiation burns. Since radiation use promised undoubted benefits in the medical field, the problem of how the potential could be safely realized became an important topic for discussion and investigation. The Second International Congress of Radiology, in addition to defining the roentgen unit, also made recommendations for protective measures; specifically, a tolerance dose of 1 R/wk was proposed to prevent the occurrence of radiation-induced erythema in radiological personnel. This congress further saw the founding of the International X-ray and Radium Protection Commission to provide a forum for work on the problem of radiation safety and to make independent recommendations on safe practices and procedures. This responsibility continued until 1950 when, in recognition of the extension of concern beyond X-rays and radium as a result of advances in nuclear physics, the Commission was reconstituted as the International Commission on Radiological Protection (ICRP) (SIEVERT, 1959). The ICRP has become the primary source of independent advice in the field of radiological protection and has been responsible for the continuing evolution of the underlying philosophy. From time to time it has published recommendations for the practical application of this philosophy to the protection of radiation workers and the general public (ICRP, 1955, 1959, 1960, 1964, 1966a, 1973, 1977). While the recommendations have no legal standing, they have been widely accepted and adapted as a basis for national legislation and codes of practice. In particular, the recommendations published in 1966 (ICRP, 1966a), together with earlier supporting data (ICRP, 1960, 1964), have found application in the development of techniques for the control of human exposure to environmental radioactivity.

(a) Basic Philosophy

The need for a balanced response to the presence of a radiation hazard in terms of the safety measures deployed has continued to be a cornerstone of radiological protection philosophy. It is explicitly recognized that the exploitation of radiation, radioactivity and nuclear energy entails some risk and that a quantification of the risk permits a rational means of resolving the problem of defining acceptable degrees of radiation exposure. The early tolerance dose was based on a single, acute biological response (i.e. skin erythema) having a more or less well-defined threshold dose beyond which the severity of the effect increased with dose in all individuals. It was soon realized that the concept of a dose below which there were no adverse effects was not supported by the evidence derived from studies of either exposed humans or experimental animals. In particular, the incidence of malignant disease, often many years after exposure, and the incidence of hereditary damage, apparently depended on the magnitude of exposure although the severity of the effect in a given individual was independent of exposure. The Commission has, therefore, adopted the conservative assumption that any exposure to radiation carries an inherent risk of the development of cancer and the induction of genetic damage expressible in subsequent generations. It has also assumed that the magnitude of the risk is linearly related to the total dose accumulated by the individual. The objectives of radiological protection are, therefore, to prevent acute radiation effects and to limit the risk of long-term damaging effects to a reasonable level while not curtailing those activities giving rise to radiation exposure and from which benefit may be derived.

Three factors are involved in an attempt to assess the level of risk which may be regarded as acceptable; (i) the relationship between radiation dose and risk; (ii) the perceived cost of the risk either to the individual or to society; and (iii) the value of the activity giving rise to the radiation exposure. None of these factors could be evaluated with any precision and the Commission accepted, therefore, that a substantial element of judgement would be involved in determining an acceptable degree of risk and setting the corresponding maximum permissible dose. Nevertheless, in view of the requirement for guidance, and on the basis of long experience with radiation and radiation effects in humans and animals, the Commission recommended maximum permissible doses which it considered would achieve the objectives of radiological protection. In the event, the Commission considered that the maximum permissible doses recommended for occupational exposure would not result in hazards exceeding those accepted in other occupations with a high standard of safety. It was also considered that the dose limits proposed for the general public would entail risks of the same order as those already regularly accepted in everyday life. The dose limits for individual members of the general public are given in Table 2-1 where it can be seen that the known variation in radiation sensitivity of the various organs and limited parts of the body is taken into account. The radiation exposure from natural background (0·1 to 0·2 rem a^{-1} (0·001 to 0·002 Sv a^{-1}) and that received, as a patient, from medical treatment are specifically excluded from the dose limits and the contributions from all other sources, both internal and external to the body, are additive within the limits.

In the case of internal exposure it is not possible to apply the dose limits directly, and the problem of assessing the exposure from internal sources of radiation is considerably

Table 2-1

Dose limits for the general public as recommended by the International Commission on Radiological Protection (ICRP, 1966a; reproduced by permission of the Pergamon Press)

Targets	Organ or tissue	Dose limit rem a^{-1} (Sv a^{-1})
Individuals	Gonads, red bone marrow, and in the case of uniform irradiation, the whole body	0·5 (0·005)
	Skin, bone	3·0 (0·03)
	Thyroid, children up to 16 a of age	1·5(0·015)
	Thyroid, persons over 16 a of age	3·0 (0·03)
	Hands and forearms, feet and ankles	7·5 (0·075)
	Other single organs	1·5 (0·015)
Populations	Gonads	5 rem (0·05 Sv) per person in 30 a, averaged over the whole population

more complex than for external exposure. Since each element is metabolized differently within the body and since the radioisotopes of an element generally emit radiations of different quality and energy and decay with different half-lives, the consequences of intake must be considered separately for each radionuclide. However, the primary criterion for protection remains an appropriate limit on the radiation exposure of either specific organs or the whole body, and this limit can be related, by means of a dosimetry model, to an organ or body burden of each radionuclide. For many elements, normal metabolism maintains a balance between intake and excretion, and there is an equilibrium body burden which may be more or less uniformly distributed or may tend to be concentrated in a specific organ or tissue. The processes involved in maintaining the equilibrium body burden are very complex and can depend on the chemical form and route of intake of the element and the age, sex, and state of health of the exposed subject. To reduce the problem to manageable proportions the ICRP introduced the concept of a 'standard man' with a defined morphology, elemental composition, and metabolism representative of the average individual. In addition, inhalation and ingestion (in food or water) are the only intake routes considered, and the question of chemical form has been reduced to an assessment of the fate of either soluble or insoluble compounds. The turnover of an element in the body or organ has been assumed to be governed by a simple exponential law, i.e. the loss per unit time is proportional to the body or organ burden; thus

$$\frac{dB}{dt} = I - \lambda_B B$$

where B is the body burden; I the intake of the element per unit time, and λ_B the elimination rate per unit time.

Integrating this expression and applying the boundary condition that $B = 0$ when $t = 0$ gives:

$$B = \frac{I}{\lambda_B}(1 - e^{-\lambda_B t})$$

For a radionuclide, which is physically decaying as well as being eliminated by metabolic processes, the expressions become:

$$\frac{dB}{dt} = I - (\lambda_B + \lambda_P)t$$

and

$$B = \frac{I}{(\lambda_B + \lambda_P)}(1 - e^{-(\lambda_B+\lambda_P)t})$$

where λ_P is the radioactive decay constant, and $(\lambda_B + \lambda_P)$ the effective elimination rate corresponding to an effective clearance half-time of $T_B T_P/(T_B + T_P)$. Thus a particular equilibrium body or organ burden (i.e. that giving the maximum permissible dose to the whole body or the critical organ) can be related to an intake of the radionuclide per unit time if the elimination rate (λ_B) is known. Hence, using the fixed respiration and drinking rates of 'standard man', maximum permissible concentrations in air and drinking water can be defined. The limiting concentrations in drinking water can be simply transformed into corresponding limits in individual food items by determining the appropriate consumption rates. These limiting concentrations in environmental materials are known as 'derived working limits'. Table 2-2 presents a selection of relevant data for 3 radionuclides which are of environmental significance.

Table 2-2

Maximum permissible concentrations of radionuclides in air and water and maximum permissible body burden for members of the general public (ICRP, 1960; reproduced by permission of the Pergamon Press)

Radionuclide	Physical half-life (d)	Chemical state	Critical organ	Biological half-life (d)	Effective half-life (d)	Maximum permissible body burden μCi (Bq)	Fraction of maximum permissible body burden in the critical organ	Maximum permissible concentration	
								Water μCi cm^{-3} (Bq cm^{-3})	Air μCi cm^{-3} (Bq cm^{-3})
Ruthenium-106	365	Soluble	Lower large intestine	0·75*				10^{-5} (0·37)	3×10^{-9} ($1·1 \times 10^{-4}$)
		Insoluble	Lung	120**					2×10^{-10} ($7·4 \times 10^{-6}$)
			Lower large intestine	0·75*				10^{-5} (0·37)	2×10^{-9} ($7·4 \times 10^{-5}$)
Caesium-137	$1·1 \times 10^{4}$	Soluble	Whole body	70	70	3·0 ($1·1 \times 10^{5}$)	1·0	2×10^{-5} (0·74)	2×10^{-9} ($7·4 \times 10^{-5}$)
		Insoluble	Lung	120**					5×10^{-10} ($1·9 \times 10^{-5}$)
			Lower large intestine	0·75*				4×10^{-5} (1·5)	8×10^{-9} ($3·0 \times 10^{-4}$)
Plutonium-239	$8·9 \times 10^{6}$	Soluble	Bone	$7·3 \times 10^{4}$	$7·2 \times 10^{4}$	4×10^{-3} (150)	0·9	5×10^{-6} (0·19)	6×10^{-14} ($2·2 \times 10^{-9}$)
		Insoluble	Lung	120**					10^{-12} ($3·7 \times 10^{-8}$)
			Lower large intestine	0·75*				3×10^{-5} (1·1)	5×10^{-9} ($1·9 \times 10^{-4}$)

*The lower large intestine does not accumulate radioactivity; it is irradiated by radioactivity passing through from food and water. The transit time is assumed to be 18 h. An effective biological half-life cannot be calculated and the body burden is equal to the daily intake.
**For insoluble particulate matter inhaled, it is assumed that 25% is immediately exhaled; 62·5% is eliminated from the lungs and swallowed within 24 h; the remaining 12·5% is cleared from the lungs into body fluids with a half-life of 120 d.
No data for the maximum permissible body burdens are given.

The assumptions concerning the dose–response relationship underlying the recommended maximum permissible doses or dose limits have 2 important consequences. Since any dose, however small, is considered to have an associated finite risk, the ICRP recommends that all unnecessary radiation exposure be avoided and that all doses be kept as far below the limits as is reasonably achievable, with due account being taken of risks, benefits, and costs. In addition, any estimate of the total radiation hazard requires, in principle, the integration of dose over the whole of the exposed population to give the collective dose in man-rem (man-sievert). Following the recommendations of the Commission, such an estimate would be required to ensure that the genetic dose limit was respected. It was also indicated that the estimate would be one of the factors necessary to make a cost–benefit analysis of any procedure resulting in radiation exposure.

The possibility of making a numerical analysis of the risks, costs, and benefits of procedures involving radiation exposure was a recurring theme in the recommendations published by the ICRP in 1966 (ICRP, 1966a). It is an attractive objective since, in principle, the cost of reducing the risk, at all levels of risk, could be determined and balanced against the benefits to be gained. This would allow an appropriate, acceptable level of risk to be set for each particular circumstance. However, it remained a possibility since the Commission conceded that the data base and methodology required were insufficiently developed. Therefore, the recommended dose limits retained their primary importance although with the added riders given above.

In the latest set of recommendations (ICRP, 1977) the emphasis has changed completely. The Commission now recommends a system of dose limitation such that: (i) no practice shall be adopted unless its introduction produces a positive net benefit; (ii) all exposures shall be kept as low as reasonably achievable, economic and social factors being taken into account; (iii) the dose equivalent to individuals shall not exceed the limits recommended for the appropriate circumstances by the Commission. The first 2 recommendations (equivalent to the riders given above) have assumed primary importance and the dose limits now simply represent the boundary beyond which the implied risks have been judged to be unacceptable. It is evident that risk–benefit and cost–benefit analyses will be required to implement these recommendations and that appropriate analytical techniques and supporting data need to be developed. For example, in addition to the collective dose equivalent discussed above, an assessment of the total risk arising from a particular activity requires the calculation of the collective dose equivalent commitment, i.e. the dose equivalent summed over the predicted lifespan of each individual in the present and future exposed population. This has particular significance for radioactivity released to the environment since, among other imponderables, it presupposes a knowledge of the future behaviour of the long-lived radionuclides. It is also clear, however, that the day-to-day assessment and monitoring of activities resulting in radiation exposure will continue to be carried out within the context of the dose limits.

The Commission has also changed the principle used to set the dose limits for certain specific organs and tissues. Previously (ICRP, 1966a) the dose limit for uniform irradiation of the whole body was equal to that recommended for certain organs or tissues (see Table 2.1) a situation which, if the intention was to operate at a single level of acceptable risk, was clearly inconsistent. For non-uniform irradiation, it is now recommended that the dose equivalents for each exposed organ or tissue should be multiplied by appropriate risk-dependent weighting factors and then summed to give an effective dose equivalent. The limit on the effective dose equivalent is thus the recommended dose equivalent limit for uniform irradiation of the whole body. Where a radionuclide deposited in the body results in significant irradiation of 2 or more organs or tissues, the previous value for the maximum permissible concentration in air or water based on a single critical organ is likely to be reduced. The detailed consequences of this change have only recently become available (ICRP, 1979a,b) and have yet to be formally incorporated into operational procedures.

The earlier recommendations of the Commission (ICRP, 1960, 1964, 1966a) have been used in the great majority of cases for the assessment and control of radioactive waste disposals into the marine environment and this approach will be adopted in the remainder of this discussion.

(b) Approaches for the Management of Waste Disposal

Due to the recognized hazards, waste management has become an increasingly important aspect of the expanding utilization of radioactivity and nuclear energy. In many cases it is neither practical nor economically feasible either to attempt to decontaminate the waste or to develop a system of permanent, controlled storage. These wastes have been arbitrarily designated low level, the implication being that the concentration of activity is low. However, the total volume of the waste, liquid or solid, can be large and, in consequence, substantial quantities of radionuclides can be involved. For such wastes, disposal is frequently the preferred option. In this context, disposal implies the release of the contaminated waste into the environment in such a way that control of its subsequent behaviour is relinquished.

The problem of the potential exposure of the general public to radioactive wastes disposed of into the environment is essentially that of ensuring that there is a very low probability of the relevant dose limits (Table 2-1) being exceeded. For liquid radioactive waste discharged into coastal waters or packaged solid wastes dumped into the deep ocean, the uncontrollable nature of the hydrographic, geochemical, and biological processes acting on the radionuclides after release means that there are only 2 points at which measures to limit exposures can be applied: i.e. the point of disposal and the point of exposure. In practice the former is the only viable option; restriction of access to contaminated beaches and restriction of commercial fishing operations would be generally unacceptable as routine control measures. Thus it is necessary to develop a means whereby the quantity of radioactivity to be disposed of can be related to the consequent exposure of individual members of the general public and the total population. Within this framework it would then be possible to limit the discharge so that radiation exposures in excess of the appropriate limits are avoided. Three quite different approaches have been employed to achieve this objective.

Point of Discharge Control

This method is the simplest and most direct in application. It depends merely on the setting of limits on the concentrations of the various radionuclides in an effluent at the point of discharge regardless of the aquatic environment into which the waste is to be introduced. The possible processes which may subsequently act to disperse or reconcentrate the radionuclides and lead to human exposure are only considered in a very superficial manner (PRESTON, 1969; IAEA, 1978a). Due allowance is made, however, for the relative radiotoxicities of the nuclides and the maximum permissible concentrations in the effluent at the discharge point are usually a set fraction of the corresponding maximum permissible drinking water concentrations for continuous occupational exposure recommended by ICRP (ICRP, 1960, 1964).

The apparent advantages of this system of control derive from the following factors: (i) a single set of maximum permissible effluent concentrations are specified which are applicable to all discharges; (ii) it is very simple for the plant operator to apply; and (iii) monitoring to ensure compliance is a straightforward task for the regulatory authorities. There are, none the less, significant and substantial deficiencies in the method as a rational approach to the control of radiation exposure. Reasonable certainty that a chance combination of restrictive environmental circumstances could not lead to overexposure from any discharge requires that large safety factors be applied to the general assumptions concerning environmental behaviour. Even so, it has been demonstrated that effluent concentration limits which have been adopted by regulatory bodies would not be adequate in all circumstances (PRESTON, 1969). In addition, the application of such large safety factors may well result in rather low values for the limiting concentrations in the effluent and may, therefore, entail investment in treatment plant which is unnecessary and unjustified in certain situations in relation to the reduction in the potential radiation hazard. Estimates of the actual radiation exposure to either individuals or populations, and hence assessment of the implied hazard, cannot be made for any site by this method (IAEA, 1978a). These points are of particular relevance in view of the most recent recommendations made by ICRP (ICRP, 1977). Finally, the method is inapplicable to the problem of solid waste dumping.

The Specific Activity Approach

The basis of this control procedure is the transformation of the recommended maximum permissible body or organ burden for each radionuclide into a specific activity (activity of radionuclide per gram of element) relative to the normal quantity of the corresponding stable element in the body or organ. On the assumption that environmental and biological processes do not differentially affect the different isotopes of an element at any point in the food chain, or ultimately, in the human body, these derived specific activities then become the limiting values for the environment receiving the waste (PRESTON, 1969; FOSTER and co-authors, 1971; IAEA, 1978a).

The relative uniformity of the chemical composition of sea water (Volume I: COLLIER, 1970, p. 66; KALLE, 1971, p. 684; Volume III: KINNE, 1976, p. 21) makes this approach appear attractive for the control of discharges of liquid radioactive wastes to the marine environment since, in principle, the derived specific activities combined with stable element concentrations in sea water can be used to set maximum permissible concentrations for each radionuclide in sea water. The rate of introduction of radionuclides, therefore, needs to be controlled so that these concentrations (and hence specific activities) are not exceeded, a procedure which only requires information as to the natural processes of dilution and dispersion in the receiving water mass. Data on the accumulation of radionuclides from sea water by edible marine organisms and the extent of their subsequent utilization as food are not required. For a point source discharge, a measure of judgement would be required to determine the volume of the receiving water

Table 2-3

Outline of the specific activity approach to the control of discharges of radioactive wastes (Based on PRESTON, 1969)

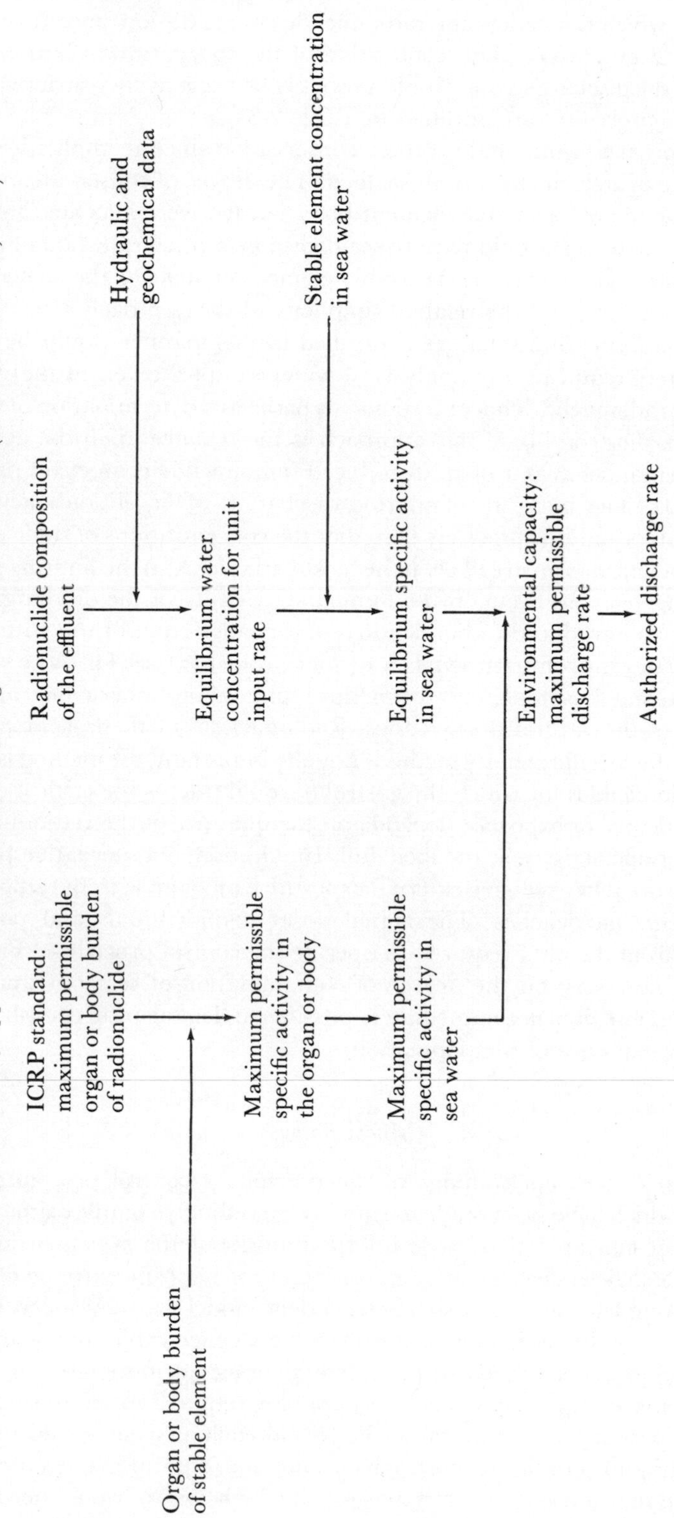

mass over which to average the radionuclide concentration since rather large gradients are likely to be present. The application of the concentration limits to the immediate vicinity of the discharge point would probably be excessively restrictive. The elements of the control procedure are outlined in Table 2-3.

The approach is inherently rather conservative since it implicitly assumes that the total intake of an element is from seafood. The degree of conservatism could be reduced if some fraction of the stable element body burden were allocated to this source. This refinement, however, would require seafood consumption rate data obtained from habits surveys and a knowledge of the stable element content of the seafoods, requirements which would confound the relative simplicity of the approach.

The monitoring of discharges controlled in this manner would be fairly straightforward, merely requiring the analysis of water samples taken in the vicinity of the discharge for radionuclide concentrations (or perhaps a determination of specific activity).

A major shortcoming of this approach is the reliance upon the assumption that the environmental behaviour of an introduced radionuclide is an exact parallel of its stable counterpart. The majority of marine discharges of liquid radioactive wastes are to coastal waters and it is precisely here that the concentrations of trace elements and their physico-chemical form are likely to be most variable. Also the limiting situation resulting in exposure tends to occur in the immediate vicinity of the discharge where it is least likely that an introduced radionuclide is in exchange equilibrium with its stable isotopes in all the relevant compartments. A further problem arises for those elements for which there is no stable isotope, e.g. technetium, plutonium, americium, and the majority of members of the natural decay series. Radionuclides of these elements cannot be controlled by the specific-activity method. Equally important, the method is not applicable for those radionuclides for which the gastrointestinal tract is the critical organ for exposure since the degree of exposure depends on the quantity of the radionuclide ingested and not the specific activity in the foodstuff. In addition, an alternative procedure is necessary to ensure that excessive external exposure cannot arise from radionuclides adsorbed to sediments on beaches. The actual doses to individuals and populations are not available from the data required to operate the control procedure. Further information would be necessary on the degree of contamination of seafoods and on consumption rates. Therefore, it is not immediately possible to demonstrate compliance with the limit on genetic exposure of the populations.

Critical Pathway Analysis

In essence, the shortcomings of the previous 2 control procedures stem from the attempts which have been made to simplify exceedingly complex situations. In each case it has been indicated that more information about the system would be required to remedy the deficiencies. Clearly, a position at the opposite extreme could be adopted in which a complete time- and space-dependent model was developed to trace the movement of each of the radionuclides through every conceivable compartment of the marine system and to assess its consequent potential for a contribution to human exposure, i.e. a total systems analysis approach. The data base required to set up such a model for each disposal operation would be rather large and each analysis would include many pathways leading to zero or, at most, trivial potential exposure. The outcome would be the identification of a few routes of exposure on which control would need to be based. This is a result which can also be obtained using another approach which occupies the middle ground between simplicity and complexity. A preliminary examination of a proposed disposal site is usually sufficient to identify the most probable combinations of radionuclides and pathways which will be important in terms of human exposure, and on which subsequent analysis should concentrate. This approach to assessment, critical pathway analysis, is outlined in Table 2-4 (ICRP, 1966b; PRESTON, 1969; FOSTER and co-authors, 1971; IAEA, 1978a).

Pathways which have been examined on occasion have included the consumption of seaweed, plankton, molluscs, crustaceans, and fish; the consumption of water obtained by desalination, and salt produced by evaporation; the inhalation of wind-suspended sediment and sea spray; and external exposure on contaminated beaches and from dredging operations, the handling of contaminated fishing gear and swimming. To determine the permissible discharge rate, the consequences of a unit input (1 Ci d^{-1}) are

Table 2-4

Outline of the critical pathway approach to the assessment of discharges of radioactive wastes (Based on PRESTON, 1969)

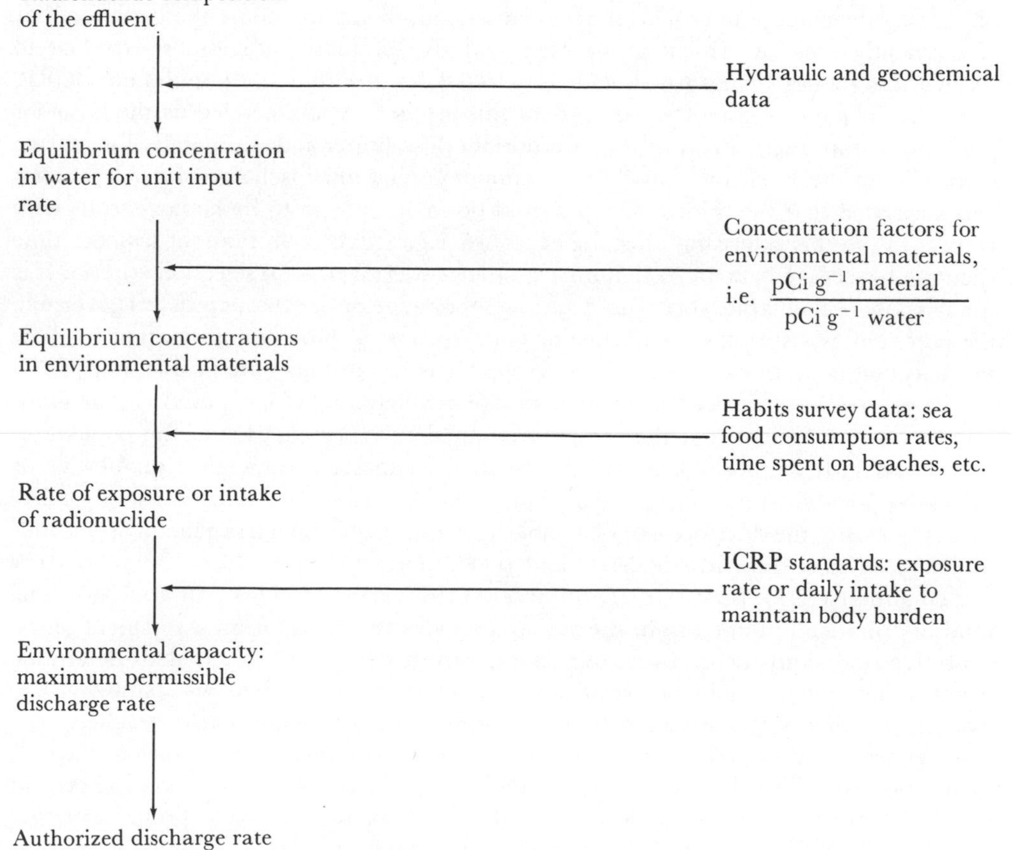

traced for each radionuclide. Site-specific hydrographic data allow the equilibrium concentrations in the receiving water mass to be predicted. These are combined with appropriate concentration factors to determine the probable concentrations of radionuclides in those components of the environment provisionally identified as being likely to engender the greatest degree of human exposure. Information concerning the consumption rates of contaminated marine products or occupancy times of contaminated beaches can then be used to calculate the daily intakes of the radionuclides or the daily exposure. The inverse ratio of these to the appropriate maximum permissible values recommended by ICRP gives an estimate of the maximum permissible daily discharge rate or, effectively, the environmental capacity to receive the waste without unacceptable consequences for human health. The major differences between this control method and the specific activity approach lie in the requirements for estimates of the degree of contamination of environmental materials and, more importantly, detailed information concerning the eating, working, and leisure habits from which exposure may ensue, in either the local or a more distant population.

Due to variations in both the environment, leading to differing degrees of contamination, and also the personal characteristics of each individual, it is to be expected that the potential exposure for a given pathway would be non-uniform throughout the population. For the same reasons it is impossible to predict the dose which might be received by each individual. The ICRP has, therefore, departed somewhat from the ideal objective of limiting the maximum dose to the individual and, to accommodate the variation, has recommended that a critical group representative of those individuals expected to receive the greatest exposure should be selected from within the population (ICRP, 1966a,b). The mean potential exposure of this group is recommended as the basis for planning within the context of the appropriate dose limits and ultimately, for marine disposals, for the determination of the maximum permissible discharge rate. Although it was suggested that the critical group should be small enough to be homogeneous with respect to the characteristics affecting exposure, e.g. age, consumption of seafood, time spent on beaches, etc, no more definitive guidance was given as to selection criteria. It is the purpose of the habits survey to determine the range of the characteristics governing the potential exposure from a particular pathway and to provide information to allow the selection of an appropriate critical group. It is usually not practicable, and by no means necessary, to survey the whole population which might be exposed via the pathway; it being sufficient that the sample examined be representative of the population and particularly of those likely to be most highly exposed. Although a qualitative or subjective identification of a critical group from habits survey data is usually fairly straightforward, the development of an objective approach which has general applicability has been more difficult (PRESTON and JEFFERIES, 1969a; PRESTON, 1971; PRESTON and co-authors, 1974; BEACH, 1975; SHEPHERD, 1975; HUNT and SHEPHERD, 1980). The total information obtained from the survey provides the basis for an assessment of the collective dose equivalent. In certain cases, care needs to be exercised to determine whether individuals could be significantly exposed via more than one pathway, e.g. fishermen eating contaminated fish and spending time on contaminated beaches.

In general, post-operational programmes of environmental monitoring have 2 main objectives. The first is to provide confirmation that the exposure via the critical pathways does not exceed the dose limits. This is normally achieved through periodic sampling and radioanalysis of the environmental materials which are directly responsible for exposure (the critical materials) and a comparison of these results with the derived working limits. In the particular case of external exposure from contaminated sediments periodic measurements of the dose rate are made. The second objective is to obtain operational data which will permit refinement of the critical pathway analyses and adjustment of the permitted discharge rates as necessary.

The limitation of human exposure is the prime objective of the 3 control procedures described. In principle, there is no reason why the limitation of the exposure of marine organisms should not be substituted as the objective provided that appropriate dose limits could be set. In practice, however, the primacy of human dose limitation is retained for the purpose of discharge control and an assessment made of the consequent exposure of, and hazard to, marine organisms at the release rates so determined (DUNSTER and co-authors, 1964).

(4) Radioactivity in the Marine Environment

Radioactivity and radiation have been pervasive factors in the terrestrial and aquatic environments since their formation, and all life has evolved in the presence of a background radiation flux of greater or lesser magnitude depending on the particular environmental niche occupied. Indeed, the earliest forms of life probably experienced higher background radiation exposures than are current today since several of the natural radionuclides have half-lives of the same order as the time over which life has been present on the earth (Table 2-5). Thus the introductions of radionuclides, either artificial, i.e. man-made, or natural, due to human activities must be assessed within the perspective of the natural background concentrations and the consequent exposure of both marine organisms and man.

(a) Concentrations of Natural Radionuclides in the Marine Environment

A summary of the concentrations of the longer-lived natural radionuclides, which have been measured in surface sea water samples from open ocean stations, is given in Table 2-5. The values quoted are broadly similar to those given in previous surveys of earlier work (MAUCHLINE and TEMPLETON, 1964; JOSEPH and co-authors, 1971). In coastal waters the thorium isotopes, in particular, can be present in concentrations greater than those shown in Table 2-5 (KOCZY and co-authors, 1957; SOMAYAJULU and GOLDBERG, 1966; MIYAKE and co-authors, 1970c) due, presumably, to an input from land runoff which is only slowly being scavenged from the water column by geochemical processes. Conversely, the concentrations of the radionuclides which are more or less conservative with sea water, e.g. potassium-40 and rubidium-87, can be lower due to the diluent action of the fresh water input. Both tritium (^{3}H) and carbon-14 are produced as a result of the interaction of high-energy cosmic rays (mainly protons with energies $\gtrsim 10^{9}$ eV) with the atmosphere. Tritium arises simply as a fragmentation product of the high-energy collisions. These collisions also produce a flux of neutrons which, after moderation, are absorbed by nitrogen nuclei in the following process:

$$^{14}N + n \rightarrow {}^{14}C + p + \text{energy}$$

The variable concentration of tritium in surface sea water arises both from its non-uniform deposition as tritiated water in rain and also from the relatively short half-life of 12·3 a which is comparable with the time scales of oceanic mixing process. The longer half-life of carbon-14 (5760 a) results in a more even distribution. Potassium-40, rubidium-87, uranium-238, uranium-235, and thorium-232 comprise a group of very long-lived radionuclides generated during the process of primordial nucleosynthesis prior to the formation of the solar system. Of this group, the uranium and thorium isotopes are the precursors of decay chains including radionuclides of the elements thorium, radium, radon, polonium, and lead which have also been measured in sea water. The different chemical properties of these elements and their interaction with marine geochemical processes result in varying degrees of radioactive disequilibrium relative to the respective parent radionuclides. The shorter-lived members of each decay chain can be assumed to be in equilibrium with the appropriate longer-lived precursor as indicated in Table 2-5.

From these data it can be seen that, among the β- and γ-active radionuclides and overall in sea water, ^{40}K is predominant, representing approximately 97% of the total activity; the α-emitting radionuclides, mainly ^{238}U and ^{234}U, make up less than 1% of the total. Since potassium is conservative in sea water and varies in direct proportion to salinity the concentration of ^{40}K would be expected to show relatively little variation throughout the ocean. For an ocean volume of $1{\cdot}37 \times 10^{21}$ l a constant concentration of 320 pCi l^{-1} represents a total activity of approximately $4{\cdot}4 \times 10^5$ MCi ($1{\cdot}6 \times 10^{22}$ Bq). For many of the other radionuclides listed in Table 2-5 there are spatial variations in concentration arising from involvement in either biological or geochemical processes, or both. For example, because the greater part of the ^{230}Th produced by the decay of ^{234}U is precipitated from the water column (activity ratio ^{230}Th/^{234}U: (0·5 to 10) $\times 10^{-4}$), the main source of ^{226}Ra is the deep ocean sediment from which the radium leaches back into the bottom water and is dispersed by the oceanic circulation (activity ratio ^{226}Ra/^{230}Th in surface water: $\approx$100) (KOCZY, 1958). The half-life of ^{226}Ra is sufficiently short that significant decay can occur before the radium-bearing bottom waters become mixed to the surface. Thus there is an increase in radium concentration with depth which may be as much as 5-fold (CHAN and co-authors, 1976). However, part of this variation is also due to the biological accumulation of radium from water in the surface layer and the sedimentation of the refractory components of dead organisms, particularly siliceous tests (KU and LIN, 1976), into deep waters where some dissolution and release of radium takes place. The total oceanic inventory of ^{226}Ra is not less than 50 MCi and may be in excess of 100 MCi ($3{\cdot}7 \times 10^{16}$ Bq).

There are relatively few data on the concentrations of natural radionuclides in the seabed, and the great majority of those available refer to measurements made in the course of geochemical and sedimentation studies in the deep ocean. In coastal and shelf waters it can be reasonably assumed that the mineral fractions of the sediment are similar to the terrestrial rocks from which, in the main, they are derived. Typical concentrations of uranium, thorium, and potassium in a number of terrestrial rocks are given in Table 2-6 (MAHDAVI, 1964; MOXHAM, 1964). The deep oceans receive very little terrigenous sediment and are characterized by very low sedimentation rates, predominantly of refractory biological detritus (e.g. skeletons of plankton and higher organisms) and material derived from geochemical processes in the water column (e.g. precipitates). HEYE (1969) has shown that the uranium and thorium contents of the sediments vary inversely with the concentration of calcium carbonate. In the uranium series, there are differing degrees of disequilibrium between the longer-lived members as a consequence of the different geochemical processes active in the water column. The most

Table 2-5

Concentration of natural radionuclides in surface seawater (Compiled from the sources listed)

Radionuclide	Half-life (y)	Daughter radionuclides assumed to be in equilibrium	Concentration $pCi\,l^{-1}$ ($mBq\,l^{-1}$)	Source
^{3}H	12·3	None	0·6–3 (22–110)	UNSCEAR (1966)
^{14}C	$5{\cdot}73 \times 10^{3}$	None	0·16–0·18 (5·9–6·7)	Broecker and co-authors (1960); UNSCEAR (1962)
^{40}K	$1{\cdot}28 \times 10^{9}$	None	320 ($1{\cdot}18 \times 10^{4}$)	
^{87}Rb	$4{\cdot}80 \times 10^{10}$	None	2·9 (107)	
^{238}U	$4{\cdot}47 \times 10^{9}$	^{234}Th, ^{234}Pa	1·2 (44)	Miyake and co-authors (1966); Miyake and co-authors (1970b)
^{234}U	$2{\cdot}45 \times 10^{5}$	None	1·3 (48)	Miyake and co-authors (1966); Miyake and co-authors (1970b)
^{230}Th	$8{\cdot}00 \times 10^{4}$	None	$[0{\cdot}6–14] \times 10^{-4}$ ($[0{\cdot}2–5{\cdot}2] \times 10^{-2}$)	Moore and Sackett (1964); Miyake and co-authors (1970c)
^{226}Ra	$1{\cdot}60 \times 10^{3}$	None	$[3{\cdot}4–8{\cdot}3] \times 10^{-2}$ (1·3–3·1)	Broecker and co-authors (1967); Szabo (1967); Ku and co-authors (1970); Broecker and co-authors (1976); Chung (1976); Ku and Lin (1976)
^{222}Rn	$1{\cdot}05 \times 10^{-2}$	^{218}Po, ^{214}Pb, ^{214}Bi, ^{214}Po	$\approx 2 \times 10^{-2}$ ($\approx 0{\cdot}7$)	Broecker and co-authors (1967)
^{210}Pb	22·3	^{210}Bi	0·01–0·14 (0·4–5·0)	Rama and co-authors (1961); Goldberg (1963); Kauranen and Miettinen (1970); Shannon and co-authors (1970); Shannon (1973); Bacon and co-authors (1976); Nozaki and co-authors (1976); Thomson and Turekian (1976); Heyraud and Cherry (1979)
^{210}Po	$3{\cdot}79 \times 10^{-1}$	None	$[0{\cdot}5–10] \times 10^{-2}$ (0·19–3·7)	Kaufman (1967); Beasley (1968); Folsom and Beasley (1968); Kauranen and Miettinen (1970); Shannon and co-authors (1970); Shannon (1973); Nozaki and co-authors (1976); Thomson and Turekian (1976); Heyraud and Cherry (1979)
^{232}Th	$1{\cdot}4 \times 10^{10}$	None	$[0{\cdot}1–7{\cdot}8] \times 10^{-4}$ ($[0{\cdot}4–29] \times 10^{-3}$)	Moore and Sackett (1964); Somayajulu and Goldberg (1966); Miyake and co-authors (1970c)
^{228}Ra	5·76	^{228}Ac	$[0{\cdot}1–10] \times 10^{-2}$ ($[0{\cdot}4–37] \times 10^{-1}$)	Kaufman and co-authors (1973)
^{228}Th	1·91	^{224}Ra, ^{220}Rn, ^{216}Po, ^{212}Pb, ^{212}Bi, ^{212}Po ^{208}Tl	$[0{\cdot}2–3{\cdot}1] \times 10^{-3}$ ($[0{\cdot}7–12] \times 10^{-2}$)	Moore and Sackett (1964); Miyake and co-authors (1970c)
^{235}U	$2{\cdot}45 \times 10^{5}$	^{231}Th	5×10^{-2} (1·9)	

Table 2-6

Concentration of natural radionuclides in beach sand, common rocks and deep ocean sediments, pCi g^{-1} (mBq g^{-1}) (Based on WOODHEAD, 1973a; IAEA, 1976; WOODHEAD and PENTREATH, 1980)

Material	U (ppm)	^{238}U	^{234}U	^{230}Th	^{226}Ra	^{235}U	Th (ppm)	^{232}Th	K (%)	^{40}K
Beach sand	3·0	1·0 (37)	Equilibrium assumed in the ^{238}U series (Beach sand to Basalt)			0·05 (1·9)	6·4	0·69 (26)	0·33	2·7 (100)
Granite	5·0	1·7 (63)				0·08 (3·0)	18	2·0 (74)	3·8	32 (1200)
Shale	3·7	1·2 (44)				0·06 (2·2)	12	1·3 (48)	1·7	14 (520)
Limestone	1·3	0·43 (16)				0·02 (0·7)	1·1	0·12 (4·4)	0·2	1·7 (63)
Sandstone	0·45	0·15 (5·6)				0·01 (0·4)	1·7	0·18 (6·7)	0·6	5·0 (190)
Basalt	0·50	0·17 (6·3)				0·01 (0·4)	2·0	0·22 (8·1)	0·5	4·2 (160)
Red clay	0·9–2·4	0·3–0·8 (11–30)	0·3–0·8 (11–30)	16–66 (590–2400)	11–47 (420–1720)	0·02–0·04 (0·7–1·5)	5·5–24	0·6–2·6 (22–96)	No data	
Globigerina ooze	0·6	0·2 (7·4)	0·2 (7·4)	4·1 (150)	2·3 (85)	0·01 (0·4)	3·3	0·4 (13)	No data	

significant is the excess of ^{230}Th (relative to the ^{234}U parent) due to the precipitation of this element from the overlying water where it is produced by the decay of ^{234}U in solution. Disequilibrium also occurs between ^{230}Th and its daughter ^{226}Ra due to the leaching of the latter element from the surface sediment. Although there is some loss of ^{222}Rn from the sediment into the bottom water to produce the known excess of radon relative to the ^{226}Ra in solution (BROECKER and co-authors, 1967; SARMIENTO and co-authors, 1976), the efflux of radon required to maintain this situation is not sufficient to disturb significantly the equilibrium between the ^{226}Ra and its daughters in the sediment. There is less information available concerning the state of equilibrium in the thorium series in deep sea sediments although circumstantial evidence suggests that equilibrium would be a reasonable assumption. The half-lives of ^{228}Ra and ^{220}Rn (5·76 a and 55·6 s respectively) are much shorter than those of their counterparts in the uranium series (^{226}Ra, 1600 a and ^{222}Rn, 3·82 d) and, therefore, any loss of the former from the sediment, together with the consequential disequilibrium, must be correspondingly less. In addition, any loss of ^{228}Ra could be balanced to some extent by the precipitation of the ^{228}Th granddaughter produced from this ^{228}Ra in solution. Data derived from deep-sea cores are given in Table 2-6 (KU, 1965).

The concentrations of natural radionuclides which have been derived for, or measured in phytoplankton, zooplankton, molluscs, crustaceans, and fish are given in Table 2-7. This grouping is necessary to provide a concise summary of the available data which refer to a wide variety of organisms; there being no instance in which there is a complete set of data for a single species or even, as is apparent, for the groups of organisms. The absence of a reference in Table 2-7 indicates that the values have been estimated.

(b) Concentrations of Artificial Radionuclides in the Marine Environment from Weapons Tests

It has been estimated that there have been 905 nuclear explosions with a cumulative energy yield of some 366 megatons of TNT equivalent in the period between the testing of the first nuclear weapon on 16 July 1945 and 30 June 1978 (CARTER and MOGHISSI, 1977; CARTER, 1979). Of this total, 390 ($\approx$288 megatons) have been atmospheric explosions and 6 devices (90 kilotons) have been detonated under water, and a majority took place at sites in the northern hemisphere prior to the partial implementation of the Limited Test Ban Treaty which came into force on 10 October 1963. The fallout pattern from an explosion depends on the yield of the device, the time of year of the explosion, the latitude of the site and whether or not the fireball intersects the surface of the earth (surface and air bursts respectively) (PETERSON, 1970). Surface bursts increase the amount of local (hours–1 day) and intermediate ($\approx$ days–$\approx$ months) fallout since an increased fraction of the radioactivity produced is retained in the troposphere in association with particulate debris. World-wide fallout derives from that proportion of the debris which is injected into the stratosphere and occurs over a period of years. The main transfer of debris from the stratosphere to the troposphere occurs in winter through the tropopause at temperate latitudes and results in the observed spring maximum deposition rate in middle latitudes. Since most of the atmospheric tests have been in the northern hemisphere the peak cumulative deposition has been in the latitude interval 30 to 60° N. However, there is some trans-equatorial air movement from the summer to winter hemisphere and a similar, though smaller, peak in deposition has been observed

Table 2-7

Concentrations of natural radionuclides in marine organisms, pCi g^{-1} (mBq g^{-1}) wet (Compiled from the sources indicated)
Part 1

Radio-nuclide	Phytoplankton	Source	Zooplankton	Source	Molluscs	Source
^{3}H	[0·5–2·7] × 10^{-3} ([0·19–1·0] × 10^{-1})	—	[0·5–2·7] × 10^{-3} ([0·19–1·0] × 10^{-1})	—	[0·5–2·7] × 10^{-3} ([0·19–1·0] × 10^{-1}	—
^{14}C	0·24–0·27 (8·9–10)	—	0·24–0·27 (8·9–10)	—	0·40–0·45 (15–17)	—
^{40}K	2·5 (93)	—	2·5 (93)	—	2·9 (110)	—
^{87}Rb	ND	—	ND	—	5 × 10^{-2} (1·9)	FUKAI and MEINKE (1959)
^{238}U	[4–5] × 10^{-2} (1·5–1·9)	MIYAKE and co-authors (1970b)	[1–2] × 10^{-2} (0·37–0·74)	MIYAKE and co-authors (1970b)	ND	—
^{234}U	[4–5] × 10^{-2} (1·5–1·9)	MIYAKE and co-authors (1970b)	[1–2] × 10^{-2} (0·37–0·74)	MIYAKE and co-authors (1970b)	ND	—
^{226}Ra	2 × 10^{-2} (0·74)	SZABO (1967)	2 × 10^{-2} (0·74)	SZABO (1967)	ND	—
^{210}Pb	0·1–0·7 (3·7–26)	SHANNON and co-authors (1970)	[1·0–25] × 10^{-2} (0·37–9·3)	BEASLEY (1968); HOLTZMAN (1969); SHANNON and co-authors (1970); HEYRAUD and CHERRY (1979)	[5–10] × 10^{-3} (0·19–0·37)	KAURANEN and MIETTINEN (1970)
^{210}Po	0·4–1·7 (15–63)	SHANNON and CHERRY (1967); SHANNON and co-authors (1970)	[5–230] × 10^{-2} (1·9–84)	SHANNON and CHERRY (1967); BEASLEY (1968); HOLTZMAN (1969); SHANNON and co-authors (1970); HEYRAUD and CHERRY (1979)	0·33–1·4 (12–52)	KAURANEN and MIETTINEN (1970); HEYRAUD and CHERRY (1979)
^{228}Th	[7–54] × 10^{-3} (0·26–2·0)	CHERRY and co-authors (1969)	[2–22] × 10^{-3} ([7·4–81] × 10^{-2})	CHERRY and co-authors (1969)	ND	—
^{235}U	2 × 10^{-2} (7·4 × 10^{-1})	—	6 × 10^{-4} (2·2 × 10^{-2})	—	ND	—

Table 2-7—*contd*

Radionuclide	Crustaceans	Source	Fish		Source
^{3}H	$[0{\cdot}5–2{\cdot}7] \times 10^{-3}$ ($[0{\cdot}19–1{\cdot}0] \times 10^{-1}$)	—		$[0{\cdot}5–2{\cdot}7] \times 10^{-3}$ ($[0{\cdot}19–1{\cdot}0] \times 10^{-1}$)	—
^{14}C	0·48–0·54 (18–20)	—		0·32–0·36 (12–13)	—
^{40}K	2·5 (93)	—		2·5 (93)	—
^{87}Rb	4×10^{-2} (1·5)	—		$2{\cdot}5 \times 10^{-2}$ (0·93)	THOMPSON and co-authors (1972)
^{238}U	ND	—	Muscle	$[0{\cdot}07–30] \times 10^{-3}$ (0·003–1·1)	ATEN and co-authors (1961); HAMILTON (1972)
			Bone	$[1{\cdot}6–80] \times 10^{-3}$ (0·06–3·0)	PENTREATH and co-authors (1980b)
^{234}U	ND	—	Muscle	$[0{\cdot}08–35] \times 10^{-3}$ (0·003–1·3)	ATEN and co-authors (1961); HAMILTON (1972)
			Bone	$[1{\cdot}8–87] \times 10^{-3}$ (0·07–3·2)	PENTREATH and co-authors (1980b)
^{226}Ra	ND	—	Muscle	$[2–51] \times 10^{-4}$ ($[0{\cdot}74–19] \times 10^{-2}$)	HOLTZMAN (1969)
^{210}Pb	$[4–7] \times 10^{-2}$ (1·5–2·6)	BEASLEY (1968); KAURANEN and MIETTINEN (1970)	Muscle	$[2–23] \times 10^{-4}$ ($[0{\cdot}7–8{\cdot}5] \times 10^{-2}$)	BEASLEY (1968); HOLTZMAN (1969); KAURANEN and MIETTINEN (1970)
			Stomach	$[1{\cdot}7–85] \times 10^{-2}$ (0·6–31)	BEASLEY (1968)

Table 2-7—*contd*

Radionuclide	Crustaceans	Source	Fish		Source
			Liver	$[1·1–2·4] \times 10^{-2}$ (0·41–0·89)	Beasley (1968); Kauranen and Miettinen (1970)
			Bone	$[9–130] \times 10^{-3}$ (0·33–4·8)	Beasley (1968); Kauranen and Miettinen (1970)
^{210}Po Whole animal	0·23–6·3 (8·5–230)	Beasley (1968); Kauranen and Miettinen (1970); Heyraud and Cherry (1979)	Muscle	$[4–1700] \times 10^{-4}$ (0·015–6·3)	Beasley (1968); Holtzman (1969); Kauranen and Miettinen (1970); Hoffman and co-authors (1974); Heyraud and Cherry (1979); Pentreath and co-authors (1980b)
Muscle	$[2·8–14] \times 10^{-2}$ (1·0–5·2)	Heyraud and Cherry (1979)	Stomach	0·2–2·6 (7·4–96)	Beasley (1968)
Hepatoprancreas	3·3–61 (120–2300)	Cherry and co-authors (1970); Heyraud and Cherry (1979)	Liver	0·2–13 (7·4–480)	Beasley (1968); Cherry and co-authors (1970); Kauranen and Miettinen (1970); Hoffman and co-authors (1974); Heyraud and Cherry (1979); Pentreath and co-authors (1980b)
			Bone	$[3·2–530] \times 10^{-3}$ (0·12–20)	Beasley (1968); Kauranen and Miettinen (1970); Hoffman and co-authors (1974); Pentreath and co-authors (1980b)
^{228}Th	ND	—		ND	—
^{235}U	ND	—	Muscle	$[0·003–1·4] \times 10^{-3}$ ($[0·1–52] \times 10^{-3}$)	Aten and co-authors (1961); Hamilton (1972)
			Bone	$[1–38] \times 10^{-4}$ ($[0·4–14] \times 10^{-2}$)	Pentreath and co-authors (1980b)

in the southern hemisphere. The limited test ban was preceded by intensive testing in 1961 and 1962 leading to a maximum annual deposition in 1963 in the northern hemisphere. Thereafter, fallout declined until 1967 when the stratospheric input from the continued atmospheric testing by France and the Peoples Republic of China more or less balanced deposition (UNSCEAR, 1972, 1977).

A substantial fraction of the atmospheric tests have taken place on small islands or coral atolls so that local and intermediate fallout entered the marine environment directly, as also has a large part of the world-wide fallout from all atmospheric tests. Thus, in principle, the whole range of fission product radionuclides and a variety of activation product radionuclides could have been detected in the marine environment. In fact, however, comprehensive data have only been obtained for ^{90}Sr, ^{137}Cs (VOLCHOK and co-authors, 1971), ^{3}H (DOCKINS and co-authors, 1967; MÜNNICH and ROETHER, 1967), ^{14}C (BIEN and SUESS, 1967; NYDAL and LÖVSETH, 1970), and ^{239}Pu (PILLAI and co-authors, 1964; MIYAKE and co-authors, 1970a; BOWEN and co-authors, 1971) in sea water; these being radionuclides of potential significance for human exposure or, since they have relatively long half-lives, as tracers of marine processes. A summary of these data is given in Table 2-8. Excepting the tritium and ^{239}Pu values for the Mediterranean Sea, the data in Table 2-8 generally cover the time period up to 1967, i.e. inclusive of the years of maximum fallout and the subsequent rapid decline in deposition rates (UNSCEAR, 1977). The ranges given, therefore, include the maximum concentrations of the 5 radionuclides in the general environment but are exclusive of the higher concentrations which might arise over limited areas from the local fallout in the immediate vicinity of oceanic test sites. Deposition rates have continued to decline slowly (UNSCEAR, 1977) and it is to be expected that the concentrations in surface sea water would be decreasing due to decay, mixing from the surface into deep water, and involvement in geochemical cycles (KAUTSKY, 1977; BOWEN and co-authors, 1980).

Through the capacity of marine organisms and sediments to accumulate radionuclides to higher concentrations than those present in sea water, a greater variety of fallout radionuclides have been detected in these materials. As for the natural radionuclides, a concise summary is only possible for general groups of organisms as shown in Tables 2-9a–c. These data represent samples of a whole variety of organisms and sediments taken at various latitudes and dates; at best, therefore, they can only be a general indication of the magnitude of the contamination by fallout radionuclides.

(c) Radionuclides from Waste Disposal

Any human activity utilizing radionuclides or their associated radiations is likely to generate some form of active waste. Many factors, including the site of production, the physical and chemical nature of the waste, the spectrum of radionuclides involved and the waste management strategies adopted, will determine whether there will be any consequential contamination of the marine environment. Of the many possible sources of waste, the nuclear fuel cycle (ore mining and uranium extraction; uranium enrichment and fuel fabrication; reactor operation; spent fuel reprocessing) is by far the most significant in terms of both the total activity involved and its concentration at the various stages. Other industrial uses (including the manufacture and eventual disposal of consumer products incorporating radionuclides) and medical and research activities are rather minor sources (see UNSCEAR, 1977 for a complete discussion of the various

Table 2-8

Concentrations of certain fallout radionuclides in surface sea water, pCi l^{-1} (mBq l^{-1}) (Compiled from WOODHEAD, 1973a and IAEA, 1976; with additions from FUKAI and co-authors, 1976; MURRAY and FUKAI, 1978 and FUKAI, pers. comm.)

	^{90}Sr		^{137}Cs		^{3}H		^{14}C		^{239}Pu
Location	Mean	Range	Mean	Range	Mean	Range	Mean	Range	Range
N Atlantic Ocean	0·13	0·02–0·50	0·21	0·03–0·80	48	31–74	0·02	0·01–0·04	[0·3–1·2] × 10^{-3}
	(4·8)	(0·74–19)	(7·8)	(1·1–30)	(1·8 × 10^3)	([1·1–2·7] × 10^3)	(0·74)	(0·37–1·5)	([1·1–4·4] × 10^{-2})
S Atlantic Ocean	0·07	0·02–0·20	0·11	0·03–0·32	19	16–22	0·03	0·02–0·04	0·2 × 10^{-3}
	(2·6)	(0·74–7·4)	(4·1)	(1·1–12)	(700)	(590–810)	(1·1)	(0·74–1·5)	(7·4 × 10^{-3})
Indian Ocean	0·10	0·02–0·15	0·16	0·03–0·24	ND		ND		ND
	(3·4)	(0·74–5·6)	(5·9)	(1·1–8·9)					
NW Pacific Ocean	0·54	0·07–3·1	0·86	0·11–5·0	29	6–70	0·03	0·02–0·03	[0·1–1·4] × 10^{-3}
	(20)	(2·6–110)	(32)	(4·1–190)	(1·1 × 10^3)	([0·22–2·6] × 10^3)	(1·1)	(0·74–1·1)	([0·37–5·2] × 10^{-2})
SW Pacific Ocean	0·08	0·01–0·20	0·13	0·02–0·32	8	0·7–22	ND		ND
	(3·0)	(0·37–7·4)	(4·8)	(0·74–12)	(300)	(26—810)			
NE Pacific Ocean	0·27	0·05–0·58	0·43	0·08–0·93	44	10–240	0·03	0–0·04	[0·1–1·3] × 10^{-3}
	(10)	(1·9–21)	(16)	(3·0–34)	(1·6 × 10^3)	([0·37–8·9] × 10^3)	(1·1)	(0–1·5)	[(0·37–4·8] × 10^{-2})
SE Pacific Ocean	0·09	0·03–0·33	0·14	0·05–0·53	8	0·3–34	0·01	0–0·03	ND
	(3·3)	(1·1–12)	(5·2)	(1·9–20)	(300)	([0·01–1·3] × 10^3)	(0·37)	(0–1·1)	
North Sea	0·50	0·31–0·97	0·80	0·50–1·55	ND		ND		ND
	(19)	(11–36)	(30)	(19–57)					
Baltic Sea	0·71	0·36–1·0	1·1	0·56–1·6	ND		ND		ND
	(26)	(13–37)	(41)	(21–59)					
Black Sea	0.47	0·07–0·78	0·75	0·11–1·25	ND		ND		ND
	(17)	(2·6–29)	(28)	(4·1–46)					
Mediterranean Sea	0·23	0·09–0·38	0·37	0·14–0·61	28	—	ND		[0·8–8·5] × 10^{-3}
	(8·5)	(3·3–14)	(14)	(5·2–23)	(1 × 10^3)	—			([0·30–3·1] × 10^{-1})

Table 2-9a

Concentrations of fallout radionuclides in marine plankton, pCi g^{-1} (mBq g^{-1}) wet (Compiled from the sources indicated)

Radio-nuclide	Phytoplankton	Source	Zooplankton	Source	Mixed plankton	Source
^{54}Mn	0·1–0·4 (3·7–15)	Folsom and co-authors (1963)	0·02–0·1 (0·74–3·7)	Folsom and co-authors (1963)	5·3 (200)	Schreiber (1967)
^{55}Fe	0·5 (19)	Palmer and Beasley (1967)	0·16–1·5 (5·9–56)	Palmer and Beasley (1967)		
^{57}Co					2·3 (85)	Seymour and Lewis (1964)
^{60}Co					0·32–16 (12–590)	Seymour and Lewis (1964)
^{63}Ni					0·1–0·4 (3·7–15)	Beasley and Held (1969)
^{90}Sr					0·02–0·34 (0·74–13)	Schreiber (1967)
$^{95}Zr/^{95}Nb$			0·5–88 (19–3300)	Osterberg (1962)	2–800 (74–$3·0 \times 10^{4}$)	Seymour and Lewis (1964); Schreiber (1967)
^{103}Ru, ^{106}Ru					0·3–30 (11–1100)	Osterberg (1962); Osterberg and co-authors (1963); Seymour and Lewis (1964)
^{125}Sb					0·9 (33)	Schreiber (1967)
^{137}Cs					0·5–36 (19–1300)	Seymour and Lewis (1964)
^{141}Ce, ^{144}Ce					0·4–480 (15–$1·8 \times 10^{4}$)	Seymour and Lewis (1964); Schreiber (1967)
^{147}Pm					3·3 (120)	Schreiber (1967)
^{155}Eu					0·4 (15)	Schreiber (1967)
^{239}Pu	$[0·3–83] \times 10^{-3}$ (0·01–3·1)	Pillai and co-authors (1964); Noshkin and co-authors (1973)	$1·1 \times 10^{-3}$ ($4·1 \times 10^{-2}$)	Pillai and co-authors (1964)		

Table 2-9b

Concentration of fallout radionuclides in marine molluscs and crustaceans, pCi g^{-1} (mBq g^{-1}) wet (Compiled from the sources indicated)

Radionuclide	Molluscs		Source	Crustaceans		Source
^{54}Mn		0·05–6 (1·9–220)	FOLSOM and co-authors (1963); CHIPMAN and THOMMERET (1970)		0·06 (2·2)	FOLSOM and co-authors (1963)
^{55}Fe		0·4–140 (15–5200)	PALMER and BEASLEY (1967)			
^{57}Co		0·05–0·44 (1·9–16)	SEYMOUR and LEWIS (1964)			
^{60}Co		0·03–0·71 (1·1–26)	SEYMOUR and LEWIS (1964)		0·65 (24)	SEYMOUR and LEWIS (1964)
^{63}Ni	Kidney	0·03–15 (1·1–560)	BEASLEY and HELD (1969)			
^{65}Zn		0.02–12 (0.74–440)	FITZGERALD and SKAUEN (1963); FOLSOM and co-authors (1963)		0.07 (2.6)	FOLSOM and co-authors (1963)
$^{95}Zr/^{95}Nb$		0·1–25 (3·7–930)	SEYMOUR and LEWIS (1964)			
^{103}Ru, ^{106}Ru		0.03–14 (1.1–520)	SEYMOUR and LEWIS (1964)			
^{108m}Ag	Hepatopancreas	0·15 (5·6)	FOLSOM and co-authors (1970)	Hepatopancreas	0·06 (2·2)	FOLSOM and co-authors (1970)
^{110m}Ag	Hepatopancreas	0·004–4·2 (0·15–160)	FOLSOM and YOUNG (1965); FOLSOM and co-authors (1970)	Hepatopancreas	1·0 (37)	FOLSOM and co-authors (1970)
^{137}Cs		0·13–0·68 (4·8–25)	SEYMOUR and LEWIS (1964)			
^{141}Ce, ^{144}Ce		0·14–49 (5·2–1800)	SEYMOUR and LEWIS (1964)			
^{239}Pu		$[9–60] \times 10^{-5}$ $([0·33–2·2] \times 10^{-2})$	PILLAI and co-authors (1964); NOSHKIN and co-authors (1973)			

Table 2-9c

Concentration of fallout radionuclides in marine fish and sediments, pCi g^{-1} (mBq g^{-1}) wet (Compiled from the sources indicated)

Radio-nuclide	Fish		Source	Sediment	Source
^{54}Mn		0·002–0·05 (0·074–1·9)	FOLSOM and co-authors (1963)	0·2–0·3 (7·4–11)	CHIPMAN and THOMMERET (1970)
^{55}Fe	Muscle	0·1–106 (3·7–1900)	PALMER and BEASLEY (1967)	[0·37–9·5] × 10^{-1}* (1.4–35)	LABEYRIE and co-authors
	Liver	1620–1860 ([6·0–6·9] × 10^{4})			
	Gonad	220–280 ([8·1–10] × 10^{3})			
^{57}Co				2·8 (100)	SEYMOUR and LEWIS (1964)
^{60}Co	Muscle	0·04–0·14 (1·5–5·2)	SEYMOUR and LEWIS (1964)		
	Liver	0·03–0·37 (1·1–14)	SEYMOUR and LEWIS (1964); FOLSOM and YOUNG (1965)		
^{65}Zn		0·05–020 (1·9–7·4)	FOLSOM and co-authors (1963)		
^{90}Sr				0·02–0·03 (0·74–1·1)	CERRAI and co-authors (1965, 1967)
^{95}Zr/^{95}Nb	Whole	0·25–0·55 (9·3–20)	SEYMOUR and LEWIS (1964)		
	Liver	0·04–7·5 (1·5–280)			
	Muscle	0·03–0·34 (1·1–13)			
^{103}Ru, ^{106}Ru	Muscle	0·04–0·68 (1·5–25)	SEYMOUR and LEWIS (1964)		
	Liver	0·04–6·6 (1·5–240)			
^{108m}Ag	Liver	0·003–0·009 (0·11–0·33)	FOLSOM and co-authors (1970)		
^{110m}Ag	Liver	0·05–0·08 (1·9–3.0)	FOLSOM and YOUNG (1965); FOLSOM and co-authors (1970)		
^{137}Cs	Muscle	0·04–0·08 (1·5–3·0)	SEYMOUR and LEWIS (1964); PRESTON (1970)	[4·1–12] × 10^{-2} (1·5–4·4)*	BOWEN and co-authors (1976)
^{141}Ce, ^{144}Ce	Whole	4·8–6·9 (180–260)	SEYMOUR and LEWIS (1964)		
	Liver	0·35–28 (13–1000)			
	Muscle	0·05–1·3 (1·9–48)			
^{147}Pm				1·5 (56)	CERRAI and co-authors (1967)
^{155}Eu				0·15 (5·6)	CERRAI and co-authors (1967)
^{239}Pu	Muscle	[0·1–18] × 10^{-5} ([0·4–67] × 10^{-4})	WONG and co-authors (1970); BOWEN and co-authors (1976)	[2·7–17] × 10^{-3} (0·10–0·63)	NOSHKIN and BOWEN (1973); BOWEN and co-authors (1976)
	Liver	[1·6–250] × 10^{-5} ([0·6–93] × 10^{-3})			

*Sediment concentrations in pCi g^{-1} (mBq g^{-1}) dry weight.

sources and an assessment of the potential human exposure). In practice, the major sources of contamination of the marine environment are reactor operation and fuel reprocessing at coastal sites and the disposal of packaged radioactive wastes from a variety of sources into the deep ocean.

In the context of a discussion of waste management and the application of control procedures it is not necessary to consider every source and just four will be used as examples: (i) the gas-cooled magnox reactor operated by the Central Electricity Generating Board at Bradwell, Essex in the United Kingdom; (ii) the boiling water reactor operated by the Jersey Central Power and Light Company at Oyster Creek, Ocean County, New Jersey in the United States; (iii) the fuel reprocessing plant operated by British Nuclear Fuels Ltd, at Windscale, Cumbria in the United Kingdom; and (iv) the disposal of packaged, solid radioactive wastes into the northeast Atlantic Ocean under the auspices of the Nuclear Energy Agency of the Organization for Economic Cooperation and Development.

At Bradwell, the primary source of liquid radioactive waste is the water in the spent-fuel storage ponds. The radionuclides present include fission products (mainly ^{134}Cs and ^{137}Cs) leached from the fuel and activation products (including ^{51}Cr, ^{55}Fe, ^{59}Fe, ^{60}Co, and ^{65}Zn) derived from impurities in the magnox canning material (PRESTON, 1968). Additional sources include the carbon dioxide gas coolant drying system where tritium, in particular, is found, and the reactor vessel cooling system. The liquid wastes are discharged into the Blackwater estuary with the condenser cooling water. The annual discharges for the years 1962 to 1971 have been published previously (WOODHEAD, 1973a; IAEA, 1976) and more detailed data for 1972 to 1979 are given in Table 2-10.

At the Oyster Creek BWR, the primary source of liquid waste is the leakage of reactor cooling water from a number of points in the coolant circulation system. This is collected via equipment drains into a storage tank which also receives spent-fuel storage pond wastes and liquid process wastes from the reactor clean-up system. All of these wastes contain both fission product radionuclides released from defective fuel elements and activation products from the zircaloy fuel cans and other reactor construction materials. Prior to discharge into Oyster Creek with the condenser cooling water the waste is filtered and passed through a mixed bed demineralizer. The laundry and the fuel shipping cask decontamination system are two minor sources of waste which are not treated before discharge to the creek (BLANCHARD and co-authors, 1976). The estimated annual discharges of activity in liquid wastes are given in Table 2-11.

Even apart from the possibility of simple mechanical failure, fuel elements have a limited effective life in a reactor. Reactor operation reduces the content of fissile ^{235}U in the fuel, a process which is only partially compensated by the generation of fissile ^{239}Pu following neutron capture in ^{238}U and subsequent double β-decay. Eventually, therefore, there would be insufficient fissile material in the reactor core to maintain the chain reaction on which sustained heat production is based. In addition, there is some loss of reactivity due to the presence of increasing quantities of fission products, some of which absorb the neutrons necessary for continuing fission. Efficient operation thus requires the periodic replacement of the spent fuel in the reactor. The objective of spent-fuel reprocessing is to separate the valuable materials, i.e. ^{239}Pu and uranium, from the accumulated fission products.

On arrival at the Windscale plant the fuel elements are retained in storage ponds to await decanning and reprocessing. The pond water is maintained at an alkaline pH to

Table 2-10

Annual discharges of radioactivity to the Blackwater estuary from the Bradwell nuclear power station, Ci(TBq) (Compiled from MITCHELL, 1975, 1977a, 1978; HETHERINGTON, 1976b; HUNT, 1979, 1980 and MAFF Directorate of Fisheries Research, unpublished)

Year	^{3}H*	^{65}Zn†	^{110m}Ag†	^{134}Cs†	^{137}Cs†	Total activity* excluding tritium
1972	251 (9·3)	0·031 (1·1 × 10^{-3})	0·11 (4·1 × 10^{-3})	6·8 (0·25)	27 (1·0)	120 (4·4)
1973	198 (7·3)	0·059 (2·2 × 10^{-3})	0·044 (1·6 × 10^{-3})	2·0 (0·07)	13 (0·48)	54 (2·0)
1974	117 (4·3)	0·14 (5·2 × 10^{-3})	0·022 (0·8 × 10^{-3})	3·0 (0·11)	24 (0·89)	90 (3·3)
1975	88 (3·3)	0·16 (5·9 × 10^{-3})	<0·028 (<1·0 × 10^{-3})	9·8 (0·36)	52 (1·9)	120 (4·4)
1976	309 (11)	0·22 (8·1 × 10^{-3})	<0·016 (<0·6 × 10^{-3})	9·1 (0·34)	37 (1·4)	65 (2·4)
1977	199 (7·4)	0·16 (5·9 × 10^{-3})	0·016 (0·6 × 10^{-3})	13 (0·48)	43 (1·6)	66 (2·4)
1978	103 (3·8)	0·19 (7·0 × 10^{-3})	7·0 × 10^{-3} (0·3 × 10^{-3})	7·5 (0·28)	35 (1·3)	54 (2·0)
1979	119 (4·4)	Analyses discontinued				43 (1·6)

*Values based on data submitted by the operator.
†Values based on analyses of quarterly bulked samples by the Fisheries Radiobiological Laboratory, Lowestoft, Suffolk, UK.

Table 2-11

Annual discharges of radioactivity to Oyster Creek from the BWR nuclear power station, Ci(TBq) (Based on BLANCHARD and co-authors, 1976 and UNSCEAR, 1977; reproduced by permission of the U.S. Environmental Protection Agency)

Radionuclide	1971	1972	1973	1974
^{3}H	21·5 (0·79)	61·6 (2·3)	36·6 (1·4)	14·1 (0·52)
^{51}Cr	0·16 ($6{\cdot}1 \times 10^{-3}$)	0·12 ($4{\cdot}4 \times 10^{-3}$)	0·49 ($1{\cdot}8 \times 10^{-2}$)	0·11 ($4{\cdot}1 \times 10^{-3}$)
^{54}Mn	0·43 ($1{\cdot}6 \times 10^{-2}$)	0·63 ($2{\cdot}3 \times 10^{-2}$)	0·17 ($6{\cdot}4 \times 10^{-3}$)	$4{\cdot}4 \times 10^{-2}$ ($1{\cdot}6 \times 10^{-3}$)
^{58}Co	0·11 ($4{\cdot}0 \times 10^{-3}$)	0·15 ($5{\cdot}7 \times 10^{-3}$)	$4{\cdot}3 \times 10^{-2}$ ($1{\cdot}6 \times 10^{-3}$)	$9{\cdot}2 \times 10^{-3}$ ($3{\cdot}4 \times 10^{-4}$)
^{60}Co	0·82 ($3{\cdot}0 \times 10^{-2}$)	1·7 ($6{\cdot}2 \times 10^{-2}$)	0·27 ($1{\cdot}0 \times 10^{-2}$)	$7{\cdot}4 \times 10^{-2}$ ($2{\cdot}7 \times 10^{-3}$)
^{59}Fe	$4{\cdot}5 \times 10^{-2}$ ($1{\cdot}7 \times 10^{-3}$)	$2{\cdot}0 \times 10^{-2}$ ($7{\cdot}4 \times 10^{-4}$)	$1{\cdot}0 \times 10^{-3}$ ($3{\cdot}7 \times 10^{-5}$)	$1{\cdot}2 \times 10^{-2}$ ($4{\cdot}4 \times 10^{-4}$)
^{65}Zn	NR	NR	$1{\cdot}0 \times 10^{-3}$ ($3{\cdot}7 \times 10^{-5}$)	$2{\cdot}0 \times 10^{-3}$ ($7{\cdot}4 \times 10^{-5}$)
^{89}Sr }	0·34 ($1{\cdot}3 \times 10^{-2}$)	0·23 ($8{\cdot}4 \times 10^{-3}$)	0·18 ($6{\cdot}7 \times 10^{-3}$)	$1{\cdot}2 \times 10^{-2}$ ($4{\cdot}4 \times 10^{-4}$)
^{90}Sr }			$2{\cdot}8 \times 10^{-2}$ ($1{\cdot}0 \times 10^{-3}$)	$6{\cdot}2 \times 10^{-3}$ ($2{\cdot}3 \times 10^{-4}$)
^{91}Sr	$5{\cdot}0 \times 10^{-2}$ ($1{\cdot}9 \times 10^{-3}$)	$6{\cdot}5 \times 10^{-2}$ ($2{\cdot}4 \times 10^{-3}$)	$2{\cdot}0 \times 10^{-3}$ ($7{\cdot}4 \times 10^{-5}$)	NR
^{99}Mo	0·13 ($4{\cdot}8 \times 10^{-3}$)	0·22 ($8{\cdot}0 \times 10^{-3}$)	0·24 ($9{\cdot}0 \times 10^{-3}$)	0·11 ($4{\cdot}1 \times 10^{-3}$)
^{99m}Tc	0·10 ($3{\cdot}7 \times 10^{-3}$)	0·20 ($7{\cdot}4 \times 10^{-3}$)	0·24 ($9{\cdot}0 \times 10^{-3}$)	NR
^{124}Sb	$3{\cdot}0 \times 10^{-3}$ ($1{\cdot}1 \times 10^{-4}$)	$3{\cdot}0 \times 10^{-3}$ ($1{\cdot}1 \times 10^{-4}$)	NR	NR
^{131}I	0·38 ($1{\cdot}4 \times 10^{-2}$)	0·45 ($1{\cdot}7 \times 10^{-2}$)	$8{\cdot}2 \times 10^{-2}$ ($3{\cdot}0 \times 10^{-3}$)	$1{\cdot}4 \times 10^{-2}$ ($5{\cdot}2 \times 10^{-4}$)
^{133}I	0·29 ($1{\cdot}1 \times 10^{-2}$)	0·41 ($1{\cdot}5 \times 10^{-2}$)	$7{\cdot}8 \times 10^{-2}$ ($2{\cdot}9 \times 10^{-3}$)	$1{\cdot}6 \times 10^{-2}$ ($5{\cdot}9 \times 10^{-4}$)
^{133}Xe	NR	0·78 ($2{\cdot}9 \times 10^{-2}$)	0·75 ($2{\cdot}8 \times 10^{-2}$)	NR
^{135}Xe	NR	2·5 ($9{\cdot}2 \times 10^{-2}$)	2·2 ($8{\cdot}2 \times 10^{-2}$)	NR
^{134}Cs	0·10 ($3{\cdot}7 \times 10^{-3}$)	2·1 ($7{\cdot}6 \times 10^{-2}$)	$8{\cdot}3 \times 10^{-2}$ ($3{\cdot}1 \times 10^{-3}$)	$2{\cdot}6 \times 10^{-2}$ ($9{\cdot}6 \times 10^{-4}$)
^{137}Cs	0·24 ($9{\cdot}0 \times 10^{-3}$)	3·0 (0·11)	8·2 × 10−2 ($3{\cdot}0 \times 10^{-3}$)	$1{\cdot}5 \times 10^{-2}$ ($5{\cdot}6 \times 10^{-4}$)
^{140}Ba–^{140}La	0·16 ($5{\cdot}9 \times 10^{-3}$)	$6{\cdot}7 \times 10^{-2}$ ($2{\cdot}5 \times 10^{-3}$)	0·15 ($5{\cdot}4 \times 10^{-3}$)	NR
^{141}Ce	NR	NR	$5{\cdot}0 \times 10^{-3}$ ($1{\cdot}9 \times 10^{-4}$)	$4{\cdot}1 \times 10^{-3}$ ($1{\cdot}5 \times 10^{-4}$)
^{144}Ce	NR	NR	$2{\cdot}0 \times 10^{-2}$ ($7{\cdot}4 \times 10^{-4}$)	$3{\cdot}1 \times 10^{-3}$ ($1{\cdot}1 \times 10^{-4}$)
^{239}Np	0·66 ($2{\cdot}4 \times 10^{-2}$)	0·68 ($2{\cdot}5 \times 10^{-2}$)	0·23 ($8{\cdot}6 \times 10^{-3}$)	NR

NR, Not reported.

Table 2-12

Annual discharges of radioactivity to the northeast Irish Sea from the BNFL fuel reprocessing plant, kCi (TBq) (Data from returns made to MAFF, Directorate of Fisheries Research, Lowestoft, England)

Year	^{3}H	^{89}Sr	^{90}Sr	^{95}Zr	^{95}Nb	^{103}Ru	^{106}Ru	^{134}Cs	^{137}Cs	^{144}Ce	^{238}Pu	$^{239+240}Pu$	^{241}Pu	^{241}Am
1972	34 (1300)	1·1 (41)	15 (560)	26 (960)	24 (890)	1·2 (44)	31 (1100)	5·8 (210)	35 (1300)	14 (520)	1.5(56)		52 (1900)	2·2 (81)
1973	20 (740)	1·5 (56)	7·4 (270)	15 (560)	28 (1000)	1·6 (59)	38 (1400)	4·5 (170)	21 (780)	15 (560)	1·8(67)		75 (2800)	3·0 (110)
1974	32 (1200)	0·28 (10)	11 (410)	2·6 (96)	7·0 (260)	0·40 (15)	29 (1100)	27 (1000)	110 (4100)	6·5 (240)	1.2(44)		46 (1700)	3·2 (120)
1975	38 (1400)	0·19 (7·0)	13 (480)	2·6 (96)	6·0 (220)	0·23 (8·5)	21 (780)	29 (1100)	141 (5200)	5·6 (210)	1·2(44)		49 (1800)	0·98 (36)
1976	33(1200)	0·06 (2·2)	10 (370)	3·1 (110)	6·0 (220)	0·33 (12)	21 (780)	20 (740)	116 (4300)	4·0 (150)	1·3(48)		35 (1300)	0·32 (12)
1977	25 (930)	0·09 (3·3)	12 (440)	2·5 (93)	5·5 (200)	0·32 (12)	22 (810)	17 (630)	121 (4500)	4·1 (150)	0·98(36)		27 (1000)	0·10 (3·7)
1978	28 (1000)	0·27 (10)	16 (590)	2·2 (81)	4·0 (150)	0·23 (8·5)	22 (810)	11 (410)	111 (4100)	2·8 (100)	0·33 (12)	1·2 (44)	48 (1800)	0·21 (7·8)
1979	32 (1200)	0·20 (7·4)	6·8 (250)	1·6 (59)	2·7 (100)	0·15 (5·6)	11 (410)	6·4 (240)	69 (2600)	2·3 (85)	0·30 (11)	1·0 (37)	40 (1500)	0·20 (7·4)

Table 2-13

Quantities of packaged solid radioactive waste dumped annually into the northeast Atlantic Ocean (OECD–NEA, 1980; reproduced by permission of OECD–NEA)

Year	Gross weight, tonnes	Approximate radioactivity, kCi (TBq)		
		α-activity	β-, γ-activity	Tritium
1967	10,900	0·25 (9·3)	7·6 (280)*	—
1969	9180	0·50 (19)	22 (810)*	—
1971	3970	0.63 (23)	11 (410)*	—
1972	4130	0.68 (25)	22 (810)*	—
1973	4350	0·74 (27)	13 (480)*	—
1974	2270	0·42 (16)	—	100 (3700)
1975	4460	0·78 (29)	31 (1100)	30 (1100)
1976	6770	0·88 (33)	33 (1200)	21 (780)
1977	5600	0·95 (35)	36 (1300)	32 (1200)
1978	8040	1·1 (41)	43 (1600)	37 (1400)
1979	5415	1·4 (52)	41 (1500)	42 (1600)

* Including tritium.

minimize the corrosion of the magnox cans but substantial quantities of caesium isotopes are still released into the water from the fuel. The concentration of caesium is controlled by circulating the water through zeolite skips immersed in the ponds. The pH and visibility in the storage ponds are maintained by continually replacing the water, and the consequent low-activity effluent passes to delay tanks to allow settlement of suspended solids before being discharged to the Irish Sea. The effluent pipeline extends 2·5 km beyond high-water mark and the waste is released at a depth of 20 m. The separation of the uranium and plutonium from the fission and activation products is achieved by a multistage solvent extraction process. The magnox cans are mechanically stripped from the fuel elements and stored on the Windscale site. The fuel rods are dissolved in nitric acid from which the uranyl and plutonium nitrates are preferentially extracted using an organic solution of the complexing agent, tributyl phosphate.The greater proportions (>99%) of the fission and activation radionuclides remain in the aqueous phase constituting a highly active waste which is retained in storage in water-cooled stainless steel tanks on the Windscale site. Further cycles of back-extraction into different strengths of nitric acid and solvent extraction give separate plutonium nitrate and uranyl nitrate streams, and produce medium- and low-activity aqueous wastes. Medium-activity wastes are stored to allow the decay of short-lived radionuclides before being discharged into the Irish Sea together with low-activity wastes. The waste streams from reprocessing are mainly acidic and are neutralized prior to discharge. The radionuclide composition of the effluent discharged into the northeast Irish Sea from the Windscale plant in recent years is given in Table 2-12; data relating to earlier years have also been published (HOWELLS, 1966; WOODHEAD, 1973a; IAEA, 1976; UNSCEAR, 1977).

The types of solid radioactive wastes which are packaged and sent for disposal into the northeast Atlantic Ocean include contaminated laboratory equipment, e.g. glove boxes, glassware, etc., process sludges, spent ion-exchange resins, and wastes arising from the industrial, medical, and research applications of radionuclides. In general, these wastes are unsuitable for land burial. The most common form of packaging is incorporation of the waste in concrete within a 640 l steel drum. The quantities of waste which have been dumped into the northeast Atlantic Ocean since 1967, when the OECD–NEA assumed a supervisory responsibility, are given in Table 2-13. The current site is located at 45°50′ to 46°10′ N, 16°00′ to 17°30′ W and has an average water depth of 4400 m (OECD–NEA, 1980).

(5) Fate of Radionuclides Discharged to the Marine Environment at 4 Representative Sites

As has already been indicated, there are many factors which influence the behaviour of radionuclides in the marine environment. From the data presented in the preceding section it can be seen that information concerning the physico-chemical form of the contaminants in the wastes is almost entirely limited to the matrix (i.e. liquid or solid) and the quantities of the major radionuclides present. Once released into sea water, tidal motions and larger scale advective processes act to dilute and disperse the radionuclides while interactions (adsorptive and/or absorptive) with marine organisms and sediments can result in accumulation to concentrations which are frequently much higher than those in the ambient sea water. For certain radionuclides in the environment, the passage of time also allows an equilibrium state of chemical speciation to develop from the form initially introduced in the waste. An understanding of these processes is a prerequisite for an accurate assessment of the consequential radiation exposure of, and hence potential hazard to, both human and marine populations.

(a) Dispersion

Blackwater Estuary

A map of the Blackwater estuary showing the position of the Bradwell nuclear power station is given in Fig. 2-5. Extensive measurements of salinity and temperature in the estuary prior to station operation showed, as expected, that the river outflow tends to follow the southern shore, but also that the estuary is well mixed both vertically and longitudinally. Further evidence of a high degree of mixing was obtained using broth cultures of the bacterium *Serratia marcesens* released as tracers in the vicinity of the cooling water outfall at various states of the tide. Within a few hours of release, volume dilutions in the region of 10^8 were observed. It has been estimated that between 1% and 5% of the high-water volume of the estuary upstream of the station is exchanged with the open sea on each tidal cycle (LOWTON and co-authors, 1966). Post-operational studies of the warm water discharged to the estuary during a complete year have provided an estimate of 9% per tidal cycle for the mean exchange between the open sea and the tidal plug, i.e. that volume of water which moves up and down the estuary on each tide. Expressed as a percentage of the high-water volume of the estuary upstream of the station, the corresponding value was 6% (LOWTON and co-authors, 1966; Central Electricity Generating Board, 1967). At full load the station requires some 9×10^7 l of water h^{-1} for condenser cooling. This is drawn from the estuary on the offshore side of a barrier wall which

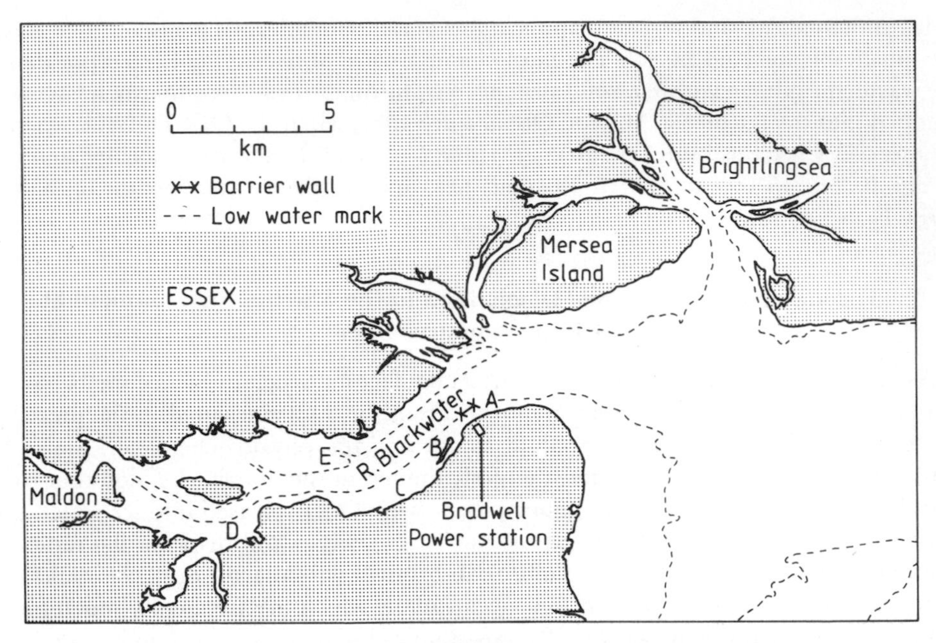

Fig. 2-5: Blackwater estuary. (After PRESTON, 1967; reproduced by permission of Pergamon Press Ltd.; Crown copyright)

prevents re-entry of the warm water discharging through an outfall positioned on the inshore side. Active wastes are added to the cooling water outflow during the first 3h of the ebb tide when maximum dispersion to the open sea is attainable (WASSON and MITCHELL, 1973).

In the pre-operational assessment of the proposed discharge from the Bradwell power station it was calculated, on the basis of assumptions concerning the hydrographic processes in the estuary, that the mean concentration of a radionuclide in the water of the tidal stretch would be approximately 150 pCi l^{-1} per Ci d^{-1} discharged (150 Bq l^{-1} per TBq d^{-1}) (see Table 2-23). Recently, a direct estimate of this value has been made. The concentration of ^{137}Cs has been measured in monthly water samples collected from the station cooling water intake (on the assumption that this is representative of the well-mixed water in the estuary) and related to the mean daily discharges of the radionuclide. The mean monthly concentration in the estuary during 1977 was found to be 58 (range 34 to 131) pCi l^{-1} per Ci d^{-1} discharged (58 Bq l^{-1} per TBq d^{-1}) (D. F. JEFFERIES, Lowestoft, pers. comm.). In view of the assumptions and uncertainties underlying these results, better agreement would not have been expected.

Oyster Creek

The site of the Oyster Creek generating station relative to the local waterways and the Atlantic Ocean is illustrated in Fig. 2-6. The hydrology of Oyster Creek and Barnegat Bay have been described by BLANCHARD and co-authors (1976). The condenser cooling water for the station is drawn from the south branch of the Forked River and discharged into Oyster Creek, and the flow, at $1{\cdot}2 \times 10^{11}$ l $month^{-1}$, is substantially greater than the

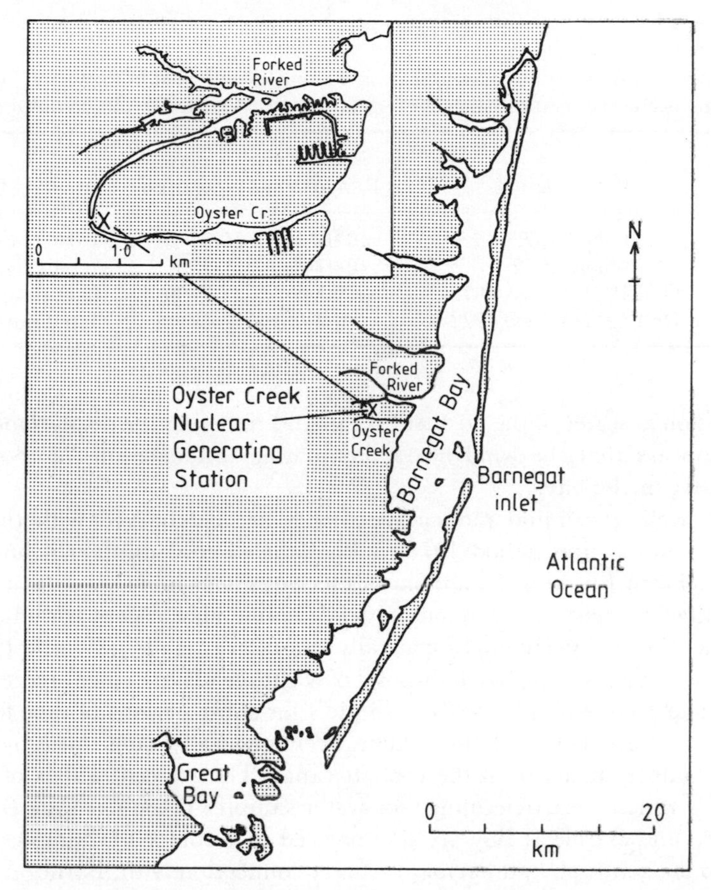

Fig. 2-6: Oyster Creek and Barnegat Bay. (After BLANCHARD and co-authors, 1976; reproduced by permission of US Environmental Protection Agency)

natural fresh water discharges and tidal movements in either of these waterways. Thus there is a forced circulation of sea water from Barnegat Bay up the Forked River and down Oyster Creek re-entering the Bay 1·8 km south of the extraction point. The monthly cooling water demand of the station is approximately half the volume of Barnegat Bay, a long, narrow and rather shallow stretch of water separated from the Atlantic Ocean by barrier beaches. There are effectively 2 direct connections between the bay and the ocean, and the most important of these, the Barnegat Inlet, is 7 km southeast of the mouth of Oyster Creek. Dye studies carried out in the bay at a time of minimal fresh water input and low to moderate winds showed that the minimum average daily exchange between the bay and the ocean is about 14% and that the half-time for exchange is about 5 d. In the absence of strong southerly winds the fresh water input to the enclosed northern basin and the salinity gradient developing from the fresh water inflow on the western shore of the bay generate a southward water flow. Water moving south from the mouth of Oyster Creek on the ebb tide is rapidly flushed out of the bay through the Barnegat Inlet. Southerly winds can produce a flow to the north producing

Table 2-14

Concentration of certain radionuclides in sea water samples collected in the vicinity of the Oyster Creek nuclear power station (Based on data from BLANCHARD and co-authors, 1976)

Radio-nuclide	Sampling period	Range of concentration	Comments
^{54}Mn	28 Sept. 1972	<0·01–2.2 (<0·37–81)	Particulate fraction only
^{60}Co	28 Sept. 1972	0·07–4·0 (2.6–150)	
^{90}Sr	18 Oct. 1971–2 Nov. 1972	<0·1–2·6 (<3·7–96)	Background from fall-out not subtracted
^{137}Cs	18 Oct. 1971–2 Nov. 1972	0·2–1·3 (7·4–48)	

an accumulation of water in the northern basin and reducing the displacement of water. Overall it appears that the wind has a dominating influence on the movement and mixing of water in the bay.

As at Bradwell, the liquid radioactive wastes are discharged with the condenser cooling water and in the period 1971 to 1973 the annual dilution volume averaged $1{\cdot}13 \times 10^{12}$ l. From this and the annual discharges given in Table 2-11 it can be seen that the average concentrations of individual radionuclides, except tritium, in the coolant outflow would be expected to be generally less than 1 pCi l^{-1} (37 mBq l^{-1}) and often very much less. However, the batch disposal of wastes leads to higher predicted values in the range 0·1 to 10 pCi l^{-1} (3·7 to 370 mBq l^{-1}) for short periods of time for the major components of the waste and these have been generally confirmed by analysis of appropriate samples taken from the coolant canal. The concentrations of ^{54}Mn, ^{60}Co, ^{90}Sr, and ^{137}Cs which were determined for water samples taken from the Forked River, Oyster Creek, and Barnegat Bay are summarized in Table 2-14. Samples collected in Great Bay to the south of, and having no direct connection with, Barnegat Bay yielded background (fallout) estimates of 0·36 to 0·50 pCi l^{-1} (13 to 19 mBq^{-1}) and 0·3 to 0·4 pCi l^{-1} (11 to 15 mBq l^{-1}) for ^{90}Sr and ^{137}Cs respectively. Thus many of the water samples collected from the vicinity of the station gave no indication of contamination attributable to the presence of these two nuclides in the discharges (BLANCHARD and co-authors, 1976).

During 1972, the average discharge rate for ^{137}Cs was $8{\cdot}35 \times 10^{-3}$ Ci d^{-1} ($3{\cdot}09 \times 10^{8}$ Bq d^{-1}); using the volume ($2{\cdot}4 \times 10^{11}$ l) and turnover half-time (5 d) for Barnegat Bay, it is estimated that the mean equilibrium concentration of ^{137}Cs in the Bay would have been 0·25 pCi l^{-1} (9·3 Bq l^{-1}). From the analytical data for water samples taken from the Bay in 1972 (BLANCHARD and co-authors, 1976), the mean concentration of ^{137}Cs can be estimated to have been 0·29 pCi l^{-1} (11 Bq l^{-1}) above the fallout background. Once again, considering the uncertainties involved, the measure of agreement between these values must be regarded as satisfactory and providing confirmation of the pre-operational estimates of local dispersion.

Windscale

Prior to the construction of the Windscale reprocessing plant it was recognized that an estimate was required of the quantity of radioactivity which could be safely discharged into the sea on a continuing basis. The available information on tidal currents suggested that, close to the shore, the water moved in an elongated ellipse with a major axis between 3 and 6 miles in length more or less parallel to the coast. There was, however, insufficient information to estimate either the degree of dilution which would be attained immediately after release and how this would vary with the distance offshore of the release point or the impact of the wind on the dispersion process. A series of experiments employing releases of fluorescein dye at distances between 1·9 and 4·8 km offshore at various states of the tide and in different weather conditions were conducted to obtain these data. It was found that the initial dilution was always at least 10^{-4} and of the order of 10^{-6} within 12 h. Measurements on the shoreline showed dilutions of 10^{-5} on the day of release and 10^{-7} within 24 h. The dye patch usually developed an eccentric shape with the longer axis parallel to the tidal current. The overall movement of the patch could be accounted for by the known tidal stream and a superimposed motion due to the wind; a northerly set was generated by onshore (westerly) winds and a southerly set by easterly or south-easterly winds. These findings were contrary to the only available theory and were an insufficient basis for developing an alternative for predicting the influence of the wind on the effluent. From the results of these experiments it was decided that the waste should be discharged through a pipeline ending 2·5 km beyond high-water mark. In these circumstances it was predicted that the equilibrium concentration at the shoreline would be 3 to 12 pCi l^{-1} per Ci d^{-1} discharged (3 to 12 Bq l^{-1} per TBq d^{-1} discharged) (SELIGMAN, 1955).

The longer-term fate of the effluent after dilution and initial movement in the tidal stream was more problematical. In agreement with the localized findings of SELIGMAN (1955), LEE (1960) concluded that the wind has a strong influence on the circulation of the northeast Irish Sea, although in calm conditions the net movement of water is southwards along the Cumbrian coast (TEMPLETON and PRESTON, 1966). In a wider context, the details of the water circulation within the Irish Sea remain a matter of debate, although there is agreement on the existence of a net flow of northeast Atlantic water in through St George's Channel in the south and out through the North Channel (RAMSTER, 1973; J. W. TALBOT, Lowestoft, unpubl.). In addition to the classical means of investigating these circulation patterns, studies of the wider dispersion of radionuclides from the Windscale discharge are also contributing to the resolution of the problem (JEFFERIES and co-authors, 1982).

The distributions of the fission products ^{95}Zr–^{95}Nb, ^{106}Ru, ^{144}Ce, and ^{137}Cs derived from the analysis of filtered (0·22 μm membrane filter) samples of surface sea water collected in the immediate vicinity of the Windscale outfall on 13–14 September 1968 are given in Figs 2-7a–d. During the sampling period the winds were light to moderate from a north-easterly quarter. The data for ^{95}Zr–^{95}Nb and ^{106}Ru clearly show the influence of the tidal flow parallel to the Cumbrian coast with an apparent net movement in a southerly direction. There is also some indication of movement offshore, possibly due to the north-easterly winds. The data for ^{144}Ce are less clear but also seem to indicate a net movement in a southerly direction. These four nuclides are more or less rapidly removed from the water column to the sediment; the distribution patterns shown in Figs 2-7a–c, therefore, reflect the influence of water movements on discharges of these nuclides immediately prior to, and during the sampling period. In contrast, ^{137}Cs is relatively conservative with water, and the rather different pattern of concentration (Fig. 2-7d) is more likely to represent the net distribution which has developed from the movement of effluent discharged over a considerably longer period of time.

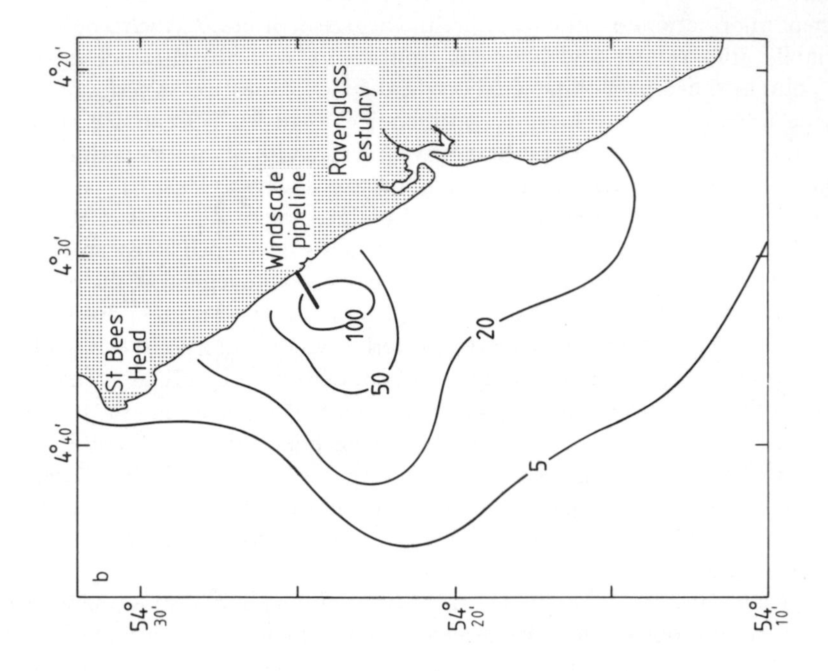
b
St Bees Head
Windscale pipeline
Ravenglass estuary
100
50
20
5
4°20'
4°30'
4°40'
54°30'
54°20'
54°10'

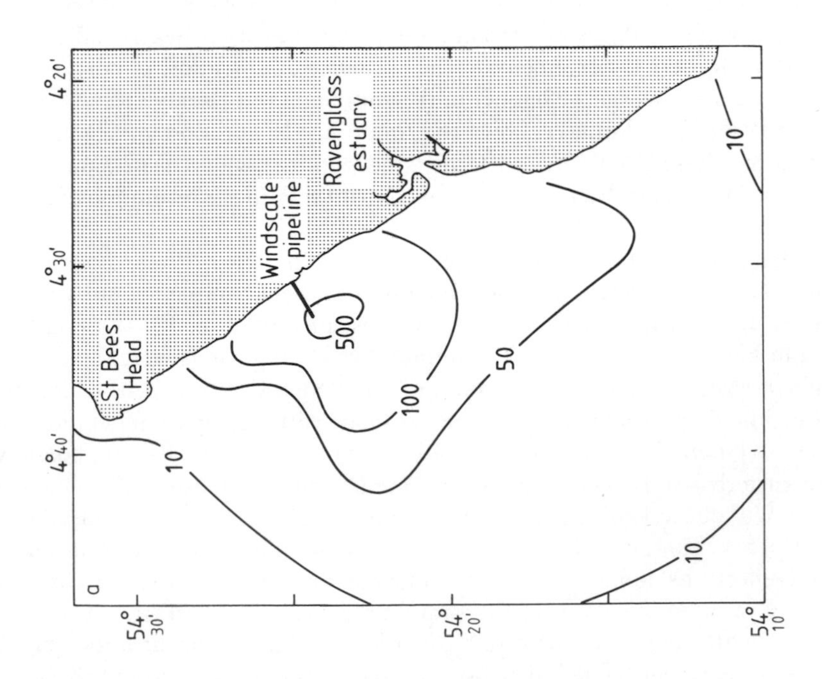
a
St Bees Head
Windscale pipeline
Ravenglass estuary
500
100
50
10
10
10
4°20'
4°30'
4°40'
54°30'
54°20'
54°10'

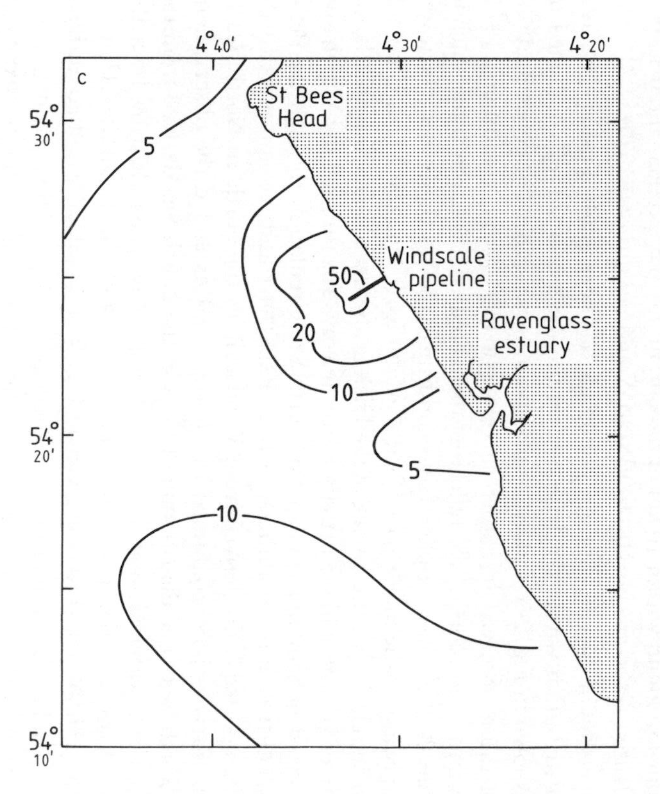

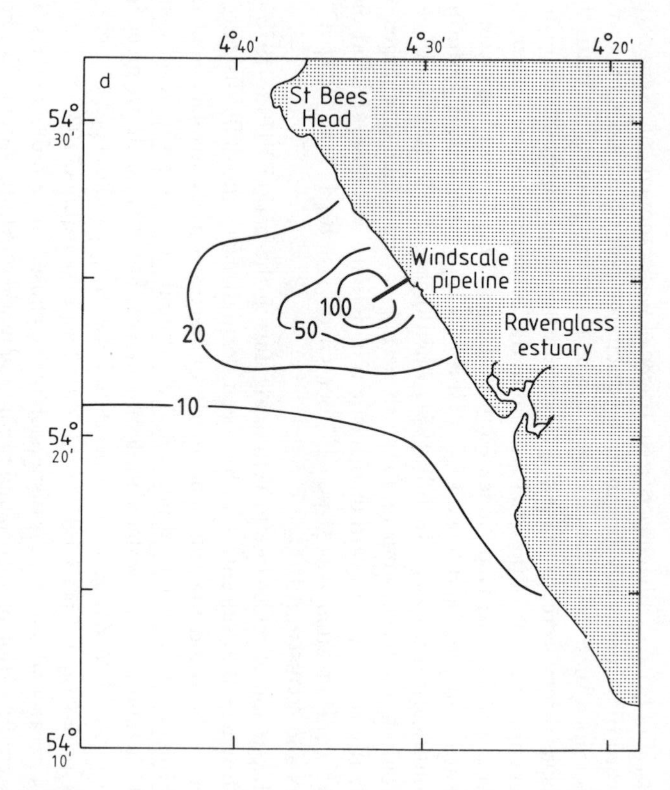

Fig. 2-7: Concentrations (pCi l^{-1}) of fission product radionuclides in surface sea water in the northeast Irish Sea in September 1968. (a) ^{95}Zr–^{95}Nb; (b) ^{106}Ru; (c) ^{144}Ce; (d) ^{137}Cs. (Ministry of Agriculture, Fisheries and Food, Directorate of Fisheries Research, unpublished)

The possibility of quite different long-term distributions due to changes in the circulation patterns in the northeast Irish Sea has been confirmed by studies of tritium which, as tritiated water (HTO), is a truly conservative tracer of water movement. Net northward and southward movements of water from the discharge point have been observed at different times. Since these observations were made after long periods of relatively steady tritium disposal they were, presumably, representative of quasi-equilibrium situations. In the immediate vicinity of the discharge point the normalized water concentrations of tritium were 4 pCi l^{-1} per Ci d^{-1} released (4 Bq l^{-1} per TBq d^{-1}) during October 1971 (net northward transport) and 5 pCi l^{-1} per Ci d^{-1} (5 Bq l^{-1} per TBq d^{-1}) during July 1973 (net southward transport). In July 1974, when disposal of tritium had been taking place for only a few weeks after an 8-month period during which discharges of tritium were virtually zero, a southward displacement of activity was observed and the normalized water concentration was 8 pCi l^{-1} per Ci d^{-1} (8 Bq l^{-1} per TBq d^{-1}) (HETHERINGTON and ROBSON, 1979).

The average dispersion of ^{137}Cs close to Windscale during the period 1970 to 1978 is shown in Fig. 2-8 where the concentration has been normalized to the daily discharge rate. In addition to the trend towards a northerly or north-westerly displacement of activity, there is also evidence for stratification. The annual average normalized concentrations of ^{137}Cs in shoreline sea water samples taken monthly at Seascale are given in Table 2-15 where it can be seen that there is variation between 2·6 and 6·3 pCi l^{-1} per Ci d^{-1} released (2·6 to 6·3 Bq l^{-1} per Ci d^{-1}) (JEFFERIES and co-authors, 1982).

Thus the post-operational data on the dispersion of the conservative radionuclides ^{3}H and ^{137}Cs provide confirmation of the original prediction made by SELIGMAN (1955). For ^{106}Ru, ^{144}Ce, and ^{239}Pu, which are not conservative with sea water, much lower normalized concentrations are found within 10 km of the outfall, i.e. 0·57, 0·063, and 0·15 pCi l^{-1} per Ci d^{-1} released respectively (HETHERINGTON and co-authors, 1975).

The wider dispersion of the effluent from Windscale has been studied using ^{137}Cs as a tracer and the distribution within the Irish Sea on the basis of a survey conducted in January 1976 is shown in Fig. 2-9 (PRESTON and co-authors, 1978). The concentrations are greater than those observed in earlier surveys (JEFFERIES and co-authors, 1973; HETHERINGTON and co-authors, 1975) reflecting the increased discharge of this nuclide since 1974 (Table 2-12). These studies have confirmed the northerly drift of water through the Irish Sea and the major part of the caesium leaves via the North Channel. There appears to be relatively little mixing of the Irish Sea water with northeast Atlantic water to the north of Ireland and the activity moves northwards through the Minch and round the north coast of Scotland to enter the North Sea circulation through the Pentland Firth (JEFFERIES and co-authors, 1973, PRESTON and co-authors, 1978). KAUTSKY (1973) reported ^{137}Cs derived from Windscale in the northern North Sea in 1971. Subsequent surveys have confirmed this finding and extended the observations throughout the North Sea. The distribution of ^{137}Cs in the North Sea derived from the analysis of filtered sea water samples obtained during May–June 1976 by both the Fisheries Radiobiological Laboratory, Lowestoft (UK), and the Deutsches Hydrographisches Institut, Hamburg (FRG) is given in Fig. 2-10 (PRESTON and co-authors, 1978). Additional studies of the shorter-lived nuclide ^{134}Cs ($t_{\frac{1}{2}}$: 2·06 yr), which is also discharged from Windscale, have indicated that the transit time for the movement of activity from the outfall to the North Channel is between 1·1 and 1·8 yr (JEFFERIES and co-authors, 1973).

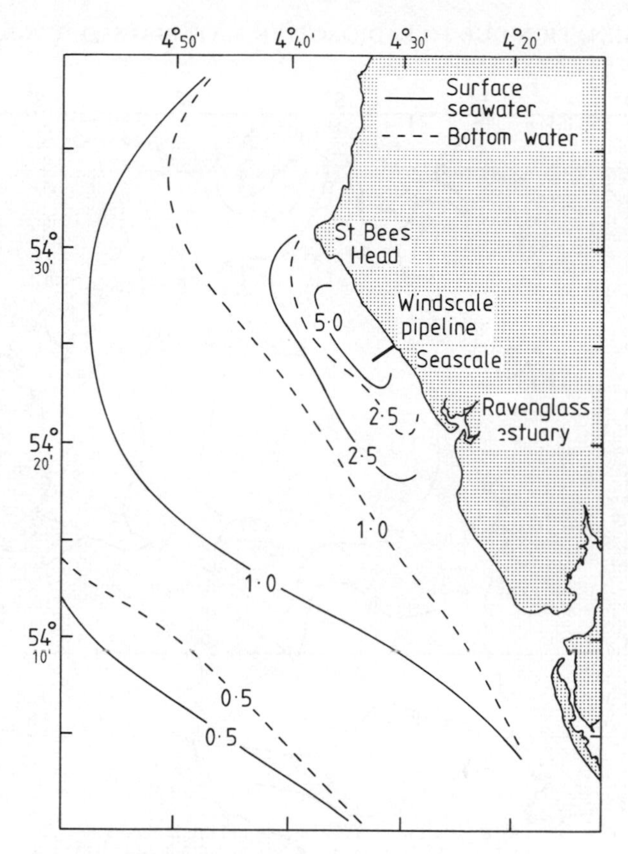

Fig. 2-8: Normalized concentrations of ^{137}Cs (pCi l^{-1} per Ci d^{-1} released) in sea water in the northeast Irish Sea, 1970–1978. (After JEFFERIES and co-authors, 1982; Crown copyright)

Table 2-15

Normalized concentration of caesium-137 in shoreline sea water at Seascale, pCi l^{-1} per Ci d^{-1} released (Bq l^{-1} per TBq d^{-1}) (MAFF, Directorate of Fisheries, Research, unpublished)

Year	Normalized concentration
1970	4·6 (4·6)
1971	6·3 (6·3)
1972	5·1 (5·1)
1973	5·5 (5·5)
1974	3·1 (3·1)
1975	4·4 (4·4)
1976	3·7 (3·7)
1977	2·8 (2·8)
1978	2·6 (2·6)

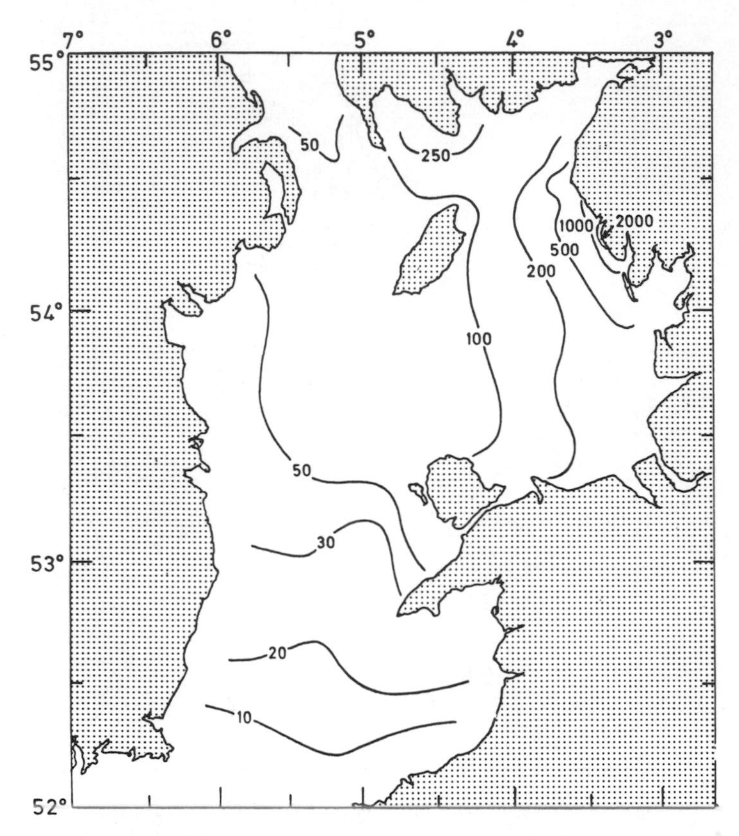

Fig. 2-9: Concentrations of ^{137}Cs (pCi l^{-1}) in filtered sea water from the Irish Sea, January 1976. (After PRESTON and co-authors, 1978; Crown copyright)

More recently, interest has become focused on the movement of the actinide elements, and in particular $^{239,240}Pu$, in the marine environment. Apart from any immediate (and rather low) radiological implications, these studies are of fundamental interest since the nuclides do not occur naturally and the northeast Irish Sea represents a labelled environment where their behaviour may be readily observed. Although a substantial fraction of the plutonium discharged from Windscale is rapidly removed from the water column to sediment, the small proportion which remains in the water behaves more or less conservatively and the concentration distribution within the Irish Sea has been determined (HETHERINGTON and co-authors, 1975; HETHERINGTON, 1976a; PENTREATH and co-authors, 1980a). As the chemical state of plutonium in sea water would be expected to have a considerable influence on the fate of the element, this aspect of its behaviour has been investigated. As discharged, it has been determined that the plutonium is almost entirely in the Pu (III + IV) oxidation states. More detailed examination of the small fraction which remains in the water column has shown that the plutonium retained by a 0·22 μm membrane filter is also predominantly Pu(III + IV) while that in the filtrate is predominantly Pu(V + VI) (NELSON and LOVETT, 1978; PENTREATH and co-authors, 1980a). Further afield, MURRAY and co-authors (1978)

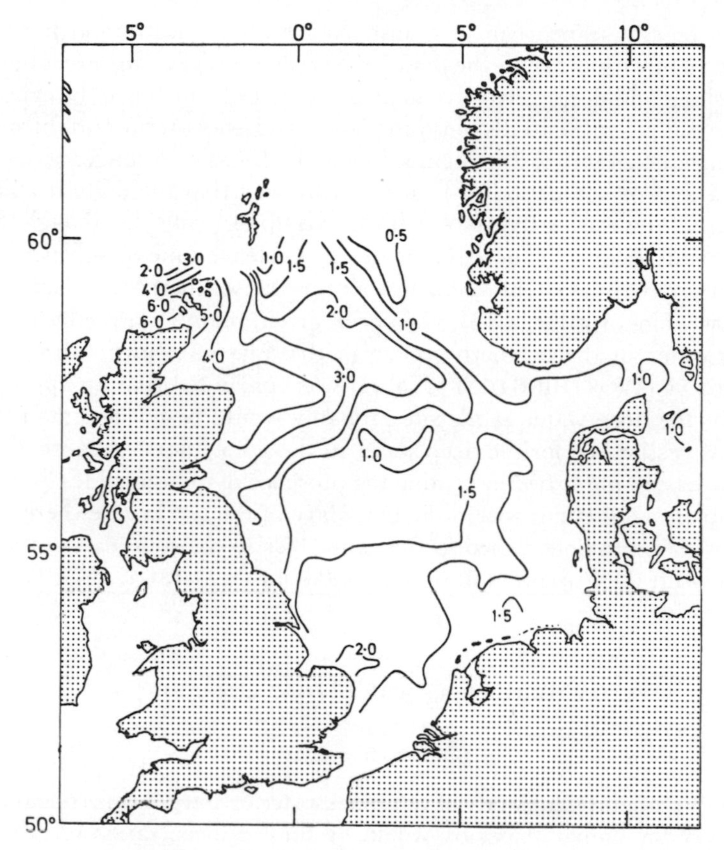

Fig. 2-10: Concentrations of ^{137}Cs (pCi l^{-1}) in filtered sea water from Scottish coastal waters and the North Sea, May–June 1976. (After PRESTON and co-authors, 1978; Crown copyright)

have concluded that $^{239,240}Pu$ derived from Windscale is present in the northern North Sea.

Northeast Atlantic Deep Ocean Disposal Site

A description of the hydrographic conditions in the vicinity of the dump site has been included as part of the hazard assessment made for a recent review of the suitability of the site for continued use (OECD–NEA, 1980). At the site the ocean is horizontally stratified and north Atlantic deep and bottom waters extend from 1500 m depth to the seabed. Above this is a layer of warm, saline water identifiable as of Mediterranean origin and extending up to about 600 m depth and horizontally out as far as the mid-Atlantic Ridge. Two further layers are present from 600 m to the surface: north Atlantic central water extending to about 100 m depth and the surface mixed layer. The presence of these well-defined layers argues that vertical mixing is rather slow compared to horizontal processes. At the seabed there is frequently a layer 50 to 100 m thick which is vertically well mixed, the so-called benthic boundary layer. The certain identification of such a layer at the dump site has not yet been made. Radionuclides released from drums resting on the seabed can be transported both vertically and horizontally by currents. Vertical transport is, in the main, a very slow process, the exceptions being in areas of upwelling and deep convective mixing, neither of which is to be expected at the site. Observations using neutrally buoyant floats and short-term current meters have shown daily mean currents of 2 to 3 cm s^{-1} near the bottom. A longer series of current meter observations over a period of 14 months showed that the daily residual current velocity at the bottom was very variable with speeds up to 4 cm s^{-1}, although the monthly average was about 2 cm s^{-1}. The estimate of the residence time of water in the deepest layer is rather approximate. The site is just within the southern boundary of the European Basin for which a value of 13 yr has been given, mainly on the basis of the large flow of deep water in the northern sector; in the adjacent North African Basin the estimate is 662 yr (WORTHINGTON, 1976). It was concluded that an appropriate residence time for the deep water at the site probably would be not less than the average value of 200 yr, estimated for the deep water in the whole north Atlantic Ocean.

No data have been published indicating the presence of radionuclides from the wastes in water samples taken either within the boundary of the site or elsewhere. Concentration profiles with depth determined for ^{90}Sr and ^{137}Cs to the southeast of the dump site are consistent with their origin in atmospheric fallout (KAUTSKY, 1977; KAUTSKY and co-authors, 1977).

(b) Interactions with Marine Organisms

Blackwater Estuary

During the pre-operational study of the Blackwater estuary it was concluded that the critical pathway for human exposure would be the reconcentration of ^{65}Zn by oysters and their consumption by the local inhabitants involved in the fishery. Subsequent investigations have, therefore, concentrated upon the interaction between the radionuclides in the station effluent and the oyster.

Fig. 2-11 shows the variation with time of the ^{65}Zn concentration in native oysters (*Ostrea edulis*) sampled 0·5 km downstream of the outfall on the southern bank of the estuary (Site A in Fig. 2-5). Also shown are the annual discharges of ^{65}Zn, and it is apparent that the relationship between the concentration of the nuclide in the oysters and the discharges has changed with time. This can be seen more clearly in Table 2-16 which gives the concentrations in the oysters normalized to the mean annual discharge in the current and previous year. Within the generally declining trend three more or less separate phases may be discerned. The reasons for this decline are not known although there are several possibilities. In 1968, the treatment plant installed to reduce the discharges of ^{137}Cs also had the effect of reducing the quantity of ^{65}Zn in the effluent (MITCHELL, 1969b); it may, in addition, have changed the chemical form of the latter nuclide and hence its biological availability. The feasibility of this explanation is reduced by the lack of a coincidence between the change in operational procedures and the first major drop in the normalized concentration of ^{65}Zn in oysters between 1970 and 1971. The stock of oysters which is sampled is known to be in decline (JEFFERIES, pers. comm.) and the causal stress may also affect the ability of the oyster to accumulate zinc. It should be noted, however, that oysters in poor condition, i.e. having low soft tissue dry weight for size, have the highest zinc content (PRESTON, 1967). The zinc concentration

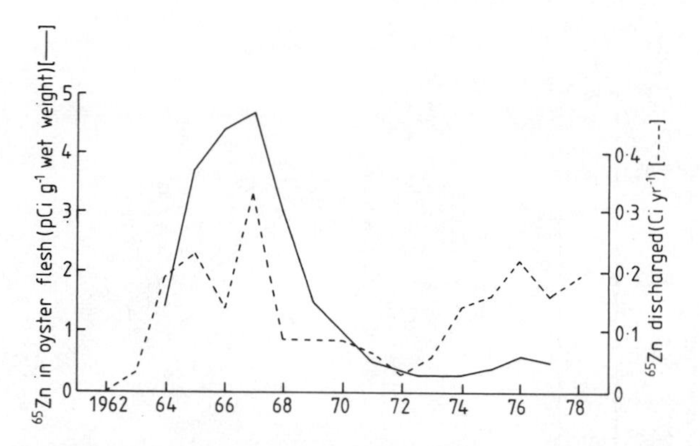

Fig. 2-11: Discharges of ^{65}Zn and concentrations of ^{65}Zn in oysters *Ostrea edulis* from the Blackwater estuary as a function of time. (Ministry of Agriculture, Fisheries and Food, Directorate of Fisheries Research, unpublished)

factor is known to be inversely related to the concentration in sea water (PRESTON, 1967), but the 75-fold increase in water concentration implied by the observed reduction in concentration factor makes this explanation rather implausible.

Although the concentrations of the radionuclides in the water are difficult to measure and often below the analytical limits of detection, their dispersion from the outfall can be traced by using oysters as indicators. Data for ^{65}Zn and ^{60}Co (April, 1966 to March, 1967) and ^{137}Cs (July to December 1965) are given for four stations (A to D) on the southern bank of the estuary in Fig. 2-12. From these it can be inferred that thc ^{137}Cs is

Table 2-16

Normalized concentration of ^{65}Zn in native oysters from the Blackwater estuary (MAFF, Directorate of Fisheries, Research, unpubl.)

Year	Normalized concentration, pCi g^{-1} per Ci y^{-1}	
1964	12·7	16·8 ± 4·3
1965	17·6	
1966	23·8	
1967	20·0	
1968	14·4	
1969	17·4	
1970	11·7	
1971	6·7	6·5 ± 0·2
1972	6·3	
1973	6·7	
1974	3·0	2·9 ± 0·3
1975	2·7	
1976	3·2	
1977	2·6	

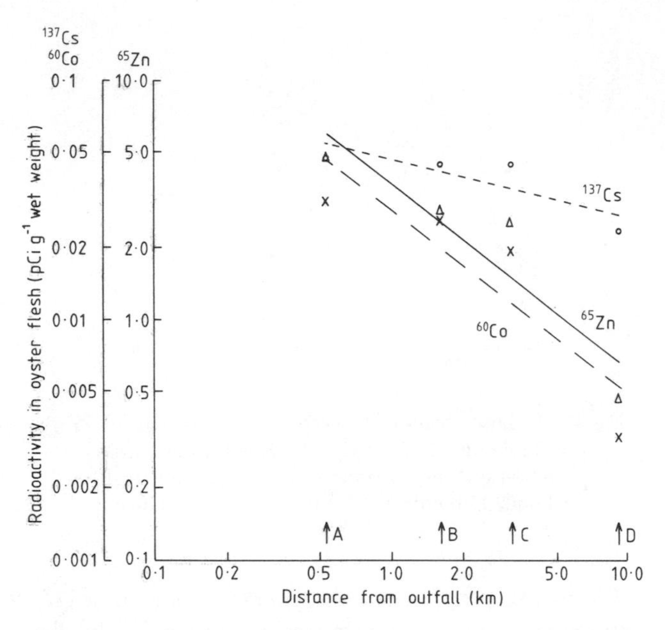

Fig. 2-12: Concentrations of ^{65}Zn, ^{60}Co, and ^{137}Cs in oysters *Ostrea edulis* from the Blackwater estuary as a function of distance from the outfall. (After PRESTON, 1968; Crown copyright)

fairly uniformly distributed within the estuary, a finding which is consistent with it behaving conservatively in the water and becoming well mixed as predicted by the hydrographic studies. In contrast, the concentrations of ^{65}Zn and ^{60}Co fall rapidly with distance indicating the presence of a sink which prevents dispersion. The concentration of ^{65}Zn in oysters from Site E, 4·4 km from the outfall but on the north shore of the estuary, were much lower than those in oysters from Site D, 9·25 km from the outfall on the southern bank. This observation could not be reconciled with the activity being transported in the well-mixed water phase, and the presumption that the activity becomes associated with silt particles was supported by the finding that Woodhead seabed drifters released at the outfall were transported along the southern shore but not across the estuary (PRESTON, 1967). Within the oyster the highest concentration of ^{65}Zn was measured in the gills which again suggested that the radionuclide was accumulated in a particulate form (PRESTON, 1968).

A summary of the concentrations of a variety of radionuclides in oysters, seaweed, fish, and sediment from the Blackwater estuary is given in Table 2-17. The data for 1968 (MITCHELL, 1969b) are included because these have previously been used for the purpose of environmental dosimetry (WOODHEAD, 1973a; IAEA, 1976). More recent data (HUNT, 1979) are given for comparison and also because measurements have been made on fish.

Oyster Creek

Although only 4 radionuclides have been positively identified in the waters of Barnegat Bay as having originated from the station, many more have been detected in

Table 2-17

Concentration of radionuclides in environmental materials from the Blackwater Estuary, pCi g^{-1} (mBq g^{-1}) wet weight (Compiled from MITCHELL, 1969b and HUNT, 1979)

Material	Year	^{32}P	^{65}Zn	^{55}Fe	^{60}Co	^{110m}Ag	^{134}Cs	^{137}Cs	^{95}Zr–^{95}Nb	^{238}Pu	$^{239\,240}Pu$	^{241}Am
Ostrea edulis	1968	0·25 (9·3)	7·6 (280)	0·28 (10)	0·28 (10)	0·38 (14)	0·6 (22)	1·4 (52)	NA	NA	NA	NA
	1977	NA	0·51 (19)	NA	ND	0·04 (1·5)	ND	0·10 (3·7)	ND	$8·6 \times 10^{-5}$ ($3·2 \times 10^{-3}$)	$3·7 \times 10^{-4}$ ($1·4 \times 10^{-2}$)	$6·8 \times 10^{-4}$ ($2·5 \times 10^{-2}$)
Fucus vesiculosus	1968	ND	ND	0·06 (2·2)	0·20 (7·4)	ND	ND	0·6 (22)	NA	NA	NA	NA
	1977	NA	ND	NA	ND	0·05 (1·9)	0·10 (3·7)	0·35 (13)	ND	NA	NA	NA
Chelon labrosus	1977	NA	ND	NA	ND	ND	$3·6 \times 10^{-2}$ (1·3)	0·24 (8·9)	ND	NA	NA	NA
Sediment (on a dry weight basis, the dry/wet ratio is ≃0·4)	1968	NA	ND	ND	0·3 (11)	ND	2·6 (96)	5·5 (200)	NA	NA	NA	NA
	1977	NA	ND	ND	0·04 (1·5)	0·20 (7·4)	0·54 (20)	2·2 (81)	0·05 (1·9)	NA	NA	NA

NA, not analysed; ND, Not detected.

Table 2-18

Concentration of radionuclides in environmental materials from Barnegat Bay, pCi g^{-1} (mBq g^{-1}) wet weight (Compiled from BLANCHARD and KAHN, 1979)

Radionuclide	Macro-algae	Fish muscle	Clam meat	Sediment (on a dry weight basis)
^{54}Mn	9×10^{-3}–0·78 (0·33–29)	0–2·4 $\times 10^{-2}$ (0–0·89)	ND	0·15–3·6 (5·6–130)
^{58}Co	0–1·6 $\times 10^{-2}$ (0–0·59)	ND	ND	ND
^{60}Co	8×10^{-3}–1·6 (0·30–59)	0–4·4 $\times 10^{-2}$ (0–1·6)	1·5 $\times 10^{-2}$–0·25 (0·56–9·3)	0·03–18·6 (1·1–690)
^{90}Sr	0–3 $\times 10^{-3}$ (0–0·11)	0–2·3 $\times 10^{-2}$ (0–0·85)	ND	ND
^{95}Zr	0–0·24 (0–8·9)	ND	ND	ND
^{95}Nb	0–0·13 (0–4·8)	ND	ND	ND
^{106}Ru	ND	0–5·8 $\times 10^{-2}$ (0–2·1)	ND	ND
^{134}Cs	0–1·4 $\times 10^{-2}$ (0–0·52)	[1·0–6·9] $\times 10^{-2}$ (0·37–2·6)	ND	0·04–0·89 (1·5–33)
^{137}Cs	0–4·6 $\times 10^{-2}$ (0–1·7)	0–0·12 (0–4·4)	ND	0·10–1·65 (3·7–61)
^{141}Ce	0–5·4 $\times 10^{-2}$ (0–2·0)	ND	ND	ND
^{144}Ce	0–0·13 (0–4·8)	ND	ND	ND

ND, Not detected.

biological materials sampled from throughout the bay and provide evidence for the widespread dispersion of the effluent. The data, with background subtracted, are summarized in Table 2-18 (BLANCHARD and KAHN, 1979).

Windscale

The magnitude of the discharge from the Windscale plant and the presence of several pathways leading to human exposure have ensured considerable investigation of the interactions between the radionuclides in the waste and the biological components of the environment. The pre-operational studies indicated that the harvesting of edible seaweed (*Porphyra umbilicalis*) from the shore close to the outfall would represent the principal constraint on the quantity of ^{106}Ru which could be discharged. Fig. 2-13 demonstrates the variation of the normalized ^{106}Ru concentration in weed sampled at Seascale and Braystones as a function of time since 1960 (JEFFERIES, pers. comm.). Data for the period 1953 to 1958 for unidentified sampling points have been published by DUNSTER (1958), and are consistent with the values given in Fig. 2-13. A relatively high mean value of 5·9 pCi g^{-1} per Ci d^{-1} (Bq g^{-1} per TBq d^{-1}) obtained for 4 sampling sites in 1959 was attributed to the prolonged calm conditions prevailing during the summer and the consequent restricted dispersion (PRESTON and JEFFERIES, 1969b). Over the continuous 1960 to 1980 period represented by Fig. 2-13 the mean value was 2·8 ± 0·9 pCi g^{-1} per Ci d^{-1} (Bq g^{-1} per TBq d^{-1}) discharged.

The variation with distance from the pipeline of the concentrations of ^{106}Ru (1961) and ^{95}Zr–^{95}Nb (1964 to 1966) in *Porphyra* is shown in Fig. 2-14 (PRESTON and JEFFERIES, 1969b; PRESTON and co-authors, 1971). The accumulation of ^{106}Ru by *Porphyra* traces the transport of this radionuclide for distances up to 160 km from the source before it becomes indistinguishable from the background due to fallout. The overall reduction in concentration with distance reflects the combined effects of dilution, removal of the radionuclides from water on to sediment, and radioactive decay. In addition, for ^{106}Ru there is some evidence for a change in the concentration factor with distance which may be attributable to a change in the chemical form of the element (PRESTON and JEFFERIES, 1969b). The cause of the sharp increase in the rate of reduction between 10 and 20 km has not been determined, but it may reflect the effective boundary between the

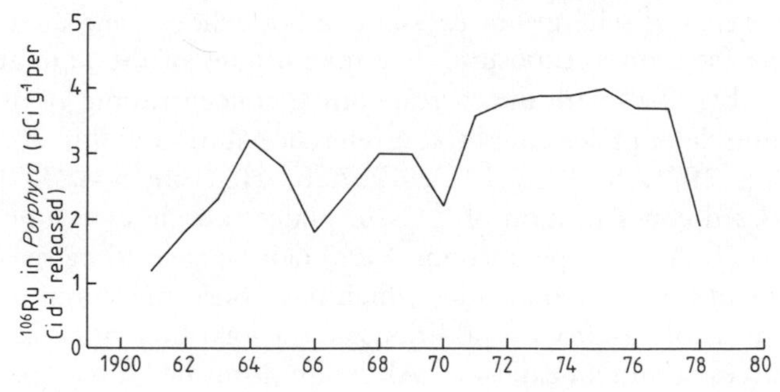

Fig. 2-13: Normalized concentration of ^{106}Ru in *Porphyra* from the Cumbrian coast as a function of time. (After PENTREATH, 1980; reproduced by permission Applied Science Publishers Ltd.; Crown copyright)

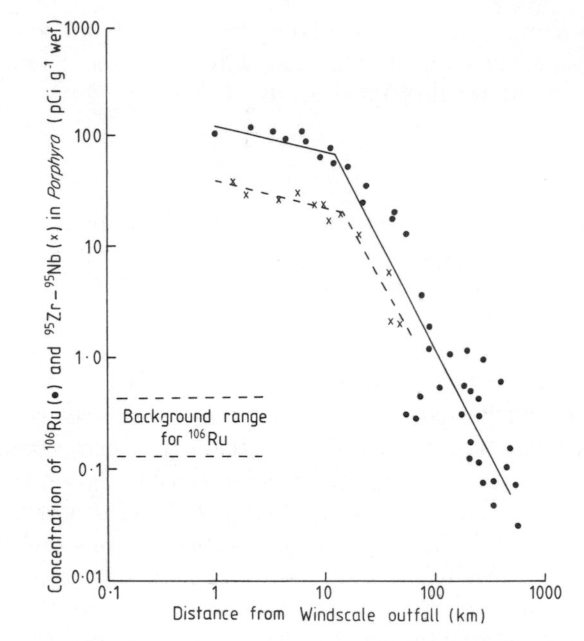

Fig. 2-14: Concentrations of ^{106}Ru and ^{95}Zr–^{95}Nb in *Porphyra* as a function of distance from Windscale. (After PRESTON and JEFFERIES, 1969b; PRESTON and co-authors. 1971; reproduced by permission of the International Atomic Energy Agency; Crown copyright)

advective transport of the nuclides along the coast in tidal currents and their more widespread dispersion by turbulent diffusion. Other radionuclides have also been measured in *Porphyra* and the data are given in Table 2-19 together with that for the brown seaweed *Fucus vesiculosus* (MITCHELL, 1969b; HUNT 1979). Where comparisons are possible, the differences in concentrations between 1968 and 1977 reflect the changes in the composition of the discharge (Table 2-12).

The presence of a small, but locally important inshore fishery, mainly for plaice, together with increasing discharges of caesium radionuclides generated another pathway of significance for human exposure. The magnitudes of the annual discharges of ^{137}Cs are plotted in Fig. 2-15 with the corresponding concentrations of the radionuclide measured in the muscle of plaice caught at a reference station within 5 km south of the outfall (MITCHELL, 1971a,b, 1973 1975, 1977a,b; HETHERINGTON, 1976b; HUNT, 1979); the normalized concentration of ^{137}Cs in plaice muscle as a function of time is shown in Fig. 2-16 (JEFFERIES, pers. comm.) and can be seen to be reasonably stable. The concentrations of other radionuclides which have been measured in plaice muscle are summarized in Table 2-20 with additional data for other organisms of potential importance as sources of human exposure. Although from the human radiological point of view it is only the activity in muscle (the edible fraction) which is important, the internal distributions, of the α-emitting radionuclides in particular, have a significant bearing on the exposure of the organisms. Data of this type for plaice are given in Table 2-21 (PENTREATH and LOVETT, 1978).

Table 2-19

Concentration of radionuclides in *Porphyra umbilicalis* and *Fucus vesiculosus* from the shoreline in the vicinity of Windscale, pCi g^{-1} wet (mBq g^{-1}) (Compiled from MITCHELL, 1969b and HUNT, 1979)

Radionuclide	*Porphyra umbilicalis*		*Fucus vesiculosus*	
	1968	1977	1968	1977
^{60}Co	NA	0·06–0·17 (2·2–6·3)	NA	0·85 (31)
$^{95}Zr–^{95}Nb$	1·2–220 (44–8·2 × 10^3)	0·70–16 (26–5·9 × 10^2)	7·0–140 (2·6 × 10^2–5·2 × 10^3)	7·8–14 (2·9 × 10^2–5·2 × 10^2)
^{106}Ru	8·2–260 (3·0 × 10^2–9·7 × 10^3)	23–230 (8·5 × 10^2–8·5 × 10^3)	2·8–40 (1·0 × 10^2–1·5 × 10^3)	31–49 (1·1 × 10^3–1·8 × 10^3)
^{134}Cs	NA	0·35–1·6 (13–59)	NA	6·2–7·0 (2·3 × 10^2–2·6 × 10^2)
^{137}Cs	NA	2·7–10 (1·0 × 10^2–3·7 × 10^2)	0·9–6·9 (33–2·6 × 10^2)	41–48 (1·5 × 10^3–1·8 × 10^3)
^{144}Ce	0·9–62 (33–2·3 × 10^3)	0–27 (0–1·0 × 10^3)	1·8–15 (67–5·6 × 10^2)	0·28–0·83 (10–31)
^{238}Pu	NA	0·34 (13)	NA	NA
$^{239,240}Pu$	NA	1·3 (48)	NA	NA
^{241}Am	NA	0·84 (31)	NA	NA
^{242}Cm	NA	2·2 × 10^{-2} (0·81)	NA	NA
$^{243,244}Cm$	NA	7·8 × 10^{-3} (0·29)	NA	NA

NA, Not analysed.

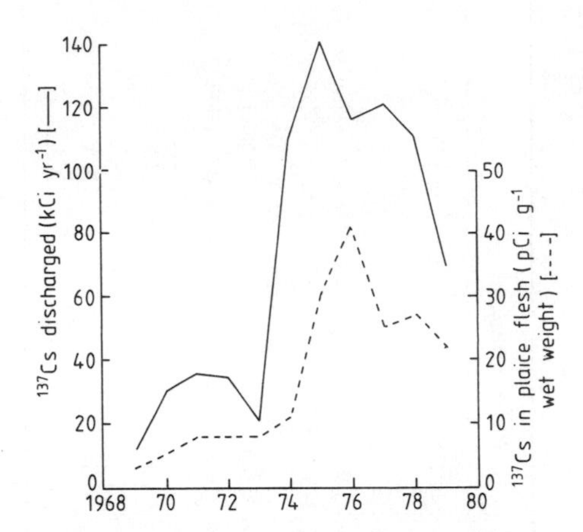

Fig. 2-15: Discharges of ^{137}Cs and concentrations of ^{137}Cs in plaice flesh as a function of time. (Original from data given in Table 2-12; MITCHELL, 1971a,b, 1973, 1975, 1977a,b; WOODHEAD, 1973a; HETHERINGTON, 1976b; HUNT, 1979, 1980 (Ministry of Agriculture, Fisheries and Food, Directorate of Fisheries Research, unpublished)

The much increased discharges of $^{134,137}Cs$ in recent years and their widespread dispersion beyond the confines of the Irish Sea into the shelf waters of northwestern Europe (p. 1162) have extended the radiological context from the limitation of individual exposure in the coastal community of Cumbria to include the assessment of the collective dose equivalent from the consumption of fish caught in these waters. This has involved the routine radiometricanalysis of fish sampled from grounds distant from Windscale (HETHERINGTON, 1976b; HUNT, 1979; MITCHELL, 1975, 1977a,b).

Initially, laboratory studies of the accumulation of radionuclides by marine organisms were made to provide estimates of the concentration factors required for critical path

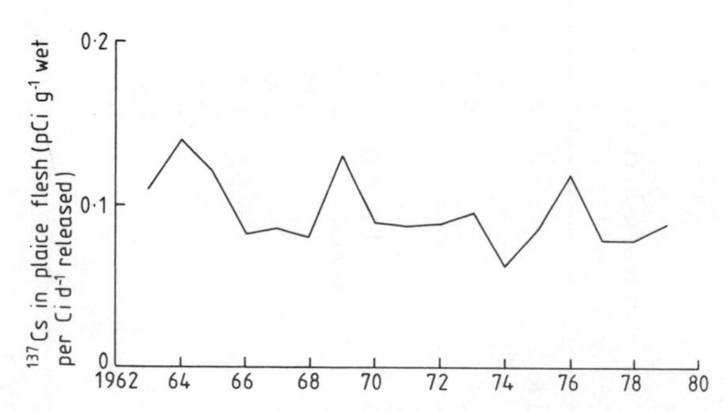

Fig. 2-16: Normalized concentrations of ^{137}Cs in plaice flesh as a function of time. (After PENTREATH, 1980; reproduced by permission of Applied Science Publishers Ltd.; Crown copyright)

Table 2-20

Concentration of radionuclides in organisms sampled close to Windscale, pCi g^{-1} wet (mBq g^{-1}) (Compiled from MITCHELL, 1969b and HUNT, 1979)

Organism	Year	^{60}Co	^{95}Zr–^{95}Nb	^{99}Tc	^{106}Ru	^{110m}Ag	^{134}Cs	^{137}Cs	^{144}Ce	^{238}Pu	$^{239,240}Pu$	^{241}Am	^{242}Cm	$^{243,244}Cm$
Plaice														
Pleuronectes platessa	1968	NA	NA	NA	NA	NA	0·2–0·7 (7·4–26)	0·7–2·0 (26–74)	NA	NA	NA	NA	NA	NA
	1977	NA	NA	NA	NA	NA	3·1 (110)	25 (930)	NA	1·9 × 10^{-4} (7·0 × 10^{-3})	7·3 × 10 × $^{-4}$ (2·7 × 10^{-2})	1·0 × 10^{-3} (3·7 × 10^{-2})	ND	ND
Cod														
Gadus morhua	1977	NA	NA	NA	NA	NA	2·6 (96)	23 (850)	NA	2·1 × 10^{-4} (7·8 × 10^{-3})	8·0 × 10^{-4} (3·0 × 10^{-2})	7·3 × 10^{-4} (2·7 × 10^{-2})	ND	ND
Crabs														
Cancer pagurus	1977	0·23 (8·5)	0·35 (13)	0·28 (10)	30 (1·1 × 10^{3})	0·92 (34)	2·2 (81)	17 (630)	0·12 (4·4)	3·8 × 10^{-2} (1·4)	0·15 (5·6)	0·44 (16)	8·0 × 10^{-3} (0·30)	3·8 × 10^{-3} (0·14)
Lobsters														
Homarus vulgaris	1977	ND	ND	NA	21 (780)	0·78 (29)	6·5 (240)	52 (1·9 × 10^{3})	ND	NA	NA	NA	NA	NA
Winkles														
Littorina littorea	1977	1·2 (44)	46 (1·7 × 10^{3})	0·11 (4·1)	540 (2·0 × 10^{4})	4·2 (160)	3·6 (130)	26 (960)	4·5 (170)	2·0 (74)	7·6 (280)	5·7 (210)	0·32 (12)	4·0 × 10^{-2} (1·5)
Mussels														
Mytilus edulis	1977	0·54 (20)	91 (3·4 × 10^{3})	NA	750 (2·8 × 10^{4})	ND	2·0 (74)	14 (520)	0·43 (16)	2·3 (85)	9·5 (350)	8·2 (300)	0·19 (7·0)	3·1 × 10^{-2} (1·1)

NA, Not analysed; ND, Not detected.

Table 2-21

Concentration of transuranic nuclides in plaice during 1975–1977, fCi g^{-1} wet (μBq g^{-1}). (After PENREATH and LOVETT; 1978; Crown copyright; reproduced by permission of Springer-Verlag)

Material	$^{239,240}Pu$	^{241}Am	^{242}Cm	$^{243,244}Cm$
Muscle	0·09–0·65 (3·3–24)	0·3–2·3 (11–85)	NA	NA
Bone	0·65–89 (24–3·3 × 10^{3})	2·1–178 (78–6·6 × 10^{3})	1·6 (59)	<0·5 (<19)
Kidney	18–88 (6·7 × 10^{2}–3·3 × 10^{3})	47–440 (1·7 × 10^{3}–1·6 × 10^{4})	0·5 (19)	<0·5 (<19)
Liver	5·0–29 (1·9 × 10^{2}–1·1 × 10^{3})	25–590 (9·3 × 10^{2}–2·2 × 10^{4})	1·8 (67)	0·5 (19)
Gill	7·9–770 (2·9 × 10^{2}–2·8 × 10^{4})	21–310 (7·8 × 10^{2}–1·1 × 10^{4})	1·1 (41)	<0·5 (<19)
Skin	1·2–4·3 (44–1·6 × 10^{2})	2·4–20 (89–7·4 × 10^{2})	NA	NA
Gut	13–72 (4·8 × 10^{2}–2·7 × 10^{3})	31–160 (1·1 × 10^{3}–5·9 × 10^{3})	NA	NA
Gut contents	540–7470 (2·0 × 10^{4}–2·8 × 10^{5})	930–7840 (3·4 × 10^{4}–2·9 × 10^{5})	90 (3·3 × 10^{3})	38 (1·4 × 10^{3})

NA, Not analysed.

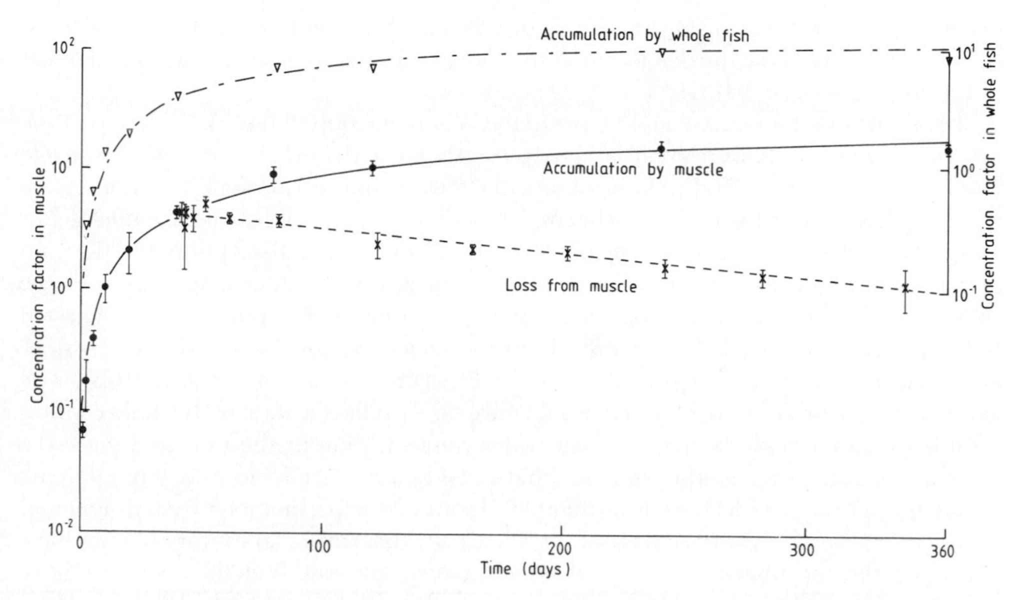

Fig. 2-17: Accumulation from sea water, and loss, of ^{134}Cs by plaice. (Original from data given in JEFFERIES and HEWETT, 1971)

analysis (DUNSTER, 1958; MORGAN, 1964; BRYAN and co-authors, 1966), but as monitoring data have accumulated, the emphasis of these investigations has shifted towards gaining a better understanding of the mechanisms underlying the environmental observations. The uptake of ^{134}Cs from water by plaice (whole fish and muscle) and its subsequent loss from muscle are illustrated in Fig. 2-17. The contamination of the fish is given in terms of the concentration factor defined as:

$$\frac{\text{radioactivity (pCi or mBq) g}^{-1}\text{ of organism or tissue}}{\text{radioactivity g}^{-1}\text{ sea water}}$$

The data have been fitted by the method of least squares with a simple exponential model (p. 1128) in which the concentration factor $[C]$ at time t is given by $[C] = [C_{ss}](1 - e^{-Kt})$ where $[C_{ss}]$ is the steady state, or equilibrium, concentration factor and K is the excretion rate per unit time (equivalent to λ_B on p. 1128). For the whole fish $[C_{ss}]$ was determined to be 10·6 but individual tissues showed values between 2·4 (blood) and 20·2 (muscle). The biological half-times (0·693/K) for plaice muscle calculated from the accumulation and excretion segments of the curves in Fig. 2-17 were 120 and 136d respectively, showing good agreement. For the whole fish the overall half-time for accumulation was somewhat shorter at 65 d, but the excretion curve (not measured) would be expected to show evidence that the loss could be described by the sum of (at least) two exponential functions corresponding to those compartments with short turnover times (gut, gills, liver, kidney, etc.) and the slower muscle compartment respectively. A comparison of these results with environmental data suggested that intake from water could only account for approximately half the steady-state body burden, the remainder being attributed to the consumption of contaminated food (JEFFERIES and HEWETT, 1971). Further investigations showed that, under laboratory conditions and using a single food organism, the intake of caesium from food accounted for approximately 42% of the body burden and that the rate of turnover in muscle was independent of the route of intake (HEWETT and JEFFERIES, 1978).

Studies of Group-I plaice caught close to the Windscale outfall have indicated that the accumulation of ^{137}Cs from food is closely correlated with the presence of polychaete worms in the gut. This food item contained the lowest concentration of ^{137}Cs among the remains identified but was the most completely digested, thus releasing the nuclide for absorption across the gut wall. The observed variations in the absorption of ^{137}Cs from both food and water, the variation in tissue concentrations due to growth, and the changes in the ambient water concentration of ^{137}Cs during the study period suggested that any value derived from the data from the concentration factor would be rather arbitrary (PENTREATH and JEFFERIES, 1971; PENTREATH and co-authors, 1973). The value obtained for an individual fish would only rarely reflect a state of dynamic equilibrium between the body burden and the water concentration at the time and place the fish was caught. These studies indicate that considerable care is necessary in applying concentration factors derived from simple laboratory experiments to environmental hazard assessments. In the context of the Windscale discharge, laboratory studies have employed the thornback ray as well as the plaice and radionuclides studied have included ^{54}Mn, ^{58}Co, ^{59}Fe, ^{65}Zn, ^{110m}Ag, and ^{237}Pu (JEFFERIES and HEWETT, 1971; HEWETT and JEFFERIES, 1978; PENTREATH, 1973a,b, 1976, 1977b, 1978a,b).

Northeast Atlantic Deep Ocean Disposal Site

Some very preliminary results from the analyses of fish sampled in 1979 at the northeast Atlantic dump site have been published by FELDT and co-authors (1979). Both ^{137}Cs (0·037 to 0·051 pCi g^{-1} dry weight) and ^{90}Sr (0·014 pCi g^{-1} dry weight) were detected in the fish and would be expected to be present from fallout. In the absence of measurements for similar fish from stations remote from the dump site it is difficult to assess the significance of these data. It should be noted that the concentration of fallout ^{137}Cs in deep ocean water from the northeast Atlantic Ocean has been determined to be 0·01 to 0·02 pCi l^{-1} (KAUTSKY, 1977) and this, together with the assumptions of a concentration factor of 50 and a dry/wet ratio of 0·2, would suggest a fish muscle concentration of $2{\cdot}5$ to $5{\cdot}0 \times 10^{-3}$ pCi g^{-1} dry weight, an order of magnitude less than the value observed. Given the assumptions in this calculation, it would be premature to attribute the observed ^{137}Cs contamination to waste disposal but the relevance of further investigations is indicated.

(c) Interaction with Sediment

Blackwater Estuary

As has already been noted (p. 1168) the relative extent of the interaction between sediment and the radionuclides in the effluent may be inferred from the distributions within the estuary of the concentration of the individual radionuclides in the oysters. Fig. 2-12 shows that the concentrations of ^{60}Co and ^{65}Zn in oysters fall much more rapidly with distance from the outfall than those for ^{137}Cs implying a scavenging mechanism in the estuary for the former. Additional data indicate that ^{55}Fe behaves similarly to ^{60}Co and ^{65}Zn while ^{32}P and ^{110m}Ag are more conservative and behave like ^{137}Cs (PRESTON, 1968; PRESTON and co-authors, 1968; PRESTON 1972). The accumulation of ^{55}Fe, ^{60}Co, and ^{65}Zn by fine particulate materials in suspension would delay their dispersion in the estuary and also increase their availability to filter feeding organisms such as the oyster. The concentrations of certain nuclides in silt sampled from a station 1·6 km upstream from the outfall are given in Table 2-17 (MITCHELL, 1969b; HUNT, 1979) from which it can be seen that, despite the inferred adsorption of ^{65}Zn to fine silt, this nuclide cannot be detected in sediment samples. In contrast, the reported presence of ^{110m}Ag in sediment is somewhat puzzling given that fact that the annual discharge has been an order of magnitude less than that of ^{65}Zn in recent years (Table 2-10) and all the other evidence for its more conservative distribution throughout the estuary as cited above. A possible explanation may be found in the fact that the identifying peak (0·658 MeV) in the analytical scheme (p. 1119 to 1121) coincides with that of ^{137}Cs (0·662 MeV) within the resolution of the sodium iodide γ-spectrometry system. The identification may, therefore, be an artefact of the computer-based spectrum analysis.

Oyster Creek

During the period October 1971 to October 1973 ^{60}Co was the only nuclide attributable to station operation which could be detected with certainty in sediment sampled from the greater part of Barnegat Bay. Concentrations decreased from maximum values in Oyster Creek to below the limit of detection ($\approx$0·03 pCi g^{-1}) at the northern and southern extremities of the bay. Manganese-54, ^{134}Cs, and ^{137}Cs were also detectable in sediment samples taken from Oyster Creek, Forked River, and the coastline of Barnegat Bay in the immediate vicinity of these two inlets. The radionuclides were found to be associated with sediments containing fine particles. The data are summarized in Table 2-18 (BLANCHARD and co-authors, 1976) where the ^{137}Cs values have been corrected for fallout background.

Windscale

The distributions of certain fission product radionuclides in the surface sediment of the northeast Irish Sea are given in Fig. 2-18a–d, based on samples collected during September 1968 from a widespread grid of 26 stations. In comparison with the corresponding concentrations in sea water (Fig. 2-7a–d) the contours for ^{95}Zr–^{95}Nb, ^{106}Ru, and ^{144}Ce show a similar elongation parallel to the coast, but are displaced to the north of the outfall in contrast to the southerly set for the water; the comparison for ^{137}Cs reveals effective identity. The distribution of the activity on the seabed is related to the nature of the sediment and illustrates the positive correlation between particle size and concentration factor (HETHERINGTON and JEFFERIES, 1974). The higher concentrations are associated with the muddy substrate to the north of the outfall whereas the lower concentrations are found in the hard sand bottom to the south and offshore. The rate of decrease in concentration with distance is greatest for ^{95}Zr–^{95}Nb and least for ^{137}Cs with ^{106}Ru and ^{144}Ce occupying an intermediate position (see also HETHERINGTON and JEFFERIES, 1974); a similar result was obtained from a study of estuarine sediments at distances up to 120 km from Windscale (PRESTON and co-authors, 1971; PRESTON, 1972). Close to the outfall this effect is mainly a reflection of the relative affinity of the

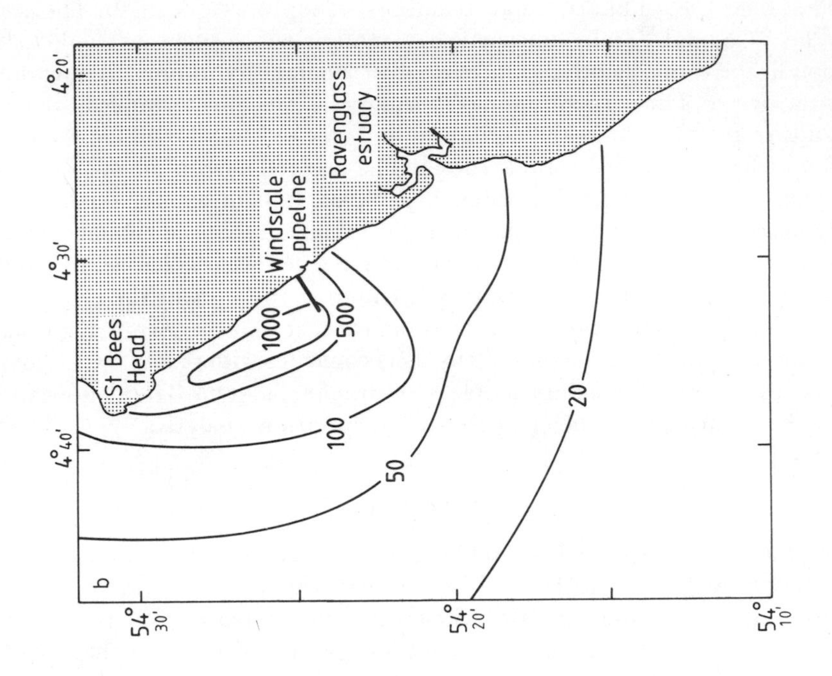
b
54° 30'
54° 20'
54° 10'
4° 40'
4° 30'
4° 20'
St Bees Head
Windscale pipeline
Ravenglass estuary
1000
500
100
50
20

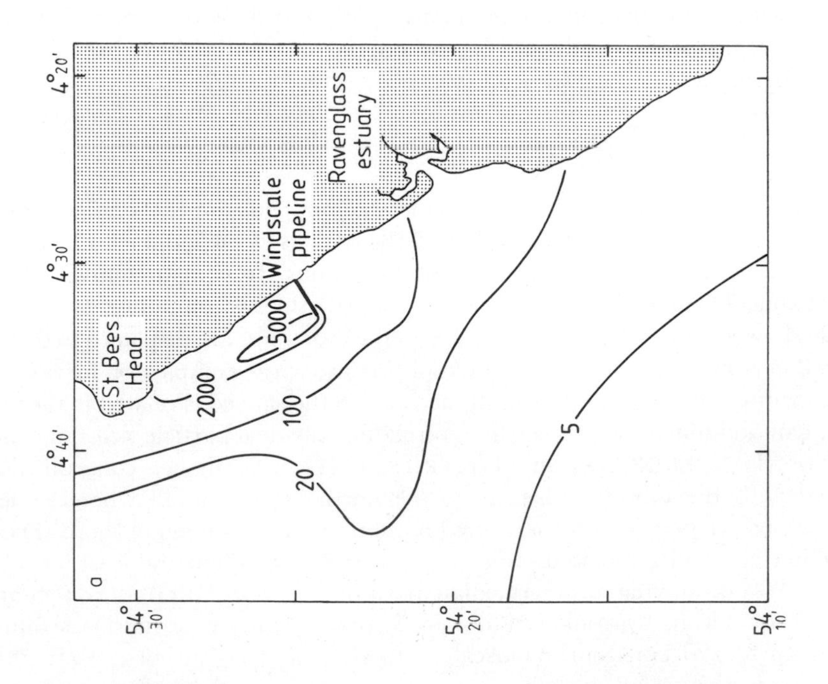
a
54° 30'
54° 20'
54° 10'
4° 40'
4° 30'
4° 20'
St Bees Head
Windscale pipeline
Ravenglass estuary
5000
2000
100
20
5

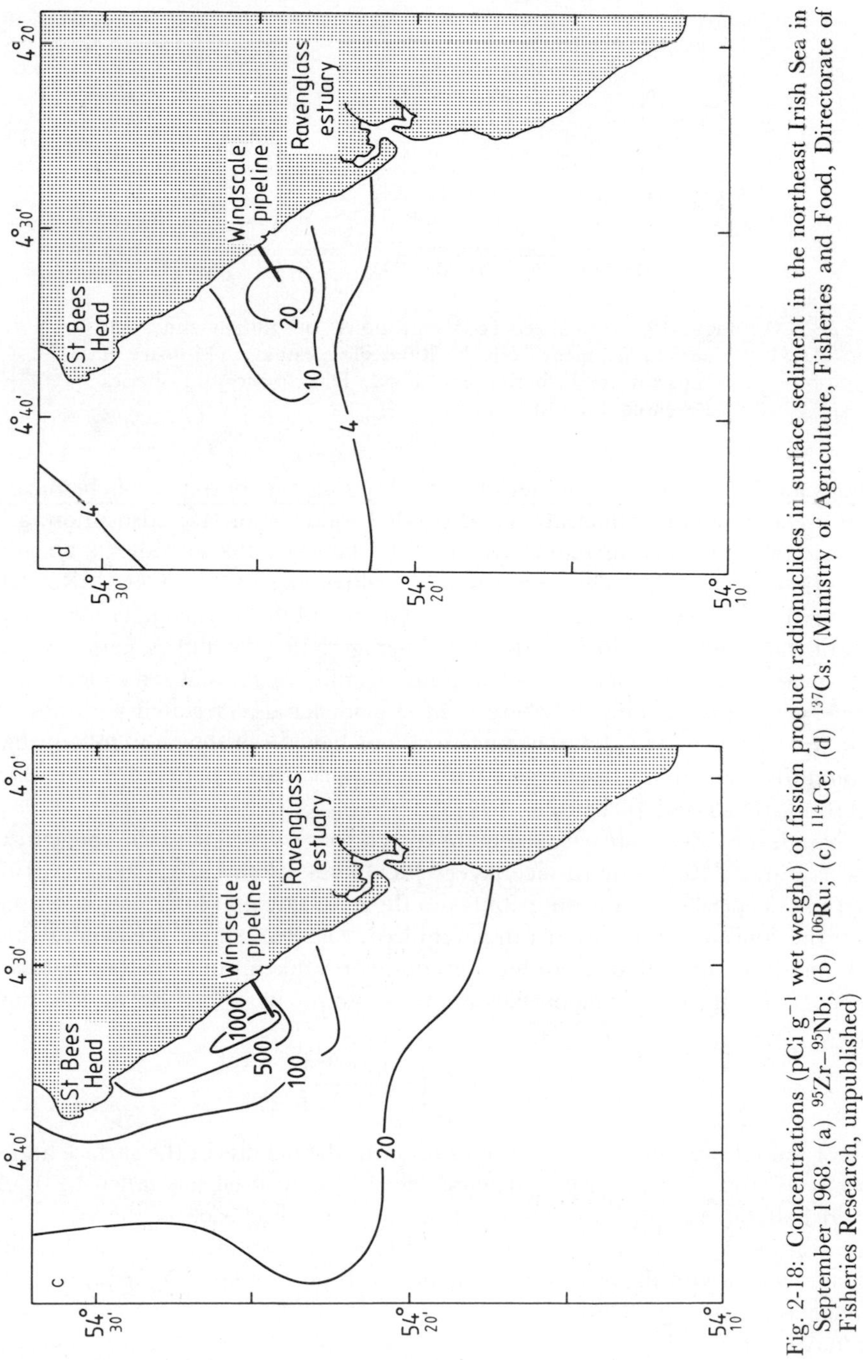

Fig. 2-18: Concentrations (pCi g^{-1} wet weight) of fission product radionuclides in surface sediment in the northeast Irish Sea in September 1968. (a) ^{95}Zr–^{95}Nb; (b) ^{106}Ru; (c) ^{114}Ce; (d) ^{137}Cs. (Ministry of Agriculture, Fisheries and Food, Directorate of Fisheries Research, unpublished)

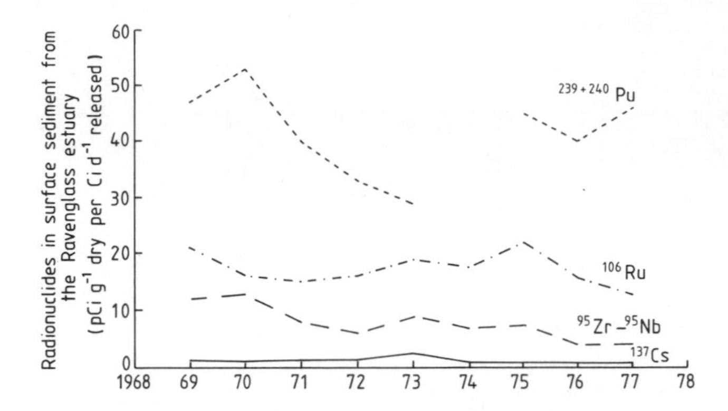

Fig. 2-19: Normalized concentrations of radionuclides in surface sediment from the Ravenglass estuary. (Ministry of Agriculture, Fisheries and Food, Directorate of Fisheries Research, unpublished)

nuclides for fine sedimentary material through either the formation of chemical species which precipitate out or nucleate on suspended solids or surface adsorption; at greater distances the difference in radioactive half-life between the nuclides is an additional factor (JEFFERIES, 1970). The normalized concentrations of ^{95}Zr–^{95}Nb, ^{106}Ru, ^{137}Cs, and the actinide nuclides $^{239+240}$Pu in surface sediments from the Ravenglass estuary 10 km south of the Windscale pipeline are given in Fig. 2-19 (JEFFERIES, pers. comm.). The greatest variation is shown by ^{95}Zr–^{95}Nb, but, overall, the normalized concentrations are relatively stable with time and there is little evidence of correlated variation. For the fission product radionuclides, the concentrations have been shown to follow changes in the quantity of activity discharged with a response time between 1 and 3 months (HETHERINGTON and JEFFERIES, 1974).

Although the γ-dose rate over sediment correlates very well with the concentrations of ^{95}Zr–^{95}Nb and ^{106}Ru in the surface layer (JEFFERIES, 1968, 1970; PRESTON, 1972), the activity in deeper layers also contributes to the dose rate and information on the variation of the concentration with depth is required. Fig. 2-20 shows the average measured depth distributions of fission products in mud cores taken from the Ravenglass estuary between 1967 and 1973. The profiles are most simply described by an equation of the form

$$A_z = A_o \, e \left(- \frac{0.693z}{z_{\frac{1}{2}}} \right)$$

where A_o and A_z are the volume concentrations of the nuclide at the surface and depth z respectively and $z_{\frac{1}{2}}$ is the depth at which the concentration has fallen to $0{\cdot}5A_o$. The derived half-value depths are tabulated in Fig. 2-20. Inclusion of the variation of the concentration of activity with depth in the sediment in a predictive model has produced good agreement with the measured γ-ray dose rates over contaminated mud flats in the Ravenglass estuary (JEFFERIES, 1970).

It may be reasonably assumed that the two caesium isotopes behave identically both in respect of diffusion and their chemical and physical interactions with sediment particles. Therefore, the difference between the two concentration profiles for these nuclides

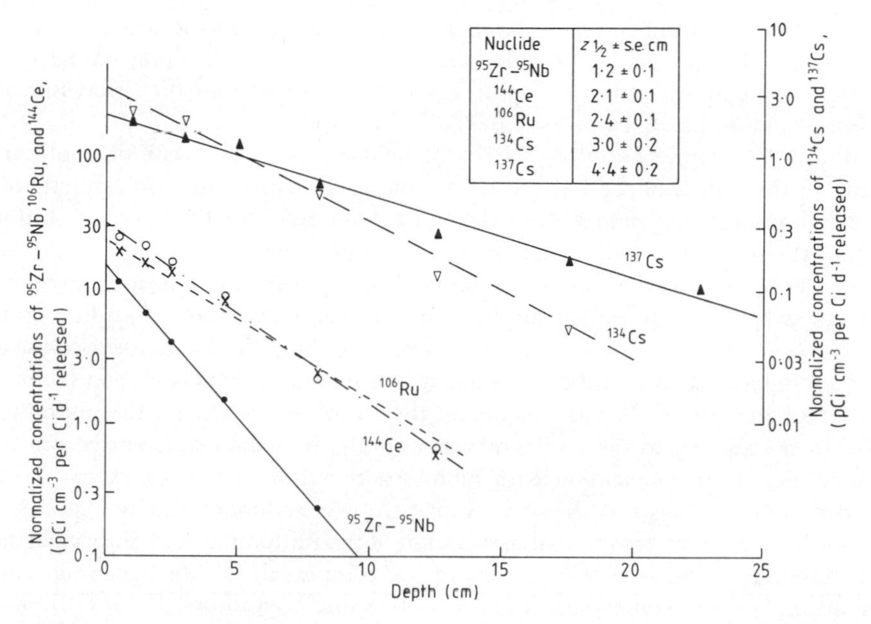

Fig. 2-20: Variation of normalized concentrations of fission product radionuclides with depth in sediment. (After HETHERINGTON and JEFFERIES, 1974; reproduced by permission of Netherlands Institute for Sea Research; Crown copyright.)

can only be ascribed to the difference between their physical half-lives and this has provided a basis for deriving an estimate of 4·2 cm yr^{-1} for the penetration rate of caesium into the mud. This has further been used to give an estimate of approximately 10^{-6} cm^2 s^{-1} for the apparent diffusion coefficient of ^{137}Cs in the sediment. This parameter represents the integrated effect of the true diffusion of the activity in the interstitial water, the retardation due to sorption on to the sediment and bioturbation. In the absence of corresponding isotope pairs the apparent diffusion coefficients of $^{95}Zr-^{95}Nb$, ^{106}Ru, and ^{144}Ce had to be evaluated from the profiles by an alternative approach. If a steady-state condition is assumed in the cores then at any point the loss of activity due to decay must be balanced by the net gain through apparent diffusion. Not unexpectedly, given the known affinity of these nuclides for sediments, the derived apparent diffusion coefficients, at 3 to 4 × 10^{-7} cm^2 s^{-1}, were somewhat lower than that for ^{137}Cs (HETHERINGTON and JEFFERIES, 1974). In addition to their use in providing a basis for understanding the variations of radiation exposure over contaminated sediments, the data on depth distributions are essential for the preparation of inventories of environmental radioactivity. When balanced against the reported discharges these provide a useful means for ensuring that all significant environmental compartments have been investigated. In the immediate vicinity of Windscale it has been estimated that more than 95% of the $^{95}Zr-^{95}Nb$, ^{106}Ru, and ^{144}Ce and approximately 60% of the ^{137}Cs are associated with the seabed. The size of this compartment emphasizes the importance of an understanding of the mechanisms which control the behaviour of the radionuclides in the sediment. For the short-lived radionuclides adsorbed on to sediment this compartment probably represents the ultimate sink, i.e. decay occurs before any hypothetical long-term processes could result in significant release back into the water column and hence the possibility of a greater potential for human exposure. This may not necessarily be true for the longer-lived nuclides and it is in this context that the behaviour of the α-emitting actinides has been investigated.

HETHERINGTON and co-authors (1975) concluded—on the basis of a plutonium inventory in the waters of the Irish Sea, the annual discharges and an estimate of one year for the turnover half-time of the water in the Irish Sea—that 96% of the plutonium input from Windscale is rapidly removed to the sediment in the immediate vicinity of the pipeline. The extent of the affinity of plutonium for sediment, as indicated by the concentration factor, is dependent on the particle size distribution, with finer-grained sediments showing higher values. After allowing for the different natures of the sediments there appeared to be little variation in the measured concentration factor with distance from the outfall. It was concluded, therefore, that although the plutonium in the effluent is removed to the sediment very rapidly, it is then transported out of the immediate vicinity in association with mobile sedimentary particles. Apart from the observed correlation between the activity per gram of sediment and the specific area there appeared to be no preferential association of the plutonium with either particular mineral fractions of the sediment or the organic component i.e. the accumulation is a surface adsorption phenomenon (HETHERINGTON and co-authors, 1975; HETHERINGTON, 1976a; 1978). Fig. 2-19 includes the normalized concentrations of $^{239+240}Pu$ in surface sediment from the Ravenglass estuary as a function of time; the overall variation is less than a factor of 2 and there is no indication of a build-up of the nuclides despite increasing rates of discharge and long physical half-lives.

The distribution of the plutonium with depth in the sediment has been investigated to determine whether the element is fixed or shows evidence of mobility. The profiles of plutonium concentration with depth have, at certain sites, been shown to be exponential (see for example Fig. 2-21a) and very similar to those of ^{137}Cs; the main difference being that the plutonium shows a slightly greater penetration. In view of the very different chemical properties which would be predicted for these two elements in the circumstances, it was concluded that the observed similarity in the sediment profiles could not be reconciled with a process of molecular diffusion and sorption–desorption reactions. It was accepted, however, that if other data indicated that this process did play a significant role in determining the distribution then it would be important evidence as to the actual (as opposed to predicted) chemical state of plutonium in the marine environment. The observed profiles are compatible with both biological reworking of the sediment and the process of continuing sedimentation. Further studies of the variation of the $^{239+240}Pu$: ^{238}Pu activity ratio with depth in the cores have provided support for the interpretation that sedimentation is the primary mechanism leading to the incorporation of these nuclides into the seabed in these areas. As fuel elements with progressively longer irradiation times have been processed, the activity ratio in the effluent has fallen. Although there was no direct information on the ratio in the discharges the change with time has been assumed to be reflected in the activity ratio measured annually in bulked surface sediment sampled monthly from the Ravenglass estuary. The data are given in Table 2-22. Therefore, it was assumed that the activity ratio at a point in a core fixed, by comparison with the data in Table 2-22, the time at which the sediment was labelled and deposited. Fig. 2-21b shows the activity ratio profile for a core and the data from Table 2-22 plotted on a linear time scale adjusted by a process of trial and error to give the best fit. From this a mean sedimentation rate of 25 mm a^{-1} over the 8 yr period was deduced

Table 2-22

Variation with time of $^{239+240}Pu$:^{238}Pu activity ratio in surface sediment from the Ravenglass estuary (Reprinted from HETHERINGTON, 1978; *Marine Science Communications* **4** (3), 239–274, by courtesy of Marcel Dekker, Inc. Crown copyright.)

Year	Activity ratio ($\pm$ 1σ counting statistics)
1966	19 ± 3
1967	15 ± 7
1968	13·5 ± 2
1969	10 ± 0·8
1970	8·6 ± 1·6
1971	5·4 ± 0·6
1972	5·1 ± 0·3
1973	5·0 ± 0·3
1974	NA
1975	4·2 ± 0·1

NA, Not analysed.

The fit could be improved further with the use of a variable time scale implying, not unreasonably, that the sedimentation rate varies with time. The preservation of the variation of the activity ratio down the core indicates that there has been relatively little disturbance of the sediment since deposition and, therefore, that biological reworking of the sediment must be of minor importance at this particular station (HETHERINGTON, 1976a, 1978).

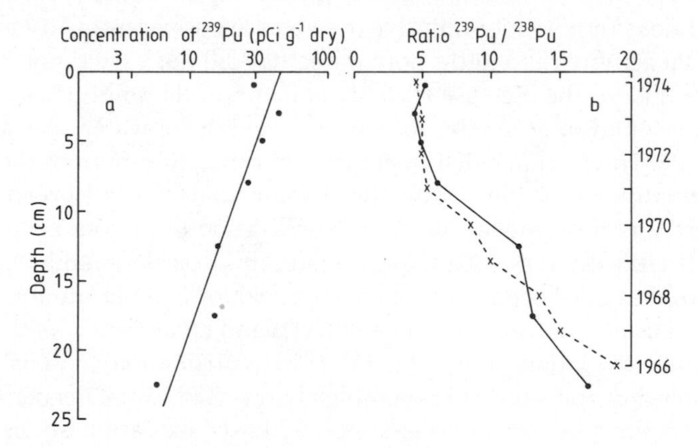

Fig. 2-21: Variation of plutonium concentration in sediment. (a) Distribution of ^{239}Pu with depth. (b) The $^{239}Pu/^{238}Pu$ ratio as a function of depth (—●—, left-hand ordinate) and as a function of time in surface sediment (—×—, right-hand ordinate). The timescale has been adjusted to give the closest coincidence between the two graphs. (After HETHERINGTON, 1976a; reproduced by permission of Ann Arbor Science Publishers, Inc.; Crown copyright)

There is, however, some evidence that this neat picture does not tell the whole story. The finding of relatively uniform activity ratio profiles in other cores (even from the same station as the example given in Fig. 2-21b) indicates mixing to a depth of at least 55 cm (HETHERINGTON, 1976a; PENTREATH and co-authors, 1980a). It was suggested that the use of a gravity barrel corer to obtain longer cores might have resulted in the spread of plutonium from the surface to the deeper segments through the coring action, and hence produced a uniform activity ratio profile (HETHERINGTON, 1976a). If it is assumed that the *in situ* activity ratio profile for the gravity core studied by HETHERINGTON (1976a) was similar to that given in Fig. 2-21b (from the same station) then the reduced activity ratio observed at depth requires a greater input of ^{238}Pu relative to $^{239+240}Pu$. Thus a mechanism must be proposed for the selective transport of ^{238}Pu and the coring action does not seem to be an acceptable explanation. In addition, the transfer of plutonium downward would increase the apparent penetration whereas the half-value depth for the gravity core is substantially less than that for the Reineck box core obtained at the same site (Fig. 2-21a). Although it is clear that undisturbed profiles are present in the sediment, the question of whether disturbed profiles are an artefact of the sampling technique or the product of natural processes, such as resuspension and sedimentation or biological reworking remains to be answered. The undisturbed profiles indicate that at least part of the plutonium discharged to the Irish Sea has remained immobilized in sediment for periods of years.

Although the preservation of the activity ratio profiles is strong circumstantial evidence for the retention of the plutonium within the sediment column in a stable state, only analyses of interstitial water can demonstrate the existence of conditions necessary for, and the extent of, movement within the column. Preliminary results have indicated that only a very small fraction ($<0{\cdot}01\%$) of the plutonium in the sediment column is present in the pore water and, therefore, available for redistribution. Apart from the surface layer, the concentrations of plutonium in interstitial water extracted from two estuarine cores showed no variation with depth within the limits of the analytical errors; thus the conditions required for diffusive transport are apparently absent. The measured $^{239+240}Pu : {}^{238}Pu$ activity ratio in the pore water also did not vary significantly with depth in marked contrast to the increase with depth found in the solid phase. The ratio in the pore water was found to be similar to that in the overlying water, as was the concentration of $^{239+240}Pu$. It was concluded, therefore, that the plutonium in the pore water was more representative of that in the overlying water than that adsorbed in the sediment and that there was no evidence for the remobilization of plutonium in these estuarine sediments (HETHERINGTON, 1978). Once again, this description may be oversimplified. Preliminary studies have been made of the chemical form of plutonium in the interstitial water obtained from a marine core (PENTREATH and co-authors, 1980a). In the surface layer the higher oxidation states, Pu (V + VI), predominate, as is the case in the overlying water, but the situation is completely reversed in the deeper layers where the plutonium is mainly in the much less mobile, lower oxidation states, Pu (III + IV) (NELSON and LOVETT, 1978). Thus it appears that, for marine, as opposed to estuarine, sediments, the plutonium in the interstitial water may not be representative of that in the overlying water. The marine geochemistry of plutonium is obviously extremely complex and the present data base is far too small to permit firm conclusions. There is still much to be learned from the northeast Irish Sea before an unqualified assessment can be made of the long-term behaviour of plutonium.

It has also been established that a substantial fraction of the ^{241}Am discharged into the northeast Irish Sea is rapidly removed to the sediment (HETHERINGTON and co-authors, 1976), and the concentration factor for curium is suspended particulate material implies that such is likely to be the case for this element (PENTREATH and co-authors, 1980a). There are no data available relating to the stability of the sediment–nuclide association for these elements in the vicinity of Windscale.

Northeast Atlantic Deep Ocean Disposal Site

Analysis of the sediment samples obtained from the dump site during 1979 has indicated the general presence of ^{137}Cs and, in two cores, traces of ^{60}Co (FELDT and co-authors, 1979). The concentrations of ^{137}Cs are such as might be expected from fallout (NOSHKIN and BOWEN, 1973); this conclusion is supported by the analysis of a single sample obtained outside and to the southeast of the disposal area. The presence of ^{60}Co is suggestive of contamination although the concentrations are very low and close to the limits of detection. Cobalt-60 was also detected in similar quantities in two samples from the area used for low-level solid waste dumping in 1967. Further sampling and specific nuclide analysis to improve the limits of detection is required to confirm these observations.

(6) Extent of Human Exposure

The 4 representative sites discussed in Sections 4c and 5 provide examples of the application of the critical pathway approach (Blackwater estuary, Windscale, and the northeast Atlantic dump site) and point of discharge control (Oyster Creek) to the management of waste disposal. These can now be examined to demonstrate the control procedure and to determine the actual extent of human exposure relative to the appropriate dose limits.

(a) Blackwater Estuary

The pre-operational investigation of the possible interactions between the radionuclides expected to be present in the effluent and other human activities in the estuary indicated that ^{65}Zn would be the critical nuclide via the consumption of oysters harvested from commercial beds along its shores. The elements of the critical pathway and the derivation of the estimated environmental capacity are given in Table 2-23. Although a value of 182 Ci a^{-1} (6·75 TBq a^{-1}) was obtained, the provisional limiting environmental capacity was set at one-tenth of this to provide a margin of safety, and required the installation of additional ion-exchange plant to remove zinc from the effluent prior to discharge. The provisional authorized maximum permissible discharge rate was set at 5 Ci a^{-1} (0·19 TBq a^{-1}) for ^{65}Zn, within a total limit of 50 Ci a^{-1} (1·9 TBq a^{-1}) for all radionuclides excluding tritium. After the station had commenced operation, the results of routine environmental monitoring showed that oysters collected from the commercial bed closest to the outfall contained $6{\cdot}2 \times 10^3$ pCi ^{65}Zn g^{-1} per Ci d^{-1} released, i.e. approximately 40% of the value estimated in the pre-operational assessment. A re-survey of eating habits yielded an increased oyster consumption rate of 75 g d^{-1}. These two data combine to give an environmental capacity (inclusive of the safety factor of 0·1) of 17 Ci a^{-1}. Although the final agreement is good, the differences in detail

Table 2-23

Pre-operational assessment of the environmental capacity of the Blackwater estuary for ^{65}Zn (Based on PRESTON, 1966; reproduced by permission of the International Atomic Energy Agency; Crown copyright)

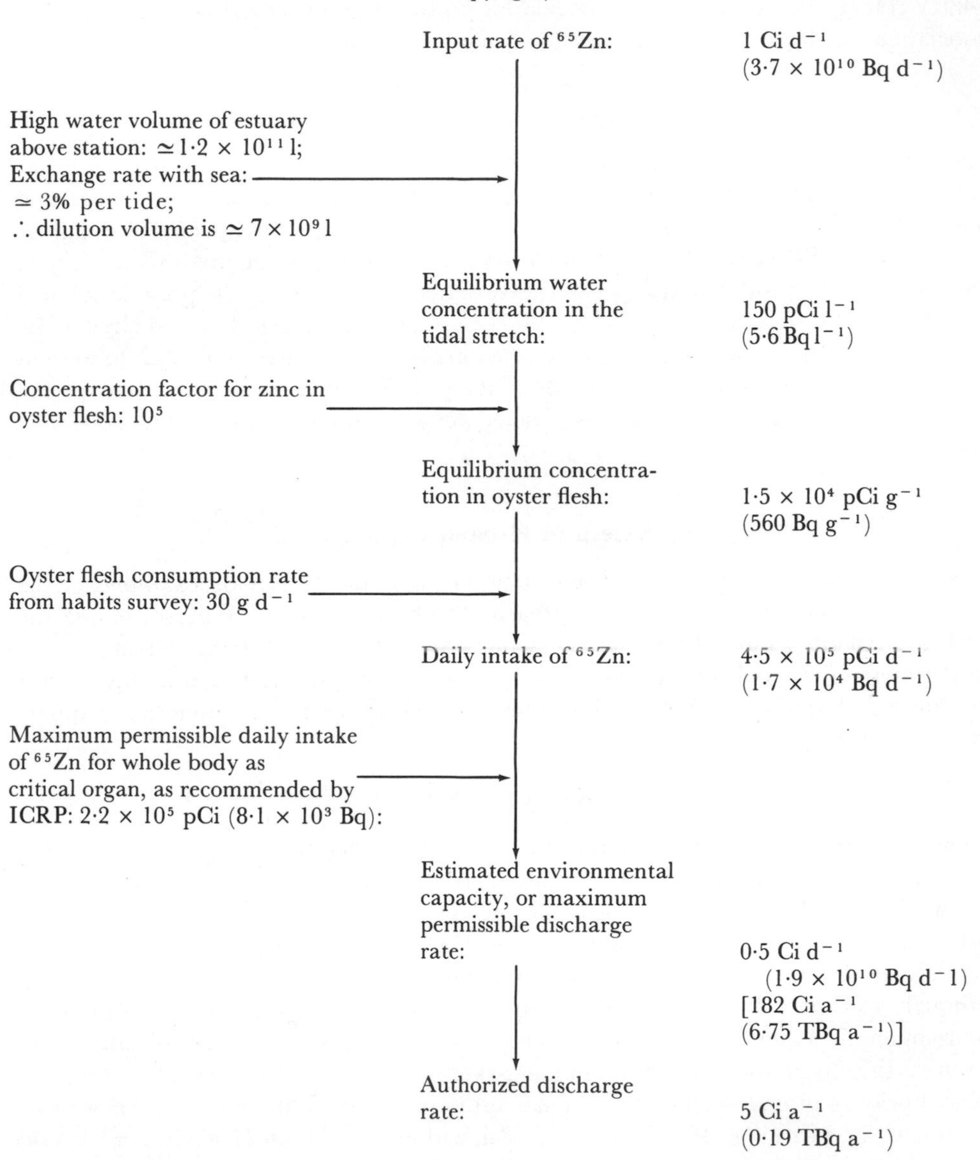

underline the advisability of setting conservative authorizations and also the need for regular re-evaluation of the data required for the assessment of the environmental capacity (PRESTON, 1966). The accumulation of monitoring data, which was not confined to the analysis of ^{65}Zn in oysters, allowed a revised authorization to be issued. The limit on ^{65}Zn was retained, but within a total limit of 200 Ci a^{-1} (7·4 TBq a^{-1}) for all radionuclides excluding tritium which was separately restricted to 1500 Ci a^{-1} (56 TBq a^{-1}). This authorization remains in force (HUNT, 1980), and—as can be seen from Table 2-10—the annual discharges to the estuary have remained well within the prescribed limits.

The critical group which has been identified for the discharge to the estuary is made up of 50 fishermen (PRESTON, 1971), and their estimated exposure based on an oyster flesh consumption rate of 75 g d^{-1} is summarized in Table 2-24. The concentrations of radionuclides in oyster flesh given in this table are those which were measured in animals harvested from the commercial bed nearest to the outfall (MITCHELL, 1969b; HUNT, 1979). The values for 1968 are less than the corresponding data given in Table 2-17 because the latter were determined for oysters specifically laid on the barrier wall adjacent to the discharge as part of a research programme and not, therefore, available for consumption. For 1968 it can be seen that ^{65}Zn only retains its position as the critical radionuclide by a very small margin relative to ^{137}Cs. The whole body remains the critical organ but there is also enhanced exposure of specific tissues; the estimated exposure of bone is given as an example. The increasing proportions of ^{134}Cs and ^{137}Cs in the effluent necessitated the installation of further ion-exchange capacity during the year to keep the total quantity of radionuclides discharged within the authorization (MITCHELL, 1969b). This measure was effective and, incidentally, also reduced the quantity of ^{65}Zn discharged. It did not, however, influence the concentration of ^{110m}Ag in the effluent and this increased until, in 1970, the presence of this radionuclide in oysters, resulting in the irradiation of the gastrointestinal tract of the fishermen, constituted a new critical pathway (MITCHELL, 1971b). During 1972 and 1973 further efforts were made to minimize the discharges of ^{65}Zn, ^{110m}Ag, and ^{137}Cs; these were successful and by 1975 the 3 nuclides, through their accumulation by oysters, had similar, very low radiological significance. Increased fishing effort in the estuary and the local consumption of the fish contaminated with ^{134}Cs and ^{137}Cs (the latter partly derived from the Windscale discharge, see Fig. 2-10) generated a further, new, critical pathway with an estimated whole body exposure in the critical group of fishermen of 2·1 mrem a^{-1} or 0·4% of the ICRP recommended limit (MITCHELL, 1978).

Overall, it can be seen that the increased radiation exposure due to the discharge is minimal and it may be concluded that the control measures applied are quite adequate to protect human health.

(b) Oyster Creek

This discharge is controlled by complying with limits on the concentrations of individual radionuclides in the effluent at the point where it enters the environment. The limits are set by the US Government (ANON, 1965; PARKER, 1965) and, in general, correspond to one-tenth of the maximum permissible concentrations in drinking water recommended by the ICRP for continuous occupational exposure (ICRP, 1960, 1964). At Oyster Creek the waste is mixed into the condenser cooling water and it is, therefore, the concentrations of the radionuclides in the discharge canal which must be maintained within the prescribed limits. A knowledge of the annual cooling water flow allows the concentration limits to be converted into total quantities of activity. Certain values of the annual discharge limits obtained for Oyster Creek, as given by BLANCHARD and co-authors (1976), are listed in Table 2-25 and may be compared with the reported annual discharges summarized in Table 2-11. This shows that, provided that the waste was released to the cooling water discharge over a reasonable period of time so as to obtain

Table 2-24

Estimated potential exposure of the critical group of fishermen eating oysters harvested from the Blackwater estuary (Original; based on data given MITCHELL 1969b and HUNT, 1979)

Organ exposed					Whole body			Bone		
ICRP dose limit (mrem a^{-1})					500			3000		
	Concentration in oyster flesh (pCi g^{-1})		Estimated daily intake (pCi)		Maximum permissible daily intake (pCi)	Estimated resultant exposure (mrem a^{-1})		Maximum permissible daily intake (pCi)	Estimated resultant exposure (mrem a^{-1})	
Radionuclide	1968	1977	1968	1977		1968	1977		1968	1977
^{32}P	$9{\cdot}7 \times 10^{-2}$	—	7·28	—	$1{\cdot}98 \times 10^{5}$	$1{\cdot}84 \times 10^{-2}$	—	$4{\cdot}40 \times 10^{4}$	$4{\cdot}96 \times 10^{-1}$	—
^{65}Zn	3·0	$5{\cdot}1 \times 10^{-1}$	225	38·3	$2{\cdot}20 \times 10^{5}$	$5{\cdot}11 \times 10^{-1}$	$8{\cdot}70 \times 10^{-2}$	$2{\cdot}20 \times 10^{6}$	$3{\cdot}06 \times 10^{-1}$	$5{\cdot}22 \times 10^{-2}$
^{55}Fe	$1{\cdot}1 \times 10^{-1}$	—	8·25	—	$4{\cdot}40 \times 10^{6}$	$9{\cdot}38 \times 10^{-4}$	—	$8{\cdot}80 \times 10^{6}$	$2{\cdot}82 \times 10^{-3}$	—
^{60}Co	$4{\cdot}3 \times 10^{-2}$	—	3·23	—	$2{\cdot}20 \times 10^{5}$	$7{\cdot}34 \times 10^{-3}$	—	—	—	—
^{110m}Ag	$2{\cdot}8 \times 10^{-1}$	$4{\cdot}0 \times 10^{-2}$	21·0	3·00	$1{\cdot}54 \times 10^{7}$	$6{\cdot}82 \times 10^{-4}$	$9{\cdot}74 \times 10^{-5}$	$4{\cdot}40 \times 10^{7}$	$1{\cdot}43 \times 10^{-3}$	$2{\cdot}04 \times 10^{-4}$
^{137}Cs	$4{\cdot}0 \times 10^{-1}$	$1{\cdot}0 \times 10^{-1}$	30·0	7·50	$3{\cdot}08 \times 10^{4}$	$4{\cdot}87 \times 10^{-1}$	$1{\cdot}22 \times 10^{-1}$	$1{\cdot}10 \times 10^{5}$	$8{\cdot}18 \times 10^{-1}$	$2{\cdot}04 \times 10^{-1}$
^{238}Pu	—	$8{\cdot}6 \times 10^{-5}$	—	$6{\cdot}45 \times 10^{-3}$	$8{\cdot}80 \times 10^{4}$	—	$3{\cdot}66 \times 10^{-5}$	$1{\cdot}10 \times 10^{4}$	—	$1{\cdot}76 \times 10^{-3}$
$^{239+240}Pu$	—	$3{\cdot}7 \times 10^{-4}$	—	$2{\cdot}78 \times 10^{-2}$	$6{\cdot}60 \times 10^{4}$	—	$2{\cdot}11 \times 10^{-4}$	$1{\cdot}10 \times 10^{4}$	—	$7{\cdot}58 \times 10^{-3}$
^{241}Am	—	$6{\cdot}8 \times 10^{-4}$	—	$5{\cdot}10 \times 10^{-2}$	$2{\cdot}20 \times 10^{4}$	—	$1{\cdot}16 \times 10^{-3}$	$1{\cdot}10 \times 10^{4}$	—	$1{\cdot}39 \times 10^{-2}$
Total estimated exposure (mrem a^{-1})						1·03	0·21		1·62	0·14
% of recommended limit						0·2	0·04		0·05	0·01

adequate dilution, there would be little likelihood of the limits being exceeded, and on average over the 4 yr the discharges were well within the limits.

In deriving the limiting concentrations and the consequential annual discharge limits for the Oyster Creek waste, no explicit or implicit account was taken of the environmental processes which might act on the radionuclides to result in human exposure. Hence, the actual annual discharge of a particular radionuclide as a proportion of the corresponding annual discharge limit cannot be interpreted as a similar proportion of some recommended dose limit. Therefore, the simple observation that the annual discharges were within the regulatory limits is insufficient to confirm their safety. To assess the potential human exposure arising from the discharge, actual or hypothetical pathways must be evaluated. The obvious pathways in respect of liquid wastes released to coastal waters are the consumption of contaminated seafood and external exposure from contaminated sediment.

The potential exposures of the whole body and bone of people eating seafood harvested from the vicinity of the station are summarized in Table 2-26, and have been developed from results obtained by BLANCHARD and KAHN (1979). The calculated values were based on estimates of the equilibrium radionuclide concentrations in the water of the discharge canal derived from the annual discharges and the cooling water throughput, concentration factor data for fish and shellfish, and notional annual consumption rates of 21 kg of fish and 5 kg of shellfish. The latter assumptions were also used to estimate the intake of radionuclides which had been actually measured in fish and shellfish sampled from the station environment (the 'measured' values). The annual intake of a particular radionuclide was then related to the radiation exposure using the maximum permissible daily intake data published by the ICRP (ICRP, 1960, 1964). The critical organ for ^{131}I is neither the whole body nor bone, but the thyroid, and the estimated exposures ('measured') are less than 0·82 and 0·18 mrem a^{-1} from fish and shellfish consumption respectively. The external exposure from ^{60}Co and ^{137}Cs contami-

Table 2-25

Derived annual radionuclide discharge limits for the Oyster Creek generating station. (After BLANCHARD and co-authors, 1976; reproduced by permission of the US Environmental Protection Agency)

Radionuclide	Discharge limit Ci a^{-1} (TBq a^{-1})	Radionuclide	Discharge limit Ci a^{-1} (TBq a^{-1})
^{3}H	3×10^6 (1×10^5)	^{99m}Tc	3×10^6 (1×10^5)
^{51}Cr	2×10^6 (7×10^4)	^{124}Sb	2×10^4 (7×10^2)
^{54}Mn	1×10^5 (4×10^3)	^{131}I	2×10^3 (70)
^{58}Co	1×10^5 (4×10^3)	^{133}I	8×10^3 (3×10^2)
^{60}Co	5×10^4 (2×10^3)	^{134}Cs	1×10^4 (4×10^2)
^{59}Fe	7×10^4 (3×10^3)	^{137}Cs	2×10^4 (7×10^2)
^{65}Zn	1×10^4 (4×10^2)	^{140}Ba–^{140}La	2×10^4 (7×10^2)
^{89}Sr	1×10^4 (4×10^2)	^{141}Ce	1×10^5 (4×10^3)
^{90}Sr	1×10^2 (4)	^{144}Ce	1×10^4 (4×10^2)
^{91}Sr	8×10^4 (3×10^3)	^{239}Np	1×10^5 (4×10^3)
^{99}Mo	4×10^4 (1×10^3)		

Table 2-26

Estimated potential exposure of the general public in the vicinity of the Oyster Creek generating station (Original; based on data given by BLANCHARD and co-authors, 1976; BLANCHARD and KAHN, 1979)

Organ exposed ICRP dose limit (mrem a^{-1})						Whole body			Bone		
						500			3000		
Source of exposure	Radio-nuclide	Concentration in the organism (pCi g^{-1}) Calculated	Measured	Estimated daily intake (pCi) Calculated	Measured	Maximum permissible daily intake (pCi)	Estimated resultant exposed (mrem a^{-1}) Calculated	Measured	Maximum permissible daily intake (pCi)	Estimated resultant exposed (mrem a^{-1}) Calculated	Measured
Internal											
Fish consumption at 21 kg a^{-1}	^{32}P	1·6	<2·0 × 10^{-1}	92	<12	1·98 × 10^{5}	2·3 × 10^{-1}	<3·0 × 10^{-2}	4·40 × 10^{4}	6·3	<8·2 × 10^{-1}
	^{55}Fe	1·5	<8·0 × 10^{-2}	86	<4·6	4·40 × 10^{6}	9·8 × 10^{-3}	<5·2 × 10^{-4}	8·8 × 10^{6}	2·9 × 10^{-2}	<1·6 × 10^{-3}
	^{60}Co	7·6 × 10^{-2}	6·0 × 10^{-3}	4·4	3·5 × 10^{-1}	2·20 × 10^{5}	1·0 × 10^{-2}	8·0 × 10^{-4}	—	—	—
	^{65}Zn	3·0 × 10^{-2}	<3·0 × 10^{-2}	1·7	<1·7	2·20 × 10^{5}	3·9 × 10^{-3}	<3·9 × 10^{-3}	2·20 × 10^{6}	2·3 × 10^{-3}	<2.3 × 10^{-3}
	^{90}Sr	6·0 × 10^{-5}	2·3 × 10^{-3}	3·5 × 10^{-3}	1·3 × 10^{-1}	1·54 × 10^{3}	1·1 × 10^{-3}	4·2 × 10^{-2}	8·80 × 10^{2}	1·2 × 10^{-2}	4·4 × 10^{-1}
	^{131}I	2·6 × 10^{-3}	<2·0 × 10^{-2}	1·5 × 10^{-1}	<1·2	4·40 × 10^{5}	1·7 × 10^{-4}	<1·4 × 10^{-3}	—	—	—
	^{137}Cs	3·3 × 10^{-2}	3·0 × 10^{-2}	1·9	1·7	3·08 × 10^{4}	3·1 × 10^{-2}	2·8 × 10^{-2}	1·10 × 10^{5}	5·2 × 10^{-2}	4·6 × 10^{-2}
Total estimated exposure (mrem a^{-1})							2·9 × 10^{-1}	<1·1 × 10^{-1}		6·4	<1·3
% recommended limit							0·06	<0·02		0·21	<0·04
Shellfish consumption at 5 kg a^{-1}	^{32}P	1·7	<0·4	23	<5·5	1·98 × 10^{5}	5·8 × 10^{-2}	<1·4 × 10^{-2}	4·40 × 10^{4}	1·6	<3·8 × 10^{-1}
	^{55}Fe	9·8	<0·1	130	<1·4	4·40 × 10^{6}	1·5 × 10^{-2}	<1·6 × 10^{-4}	8·80 × 10^{6}	4·4 × 10^{-2}	<4·8 × 10^{-4}
	^{60}Co	7·6 × 10^{-1}	1·8 × 10^{-1}	10	2·5	2·20 × 10^{5}	2·3 × 10^{-2}	5·7 × 10^{-3}	—	—	—
	^{65}Zn	4·2	<2·5 × 10^{-2}	58	<3·4 × 10^{-1}	2·20 × 10^{5}	1·3 × 10^{-1}	<7·7 × 10^{-4}	2·20 × 10^{6}	7·9 × 10^{-2}	<4·6 × 10^{-4}
	^{90}Sr	6·0 × 10^{-4}	<1·2 × 10^{-2}	8·2 × 10^{-3}	<1·6 × 10^{-1}	1·54 × 10^{3}	2·7 × 10^{-3}	<5·2 × 10^{-2}	8·80 × 10^{2}	2·8 × 10^{-2}	<5·5 × 10^{-1}
	^{131}I	1·3 × 10^{-2}	<2·0 × 10^{-2}	1·8 × 10^{-1}	<2·7 × 10^{-1}	4·40 × 10^{5}	2·0 × 10^{-4}	<3·1 × 10^{-4}	—	—	—
	^{137}Cs	2·0 × 10^{-2}	3·0 × 10^{-2}	2·7 × 10^{-1}	4·1 × 10^{-1}	3·08 × 10^{4}	4·4 × 10^{-3}	6·7 × 10^{-3}	1·10 × 10^{5}	7·4 × 10^{-3}	1·1 × 10^{-2}
Total estimated exposure (mrem a^{-1})							2·3 × 10^{-1}	7·9 × 10^{-2}		1·7	9·3 × 10^{-1}
% recommended limit							0·05	0·02		0·06	0·03

nation of sediment was estimated to be 0·02 to 0·1 mrem a^{-1} on the assumption that the maximum exposed individual spent 67 h a^{-1} on the foreshore.

In most cases the exposures derived from the measured concentrations of radionuclides in fish and shellfish are less than, or of the same order as, the calculated values; the notable exception being ^{90}Sr. On the basis of the calculated values, ^{32}P would be predicted to be the critical radionuclide, but in practice both ^{90}Sr and ^{131}I appear to be of similar significance. The actual relative significance of the radionuclides is difficult to judge because the majority of the results for the measurements of radionuclide concentrations are 'less than' values. It is pertinent to note that the discharge of ^{131}I at rates approaching the regulatory limit given in Table 2-25 could lead to thyroid exposure at a substantial fraction of the recommended dose equivalent limit. This is a problem with the point of discharge control procedure which has been identified previously (PRESTON, 1969). It is interesting to note that what essentially amounts to the critical pathway approach was recommended as a more suitable means for assessing and controlling the radiation exposure of the general public arising from this discharge (BLANCHARD and co-authors, 1976; BLANCHARD and KAHN, 1979).

(c) Windscale

During the design stage of the first fuel reprocessing plant at Windscale it was realized that it would be necessary to discharge large volumes of low-activity liquid wastes into the sea. At that time (1947 to 1952) there was virtually no prior experience to indicate how the problem of the potential radiation exposure of the general public might be managed although it was recognized that there would be limits to the quantities of radionuclides which could be safely discharged. The assessment and control procedure eventually developed contained all the elements of what is now known as the critical pathway approach (DUNSTER, 1958). The preliminary investigations considered many pathways by which the radionuclides dispersed into coastal waters could result in human exposure but it was quickly shown that there was just 3 which would be likely to be significant. These were the accumulation of radionuclides in edible seaweed and fish, and external exposure from contaminated sediments deposited on beaches (DUNSTER, 1958).

The short-term hydrographic processes in the area were investigated by releases of a fluorescent dye at different states of the tide and under various weather conditions. The results enabled estimates to be made of both the initial dilution to be expected and the longer-term standing concentrations at various distances from the outfall. In addition to providing data for the radiological assessment the experiments also provided a basis for the design of the effluent pipeline (SELIGMAN, 1955). Laboratory studies of the accumulation by marine organisms of the radionuclides present in the effluent were started to obtain concentration factor values although it was recognized that these might not be completely representative of the environmental situation (DUNSTER, 1958). Habit surveys were conducted to obtain estimates of the consumption rates of the edible seaweed (*Porphyra umbilicalis*) in the form of laverbread by a population in South Wales and of fish caught in the vicinity of the outfall; the values employed in the preliminary analysis were 75 g d^{-1} and 25 g d^{-1} respectively. All these data, together with the early recommended values of maximum permissible daily intake of radionuclides (ICRP, 1955) allowed tentative evaluations of the environmental capacities to which a safety factor of 10 was

Table 2-27

Constraints on the Windscale discharge imposed by the *Porphyra*/laverbread critical pathway (Original)

Organ or tissue exposed	Radionuclide	a Concentration in seaweed for a discharge of 10^3 Ci month^{-1} (pCi g^{-1})	b Maximum permissible daily intake (ICRP, 1960, 1964) (pCi)	c Maximum permissible concentration in seaweed, (b/80) (pCi g^{-1})	Environmental capacity or maximum permissible discharge rate ($c/a \times 1{\cdot}2 \times 10^4$) (Ci a^{-1})	Authorized discharge rate (Ci a^{-1})	
Whole body	^{137}Cs	6·2	$3{\cdot}08 \times 10^4$	385	$7{\cdot}5 \times 10^5$		Total β-activity limited to $3{\cdot}0 \times 10^5$ (for ^{137}Cs to ^{90}Sr)
Lower large intestine	^{106}Ru	57·1	$2{\cdot}20 \times 10^4$	275	$5{\cdot}8 \times 10^4$	$6{\cdot}0 \times 10^4$	
	^{103}Ru	12	$1{\cdot}76 \times 10^5$	$2{\cdot}2 \times 10^3$	$2{\cdot}2 \times 10^6$		
	^{95}Zr–^{95}Nb (at activity ratio 1·0)	9·5	$1{\cdot}76 \times 10^5$	$2{\cdot}2 \times 10^3$	$2{\cdot}8 \times 10^6$		
	^{91}Y+RE*	11·3	$6{\cdot}60 \times 10^4$	825	$8{\cdot}8 \times 10^5$		
	^{144}Ce	8·8	$2{\cdot}20 \times 10^4$	275	$3{\cdot}8 \times 10^5$		
Bone	^{90}Sr	0·24	$8{\cdot}80 \times 10^2$	11	$5{\cdot}5 \times 10^5$	$3{\cdot}0 \times 10^4$	
	Total α-activity (as ^{239}Pu)	23	$1{\cdot}10 \times 10^4$	138	$7{\cdot}2 \times 10^4$	$6{\cdot}0 \times 10^3$	

*RE represents the rare earth radionuclides, principally ^{147}Pm; the maximum permissible daily intake is that for ^{91}Y, the more restrictive of the two.

applied to allow for uncertainty in many of the data. It was concluded that discharges of 100 Ci d^{-1} of β-activity and 0·1 Ci d^{-1} of α-activity could be made with safety (DUNSTER, 1958).

By 1952 sufficient low-level liquid waste was being generated within the plant to require continuing discharges within these limitations. A parallel programme of environmental monitoring was developed to establish not only the safety of the discharges but also to obtain more reliable values for the parameters employed in determining the maximum permissible discharge rates. The data showed that the provisional assessment had been conservative and allowed upward revision of the estimated environmental capacities and the possibility of an increase in the authorized discharge rates. The contamination of the seaweed *Porphyra umbilicalis*, principally by ^{106}Ru, and its consumption in the form of laverbread was identified as the controlling, or critical, pathway for the discharges. The details of this pathway are given in Table 2-27 which shows how the environmental capacities for certain radiologically significant components of the waste have been determined and the resultant constraints which have been placed on the discharge.

The data in Column 3 of Table 2-27 are empirical values obtained from the post-operational monitoring programme (JEFFERIES, pers. comm.) and replace both the dilution factors obtained from hydrographic studies and the concentration factors in environmental materials which were used in the provisional assessment. It can be seen that ^{106}Ru, due to irradiation of the gut, has the lowest environmental capacity, closely followed by the total α-activity (assumed to be ^{239}Pu) for which bone is the critical tissue. On this basis alone, these 2 nuclides would not necessarily become of critical importance; this designation depends on the relative quantities of the various nuclides in the discharge and the fraction of the corresponding environmental capacity (equivalent to potential radiation exposure) which these represent. As it happens, ^{106}Ru was the predominant radionuclide in the early discharges (IAEA, 1976) and still constitutes a significant, although much smaller, fraction of the total activity in the effluents (Table 2-12). It became, therefore, the critical radionuclide. The environmental capacity for ^{106}Ru has been calculated on the assumption that the laverbread, as consumed, is made entirely from *Porphyra* of Windscale origin and the authorized discharge rate set accordingly. This provides a measure of conservatism in the assessment since it is known that weed from other sources is utilized and almost always results in dilution (PRESTON and JEFFERIES, 1967).

As well as ^{106}Ru, all the other radionuclides which are accumulated by the seaweed also contribute to the irradiation of the gut. Therefore, the total activity discharged cannot be the sum of the permissible discharges for the individual radionuclides. This circumstance can be controlled by requiring conformity with an inequality expression of the form

$$\sum_i \frac{D_i}{D_{i_{max}}} \ngtr 1$$

where D_i and $D_{i_{max}}$ are, respectively, the projected, and the maximum permissible, discharges of the *i*th radionuclide in a given period. In theory each radionuclide should be specifically included in the summation, but in practice it has been found that, apart from ^{106}Ru, ^{144}Ce is the only nuclide requiring individual consideration and that the remainder can be grouped together into a 'total β-activity' term. This allows considerable simplification of the expression, and the limitation imposed on the discharge over a period of a calender quarter is:

$$\frac{\text{Ci } ^{106}\text{Ru discharged}}{1.5 \times 10^4} + \frac{\text{Ci } ^{144}\text{Ce discharged}}{9 \times 10^4} + \frac{\text{Ci total } \beta\text{-activity discharged}}{3 \times 10^5} \ngtr 1$$

(PRESTON, 1971). For ^{106}Ru and ^{144}Ce the denominators are (with rounding) one-quarter of the annual environmental capacities given in Column 6 of Table 2-27. Of the remaining β-emitting radionuclides, ^{91}Y + RE has the smallest environmental capacity based on irradiation of the gut, i.e. it is the most restrictive; therefore, a completely conservative approach would dictate the adoption of $2{\cdot}2 \times 10^5$ for the denominator of the total β-activity term. However, it is known that ^{91}Y + RE constitute only a small proportion of the activity in the waste (IAEA, 1976) and the value of 3×10^5 employed is more than adequate to provide protection from ^{91}Y + RE and is conservative for the other radionuclides. In a sense, discussion of this value is academic because the total β-activity is the subject of a separate, and lower limit due to the accumulation of the radionuclides, particularly ^{95}Zr–^{95}Nb, by sediment and the consequent exposure of people on the mud banks and beaches of the Ravenglass Estuary. The authorized limit is 3×10^5 Ci a^{-1} based on ^{95}Zr–^{95}Nb (and is conservative for the other radionuclides) and means that the gut dose from the total β-activity could not be greater than a fraction of the appropriate limit even in the unlikely event of all the exposure being due to these nuclides being discharged at the limiting rates.

Although the discharge at Windscale has been restricted primarily on the basis of the *Porphyra*/laverbread pathway and secondarily through considerations of external exposure over mud banks in the Ravenglass estuary, there are two further pathways, involving the consumption of fish and shellfish, which lead to exposure and which are routinely monitored. The *Porphyra*/laverbread pathway is now of only potential importance since there has been no harvesting of the seaweed from the Cumbrian coast since 1972. The last data available concerning exposure via this pathway, therefore, relate to 1971. As mentioned above, it is known that the Cumbrian coast is not the sole source of weed for laverbread manufacture: a large and variable proportion being obtained from other, uncontaminated areas (PRESTON and JEFFERIES, 1967). Thus the simplest means of monitoring the possible exposure entails the routine sampling of the foodstuff as marketed in South Wales, and measuring the radionuclide content. From these data it was estimated that, depending on the manufacturer, the radiation exposure of the lower large intestine of the critical group ranged between 1% and 33% of the ICRP-recommended dose limit, with a weighted mean value of 8% (equivalent to 120 mrem a^{-1}), mainly due to ^{106}Ru (MITCHELL, 1973). However, since the discharge assessment and the consequent controls are based on the assumption that Windscale would be the sole source of *Porphyra* for South Wales, and since it is possible that laverbread could occasionally have been manufactured from undiluted Windscale weed, a demonstration of the effective control of the discharge by the constraints applied requires that the potential exposure from *Porphyra* of Windscale origin be estimated. The relevant data are given in Table 2-28. It can be seen that the lower large intestine was the critical organ with the β, γ-emitters ^{106}Ru, ^{144}Ce, and ^{95}Zr–^{95}Nb contributing respectively 88·6%, 9·3%, and 1·8% of the total dose and ^{90}Sr and the α-emitters delivering the remaining 0·3%. The total estimated dose equivalent rate of $1{\cdot}4 \times 10^3$ mrem a^{-1} (14 mSv a^{-1}) is

91% of the relevant ICRP limit and confirms the effective control of the discharge. The estimated potential exposure of bone (mainly from ^{90}Sr and $^{239-240}$Pu) and the whole body (mainly from ^{90}Sr, ^{137}Cs, and ^{106}Ru) were of much less significance.

Radiation exposure following the consumption of contaminated fish and shellfish is of greatest significance in the local fishing community. A survey of eating habits showed that, for 1977, there were 2, essentially independent, pathways, i.e. people common to both critical groups were not found. In the first case, the maximum consumption rate of fish (224 g d^{-1}), mainly the plaice (*Pleuronectes platessa*), and crabs (41 g d^{-1}) leads to an estimated whole body exposure of 120 mrem (1·2 mSv) during 1977, or 24% of the recommended limit. The exposure derives almost entirely from ^{137}Cs and ^{134}Cs. The irradiation of the lower large intestine, mainly due to ^{106}Ru, and of bone by ^{137}Cs and ^{134}Cs are much smaller fractions of the corresponding recommended limits. Contaminated winkles *Littorina littorea* constitute the second pathway and the maximum consumption rate of 5 g d^{-1} results in an estimated dose equivalent rate to the lower large intestine of 190 mrem a^{-1} (1·9 mSv a^{-1}), again mainly from ^{106}Ru; the concomitant exposures of bone and the whole body are of much lesser significance.

The final pathway arises from the external irradiation of fishermen by the contaminated fine grain sediments in the Ravenglass estuary to the south of Windscale. The dose rate over the sediment is measured periodically throughout the year and these data, in combination with a maximum occupancy time of 350 h a^{-1} during 1977, provide an estimate of 21 mrem a^{-1} (0·21 mSv a^{-1}) for the whole body exposure.

The variations with time of the estimated annual dose equivalents for each of the 3 main pathways at Windscale are shown in Fig. 2-22. (MITCHELL, 1967, 1968, 1969b; 1971a,b, 1973, 1975, 1977a,b; PRESTON and MITCHELL, 1973; HETHERINGTON, 1976b; HUNT, 1979, 1980). It is apparent that the consumption of fish and shellfish represents the critical pathway at the present time, although all the pathways are of sufficient potential or actual significance to warrant continued monitoring. The increasing quantities of caesium radionuclides in the discharge have also raised a wider problem than the irradiation of the local fishing community. The combination of relatively long half-life and conservative behaviour in sea water has resulted in the widespread dispersion of these nuclides, albeit at low concentrations, in the shelf waters of northwestern Europe (Figs 2-9 and 2-10). These waters support large stocks of commercially important fish which have become slightly contaminated and which are eaten both in the United Kingdom and in many countries in northwestern Europe. In these circumstances the control of the discharge cannot be based solely upon the consideration of the most highly exposed groups which are, in general, rather small. The potential integrated collective dose equivalent to the whole fish-eating population must also be assessed to ensure compliance with the limitation on genetic (gonad) dose (JEFFERIES and co-authors, 1977). The estimated annual collective and *per capita* dose equivalents for the populations of the United Kingdom and northwestern Europe are given in Table 2-29 (MITCHELL, 1971b, 1973, 1975, 1977a; HUNT, 1979, 1980). The ICRP recommends a limit of 5 rem person^{-1} in 30 a, averaged over the whole population. In the United Kingdom one-fifth of this has been allocated to waste disposal corresponding to a mean *per capita* dose equivalent rate of 33 mrem a^{-1}. It can be seen from Table 2-29 that the estimated values are well within this limit. The rise to prominence of ^{137}Cs and, to a lesser extent, ^{134}Cs in fish and shellfish and their continuing significance make it likely that these radionuclides will be the subject of specific limitation in any future, amended discharge authorization.

Table 2-28

Estimated potential exposure of the critical populations due to the Windscale discharge (Original; based on data provided by PRESTON, 1971; MITCHELL, 1973 and HUNT, 1979)

Organ exposed ICRP dose limit (mrem a^{-1}) Sources of exposure	Exposed population group	Daily consumption or annual occupancy	Radionuclide	Concentration in the material (pCi g^{-1})	Estimated daily intake (pCi)
Porphyra (seaweed)	Laverbread consumers in S. Wales; critical group: 100 persons during 1971	160 g laver-bread (≃80 g seaweed) d^{-1}	^{106}Ru	10–800	$8{\cdot}0 \times 10^{2}$–$6{\cdot}4 \times 10^{4}$
				224	$1{\cdot}8 \times 10^{4}$
			^{144}Ce	1·1–130	$8{\cdot}8 \times 10^{1}$–$1{\cdot}0 \times 10^{4}$
				24·3	$1{\cdot}9 \times 10^{3}$
			^{95}Zr–^{95}Nb	1–280	$8{\cdot}0 \times 10^{1}$–$2{\cdot}2 \times 10^{4}$
				37·0	$3{\cdot}0 \times 10^{3}$
			^{137}Cs	0·6–14	$4{\cdot}8 \times 10^{1}$–$1{\cdot}1 \times 10^{3}$
				2·52	$2{\cdot}0 \times 10^{2}$
			$^{239-240}Pu$	0·8–2·0	$6{\cdot}4 \times 10^{1}$–$1{\cdot}6 \times 10^{2}$
				1·39	$1{\cdot}1 \times 10^{2}$
			^{241}Am	0·3–1·1	$2{\cdot}4 \times 10^{1}$–$8{\cdot}8 \times 10^{1}$
				0·63	$5{\cdot}0 \times 10^{1}$
			^{90}Sr	0·05–0·01	$4{\cdot}0 \times 10^{0}$–$8{\cdot}1 \times 10^{1}$
				0·42	$3{\cdot}4 \times 10^{1}$
Total estimated exposure (mrem a^{-1})					
% recommended limit					
Fish and crabs *Pleuronectes platessa* and *Cancer pagurus*	Local fishing community (1977)	Maximum consumption Fish: 224 g d^{-1}	^{137}Cs	25	$5{\cdot}6 \times 10^{3}$
			^{134}Cs	3·1	$6{\cdot}9 \times 10^{2}$
			^{238}Pu	$1{\cdot}9 \times 10^{-4}$	$4{\cdot}3 \times 10^{-2}$
			$^{239+240}Pu$	$7{\cdot}3 \times 10^{-4}$	$1{\cdot}6 \times 10^{-1}$
			^{241}Am	$1{\cdot}0 \times 10^{-3}$	$2{\cdot}2 \times 10^{-1}$
		Crab: 41 g d^{-1}	^{60}Co	$2{\cdot}3 \times 10^{-1}$	9·4
			^{95}Zr–^{95}Nb	$3{\cdot}5 \times 10^{-1}$	14
			^{99}Tc	$2{\cdot}8 \times 10^{-1}$	11
			^{106}Ru	30	$1{\cdot}2 \times 10^{3}$
			^{110m}Ag	$9{\cdot}2 \times 10^{-1}$	38
			^{134}Cs	2·2	90
			^{137}Cs	17	$7{\cdot}0 \times 10^{2}$
			^{144}Ce	$1{\cdot}2 \times 10^{-1}$	4·9
			^{238}Pu	$3{\cdot}8 \times 10^{-2}$	1·6
			$^{239+240}Pu$	$1{\cdot}5 \times 10^{-1}$	6.2
			^{241}Am	$4{\cdot}4 \times 10^{-1}$	18
			^{242}Cm	$8{\cdot}0 \times 10^{-3}$	$3{\cdot}3 \times 10^{-1}$
			$^{243+244}Cm$	$3{\cdot}8 \times 10^{-3}$	$1{\cdot}6 \times 10^{-1}$
Total estimated exposure (mrem a^{-1})					
% recommended limit					

Table 2-28—*contd*

Lower large intestine		Bone		Wholebody	
1500		3000		500	
Maximum permissible daily intake (pCi)	Estimated resultant exposure (mrem a^{-1})	Maximum permissible daily intake (pCi)	Estimated resultant exposure (mrem a^{-1})	Maximum permissible daily intake (pCi)	Estimated resultant exposure (mrem a^{-1})
2.2×10^{4}	5.5×10^{1}–4.4×10^{3}	2.2×10^{6}	1.1×10^{0}–8.7×10^{1}	4.4×10^{6}	9.1×10^{-2}–7.3×10^{0}
	1.2×10^{3}		2.5×10^{1}		2.0×10^{0}
2.2×10^{4}	6.0×10^{0}–6.8×10^{2}	1.8×10^{7}	1.5×10^{-2}–1.7×10^{0}	6.6×10^{7}	6.7×10^{-4}–7.6×10^{-2}
	1.3×10^{2}		3.2×10^{-1}		1.4×10^{-2}
1.8×10^{5}	6.7×10^{-1}–1.8×10^{2}	9.9×10^{8}	2.4×10^{-4}–6.7×10^{-2}	5.5×10^{8}	7.3×10^{-5}–2.0×10^{-2}
	2.5×10^{1}		9.1×10^{-3}		2.7×10^{-3}
—	—	1.1×10^{5}	1.3×10^{0}–3.0×10^{1}	3.1×10^{4}	7.7×10^{-1}–1.8×10^{1}
	—		5.5×10^{0}		3.2×10^{0}
6.6×10^{4}	1.5×10^{0}–3.6×10^{0}	1.1×10^{4}	1.7×10^{1}–4.4×10^{1}	6.6×10^{4}	4.8×10^{-1}–1.2×10^{0}
	2.5×10^{0}		3.0×10^{1}		8.3×10^{-1}
6.6×10^{4}	5.5×10^{-1}–2.0×10^{0}	1.1×10^{4}	6.5×10^{0}–2.4×10^{1}	2.2×10^{4}	5.5×10^{-1}–2.0×10^{0}
	1.1×10^{0}		1.4×10^{1}		1.1×10^{0}
1.1×10^{5}	5.5×10^{-2}–1.1×10^{0}	8.8×10^{2}	1.4×10^{1}–2.8×10^{2}	1.5×10^{3}	1.3×10^{0}–2.7×10^{1}
	4.6×10^{-1}		1.2×10^{2}		1.1×10^{1}
	1.4×10^{3}		1.9×10^{2}		1.9×10^{1}
	91		6.3		3.7
—	—	1.1×10^{5}	1.5×10^{2}	3.1×10^{4}	9.0×10^{1}
—	—	1.5×10^{5}	1.4×10^{1}	2.0×10^{4}	1.7×10^{1}
6.6×10^{4}	9.8×10^{-4}	1.1×10^{4}	1.2×10^{-2}	8.8×10^{4}	2.4×10^{-4}
6.6×10^{4}	3.6×10^{-3}	1.1×10^{4}	4.4×10^{-2}	6.6×10^{4}	1.2×10^{-3}
6.6×10^{4}	5.0×10^{-3}	1.1×10^{4}	6.0×10^{-2}	2.2×10^{4}	5.0×10^{-3}
1.1×10^{5}	1.3×10^{-1}	—	—	2.2×10^{5}	2.1×10^{-2}
1.8×10^{5}	1.2×10^{-1}	9.9×10^{8}	4.2×10^{-5}	5.5×10^{8}	1.3×10^{-5}
6.6×10^{5}	2.5×10^{-2}	6.6×10^{7}	5.0×10^{-4}	2.2×10^{7}	2.5×10^{-4}
2.2×10^{4}	8.2×10^{1}	2.2×10^{6}	1.6×10^{0}	4.4×10^{6}	1.4×10^{-1}
6.6×10^{4}	8.6×10^{-1}	4.4×10^{7}	2.6×10^{-3}	1.5×10^{7}	1.3×10^{-3}
—	—	1.5×10^{5}	1.8×10^{0}	2.0×10^{4}	2.3×10^{0}
—	—	1.1×10^{5}	1.9×10^{1}	3.1×10^{4}	1.1×10^{1}
2.2×10^{4}	3.3×10^{-1}	1.8×10^{7}	8.2×10^{-4}	6.6×10^{7}	3.7×10^{-5}
6.6×10^{4}	3.6×10^{-2}	1.1×10^{4}	4.4×10^{-1}	8.8×10^{4}	9.1×10^{-3}
6.6×10^{4}	1.4×10^{-1}	1.1×10^{4}	1.7×10^{0}	6.6×10^{4}	4.7×10^{-2}
6.6×10^{4}	4.1×10^{-1}	1.1×10^{4}	4.9×10^{0}	2.2×10^{4}	4.1×10^{-1}
4.4×10^{4}	1.1×10^{-2}	4.4×10^{5}	2.3×10^{-3}	1.1×10^{6}	1.5×10^{-4}
4.4×10^{4}	5.5×10^{-3}	1.1×10^{4}	4.4×10^{-2}	4.4×10^{4}	1.8×10^{-3}
	8.4×10^{1}		2.0×10^{2}		1.2×10^{2}
	5.6		6.5		24

Table 2-28—*contd*

Source of exposure	Exposed population group	Daily consumption or annual occupancy	Radionuclide	Concentration in the material ($pCi\,g^{-1}$)	Estimated daily intake (pCi)
Organ exposed ICRP dose limit ($mrem\ a^{-1}$)					
Winkles *Littorina littorea*	Local fishing community (1977)	Maximum consumption $5\ g\ d^{-1}$	^{60}Co	1·2	6·0
			^{95}Zr–^{95}Nb	46	$2{\cdot}3 \times 10^{2}$
			^{99}Tc	$1{\cdot}1 \times 10^{-1}$	$5{\cdot}5 \times 10^{-1}$
			^{106}Ru	$5{\cdot}4 \times 10^{2}$	$2{\cdot}7 \times 10^{3}$
			^{110m}Ag	4·2	21
			^{134}Cs	3·6	18
			^{137}Cs	26	$1{\cdot}3 \times 10^{2}$
			^{144}Ce	4·5	23
			^{238}Pu	2·0	10
			$^{239+240}Pu$	7·6	38
			^{241}Am	5·7	29
			^{242}Cm	$3{\cdot}2 \times 10^{-1}$	1·6
			$^{243+244}Cm$	$4{\cdot}0 \times 10^{-2}$	$2{\cdot}0 \times 10^{-1}$
Total estimated exposure ($mrem\ a^{-1}$)					
% recommended limit					
Estuarine silt	Fishermen: critical group: 10 persons (1977)	$350\ h\ a^{-1}$	^{95}Zr–^{95}Nb		
			^{106}Ru		
			^{134}Cs	—	—
			^{137}Cs		
			^{144}Ce		

(d) Northeast Atlantic Deep Ocean Disposal Site

Direct estimates of the possible human exposure due to disposals at this site are not possible since specific pathways which could lead to such exposure have not yet been identified. In addition, measurements of radionuclides in environmental materials taken from the site have not detected any significant contamination which can be attributed with certainty to the wastes. This difficulty was recognized by the authors of a recent report which reviewed the continued suitability of the site for radioactive waste disposal (OECD–NEA, 1980). It was, however, necessary to make estimates of the potential human exposure and a theoretical approach was adopted. The International Atomic Energy Agency has developed a generalized model of the oceanographic and radiological factors controlling the extent of potential human exposure from radiounclides released into the deep ocean (IAEA, 1978b,c). This has been used to define radioactive wastes which are unsuitable for dumping at sea for the purposes of the London Dumping Convention, but may be used in the present context. With appropriate qualifications, it

Table 2-28—*contd*

Lower large intestine		Bone		Wholebody	
1500		3000		500	
Maximum permissible daily intake (pCi)	Estimated resultant exposure (mrem a^{-1})	Maximum permissible daily intake (pCi)	Estimated resultant exposure (mrem a^{-1})	Maximum permissible daily intake (pCi)	Estimated resultant exposure (mrem l^{-1})
$1{\cdot}1 \times 10^5$	$8{\cdot}2 \times 10^{-2}$	—	—	$2{\cdot}2 \times 10^5$	$1{\cdot}4 \times 10^{-2}$
$1{\cdot}8 \times 10^5$	$1{\cdot}9 \times 10^0$	$9{\cdot}9 \times 10^8$	$7{\cdot}0 \times 10^{-4}$	$5{\cdot}5 \times 10^8$	$2{\cdot}1 \times 10^{-4}$
$6{\cdot}6 \times 10^5$	$1{\cdot}3 \times 10^{-3}$	$6{\cdot}6 \times 10^7$	$2{\cdot}5 \times 10^{-5}$	$2{\cdot}2 \times 10^7$	$1{\cdot}3 \times 10^{-5}$
$2{\cdot}2 \times 10^4$	$1{\cdot}8 \times 10^2$	$2{\cdot}2 \times 10^6$	$3{\cdot}7 \times 10^0$	$4{\cdot}4 \times 10^6$	$3{\cdot}1 \times 10^{-1}$
$6{\cdot}6 \times 10^4$	$4{\cdot}8 \times 10^{-1}$	$4{\cdot}4 \times 10^7$	$1{\cdot}4 \times 10^{-3}$	$1{\cdot}5 \times 10^7$	$7{\cdot}0 \times 10^{-4}$
—	—	$1{\cdot}5 \times 10^5$	$3{\cdot}6 \times 10^{-1}$	$2{\cdot}0 \times 10^4$	$4{\cdot}5 \times 10^{-1}$
—	—	$1{\cdot}1 \times 10^5$	$3{\cdot}5 \times 10^0$	$3{\cdot}1 \times 10^4$	$2{\cdot}1 \times 10^0$
$2{\cdot}2 \times 10^4$	$1{\cdot}6 \times 10^0$	$1{\cdot}8 \times 10^7$	$3{\cdot}8 \times 10^{-3}$	$6{\cdot}6 \times 10^7$	$1{\cdot}7 \times 10^{-4}$
$6{\cdot}6 \times 10^4$	$2{\cdot}3 \times 10^{-1}$	$1{\cdot}1 \times 10^4$	$2{\cdot}7 \times 10^0$	$8{\cdot}8 \times 10^4$	$5{\cdot}7 \times 10^{-2}$
$6{\cdot}6 \times 10^4$	$8{\cdot}6 \times 10^{-1}$	$1{\cdot}1 \times 10^4$	$1{\cdot}0 \times 10^1$	$6{\cdot}6 \times 10^4$	$2{\cdot}9 \times 10^{-1}$
$6{\cdot}6 \times 10^4$	$6{\cdot}6 \times 10^{-1}$	$1{\cdot}1 \times 10^4$	$7{\cdot}9 \times 10^0$	$2{\cdot}2 \times 10^4$	$6{\cdot}6 \times 10^{-1}$
$4{\cdot}4 \times 10^4$	$5{\cdot}5 \times 10^{-2}$	$4{\cdot}4 \times 10^5$	$1{\cdot}1 \times 10^{-2}$	$1{\cdot}1 \times 10^6$	$7{\cdot}3 \times 10^{-4}$
$4{\cdot}4 \times 10^4$	$6{\cdot}8 \times 10^{-3}$	$1{\cdot}1 \times 10^4$	$5{\cdot}5 \times 10^{-2}$	$4{\cdot}4 \times 10^4$	$2{\cdot}3 \times 10^{-3}$
	$1{\cdot}9 \times 10^2$		$2{\cdot}9 \times 10^1$		3·9
	13		0·96		0·78
—	—	—	—	—	21 mrem a^{-1} (measured) 4·2% of recommended limit

was estimated that the average annual dumping rates over the period 1975 to 1979 could not result in exposures greater than 0·3%, and very probably much less than 0·1%, of the relevant ICRP limits. It was also recommended that a specific model of the oceanographic, chemical, and biological processes operative at the site be developed to permit a more realistic assessment to be made in the future.

(7) Effects of Radiation on Marine Organisms

It has already been made abundantly clear that dosimetry is a necessary prerequisite for an assessment of the hazards of radiation to marine organisms (see Sections 1 and 2d). There are several difficulties in the way of making *in situ* measurements of the radiation dose rate to marine organisms from external sources: first, the expected dose rates in most situations are rather low; second, the sea is hostile to the sensitive detectors which would need to be employed and to the associated high voltages and electronic equipment; finally, one environment of particular interest, the seabed, is relatively

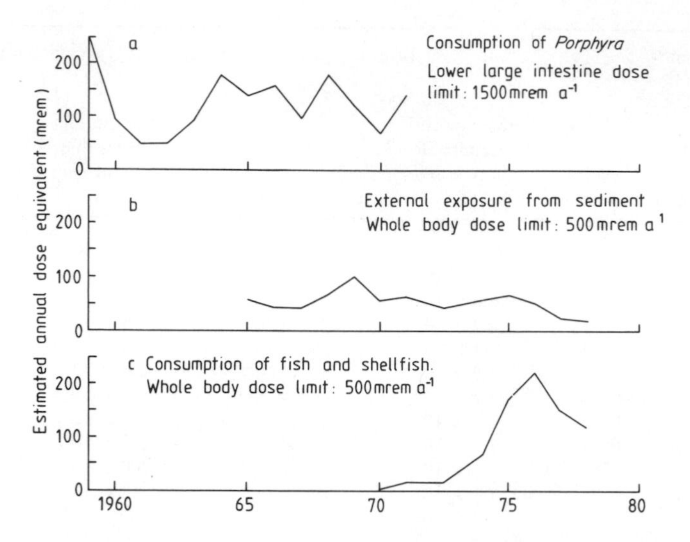

Fig. 2-22: Variation with time of the estimated annual dose equivalent due to the three pathways of importance for the Windscale discharge (a) After PRESTON and MITCHELL, 1973; reproduced by permission of the International Atomic Energy Agency; Crown copyright. (b) (c) originals from data in MITCHELL, 1967, 1968, 1969b, 1971a,b, 1973, 1975, 1977a,b; HETHERINGTON, 1976b; HUNT, 1979, 1980)

Table 2-29

Collective and mean *per capita* whole body dose equivalent rates due to the radiocaesium discharges from Windscale (Compiled from MITCHELL, 1971b, 1973, 1975, 1977a, and HUNT, 1979, 1980)

	United Kingdom $5 \cdot 5 \times 10^7$ persons		Northwestern Europe $1 \cdot 4 \times 10^8$ persons	
Year	Collective dose equivalent rate (man-rem a^{-1})	*Per capita* dose equivalent rate (mrem a^{-1})	Collective dose equivalent rate (man-rem a^{-1})	*Per capita* dose equivalent rate (mrem a^{-1})
1970	$1 \cdot 5 \times 10^2$	$2 \cdot 7 \times 10^{-3}$	No data	No data
1971	$1 \cdot 0 \times 10^2$	$1 \cdot 8 \times 10^{-3}$	No data	No data
1972, 1973	$1 \cdot 5 \times 10^3$	$2 \cdot 7 \times 10^{-2}$	No data	No data
1974	$4 \cdot 8 \times 10^3$	$8 \cdot 7 \times 10^{-2}$	$3 \cdot 8 \times 10^3$	$2 \cdot 7 \times 10^{-2}$
1975	$8 \cdot 3 \times 10^3$	$1 \cdot 5 \times 10^{-1}$	$5 \cdot 7 \times 10^3$	$4 \cdot 1 \times 10^{-2}$
1976	$1 \cdot 4 \times 10^4$	$2 \cdot 5 \times 10^{-1}$	$1 \cdot 2 \times 10^4$	$8 \cdot 6 \times 10^{-2}$
1977	$8 \cdot 9 \times 10^3$	$1 \cdot 6 \times 10^{-1}$	$8 \cdot 0 \times 10^3$	$5 \cdot 7 \times 10^{-2}$
1978	$1 \cdot 3 \times 10^4$	$2 \cdot 4 \times 10^{-1}$	$1 \cdot 1 \times 10^4$	$7 \cdot 9 \times 10^{-2}$

inaccessible. The latter two problems can, in principle, be circumvented by enclosing the equipment in a watertight pressure vessel, but this leads to problems in the interpretation of the response of the detector. Therefore, with two exceptions (LAPPENBUSCH and co-authors, 1971; WOODHEAD, 1973b), dose rates have been calculated from the measured concentrations of radionuclides in the various components of the environment together with suitable models. In addition, this is the only approach which can be adopted for determining the exposure from internal sources of radiation.

(a) Dose Rate Regimes in the Marine Environment

FOLSOM and HARLEY (1957) were the first to consider the problem of the natural background exposure in the aquatic environment and although their estimates, based on limited data, have been largely superseded, several important points were made. It was indicated that the size of the organism would be a significant determinant of the relative contributions from internal and external sources and that, for small organisms, the macroscopic concept of absorbed dose would probably be an inadequate means of describing the background radiation exposure. In addition, it was concluded that organisms inhabiting certain deep ocean sediments might be the most highly exposed in the marine environment.

The much greater quantity of data on environmental radioactivity which have become available since 1957 have been used to make a fairly comprehensive re-examination of the dose rates to different organisms from a variety of sources (IAEA, 1976; WOODHEAD, 1973a; 1974). Simple geometrical shapes were adopted to represent the groups of organisms as follows:

phytoplankton:	sphere, 50 μm diameter;
zooplankton:	cylinder, 0·5 cm long × 0·2 cm diameter;
molluscs:	flat cylinder, 1·0 cm high × 4·0 cm diameter;
crustaceans:	cylinder, 15 cm long × 6·0 cm diameter;
fish:	cylinder, 50 cm long × 10 cm diameter.

A tissue density of unity was assumed for all the organisms and, except for certain nuclides in crustaceans and fish, it was assumed that the radioactivity is uniformly distributed throughout the volume.

The limited size of the sphere representing phytoplankton relative to the ranges of the α-particles emitted by the natural and contaminant radionuclides means that a significant fraction of the total α-energy derived from incorporated activity is dissipated in the surrounding water. On the very simplified assumption that the average range of α-particles is 60 μm in water at constant linear energy transfer, it was concluded that approximately 30% of the energy of α-particles emitted within the sphere would be absorbed within the volume, i.e.

$$D_\alpha \approx 0.3 D_\alpha(\infty)$$

where

$$D_\alpha(\infty) = 2.13 \bar{E}_\alpha C \mu\text{rad h}^{-1} \; (0{\cdot}576 \, \bar{E}_\alpha C \; \mu\text{Gy h}^{-1})$$

and is the dose rate in an infinite volume containing a uniform distribution of C pCi g^{-1}

(C Bq g^{-1}) (C being the concentration of activity in the tissue) of a radionuclide emitting α-particles of mean energy $\bar{E}_\alpha$ MeV per disintegration. For the other 4 groups of organisms the range of the α-particles is so short in comparison with the assumed size that it was reasonable to conclude that the mean absorbed dose rate from internally deposited α-emitting radionuclides would be equal to D_α (∞).

In unit density tissue, the β-particles emitted by radionuclides encountered in the marine environment have ranges up to 2 cm. Thus, for both phytoplankton and zooplankton, a major fraction of the energy of the β-particles emitted by internal activity is not absorbed by the organism. The point source β-dose function developed empirically by LOEVINGER (LOEVINGER and co-authors, 1956b) was used to estimate, as a function of the maximum β-ray energy, the mean dose rate in the sphere representing the phytoplankton and the dose rate at the centre of the cylinder representing zooplankton. For the molluscs, crustaceans, and fish the dimensions of the models were sufficiently large that the maximum dose rate could be assumed to be D_β (∞) to a very close approximation.

The absorption coefficient of γ-radiation is so low that it was assumed that γ-ray emission from internal radionuclides would not make a significant contribution to the exposure of phytoplankton. For the other 4 groups of organisms geometrical factors had also to be considered and the average dose rate from internal sources was obtained from the expression

$$\bar{D}_\gamma = \Gamma C \rho \bar{g} \times 10^{-3} \ \mu\text{rad h}^{-1} \ (\Gamma C \rho \bar{g} \times 10^{-3} \ \mu\text{Gy h}^{-1}$$

where Γ is the specific γ-ray constant in cm^2 rad h^{-1} mCi^{-1} (cm^2 Gy h^{-1} Gbq^{-1}); C the radionuclide concentration in pCi g^{-1} (Bq g^{-1}); ρ the tissue density and taken to be 1 g cm^{-3}; and $\bar{g}$ the mean geometrical factor in cm. The values of $\bar{g}$ interpolated from published data for cylinders (LOEVINGER and co-authors, 1956a) are 1, 10, 25, and 41 cm for zooplankton, molluscs, crustaceans, and fish respectively.

The dose rates to phytoplankton and zooplankton from α- and β-activity in the water were simply calculated from the general reciprocity relation:

$$D(\text{external}) = D(\infty) - D(\text{internal})$$

where D (internal) was calculated as above for the particular organism geometry and particle energy, but evaluated at the radionuclide concentration appropriate to water as also was D (∞). The α- and β-dose rates to the molluscs, crustaceans, and fish from activity in the water were taken to be zero. For all the groups of organisms the γ-ray dose rate from activity in the water was taken to be D_γ (∞).

For the natural radionuclides in coastal sediments it was assumed that the seabed represented an infinitely thick, plane source of uniformly distributed activity so that the dose rate at the sediment–water boundary could be taken as 0·5 D (∞) for each radiation type. For the radionuclides discharged from Windscale and for certain of the natural radionuclides in deep-sea sediments, the finite thickness of the source and the variation in the concentration of activity with depth were taken into account in calculating the γ-ray dose rate at the sediment surface. For the Blackwater estuary and Barnegat Bay, for which equivalent data are unavailable but where similar variations are almost certainly present, it has been assumed that these factors, as at Windscale, result in the γ-radiation dose rate at the sediment surface being approximately 0·25 D_γ (∞).

The dose rates to marine organisms calculated on the basis of these very simple models and the concentrations of radionuclides in the environment as given in Tables

2-5, 2-6, 2-7, 2-8, 2-9, 2-14, 2-17, 2-18, 2-20, and 2-21 are summarized in Table 2-30 (for details, see WOODHEAD, 1973a, 1974; IAEA, 1976).

The depths indicated for each of the different groups of organisms have been chosen arbitrarily and only affect the contributions from cosmic radiation which have been estimated from the data given by FOLSOM and HARLEY (1957). The greater part of the natural exposure for organisms which live away from the seabed is due to radionuclides accumulated within the body. The α-emitting radionuclides, principally ^{210}Po, are the main source as is indicated by the generally much higher values of biologically effective dose included in parentheses in Table 2-30; ^{10}K contributes most of the remainder. The apparent importance of α-radiation from accumulated radionuclides demonstrates that the assumption of a uniform distribution of activity throughout the organism can lead to underestimates of the dose rates to particular tissues from this short-range radiation where the measured whole body burdens of the α-emitting nuclides are due, in part, to higher degrees of accumulation in these tissues. For example, the biologically effective dose rates to the crustacean hepatopancreas and to the fish liver from ^{210}Po are in the ranges 750–1·38 × 10^4 μrem h^{-1} (7·5 to 138 μSv h^{-1}) and 40 to 2940 μrem h^{-1} (0·4 to 29·4 μSv h^{-1}) respectively, and higher than the whole body dose rates given in Table 2-30. For benthic organisms, the natural γ-radiation from the seabed is an equally important source of exposure. The magnitudes of the α- and β-radiation dose rates at the sediment surface indicate that these could be important but the significance for marine organisms is very dependent on the geometrical relationship between the source and the target. These radiations are clearly of consequence for the exposure of the gut epithelium of sediment browsers and for the micro- and meio-fauna inhabiting the sediment.

The estimated doses from fallout are very variable reflecting the variation in the measured concentrations of radionuclides in organisms due to spot sampling. The data serve mainly to indicate the general magnitude of the exposure which may have been experienced by the organisms over relatively short periods of time in the past. The incorporated activity was, and remains, the only significant source of exposure, and in the long term ^{137}Cs is the major contributor to the dose rate to all groups of organisms; ^{239}Pu becomes of some significance in the case of phytoplankton in terms of the biologically effective dose. In general, the dose rates from this source have been of the same order as, or less than, that due to the natural background.

In the Blackwater estuary the activity accumulated by the organisms and the sediment represent the major sources of exposure. The decreased quantities of ^{110m}Ag, ^{134}Cs, and ^{137}Cs discharged in 1977 compared with 1968 and the apparent change in the biological availability of ^{65}Zn (see Section 5b) are reflected in the lower estimates of the dose rates in 1977. From the annual discharge data it seems unlikely that the additional radiation exposure of organisms in the Blackwater estuary due to the wastes has ever been more than a fraction of natural background.

In Oyster Creek and Barnegat Bay the greatest exposure arises from contaminated sediment with the highest dose rates restricted to the creek and being mainly due to ^{60}Co. Radionuclides in the water and accumulated by the organisms are very minor sources of exposure and apart from limited areas in Oyster Creek the total dose rates are much less than the natural background.

As might have been predicted on the basis of the relative magnitudes of the annual discharges of radioactivity, the situation in the northeast Irish Sea is quite different. For all groups of organisms the estimated dose rates are greater than those from the natural

Table

Summary of the radiation exposure of marine organisms from different sources of radioactivity, by a factor of 10^{-2} (Compiled from WOODHEAD, 1973a,

Source	Phytoplankton 2 m depth, remote from the seabed	Zooplankton 2 m depth, remote from the seabed	Molluscs 5 m depth, on the seabed
Natural background			
Cosmic radiation	2·2	2·2	1·6
Internal activity	1·9–7·3 (37–140)	2·5–30 (23–560)	6·6–19 (78–320)
Water activity	0·43	0·24	0·05
Sediment γ	(0·79)	—	1·5–16
β	—	—	1·6–21
α	—	—	14–170 (280–3300)
Total:*	4·5–9·9 (40–150)	5·0–32 (26–59)	9·8–36 (81–340)
Fallout			
Internal activity			
^{3}H, ^{14}C, $^{90}Sr-^{90}Y$, ^{137}Cs, ^{239}Pu	0·01–0·88 (0·03–6·0)	0·22–14 (0·45–14)	$5·9 \times 10^{-2}$–0·32 ($7·8 \times 10^{-2}$–0·44)
Other radionuclides	0·25–25	1·2–130	0·04–7·7
Water activity	$5·8 \times 10^{-5}$–$1·6 \times 10^{-2}$ ($6·5 \times 10^{-5}$–$1·7 \times 10^{-2}$)	$4·0 \times 10^{-5}$–$1·1 \times 10^{-2}$	$1·3 \times 10^{-5}$–$3·2 \times 10^{-3}$
Total:*	0·26–26 (0·28–31)	1·4–150 (1·7–150)	$9·9 \times 10^{-2}$–8·0 (0·12–8·2)
Waste disposal			
Blackwater estuary			
Internal activity 1968	—	—	1·4–1·8
1977	—	—	$8·6 \times 10^{-2}$ (0·34)
Sediment activity 1968			
γ	—	—	1·7
β	—	—	1·3
1977			
γ	—	—	0·62
β	—	—	0·22
Total:* 1968	—	—	3·1–3·5
1977	—	—	0·70 (0·95)
Barnegat Bay			
Internal activity	—	—	$5·1 \times 10^{-3}$–$8·5 \times 10^{-2}$
Water activity	$9·6 \times 10^{-4}$–$3·4 \times 10^{-2}$	$7·9 \times 10^{-4}$–$3·0 \times 10^{-2}$	$3·2 \times 10^{-4}$–$1·3 \times 10^{-2}$
Sediment activity γ	—	—	$7·1 \times 10^{-2}$–11
β	—	—	$1·2 \times 10^{-2}$–0·98
Total:*	$9·6 \times 10^{-4}$–$3·4 \times 10^{-2}$	$7·9 \times 10^{-4}$–$3·0 \times 10^{-2}$	$7·6 \times 10^{-2}$–11
Windscale			
Internal activity	200—2100†	530–6900†	1740–2430 (5140–6900)
Water activity	0·2–3·3	0·2–3·0	$5·0 \times 10^{-2}$–1·2
Sediment activity γ	—	—	36–3340
β	—	—	210–5380
α	—	—	31–570 (620–$1·14 \times 10^{4}$)
Total:*	200–2100	530–6900	1780–5770 (5180–$1·02 \times 10^{4}$)

*Possible contributions to the total from α- and β-radiation from sediment have not been included because the magnitude is very dependent on the precise geometry.
†Estimates based on water concentration and published values for concentration factor.

μrad h^{-1} (μrem h^{-1}). To convert to SI units of μGy h^{-1} (μSv h^{-1}) the values given must be decreased
1974; IAEA, 1976; WOODHEAD and PENTREATH, 1980)

Crustaceans 10 m depth, on the seabed	Fish 20 m depth, remote from the seabed	 50 m depth, on the seabed	 4500 m depth on the seabed
1·1	0·5	0·2	—
5·2–74	2·6–5·3	2·6–4·3	2·6–6·0
(55–1420)	(3·3–56)	(2·8–36)	(4·3–70)
0·05	0·11	0·05	0·05
1·5–16	—	1·5–16	5·3–86
1·6–21	—	1·6–21	3·5–65
14—170	—	14–170	110–1910
(280–3300)		(280–3300)	(2210–3·82 × 10^{4})
7·8–91	3·2–5·9	4·3–21	7·9–92
(57–1440)	(3·3–57)	(4·5–52)	(9·5–150)
4·3 × 10^{-3}–9·7 × 10^{-2}	2·4 × 10^{-2}–5·7 × 10^{-2}	2·4 × 10^{-2}–5·7 × 10^{-2}	—
(4·3 × 10^{-3}–0·10)	(2·4 × 10^{-2}–9·5 × 10^{-2})	(2·4 × 10^{-2}–9·5 × 10^{-2})	
0·36	0·12–1·7	0·12–1·7	—
1·3 × 10^{-5}–3·2 × 10^{-3}	2·6 × 10^{-5}–6·5 × 10^{-3}	1·3 × 10^{-5}–3·2 × 10^{-3}	—
0·36–0·46	0·12–1·8	0·12–1·8	—
(0·36–0·47)	(0·12–1·8)	(0·12–1·8)	—
—	—	—	—
	0·15	0·15	
1·7	—	1·7	—
1·3	—	1·3	—
0·62	—	0·62	—
0·22	—	0·22	—
1·7	—	1·7	—
0·62	0·15	0·76	—
—	7·0 × 10^{-3}–0·37	7·0 × 10^{-3}–0·37	—
3·2 × 10^{-4}–1·3 × 10^{-2}	6·4 × 10^{-4}–2·7 × 10^{-2}	3·2 × 10^{-4}–1·3 × 10^{-2}	—
7·1 × 10^{-2}–11	—	7·1 × 10^{-2}–11	—
1·2 × 10^{-2}–0·98	—	1·2 × 10^{-2}–0·98	—
7·1 × 10^{-2}–11	7·6 × 10^{-3}–0·40	7·8 × 10^{-2}–11	—
75–110	9·0–9·9	9·0–9·9	—
(220–260)	(9·3–10·4)	(9·3–10·4)	—
5·0 × 10^{-2}–1·2	9·0 × 10^{-2}–2·4	5·0 × 10^{-2}–1·2	—
36–3340	—	36–3340	—
210–5380	—	210–5380	—
31–570	—	31–570	-
(620–1·14 × 10^{4})		(620–1·14 × 10^{4})	
110–3450	9·0–12·3	45–3350	—
(250–3600)	(9·4–12·8)	(46–3350)	

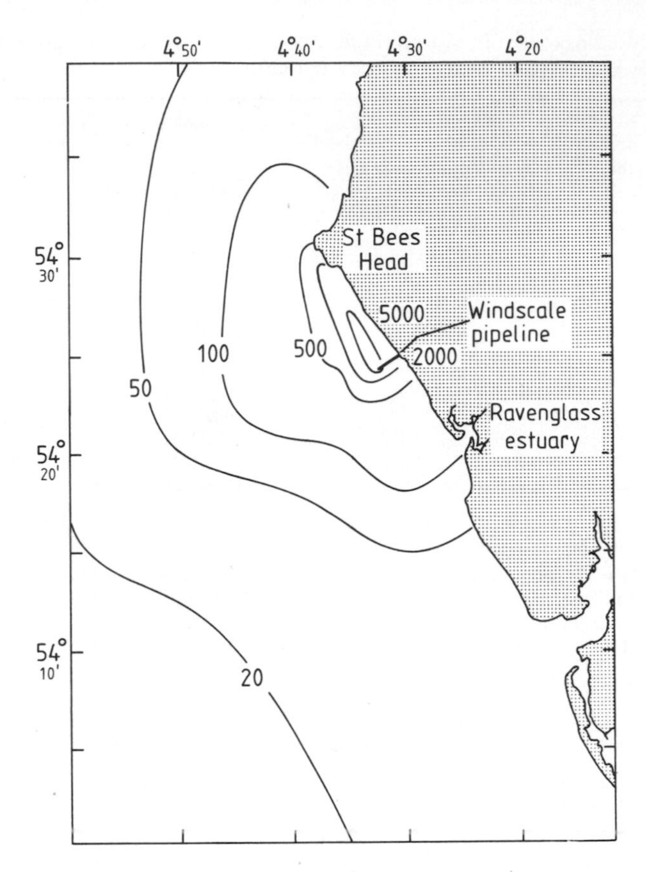

Fig. 2-23: Dose rate (μrad h^{-1}) contours at the seabed surface in the northeast Irish Sea, September 1968. (After WOODHEAD, 1973b; reproduced by permission of Pergamon Press Ltd.; Crown copyright)

background and in most cases by a very substantial margin. The activity in the sea water is unimportant as a source of exposure compared with the radionuclides accumulated by the organisms or associated with the sediment. The values given for phytoplankton and zooplankton must be treated with some reservations since the estimates are based on the radionuclide concentrations in the water and generalized concentration factors (LOWMAN and co-authors, 1971). The values given are probably maxima and are likely to be overestimates to an extent which depends on the fraction of the equilibrium organism burden implied by the concentration factor which can be attained before dilution and dispersion of the activity can occur. For pelagic fish, the accumulated activity is the most important source of exposure and the magnitude of the absorbed dose rate is of the same order as the natural background, but in terms of biologically effective dose it is somewhat less. For benthic organisms the underlying sediment is by far the most important source. Fig.2-23 shows the dose rate contours at the seabed–sea water interface as calculated from the radionuclide distributions shown in Fig. 2-18.

The estimated dose rates and the area over which these exceeded natural background ($\approx$2000 km^2) were sufficient to make it feasible to attempt *in situ* dosimetry provided that

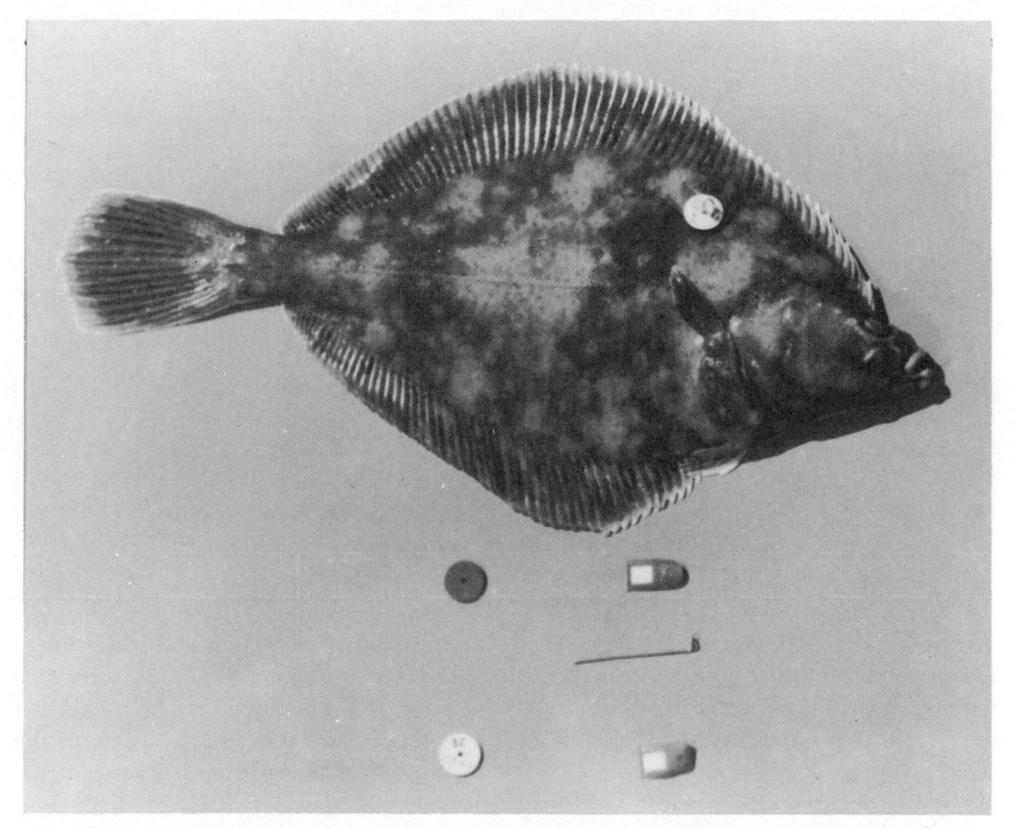

Fig. 2-24: Lithium fluoride thermoluminescent dosimeter components and a tagged plaice. (After WOODHEAD, 1973b; reproduced by permission of Pergamon Press Ltd.; Crown copyright)

a suitable dosimeter could be developed. The dosimeter design finally adopted is shown in Fig. 2-24. It consists of 32 mg of thermoluminescent grade lithium fluoride crystals encapsulated in a flat plastic sachet fabricated from 25 mg cm^{-2} thick PVC sheet. The Petersen disc tag was used to attach the dosimeters to the plaice *Pleuronectes platessa* through the skeletal muscle as shown in Fig. 2-24. One dosimeter was situated on each of the upper and lower surfaces to provide an estimate of both the γ-ray and β-ray components of the total dose from the sediment by using the body of the fish to shield the upper dosimeter from the β-radiation. On the basis of the known radionuclide compositions and distributions, the β-radiation contributes between 60% and 90% of the total dose rate as indicated by the contours shown in Fig. 2-23. Thus, provided that the upper dosimeter is in fact shielded from β-radiation, the average ratio of the responses of the upper and lower dosimeters would be expected to be 0·2. In June 1967, 2488 plaice (25–35 cm long) were marked with the combined tag–dosimeter and released 0·4 km south of the pipeline outfall. Because the recaptures of the marked fish by commercial fishermen and their subsequent return to the laboratory during the remainder of 1967 was somewhat lower than expected, a further 1092 plaice were marked and released at the same point in April 1968. Altogether 1053 (29%) fish were recaptured up to $2\frac{1}{2}$ yr after release and the dosimeters from 969 yielded useful data. The distribution of aver-

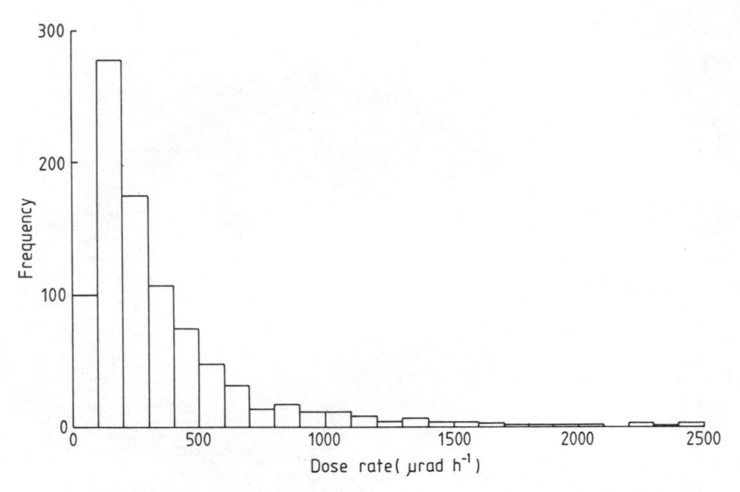

Fig. 2-25: Distribution of measured dose rates to plaice. (After WOODHEAD, 1973b; reproduced by permission of Pergamon Press Ltd.; Crown copyright)

age dose rates experienced between release and recapture as indicated by the lower dosimeter is illustrated in Fig. 2-25. The mean dose rate was 350 μrad h^{-1} (3·5 μGy h^{-1}) with occasional dosimeters registering up to 2·5 mrad h^{-1} (25 μGy h^{-1}). The maximum dose rates measured were of the same order as those calculated on the basis of the concentrations of radionuclides in the seabed but it is clear that the natural movements of the fish substantially reduced the mean dose rate experienced. Analysis of the dose rate distributions as a function of the time at liberty showed little correlation and it was concluded that the data give a fairly accurate indication of the radiation experience of the plaice population in this size range in the northeast Irish Sea. The mean ratio of the responses of the upper and lower dosimeters was 0·68 with a range of 0·13 to 1·79, i.e. values substantially greater than anticipated. It was concluded that there were two factors related to the behaviour of the fish which could account for the discrepancy. First, the resting fish tends to bury itself in the sediment thus exposing the upper dosimeter to β-radiation; second, when swimming just clear of the seabed in search of food, even a thin intervening layer of water would substantially reduce the β-radiation dose rate to the lower dosimeter. In both instances the γ-ray dose rate to both dosimeters would be almost identical. From the relative responses of the two dosimeters and assumptions concerning the behaviour of the fish it was concluded that the mean dose rate to the gonads was 210 μrad h^{-1} (2·1 μGy h^{-1}) (WOODHEAD, 1973b).

For the northeast Atlantic Ocean disposal site there are insufficient data yet available on which to base estimates of the increments of radiation exposure due to dumping. All the indications are, however, that the general increases in radiation exposure at current dumping rates are likely to be of the order of, or less than, natural background. A few organisms in the immediate vicinity of a very small proportion of the waste drums may experience higher external dose rates from the γ-emitting radionuclides contained within the drums.

The computational approach underlying the dosimetry models on which the data in Table 2-30 are based has recently been generalized (WOODHEAD, 1979) to allow the exposure of any aquatic organism (or constituent organ or tissue) to be estimated provided that the distributions of the radionuclides, both internal and external, are known (e.g. HOPPENHEIT and co-authors, 1980). The methods were exemplified by application to the case of fish eggs developing in contaminated water and will be considered in greater detail within the same context further on (pp. 1240–1242).

(b) Effects of Radiation on Aquatic Organisms

The magnitudes of the dose rates to marine organisms from the natural background, global fallout and waste radionuclides, as given in Table 2-30, provide a very necessary perspective within which the possible harmful effects of the increased radiation exposure of marine organisms must be considered. The estimates of the incremental exposure from waste radionuclides provide the only valid bases for an assessment of their potential impact in aquatic environments. The dose rates in the marine environment due to radionuclide inputs arising from human activities range from less than the natural background for typical nuclear power stations in routine operation up to a few tens of mrem h^{-1} for the rather exceptional case of the Windscale discharge into the northeast Irish Sea. Ideally, therefore, firm conclusions concerning the potential impact of waste disposal operations on the marine environment should be derived from the results of chronic irradiation experiments employing dose rates in this range, a point which has recently been emphasized (IAEA, 1979). However, the great majority of the data relating to the effects of radiation on aquatic organisms have been obtained at total doses and dose rates several orders of magnitude greater than even the highest of those estimated for contaminated environments. This situation reflects the difficulties involved in detecting radiation effects at very low doses and dose rates; the gross effects, such as mortality, do not occur or are only apparent as very slight changes in normal life expectancy, and the techniques for detecting the more subtle effects, such as chromosome damage or changes in the cell-type distributions in the gonads are only slowly being developed. The relevant experiments are frequently expensive in terms of the time required for detectable damage to develop and the number of organisms required to obtain statistically valid indications of a differential response with respect to dose: factors which have tended to discourage investigation. Thus the conclusions drawn about the incidence of environmental effects have largely rested on more or less extended extrapolation.

Such extrapolation almost always involves the explicit or implicit assumption of a linear extension to low radiation doses of the relationship between dose and the consequent effect observed at high doses. This has the corollary that any radiation dose, however small, carries a finite risk of harm. Discussion of the validity of this procedure has been confined almost entirely to the field of human radiological protection where ethical considerations require that the risk of harm to an individual in either the present generation (somatic effects), or in subsequent generations (genetic effects) be acceptably small. In this instance the exact form of the dose–response relationship at very low doses obviously has a substantial bearing on the assessment of the consequential risk from exposure, and hence on the setting of appropriate dose limits. As indicated above for aquatic organisms, much work remains to be done to elucidate the complexities of the relationships between very low doses received by individuals and a variety of possible biological responses. However, on the basis of the available data, and on theoretical grounds, the ICRP has concluded that the linear relationship is an acceptable, and perhaps, conservative simplifying assumption for stochastic effects at the very low doses and dose rates which are the predominant concern for public health. Stochastic effects are those for which it is the probability of occurrence rather than the severity of effect which increases with the total absorbed dose. This class of response includes both somatic effects, of which the most important in human terms is tumour induction, and mutations in germ cells. Non-stochastic effects are those for which the severity of the effect increases with the total dose received by an individual, and there is frequently an effective threshold dose below which the response, although present, is not considered to be detrimental. The lens of the eye, the skin, haematopoiesis and gametogenesis are the tissues and cell systems of particular significance for human exposure, and of these, gametogenesis exhibits the greatest sensitivity to irradiation (ICRP, 1977). A linear extrapolation of the dose–response (severity) relationship to very low doses can be seen to be conservative if there is an effective threshold. Indeed, there may be a true threshold if there exist effective mechanisms for repairing damage which are saturated at the high doses and dose rates at which the effects are most usually scored in laboratory experiments. Such repair mechanisms may be intracellular or operate through the dynamic process of cell turnover and differentiation at the tissue level; there is some evidence for their existence in aquatic organisms (OPHEL and co-authors, 1976).

In direct contrast to the situation for humans, the fate of any given individual in a population of marine organisms is unimportant; the total population is the significant entity and, therefore, it is the capacity of the population to maintain itself under the impact of external stresses which must be considered. Thus, although enhanced radiation exposure of marine organisms due to waste disposal may, on the linear hypothesis, be expected to induce detrimental effects in individuals, these will be unimportant if there are no concomitant effects on subsequent recruitment to the population or significant changes in population structure.

There are very few data on the effects of radiation on aquatic populations. In general, therefore, it is necessary to assemble the available data on the effects of radiation on the lifespan, fertility, fecundity and genome of individual organisms, placing the greatest weight on results from experiments which have employed chronic exposure, and attempt to interpret them in terms of a response at the population level. However, the latter step is only required where it seems possible that the estimated dose rates in the environment could cause significant effects in the individual since it is inconceivable that an exposure which does not produce any effect in individuals could have any significant impact on the population.

In addition to the survey in Volume I of *Marine Ecology* (CHIPMAN, 1972) there have been several other more or less comprehensive reviews of published data on the effects of radiation on aquatic organisms (POLIKARPOV, 1966; TEMPLETON and co-authors, 1971; OPHEL and co-authors, 1976; BLAYLOCK and TRABALKA, 1978; EGAMI and IJIRI, 1979). Thus it is not necessary to give a further, detailed account of the entire literature. In most respects the conclusions and generalizations developed by these authors, together with relevant data which have been obtained subsequently, provide a satisfactory basis for an assessment of potential environmental damage. The major deficiency which has been identified concerns the substantial, and often contradictory, literature on the effects of radiation on developing fish embryos and, hence, recruitment. An effort has been made to analyse these data critically and to determine how far they provide a sound foundation for predicting the degree of effect in contaminated environments.

Effects of Radiation on Mortality rate

In a majority of investigations the lethal effects of radiation on aquatic organisms have been evaluated in terms of their short-term survival after acute exposures and most often by determining the median lethal dose at 30 d ($LD_{50/30}$). On this basis it appears that the radiosensitivity of aquatic organisms increases with increasing biological complexity, i.e. the higher the phylogenetic position the lower the $LD_{50/30}$. Thus fish show the greatest sensitivity and appear to be somewhat less radiosensitive than mammals. The 30-d time interval for scoring mortality is a carryover from the extensive studies with small mammals where it has been found that essentially all the acute lethality is expressed within this time and that survivors have a subsequent long-term pattern of mortality which is little different from that of the controls. However, it has been pointed out by WHITE and ANGELOVIC (1966) that the 30-d period is inappropriate for aquatic organisms since there is often continuing, radiation-induced mortality beyond this time. It was demonstrated for a number of organisms that the median lethal dose first decreases with time after irradiation before becoming more or less independent of time after all the acute damage has been expressed. At 30 d after irradiation the LD_{50} for 6 species of fish showed an overall variation by a factor of 5·3; by 40 d this had fallen to a factor of 3·3 and at 50 d the LD_{50} values (extrapolated for 2 species) were in the range 925 to 2250 rad (a factor of 2·4) and still falling for 2 of the species. For invertebrates the reduction in median lethal dose with time after irradiation was often even more pronounced. For the oyster *Crassostrea virginica* for example, the $LD_{50/30}$ was estimated to be greater than 180 krad, but at 60 d, when the greater part of the acute response had become apparent, the LD_{50} was estimated to be approximately 6·6 krad. These findings do not alter the general conclusion given above regarding the relative sensitivities of aquatic organisms, but the differences of radiosensitivity in comparision with terrestrial vertebrates and invertebrates become much less marked.

There are very few data concerning the lethal effects of chronic irradiation on aquatic organisms. The mortality of freshwater snails *Physa heterostropha* exposed to chronic ^{60}Co γ-irradiation did not become significantly greater than that of the controls until the accumulated doses reached 57 krad at 25 rad h^{-1} and 28 krad at 10 rad h^{-1}. At 1 rad h^{-1} (total dose 4·5 krad) there was no significant differential mortality over the 24-wk lifespan of the snail under laboratory conditions (COOLEY and MILLER, 1971). Juvenile marine clams *Mercenaria mercenaria* and scallops *Argopecten irradians* were also found to be resistant to ^{60}Co γ-radiation at cumulative annual doses of 8 to 8·8 krad (0·9 to 1·0 rad h^{-1}) as assessed by survival (BAPTIST and co-authors, 1976). The age-specific survival rate of *Daphnia pulex* maintained on a constant *per capita* food supply showed very little influence of ^{60}Co γ-irradiation at a dose rate of 22·3 rad h^{-1} (total dose over the 28-d median lifespan of approximately 12 krad), and higher dose rates up to 76 rad h^{-1} produced only relatively minor effects on the life expectancy at birth (MARSHALL, 1962). In a second study using populations of *Daphnia* competing for a fixed food supply the observed post-natal death rate was insignificantly increased relative to the controls at a mean dose rate of 3·5 to 4·5 rad h^{-1}; the calculated total death rate (including pre-natal mortality) was, however, increased (MARSHALL, 1966). Over a period of 70 d the survival of blue crabs *Callinectes sapidus* was unaffected by continuous irradiation from ^{60}Co at either 3·2 or 7·3 rad h^{-1}; at 29 rad h^{-1} the mortality exceeded that of the controls after 50 d at which time the cumulative dose was approximately 33 krad (ENGEL, 1967). Exposure of the mosquito fish *Gambusia affinis* to ^{60}Co γ-rays at dose rates of 1·43, 2·45, and 5·43 rad h^{-1}, at either 15 °C or 25 °C for a period of 40 d (total doses 1200, 2300, and 5000 rad), elicited no increased mortality compared with the controls (COSGROVE and BLAYLOCK, 1973). Pairs of guppies *Poecilia reticulata* exposed to ^{137}Cs γ-radiation at a mean dose rate of 1·27 rad h^{-1} accumulated 7200 rad over a period of 238 d with no increased mortality (WOODHEAD, 1977). At each of 2 lower dose rates (0·17 rad h^{-1} for 974 d and 0·40 rad h^{-1} for 771 d) there was some indication of increased mortality relative to the controls. The data (Fig. 2-26) were obtained incidentally to the main objectives of the investigation and, because only small numbers of fish were involved, have relatively limited significance. They are included here, however, because they represent the only data available which concern the effects of chronic irradiation over essentially the whole lifespan of a fish (WOODHEAD, unpubl.).

From these limited data it can be reasonably concluded that there would not be any significant reduction in the life expectancy of aquatic organisms irradiated at the dose rates which have been estimated to prevail in contaminated environments (Table 2-30). It also seems likely that fish would be most susceptible to damage.

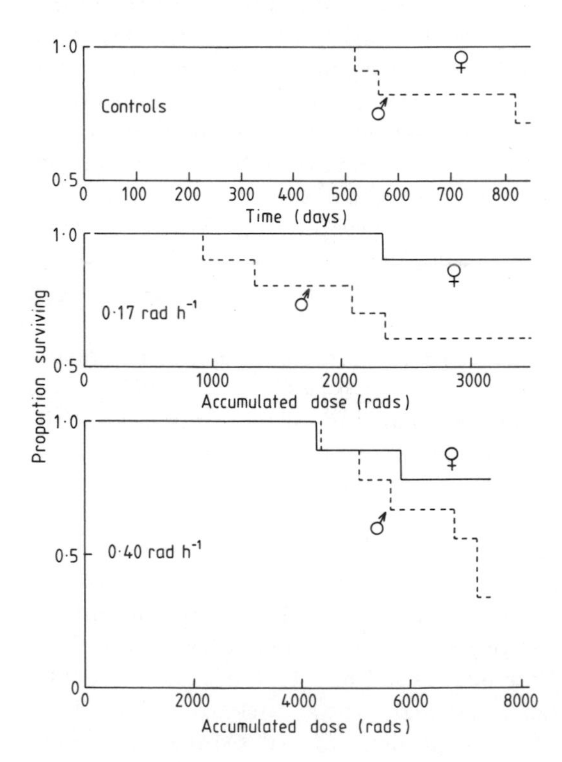

Fig. 2-26: Survival of guppies *Poecilia reticulata* under chronic γ-irradiation. (Ministry of Agriculture, Fisheries and Food, Directorate of Fisheries Research, unpublished)

Effects of Radiation on Fecundity

The gonads have been found to be one of the more sensitive of cell systems (CARLSON and GASSNER, 1964; BLAYLOCK and TRABALKA, 1978), and a detailed account of the effects of radiation on the gonads of teleost fish has been given by EGAMI and IJIRI (1979).

For the freshwater snail, *Physa acuta* an acute 2 kR X-ray exposure reduced the rate of egg production per adult to 74% of the control value (RAVERA, 1967). In two separate experiments (COOLEY and MILLER, 1971; COOLEY, 1973a) exposure of *P. heterostropha* to ^{60}Co γ-rays at 1·1 rad h^{-1} produced a significant reduction in egg production rate over 12 (total dose 2·2 krad) and 14 (total dose 2·6 krad) wk respectively, although in the first series recovery to control values occurred between 12 and 24 wk. Dose rates of approximately 2, 5, 10, and 25 rad h^{-1} were progressively more damaging.

Exposure of young adult male brine shrimp *Artemia sp.* to 2 kR or more of ^{137}Cs γ-rays had a severe effect on the testes as evidenced by their subsequent reduced ability to produce offspring when mated with unirradiated females (SQUIRE and GROSCH, 1970). Similarly, exposures greater than 1 kR of young adult female brine shrimp had severe effects on the ovary as indicated by the total number of nauplii and encysted embryos produced per individual in matings with unirradiated males (SQUIRE, 1970). HOLTON and co-authors (1973a) irradiated both male and female brine shrimps as juveniles (mainly 10th instar) with ^{60}Co γ-rays and found that 900 rad were sufficient to produce a reduction in the number of broods per pair in subsequent matings and 2400 rad effectively sterilized the animals. A second experiment in which initial populations, consisting of 300 7th instar to adult individuals, were given acute doses of 500, 1500, and 3000 rad of ^{60}Co γ-rays and then examined at intervals over 20 wk provided some contrasting results (HOLTON and co-authors, 1973b). The populations which received 3000 rad did not become extinct as might have been predicted on the basis of the results of the first experiment. This was attributed to the presence of adults in the initial population whose brood production capacity was relatively less radiosensitive than that of the 10th instar juveniles examined in the first experiment. Although the number of broods produced by each pair was significantly reduced throughout the experiment, sufficient young were born to allow the population size to recover to control numbers by 20 wk. Since the 20 wk of observations covered several generations the persisting reduced mean numbers of broods produced by each pair sampled from the irradiated populations implies the presence of inherited damage. Even though the continuing fecundity was demonstrably affected by the irradiation of the initial populations, only a small fraction of the available reproductive potential was realizable under the culture conditions employed and by 20 wk the effects of radiation had become a relatively minor factor in determining the population size.

IWASAKI (1973) showed from histological preparations that the radiosensitivity of the ovary of the brine shrimp to acute irradiation from ^{60}Co γ-rays depended on the age of the animal. The encysted embryo is arrested at the gastrula stage and the ovary is not present as an identifiable entity. Irradiation at this stage with less than 10 krad produces no histologically detectable damage in the adult ovary. Mitotically active oogonia are present in juvenile shrimp and doses of 2 krad or more produce immediate and severe depletion of the cell population. At the lowest dose recovery was evident at 10 to 14 d post-irradiation. In the adults all stages of oogonia and oocytes are present and at least 5 krad were necessary to produce significant, histologically detectable damage to the most sensitive, pre-meiotic oocyte stage. Acute irradiation of adult female amphipods *Gammarus duebeni* with single exposures between 220 and 740 R progressively reduced the cumulative egg production per female over the ensuing 23 wk (HOPPENHEIT, 1973). In populations of *Daphnia pulex* exposed to chronic γ-irradiation from ^{60}Co and provided with constant *per capita* food supply the progressive reduction in the instantaneous population birth rate with increasing dose rate above 22·3 rad h^{-1} was attributed almost entirely to reduced fecundity (MARSHALL, 1962). In the second study in which the individuals competed for a fixed food supply to the population, the mean brood size (taken as an indicator of fecundity) showed a slight tendency to increase up to the maximum dose rate of approximately 18 rad h^{-1} which could apparently be tolerated more or less indefinitely. This was interpreted as being due to the increased food supply available to each individual in the progressively smaller populations maintained at the increasing dose rates and which tended to compensate for the damaging effects of radiation on fecundity (MARSHALL, 1966).

SOLBERG (1938a) found a 65% decrease in the fecundity of female medaka *Oryzias latipes* 7 d after an X-ray exposure of 1980 R with recovery of egg production by Day 20. These results were confirmed by subsequent studies which showed the partially mature oocytes to be the most radiosensitive (EGAMI and HYODO, 1965; EGAMI and IJIRI, 1979). Exposure of female loach *Misgurnus anguillicaudatus* to less than 1 kR produced no marked changes in the ovary over a period of 2 months, but exposures of 2 kR and more again caused degeneration of the partially mature oocytes (EGAMI and AOKI, 1966). At least part of the observed damage to the oocytes has been attributed to radiation-induced disturbances of the hormone system governing gonad activity since 2 kR exposures of the head alone (EGAMI and AOKI, 1966) or the whole body with the ovary shielded (EGAMI and HYODO, 1965) produced essentially the same ovarian response. This suggestion is supported by the finding that 10 to 15 d after an exposure of 2 kR there was an increased accumulation of gonadotropic secretion granules in cells of the pituitary gland of the medaka relative to controls (YAMAMOTO and EGAMI, 1974). The immature ovaries in developing embryos and newly hatched fry also appear to be relatively radiosensitive. Exposure of medaka embryos to 1 kR did not significantly change the number of cells present in the developing ovary but delayed the onset of meiosis until 12 d after hatching.

At hatching, when meiosis normally commences, and in the immediate post-hatch period there is a profound change in the radiosensitivity of the cells and in the capacity of the survivors of irradiation to repopulate the ovary. During this period, and exposure of 1 kR kills most of the oogonia and diplotene oocytes immediately; many of the remaining oogonia and apparently less sensitive cells in earlier stages of meiosis only survive until the transformation from pachytene to diplotene (HAMAGUCHI and EGAMI, 1975; HAMAGUCHI, 1976, 1980). KONNO (1980) found that exposure of rainbow trout *Salmo gairdneri irideus* embryos at the late-eyed stage to 600 and 800 R of ^{60}Co γ-rays produced a high incidence of sterility in surviving fry of both sexes; at lower X-ray exposures up to 200 R, however, no significant effects were observed on the fecundity (WELANDER and co-authors, 1971). X-irradiation of rainbow trout fry at an age of 130 d to an exposure of 500 R caused a significant reduction in the ovarian weight as a proportion of body weight which persisted for at least 310 d; a lower exposure of 100 R produced a marginally significant reduction in ovarian weight 100 to 120 d after irradiation with subsequent recovery (KOBAYASHI and MOGAMI, 1958). Exposures in excess of 500 R have been found to have a marked effect on the testes of the medaka both histologically (SOLBERG, 1938a) and in terms of the gonado-somatic index (MICHIBATA, 1976); in the latter case, recovery was essentially complete after 60 d. An exposure of 2 kR to male medaka reduced the number of fertilized eggs produced in matings with unirradiated females in the period 2 to 6 wk after irradiation (EGAMI and IJIRI, 1979).

For the medaka, the guppy *Poecilia reticulata* and the marine goby *Chasmichthys glosus* the effect of exposures between 100 and 1000 R were broadly comparable. Spermatogonia (Ib and II) showed an immediate reduction even at the lowest dose; spermatocytes were also very sensitive but the complete response took longer to develop. Spermatids and spermatozoa were relatively more resistant. Recovery and repopulation of the testes took place through the multiplication and differentiation of a small subpopulation of primary spermatogonia (Ia) which is not normally mitotically active. For the guppy, the recovery in each cell type was bimodal and it was suggested that the first, short-term, increase was the result of recovery from radiation-induced mitotic delay, while the more enduring recovery was due to repopulation from the less sensitive stem cell spermatogonia (Ia) (KOBAYASHI and YAMAMOTO, 1971; HYODO-TAGUCHI and EGAMI, 1971, 1976). Exposure of medaka to 1000 R just before or after hatching caused complete destruction of the gonad even though differentiation of the precursor cells into adult spermatogonia had not commenced at that time (HAMAGUCHI, 1980).

Overall, for both male and female fish, it is evident that acute sublethal irradiation causes severe damage to the gonads and that the testes show the greater radiosensitivity with doses of as little as 100 rad producing a readily observable response.

NELSON and co-authors (1976) attempted to sterilize sexually undifferentiated fry of *Tilapia zilli* with chronic ^{60}Co γ-irradiation at dose rates of 1·7, 2·5, and 4·0 rad h^{-1} for total doses of 1·4, 2·1, and 3·3 krad respectively. At the end of the 35-d exposure period histological examination of the gonads showed no significant effects. A similar study with mosquito fish *Gambusia affinis* at ^{60}Co γ-ray dose rates of 1·3, 2·5, and 5·4 rad h^{-1} over 47 d revealed no histological evidence of damage to the ovaries, but the testes showed severe damage at all dose rates from 18 d onwards (COSGROVE and BLAYLOCK, 1973). The gonads of 6 pairs of guppies (initial age 0 to 3 d) maintained for 238 d under chronic ^{137}Cs irradiation at a mean dose rate of 1·27 rad h^{-1} were completely destroyed by an accumulated dose not greater than 7200 rad. Four of the 6 pairs produced one small brood after accumulated doses between 3·29 and 4·07 krad. At 2 lower dose rates (0·17 and 0·40 rad h^{-1}) the fecundity was also reduced as a result of smaller actual brood sizes, an increased incidence of temporary sterility and the earlier onset of permanent sterility. At the lowest dose rate the pair producing the last brood (Day 830) had accumulated 3·47 krad. The testes of 6 fish were examined histologically and appeared normal after total doses between 2·50 and 4·6 krad. Nine females were also examined, and of these 6 appeared normal (2·86–4·06 krad), 1 lacked maturing oocytes although oogonia were present (3·69 krad) and the ovaries of the remaining 2 showed no signs of gametogenesis (2·86 and 3·23 krad). The testes of 6 fish irradiated at 0·4 rad h^{-1} appeared histologically normal after 4·36 to 7·37 krad, as did the ovaries of 6 females which had accumulated 5·81 to 7·37 krad; however, the ovaries of a further 2 females were devoid of germ cells after 6·52 and 7·37 krad. At this dose rate the last brood was produced on Day 749 (7·19 krad). It was concluded that the reduced fecundity was not necessarily dependent upon histologically detectable damage to the gonads and that the ovary was, in gross terms, more radiosensitive than the testis. It is also apparent that an accumulated gonad dose of several krad does not preclude successful reproduction in the first generation (WOODHEAD, 1977). A rather more detailed study of the effects of chronic irradiation on spermatogenesis has been made with the medaka by HYODO-TAGUCHI (1980) at ^{137}Cs γ-ray dose rates of 1·3 to 84·3 rad d^{-1}. After 30 d of exposure the gonado-somatic index was significantly reduced at all dose rates greater than 2·9 rad

d^{-1} (approximately 0·12 rad h^{-1}). Between 30 and 120 d more or less complete recovery took place at dose rates less than 15·6 rad d^{-1} ($\approx$0·65 rad h^{-1}). Histological examination of the testes showed marked reductions in the numbers of primary spermatogonia (Ib) present after 30-d exposure at dose rates greater than 6·8 rad d^{-1}. Between 30 and 120 d the cell population recovered at dose rates less than 15·6 rad d^{-1} ($\approx$0·65 rad h^{-1}). At 15·6 and 24·2 rad d^{-1} the cells attained lower densities which appeared to represent new homeostatic equilibria while 84·3 rad d^{-1} almost completely destroyed the cell population. The effect on fecundity of irradiation for 60 d at these dose rates was tested in matings with unirradiated females. Significant and progressively greater reductions in fecundity were observed at dose rates greater than 6·8 rad d^{-1} (0·28 rad h^{-1}); the proportion of completely infertile males also increased progressively with dose rate.

Several studies have been made of the effects on the gonads of radionuclides present in the body of the fish. Cobalt-60 injected into catfish *Heteropneustes fossilis* damaged the ovary (SRIVASTAVA and RATHI, 1967), and $^{90}Sr-^{90}Y$ accumulated from the water damaged the ovary of a marine goby (HYODO-TAGUCHI and co-authors, 1971), as well as the testes of the medaka (YOSHIMURA and co-authors, 1969). However, these results are only qualitative since no estimates of the doses or dose rates were given, nor were sufficient data provided to allow independent dose estimates to be made. Mature loach *Misgurnus fossilis* received estimated doses of 0·02 and 0·2 rad to the ovary from $^{90}Sr-^{90}Y$ accumulated over 3 months from ambient water concentrations of $3{\cdot}11 \times 10^{-10}$ and $1{\cdot}32 \times 10^{-6}$ Ci l^{-1} respectively, and greater doses of pituitary extract were required to stimulate egg production in these fish than in the controls (SHEKHANOVA and PECHKURENKOV, 1969). Exposure of male medaka to various concentrations of tritiated water caused detectable damage to the testes. After 10 d exposure, the gonado-somatic index was significantly reduced at a concentration of 10^{-1} Ci l^{-1} (dose rate $\approx$0·87 rad h^{-1}), as were the number of primary spermatogonia (Ib) at concentrations greater than 10^{-2} Ci l^{-1} ($\approx$0·085 rad h^{-1}). At 30 d of exposure the reduction in gonado-somatic index had become significant at the lower concentration of 5×10^{-2} Ci l^{-1} ($\approx$0·43 rad h^{-1}), while some recovery was apparent in the numbers of primary spermatogonia (Ib) at 10^{-2} Ci l^{-1}. It should be noted, however, that there was progressive decrease in the numbers of spermatogonia with increasing concentration so that the reduction at the lowest concentration of 5×10^{-3} Ci l^{-1} ($\approx$0·042 rad h^{-1}) while not significant in isolation might nevertheless have been indicative of a real effect (HYODO-TAGUCHI and EGAMI, 1977). BONHAM and DONALDSON (1972) showed that a dose rate of 0·42 rad h^{-1} during the 80-d egg incubation period of chinook salmon *Oncorhynchus tshawytscha* markedly affected the gonad development in the fry.

It is clear from the preceding summary that fish are the most radiosensitive aquatic organisms in the context of reduced fecundity. There is considerable apparent variation between species but in general terms it may be concluded that minor damage can be expected in fish gonads at dose rates greater than 0·1 rad h^{-1}. This is substantially greater than the maximum dose rate which has been estimated for benthic fish in contaminated environments (Table 2-30), and significant reductions in fecundity due to the increased radiation exposure can be discounted.

Critical Reassessment of the Influence of Radiation on Developing Fish Embryos

Aside from fecundity, the factor likely to have the greatest influence on potential recruitment is mortality during embryonic development. Since POLIKARPOV and IVANOV

(1961, 1962b) found apparent radiation-related damage affecting the embryos of a variety of fish developing in sea water contaminated with low concentrations of $^{90}Sr-^{90}Y$, and from these results proceeded to predict considerable adverse consequences for commercial fish stocks at the then-prevailing environmental concentrations (POLIKARPOV, 1966), this has been the most contentious aspect of aquatic radiobiology (GARROD, 1966; TEMPLETON, 1966).

The fact that the vertebrate embryo was at least as radiosensitive as the adult, and probably more so, had been recognized quite early in radiobiological studies (for reviews see RUSSELL, 1954; ELLINGER, 1957; RUGH, 1960). This conclusion could also have been predicted on the basis of the 'law' of BERGONIÉ and TRIBONDEAU (1906) which was deduced mainly from studies of irradiated testes and also from consideration of other investigations. This 'law' states that 'the radiosensitivity of cells and tissues is proportional to their reproductive activity and inversely proportional to their degree of differentiation'.

The developing eggs of an aquatic vertebrate were employed in a radiation experiment very soon after the discovery of X-rays (TARKHANOV, 1896). It was found that when artificially fertilized eggs of the lamprey were exposed to X-rays for 15 min once or twice daily, embryonic development was inhibited although control eggs produced normal larvae. In this case, as with a great many of the early experiments with other organisms it is impossible to estimate the dose rate or the absorbed dose. The early recognition of the importance of dosimetry to the safe realization of the diagnostic and therapeutic properties of radiations has already been mentioned (see Section 1), and the fundamental significance of radiation dose in radiological protection will have become apparent in Section 3. The absolute necessity for dosimetry in aquatic radiobiological studies if the results are to have any meaning will be a recurring theme in this section.

In view of the probable greater sensitivity of the embryonic phase of the life cycle, it was natural that in the context of contaminated aquatic environments attention should have become focused on developing fish eggs since fish have very significant food and amenity value. The importance of such investigations was underlined by the finding that fallout radioactivity extracted from rainwater after the test series in the Pacific in 1954 had a damaging effect on the developing eggs of the zebra fish (HIBIYA and YAGI, 1956; MIKAMI and co-authors, 1956, 1965). The interpretation of these results, however, is hampered by the lack of information on the radionuclide composition and chemical form of the residue (OPHEL and co-authors, 1976).

Several more or less complete reviews or summaries of the published data on the effects of radiation on developing fish embryos have recently been published. DABROWSKI (1975) has given a very brief summary of some of the main results which had been published up until 1973; this has the virtue of including many previously unconsidered Russian data. However, no critical comparisons of the data were made and the problem of dosimetry was ignored. A short review of selected data has been given by ETOH (1976) as an introduction to his own results. Although the dose rates to the developing embryo from radionuclides in the water and on and in the egg are considered, the discussion and comparison of the data are given in terms of the radionuclide concentrations in the water. In an attempt to give a definitive assessment of the possible adverse biological consequences in contaminated aquatic environments, estimates of the likely dose rates to a variety of organisms from a variety of sources have been made and a wide-ranging review of the data on the effects of irradiation on aquatic organisms has been given (IAEA, 1976). With respect to fish embryos developing in water contaminated with radionuclides the problem of dosimetry was not addressed and once again the data were compared and discussed on the basis of the radionuclide concentrations in the water. It is apparent, though not explicitly stated, that many of the data could not be used as a basis for the assessment. BLAYLOCK and TRABALKA (1978) emphasized the fundamental importance of dosimetry to radiobiological studies of developing fish embryos. They also recognized that some of the apparent inconsistencies between the results of different investigators arose from the differential accumulation of the various radionuclides by the eggs, the variety of species studied, and the range of biological end-points employed as criteria of effect. Despite these insights, the results of the experiments were once again discussed in terms of the concentration of the radionuclide in the water.

It seems clear that the authors of the latter three reviews appreciated that there were many problems involved in interpreting the results of much of the experimental work either in an environmental context or for the purposes of assessment and in the absence of any obvious means of resolving the problems confined themselves to a straightforward statement of the results. It is the objective of this section to provide a critical reappraisal of the published data on the effects of radiation on the developing fish embryo and to determine its utility as a basis for the assessment of the possible effects in contaminated aquatic environments.

Effects arising from external irradiation of the developing embryo with X-rays or with γ-rays from sealed sources

Acute irradiation The response of developing fish embryos to acute, external irradiation is of limited direct relevance to the main problem of environmental concern, i.e. chronic, low dose rate exposure. However, some of the results which have been obtained can provide a useful insight into the mechanisms which may underlie the response in the latter case. A number of experiments have shown that the radiosensitivity of the developing embryo changes with increasing development. BONHAM and WELANDER (1963) determined the median lethal dose (LD_{50}) at hatching and at 150 d after fertilization (larval age approximately 90 d) for the silver salmon. The minimum values were calculated to be 30 R and 16 R respectively for irradiation early in the one-cell stage. For both criteria the early phases of development (up to 32 cells) showed cyclic variations in radiosensitivity. During the later stages, which were less intensively investigated, the LD_{50} at hatching showed a steady decrease in radiosensitivity with a final value of 1871 R for irradiation at the late-eyed stage, while the LD_{50} for the larvae continued to give evidence of a number of more sensitive periods of embryonic development against a background of increasing radioresistance (an LD_{50} of 874 R for irradiation of late-eyed embryos). Such variations in the radiosensitivity of the very early phases of embryogenesis have been noted on a number of occasions (SOLBERG, 1938b; NEIFAKH and ROTT, 1958; BELYAEVA and POKROVSKAYA, 1959; PROKOF'YEVA-BEL'GOVSKAYA, 1961; KULIKOV, 1970a,b) and the synchrony of early cleavage divisions and the differential sensitivity of the phases of the mitotic cycle have been implicated as the underlying cause. The trend of decreasing radiosensitivity with increasing development is a general finding (SOLBERG, 1938b; WELANDER, 1954; MCGREGOR and NEWCOMBE, 1968; FRANK and BLAYLOCK, 1971; WADLEY and WELANDER, 1971; EGAMI and HAMA, 1975). WELANDER (1954) determined the LD_{50} at hatching and at the end of the yolk sac stage for trout, and obtained the low values of 78·3 R and 54·8 R respectively for irradiation of the one-cell stage, and 735 R and 454 R respectively for irradiation of late germ ring embryos. For late-eyed embryos the LD_{50} at the end of the yolk sac stage increased to 904 R. The developing embryos of the carp seem to be substantially less radiosensitive than salmonid embryos. BLAYLOCK and GRIFFITH (1971) found the LD_{50} at hatching to be 601 rad and 501 rad for exposure of single cell eggs to γ-radiation and β-radiation respectively. FRANK and BLAYLOCK (1971) further showed that the LD_{50} at hatching reached 16 krad for irradiation of early-eyed carp embryos. The only datum which has apparently been obtained for a marine fish is the value of 90 rad for the LD_{50} at metamorphosis for plaice larvae which had been irradiated at the blastula stage of embryogensis (WARD and co-authors, 1971).

Two sets of data (WELANDER, 1954; BONHAM and WELANDER, 1963) have indicated that lower values of LD_{50} are obtained if the time interval between irradiation and the assessment of the damage inflicted during embryogenesis is increased.

It has been noted that an acute radiation dose may result in the mortality of the eggs at a specific stage later in development. Thus, irradiated rainbow trout embryos experienced much increased mortality during the late cleavage stage and during the development of the eye, brain, and caudal region (WELANDER, 1954); for salmon embryos the increased mortality occurred at gastrulation and at hatching (WADLEY and WELANDER, 1971), while for loach and sturgeon embryos a high dose (10 kR) at any time between fertilization and early blastula development prevented successful gastrulation (NEIFAKH and ROTT, 1958). In the latter case it appeared that the genetic information necessary to control development as far as gastrulation had been released before fertilization and was in a form that was insensitive to the effects of radiation. Only when new information was required to govern further development did the damage to the genetic apparatus become apparent. Other data, obtained after even higher radiation exposures, have demonstrated radiation damage to the processes of gene activation and repression at specific times during the development of loach embryos (NEIFAKH and DONTSOVA, 1962; BELITSINA and co-authors, 1963; UMANSKII, 1968). Clearly, at these high doses the damage is unconditionally lethal, but at lower doses it is to be expected that damage to these processes to different extents in differing proportions of cells in each embryo could generate the observed probability distributions of mortality and abnormal development.

The most usual criteria of radiation effect which have been employed are embryo mortality as indicated by hatching success (WELANDER, 1954; KONNO and co-authors, 1955; NEIFAKH and ROTT, 1958; BELYAEVA and POKROVSKAYA, 1959; BONHAM and WELANDER, 1963; WHITE, 1964; KULIKOV and co-authors, 1968; MCGREGOR and NEWCOMBE, 1968; HYODO-TAGUCHI and EGAMI, 1969; KULIKOV and co-authors, 1969, 1970b, AL'SHITS and co-authors, 1970; KULIKOV, 1970a; BLAYLOCK and GRIFFITH, 1971; FRANK and BLAYLOCK, 1971; KULIKOV and co-authors, 1971a; WADLEY and WELANDER, 1971), and larval mortality (WELANDER, 1954; KONNO and co-authors, 1955; BONHAM and WELANDER, 1963; KULIKOV and co-authors, 1968; HYODO-TAGUCHI and EGAMI, 1969; KULIKOV and co-authors, 1969, 1970b; KULIKOV, 1970a; KULIKOV and co-authors, 1971a; WARD and co-authors, 1971). These parameters have the advantage of being relatively easy to quantify in relation to the absorbed dose and can be used to derive a reasonably unambiguous indication of radiosensitivity (i.e. the LD_{50} at some specified time after irradiation). However, mortality is the final and most easily identifiable expression of damage to biological functions, tissues, and organs which are developing in the embryo. Any or all of these may be damaged to a greater or lesser degree in a moribund embryo or larva and similar, although less extensive, damage may be present in apparently viable individuals. Another criterion of damage which is frequently employed is the incidence of abnormal embryos and larvae. In severe cases this is the equivalent of mortality, but milder, more or less viable forms of anomaly have been observed. Due to the rather more subjective nature of the assessment and also to the frequent existence of multiple anomalies in a given individual, particularly at high doses, it is less easy to quantify than mortality.

WELANDER (1954) scored visible anomalies such as microphthalmia, abdominal and caudal foreshortening, abdominal lordosis and abnormal fin structures in rainbow trout. An increased incidence of abnormal fish with increasing absorbed dose was only detected with certainty for those embryos irradiated at the 32 cell and the early-eyed stages. The relatively minor changes in the other characters scored (standard length, eye diameter, head length, pigmentation, and fin ray numbers) probably represent, in a mild form, the same type of damage which was scored as abnormality. The late cleavage stage of rainbow trout embryos was found to be significantly more sensitive to an absorbed dose of 1000 rad than the early cleavage, blastula, and germ ring stages by MCGREGOR and NEWCOMBE (1968), who scored major malformations of the eyes and body. In the absence of comparable absorbed doses and stages at irradiation it is not possible to make a close comparison of the results of these two experiments. KONNO and co-authors (1955) noted skeletal deformities and hypertrophy of the yolk sac and pericardium in goldfish which increased in incidence with absorbed dose. FRANK and BLAYLOCK (1971) scored as abnormal those carp embryos which showed crooked tails and bodies, enlarged abdomens and reduced overall size; again, for irradiation at a given stage of development, the subsequent incidence of abnormal larvae increased with absorbed dose and the sensitivity to irradiation declined with increasing development. Irradiation of the embryos of the euryhaline fish *Fundulus heteroclitus* mainly produced anomalies in the development of the brain and a general stunting of growth (RUGH and CLUGSTON, 1955; RUGH and GRUPP, 1959).

ALLEN and MULKAY (1960) made an extensive histological investigation of the effects on embryos of the paradise fish *Macropodus opercularis,* of an X-ray exposure of 1000 R at different stages of development. Twelve tissue or organ systems were investigated and were found to have the following approximate order of decreasing sensitivity: (1) blood and haemopoietic tissue; (2) eye; (3) central nervous system; (4) germ cells; (5) muscles; (6) gut; (7) heart; (8) ear and lateral line; (9) pronephros; (10) olfactory organ; (11) notochord; (12) pigment cells. If damage to muscles can be assumed to be, in part at least, responsible for deformations of the body, then the abnormalities which are most easily scored also correspond to damage to some of the most sensitive tissues. For all tissues and organs it was found that specific types of damage could be elicited by irradiation prior to the appearance of any identifiable precursor cells, i.e. up to and including at least the late blastula stage and for some tissues (germ cells, heart, pronephros, and olfactory organ) much later. This finding is in agreement with both the stated conclusions (WELANDER, 1954; RUGH and GRUPP, 1959; MCGREGOR and NEWCOMBE, 1968) and the results of other workers (KULIKOV and co-authors, 1969; AL'SHITS and co-authors, 1970; KULIKOV, 1970b; FRANK and BLAYLOCK, 1971). The process of differentiation of a tissue and any period of increased mitotic activity in a tissue or organ were found to result in increased radiosensitivity, thus confirming the correlation suggested by WELANDER (1954). It was also found that regeneration only occurred in those tissues in which there was significant mitotic activity to be expected subsequent to the stage of development at irradiation.

After an exposure of 500 R at different times during the first cleavage division, a correlation was found between the incidence of dead and abnormal embryos and the incidence of cells with chromosome aberrations. The greatest sensitivity was found at mitotic metaphase; but at an exposure of 50 R the correlation was much less convincing (BELYAEVA and POKROVSKAYA, 1959). At the higher dose it is possible that the high incidence of chromosome anomalies represented a degree of 'overkill' and studies at intermediate doses would be required to obtain a better definition of the dose–response relation. KULIKOV and co-authors (1970b, 1971a) have shown that, for irradiation prior to the first cleavage division, the incidence of aberrant anaphase figures is directly proportional to exposure in the range 50 to 800 R, reaching 73% at the highest dose. PROKOF'YEVA-BEL'GOVSKAYA (1961) also found an incidence of 67% to 83% aberrant mitoses after an exposure of 800 R at the 8th blastomere stage. More extensive studies of the dose–response relation for the early cleavage divisions as a function of low doses (< 50 rad) would be very useful in indicating the sensitivity at expected environmental dose rates.

HYODO-TAGUCHI and EGAMI (1969) found that, as embryos of *Oryzias latipes* developed, a dose-independent segment of the dose–survival time curve became apparent. The range of exposures (4 to 64 kR) over which this response extended was the same as that which, in adult fish, corresponds to mortality from intestinal failure. It was suggested, therefore, that the appearance of this response in the pre-hatching embryo arose as a consequence of the differentiation and functioning of the intestinal epithelium at that stage of development. This conclusion is not too convincing because, from the data given in both numerical and graphical form, the dose-independent response appears to be present from the late germ ring stage onward. KULIKOV and co-authors (1969, 1970b) noted that exposure of tench eggs to 25 to 100 R of ^{60}Co γ-rays prior to the first cleavage division appeared to provide protection against a subsequent exposure of 4 kR delivered to young larvae. Further observations of the irradiated larvae over longer periods of time would have provided more convincing evidence of the significance of this apparent effect. The consequences of irradiation during embryonic development on the growth of the resulting larvae are quite complicated. High exposures (> 600 R) at any stage of development can be expected to cause a retardation of growth (WELANDER, 1954; KONNO and co-authors, 1955) whereas lower doses may cause no effect or a retardation if delivered at late stages of development (WELANDER, 1954; WELANDER and co-authors, 1971) or may stimulate or retard growth if exposure occurs at early stages of development (WELANDER, 1954; WADLEY and WELANDER, 1971). It has been pointed out that, when a comparison of growth is made between a series of irradiated groups and a sham-treated control, the presence of differential mortality can generate a false indication of growth stimulation if the effect of radiation is selectively to eliminate the smaller and weaker individuals (WADLEY and WELANDER, 1971). In any one series of experiments it may not be possible to determine a consistent trend of responses (WELANDER, 1954) and for the same experimental organism the results of different experiments may be, in part, contradictory (WELANDER, 1954; WELANDER and co-authors, 1971). HAMAGUCHI (1976) showed that the exposure of *Oryzias latipes* to 1000 R on the day prior to hatching delayed the proliferation of oogonia in the developing gonad, but that recovery was in progress at a larval age of 12 d. Irradiation of rainbow trout with exposures up to 203 R at the eyed-embryo stage produced no significant effect on either the subsequent fecundity of the females or on the survival to the eyed-embryo stage of eggs from crosses between similarly irradiated parents (WELANDER and co-authors, 1971).

WHITE (1964) investigated the effect of fractionated exposure (daily exposures in the range 100 to 2000 R on the first 6 d of development) on embryos of *Fundulus heteroclitus*. At all exposures there was a reduction in the rate of development and a delay in hatching, and except at 100 R d^{-1} (total exposure 600 R), there was a decline in the hatch rate and a lower survival of embryos and larvae. In the absence of corresponding data for a single acute exposure it is not possible to determine whether fractionation of a given dose ameliorates the damage produced. EGAMI and HAMA (1975) have shown that 2 kR delivered at early stages of development produce a similar response in terms of reduced hatching rate and lengthened incubation period when the exposure rate is either 250 R min^{-1} or 33·3 R min^{-1}, but that at 1·7 R min^{-1} there is little change from control values. In the latter half of the incubation period 2 kR appeared to have little effect on these parameters regardless of dose rate. Lowering the temperature of the eggs during exposure eliminated the sparing effect of either reduced dose rate or dose fractionation.

Chronic irradiation Compared with the number of studies which have utilized radionuclides in solution as the source of chronic radiation exposure (see pp. 1242–1258), there are relatively few investigations which have adopted the simpler procedure of using an external, sealed source of γ-rays, e.g. ^{60}Co or ^{137}Cs. The obvious advantage of the latter approach lies in the ease with which unambiguous estimates of the dose rate and total absorbed dose to the fish embryo can be obtained by employing conventional dosimetric techniques, e.g. film, ionization chambers or thermoluminescent dosimeters. Thus, the biological response of the developing embryo to irradiation can be directly related to the primary cause of damage, i.e. the absorbed energy.

Concern as to the possible effects of the radioactive effluent from the plutonium production reactors on the Hanford reservation in Washington State on the natural salmon runs up the Columbia River prompted early and detailed investigations. The studies which have been made are unique in that they have utilized the salmon runs which have been developed to return to the ponds maintained by the College of Fisheries on the University of Washington campus. There, normal fish hatchery procedures have been employed to rear populations of both irradiated and control embryos to migratory size. These fish have been differentially marked and then released to return to the sea so that an assessment of the effects of irradiation can be made under essentially natural conditions (DONALDSON and BONHAM, 1964, 1970; BONHAM and DONALDSON, 1966, 1972; DONALDSON, 1970, 1971; DONALDSON and co-authors, 1972; HERSHBERGER and co-authors, 1973, 1974, 1976, 1978). The initial level of irradiation was arbitrarily chosen to be approximately 0·5 R d^{-1} or a total exposure of some 40 R between the fertilization of the salmon eggs and the attainment of free feeding after the absorption of the yolk sac. This exposure was greater than the 1 R which it was estimated might obtain under the contaminated conditions in the Columbia River, and of the same order as the acute exposures which had been shown to have significant effects at different stages of the development of rainbow trout and coho and silver salmon embryos (WELANDER and co-authors 1948; WELANDER, 1954; BONHAM and WELANDER, 1963). Subsequent experiments employed increased exposure rates of 1·3, 2·8, 5·0, 10·0, 20·0, and 17 to 50 R d^{-1}.

The main results of these studies are summarized in Table 2-31. From the data it can be seen that consistently deleterious effects in the irradiated fish did not become apparent until the exposure rate was 10 R d^{-1} or greater, when gonad development was affected in the smolts, the survival of the fish during the marine phase of the life cycle (as indicated by the numbers of irradiated fish returning to spawn) is significantly reduced, and male sterility occurs. In so far as the migratory drive may be dependent on the attainment of sexual maturity, these effects may be interrelated. Damage significantly affecting the survival potential of first generation offspring may be induced at the lower exposure rate of 5 R d^{-1}.

BROWN and TEMPLETON (1964) and TEMPLETON (1966) used the fact that healthy eggs of the plaice *Pleuronectes platessa* float on the surface of sea water of normal salinity (34·3‰) to irradiate the developing embryos more or less uniformly with sealed sources of ^{137}Cs positioned above the rearing tanks. Four dose rates between 1·4 mrad h^{-1} and 1·08 rad h^{-1} were employed over the 18 to 19-d development period (total doses 0·6 to 493 rad). The data obtained from two separate experiments were rather variable and it was not possible to detect any significant effect of irradiation on the proportion of eggs hatching, the incidence of abnormal larvae or the larval length. The amount of variation present suggests that greater replication would be necessary to separate the variation within treatments from any radiation-induced differences between treatments.

PECHKURENKOV (1976, 1977) has shown that the radiation-induced chromosome damage which can be scored as aberrant anaphase configurations in cells from developing loach *Misgurnus fossilis* embryos is repaired much more efficiently during early cleavage than during organogenesis. Using EDTA (ethylenediaminetetraacetic acid) to inhibit the repair process it was also shown that very little repair took place in the later stages of embryogenesis, even in the absence of EDTA, at an exposure rate of 10 R h^{-1}, whereas at 0·9 R h^{-1} the major part of the damage was repaired. Thus it appears that the radiosensitivity (using the incidence of aberrant anaphases as the criterion of damage) is dependent upon the relative rates of damage induction and repair, rather than the total accumulated dose. At the higher exposure rate over a period of 24 h, and also at 4 R h^{-1} during the whole of embryogenesis, the incidence of aberrant anaphase configurations reached equilibrium levels due to the selective elimination of critically damaged cells.

The effects of chronic irradiation on the overall breeding performance have been studied with the guppy *Poecilia reticulata* (WOODHEAD, 1977). Although the fertilized eggs are retained within the female (ovoviviparity) the results do provide some information concerning the radiosensitivity of developing embryo. The mean dose rates employed were 0·17, 0·40, and 1·27 rad h^{-1} giving total absorbed doses of 94, 215, and 670 rad during an assumed embryonic development time of 22 d. The lifetime fecundity was significantly reduced at all dose rates due to both an increased incidence of temporary and permament sterility and a reduced mean actual brood size (i.e. broods consisting of one or more young). Somatic damage to the developing embryos was not exposed

Table 2-31

Effects of chronic γ-irradiation on embryonic development and subsequent life history of Chinook salmon (Compiled from DONALDSON and BONHAM, 1964, 1970; BONHAM and DONALDSON, 1966, 1972; DONALDSON, 1970, 1971; DONALDSON and co-authors, 1972; HERSHBERGER and co-authors, 1973, 1974, 1976, 1978)

Brood year	1960	1962	1963	1964	1965	1966	1967	1968	1969	1970
Dose rate, R d^{-1}	0·54	Bg*	Bg*	0·53	1·3	2·8	5·0	10·0	20·0	50·0†
Total dose, R	33–37	≈0·2	≈0·2	40–42	95–100	230	355	810	1720	3800–4000
Stock	Unirradiated chinook salmon	Irradiated and control chinook salmon from the 1960 brood year mated Irr × Irr and Con × Con (1962–1964)			← Unirradiated chinook salmon → (1965–1970)					
Mortality	0	0	0	0					0	E
Smolt weight	++‡			0	0	0	0	–	– –	– –
Smolt length	0	– – –	+++	0	0	0	0	– –	0	– –
Growth rate		0	0	0					0	0
Abnormalities: Number of vertebrae		–	0	– – –						
Vertebral fusions		0	0§	0						
Gross abnormalities				0					E	E
Gonad development			0	0	0	0	0	– – –	– – –	– – –
Returns: Age 1 yr ♂	0	0	0	0	0	0	0	– – –	0	0
♀	0	0	0	0	0	0	0	0	0	0
2 yr ♂	0	0	0	– – –	– –	0	0	– – –	– – –	– – –
♀	0	0	0	0	0	0	0	0	0	0
3 yr ♂	+	+++	++	0	0	0	0	– – –	– – –	– – –
♀	0	0	0	0	0	0	0	– – –	– – –	– – –
4 yr ♂	0	++	+++	0	0	0	0	0	– – –	– – –
♀	+	0	+	0	0	0	0	0	– – –	– – –
5 yr ♂	0	0	0	0	0	0	0	+	0	0
♀	0	0	++	0	0	0	0	–	0	– –

Total ♂	0	+++	+++	−	−−	0	0	−−−	−−−	−−−
Total ♀	+	0	+++	0	0	0	0	−−−	−−−	−−−
Total ♂ + ♀	+	+++	+++	0	−	0	0	−−−	−−−	−−−
Relative fecundity: Age 3 yr					0	0	+	0	0	0
4 yr				−−	0	0	0	−−	−−	0
5 yr				0	0	0	0			
F_1 crosses (unirradiated)										
Mortality						0	0	0	0	
Smolt weight										
Parental age 3 yr E♀ × E♂								−		
E♀ × C♂								0		
C♀ × E♂								0		
4 yr E♀ × E♂							−	−		
E♀ × C♂							−	−		
C♀ × E♂							−	0		
Smolt length										
Parental age 3 yr E♀ × E♂	+++							−		
E♀ × C♂								0		
C♀ × E♂								0		
4 yr E♀ × E♂							+	−		
E♀ × C♂							0	−		
C♀ × E♂							0	0		

*Bg: natural background.

†12 of the 16 irradiated lots received an exposure rate of 17 R d^{-1} for the first 5 d of development and 50 R d^{-1} for the remainder; the remaining 4 lots received 50 R d^{-1} for the whole of the incubation period.

‡In HERSHBERGER and co-authors (1978) the same data for weights as given in DONALDSON and BONHAM (1964) for irradiated and control smolts are said not to be significantly different.

§Data given for two observers: the first finds no significant ($P > 0{\cdot}25$) increase in vertebral fusions, the second, a significant ($P < 0{\cdot}01$) increase in incidence in irradiated smolts.

Plus signs indicate that irradiated fish outperformed controls at the following levels of significance: + $P < 0{\cdot}05$; ++ $P < 0{\cdot}01$, and +++ $P < 0{\cdot}001$.

Minus signs indicate that controls outperformed irradiated fish at the following levels of significance: − $P < 0{\cdot}05$; −− $P < 0{\cdot}01$, and −−− $P < 0{\cdot}001$.

0 indicates no significant effect either way.

E indicates a deleterious effect with no attached degree of significance.

either as an increased incidence of dead or abnormal embryos or as an increased post-natal mortality of the young. It was, therefore, concluded that the reduced mean actual brood size could more probably be attributed to a reduced production of viable gametes than to increased embryonic mortality. Thus chronic irradiation at these dose rates did not appear to have any significant effect on the developing embryo.

Effects arising from irradiation of the developing embryo by radionuclides in the water and by radionuclides adsorbed and absorbed by the eggs

Fish eggs developing in a contaminated environment will be irradiated not only by radionuclides associated with external components of the environment, e.g. water, sea-bed, etc., but also by radionuclides which are adsorbed on to the surface of, or absorbed into, the eggs themselves. The possibility that such accumulations might produce enhanced radiation fields in the vicinity of the embryo has prompted investigations of the response to irradiation of developing fish eggs under conditions which attempt to parallel reasonably closely those which might obtain in contaminated environments, that is, the eggs have been incubated in water labelled with increasing concentrations of the radionuclide being studied. It seems, therefore, that there are three aspects of the problem which require detailed study if the experimental data are to have any relevance to environmental concerns, viz. the kinetics of radionuclide accumulation by developing fish eggs, dosimetry, and the quantification of the radiation response as a function of dose.

Accumulation of radionuclides by developing fish eggs. The accumulation of radionuclides on the surface, and in the contents of developing fish eggs to bulk concentrations (i.e. activity per unit weight of egg) above that of the surrounding water would, in general, be expected to have some significance in respect of an increased radiation field in the vicinity of the embryo. The concentration factor, defined as

$$\mathrm{CF} = \frac{\text{quantity of radionuclides g}^{-1}\text{ egg}}{\text{quantity of radionuclides g}^{-1}\text{ water}}$$

has been used as a convenient basis on which the results of accumulation experiments conducted under laboratory conditions can be either compared, or applied in other situations: in particular in contaminated environments. That this basis is not completely sound in the case of fish has recently been pointed out by PENTREATH (1977a). For fish eggs the situation is somewhat simpler in that there is no food input term, and the accumuated radionuclides are directly derived from solution or from suspended material. However, there remain the reservations as to the utility of a single CF value in the absence of a precise definition of the physico-chemical form of the radionuclide under the conditions of the measurement.

Much of the available information concerning the accumulation of radionuclides by fish eggs is summarized in Table 2-32. With the possible exception of the values reported by GUS'KOVA and co-authors (1975) all the data are derived from laboratory studies. In 17 of the 41 references cited the data were collected in the course of experiments investigating the effects of water contamination on the developing fish embryo. Presumably, therefore, these data were intended to provide a basis for estimates of dose rate or absorbed dose; their utility in this respect will be considered in detail later.

Many authors omit any mention of the chemical form in which the radionuclide was added to the incubation system (BROWN and TEMPLETON, 1964; IVANOV, 1965b; POLIKARPOV and GAMEZO, 1966; TEMPLETON, 1966; KULIKOV and co-authors, 1970a; WOODHEAD, 1970; GUS'KOVA and co-authors, 1971; KOSHELEVA, 1973a; KULIKOV and co-authors, 1971b; MIGALOVSKIY, 1973a,b; IVANOV, 1972; OLSON, 1973; DOKHOLYAN and co-authors, 1974; GUS'KOVA and co-authors, 1975; KULIKOV and OZHEGOV, 1975), or insufficient information is given to permit the calculation of the concentration factor (BROWN, 1962; SHEKHANOVA, 1970; GUS'KOVA and co-authors, 1971; MIGALOVSKIY, 1973a,b; PATIN and co-authors, 1971; OLSON, 1973; STRAND and co-authors, 1973a,b; TILL, 1976; TILL and FRANK, 1977), two factors which immediately limit the usefulness of the data in any other context. Approximately 40% of the data concern the accumulation of strontium and yttrium. Where the radionuclide pair $^{90}Sr-^{90}Y$ has been used it is sometimes impossible to determine whether the values given refer either to measurements made immediately after sampling, when unsupported ^{90}Y is likely to be present, or to a situation of secular equilibrium when they are CF values for strontium alone (FEDOROVA, 1963a; BROWN and TEMPLETON, 1964; TEMPLETON, 1966; SHEKHANOVA, 1970; KOSHELEVA, 1973a; MIGALOVSKIY, 1973b). A similar criticism can be levelled at the data for $^{95}Zr-^{95}Nb$ (IVANOV, 1965a; KULIKOV and co-authors, 1970a) and $^{140}Ba-^{140}La$ (GUS'KOVA and co-authors, 1971). The concentration factor for strontium in eggs spawned in fresh water ranges from 1·7 to 37 and for marine eggs the range is 0·1 to 5·4. Where data are given for the differential accumulation by the components of the eggs, large variations are not found. The corresponding data for yttrium are 1 to 80 and 5 to 233 with the major part of the element being associated with the chorion or egg membrane. A proportion of the variation within each of these ranges of concentration factors is probably attributable to species differences in both incubation time and egg size. For an element such as strontium present in ionic form in water, the former factor will be important if diffusion across the chorion controls the rate of accumulation. The latter factor is probably of particular importance for elements like yttrium, where surface adsorption seems to play a major role in the accumulation process. Of particular interest among the data for ^{90}Sr and ^{90}Y are those reported by FEDOROVA (1963a), DOKHOLYAN and co-authors (1974), and NILOV (1974) for fresh water fish eggs. These show concentration factors for both elements to be inversely related to the water concentrations of the radionuclides. Both FEDOROVA (1963a) and NILOV (1974) maintained the eggs in Petri dishes at densities of 100 and 300 to 400 eggs per dish respectively. Neither author appears to have changed the solutions to maintain a constant concentration of radioactivity, and no data are given relating to changes in the concentrations of the radionuclides in the water during the experiment. In these circumstances the possibility that the high apparent concentration factors measured at the low water concentrations might be due more to the depletion of the radionuclides from the water than to the accumulation of the radioactivity by the eggs cannot be ruled out. The data of DOKHOLYAN and co-authors (1974) may be more dependable since the eggs were apparently incubated in larger vessels and the water was changed to ensure good rearing conditions. This question clearly requires further investigation since it has implications both for the results of radiation effect studies in the laboratory where the radiation source is $^{90}Sr-^{90}Y$ in a fresh water system, and for the problem of extrapolating the results of laboratory studies to the environment. FEDOROVA (1965a) also noted a variation in the relative accumulation of ^{14}C-labelled sodium acetate from solutions of two different concentrations by fresh water eggs. No data of a similar nature have been

Table 2-32

Radionuclide accumulation by fresh water and marine fish eggs (Original; compiled from the sources indicated)

Species	Radionuclide	Concentration factor*	Other details given of the accumulation process	Source
Fresh water spawning fish				
Salmo salar	^{90}Sr-nitrate	37 ($t_h = 60$)	A,B†	BROWN (1962)
	^{90}Y-nitrate	ID‡	B	BROWN (1962)
	^{90}Sr–^{90}Y	8–10 ($t_{eq} = 9$)	B	KOSHELEVA (1973a)
	^{90}Sr–^{90}Y	ID	C	MIGALOVSKIY (1973b)
	^{137}Cs	1–2 ($t_{eq} = 10$)	B	KOSHELEVA (1973a)
Oncorhynchus kisutch	^{45}Ca	ID	A,B	OLSON (1973)
Salmo trutta	^{59}Fe-citrate	16–18 ($t_h = 51$)	A,B	DABROWSKI and co-authors (1975)
	^{90}Sr-nitrate	25 ($t_h = 57$)	A,B	BROWN (1962)
	^{90}Sr–^{90}Y	25 ($t_h = 58$)	None	BROWN and TEMPLETON (1964); TEMPLETON (1966)
	^{90}Y-nitrate	ID	B	BROWN (1962)
	^{90}Sr	4·5 at $1{\cdot}5 \times 10^{-8}$ Ci l^{-1} ($t_h = ?$)	A	DOKHOLYAN and co-authors (1974)
	^{90}Sr	2·1 at $1{\cdot}75 \times 10^{-6}$ Ci l^{-1} ($t_h = ?$)	A	DOKHOLYAN and co-authors (1974)
	^{90}Y	57·5 at $1{\cdot}5 \times 10^{-8}$ Ci l^{-1} ($t_h = ?$)	A	DOKHOLYAN and co-authors (1974)
	^{90}Y	26·8 at $1{\cdot}75 \times 10^{-6}$ Ci l^{-1} ($t_h = ?$)	A	DOKHOLYAN and co-authors (1974)
Salmo gairdneri	^{3}H-water	ID	A,B	STRAND and co-authors (1973b)
(≡*Salmo irideus*)	^{14}C-sodium carbonate	ID	A,B	SHEKHANOVA (1970)
	^{32}P-disodium hydrogen phosphate	ID	A,B	SHEKHANOVA (1970)
	^{45}Ca-chloride	ID	A,B	SHEKHANOVA (1970)
	^{60}Co-chloride	7 ($t_h = 30$)	A,B	KIMURA and HONDA (1977)
	^{89}Sr-chloride	ID	A,B	SHEKHANOVA (1970)
	^{90}Sr–^{90}Y chloride	ID	B	SHEKHANOVA (1970)
	^{91}Y-chloride	ID	B	SHEKHANOVA (1970)

	^{106}Ru (chloro-complex in 10 N HCl)	30 ($t_{eq} = 6$)	A	HONDA and co-authors (1972)
	^{106}Ru (NO-nitro complex in water)	11 ($t_h = 30$)	A,B	KIMURA and HONDA (1977)
	^{106}Ru (NO-nitrato complex in 0·01 N HNO_3)	15 ($t_h = 30$)	A,B	KIMURA and HONDA (1977)
	^{106}Ru (NO-binuclear complex in water)	7 ($t_h = 30$)	A,B	KIMURA and HONDA (1977)
	^{131}I-sodium iodide	0·4 ($t_h = 30$)	A,B	KIMURA and HONDA (1977)
	^{137}Cs-chloride	0·4 ($t_h = 30$)	A,B	KIMURA and HONDA (1977)
	^{144}Ce-chloride	120 ($t_h = 30$)	A,B	KIMURA and HONDA (1977)
Salvelinus lepechini	^{14}C-sodium acetate	10·8 at 2×10^{-5} Ci l^{-1} ($t = 70$)	A	FEDOROVA (1965)
	^{14}C-sodium acetate	3·5 at 2×10^{-4} Ci l^{-1} ($t = 70$)	A,B	FEDOROVA (1965)
Coregonus peled	^{14}C-sodium acetate	26·5 ($t = 50$)	A,B	FEDOROVA (1965a)
	^{14}C-sodium acetate	49 ($t_h = 80$)	B	FEDOROVA (1972b)
	^{106}Ru	50 ($t_{max} = 9$–11)	None	GUS'KOVA and co-authors (1975)
	^{131}I	22 ($t_{max} = 2$–5)	None	GUS'KOVA and co-authors (1975)
	^{140}Ba–^{140}La	ID	A,B	GUS'KOVA and co-authors (1971)
Coregonus lavaretus	^{60}Co	20 ($t = 16$)	A	KULIKOV and OZHEGOV (1975)
	^{90}Sr-chloride	3 ($t_h = 120$)	A	KULIKOV and OZHEGOV (1976)
	^{90}Y-chloride	80 ($t_h = 120$)	A	KULIKOV and OZHEGOV (1976)
	^{90}Sr-^{90}Y-nitrate	10^3 at 2×10^{-9} Ci l^{-1} ($t = 25$)	A	FEDOROVA (1963†)
	^{90}Sr–^{90}Y-nitrate	80 at 2×10^{-7} Ci l^{-1} ($t_h = 30$)	A	FEDOROVA (1963†)
	^{90}Sr–^{90}Y-nitrate	20 at 2×10^{-5} Ci l^{-1} ($t_h = 30$)	A	FEDOROVA (1963†)
Ctenopharyngodon idella	^{90}Sr-chloride	18 at 8×10^{-10} Ci l^{-1} ($t_h = 1·5$)	A	NILOV (1974)
	^{90}Sr-chloride	4 at $3·2 \times 10^{-8}$ Ci l^{-1} ($t_h = 1·5$)	A	NILOV (1974)
	^{90}Sr-chloride	1·7 at $1·1 \times 10^{-6}$ Ci l^{-1} ($t_h = 1·5$)	A	NILOV (1974)
	^{90}Sr-chloride	1·8 at $1·39 \times 10^{-4}$ Ci l^{-1} ($t_h = 1·5$)	A	NILOV (1974)

Table 2-32—*contd*

Species	Radionuclide	Concentration factor*	Other details given of the accumulation process	Source
	^{90}Y-chloride	24 at 8×10^{-10} Ci l^{-1} ($t_h = 1{\cdot}5$)	A	NILOV (1974)
	^{90}Y-chloride	9 at $3{\cdot}2 \times 10^{-8}$ Ci l^{-1} ($t_h = 1{\cdot}5$)	A	NILOV (1974)
	^{90}Y-chloride	2 at $1{\cdot}1 \times 10^{-6}$ Ci l^{-1} ($t_h = 1{\cdot}5$)	A	NILOV (1974)
	^{90}Y-chloride	1 at $1{\cdot}39 \times 10^{-4}$ Ci l^{-1} ($t_h = 1{\cdot}5$)	A	NILOV (1974)
Rutilus rutilus	^{14}C-acetic acid	2·9 ($t_h = 4$)	A	FEDOROVA (1963b)
Acerina cernua	^{14}C-acetic acid	2·3 ($t_h = 4$)	None	FEDOROVA (1963b)
Alburnus alburnus	^{14}C-acetic acid	2 ($t_h = 4$)	None	FEDOROVA (1963b)
Carassius carassius	^{14}C-acetic acid	0·7 ($t = 4$)	None	FEDOROVA (1963b)
	^{90}Sr	5 at 10 °C and 20 °C ($t = 4$)	A	KULIKOV and co-authors (1971b)
	^{90}Y	24 at 10 °C ($t = 8$)	A	KULIKOV and co-authors (1971b)
	^{90}Y	66 at 20 °C ($t = 4{\cdot}5$)	A	KULIKOV and co-authors (1971b)
Perca fluviatilis	^{60}Co	32 ($t = 14$)	A	KULIKOV and OZHEGOV (1975)
	^{90}Sr	8 ($t = 7$)	A	KULIKOV and co-authors (1971b)
	^{90}Y	31 ($t = 7$)	A	KULIKOV and co-authors (1971b)
	^{95}Zr–^{95}Nb	4 at 10 °C ($t_h = 12$)	A	KULIKOV and co-authors (1970a)
	^{95}Zr–^{95}Nb	8 at 20 °C ($t_h = 6$)	A	KULIKOV and co-authors (1970a)
	^{106}Ru	6 at 10 °C ($t_h = 12$)	A	KULIKOV and co-authors (1970a)
	^{106}Ru	14 at 20 °C ($t_h = 6$)	A	KULIKOV and co-authors (1970a)
	^{137}Cs	2 at 10 °C ($t_h = 12$)	A	KULIKOV and co-authors (1970a)
	^{137}Cs	3 at 20 °C ($t_h = 6$)	A	KULIKOV and co-authors (1970a)
	^{144}Ce	18 at 10 °C ($t_h = 12$)	A	KULIKOV and co-authors (1970a)
	^{144}Ce	26 at 20 °C ($t_h = 6$)	A	KULIKOV and co-authors (1970a)
Esox lucius	^{60}Co	29 at 10 °C ($t = 12$)	A	KULIKOV and OZHEGOV (1975)
	^{60}Co	36 at 20 °C ($t = 4$)	A	KULIKOV and OZHEGOV (1975)
	^{90}Sr	2 at 10 °C ($t = 4$)	A	KULIKOV and co-authors (1971b)

	^{90}Sr	2 at 20 °C ($t = 4$)	A	KULIKOV and co-authors (1971b)
	^{90}Y	53 at 10 °C ($t = 8$)	A	KULIKOV and co-authors (1971b)
	^{90}Y	102 at 20 °C ($t = 4$)	A	KULIKOV and co-authors (1971b)
Tinca tinca	^{90}Sr	4 ($t = 1{\cdot}5$)	A	KULIKOV and co-authors (1971b)
	^{90}Y	10 ($t = 1{\cdot}5$)	A	KULIKOV and co-authors (1971b)
	^{95}Zr–^{95}Nb	23 ($t_h = 2$)	A	KULIKOV and co-authors (1970a)
	^{106}Ru	25 ($t_h = 2$)	A	KULIKOV and co-authors (1970a)
	^{137}Cs	1 ($t_h = 2$)	A	KULIKOV and co-authors (1970a)
	^{144}Ce	41 ($t_h = 2$)	A	KULIKOV and co-authors (1970a)
Misgurnus fossilis	^{90}Sr-chloride	5·5 ($t = ?$)	A,B	SHEKHANOVA and PECHKURENKOV (1968)
	^{90}Y-chloride	15·8 ($t = ?$)	A,B	SHEKHANOVA and PECHKURENKOV (1968)
	^{239}Pu-nitrate	ID	A	PATIN and co-authors (1971)
Cyprinus carpio	^{238}Pu-citrate	12·6 ($t = 1{\cdot}9$)	A,B	TILL (1976); TILL and FRANK (1977; TILL (1978)
	^{241}Am-citrate	ID	A,B	TILL and FRANK (1977)
	^{244}Cm-citrate	ID	A,B	TILL and FRANK (1977)
	^{233}U-citrate	0·2 ($t = 2$)	A,B	TILL (1976)
Pimephales promelas	^{233}U-citrate	ID	A,B	TILL (1976)
Oryzias latipes	^{3}H-water	0·6 (t_{eq})	A,B	UENO (1974)
Marine spawing fish				
Rhombus maeoticus	^{54}Mn-chloride	3·3 ($t_h = 3$)	A,B	IVANOV (1969)
(≡*Scophthalmus*	^{60}Co-chloride	4·94 ($t_h = 3$)	A	IVANOV (1969)
maeoticus)	^{65}Zn-chloride	31 ($t_h = 4{\cdot}5$)	A	IVANOV (1970)
	^{90}Sr	1·6 ($t_h = 5{\cdot}6$)	A,B	IVANOV (1965a)
	^{90}Y	57 ($t_h = 5{\cdot}6$)	A,B	IVANOV (1965a)
	^{137}Cs	4 ($t_h = 5{\cdot}6$)	B	IVANOV (1965a)
	^{137}Cs	8·8 ($t = ?$)	None	IVANOV (1972)
	^{144}Ce	308 ($t_h = 5{\cdot}6$)	A,B	IVANOV (1965a)
	^{185}W-sodium tungstate	9·45 ($t_h = 3$)	A,B	IVANOV (1969)
Odontogadus merlangus	^{35}S-sodium sulphate	1 ($t_h = ?$)	B	IVANOV (1965b)
ponticus	^{59}Fe-chloride	169 ($t_h = ?$)	A	IVANOV (1965b)
(≡*Gadus merlangus*	^{90}Sr-chloride	4·4 ($t_h = ?$)	None	IVANOV (1965b)
euxinus)	^{90}Y-chloride	73 ($t_h = ?$)	None	IVANOV (1965b)
	^{106}Ru-chloride	17·8 ($t_h = ?$)	A,B	IVANOV (1965b)
	^{144}Ce-chloride	152 ($t_h = ?$)	A,B	IVANOV (1965b)

Table 2-32—*contd*

Species	Radionuclide	Concentration factor*	Other details given of the accumulation process	Source
Belone belone euxini	^{32}P-phosphoric acid	5·5 (t = 4)	A,B	Zesenko and Ivanov (1966)
	^{90}Sr	1 (t = ?)	None	Ivanov (1965a)
	^{90}Y	86 (t = ?)	None	Ivanov (1965a)
	^{95}Zr	24 (t = 4)	B	Ivanov (1972)
	^{95}Nb	106 (t = 7·1)	B	Ivanov (1972)
	^{106}Ru	21 (t = ?)	B	Ivanov (1965a)
	^{144}Ce	21·4 (t = ?)	B	Ivanov (1965a)
Engraulis encrasicholus ponticus	^{32}P-phosporic acid	6·7 (t = 1·25)	A,B	Zesenko and Ivanov (1966)
	^{90}Sr	0·8 (t = ?)	None	Ivanov (1965a)
	^{90}Y	100 (t = ?)	None	Ivanov (1965a)
	^{91}Y	233 (t_h = 1·33)	A,B	Ivanov (1965a)
	^{95}Zr–^{95}Nb	14·5 (t_h = 1·25)	A,B	Ivanov (1965a)
	^{106}Ru	12 (t_h = 1·25)	A,B	Ivanov (1965a)
	^{137}Cs	9 (t_h = 1·33)	A,B	Ivanov (1965a)
Trachurus mediterraneus ponticus	^{90}Sr	1·3 (t = ?)	B	Ivanov (1965a)
	^{90}Sr	0·9 (t = ?)	B	Ivanov (1972)
	^{90}Y	191 (t = ?)	B	Ivanov (1965a)
	^{91}Y	196 (t = ?)	B	Ivanov (1965a)
	^{144}Ce	495 (t = ?)	A,B	Ivanov (1965a)
Mullus barbatus ponticus	^{90}Sr	0·8 (t = ?)	None	Ivanov (1965a)
	^{90}Y	83 (t = ?)	None	Ivanov (1965a)
Scorpaena porcus	^{14}C	30 (t = ?)	None	Ivanov (1972)
	^{89}Sr	1·8 (t = ?)	B	Ivanov (1972
	^{89}Sr	0·83 (t = 1·67)	A	Polikarpov and Gamezo (1966)
Serranus scriba	^{14}C	25 (t = ?)	None	Ivanov (1972)
	^{89}Sr	0·8 (t = ?)	B	Ivanov (1972)
Uranoscopus scaber	^{95}Zr	35 (t = ?)	B	Ivanov (1972)
Pleuronectes platessa	^{51}Cr (III)	30 (t_h = 15)	B	Pentreath (1977a, pers. comm.)
	^{51}Cr (VI)	30 (t_h = 15)	B	Pentreath (1977a, pers. comm.)
	^{54}Mn-chloride	1·5 (t_h = 15)	A,B	Pentreath (1976)

Species	Radionuclide	Concentration factor*	Details†	Reference
	^{65}Zn-chloride	30 (t_h = 15)	A,B	PENTREATH (1976)
	^{90}Sr-nitrate	0·1 (t_h = 12)	A,B	WOODHEAD (1970§)
	^{90}Sr	0·35 (t_h = 18)	None	BROWN and TEMPLETON (1964); TEMPLETON (1966)
	^{90}Y-nitrate	5 (t_h = 12)	A,B	WOODHEAD (1970§)
	^{90}Y	10·5 (t_h = 18)	None	BROWN and TEMPLETON (1964); TEMPLETON (1966)
	^{95}Zr-oxalate complex	10 (t_h = 12)	A,B	WOODHEAD (1970§)
	^{95}Nb-oxalate complex	4·8 (t_h = 12)	A,B	WOODHEAD (1970§)
	^{106}Ru (nitrosyl Ru nitrato complex in 8 N HNO_3)	10 (t_h = 12)	A,B	WOODHEAD (1970§)
	^{110m}Ag-nitrate	1600 (t_h = 17)	A,B	PENTREATH (1977a,b)
	^{137}Cs-chloride	0·7 (t_h = 12)	A,B	WOODHEAD (1970§)
	^{144}Ce-chloride	10 (t_h = 12)	A,B	WOODHEAD (1970§)
	^{239}Pu in 0·5 M nitric acid	5·8 (t_h = 11·5)	A,B (See Fig. 2-27a, b, c)	HETHERINGTON and co-authors (1976§)
	^{239}Pu in 0·5 M nitric acid	35 (t_h = 13)	A,B (See Fig. 2-27a, b, c)	HETHERINGTON and co-authors (1976§)
	^{241}Am in 0·5 M hydrochloric acid	24 (t_h = 11·5)	A,B (See Fig. 2-27a, b, c)	WOODHEAD (unpubl.)
Neogobius melanostomus	^{14}C-sodium carbonate	ID	A	SHEKHANOVA (1970)
	^{45}Ca-chloride	ID	A	SHEKHANOVA (1970)
	^{89}Sr-chloride	ID	A	SHEKHANOVA (1970)

*Where possible the concentration factor has been given at, or just prior to hatching and is indicated by (t_h = incubation time in d). For other data the time of the concentration factor measurement is indicated by (t = time in d); (t = ?) indicates that the time of measurement either after fertilization or after immersion in contaminated water is not known.

†Categories of details given:

A. Tabulated or graphical data for concentration factor or accumulation process are given for the duration of exposure to the radionuclide; this may cover the whole incubation period.

B. Information is given concerning the differential distribution of the radionuclide through the egg.

C. Data are given in terms of radioactivity units per egg and insufficient additional data are provided to permit calculation of the concentration factor.

‡Insufficient information is provided to determine the concentration factor.

§Chemical form of radionuclide as added to the sea water not given in original publication.

reported for the accumulation of radionuclides by the eggs of marine species. The presence in sea water of much higher concentrations of ionic and/or colloidal forms of the majority of the elements can be expected to eliminate this effect on the accumulation process at the low gravimetric concentrations of the majority of radionuclides.

A further 25% of the data in Table 2-32 refer to other fission product radionuclides. Of these, ^{144}Ce and, to a lesser extent, ^{106}Ru are significantly accumulated by fish eggs reaching concentration factors in the range 18 to 495 and 6 to 50 respectively, with the major part of the radioactivity being associated with the chorion.

A few studies have been made of the accumulation of the actinide elements by developing fish eggs. These might be of importance in two respects: firstly, there are no stable isotopes of the elements to have a diluent effect on the accumulation process leading to the possibility of high concentration factors and, secondly, many of the nuclides decay by α-particle emission. The short range and relatively high energies of the α-particles mean that if there is a substantial accumulation of the radionuclides by the egg contents, and particularly if localization occurs in the developing embryo, a significant increase in radiation exposure could be expected. Figs 2-27 a, b, and c show the accumulation curves for ^{239}Pu (2) and ^{241}Am. In each case for the plutonium study approximately 1000 eggs at the gastrulation stage were placed in 2·5 l of filtered sea water (0·22 μm filter) labelled with a solution of ^{239}Pu in 0·5 M nitric acid to a concentration of 0·1 μCi l^{-1}. The development of the eggs proceeded normally with no significant mortality and normal, healthy larvae were obtained at hatching. However, as can be seen from Figs 2-27 a and b the kinetics of accumulation were quite different yielding respectively linear and power law functions of time with quite different concentration factors for the whole eggs just prior to hatching. Low-level bacterial contamination of the incubation system in the second case has been implicated as a possible cause of the difference (HETHERINGTON and co-authors, 1976). Analysis of the newly hatched larvae indicated that more than 90% of the accumulated radioactivity was associated with the chorion. The accumulation curve for americium (original solution ^{241}Am in 0·5 M hydrochloric acid) obtained under similar experimental conditions shows that the rate of absorption/adsorption of the nuclide decreases with time and that the concentration factor tends towards an asymptotic value. Again it was determined that the majority of the radioactivity was associated with the chorion.

Among the remaining data, the very high concentration factor of 1600 for ^{110m}Ag by plaice eggs and individual values of 169 for ^{59}Fe, 49 for ^{14}C, and 36 for ^{60}Co are notable. For both silver and cobalt the greater part of the accumulation was found to be by the chorion.

Where it has been indicated in Column 4 of Table 2-32 that data have been obtained regarding the differential distribution of the radionuclide within the egg, this most often refers to measurements of the concentration factor in newly hatched larvae. Such measurements give an approximate indication of the distribution of the radionuclide between the egg shell and the contents. That this may not be a completely adequate measure of the relative distributions of the nuclides during the whole of the incubation period is shown by the data reported by SHEKHANOVA (1970).

IVANOV and PARCHEVSKAYA (1974) have shown that there is a distribution of concentration factors among individual eggs sampled at a given time and also that, in the case of ^{91}Y, an aged solution generates a different and much broader distribution. Thus a series of measurements of the accumulation of a radionuclide during incubation based

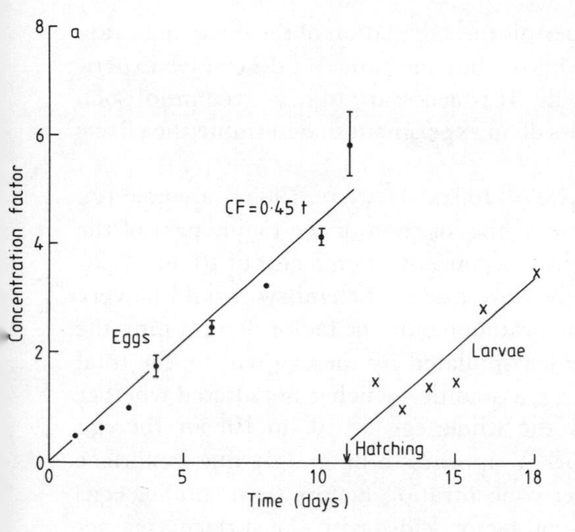

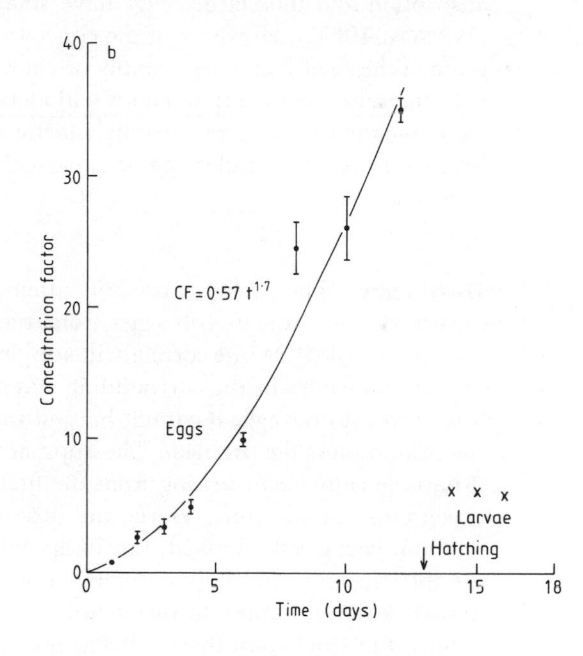

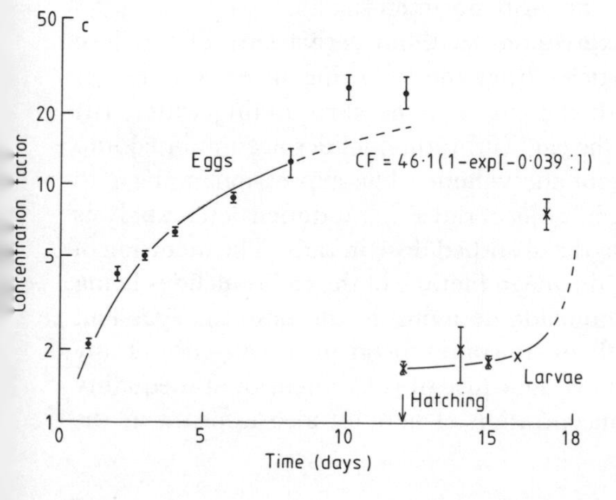

Fig. 2-27: Accumulation by developing plaice eggs of (a), (b) ^{239}Pu; (c) ^{241}Am. (Ministry of Agriculture, Fisheries and Food, Directorate of Fisheries Research, unpublished)

on replicate samples of perhaps 20 eggs may permit the calculation of the mean dose rate and the mean total absorbed dose to the embryos, but the range of dose rates experienced by individual embryos may be quite wide. It is necessary to take account of such factors when attempting to interpret the results of an experiment to determine the effects of radiation from accumulated radionuclides.

In the early studies (POLIKARPOV and IVANOV, 1962a; IVANOV, 1965a) a whole egg concentration factor of 100 for ^{90}Y which, due to the location of the major part of the radionuclide on the egg surface, was equivalent to a concentration factor of 10^3 to 10^4 for the egg membrane, was taken to mean that the dose rate to the embryo would be very high. This assumption is simplistic. For a given radionuclide the factor determining the dose rate to the embryo from a radionuclide accumulated by the egg will be the total quantity of radionuclide present on or in the egg, a quantity which is not altered whether the concentration factor is given as 100 for the whole egg or 10^3 to 10^4 for the egg membrane. The high concentration factor for ^{90}Y appears to be mainly due to surface adsorption and thus larger eggs have smaller concentration factors than smaller eggs (IVANOV, 1965a). However, if the concentration factor scales with the surface area per gram of the eggs, it can very simply be shown that if the egg radius is reduced by a factor of 2, then the concentration factor will increase by a factor of 2, but that the quantity of radionuclide per egg is reduced by a factor of 4. Thus, despite the increase in concentration factor for the smaller egg, it is clear that the dose rate to an isolated egg could be reduced.

Dosimetry The earliest apparent attempt to treat the problem of determining the absorbed dose rate to fish eggs from radionuclides in their environment is that of FEDOROV (1965). While correctly identifying some of the important sources of radiation, e.g. radionuclides in the surrounding water, adsorbed on to the surface of the egg and absorbed into the egg, it cannot be said that the treatment contributed significantly to the resolution of the problem. The approach utilized to determine the dose rate from the β-particle component arising from the first 2 sources is erroneous in both concept and execution. For the third source the dose rate in an infinite, uniformly contaminated medium is correctly derived, but in applying this result to fish eggs, correction factors are introduced without either explanation of their necessity or derivation of their form. Similar strictures apply to the treatment of the γ-ray dosimetry.

KULIKOV and co-authors (1970a) give expressions (without derivation) for the dose rate at the centre of the egg due to β-particles from the following three sources: (i) radionuclides uniformly distributed through the egg and the surrounding water; (ii) radionuclides adsorbed on to the surface of the egg; (iii) radionuclides accumulated into the egg and uniformly distributed throughout the volume. The expressions appear to provide a means of calculating the dose rate in units of rad s^{-1}, but dimensional analysis shows that the quantity determined is simply the absorbed dose in rads. The inclusion of 2 multiplying factors in the expressions, i.e. the mean lifetime of the radionuclides being considered, and the proportion of the radionuclide decaying in the time the eggs had been exposed to contaminated water, is difficult to comprehend in the absence of any detailed derivation. The first of these factors is the cause of the dimensional inequality and the second is redundant since the concentration of activity also appears in the equations.

PODYMAKHIN (1973b) provides some insight into the origins of certain of the formulae given by FEDOROV (1965) and KULIKOV and co-authors (1970a). A point source dose distribution function is used to determine the dose rate due to complex source geometries. The function is of the form

$$D(r) = \frac{ke^{-\mu r}}{r^2} \text{ rad dis}^{-1} \tag{1}$$

where k is a normalizing constant defined by the requirement that the total energy adsorbed in a very large sphere about the point source must be the average β-particle energy per disintegration, $\bar{E}_\beta$;

$$\text{thus } k = \frac{1.27 \times 10^{-9} \mu \bar{E}_\beta}{\rho} \tag{2}$$

where μ cm^2 g^{-1} is the mass coefficient of adsorption of the β-radiation and ρ is the density of the medium. For an egg of radius r uniformly contaminated throughout with a β-active radionuclide, the resultant dose rate at the centre obtained by integrating Equation 1 over the volume of the egg, assumed to be spherical, is

$$D(0) = \frac{A}{m} D_\beta(\infty)(1 - e^{-\mu r}) \text{ rad s}^{-1}$$

where A is the activity per egg in μCi, m is the mass of the egg in g and $D_\beta(\infty)$ is the equilibrium dose rate in an infinite, uniformly contaminated volume in rad s^{-1} (μCi g^{-1})$^{-1}$. The term $(1 - e^{-\mu r})$ is common to formulae given by both FEDOROV and KULIKOV and co-authors and represents the geometric factor allowing for the average attenuation of the radiation flux between the elements of the source geometry and the point at which the dose rate is to be determined. PODYMAKHIN (1973b) goes on the state that the geometric factor at the surface of the sphere is half that at the centre without adding the qualification that this is only true for a dose distribution function of the form $D(r) = k/r^2$ where the absorption of the radiation is neglected; the use of 0·75 times the geometric factor at the centre of a sphere as the mean geometric factor for the sphere (i.e. to obtain the mean dose rate throughout the sphere) is subject to the same qualification. When the attenuation of the radiation is included in the calculations, the relationships between the various geometric factors become complicated functions of μr and only approach the simpler relationships for certain extreme values of μ, r, and μr. PODYMAKHIN does state, however, that the expression

$$D = 0.75 \frac{A}{m} D_\beta(\infty)(- e^{-\mu r}) \text{ rad s}^{-1}$$

for the average dose rate is subject to errors not exceeding 20% depending on the value of μr. This expression is that given without supporting details by FEDOROV (1965).

TELYSHEVA and SHULYAKOVSKIY (1974) give a formula for the dose rate to the embryo from radioactivity accumulated by the egg which is derived from the dose rate in an infinite, uniformly contaminated medium. Two modifying factors are introduced.

The first, containing two additive terms, appears to be intended to take account of the accumulation of radionuclides both in and on the egg. The definitions of the two terms are not all all clear and it seems at least possible that they could have different units. No explanation is given as to why it is necessary to consider only half the amount of radionuclide deposited on the membrane. The second modifying factor is to allow for the fractional energy deposition in the egg if the particle range is greater than the diameter of the egg. There are two errors implicit here: firstly, this factor would not, in general, have the same value for particles originating from radioactivity uniformly distributed over the egg surface as for particles originating from radioactivity uniformly distributed throughout the egg; secondly, for lower energy particles where the range is less than the diameter of the egg, the factor would not have the implied value of unity. Thus this formula does not provide a sound basis for dosimetric calculations.

The main shortcoming of these approaches lies in the form of the dose distribution function employed which, while it is not grossly in error, does not accord closely with measured and theoretical β-ray dose distributions (BERGER, 1971; BOCHKAREV and co-authors, 1972; MURTHY and co-authors, 1973). ADAMS (1968) gave estimates of the dose rates received by contaminated fish eggs using the simple inverse square law dose distribution function and, in all cases considered, showed these to be upper limits to the dose rates which could be calculated using the empirical dose distribution function developed by LOEVINGER and co-authors (1956b). This distribution function has two adjustable constants which permit a good fit between the calculated dose rates and those measured for a variety of β-particle sources. The function has the form:

$$D_\beta(r) = \frac{k}{(\rho v r)^2}\left\{c\left[1 - \frac{\rho v r}{c}\exp\left(1 - \frac{\rho v r}{c}\right)\right] + \rho v r \exp(1 - \rho v r)\right\} \text{ rad h}^{-1}\text{ mCi}^{-1}$$

where

$$\left[1 - \frac{\rho v r}{c}\exp\left(1 - \frac{\rho v r}{c}\right)\right] \equiv 0 \text{ for all } r \geqslant \frac{c}{\rho v}$$

and the normalizing content

$$k = \frac{170\rho^2 v^3 \bar{E}_\beta}{3c^2 - (c^2 - 1)e} \text{ rad h}^{-1}\text{ mCi}^{-1}$$

The two adjustable parameters are v, the apparent mass absorption coefficient, and c.

PARCHEVSKAYA (1969, 1972) used the Loevinger dose distribution function to determine the dose rate at the centre of the egg from β-active radionuclides uniformly distributed throughout the egg, uniformly distributed over the surface of the egg, and in the surrounding water. The results were given in graphical form for a number of different radii as a function of the mean energy of emission of the β-particles. These data would allow an estimate to be made of the absorbed dose to a developing embryo during incubation in contaminated water provided that the time dependence of the accumulation process, and the differential distribution of the radionuclide between the components of the egg, were determined. Independently, the same approach had been used to estimate the dose rate to developing plaice embryos at a point 0·01 cm from the surface of the egg (diameter = 0·2 cm) from fission products in a contaminated environment (WOODHEAD, 1970). The accumulation as a function of time and the distribution between egg shell and egg contents were determined for each radionuclide. Estimates of the γ-ray dose rate to the developing embryos were also made using the simple inverse square law dose distribution function.

A complete suite of methods for calculating the dose rates to aquatic organisms from radionuclides in their environment has been given (WOODHEAD, 1979). Fish eggs were selected as an example to demonstrate the application of the methods and the dose rates from α-, β-, and γ-emitting radionuclides in the external environment, and on and in the eggs, are given. The basic expression for estimating the dose rate from α-particles is

$$D_\alpha(r) = \frac{1{\cdot}70 \times 10^{-7}}{\rho r^2}(A + Br^2) \text{ rad h}^{-1} \text{ pCi}^{-1}$$

where ρ g cm^{-3} is the density of the medium; A MeV μm^{-1} is the stopping power of the medium at the α-particle emission energy E_α MeV, and B is defined by the requirement that the energy absorbed within a sphere of radius $R(E_\alpha)$ about the point source must equal the energy emitted by the source, thus

$$B = 3\left(\frac{E_\alpha - AR(E_\alpha)}{R(E_\alpha)^3}\right) \text{ MeV } \mu\text{m}^{-3},$$

$R(E_\alpha)$ μm being the nominal residual range of the α-particle at the emission energy. The dose rate at the point of interest is obtained by integrating this point source dose function over the appropriate source geometry. For β-radiation a slightly modified version of Loevinger's original point source dose function, containing three adjustable constants, was used to obtain improved fits to the extensive tabulations of theoretical dose distributions given by BERGER (1971). The expression is

$$D_\beta(r) = \frac{k}{(\rho v r)^2}\left\{a\left[1 - \frac{\rho v r}{c}\exp\left(1 - \frac{\rho v r}{c}\right)\right] + \rho v r \exp(1 - \rho v r)\right\} \text{ rad h}^{-1} \text{ pCi}^{-1}$$

where

$$\left[1 - \frac{\rho v r}{c}\exp\left(1 - \frac{\rho v r}{c}\right)\right] \equiv 0 \text{ for all } r \geqslant \frac{c}{\rho v}$$

and

$$k = \frac{1{\cdot}70 \times 10^{-7}\rho^2 v^3 \bar{E}_\beta n_\beta}{ac(3 - e) + e} \text{ rad h}^{-1} \text{ pCi}^{-1}$$

Values for the three constants a, c, and v were obtained from the tabulated dose distributions for 41 radionuclides for which greater than 98% of the emitted electrons (or positrons) result from a single transition. The dependence of each of the 3 constants on the maximum emission energy was determined to provide a basis for estimating the parameters required to calculate the dose distribution for radionuclides not included in the tabulation given by Berger and for radionuclides emitting complex spectra. For organisms of the size of developing fish eggs, the simple inverse square law, neglecting absorption and scattering, is an adequate description of the γ-ray flux emitted by the accumulated, γ-active radionuclides. The use of this function to describe the γ-ray dose field about a point source implies the complete absorption of the energy at the site of the γ-ray–electron interaction although it is dissipated along the secondary electron track. Thus the inverse square law tends to overestimate the dose rate to a small contaminated organism which has dimensions less than the mean range of the secondary electrons. The exact derivation of the factor to take account of the build-up of ionization intensity is difficult (BERGER, 1968); however, a factor of the form

$$\left[1 - \exp\left(\frac{-2{\cdot}30r}{r(0{\cdot}3E_\gamma)}\right)\right]$$

which implies that intensity reaches 90% of the value predicted at a distance equal to the range of an electron with energy $0{\cdot}3E_\gamma$ MeV, would serve to reduce the discrepancy. Thus the point source dose distribution function applicable to small, contaminated organisms is

$$D_\gamma(r) = \frac{1{\cdot}70\mu n_\gamma E_\gamma}{\rho r^2}\left[1 - \exp\left(\frac{2{\cdot}30r}{r(0{\cdot}3E_\gamma)}\right)\right] \text{ rad h}^{-1}\text{ pCi}^{-1}$$

where μ/ρ cm^2 g^{-1} is the true mass energy absorption coefficient, at energy E_γ MeV, of the tissue being irradiated and n_γ is the number of photons of energy E_γ MeV emitted per disintegration.

Effects observed in developing fish eggs under experimental conditions A considerable number of papers have been published on the effects of radiation from radionuclides in the water on the development of both fresh water and marine fish eggs. An attempt has been made to summarize the main results described by the various authors and this is presented in Table 2-33. Only those types of damage which have been most frequently used as evidence of radiation-induced effects have been included so as to keep the summary reasonably concise. Other types of possible radiation-induced damage which have been studied include changes in mitotic index (IVANOV, 1967; DOKHOLYAN and co-authors, 1974; PECHKURENKOV and co-authors, 1974; SHEKHANOVA and co-authors, 1974; TSYTSUGINA, 1975), larval survival (FEDOROVA, 1963a, 1964; KOSHELEVA, 1973a; PODYMAKHIN, 1973a), embryonic and larval growth rates (BROWN and TEMPLETON, 1964; NEUSTROEV and PODYMAKHIN, 1966b; TEMPLETON, 1966; ERICKSON, 1973; MIGALOVSKIY, 1973a,b; STRAND and co-authors, 1973b; KASATKINA and co-authors, 1974), larval eye diameter (WALDEN, 1973; ICHIKAWA and SUYAMA, 1974), haemoglobin production (SHEKHANOVA and co-authors, 1970), RNA and DNA content of developing embryos (SHEKHANOVA and co-authors, 1970), gonad formation (HYODO-TAGUCHI and EGAMI, 1970; MIGALOVSKIY, 1973b; KASATKINA and co-authors, 1974), the function of the hatching gland

(KASATKINA, 1973; KASATKINA and co-authors, 1974), duration of hatching (KOSHELEVA, 1973a; MIGALOVSKIY, 1973a,b), respiration rate (NEUSTROEV and PODYMAKHIN, 1966a; SHEKHANOVA and co-authors, 1970; KLYASHTORIN and co-authors, 1972; SHEKHANOVA and co-authors, 1974) changes in peripheral blood (KOSHELEVA, 1973b), interaction of incubation temperature with irradiation (IVANOV, 1966a), variation of radiosensitivity with development stage (IVANOV, 1966b; FEDOROVA, 1972a,b), the functional state of the hypophysis and the thyroid (OGANESYAN, 1973), thermal tolerance (STRAND and co-authors, 1973b), and susceptibility to predation (STRAND and co-authors, 1973b).

It will be noted that the insult is given in terms of the concentration of radioactivity in the water; in all cases, as far as can be ascertained from the publications, this represents the concentration of the radionuclide in the water at the beginning of the experiment. The tabulation has been assembled on this basis because these data represent the only common points of reference between publications. In only 18 of the 59 references cited is there any attempt to assess either the dose rate to, or the absorbed dose received by, the embryo during development. These 2 quantities are the fundamental parameters required for an analysis of the response of a biological entity to radiation. They are only distantly related to the concentration of radioactivity in the water at the beginning of exposure; other contributory variables include the radionuclide employed, the variation of the concentration of radioactivity in the water during incubation, the differential accumulation of the radioactivity on and in the eggs and in other components of the incubation system, the geometry of the individual eggs (i.e. species) and their geometrical relationship to each other in the incubation system and the total development time. The possible importance of the accumulation of radioactivity by the egg as a significant determinant of the response of developing fish embryos to irradiation was first suggested by POLIKARPOV and IVANOV (1961). Although no mention of the consequent absorbed dose was made, a clear inference can be drawn that the enhanced radiation field due to the accumulated radioactivity was the cause of the effects observed at rather low water concentrations of $^{90}Sr-^{90}Y$. Subsequent studies would have been expected to concentrate on elucidating the relationship between the observed effect on the developing embryos and the accumulation of radionuclides by the eggs and the resultant dose rate.

Of the 59 publications cited, only 18 give any information concerning either the variation of the concentration of the radionuclide in the water during the experiment (BROWN, 1962; STRAND and co-authors, 1973a; DABROWSKI and co-authors, 1975) or the accumulation of radioactivity on and in the eggs. In the latter case the data vary from a single value for the radioactivity per egg (MIGALOVSKIY, 1973b) or for the accumulation coefficient (presumably at equilibrium) (BROWN and TEMPLETON, 1964; FEDOROV and co-authors, 1964; TEMPLETON 1966; FEDOROVA, 1972b; KOSHELEVA, 1973a; GUS'KOVA and co-authors, 1975) through descriptions of the accumulation process in the whole egg as a function of time during incubation (BROWN, 1962; FEDOROVA, 1963a; POLIKARPOV and GAMEZO, 1966; NILOV, 1974) or, less usefully, as a function of developmental stage (SHEKHANOVA and PECHKURENKOV, 1968; DOKHOLYAN and co-authors, 1974) to more or less comprehensive details of the differential distribution of the accumulated radioactivity between the components of the egg (e.g. egg membrane, egg contents, perivitelline fluid, yolk, and embryo) as a function of incubation time (GUS'KOVA and co-authors, 1971; KOSHELEVA, 1973a; STRAND and co-authors, 1973a;

Table 2-33

Effects of radionuclides in the water on the development of fish embryos as a function of concentration (Original; compiled from the sources indicated)

Radio-nuclide	Species	Biological effect scored	Concentration of radionuclide, $Ci\,l^{-1}$ 10^{-12} 10^{-10} 10^{-8} 10^{-6} 10^{-4} 10^{-2} 10^{0} 10^{2}	Source
^{3}H	*Paralichthys olivaceous*	Embryo survival		ICHIKAWA and SUYAMA (1974)
	Fugu niphobles	Embryo survival		ICHIKAWA and SUYAMA (1974)
	Salmo gairdneri	Embryo survival		STRAND and co-authors (1973b)
		Abnormal larvae		STRAND and co-authors (1973b)
		Immune response		STRAND and co-authors (1973a); STRAND and co-authors (1977)
^{14}C	*Coregonus peled*	Incubation time		FEDOROVA (1968)
		Incubation time		FEDOROVA (1972a,b)
		Larval abnormality		FEDOROVA (1972a,b)
		Embryo survival		FEDOROVA (1972b)
	Alburnus alburnus	Embryo survival		FEDOROVA (1964)
		Incubation time		FEDOROVA (1964)
		Mortality during hatching		FEDOROVA (1964)
		Abnormal larvae		FEDOROVA (1964)
	Acerina cernua	Embryo survival		FEDOROVA (1964)
		Incubation time		FEDOROVA (1964)
		Mortality during hatching		FEDOROVA (1964)

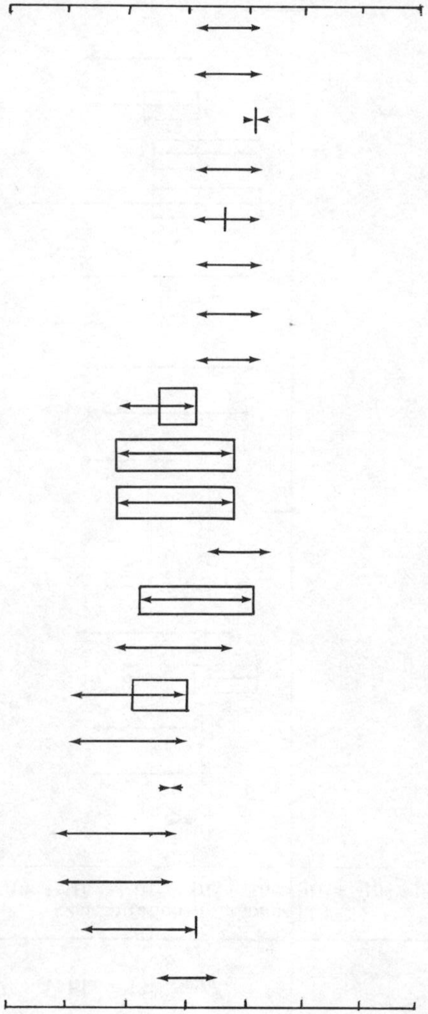

Radionuclide	Species	Effect		Reference
^{14}C		Abnormal larvae		FEDOROVA (1964)
	Carassius carassius	Embryo survival		FEDOROVA (1964)
		Incubation time		FEDOROVA (1964)
		Mortality during hatching		FEDOROVA (1964)
	Rutilus rutilus	Embryo survival		FEDOROVA (1964)
		Incubation time		FEDOROVA (1964)
		Mortality during hatching		FEDOROVA (1964)
		Abnormal larvae		FEDOROVA (1964)
	Scorpaena porcus	Chromosome aberrations		TSYTSUGINA (1971, 1972, 1975)
^{45}Ca	*Neogobius melanostomus*	Embryo abnormality		MOSKALKOVA (1970)
		Embryo survival		MOSKALKOVA (1970)
^{59}Fe	*Salmo trutta*	Embryo survival		DABROWSKI and co-authors (1975)
^{89}Sr	*Rhombus maeoticus*	Chromosome aberrations		TSYTSUGINA (1971, 1972, 1973)
	Salmo irideus	Embryo survival		SHEKHANOVA and co-authors (1970)
$^{90}Sr–^{90}Y$	*Salmo salar*	Embryo survival		NEUSTROEV and PODYMAKHIN (1966b)
		Embryo survival		PODYMAKHIN (1973a)
		Embryo survival		BROWN (1962)
	Oncorhynchus gorbuscha	Embryo survival		SHEKHANOVA and co-authors (1970)
	Salmo irideus	Embryo survival		SHEKHANOVA and co-authors (1970)
	Salmo trutta	Embryo survival		DOKHOLYAN and co-authors (1974)
		Embryo survival		BROWN and TEMPLETON (1964); TEMPLETON (1966)

Table 2-33—*contd*

Radio-nuclide	Species	Biological effect scored	Concentration of radionuclide, Ci l^{-1} (10^{-12} 10^{-10} 10^{-8} 10^{-6} 10^{-4} 10^{-2} 10^{0} 10^{2})	Source
^{90}Sr–^{90}Y	*Salmo trutta*	Embryo survival		Brown (1962)
	Huso huso	Embryo survival		Dokholyan and co-authors (1974)
	Aristichthys nobilis	Embryo survival		Dokholyan and co-authors (1974)
	Ctenopharyngodon idella	Embryo survival		Nilov (1974)
	Tinca tinca	Embryo survival		Kulikov (1973)
	Esox lucius	Embryo survival		Timofeeva and Al'shits (1970); Kulikov (1973)
		Embryo survival		Pitkyanen (1971)
	Neogobius melanostomus	Embryo survival		Shekhanova and co-authors (1970)
	Misgurnus fossilis	Embryo survival		Pechkurenkov and co-authors (1972a,b)
		Embryo survival		Shekhanova and Pechkurenkov (1968)
	Pleuronectes platessa	Embryo survival		Fedorov and co-authors (1964)
		Embryo survival		Brown and Templeton (1964); Templeton (1966)
	Engraulis encrasicholus	Embryo survival		Polikarpov and Ivanov (1961)
	Serranus scriba	Embryo survival		Polikarpov and Ivanov (1961)
	Salmo salar	Abnormal larvae		Neustroev and Podymakhin (1966b)
		Abnormal larvae		Podymakhin (1973a)
	Salmo trutta	Abnormal larvae		Brown and Templeton (1964); Templeton (1966)

^{90}Sr–^{90}Y	*Ctenopharyngodon idella*	Abnormal larvae	NILOV (1974)
	Tinca tinca	Abnormal larvae	KULIKOV (1973)
	Esox lucius	Abnormal larvae	KULIKOV (1973)
		Abnormal larvae	PITKYANEN (1971)
	Misgurnus fossilis	Abnormal larvae	PECHKURENKOV and co-authors (1972a,b)
		Abnormal larvae	SHEKHANOVA and PECHKURENKOV (1968)
	Pleuronectes platessa	Abnormal larvae	FEDOROV and co-authors (1964)
		Abnormal larvae	BROWN and TEMPLETON (1964); TEMPLETON (1966)
	Engraulis encrasicholus	Abnormal larvae	POLIKARPOV and IVANOV (1961)
		Abnormal larvae	POLIKARPOV and IVANOV (1962b)
	Mullus barbatus	Abnormal larvae	POLIKARPOV and IVANOV (1962b)
	Trachurus trachurus	Abnormal larvae	POLIKARPOV and IVANOV (1962b)
	Crenilabrus hybrid	Abnormal larvae	POLIKARPOV and IVANOV (1962b)
	Scorpaena porcus	Abnormal larvae	POLIKARPOV and GAMEZO (1966)
	Sarguis annularis	Abnormal larvae	POLIKARPOV and GAMEZO (1966)
	Salmo salar	Chromosome aberrations	MIGALOVSKAYA (1973)
	Salmo trutta	Chromosome aberrations	DOKHOLYAN and co-authors (1974)
	Aristichthys nobilis	Chromosome aberrations	DOKHOLYAN and co-authors (1974)
	Misgurnus fossilis	Chromosome aberrations	PECHKURENKOV (1970)
	Scorpaena porcus	Chromosome aberrations	POLIKARPOV (1973); TSYTSUGINA (1971, 1972, 1975)

Table 2-33—*contd*

Radio-nuclide	Species	Biological effect scored	Concentration of radionuclide, $Ci\,l^{-1}$ (10^{-12} 10^{-10} 10^{-8} 10^{-6} 10^{-4} 10^{-2} 10^{0} 10^{2})	Source
^{90}Sr–^{90}Y	*Esox lucius*	Chromosome aberrations		TIMOFEEVA and AL'SHITS (1970); TIMOFEEVA and co-authors (1971)
	Salmo salar	Incubation time		MIGALOVSKIY (1973a)
		Incubation time		NEUSTROEV and PODYMAKHIN (1966b)
	Engraulis encrasicholus	Incubation time		POLIKARPOV and IVANOV (1961)
	Esox lucius	Incubation time		PITKYANEN (1971)
^{91}Y	*Scorpaena porcus*	Chromosome aberrations		TSYTSUGINA (1971, 1975)
	Scopthalmus maeoticus	Chromosome aberrations		TSYTSUGINA (1971, 1972, 1973, 1975)
^{106}Ru	*Coregonus peled*	Embryo survival		GUS'KOVA and co-authors (1975)
		Abnormal larvae		GUS'KOVA and co-authors (1975)
^{137}Cs	*Trichiurus lepturus*	Embryo survival		TELYSHEVA and SHULYAKOVSKIY (1974)
	Dentex canadiensis	Embryo survival		TELYSHEVA and SHULYAKOVSKIY (1974)
	Trachurus trachurus	Embryo survival		TELYSHEVA and SHULYAKOVSKIY (1974)
	Pleuronectes platessa	Embryo survival		FEDOROV and co-authors (1964)
	Salmo salar	Embryo survival		PECHKURENKOV and co-authors (1972a,b)
	Esox lucius	Embryo survival		PECHKURENKOV and co-authors (1972a,b)
	Dentex canadiensis	Abnormal larvae		TELYSHEVA and SHULYAKOVSKIY (1974)
	Trachurus trachurus	Abnormal larvae		TELYSHEVA and SHULYAKOVSKIY (1974)
	Pleuronectes platessa	Abnormal larvae		FEDOROV and co-authors (1964)
	Salmo salar	Abnormal larvae		PECHKURENKOV and co-authors (1972a,b)

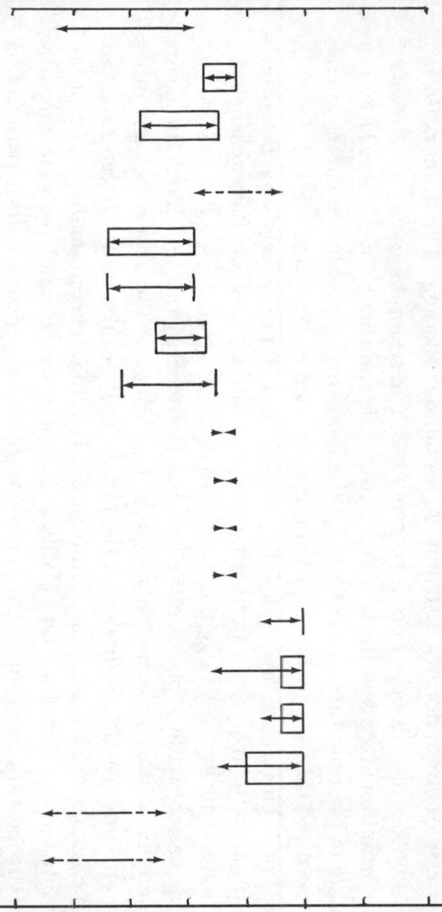

^{137}Cs	*Esox lucius*	Abnormal larvae	PECHKURENKOV and co-authors (1972a,b)
	Scopthalmus maeoticus	Chromosome aberrations	TSYTSUGINA (1975)
	Salmo salar	Incubation time	KASATKINA and co-authors (1974)
^{140}Ba–^{140}La	*Coregonus peled*	Embryo survival	GUS'KOVA and co-authors (1971)
^{144}Ce	*Salmo salar*	Embryo survival	PODYMAKHIN (1973a)
		Abnormal larvae	PODYMAKHIN (1973a)
		Incubation time	KASATKINA and co-authors (1974)
		Incubation time	MIGALOVSKIY (1973a)
	Trichiurus lepturus	Embryo survival	TELYSHEVA and SHULYAKOVSKIY (1974)
	Trachurus trachurus	Embryo survival	TELYSHEVA and SHULYAKOVSKIY (1974)
	Trichiurus lepturus	Abnormal larvae	TELYSHEVA and SHULYAKOVSKIY (1974)
	Trachurus trachurus	Abnormal larvae	TELYSHEVA and SHULYAKOVSKIY (1974)
^{238}Pu	*Cyprinus carpio*	Embryo survival	TILL (1976, 1978)
	Pimephales promelas	Embryo survival	TILL (1976, 1978)
	Cyprinus carpio	Abnormal larvae	TILL (1976, 1978)
	Pimephales promelas	Abnormal larvae	TILL (1976, 1978)
^{239}Pu	*Misgurnus fossilis*	Embryo survival	PECHKURENKOV and co-authors (1972a,b)
		Abnormal larvae	PECHKURENKOV and co-authors (1972a,b)

←→ Range of concentrations over which embryos were exposed; no significant effects noted.
→← Experiment conducted at a single concentration with no significant effect.
←[box]→ Embryos exposed to a range of radionuclide concentrations and significant effects noted over the range included by the box.
→|← Embryos exposed at a single concentration at which a significant effect was noted.
←- - —→ Dashed part of line indicates uncertainty as to the concentrations employed.

DABROWSKI and co-authors, 1975; TILL, 1976, 1978). In no instance are any data given concerning the accumulation of radioactivity by the different components of the incubation systems which might constitute significant sources of external exposure. These data must be judged in respect of their utility for providing reasonably accurate estimates of the dose rates and absorbed doses received by the developing embryos.

STRAND and co-authors, (1973a) give data on the accumulation of tritium (as tritiated water, HTO) by eggs of the rainbow trout in terms of a relative count rate as a function of time for both a volatile fraction and a 'bound' fraction. The accumulation of the tritiated water reached a peak value after 2 d and then declined slowly. It was noted, without giving details, that the concentration of the radionuclide in the incubation system decreased with time and this presumably accounts for the slow decline in the tritium content of the eggs. Preliminary results suggested that the equilibrium concentration factor had a value between 0·92 and 1·1, and autoradiographs showed that the tritium was essentially uniformly distributed throughout the egg. The authors do not give any estimate of the absorbed doses received by the developing embryos. However, using the information given and the fact that at the emission energy of tritium $D_\beta(\infty)$ is a valid estimate of the dose rate to the embryo, the following approximate values for the total doses received during the 28-d incubation period can be obtained: at 0·01, 0·1, 1·0, and 10·0 μCi ml^{-1} initial concentration, 0·04, 0·4, 4·0, and 40 rad respectively. Because of the rapid uptake of the tritiated water by the eggs the dose rates would have reached the maximum values during the early part of embryogenesis. Such estimates are easy to obtain in the case of tritium because the low emission energy of the β-particles means that radioactivity in the egg is the only significant source. In the absence of any other data, rapid and uniform accumulation of the radioactivity to the same concentration as the water (i.e. a concentration factor of unity) would be an unexceptionable assumption. Thus, the only data required for reasonably accurate dose estimates relate to the variation of the concentration of radioactivity in the water during the incubation period. It is only in the case of exposure from tritiated water that the problem is so readily solved.

DABROWSKI and co-authors (1975) investigated the accumulation kinetics of ^{59}Fe by trout eggs and also noted the cumulative mortality in control and treated groups during incubation. The variation in the concentration of radioactivity in the water throughout the 60-d incubation period of the eggs is given, as also is the accumulation factor for the whole egg, egg membrane, yolk, perivitelline fluid, and embryo. If it is assumed that the eggs are spherical and of unit density, then it would be possible to develop from the given weights of the egg components a geometrical model as a basis for estimating the dose rate from the accumulated radioactivity. Trout eggs normally adhere to one another and to the substrate, and information about the relative positions of the eggs and possible adsorption of radioactivity to the substrate would be necessary to make reasonable estimates of the dose rate to a given egg from external sources. Such information is lacking. Thus it appears that, in this case, only a partial description of the radiation exposure of the embryos would be possible on the basis of the data given.

When, as in the experiments with salmon and trout eggs conducted by BROWN (1962), the radionuclide pair ^{90}Sr–^{90}Y is the source of radiation, the differential accumulation of the two nuclides by the eggs complicates the situation. Data are given on the variation of the radioactivity (at secular equilibrium) in the water and in the eggs during the incubation period. Single observations are reported of the differential uptake of ^{90}Sr and ^{90}Y in whole eggs and also in separated egg membranes and egg contents; for dose estimation these data would be assumed to apply to the whole incubation period. Although a salmon egg was found to be 50% heavier than a trout egg, the surface areas were shown to be almost identical and it is not possible, from the information given, to generate a geometrical model of the eggs. No information is given on the relative positions of the eggs during incubation or on the possible adsorption of radioactivity to the substrate. From the available data it is not possible to make realistic estimates of the total doses received by the embryos during incubation. The author adopted a very simple model to obtain the absorbed dose values given in the paper, but mistakenly multiplied these by the weights of the individual eggs to obtain 'the average dose received per egg'. This error also appears to have been committed by TELYSHEVA and SHULYAKOVSKIY (1974).

In view of what has been written in respect of the foregoing 3 papers it is clear that a single value of the accumulation coefficient does not provide a secure basis for realistic estimates of the absorbed dose to embryos during incubation. Thus the calculated dose rates given by FEDOROV and co-authors (1964) and the total absorbed doses given by BROWN and TEMPLETON (1964) and TEMPLETON (1966) must be treated with caution. The dose estimates given by FEDOROVA (1972b) may be more reliable since they appear to take account of the time dependence of the accumulation process; in addition the low energy of the β-particle from ^{14}C decay eliminates the necessity to consider any source except the radioactivity accumulated by the egg contents. No data are given, however, with which to make an assessment of the validity of the estimates. GUS'KOVA and co-authors (1975), KOSHELEVA (1973a), and MIGALOVSKIY (1973b) did not attempt to make estimates of the dose rate or absorbed dose.

The uptake curve for ^{89}Sr by the eggs of *Scorpaena porcus* given by POLIKARPOV and GAMEZO (1966) could only be used as a basis of reasonable dose estimates for ^{90}Sr if assumptions were made about the egg geometry and the distribution of radioactivity on and in the egg mass; also there is no information given on which to base estimates of the dose rate to a given egg from external sources. In any event, the resulting values would, in isolation, be of little help in interpreting the data given in the paper on the effects of ^{90}Sr–^{90}Y on the developing embryos of this fish. NILOV (1974) gives the accumulation curves for both ^{90}Sr and ^{90}Y as a function of time for the eggs of *Ctenopharyngodon idella* incubated in solutions of 4 different concentrations. The necessary additional data required for realistic estimates of the absorbed doses from both internal and external sources is not given. NILOV did, however, make estimates of the absorbed dose with what is described as a 'generally accepted method'. In the absence of a reference and any other details, and in view of the fact that the results are expressed in units of rad g^{-1}, their validity must be open to question. The data given by FEDOROVA (1963a) do not distinguish the differential accumulations of ^{90}Sr and ^{90}Y and thus cannot be used as a basis of dosimetry calculations. An interesting point arises from the last two papers inasmuch as the data indicate that the accumulation factors for ^{90}Sr and ^{90}Y may depend on the water concentrations of the radionuclides. This observation clearly has implications for the problem of dosimetry in such experiments.

DOKHOLYAN and co-authors (1974) give the accumulation factors for ^{90}Sr and ^{90}Y for eggs of the Caspian salmon incubated in two different concentrations of the radionuclides. The data are related to the development stages of the embryos. This factor would make it impossible to obtain estimates of the total absorbed doses to the embryo during development even if all the other necessary data were available. Again it is notable that the magnitude of the concentration factors appears to be dependent on the concentration of the radionuclides in the water. SHEKHANOVA and PECHKURENKOV (1968) also relate the uptake of ^{90}Sr and ^{90}Y by eggs of the loach *Misgurnus fossilis* to the development stage of the embryos. Although there is insufficient information on which to base independent estimates, the authors did use the data on accumulation kinetics to obtain values for the dose rate and total absorbed dose to the embryos. No details of the method are given and no mention is made of the contribution which might be derived from external sources; the estimates must, therefore, be treated with caution.

The accumulation of ^{140}B–^{140}La on the membrane and in the contents of the developing eggs of *Coregonus peled* is described by GUS'KOVA and co-authors (1971). Apparently no data were obtained concerning the differential uptake of the two radionuclides and the water concentrations employed in the experiment are not defined precisely ($n \times 10^{-4}$ and $n \times 10^{-6}$ Ci l^{-1}). Thus it is not possible to make realistic, independent estimates of dose rates or absorbed doses from the data given. In view of the relatively high β-ray energies and the intense γ-radiation ($\Gamma = 12$ cm^2 r(mCi h)$^{-1}$ for ^{140}La alone) following decay of these nuclides it is essential that possible sources of radiation external to the eggs be adequately considered. Values of the total absorbed dose as a function of the development time are given but there are no details of the derivation or of the sources considered. Once more, therefore, the data must be considered to be of doubtful validity.

TILL has determined the accumulation kinetics of ^{238}Pu (1976, 1978) and ^{233}U (1976) by developing eggs of the carp and has investigated the differential distribution of the radionuclides between the egg membrane and the egg contents. Autoradiographic studies indicated that the plutonium penetrating the chorion became uniformly distributed throughout the contents, but that the uranium was largely associated with the yolk. In corresponding investigations of the radiotoxicity of ^{238}Pu and ^{232}U to developing carp and fathead minnow embryos these data were used as a basis of calculations of the absorbed doses. As ^{238}Pu decays solely through α-emission, and because there is no significant build-up in the activity of daughter nuclides with other decay modes, the only source of radiation that needed to be considered in the dose calculations was the plutonium distributed through the content of the eggs. Thus the derived estimates of the total absorbed dose to the carp embryos during development can be accepted with confidence. The extrapolation of the accumulation data for carp eggs to developing minnow eggs must introduce some uncertainty as to the accuracy of the absorbed dose estimates for the latter. Due to the build-up of daughter radionuclides of ^{232}U, the dosimetry problem in these experiments was much more complex. At the time of the experiment with the carp eggs, 35 d after the separation of pure ^{232}U, some 15% of the total α-activity of the solution was due to daughter radionuclides; for the experiment with minnow eggs the corresponding data were 180 d and 45%. Thus, although the tracer experiment with ^{233}U undoubtedly provides a reasonable basis for the calculation of the dose rate from ^{232}U to carp embryos, and to a lesser extent (due to the extrapolation involved) to minnow embryos, it has no relevance at all to the dose rate from the daughter radionuclides. It is not sufficient, as was done, to adjust the value of the mean energy of the α-particles used in the calculation to take account of the increasing mean energy of the α-particles derived from the daughter nuclides. A dose rate calculated on the basis of the α-irradiation arising from the accumulation of ^{232}U alone would be a minimum estimate; in addition, there could be contributions from the accumulation of each of the daughter radionuclides, and from the β- and γ-active daughter radionuclides in the solution. By far the largest of these additional contributions is likely to arise from α-active daughters distributed within the egg. Of the daughter nuclides, ^{228}Th and ^{224}Ra have sufficiently long half-lives for the possible accumulation within the egg to become significant and, depending on the rate of diffusion, ^{220}Rn may become more or less uniformly distributed through the egg–water system regardless of its original point of production. Clearly the problem of the dosimetry of the daughter radionuclides is extremely complex, and no information is given which would permit reasonable estimates to be made. In view of the fact that the absorbed doses from ^{232}U plus daughters have probably been underestimated, it is interesting to note that the estimated absorbed doses at which effects were found on hatching, survival, and the induction of abnormalities in carp and minnow embryos were always lower than the corresponding values for ^{238}Pu (Table 11: TILL, 1976).

A further 8 papers include estimates of the dose rates or total absorbed doses to the embryo during development although no data on the accumulation of the radionuclide(s) on which the estimates may have been based are given (PECHKURENKOV, 1970; PODYMAKHIN, 1973a; WALDEN, 1973; ICHIKAWA and SUYAMA, 1974; SHEKHANOVA and co-authors, 1974; TELYSHEVA and SHULYAKOVSKIY, 1974; TSYTSUGINA, 1975; STRAND and co-authors, 1977).

In the case of tritiated water (WALDEN, 1973; ICHIKAWA and SUYAMA, 1974; STRAND and co-authors, 1977) the explicit or implicit assumptions of rapid uptake to an equilibrium concentration factor of unity, and uniform distribution throughout the eggs, are unexceptionable and there is no problem with radiation sources external to the eggs. Thus the derived absorbed doses are likely to be fairly accurate. TSYTSUGINA (1975) utilized the dosimetry models developed by PARCHEVSKAYA (1969, 1972) to calculate the total absorbed doses received by the embryos during the exposure to the radionuclides in the water. It appears that the accumulation and differential distribution of the radionuclides in the eggs and the contribution from radioactivity in the water were all taken into account. However, none of the relevant data forming the basis of the calculations are given. Values for absorbed dose rates or absorbed doses are quoted by PODYMAKHIN (1973a), PECHKURENKOV (1970), and SHEKHANOVA and co-authors (1974) without giving any details of the methods used to derive them, the data forming the basis of the calculations, or the sources of radiation considered. Such estimates are effectively worthless since there is no means by which their validity may be assessed. As already discussed, the formula given by TELYSHEVA and SHULYAKOVSKIY (1974) does not provide an adequate basis for dose calculations and no significance can be attached to the dose rate and absorbed dose values given. The treatment of the estimates of absorbed dose also indicates a misunderstanding of the definition of the quantity.

For those papers where it is reasonable to conclude that the absorbed dose estimates are fairly realistic it is possible to attempt some comparison between the data on the effects of irradiation. The relevant data are summarized in Table 2-34 relative to the total absorbed dose during the exposure time of the embryos. A similar plot of the data relative to the average dose rate during the exposure period does not alter the range of the independent variable, and produces only minor changes in the relative positions of the ranges over which a given effect is found. It should be noted that the total doses which have been employed in the experiments span 9·2 orders of magnitude and that the lowest average dose rate is 0·1 μrad h^{-1} (corresponding to approximately 1% of the

Table 2-34

Effects of radionuclides in the water on the development of fish embryos as a function of concentration (Original; compiled from the sources indicated)

Biological effect scored	Radio-nuclide	Species	Duration of exposure	Estimated dose (rad) 10^{-5} 10^{-3} 10^{-1} 10^{1} 10^{3} 10^{5}	Source
Embryo survival	^{3}H	*Paralichthys olivaceous*	*F* + 4 h F + 96 h		Ichikawa and Suyama (1974)
		Fugu niphobles	F + 1·3 h F + 180 h		Ichikawa and Suyama (1974)
		Salmo gairdneri	F + 6 h F + 28·5 d		Strand and co-authors (1973b)
			F + 6 h F + 21 d		Strand and co-authors (1973b)
	^{14}C	*Coregonus peled*	F + ? F + 80 d		Fedorova (1972b)
			F + ? F + 15 d		Fedorova (1972b)
			F + 7 d F + 70 d		Fedorova (1972b)
			F + 7 d F + 22 d		Fedorova (1972b)
	^{238}Pu	*Cyprinus carpio*	F F + 72 h		Till (1976, 1978)
Abnormal embryos	^{3}H	*Salmo gairdneri*	F + 6 h F + 28·5 d		Strand and co-authors (1973b)
			F + 6 h F + 28·5 d		Strand and co-authors (1973b)
	^{238}Pu	*Cyprinus carpio*	F F + 72 h		Till (1976, 1978)

Effect	Nuclide	Species	Start	End	Dose-rate range	Reference
Larval survival	^{238}Pu	*Cyprinus carpio*	F	F + 72 h		Till (1976, 1978)
Larval growth rate	^{3}H	*Salmo gairdneri*	F + 6 h	F + 28·5 d		Strand and co-authors (1973b)
			F + 6 h	F + 20 d		Strand and co-authors (1973b)
Larval eye diameter	^{3}H	*Fugu niphobles*	F + 19 h	F + 180 h		Ichikawa and Suyama (1974)
		Gasterosteus aculeatus	F + ?	F + 7 d		Walden (1973)
Mitotic activity	^{90}Sr–^{90}Y	*Scorpaena porcus*	F + ?	F + ?		Tsytsugima (1975)
	^{91}Y	*Scorpaena porcus*	F + ?	F + ?		Tsytsugina (1975)
Chromosome aberrations	^{14}C	*Scorpaena porcus*	F + ?	F + 22 h		Tsytsugina (1971, 1972, 1975)
	^{90}Sr–^{90}Y	*Scorpaena porcus*	F + ?	F + 22 h		Tsytsugina (1971, 1972, 1973, 1975)
	^{91}Y	*Scorpaena porcus*	F + ?	F + 22 h		Tsytsugina (1971, 1975)
		Scopthalmus maeoticus	F + ?	V + 25 h		Tsytsugina (1971, 1975)
	^{137}Cs	*Scopthalmus maeoticus*	F + ?	F + 25 H		Tsytsugina (1975)
Immune response	^{3}H	*Salmo gairdneri*	F + ?	F + 20 d		Strand and co-authors (1977)
Susceptibility to predation	^{3}H	*Salmo gairdneri*	F + 6 h	F + 20 d		Strand and co-authors (1973b)
Thermal tolerance	^{3}H	*Salmo gairdneri*	F + 6 h	F + 20 d		Strand and co-authors (1973b)

Symbols as for Tables 2-33 except for dashed lines; here these represent the extrapolation of dose rates given by Tsytsugina (1975) to results presented in earlier publications (Tsytsugina, 1971, 1972, 1973) for what appears to be the same series of experiments.

expected natural background dose rate), and that the highest is 220 rad h^{-1}. It is immediately apparent from Table 2-34 that the normalization of the radiation effects data on the basis of absorbed dose has not eliminated all the inconsistencies between the various data.

Using the criterion of embryo survival (as measured by hatching rate) after irradiation for the major part of embryogenesis, the highest total absorbed dose at which no effect has been observed is 8600 rad (TILL, 1976, 1978) while the lowest at which an effect has been noted is 0·3 rad (STRAND and co-authors, 1973b). This discrepancy may well be much greater since the former exposure was derived from ^{238}Pu while the latter was due to tritium, and the relative biological effectiveness of the α-radiation would be expected to be greater than that of the β-radiation. There are 2 points which generate some reservations about the validity of the lower limit. STRAND and co-authors (1973b) in fact report 2 experiments in which the hatching success was measured, and 2 concentrations of tritium are duplicated between these experiments. In the first experiment no response was noted in respect of hatching success at these two concentrations, whereas in the second, such an effect was observed. Thus there can be some doubt as to whether a genuine radiation-induced effect was being produced. This conclusion is supported by the fact that the decline in hatching success was not found to be monotonically dependent on increasing dose rate; the highest dose produced less effect on hatching success than either of the 2 lower doses. If these data are excluded on the basis of these apparent internal inconsistencies, then the remaining doses at which an effect has been observed are 2160 rad (ICHIKAWA and SUYAMA, 1974) and 750 rad (FEDOROVA, 1972b). Since FEDOROVA (1972b) only gives data for a single concentration of ^{14}C, it is not possible to be certain that 750 rad represents the lower limit for a significant effect on the hatching success of *Coregonus peled*. From the data given in the paper it can also be concluded that the actual dose to the embryo could be as much as 50% greater, i.e. 1130 rad. FEDOROVA'S data also provide confirmation that a significant part of the total mortality is caused by a small fraction of the total dose delivered in the early part of embryogenesis, and also that the embryo becomes less sensitive during the course of development.

There are only 2 sets of data on the incidence of embryo abnormality (STRAND and co-authors, 1973b; TILL, 1976, 1978). Once again, the data of STRAND and co-authors, (1973b) can only be accepted with reservations due to the clear lack of consistency within and between the 2 experiments reported; the embryos receiving the highest absorbed dose yielded a lower proportion of anomalous individuals than any other treatment, including the 2 control populations. TILL (1976, 1978) observed that the lowest dose at which a significantly increased proportion of abnormal embryos was produced was 3400 rad.

The survival of larvae, irradiated as embryos, does not appear to be affected until the total absorbed dose reaches 8600 rad (TILL, 1976, 1978) and the data for larval growth rate as indicated by body length are inconclusive (STRAND and co-authors, 1973b); however, the larval eye diameter shows a significant reduction at total absorbed doses to the embryo greater than 1960 rad (WALDEN, 1973; ICHIKAWA and SUYAMA, 1974).

TSYTSUGINA (1975) has investigated certain of the cellular factors which may underlie the effects scored at the level of the whole embryo, i.e. mortality, abnormality, etc. Changes in the mitotic index and the incidence of chromosome anomalies in the embryo were the chosen criteria of effect. Two experiments investigating the effects of ^{90}Sr–^{90}Y

and ^{91}Y on the mitotic index in *Scorpaena porcus* embryos are inconsistent until a total absorbed dose of 0·6 rad is reached (average dose rate approximately 3×10^{-3} rad h^{-1}). For chromosome anomalies, the data given indicate that the incidence can be expected to increase at total absorbed doses greater than (4 to 20) $\times 10^{-3}$ rad. The threshold absorbed dose for a significant effect was not determined for either *Scorpaena porcus* eggs in solutions of ^{90}Sr–^{90}Y and ^{91}Y, or *Scopthalmus maeoticus* eggs in solutions of ^{91}Y. However, additional data (from what appears to be the same series of experiments), represented in earlier papers (TSYTSUGINA, 1971, 1972, 1973) can be used to deduce the following threshold absorbed doses respectively: $1{\cdot}7 \times 10^{-4}$, $7{\cdot}8 \times 10^{-5}$, and $4{\cdot}0 \times 10^{-5}$ rad. These deductions are based on the assumption that the absorbed dose scales with the concentration of radionuclide in the water. That this assumption might not be valid can be concluded from the data cited above (FEDOROVA, 1963a; DOKHOLYAN and co-authors, 1974; NILOV, 1974) which indicate that the concentration factors for ^{90}Sr and ^{90}Y may be functions of the water concentrations of the radionuclides. This lowering of the thresholds for a significant effect introduces the problem of a lack of consistency. For both the data on the mitotic index and the incidence of chromosome anomalies, the lack of consistent effects for a given species of embryo over certain ranges of total dose must raise doubts as to the validity of the absorbed dose values. The same concern arises from the fact that effects are indicated at total absorbed doses which correspond to dose rates of 8, 4, and 0·4 μrad h^{-1}, i.e. of the order of, or less than, the natural background dose rate which the eggs might be expected to experience in the laboratory environment. Other questions also have to be considered in relation to these results. Firstly, the magnitude of the change in the incidence of chromosome anomalies at a give dose (10^{-1} rad) varies by a factor of 5; secondly, on the basis of the data given by TSYTSUGINA (1975) there is no indication of an increasing incidence of chromosome anomalies with increasing dose; and thirdly, the control values for each of the 2 species indicate considerable heterogeneity in the population. All these points require adequate investigation and explanation before the results can be accepted.

STRAND and co-authors (1973a, 1977) also investigated a radiation effect arising at the cellular level, i.e. the suppression by embryonic irradiation of the primary immune response in juvenile rainbow trout to an injected antigen. By the 9th and 11th wk following vaccination an analysis of variance of the results from replicated samples demonstrated that there were significant differences between the 5 treatment groups. At 9 wk the threshold dose eliciting a significant suppression of antibody production was 0·4 rad and there was a clear trend of decreased response with increasing dose. By 11 wk, however, the threshold had risen to 4 rad and the smooth trend as a function of dose was much less apparent. Since the experiment was terminated before the peak response was achieved the minimum dose for a significant suppression of immune response may not have been determined. The overall variability of the data prevented the derivation of a satisfactory dose–response relation. Irradiation of rainbow trout embryos with total absorbed doses of 3 rad was not found to influence the thermal tolerance of the juvenile fish or their susceptibility to predation (STRAND and co-authors, 1973b).

Thus it appears that few of the data which have been published on the effects on developing fish embryos of radiation from radionuclides accumulated by the eggs from the surrounding water are particularly informative. In the majority of cases the only conclusion that can be drawn is that the contamination of the eggs with radionuclides may (or may not) have had an effect on the embryos under the particular experimental conditions employed. There is no way in which either valid intercomparisons can be made between the data, or predictive extrapolations can be made to conditions as they might exist in a contaminated environment. Therefore, despite the large number of experiments which have been carried out, it may be concluded that there has been relatively little real advance from the original rather limited observations of POLIKARPOV and IVANOV (1961).

Assessment of radiation effects in contaminated environments

It is now apparent that an assessment of the effect of contaminant radionuclides on developing fish embryos in the natural environment requires an estimate of the incremental radiation exposure. This represents the only secure basis for an extrapolation of data on the effects of radiation obtained under controlled conditions in the laboratory to a contaminated natural environment.

DONALDSON and BONHAM (1964), using concentrations of radionuclides measured in the Columbia River, estimated that the total exposure of the gametes and young salmon up to the time of their migration to the sea would not exceed 1 R. As no details were given of the derivation of the estimate, it is not possible to assess its accuracy. However, it represented 1/40 of the lowest exposure employed in the long-term experiments with Chinook salmon embryos and 1/800 of the total exposure at which consistently deleterious effects were noted (Table 2-31). Thus it can be reasonably concluded that the degree of contamination of the Columbia River did not have any significant effects on the long-term viability of the Chinook salmon population of the river.

The potential exposure of the developing plaice embryos in the northeast Irish Sea due to the fission product radionuclides (^{90}Sr–^{90}Y, ^{95}Zr–^{95}Nb, ^{106}Ru, ^{137}Cs, and ^{144}Ce) discharged from the fuel reprocessing plant at Windscale has been estimated to be 0·085 μrad h^{-1} (WOODHEAD, 1970). This value was derived from the concentrations of radionuclides measured in sea water and from accumulation factors determined under laboratory conditions. The natural background exposure of a developing plaice embryo has 2 main components: the natural potassium-40 in both sea water and the egg which contributes 0·7 μrad h^{-1} and the cosmic radiation dose rate which varies from 4 μrad h^{-1} at the sea surface to 0·5 μrad h^{-1} at 20 m depth. The total exposure can, therefore, vary between 1·2 and 4·7 μrad h^{-1} depending on the position of the egg in the water column. Thus, provided that the extrapolation of accumulation-factor data from the laboratory to the natural environment is valid, the exposure due to the contaminant radioactivity can be seen to be much less than the variation in the natural background likely to be experienced by the developing egg. The total dose received by the embryo up to hatching was approximately 31 μrad; this is much lower than the total doses during embryonic development at which effects have been noted (Table 2-34) if the uncertain data due to TSYTSUGINA are discounted. Thus it can be concluded that the fission product radionuclide concentrations in sea water arising from the controlled disposal of low-level liquid radioactive wastes into the northeast Irish Sea do not inflict any significant damage on developing plaice embryos.

The dose rate to the developing plaice embryo in the vicinity of the Windscale discharge from plutonium-239 has also been investigated (HETHERINGTON and co-authors,. 1976). Once again, accumulation factors derived from laboratory studies (Figs 2-27a,b) have been combined with measurements of the concentration of the nuclide in the environment (1 pCi l^{-1}) to estimate the dose rate (0·09 or 0·47 μrad h^{-1} depending on the accumulation factor used). Although these values are within the range defined by the dose rate from the fission product radionuclides and that from the natural background, closer examination of the discrete nature of the processes involved indicates that caution is needed in the interpretation of the data. The quantity of plutonium associated with a given egg is so low that Poisson statistics are applicable. It has been shown that a considerable proportion of a population of plaice eggs could complete embryonic development without being irradiated by even a single α-particle from the decay of plutonium-239 adsorbed on to the egg and, at most, only 3 α-particles would penetrate the egg during the development period. Under these conditions, the assumptions underlying the calculation of the absorbed dose rate clearly become invalid. Even the inclusion of the contributions from plutonium-238 (0·25 pCi l^{-1}; 0·13 μrad h^{-1} maximum) and americium-241 (see Fig. 2-27c; 0·5 pCi l^{-1}; 0·17 μrad h^{-1}) does not alter this conclusion. In any event, significant effects in the developing embryos are extremely unlikely.

Much of the radioactivity discharged into the sea from the Windscale plant rapidly becomes associated with the sediment and there produces an environment in which the increase in radiation exposure is much greater than that due to water-borne radionuclides (WOODHEAD, 1973a,b; IAEA, 1976). In the immediate vicinity of the outfall the combined β- and γ-ray dose rate at the sediment surface may reach 10 mrad h^{-1}. Thus fish eggs which develop on the seabed will be exposed to much higher dose rates from external sources than pelagic eggs; although even in this instance the highest dose rates would be experienced by only a small proportion of the total population, and are generally less than those at which significant effects have been noted.

Accumulation curves obtained under laboratory conditions have also been used to estimate the exposure of developing fish eggs in fresh waters contaminated with plutonium-239 (TILL and FRANK, 1977; TILL, 1978). The estimated averaged dose rates over the development period ranged from 3×10^{-5} μrad h^{-1} for carp eggs exposed, essentially, to fallout plutonium-239 in Lake Michigan, to 4×10^{-3} μrad h^{-1} for fathead minnow eggs developing in White Oak Lake. The further extrapolations made from the laboratory data for carp and minnow eggs to the eggs of other species with different dimensions and development times must be of doubtful validity. The caveat given above concerning the estimates of the α-particle dose rates to plaice eggs applies with even greater force to these much lower values, and it is only reasonable to conclude that no significant damage to the developing embryos can result from this source of radiation.

Conclusions

The major part of the published literature on the effects of radiation on developing fish embryos has been critically reviewed to determine its utility as a basis for assessing the potential impact of radioactive waste disposal on fish populations. Since, in these circumstances, chronic, low dose rate exposure would be expected, the most detailed consideration has been given to data which have been obtained from experiments employing these irradiation conditions. A limited examination of the results of acute irradiation experiments has been made where these provide information about the mechanisms which underlie the radiation response. It has been emphasised that absorbed dose or dose rate are the only parameters which permit either a valid comparison of the results of different experiments, or the valid extrapolation of the data obtained in laboratory experiments to a contaminated aquatic environment.

The most significant conclusion to be drawn from the results of experiments in which the developing fish embryos have been exposed to acute irradiation concerns the fact that the radiosensitivity varies during embryogenesis about a generally declining trend. Thus it is important that studies of the effects of chronic irradiation commence as soon as possible after fertilization so as to include the most sensitive phases of development.

There are few studies of the effects of chronic irradiation using external sealed sources. It seems that 0·21 rad h^{-1} is the lowest dose rate at which long-term deleterious effects have been noted (Table 2-31), although other studies have concluded that there are no significant effects at dose rates up to 1·27 rad h^{-1}.

The addition of radionuclides to the water in the incubation system has been the most usual way of irradiating the developing fish embryo, probably because it resembles to some degree the circumstances which would exist in a contaminated environment. However, there appears, in the majority of cases, to be little appreciation of the additional complexity of the experimental procedures required to obtain interpretable results, particularly if it is accepted that dosimetry is an absolute necessity. Generally, the induced radiation effect has been related to the concentration of the radionuclide in the water at the start of the experiment, and often, so few experimental and operational details are provided that it would be difficult to attempt, with confidence, the verifications of a given observation. To avoid the possibility of additional conflicting and uninterpretable results, and to raise their level of scientific credibility, future investigations in this field have to give much greater attention to the following aspects of the problem:

(i) Full details must be given of the differential accumulation of the radionuclide(s) by the eggs, and the components of their environment, as a function of time at each radionuclide concentration.

(ii) Estimates of the absorbed dose to the embryo and its variation during development must be made and full details of the derivation given.

(iii) Sufficient replicates must be made to permit the relative statistical significance of any observation to be determined.

(iv) A range of radionuclide concentrations must be employed such as to allow the construction of a reliable dose-effect curve over the range of 0 to 100% response.

Experiments performed with this degree of attention to detail will not be easy or straightforward although guidance on methods to be adopted has been given (PECHKURENKOV and co-authors, 1972a,b; NISHIWAKI and co-authors, 1979; POLIKARPOV, 1979; WOODHEAD, 1979). Only when data become available from a number of studies of this type can the hypothesis of a high sensitivity of developing fish eggs to a low-level contamination of their environment by radionuclides be proven true or false.

On the basis of the limited data summarised in Table 2-34, the following provisional conclusions can be drawn concerning the effects of radiation from radionuclides both accumulated by the egg and in the surrounding environment:

(i) Developing fish embryos do not appear to be particularly radiosensitive when the criteria of embryo mortality, the incidence of embryo abnormality, larval survival and larval growth rate are employed.

(ii) Those effects which have been studied at the cellular level (e.g. mitotic index, chromosome aberrations, immune response) indicate a potential for being much more sensitive indicators of radiation damage.

However, much further work is required to provide a sounder basis for these tentative conclusions

The dose rates to developing fish eggs in contaminated environments generally appear to be low. On the assumption that radioactive waste disposal to aquatic environments will continue to be controlled so that the consequent exposure of human populations is minimized, it seems extremely unlikely that dose rates in excess of those currently existing in the immediate vicinity of, for example, the Windscale discharge will be experienced by developing fish embryos. Using the exposure from natural background as a baseline, and the results of experiments with external sealed sources of γ-rays, it has been concluded that there is no hazard to fish populations through the induction of damage in developing embryos. The results of the experiments with radionuclides added to the water as the source of radiation do not, as yet, have sufficient internal consistency for hazard assessment. There is a need for rigorously controlled experiments, as discussed above, to eliminate this last, small uncertainty.

Genetic Effects of Radiation in Aquatic Organisms

There have been relatively few studies of the mutagenic effects of radiation employing aquatic organisms and the available data only allow rather qualitative comparisons to be made with the extensive information which has been obtained for *Drosophilia* and the mouse. Exposure of prophase oocytes of parthenogenetic diploid and tetraploid strains of brine shrimp *Artemia* sp. to 1000 R of X-rays produced a decline in hatchability which can be interpreted as a measure of the induction of dominant lethal mutations (METALLI and BALLARDIN, 1962). For the diploid strain the reduction in hatching success in the first generation provides an estimate of $3{\cdot}9 \times 10^{-4}$ gamete^{-1} rad^{-1} for the mutation rate; the corresponding value for the tetraploid strain is $3{\cdot}8 \times 10^{-5}$ gamete^{-1} rad^{-1}. In the second generation the diploid strain experienced a further reduction in hatching success which was interpreted as being due to the expression of recessive lethals although, given the particular nature of the meiotic process in the developing oocytes, the mechanism of expression could not be adduced. For the tetraploid strain no further decline in viability was found in the second generation. It was suggested that the tetraploid genotype is more effective than the diploid in ameliorating the expression of lethal damage of both dominant and recessive forms, a conclusion more fully supported by later studies (BALLARDIN and METALLI, 1965; METALLI and BALLARDIN, 1972). The effects of acute radiation exposure of successive generations on the continuing fitness of populations of diploid parthenogenetic brine shrimp have also been investigated (BALLARDIN and METALLI, 1968). The fitness was defined as the number of adult progeny produced per female from eggs developing within a 20-d period. X-ray exposures of 500 and 1000 R were given to juvenile females, i.e. the oogonial stem cells were irradiated. No effects were seen until the 8th generation when the fitness dropped to 48% and 25% of the control respectively. The fact that an effect suddenly became apparent in the 8th generation could not be explained, although overall it was concluded that *Artemia* appeared to be more radiosensitive than *Drosophila* by the criterion of reduced fitness. BLAYLOCK (1971, 1973) has investigated the cytogenetic effects of radiation in the fresh water midge *Chironomus riparius*. Acute (2000 rad) and chronic (1650 rad at 3 rad h^{-1}) ^{60}Co γ-ray exposures were employed and in a separate experiment individuals were allowed to complete their 20-d development (egg to mature adult) in water contaminated with 125, 250, and 500 μCi ml^{-1} of tritium. For the latter the total doses were 760 rad (1·6 rad h^{-1}), 1525 rad (3·2 rad h^{-1}), and 3050 rad (6·4 rad h^{-1}) respectively.

Somatic damage was scored in chromosome preparations made from the salivary glands of 4th instar larvae sampled from the generation chronically irradiated with tritium β-rays and genetic damage was scored for all exposure conditions in similar preparations from individuals sampled from the first post-irradiation generation. For the acute exposure the chromosome aberration induction rate was determined to be $1{\cdot}8 \times 10^{-4}$ gamete^{-1} rad^{-1} while the chronic exposures yielded estimates of (1 to 2) $\times 10^{-5}$ gamete^{-1} rad^{-1}. The more realistic conditions (in environmental terms) of chronic exposure indicated a substantially lower sensitivity to the induction of heritable chromosome aberrations as compared with acute exposure. There are several factors, in addition to dose rate, which may be contributing to the observed difference. Only mature gametes in adult male midges were given the acute exposure before mating to virgin unirradiated females whereas both males and females received chronic exposure and the gametogenic cells were irradiated throughout the life cycle. The basic cause of the difference is probably the different degrees of repair which are possible in the 2 cases. Accurate estimation of the induction rate for somatic chromosome aberrations is not possible since the total doses for the tritium exposure from developing embryo to 4th instar larva are not given, but it is reasonable to conclude that it is unlikely to be greater than 5×10^{-6} cell^{-1} rad^{-1}.

SCHRÖDER (1973, 1979) has provided useful summaries of the genetic experiments which have employed teleost fish. Estimates of the induced mutation rate per gamete and per locus have been obtained for the guppy *Poecilia reticulata* for a variety of types of mutation after acute irradiation. The values are in the ranges (0·38 to 10·9) $\times 10^{-5}$ gamete^{-1} rad^{-1} and (2·4 to 4·7) $\times 10^{-7}$ locus^{-1} rad^{-1} (PURDOM, 1966; SCHRÖDER, 1969a; PURDOM and WOODHEAD, 1973). In each case it was concluded that the sensitivity of the fish was somewhat less than that of the mouse. The reduction in brood size has been used to obtain estimates of (3·1 to 8·7) $\times 10^{-4}$ gamete^{-1} rad^{-1} for the induction rate of dominant lethal mutations in pairs of guppies subjected to chronic ^{137}Cs γ-irradiation (WOODHEAD, 1977). In this case the much-qualified conclusion was drawn that the guppy had a radiosensitivity of the same order as that of the mouse. SCHRÖDER and HOLZBERG (1972) have shown that the random recessive mutations produced by irradiation in the genome of 'wild type' guppies can yield apparently different estimates of the mutation rates depending on the genotype with which the test matings are made. Segregation ratios in the F_2 generation after matings with a stock homozygous for 3 recessive genes situated on non-homologous chromosomes were markedly different from expectation. Ancestral irradiation has effects on quantitative characters, such as the number of vertebrae and body dimensions (SCHRÖDER, 1969b,c,d), the degree of expression of other genes (ANDERS and co-authors, 1971), and behavioural traits, such as male aggressiveness (HOLZBERG, 1973). All these characters are considered to be under polygenic control and the observed changes are undoubtedly of mutational origin, but the relative significance of these effects in comparison with the specific locus, recessive lethal and dominant lethal mutations which have been more commonly investigated has yet to be determined (SCHRÖDER, 1980).

The radiation induction of chromosome aberrations has been investigated both *in vivo* and *in vitro*. In cells of the gut epithelium of the mud minnow *Umbra limi* an X-ray dose of approximately 350 rad produced a total chromosome aberration rate of $1{\cdot}0 \times 10^{-3}$ cell^{-1} rad^{-1} (KLIGERMAN and co-authors, 1975), while cells of *Ameca splendens* in tissue culture exposed to 290 rad of ^{60}Co γ-rays yielded $1{\cdot}2 \times 10^{-3}$ aberrations cell^{-1} rad^{-1}

(WOODHEAD, 1976). This fairly good agreement must be regarded as fortuitous since quite different aberrations were observed in the 2 experiments. *In vivo* only chromatid gaps and breaks were scored, while *in vitro* the aberrations were primarily exchanges and acentric fragments. SCHRÖDER (1969e) found an increased incidence of exchanges between the X and Y chromosomes of the guppy after spermatogonial irradiation. The induction rate was 5×10^{-5} cell^{-1} rad^{-1} and is much lower than the 2 values given above because only 2 chromosomes out of the total diploid complement were being scored. Major congenital anomalies induced by ^{60}Co γ-ray exposure of mature ova and sperm of the rainbow trout *Salmo gairdneri*, and attributed to chromosome damage, have been used to obtain estimates of 26 rad for the dose required to double the control incidence and 3×10^{-4} gamete^{-1} rad^{-1} for the induction rate (NEWCOMBE and MCGREGOR, 1967). Later experiments determined the dose–response relationships for the induction of eye malformations and body deformities at low doses to sperm. For the former, the response was linear with dose giving an induction rate of $8{\cdot}4 \times 10^{-5}$ gamete^{-1} rad^{-1} (MCGREGOR and NEWCOMBE, 1972a), and for the latter, the response was a non-linear (quadratic) function of the dose and the induction rate at 25 rad was $1{\cdot}2 \times 10^{-4}$ gamete^{-1} rad^{-1} (NEWCOMBE and MCGREGOR, 1975). This series of experiments also produced 2 surprising results. The 2 lower doses employed (25 and 50 rad) yielded greater proportions of fertilized eggs and improved embryo survival relative to the controls (MCGREGOR and NEWCOMBE, 1972b; NEWCOMBE and MCGREGOR, 1972). These apparently 'beneficial' effects of sperm irradiation would have compensated to some extent the damaging effect on the total fertility due to the incidence of inviable, malformed embryos (NEWCOMBE, 1973).

From these limited data it may be concluded that aquatic organisms show a similar sensitivity to the induction of mutations by radiation as do other, more intensively studied, species, and a mutation rate in the range (10^{-3} to 10^{-4}) gamete^{-1} rad^{-1} seems to be a reasonable assumption. At the dose rates prevailing in contaminated environments, this would not lead to detrimental effects (WOODHEAD, 1974; TEMPLETON and co-authors, 1976).

(c) Responses of Natural Populations to Increased Radiation Exposure

In addition to laboratory studies concerning effects of radiation on aquatic organisms, there have been several investigations of the possible effects in contaminated environments. A population of the snails *Physa heterostropha* inhabits a small delay pond on a stream which received some seepage from intermediate liquid radioactive waste disposal pits on the site of the Oak Ridge National Laboratory (COOLEY, 1973b). The dose rate to the snails at the time of the study (1969 to 1970) was estimated to be 27 mrad h^{-1}, and had probably been much higher in previous years. After eliminating the interfering influence of temperature differences, it was found that individuals from the irradiated population produced significantly fewer egg capsules per snail than a control population. However, the irradiated individuals produced significantly more eggs per capsule so that there were no significant differences in overall fecundity (eggs per snail). It was suggested that the increased number of eggs per capsule produced by the irradiated population was an adaptive response to the rather more severe environmental conditions (low water levels, high water temperatures, rapid temperature fluctuations, etc.) experienced as compared with the controls. Nevertheless, it was clear that any effect of the radiation was of negligible overall significance in terms of the fecundity of the population.

BLAYLOCK has investigated the incidence of chromosome anomalies in the salivary glands sampled from natural populations of larvae of the midge *Chironomus tentans*. An irradiated population exists in White Oak Lake and White Oak Creek which receive low-level wastes from the Oak Ridge National Laboratory; the dose rate has been estimated to be 26 mrad h^{-1}, or about $10^3 \times$ natural background. In total, 17 different chromosome anomalies were identified. Of these, 6 were common to both irradiated and control populations, with 3 occurring at relatively high frequency (0·09 to 0·21). On the basis of an examination of their relative frequencies between populations and over time, it was concluded that these 6 inversions were of natural origin and endemic in the population. The remaining 11 aberrations were only found in the irradiated population and in 10 cases only 1 example was seen. The 11th aberration, a simple paracentric inversion, was recorded 5 times and always in association with the same example of an endemic inversion. The 5 larvae concerned were collected at 1 site over a 6-week period, and it was suggested that the inversion had initially occurred in a gonial cell in a single individual and was subsequently lost from the population through the action of selection or genetic drift. The unique occurrence of each of the other aberrations also implies low carrier viability. It was concluded that chromosome aberrations induced by the increased radiation exposure of 26 mrad h^{-1} were present at a frequency of 0·02. It was noted that these aberrations represented 2-hit events and that 1-hit or point mutations would, therefore, have been present at a higher frequency. It was also concluded that although the observed aberrations were apparently lethal in overall effect, there was no observable response at the population level (NELSON and BLAYLOCK, 1963; BLAYLOCK 1965, 1966a,b).

It is instructive to compare these data with those obtained from laboratory studies with the related species *Chironomus riparius* (p. 1261). From the latter, a value of $(1 \text{ to } 2) \times 10^{-5}$ $gamete^{-1}$ rad^{-1} was derived for the induction rate of chromosome aberrations in this species by chronic irradiation. If the probable influence of the large differences in dose rate is ignored, then the implied aberration frequency at a dose rate of 26 mrad h^{-1} over a natural lifespan of 46 to 60 days would be in the range $(3 \text{ to } 8) \times 10^{-4}$, i.e. substantially less (by 1 to 2 orders of magnitude) than that actually observed. Since the aberrations are the result of 2-hit damage, the higher dose rates used in the laboratory experiments would have been expected to yield the higher frequency, and not the reverse as observed. The resolution of this problem may, in part, lie with the much higher total doses acccumulated in the laboratory experiments and the consequent extent of cell-killing leading to a loss of cells carrying aberrant chromosomes. It may be, of course, that not all the chromosome aberrations observed in environmental samples are attributable to increased radiation exposure. Radionuclides are not the only contaminants in the wastes discharged to the system and there may be mutagenic chemicals which are producing an effect in addition to that predicted to result from the increased radiation exposure. However, further work in both the laboratory and the contaminated environment would be required to resolve satisfactorily the apparent discrepancies. In any event, these observations emphasise the great difficulties involved in extrapolating the results obtained at high doses and dose rates in the laboratory to the much lower dose rates characteristic of contaminated environments.

BLAYLOCK and his colleagues have also investigated the effects of increased radiation exposure on the mosquito fish *Gambusia affinis* in White Oak Lake. Initially, it was estimated that the dose rate during 1965 was some 450 mrad h^{-1} (BLAYLOCK, 1969), but later studies indicated that this was in error (TRABALKA and ALLEN, 1977). The re-evaluation showed that the dose rate had probably declined from 15 mrad h^{-1} in 1965 to 7·5 mrad h^{-1} in 1971 and 2·5 mrad h^{-1} in 1975 as a consequence of reduced discharges and burial of the radioactivity by sediment with a lesser degree of contamination. From samples of fish collected during 1965 it was demonstrated that the irradiated population produced more viable offspring per female after correction for maternal size than the controls and that there was also an increased incidence of dead and deformed embryos. It was suggested that the former could be an adaptive, genetic response to increased radiation exposure while the latter was attributed to the presence of radiation-induced recessive lethal mutations in the genome. Further studies (BLAYLOCK and MITCHELL, 1969) showed that the irradiated fish from White Oak Lake tended to be marginally more sensitive to the effects of acute radiation exposure than controls; this response was attributed to the presence of some sublethal damage due to the accumulated chronic exposure. Investigations by TRABALKA and ALLEN (1977) confirmed the increased fecundity of irradiated fish when sampled directly from White Oak Lake, but when irradiated and control fish were reared under identical laboratory conditions, the F_1 generations showed no differential effect. It was concluded that the field observations were a consequence, not of genetic adaptation, but of increased productivity under the more eutrophic conditions of White Oak Lake as compared with the control sites. However, the increased incidence of dead and deformed embryos persisted under laboratory conditions and it was concluded that this was indeed due to the presence of recessive lethal mutations in the genome. Further, a study of the thermal tolerance of male fish from White Oak Lake showed a lower mean value and greater variance as compared with the controls. This was interpreted as being due to the semi-dominant expression of the heterozygous recessive lethals. Notwithstanding the problems regarding dosimetry and the difficulties of interpretation of the observations it seems certain that there is a significantly increased genetic load consequent upon the increased irradiation of the White Oak Lake population. However, it was also concluded that this was of very little, and probably no significance at the population level (TRABALKA and ALLEN, 1977).

Indications of a mutagenic consequence of increased environmental radiation have also been observed in several populations of the goldfish *Carassius auratus gibelio,* a strain which reproduces gynogenetically. The development of this fish takes place ameiotically with the offspring of a given female forming a clone, and in the reservoirs studied each population was assumed to have been derived from relatively few individuals. Under these circumstances the initial genetic variability would have been low and it was assumed that much of the protein variation demonstrated by electrophoresis was due to mutation. Insufficient data are given to determine either the dose rate or the total dose accumulated by the genome, but there is an apparent increase in the number of new and rare protein variants in the 5 irradiated populations (DUBININ and co-authors, 1975). Properly controlled laboratory experiments with this fish would be of great interest as has been recently suggested for another similar species, *Poecilia formosa* (IAEA, 1979).

There are several ponds and streams on the Hanford Reservation which have been contaminated with low-level liquid radioactive wastes, and these have been investigated for the presence of possible radiation effects (EMERY and MCSHANE, 1980). The dose rates at the sediment–water interface were measured with lithium fluoride thermoluminescent dosimeters and mostly ranged from background up to approximately 3·8 mrad h^{-1}, but 1 system (the 100-N trench) was highly contaminated and gave dose rates between 2·2 and 4·4 rad h^{-1}. The biological properties investigated included the rates of periphyton (measured as chlorophyll *a* production) and invertebrate colonization of artificial substrates and diversity indices for both the macro-algae and the invertebrates. It was not possible to detect any correlation between the degree of contamination of the several sites and variations in any of these parameters. It was concluded that these properties, which are all indicators at the population and community level, were not responsive to the present degree of radiation exposure.

From these environmental studies it appears to be fairly well established that genetic effects arising from the increased radiation exposure of 1 or more generations of aquatic organisms can be discerned if a suitable detection system is available. It is equally apparent that this response has had no detrimental, overall effect at the population level. These results support Purdom's conclusion that:

> 'at the population level genetic damage is reparable by the process of natural selection. . . . It would seem likely that the genetic response of populations is relatively unimportant and that general mortality and infertility would be the limiting factors in the extent to which populations may overcome radiation exposure.' (PURDOM, 1966, p. 862)

(8) Conclusions

From the above review of the effects of increased environmental radiation on the mortality, fecundity, fertility, and genome of aquatic organisms and from the studies carried out in contaminated environments, it can be concluded that very minor effects could be expected at the maximum dose rates which have been estimated for the Windscale discharge to the northeast Irish Sea and which only affect small proportions of the populations of the various organisms; at the other 3 disposal sites considered the dose rates are far too low for it to be reasonable to expect detectable responses. With the possible exception of chromosome aberrations in a suitable indicator species (in the northeast Irish Sea the dog whelk *Nucella lapillus* and the mussel *Mytilus edulis* are possible candidates) it appears that these effects would not be easy either to detect or to differentiate from both natural variability and the responses to other environmental variables. The overwhelming weight of the evidence indicates that even if there were minor effects in individual organisms, these would not become apparent at the population, community or ecosystem level of complexity.

It is apparent, therefore, that the controls which are applied to the discharge of radioactive wastes into the marine environment to limit the potential exposure of human populations also provide quite adequate protection for populations of marine organisms. The critical pathway system of control has been confirmed as a reliable means of managing discharges of radioactive wastes to, and for the protection of, the marine environment.

Acknowledgements

It is a pleasure to acknowledge the help of my colleagues at the Fisheries Radiobiological Laboratory upon whose work much of this review is based. I am particularly grateful to Mr D. F. Jefferies for making available previously unpublished data. Thanks are also due to Professor G. G. Polikarpov of the Institute of Biology of the Southern Seas, Ukrainian SSR Academy of Sciences, Sevastopol, for his help in collecting together and interpreting much of the Russian literature on the effects of radiation on developing fish embryos.

Literature Cited (Chapter 2)

ADAMS, N. (1968). Doserate distributions from spherical and spherical-shell radiation sources with special reference to fish eggs in radioactive media. *Rep. U.K. atom. Energy Auth., A.H.S.B. Rep.*, **R87**, 1–19.

ALLEN, A. L. and MULKAY, L. M. (1960). X-ray effects on embryos of the paradise fish, with notes on the normal stages. *Growth*, **24**, 131–168.

AL'SHITS, L. K., TIMOFEEVA, N. A., and KULIKOV, N. V. (1970). Influence of γ-rays from ^{60}Co upon the embryonic development of pike (*Esox lucius*, L). *Trudy Instituta ekologii rasteniji i zhivotnykh (Sverdlovsk)*, **74**, 12–16.

ANDERS, A., ANDERS, F., and PURSGLOVE, D. (1971). X-ray induced mutations of the genetically-determined melanoma system of Xiphophorin fish. *Experientia*, **27**, 931–932.

ANONYMOUS (1965) *Standards for Protection Against Radiation* (United States Atomic Energy Commission, Title 10, Code of Federal Regulations, Part 20), U.S. Government Printing Office, Washington.

ATEN, A. H. W., DALENBERG, J. W., and BAKKUM, W. C. M. (1961). Concentration of uranium in sea fish. *Hlth Phys.*, **5**, 225–226.

ATTIX, F. H. and ROESCH, W. C. (Eds) (1966). *Radiation Dosimetry* (2nd ed.), Vol. II, Academic Press, New York.

ATTIX, F. H. and ROESCH, W. C. (Eds) (1968). *Radiation Dosimetry* (2nd ed.), Vol. I, Academic Press, New York.

ATTIX, F. H. and TOCHILIN, E. (Eds) (1969). *Radiation Dosimetry* (2nd ed.), Vol. III, Academic Press, New York.

BACON, M. P., SPENCER, D. W., and BREWER, P. G. (1976). ^{210}Pb/^{226}Ra and ^{210}Po/^{210}Pb disequilibria in seawater and suspended particulate matter. *Earth Planet. Sci. Lett.*, **32**, 277–296.

BACQ, Z. M. and ALEXANDER, P. (1961). *Fundamentals of Radiobiology* (2nd ed.), Pergamon Press, Oxford.

BALLARDIN, E. and METALLI, P. (1965). Sulla relazione fra poliploidia e radiosensibilità in ovociti di *Artemia salina*, Leach. *Boll. Zool.*, **32**, 613–618.

BALLARDIN, E. and METALLI, P. (1968). Stima di alcune componenti della 'fitness' in *Artemia salina* partenogenetica diploide irradiata per piu' generazioni. *Atti Ass. genet. ital.*, **13**, 341–345.

BAPTIST, J. P., WOLFE, D. A., and COLBY, D. R. (1976). Effects of chronic gamma-radiation on the growth and survival of juvenile clams (*Mercenaria mercenaria*) and scallops (*Argopecten irradians*). *Hlth Phys.*, **30**, 79–83.

BEACH, S. A. (1975). The identification of a homogeneous critical group using statistical extreme value theory: application to laverbread consumers and the Windscale liquid effluent discharges. *Hlth Phys.*, **29**, 171–180.

BEASLEY, T. M. (1968). *Lead-210 in Selected Marine Organisms*, Ph.D. Thesis, Laboratory of Radiation Ecology, College of Fisheries, Univeristy of Washington, Seattle.

BEASLEY, T. M. and HELD, E. E. (1969). Nickel-63 in marine and terrestrial biota, soil and sediment. *Science, N.Y.*, **164**, 1161–1163.

BELITSINA, N. V., GAVRILOVA, L. P., NEIFAKH, A. A., and SPIRIN, A. S. (1963). The effect of the radiation inactivation of nuclei on the embryos of groundling fish (*Misgurnus fossilis*). *Dokl. Akad. Nauk SSSR*, **153**, 1204–1206.

BELYAEVA, V. N. and POKROVSKAYA, G. L. (1959). Changes in the radiation sensitivity of loach spawn during the first embryonic mitoses. *Dokl. Akad. Nauk SSSR*, **125**, 632–635.

BERGER, M. J. (1968). Energy deposition in water by photons from point isotropic sources. *J. nucl. Med.*, **9**, (Suppl. 1), 15–25.

BERGER, M. J. (1971). Distribution of absorbed dose around point sources of electrons and β-particles in water and other media. *J. nucl. Med.*, **12** (Suppl. 5), 5–23.

BERGONIÉ, J. and TRIBONDEAU, L. (1906). Interprétation de quelques resultats de la radiothérapie et essai de fixation d'une technique rationelle. *C.r. hebd. Séanc. Acad. Sci., Paris*, **143**, 983–985.

BIEN, G. S. and SUESS, H. E. (1967). Transfer and exchange of ^{14}C between the atmosphere and the surface water of the Pacific Ocean. In *Radioactive Dating and Methods of Low-Level Counting*, Proc. Symp. Monaco, March 1967. International Atomic Energy Agency, Vienna. pp. 105–114.

BLANCHARD, R. L., BRINCK, W. L., KOLDE, H. E., KRIEGER, H. L., MONTGOMERY, D. M., GOLD, S., MARTIN, A., and KAHN, B. (1976). Radiological surveillance studies at the Oyster Creek BWR nuclear generating station. *Rep. U.S. environ. prot. Ag.*, **EPA-520/5–76–003**. United States Environmental Protection Agency, Cincinnati, Ohio.

BLANCHARD, R. L. and KAHN, B. (1979). Abundance and distribution of radionuclides discharged from a BWR nuclear power station into a marine bay. *Nucl. Saf.*, **20**, 190–204.

BLAYLOCK, B. G. (1965). Chromosomal aberrations in a natural population of *Chironomus tentans* exposed to chronic low-level radiation. *Evolution, Lancaster, Pa.*, **19**, 421–429.

BLAYLOCK, B. G. (1966a). Chromosomal polymorphism in irradiated natural populations of *Chironomus*. *Genetics, Princeton*, **53**, 131–136.

BLAYLOCK, B. G. (1966b). Cytogenetic study of a natural population of *Chironomus* inhabiting an area contaminated by radioactive waste. In *Disposal of Radioactive Wastes into Seas, Oceans and Surface Waters*, Proc. Symp. Vienna, May 1966. International Atomic Energy Agency, Vienna. pp. 835–846.

BLAYLOCK, B. G. (1969). The fecundity of a *Gambusia affinis affinis* population exposed to chronic environmental radiation. *Radiat. Res.*, **37**, 108–117.

BLAYLOCK, B. G. (1971). The production of chromosome aberrations in *Chironomus riparius* (Diptera: Chironomidae) by tritiated water. *Can. Ent.*, **103**, 448–453.

BLAYLOCK, B. G. (1973). Chromosome aberrations in *Chironomus riparius* developing in different concentrations of tritiated water. In D. J. Nelson (Ed.), *Radionuclides in Ecosystems*, Vol. 2, Proc. Symp. Oak Ridge, Tennessee, May 1971. United States Atomic Energy Commission, Oak Ridge. pp. 1169–1173.

BLAYLOCK, B. G. and GRIFFITH, N. A. (1971). Effects of acute beta and gamma radiation on developing embryos of carp (*Cyprinus carpio*). *Radiat. Res.*, **46**, 99–104.

BLAYLOCK, B. G. and MITCHELL, T. J. (1969). The effect of temperature on the dose response of *Gambusia affinis affinis* from two natural populations. *Radiat. Res.*, **40**, 503–511.

BLAYLOCK, B. G. and TRABALKA, J. R. (1978). Evaluating the effects of ionizing radiation on aquatic organisms. *Adv. Radiat. Biol.*, **7**, 103–152.

BOCHKAREV, V. V., RADZIEVSKY, G. B., TIMOFEEV, L. V., and DEMIANOV, N. A. (1972). Distribution of absorbed energy from a point beta-source in a tissue-equivalent medium. *Int. J. appl. Radiat. Isotopes*, **23**, 493–504.

BONHAM, K. and DONALDSON, L. R. (1966). Low-level chronic irradiation of salmon eggs and alevins. In *Disposal of Radioactive Wastes into Seas, Oceans and Surface Waters*, Proc. Symp. Vienna, May 1966. International Atomic Energy Agency, Vienna. pp. 869–883.

BONHAM, K. and DONALDSON, L. R. (1972). Sex ratios and retardation of gonadal development in chronically gamma-irradiated chinook salmon smolts. *Trans. Am. Fish. Soc.*, **101**, 428–434.

BONHAM, K. and WELANDER, A. D. (1963). Increase in radioresistance of fish to lethal doses with advancing embryonic development. In V. Schultz and A. W. Klement (Eds), *Radioecology*, Proc. 1st Nat. Symp. on Radioecology, Fort Collins, Colorado, September, 1961. Reinhold Publishing Corp., New York. pp. 353–358.

BOWEN, V. T., LIVINGSTON, H. D., and BURKE, J. C. (1976). Distributions of transuranium nuclides in sediment and biota of the North Atlantic Ocean. In *Transuranium Nuclides in the Environment*, Proc. Symp. San Francisco, November 1975. International Atomic Energy Agency, Vienna. pp. 107–120.

BOWEN, V. T., NOSHKIN, V. E., LIVINGSTON, H. D., and VOLCHOK, H. L. (1980). Fallout radionuclides in the Pacific Ocean. *Earth Planet. Sci. Lett.*, **49**, 411–434.

BOWEN, V. T., WONG, K. M., and NOSHKIN, V. E. (1971). Plutonium-239 in and over the Atlantic Ocean. *J. mar. Res.*, **29**, 1–10.

BROECKER, W. S., GERARD, R., EWING, M., and HEEZEN, B. C. (1960). Natural radiocarbon in the Atlantic Ocean. *J. geophys. Res.*, **65**, 2903–2931.

BROECKER, W. S., GODDARD, J., and SARMIENTO, J. L. (1976). The distribution of ^{226}Ra in the Atlantic Ocean. *Earth Planet. Sci. Lett.*, **32**, 220–235.

BROECKER, W. S., LI, Y. H., and CROMWELL, J. (1967). Radium-226 and radon-222: concentration in Atlantic and Pacific Oceans. *Science, N.Y.*, **158**, 1307–1310.

BROWN, V. M. (1962). The accumulation of strontium-90 and yttrium-90 from a continuously flowing natural water by eggs and alevins of the Atlantic salmon and the sea trout. *Rep. U.K. atom. Energy Auth.*, **PG-288(W)**, 1–16.

BROWN, V. M. and TEMPLETON, W. L. (1964). Resistance of fish embryos to chronic irradiation. *Nature, Lond.*, **203**, 1257–1259.

BRYAN, G. W., PRESTON, A. and TEMPLETON, W. L. (1966). Accumulation of radionuclides by aquatic organisms of economic importance in the United Kingdom. In *Disposal of Radioactive Wastes into Seas, Oceans and Surface Waters*, Proc. Symp. Vienna, May 1966. International Atomic Energy Agency, Vienna. pp. 623–637.

CARLSON, W. D. and GASSNER, F. X. (Eds) (1964). *Effects of Ionizing Radiation on the Reproductive System.* Proc. Symp. Fort Collins, Colorado, 1962, Pergamon Press, Oxford.

CARTER, M. W. (1979). Nuclear testing. *Hlth Phys.*, **36**, 432–437.

CARTER, M. W. and MOGHISSI, A. A. (1977). Three decades of nuclear testing. *Hlth Phys.*, **33**, 55–71.

CENTRAL ELECTRICITY GENERATING BOARD (1967). *Hydrobiological Studies in the River Blackwater in Relation to the Bradwell Nuclear Power Station*, Central Electricity Generating Board, London.

CERRAI, E., SCHREIBER, B., and TRIULZI, C. (1965). Strontium-90 in upper layers of coastal sediments of the Ligurian Sea and contribution of some nuclides to their radioactivity. *Energia nucl., Milano*, **12**, 549–552.

CERRAI, E., SCHREIBER, B., and TRIULZI, C. (1967). Vertical distribution of ^{90}Sr, ^{144}Ce, ^{147}Pm, and ^{155}Eu in coastal marine sediments. *Energia nucl. Milano.*, **14**, 586–592.

CHAN, L. H., EDMOND, J. M., STALLARD, R. F., BROECKER, W. S., CHUNG, Y. C., WEISS, R. F., and KU, T. L. (1976). Radium and barium at GEOSECS stations in the Atlantic and Pacific. *Earth Planet. Sci. Lett.*, **32**, 258–267.

CHERRY, R. D., GERICKE, I. H., and SHANNON, L. V. (1969). ^{228}Th in marine plankton and seawater. *Earth Planet. Sci. Lett.*, **6**, 451–456.

CHERRY, R. D., SHAY, M. M., and SHANNON, L. V. (1970). Natural α-radioactivity concentrations in bone and liver from various animal species. *Nature, Lond.*, **228**, 1002–1003.

CHIPMAN, W. A. (1972). Ionizing radiation. In O. Kinne (Ed.), *Marine Ecology*, Vol. I, Environmental Factors, Part 3. Wiley, London. pp. 1579–1657.

CHIPMAN, W. A. and THOMMERET, J. (1970). Manganese content and the occurrence of fallout ^{54}Mn in some marine benthos of the Mediterranean. *Bull. Inst. océanogr. Monaco*, **69** (1402), 1–15.

CHUNG, Y. C. (1976). A deep ^{226}Ra maximum in the northeast Pacific. *Earth Planet. Sci. Lett.*, **32**, 249–257.

COLLIER, A. W. (1970). Oceans and coastal waters as life-supporting environments. In O. Kinne (Ed.), *Marine Ecology*, Vol. I, Environmental Factors, Part 1. Wiley, London. pp. 1–88.

COOLEY, J. L. (1973a). Effects of temperature and chronic irradiation on populations of the aquatic snail, *Physa heterostropha*. In D. J. Nelson (Ed.), *Radionuclides in Ecosystems*, Vol. 1, Proc. Symp. Oak Ridge, Tennessee, May 1971. United States Atomic Energy Commission, Oak Ridge. pp. 585–590.

COOLEY, J. L. (1973b). Effects of chronic environmental radiation on a natural population of the aquatic snail, *Physa heterostropha. Radiat. Res.*, **54**, 130–140.

COOLEY, J. L. and MILLER, F. L. (1971). Effects of chronic irradiation on laboratory populations of the aquatic snail, *Physa heterostropha. Radiat. Res.*, **47**, 716–724.

COSGROVE, G. E. and BLAYLOCK, B. G. (1973). Acute and chronic irradiation effects in mosquito fish at 15 or 20 °C. In D. J. Nelson (Ed.), *Radionuclides in Ecosystems*, Vol. 1, Proc. Symp. Oak

Ridge, Tennessee, May 1971. United States Atomic Energy Commission, Oak Ridge. pp. 579–584.

DABROWSKI, K. (1975). The effect of ionizing radiation and radionuclides on the embryonal development of fish. *Wiad. Ekol.*, **11**, 123–132.

DABROWSKI, K., TUCHOLSKI, S., and CZARNOCKI, J. (1975). The effect of radioactive iron, ^{59}Fe, on the embryogeny of sea trout (*Salmo trutta*, L). *Polskie Arch. Hydrobiol.*, **22**, 577–592.

DOCKINS, K. O., BAINBRIDGE, A. E., HOUTERMANS, J. C., and SUESS, H. E. (1967). Tritium in the mixed layer of the north Pacific Ocean. In *Radioactive Dating and Methods of Low-level Counting*, Proc. Symp. Monaco, March 1967. International Atomic Energy Agency, Vienna. pp. 129–140.

DOKHOLYAN, V. K., KOSTROV, B. P., and BITYUTSKAYA, G. S. (1974). Accumulation and effect of strontium-90 and yttrium-90 dissolved in water on the embryonic development of different Caspian Sea fish species. In G. P. Andrushaytis, G. G. Polikarpov, R. M. Aleksakhin, D. G. Fleishman, and M. P. Leinert (Eds), *Radioecology of Water Organisms*, Vol. 3, The Effects of Ionizing Radiation on Water Organisms (USAEC translation from the Russian, **AEC-tr-7592**). United States Atomic Energy Commission, Oak Ridge, Tennessee. pp. 19–24.

DONALDSON, L. R. (1970). Low-level chronic irradiation of salmon. Annual report to June 1970. *Rep. Coll. Fish. Univ. Wash.*, **RLO-2225-T2-1**, 1–22.

DONALDSON, L. R. (1971). Low-level chronic irradiation of salmon. Annual report to June 1971. *Rep. Coll. Fish. Univ. Wash.*, **RLO-2225-T2-2**, 1–19.

DONALDSON, L. R. and BONHAM, K. (1964). Effects of low-level chronic irradiation of chinook and coho salmon eggs and alevins. *Trans. Am. Fish. Soc.*, **93**, 333–341.

DONALDSON, L. R. and BONHAM, K. (1970). Effects of chronic exposure of chinook salmon eggs and alevins to γ-irradiation. *Trans. Am. Fish. Soc.*, **99**, 112–119.

DONALDSON, L. R., BONHAM, K., and HERSHBERGER, W. K. (1972). Low-level chronic irradiation of salmon. Annual report to June 1972. *Rep. Coll. Fish., Univ. Wash.*, **RLO-2225-T2-3**, 1–35.

DUBININ, N. P., ALTUKHOV, Y. P., SALMENKOVA, E. A., MILISHNIKOV, A. N., and NOVIKOVA, T. A. (1975). Analysis of monomorphic markers of genes in populations as a method of evaluating the mutagenicity of the environment. *Dokl. Akad. Nauk. SSSR*, **225**, 693–696.

DUNSTER, H. J. (1958). The disposal of radioactive liquid wastes into coastal waters. In *Peaceful Uses of Atomic Energy*, Vol. 18, Waste Treatment and Environmental Aspects of Atomic Energy, Proc. 2nd Int. Conf. Geneva, September 1958. U.N. Geneva. pp. 390–399.

DUNSTER, H. J., GARNER, R. J., HOWELLS, H., and WIX, L. F. U. (1964). Environmental monitoring associated with the discharge of low activity radioactive waste from Windscale works to the Irish Sea. *Hlth Phys.*, **10**, 353–362.

DUTTON, J. W. R., MITCHELL, N. T., REYNOLDS, E., and WOOLNER, L. (1974). Analytical systems applied to monitoring the aquatic environment in the control of radioactive waste disposal. In *Environmental Surveillance Around Nuclear Installations*, Vol. 1, Proc. Symp. Warsaw, November 1973. International Atomic Energy Agency, Vienna. pp. 155–167.

EGAMI, N. and AOKI, K. (1966). Effects of X-irradiation of a part of the body on the ovary of the loach, *Misgurnus anguillicaudatus*. *Annotnes zool. jap.*, **39**, 7–15.

EGAMI, N. and HAMA, A. (1975). Dose rate effects on the hatchability of irradiated embryos of the fish, *Oryzias latipes*. *Int. J. Radiat. Biol.*, **28**, 273–278.

EGAMI, N. and HYODO, Y. (1965). Effect of X-irradiation on the oviposition of the teleost, *Oryzias latipes*. *Annotnes zool. jap.*, **38**, 171–181.

EGAMI, N. and IJIRI, K. I. (1979). Effect of irradiation on germ cells and embryonic development in teleosts. *Int. Rev. Cytol.*, **59**, 195–248.

ELKIND, M. M. and WHITMORE, G. F. (1967). *The Radiobiology of Cultured Mammalian Cells*, Gordon and Breach, New York.

ELLINGER, F. (1957). *Medical Radiation Biology*, Charles C. Thomas, Springfield, Illinois.

EMERY, R. M. and MCSHANE, M. C. (1980). Nuclear waste ponds and streams on the Hanford site: an ecological search for radiation effects. *Hlth Phys.*, **38**, 787–809.

ENGEL, D. W. (1967). Effect of single and continuous exposure of gamma radiation on the survival and growth of the blue crab, *Callinectes sapidus*. *Radiat. Res.*, **32**, 685–691.

ERICKSON, R. C. (1973). Effects of chronic irradiation by tritiated water on *Poecilia reticulata*, the guppy. In D. J. Nelson (Ed.), *Radionuclides in Ecosystems*, Vol. 2. Proc. Symp. Oak Ridge, Tennessee, May 1971. United States Atomic Energy Commission, Oak Ridge. pp. 1091–1099.

ETOH, H. (1976). Effects of radionuclides on the embryogenesis and gonads of fish. *Hoken Butsuri*, **11**, 183–191.

FEDOROV, A. F. (1965). Radiation doses for some types of marine biota under present day conditions. *Bull. Inst. océanogr. Monaco*, **64**, (1334), p. 1–28.

FEDOROV, A. F., PODYMAKHIN, V. N., SHCHITENKO, N. T., and CHUMACHENKO, V. V. (1964). The influence of low radioactive contamination of water on the development of *Pleuronectes platessa*, L. *Vop. Ikhtiol.*, **4**, 579–585.

FEDOROVA, G. V. (1963a). Effects of strontium-90 on roe and larvae of Ludoga white-fish. *Vest. leningr. gos. Univ. (Ser. Biol.)*, **3**, 48–53.

FEDOROVA, G. V. (1963b). Experimental studies of ^{14}C uptake by developing spawn and young of freshwater fish. *Radiobiologiya*, **3**, 677–681.

FEDOROVA, G. V. (1964). On the effect of ^{14}C on developing ova and larvae of freshwater fish. *Vop. Ikhtiol.*, **4**, 723–728.

FEDOROVA, G. V. (1965). Entrance of ^{14}C into roe and larvae of autumn spawning fish. *Radiobiologiya*, **5**, 690–692.

FEDOROVA, G. V. (1968). Effect of ^{14}C on hatching of white-fish larvae from roe. *Vest. leningr. gos. Univ. (Ser. Biol.)* **4**, 45–49.

FEDOROVA, G. V. (1972a). The biological effect of ^{14}C on fish in the early development stages. *Vop. Ikhtiol.*, **12**, 198–201.

FEDOROVA, G. V. (1972b). Biological effect of ^{14}C incorporated in fish eggs. *Radiobiologiya*, **12**, 561–565.

FELDT, W., KANISCH, G., BÜHRINGER, H., and LAUER, R. (1979). *Preliminary Report on the 'Anton Dohrn' cruise, 1979, to the NEA Radioactive Waste Dumping Areas*, Isotopenlaboratorium der Bundesforschungsanstalt für Fisherei, Hamburg.

FITZGERALD, B. W. and SKAUEN, D. M. (1963). Zinc-65 in oysters in Fishers Island Sound and its estuaries. In V. Schultz and A. W. Klement (Eds), *Radioecology*, Proc. 1st Nat. Symp. on Radioecology, Fort Collins, Colorado, September, 1961. Reinhold Publishing Corp. New York. pp. 159–162.

FOLSOM, T. R. and BEASLEY, T. M. (1968). *The Contribution of α-Emitters to the Natural Radiation Environment of Marine Organisms*, Manuscript submitted to the Committee on Effects of Atomic Radiation on Oceanography and Fisheries of the National Academy of Sciences (unpubl.)

FOLSOM, T. R., GRISMORE, R., and YOUNG, D. R. (1970). Long-lived γ-ray emitting nuclide silver-108m found in Pacific marine organisms and used for dating. *Nature, Lond.*, **227**, 941–943.

FOLSOM, T. R. and HARLEY, J. H. (1957). Comparison of some natural radiations received by selected organisms. In The Effects of Atomic Radiation on Oceanography and Fisheries. *Publs natn. Res. Coun., Wash.*, (511), 28–33.

FOLSOM, T. R. and YOUNG, D. R. (1965). Silver-110m and cobalt-60 in oceanic and coastal organisms. *Nature, Lond.*, **206**, 803–806.

FOLSOM, T. R., YOUNG, D. R., JOHNSON, J. N., and PILLAI, K. C. (1963). Manganese-54 and zinc-65 in coastal organisms of California. *Nature, Lond.*, **200**, 327–329.

FOSTER, R. F., OPHEL, I. L., and PRESTON, A. (1971). Evaluation of human radiation exposure. In *Radioactivity in the Marine Environment*. National Academy of Sciences, Washington. pp. 240–260.

FRANK, M. L. and BLAYLOCK, B. G. (1971). Effects of acute ionizing radiation on carp (*Cyprinus carpio*, L) embryos. *Rep. Oak Ridge Nat. Lab.*, **ORNL-TM-3346**, 1–48.

FUKAI, R., BALLESTRA, S., and HOLM, E. (1976). ^{241}Am in Mediterranean surface waters. *Nature, Lond.*, **264**, 739–740.

FUKAI, R. and MEINKE, W. W. (1959). Trace analysis of marine organisms: a comparison of activation analysis and conventional methods. *Limnol. Oceanogr.*, **4**, 398–408.

GARROD, D. J. (1966). Discussion of the paper by Templeton (1966). In *Disposal of Radioactive Wastes into Seas, Oceans and Surface Waters*, Proc. Symp., Vienna, May, 1966. International Atomic Energy Agency, Vienna. pp. 859–860.

GOLDBERG, E. D. (1963). Geochronology with lead-210. In *Radioactive Dating*, Proc. Symp. Athens, November 1962. International Atomic Energy Agency, Vienna. pp. 121–130.

GUS'KOVA, V. N., BRAGINA, A. N., KUPRIYANOVA, V. M., MASHNEVA, N. I., RODIONOVA, L. F., and SUKAL'SKAYA, S. Y. (1971). Kinetics of ^{140}Ba–^{140}La accumulation by some hydrobionts and influence of these nuclides on fish eggs and the biochemical processes characterizing the self-cleaning of water bodies. *Trudy Instituta ekologii rasteniji i zhivotnykh (Sverdlovsk)*, **78**, 160–167.

GUS'KOVA, V. N., MASHNEVA, N. I., RODIONOVA, L. F., KUPRIYANOVA, V. M., ZAZEDATELEV, A. A., and SUKAL'SKAYA, S. Y. (1975). Biological effect of ^{106}Ru and ^{131}I on water organisms. In G. P. Andrushaytis, G. G. Polikarpov, R. M. Aleksakhin, D. G. Fleishman, and M. P. Leinert (Eds), *Radioecology of Water Organisms*, Vol. 2, Distribution and Migration of Radionuclides in Freshwater and Seawater Biocenoses (USAEC translation from the Russian, **AEC-tr-7606**). United States Energy Research and Development Administration, Oak Ridge, Tennessee. pp. 110–114.

HAMAGUCHI, S. (1976). Change in the radiation responses of oogonia in the embryos and fry of the fish, *Oryzias latipes. Int. J. Radiat. Biol.*, **29**, 565–570.

HAMAGUCHI, S. (1980). Differential radiosensitivity of germ cells according to their development stages in the teleost, *Oryzias latipes*. In N. Egami (Ed.), *Radiation Effects on Aquatic Organisms*, Proc. Symp. Kanagawa, Japan, May, 1979. Japan Scientific Societies Press, Tokyo/University Park Press, Baltimore. pp. 119–128.

HAMAGUCHI, S. and EGAMI, N. (1975) Post-irradiation changes in oocyte populations in the fry of the fish, *Oryzias latipes. Int. J. Radiat. Biol.*, **28**, 279–284.

HAMILTON, E. I. (1972). The concentration of uranium in man and his diet. *Hlth Phys.*, **22**, 149–153.

HERSHBERGER, W. K., BONHAM, K., and DONALDSON, L. R. (1978). Chronic exposure of chinook salmon eggs and alevins to gamma irradiation: effects on their return to freshwater as adults. *Trans. Am. Fish. Soc.*, **107**, 622–631.

HERSHBERGER, W. K., DONALDSON, L. R., BONHAM, K., and BRANNON, E. L. (1973). Low-level chronic irradiation of salmon. Annual report to June 1973. *Rep. Coll. Fish., Univ. Wash.*, **RLO-2225-T2-5**, 1–33.

HERSHBERGER, W. K., DONALDSON, L. R., BONHAM, K., and BRANNON, E. L. (1974). Low-level chronic irradiation of salmon. Annual report to July 1974. *Rep. Coll. Fish., Univ. Wash.*, **RLO-2225-T2-6**, 1–25.

HERSHBERGER, W. K., DONALDSON, L. R., BONHAM, K., and BRANNON, E. L. (1976). Low-level chronic irradiation of salmon. Annual report to June 1976. *Rep. Coll. Fish., Univ. Wash.*, **RLO-2225-T2-8**, 1–13.

HETHERINGTON, J. A. (1976a). The behaviour of plutonium nuclides in the Irish Sea. In M. W. Miller and J. N. Stanndard (Eds), *Environmental Toxicity of Aquatic Radionuclides: Models and Mechanisms*. Ann Arbor Science Publishers Inc., Ann Arbor. pp. 81–106.

HETHERINGTON, J. A. (1976b). Radioactivity in surface and coastal waters of the British Isles, 1974. *Tech. Rep. Fish. Radiobiol. Lab., MAFF Direct. Fish. Res., Lowestoft*, **FRL 11**, 1–35.

HETHERINGTON, J. A. (1978). The uptake of plutonium nuclides by marine sediments. *Mar. Sci. Commun.*, **4**, 239–274.

HETHERINGTON, J. A. and JEFFERIES, D. F. (1974). The distribution of some fission product radionuclides in sea and estuarine sediments. *Neth. J. Sea Res.*, **8**, 319–338.

HETHERINGTON, J. A. JEFFERIES, D. F., and LOVETT, M. B. (1975). Some investigations into the behaviour of plutonium in the marine environment. In *Impacts of Nuclear Releases into the Aquatic Environment*, Proc. Symp. Otaniemi, June–July 1971. International Atomic Energy Agency, Vienna. pp. 193–212.

HETHERINGTON, J. A. JEFFERIES, D. F., MITCHELL, N. T., PENTREATH, R. J., and WOODHEAD, D. S. (1976). Environmental and public health consequences of the controlled disposal of transuranic elements to the marine environment. In *Transuranium Nuclides in the Environment*, Proc. Symp. San Francisco, November 1975. International Atomic Energy Agency, Vienna. pp. 139–154.

HETHERINGTON, J. A. and ROBSON, J. C. (1979). An assessment of the radiological impact of tritium released to sea from the Windscale fuel element reprocessing plant. In *Behaviour of Tritium in the Environment*, Proc. Symp. San Francisco, October 1978. International Atomic Energy Agency, Vienna. pp. 283–301.

HEWETT, C. J. and JEFFERIES, D. F. (1978). The accumulation of radioactive caesium from food by the plaice (*Pleuronectes platessa*) and the brown trout (*Salmo trutta*). *J. Fish Biol.*, **13**, 143–153.

HEYE, D. (1969). Uranium, thorium and radium in ocean water and deep sea sediments. *Earth Planet. Sci. Lett.*, **6**, 112–116.

HEYRAUD, M. and CHERRY, R. D. (1979). Polonium-210 and lead-210 in marine food chains. *Mar. Biol.*, **52**, 227–236.

HIBIYA, T. and YAGI, T. (1956). Effects of fission materials upon the development of aquatic animals. In *Research in the Effects and Influences of the Nuclear Bomb Explosions*, Vol. 2. Japanese Society for the Promotion of Science, Tokyo. pp. 1219–1224.

HOFFMAN, F. L., HODGE, V. F., and FOLSOM, T. R. (1974). ^{210}Po radioactivity in organs of selected tunas and other marine fish. *J. Radiat. Res.*, **15**, 103–106.

HOLTON, R. L., FORSTER, W. O., and OSTERBERG, C. L. (1973a). The effect of gamma irradiation on the reproductive performance of *Artemia* as determined by individual pair matings. In D. J. Nelson (Ed.), *Radionuclides in Ecosystems*, Vol. 2, Proc. Symp. Oak Ridge, Tennessee, May 1971. United States Atomic Energy Commission, Oak Ridge. pp. 1191–1197.

HOLTON, R. L., FORSTER, W. O., and OSTERBERG, C. L. (1973b). The effects of gamma irradiation on the maintenance of population size in the brine shrimp, *Artemia*. In D. J. Nelson (Ed.), *Radionuclides in Ecosystems*. Vol. 2, Proc. Symp. Oak Ridge, Tennessee, May, 1971. United States Atomic Energy Commission, Oak Ridge. pp. 1198–1205.

HOLTZMAN, R. B. (1969). Concentrations of the naturally-occurring radionuclides radium-226, lead-210 and polonium-210 in aquatic fauna. In D. J. Nelson and F. C. Evans (Eds), *Proceedings of the 2nd National Symposium on Radioecology*, Ann Arbor, Michigan, May 1967. United States Atomic Energy Commission, Washington. pp. 535–546.

HOLZBERG, S. (1973). Change in aggressive readiness in post-irradiation generations of the convict cichlid fish, *Cichlasoma nigrofasciatum*. In J. H. Schröder (Ed.), *Genetics and Mutagenesis of Fish*. Springer-Verlag, Berlin. pp. 173–176.

HONDA, Y., KIMURA, Y., TAMURA, Y., and TANAKA, C. (1972). Uptake of ^{106}Ru by eggs and fry of rainbow trout. *J. Radiat. Res.*, **13**, 95–99.

HOPPENHEIT, M. (1973). Effects on fecundity and fertility of single sublethal X-irradiation of *Gammarus duebeni* females. In *Radioactive Contamination of the Marine Environment*, Proc. Symp. Seattle, July 1972. International Atomic Energy Agency, Vienna. pp. 479–486.

HOPPENHEIT, M., MURRAY, C. N., and WOODHEAD, D. S. (1980). Uptake and effects of americium-241 on a brackish water amphipod. *Helgoländer Meeresunters*, **33**, 138–152.

HOWELLS, H. (1966). Discharges of low-activity, radioactive effluent from the Windscale works into the Irish Sea. In *Disposal of Radioactive Wastes into Seas, Oceans and Surface Waters*, Proc. Symp. Vienna, May 1966. International Atomic Energy Agency, Vienna. pp. 769–785.

HUNT, G. J. (1979). Radioactivity in surface and coastal waters of the British Isles, 1977. *Aquat. Environ. Monit. Rep., MAFF Direct. Fish. Res., Lowesoft*, **3**, 1–37.

HUNT, G. J. (1980). Radioactivity in surface and coastal waters of the British Isles, 1978. *Aquat. Envrion. Monit. Rep., MAFF Direct. Fish. Res., Lowestoft*, **4**, 1–36.

HUNT, G. J. and SHEPHERD, J. G. (1980). The identification of critical groups. In A. Eisenberg (Ed.), *Radiation Protection: A Systematic Approach to Safety*, Proc. 5th Int. Congr. IRPA Jerusalem, March 1980. Pergamon Press, London. Vol. 1, pp. 141–145.

HYODO-TAGUCHI, Y. (1980). Effects of chronic γ-irradiation on spermatogenesis in the fish *Oryzias latipes* with special reference to the regeneration of testicular stem cells. In N. Egami (Ed.), *Radiation Effects on Aquatic Organisms*, Proc. Symp. Kanagawa, Japan, May 1979. Japan Scientific Societies Press, Tokyo/University Park Press, Baltimore. pp. 91–104.

HYODO-TAGUCHI, Y. and EGAMI, N. (1969). Change in dose–survival time relationship after X-irradiation during embryonic development in the fish, *Oryzias latipes*. *J. Radiat. Res.*, **10**, 121–125.

HYODO-TAGUCHI, Y. and EGAMI, N. (1970). Inhibitory effects of β-rays from ^{90}Sr–^{90}Y on gonad formation in embryos of *Oryzias latipes*. *Zool. Mag., Tokyo*, **79**, 185–187.

HYODO-TAGUCHI, Y. and EGAMI, N. (1971). Notes on the X-ray effects on the testis of the goby, *Chasmichthys glosus*. *Annotnes zool. jap.*, **44**, 19–22.

HYODO-TAGUCHI, Y. and EGAMI, N. (1976). Effect of X-irradiation on spermatogonia of the fish, *Oryzias latipes*. *Radiat. Res.*, **67**, 324–331.

HYODO-TAGUCHI, Y. and EGAMI, N. (1977). Damage to spermatogenic cells in fish kept in tritiated water. *Radiat. Res.*, **71**, 641–652.

HYODO-TAGUCHI, Y., EGAMI, N., and MORI, N. (1971). Histological effects of β-rays from ^{90}Sr–^{90}Y on the ovary of the marine goby, *Chasmichthys glosus. J. Fac. Sci. Tokyo Univ. (Sec. IV.)*, **12**, 337–344.

HYODO-TAGUCHI, Y., ETOH, H., and EGAMI, N. (1973). RBE of fast neurons for inhibition of hatchability in fish embryos irradiated at different developmental stages. *Radiat. Res.*, **53**, 385–391.

HYODO-TAGUCHI, Y. and MARUYAMA, T. (1977). Effects of fast neutrons on spermatogenesis of the fish, *Oryzias latipes. Radiat. Res.*, **70**, 345–354.

IAEA (1970). Reference methods for marine radioactivity studies. *Tech. Rep. Int. atom. Energy Ag.*, (**118**), International Atomic Energy Agency, Vienna.

IAEA (1975). Reference methods for marine radioactivity studies II. *Tech. Rep. Int. atom. Energy Ag.*, (**169**), International Atomic Energy Agency, Vienna.

IAEA (1976). Effects of ionizing radiation on aquatic organism and ecosystems. *Tech. Rep. Int. atom. Energy Ag.*, (**172**), International Atomic Energy Agency, Vienna.

IAEA (1978a). Principles for establishing limits for the release of radioactive materials into the environment. *Saf. Ser. Int. atom. Energy Ag.*, (**45**), International Atomic Energy Agency, Vienna.

IAEA (1978b). *The Oceanographic Basis of the IAEA Revised Definition and Recommendations Concerning High-level Radioactive Waste Unsuitable for Dumping at Sea.*, IAEA-210, International Atomic Energy Agency, Vienna.

IAEA (1978c). *The Radiological Basis of the IAEA Revised Definition and Recommendations Concerning High-level Radioactive Waste Unsuitable for Dumping at Sea*, IAEA-211, International Atomic Energy Agency, Vienna.

IAEA (1979). Methodology for assessing impacts of radioactivity on aquatic ecosystems. *Tech. Rep. Int. atom. Energy Ag.*, (**190**), International Atomic Energy Agency, Vienna.

ICHIKAWA, R. and SUYAMA, I. (1974). Effects of tritiated water on the embryonic development of two marine teleosts. *Bull. Jap. Soc. scient. Fish.*, **40**, 819–824.

ICRP (1955). Recommendations of the International Commission on Radiological Protection (Revised December 1, 1954). *Br. J. Radiol.*, (**Suppl. 6**), 1–92.

ICRP (1959). *Recommendations of the International Commission on Radiological Protection* (Adopted 9 September 1958), Pergamon Press, London.

ICRP (1960). *Recommendations of the International Commission on Radiological Protection* (Report of Committee II on Permissible Dose for Internal Radiation. ICRP Publication, 2), Pergamon Press, London.

ICRP (1964). *Recommendations of the International Commission on Radiological Protection* (As amended 1959 and revised 1962. ICRP Publication, 6), Pergamon Press, London.

ICRP (1966a). *Recommendations of the International Commission on Radiological Protection* (Adopted 17 September, 1965. ICRP Publication, 9), Pergamon Press, London.

ICRP (1966b). *Principles of Environmental Monitoring Related to the Handling of Radioactive Materials* (A Report by Committee 4 of the International Commission on Radiological Protection. ICRP Publication, 7), Pergamon Press, London.

ICRP (1973). *Recommendations of the International Commission on Radiological Protection* (Implications of Commission Recommendations that Doses be Kept as Low as Readily Achievable. ICRP Publication, 22), Pergamon Press, London.

ICRP (1977). Recommendations of the International Commission on Radiological Protection. *Annals of the ICRP*, **1**, 1–53 (ICRP Publication, 26).

ICRP (1979a). Part 1. Limits for intakes of radionuclides by workers. *Annals of the ICRP*, **2**, 1–116 (ICRP Publication, 30).

ICRP (1979b). Supplement to Part 1. Limits for intakes of radionuclides by workers. *Annals of the ICRP*, **3**, 1–555 (ICRP Publication, 30).

IVANOV, V. N. (1965a). Accumulation of fission product radionuclides by eggs of Black Sea fishes. *Radiobiologiya*, **5**, 295–300.

IVANOV, V. N. (1965b). Accumulation of ^{90}Sr, ^{90}Y, ^{144}Ce, ^{106}Ru, ^{59}Fe, and ^{35}S in the developing Black Sea haddock roe. *Radiobiologiya*, **5**, 57–60.

IVANOV, V. N. (1966a). Influence of temperature on the process of radiation damage of Black Sea fish eggs. *Gidrobiol. Zh.*, **2**, 79–84.

IVANOV, V. N. (1966b). Changes in radiosensitivity of fish eggs in the process of development. *Gidrobiol. Zh.*, **2**, 79–84.

IVANOV, V. N. (1967). Effect of radioactive substances on the embryonic development of fish. In *Voprosy Bio-okeanograffi. Materially II Mezhdunarodnogo Okeanograficheskogo Kongressa*, Moscow, Maya-Iyunya, 1966. Naukova Dumka, Kiev. pp. 185–190 (Available as USAEC translation, **AEC-tr-6940**; pp. 47–51).

IVANOV, V. N. (1969). Accumulation of ^{54}Mn, ^{60}Co, and ^{185}W in the roe of Black Sea flounder. *Gidrobiol. Zh.*, **5**, 58–59.

IVANOV, V. N. (1970). The accumulation of ^{65}Zn by the eggs of the Black Sea flounder, *Scophthalmus maeocticus* (Pallas). *Vop. Ikhtiol.*, **10**, 1129–1131.

IVANOV, V. N. (1972). Accumulation of radionuclides by eggs and larvae of Black Sea fishes. In G. G. Polikarpov (Ed.), *Marine Radioecology* (USAEC translation from the Russian, **AEC-tr-7299**). United States Atomic Energy Commission, Oak Ridge, Tennessee. pp. 147–157.

IVANOV, V. N. and PARCHEVSKAYA, D. S. (1974). Probability nature of radionuclide accumulation and action in seawater. *Dokl. Akad. Nauk SSSR*, **218**, 215–218.

IWASAKI, T. (1973). The differential radiosensitivity of oogonia and oocytes at different developmental stages of the brine shrimp, *Artemia salina*. *Biol. Bull. mar. biol. Lab., Woods Hole*, **144**, 151–161.

JEFFERIES, D. F. (1968). Fission-product radionuclides in sediments from the northeast Irish Sea. *Helgoländer Meeresunters.*, **17**, 280–290.

JEFFERIES, D. F. (1970). Exposure to radiation from gamma-emitting fission-product radionuclides in estuarine sediments from the northeast Irish Sea. In W. C. Reinig (Ed.), *Environmental Surveillance in the Vicinity of Nuclear Facilities*. Charles C. Thomas, Springfield, Illinois. pp. 205–216.

JEFFERIES, D. F. and HEWETT, C. J. (1971). The accumulation and excretion of radioactive caesium by the plaice (*Pleuronectes platessa*) and the thornback ray (*Raja clavata*). *J. mar. biol. Ass. U.K.*, **51**, 411–422.

JEFFERIES, D. F., MITCHELL, N. T., and HETHERINGTON, J. A. (1977). Collective population radiation exposure from waste disposal from a fuel reprocessing plant. *Proc. 4th Int. Congr. Int. Radiat. Prot. Ass. Paris, April 1977*, **3**, 929–939.

JEFFERIES, D. F., PRESTON, A., and STEELE, A. K. (1973). Distribution of caesium-137 in British coastal waters. *Mar. Pollut. Bull., N.S.*, **4**, 118–122.

JEFFERIES, D. F., STEELE, A. K., and PRESTON, A. (1982). Further studies on the distribution of ^{137}Cs in British coastal waters. Part I. Irish Sea. *Deep Sea Res.*, **29**, 713–738.

JOSEPH, A. B., GUSTAFSON, P. F., RUSSELL, I. R., SCHUERT, E. A., VOLCHOK, H. L., and TAMPLIN, A. (1971). Sources of radioactivity and their characteristics. In *Radioactivity in the Marine Environment*. National Academy of Sciences, Washington. pp. 6–41.

KALLE, K. (1971). Salinity: general introduction. In O. Kinne (Ed.), *Marine Ecology*, Vol. I, Environmental Factors, Part 2. Wiley, London. pp. 683–688.

KASATKINA, S. V. (1973). Effect of ^{90}Sr–^{90}Y on the development and functioning of the hatching glands in Atlantic salmon. In B. P. Sorokin (Ed.), *Effect of Ionizing Radiation on the Organism* (USAEC translation from the Russian, **AEC-tr-7418**). United States Atomic Energy Commission, Oak Ridge, Tennessee. pp. 66–74.

KASATKINA, S. V., KOSHELEVA, V. V., MIGALOVSKAYA, V. N., MIGALOVSKIY, I. P., and OGANESYAN, S. A. (1974). Chronic effect of ^{144}Ce and ^{137}Cs radionuclides dissolved in water on the embryonic development of Atlantic salmon. In G. P. Andrushaytis, G. G. Polikarpov, R. M. Aleksakhin, D. G. Fleishman, and M. P. Leinert (Eds), *Radioecology of Water Organisms*, Vol. 3, The Effects of Ionizing Radiation on Water Organisms (USAEC translation from the Russian, **AEC-tr-7592**). United States Atomic Energy Commission, Oak Ridge, Tennessee. pp. 13–18.

KAUFMAN, A. (1967). Uranium and thorium series isotope programme. *Rep. U.S. Atom. Energy Commn*, **CU-3139-1** (*App. A*), 1–16.

KAUFMAN, A., TRIER, R. M., BROECKER, W. S., and FEELY, H. W. (1973). Distribution of ^{228}Ra in the World Ocean. *J. geophys. Res.*, **78**, 8827–8848.

KAURANEN, P. and MIETTINEN, J. K. (1970). Polonium and radiolead in some aqueous ecosystems in Finland. Helsinki University Department of Radiochemistry. *Annual Report, August 1969–August 1970*, **NY-3446-14**, Paper 27.

KAUTSKY, H. (1973). The distribution of the radionuclide caesium-137 as an indicator of North Sea watermass transport. *Dt. Hydrogr. Z.*, **26**, 242–246.

KAUTSKY, H. (1977). Distribution of radioactive fallout products in Atlantic water between 10° S and 81° N in the years 1969 and 1972. *Dt. Hydrogr. Z.*, **30**, 216–227.

KAUTSKY, H., KOLTERMANN, K. P., and PRAHM, G. (1977). Iberische Tiefsee: Hydrographische und radiologische Untersuchungen. *Rep. Dt. Hydrogr. Inst.*, (**2149/17**), 1–244.

KIMURA, Y. and HONDA, Y. (1977). Uptake and elimination of some radionuclides by eggs and fry of rainbow trout. *J. Radiat. Res.*, **18**, 170–181.

KINNE, O. (1976). Cultivation of marine organisms: water-quality management and technology. In O. Kinne (Ed.), *Marine Ecology*, Vol. III, Cultivation, Part 1. Wiley, London. pp. 19–300.

KLIGERMAN, A. D., BLOOM, S. E., and HOWELL, W. M. (1975). *Umbra limi*: a model for the study of chromosome aberrations in fishes. *Mutat. Res.*, **31**, 225–233.

KLYASHTORIN, L. B., SHEKHANOVA, I. A., and YARZHOMBEK, A. A. (1972). On the effect of radioactive strontium on the respiration rate of the embryos of loach. *Trudy vses. nauchno-issled. Inst. morsk. rýb. Khoz. Okeanogr.*, **85**, 27–30.

KOBAYASHI, S. and MOGAMI, M. (1958). Effects of X-irradiation upon rainbow trout (*Salmo irideus*). III. Ovary growth in the stages of fry and fingerling. *Bull. Fac. Sci. Hokkaido Univ.*, **9**, 89–94.

KOBAYASHI, S. and YAMAMOTO, T. (1971). Effects of irradiation on spermatogenesis in the guppy, *Lebistes reticulatus*. *Tokushima J. exp. Med.*, **18**, 21–27.

KOCZY, F. F. (1958). Natural radium as a tracer in the ocean. In *Peaceful Uses of Atomic Energy*, 18, Waste Treatment and Environmental Aspects of Atomic Energy, Proc. 2nd Int. Conf. Geneva, September 1958. U.N., Geneva. pp. 351–357.

KOCZY, F. F., PICCIOTTO, E., POULAERT, G., and WILGAIN, S. (1957). Mesure des isotopes du thorium dans l'eau de mer. *Geochim. cosmochim. Acta*, **11**, 103–129.

KONNO, K. (1980). Effects of γ-irradiation on the gonads of the rainbow trout, *Salmo gairdnerii irideus*, during embryonic stages. In N. Egami (Ed.), *Radiation Effects on Aquatic Organisms*, Proc. Symp. Kanagawa, Japan, May 1979. Japan Scientific Societies Press, Tokyo/University Park Press, Baltimore. pp. 129–133.

KONNO, K., KIKUCHI, T., OSAKABE, I. and OKADA, I. (1955). On the influence of γ-ray radiation on the aquatic animals. I. On the influence in the early development of goldfish (*Carassius auratus* L). *J. Tokyo Univ. Fish.*, **41**, 163–168.

KOSHELEVA, V. V. (1973a). Accumulation of radioactive isotopes by the developing eggs of the Atlantic salmon. In B. P. Sorokin (Ed.), *Effect of Ionizing Radiation on the Organism* (USAEC translation from the Russian, **AEC-tr-7418**). United States Atomic Energy Commission, Oak Ridge, Tennessee. pp. 7–15.

KOSHELEVA, V. V. (1973b). Change in the peripheral blood in the embryos and larvae of Atlantic salmon under the influence of radioactive contamination of the water, and X-rays. In B. P. Sorokin (Ed.), *Effect of Ionizing Radiation on the Organism* (USAEC translation from the Russian, **AEC-tr-7418**). United States Atomic Energy Commission, Oak Ridge, Tennessee. pp. 75–88.

KU, T. L. (1965). An evaluation of the $^{234}U/^{238}U$ method as a tool for dating pelagic sediments. *J. geophys. Res.*, **70**, 3457–3474.

KU, T. L., LI, Y. H., MATHIEU, G. G., and WONG, H. K. (1970). Radium in the Indian-Antarctic Ocean south of Australia. *J. geophys. Res.*, **75**, 5286–5292.

KU, T. L. and LIN, M. C. (1976). ^{226}Ra distribution in the Antarctic Ocean. *Earth Planet. Sci. Lett.*, **32**, 236–248.

KULIKOV, N. V. (1970a). Radiosensitivity of tench (*Tinca tinca*, L) embryos at early stages of development. *Radiobiologiya*, **10**, 127–130.

KULIKOV, N. V. (1970b). Radiosensitivity of roe of pike (*Esox lucius*) during fertilization and early cleavage. *Radiobiologiya*, **10**, 768–770.

KULIKOV, N. V. (1973). Radioecology of freshwater plants and animals. In V. M. Klechkovskii, G. G. Polikarpov, and R. M. Aleksakhin (Eds), *Radioecology*. Wiley, New York. pp. 323–337.

KULIKOV, N. V., AL'SHITS, L. K., and TIMOFEEVA, N. A. (1971a). Influence of γ-rays from ^{60}Co upon the embryonic development of freshwater fishes (pike, *Esox lucius*, L and tench, *Tinca tinca*, L.). *Trudy Instituta ekologii rasteniji i zhivotnykh (Sverdlovsk)*, **78**, 189–194.

KULIKOV, N. V., BEZEL, V. S., and OZHEGOV, L. N. (1970a). Accumulation of radioisotopes by developing roe of tench (*Tinca tinca*, L) and perch (*Perca fluviatilis*, L). *Ekologiya*, **5**, 73–77.

KULIKOV, N. V., BEZEL, V. S., and OZHEGOV, L. N. (1971b). The accumulation of ^{90}Sr–^{90}Y by the eggs of freshwater fish and the calculation of absorbed dose. *Trudy Instituta ekologii rastenij i zhivotnykh (Sverdlovsk)*, **78**, 135–142.

KULIKOV, N. V. and OZHEGOV, L. N. (1975). Accumulation of ^{60}Co in the developing spawn of *Esox lacustris, Perca fluviatilis* and *Coregonus lavaretus*. *Ekologiya*, **4**, 100–103.

KULIKOV, N. V. and OZHEGOV, L. N. (1976). Accumulation of ^{90}Sr–^{90}Y by the developing spawn and larvae of *Coregonus lavaretus*, L. *Ekologiya*, **2**, 97–99.

KULIKOV, N. V., TIMOFEEVA, N. A., and AL'SHITS, L. K. (1969). Decrease in the radiosensitivity of tench embryos (*Tinca tinca*, L) as a result of preliminary irradiation. *Radiobiologiya*, **9**, 637–639.

KULIKOV, N. V., TIMOFEEVA, N. A., and AL'SHITS, L. J. (1970b). Action of prior irradiation on the subsequent radiosensitivity of tench (*Tinca tinca*, L) larvae. *Trudy Instituta ekologii rastenij i zhivotnykh (Sverdlovsk)*, **74**, 17–20.

KULIKOV, N. V., TIMOFEEVA, N. A., and SHISHENKOVA, L. K. (1968). Radiosensitivity of developing tench embryos (*Tinca tinca*, L). *Radiobiologiya*, **8**, 391–395.

LABEYRIE, L. D., LIVINGSTON, H. D., and BOWEN, V. T. (1976). Comparison of the distributions in marine sediments of the fallout derived nuclides ^{55}Fe and $^{239,240}Pu$. In *Transuranic Nuclides in the Environment*, Proc. Symp. San Francisco, November 1975. International Aomic Energy Agency, Vienna. pp. 121–137.

LAPPENBUSCH, W. L., WATSON, D. G., and TEMPLETON, W. L. (1971). *In situ* measurement of radiation dose in the Columbia River. *Hlth Phys*., **21**, 247–251.

LEDERER, C. M. and SHIRLEY, V. S. (Eds) (1978). *Table of Isotopes* (7th ed.), Wiley, New York.

LEE, A. (1960). Hydrographical observations in the Irish Sea, January–March 1953. *Fishery Invest., Lond*. (*Ser. II*), **23**, 1–25.

LOEVINGER, R., HOLT, J. G., and HINE, G. J. (1956a). Internally adminstered radioisotopes. In G. J. Hine and G. L. Brownell (Eds), *Radiation Dosimetry*. Academic Press, New York. pp. 801–873.

LOEVINGER, R., JAPHA, E. M., and BROWNELL, G. L. (1956b). Discrete radioisotope sources. In G. J. Hine and G. L. Brownell (Eds), *Radiation Dosimetry*. Academic Press, New York. pp. 693–799.

LOWMAN, F. G., RICE, T. R., and RICHARDS, F. A. (1971). Accumulation and redistribution of radionuclides by marine organisms. In *Radioactivity in the Marine Environment*. National Academy of Sciences, Washington. pp. 161–199.

LOWTON, R. J., MARTIN, J. H., and TALBOT, J. W. (1966). Dilution, dispersion and sedimentation in some British estuarines. In *Disposal of Radioactive Wastes into Seas, Oceans and Surface Waters*, Proc. Symp. Vienna, May, 1966. International Atomic Energy Agency, Vienna. pp. 189–206.

MCGREGOR, J. F. and NEWCOMBE, H. B. (1968). Major malformations in trout embryos irradiated prior to active organogenesis. *Radiat. Res*., **35**, 282–301.

MCGREGOR, J. F. and NEWCOMBE, H. B. (1972a). Dose–response relationship for yields of major eye malformations following low doses of radiation to trout sperm. *Radiat. Res*., **49**, 155–169.

MCGREGOR, J. F. and NEWCOMBE, H. B. (1972b). Decreased risk of embryo mortality following low doses of radiation to trout sperm. *Radiat. Res*., **52**, 536–544.

MAHDAVI, A. (1964). The thorium, uranium and potassium contents of Atlantic and Gulf coast beach sands. In J. A. S. Adams and M. Lowder (Eds), *The Natural Radiation Environment*. University of Chicago Press, Chicago. pp. 87–114.

MARSHALL, J. S. (1962). The effect of continuous γ-radiation on the intrinsic rate of natural increase of *Daphnia pulex*. *Ecology*, **43**, 598–607.

MARSHALL, J. S. (1966). Population dynamics of *Daphnia pulex* as modified by chronic radiation stress. *Ecology*, **47**, 561–571.

MAUCHLINE, J. and TEMPLETON, W. L. (1964). Artificial and natural radioisotopes in the marine environment. *Oceanogr. mar. Biol. A. Rev.* **2**, 229–279.

METALLI, P. and BALLARDIN, E. (1962). First results on X-ray-induced genetic damage in *Artemia salina* Leach. *Atti Ass. genet. ital.*, **7**, 219–231.

METALLI, P. and BALLARDIN, E. (1972). Radiobiology of *Artemia*: radiation effects and ploidy. *Curr. Top. Radiat. Res. Q.*, **7**, 181–240.

MICHIBATA, H. (1976). The role of spermatogonia in the recovery process from temporary sterility induced by gamma-ray irradiation in the teleost, *Oryzias latipes*. *J. Radiat. Res.*, **17**, 142–153.

MIGALOVSKAYA, V. N. (1973). Chronic effect of strontium-90 and yttrium-90 on the frequency of chromosomal aberrations in the embryonal cells of the Atlantic salmon. In B. P. Sorokin (Ed.), *Effect of Ionizing Radiation on the Organism* (USAEC translation from the Russian, **AEC-tr-7418**). United States Atomic Energy Commission, Oak Ridge, Tennessee. pp. 89–99.

MIGALOVSKIY, I. P. (1973a). Development of Atlantic salmon eggs under conditions of radioactive contamination of water by strontium-90–yttrium–90 and cerium–144. In B. P. Sorokin (Ed.), *Effect of Ionizing Radiation on the Organism* (USAEC translation from the Russian, **AEC-tr-7418**). United States Atomic Energy Commission, Oak Ridge, Tennessee. pp. 16–30.

MIGALOVSKIY, I. P. (1973b). Development of fish eggs and the early period of gametogenesis in the embryos and larvae of the Atlantic salmon under conditions of radioactive contamination of the water. In B. P. Sorokin (Ed.), *Effect of Ionizing Radiation on the Organism* (USAEC translation from the Russian, **AEC-tr-7418**). United States Atomic Energy Commission, Oak Ridge, Tennessee. pp. 36–52.

MIKAMI, Y., MUTO, T., and AGAWA, Y. (1965). Effects of rainwater radioactivity on organisms. *Mie Kaibo Gyosekishu*, **16**, 1–27.

MIKAMI, Y., WATANABE, H. and TAKANO, K. (1956). The influence of radioactive rainwater on the growth and differentiation of a tropical fish, *Zebra danio*. In *Research in the Effects and Influences of the Nuclear Bomb Explosions*, Vol. 2. Japanese Society for the Promotion of Science, Tokyo. pp. 1225–1229.

MITCHELL, N. T. (1967). Radioactivity in surface and coastal waters of the British Isles. *Tech. Rep. Fish. Radiobiol. Lab., MAFF Direct. Fish. Res., Lowestoft*, **FRL 1**, 1–45.

MITCHELL, N. T. (1968). Radioactivity in surface and coastal waters of the British Isles, 1967. *Tech. Rep. Fish. Radiobiol. Lab., MAFF Direct. Fish. Res., Lowestoft*, **FRL 2**, 1–41.

MITCHELL, N. T. (1969a). Monitoring of the aquatic environment of the United Kingdom and its application to hazard assessment. In *Environmental Contamination by Radioactive Materials*, Proc. Seminar, Vienna, March 1969. International Atomic Energy Agency, Vienna. pp. 449–462.

MITCHELL, N. T. (1969b). Radioactivity in surface and coastal waters of the British Isles, 1968. *Tech. Rep. Fish. Radiobiol. Lab., MAFF Direct. Fish. Res., Lowestoft*, **FRL 5**, 1–39.

MITCHELL, N. T. (1971a). Radioactivity in surface and coastal waters of the British Isles, 1969. *Tech. Rep. Fish. Radiobiol. Lab., MAFF Direct. Fish. Res., Lowestoft*, **FRL 7**, 1–33.

MITCHELL, N. T. (1971b). Radioactivity in surface and coastal waters of the British Isles, 1970. *Tech. Rep. Fish. Radiobiol. Lab., MAFF Direct. Fish. Res., Lowestoft*, **FRL 8**, 1–35.

MITCHELL, N. T. (1973). Radioactivity in surface and coastal waters of the British Isles, 1971. *Tech. Rep. Fish. Radiobiol. Lab., MAFF Direct. Fish. Res., Lowestoft*, **FRL 9**, 1–34.

MITCHELL, N. T. (1975). Radioactivity in surface and coastal waters of the British Isles, 1972–1973. *Tech. Rep. Fish. Radiobiol. Lab., MAFF Direct. Fish. Res., Lowestoft*, **FRL 10**, 1–40.

MITCHELL, N. T. (1977a). Radioactivity in surface and coastal waters of the British Isles, 1975. *Tech. Rep. Fish. Radiobiol. Lab., MAFF Direct. Fish. Res., Lowestoft*, **FRL 12**, 1–33.

MITCHELL, N. T. (1977b). Radioactivity in surface and coastal waters of the British Isles, 1976. Part I. The Irish Sea and its environs. *Tech. Rep. Fish. Radiobiol. Lab., MAFF Direct. Fish. Res., Lowestoft*, **FRL 13**, 1–16.

MITCHELL, N. T. (1978). Radioactivity in surface and coastal waters of the British Isles, 1976. Part 2. Areas other than the Irish Sea and its environs. *Tech. Rep. Fish. Radiobiol. Lab., MAFF Direct. Fish. Res., Lowestoft*, **FRL 14**, 1–21.

MIYAKE, Y., KATSURAGI, Y., and SUGIMURA, Y. (1970a). A study on plutonium fallout. *J. geophys. Res.*, **75**, 2329–2330.

MIYAKE, Y., MAYEDA, M., and SUGIMURA, Y. (1970b). Uranium content and the activity ratio $^{234}U/^{238}U$ in marine organisms and seawater in the western north Pacific. *J. oceanogr. Soc. Japan*, **26**, 123–129.

MIYAKE, Y., SUGIMURA, Y., and UCHIDA, T. (1966). Ratio $^{234}U/^{238}U$ and the uranium concentration in seawater in the western north Pacific. *J. geophys. Res.*, **71**, 3083–3087.

MIYAKE, Y., YASUJIMA, T., and SUGIMURA, Y. (1970c). Thorium concentration and the activity ratios $^{230}Th/^{232}Th$ and $^{228}Th/^{232}Th$ in seawater in the western north Pacific. *J. oceanogr. Soc. Japan*, **26**, 130–136.

MOORE, W. S. and SACKETT, W. M. (1964). Uranium and thorium series inequilibrium in seawater. *J. geophys. Res.*, **69**, 5401–5405.

MORGAN, F. (1964). The uptake of radioactivity by fish and shellfish. I. 134-caesium by whole animals. *J. mar. biol. Ass. U.K.*, **44**, 259–271.

MOSKALKOVA, K. I. (1970). On the development of *Neogobius melanostomus* Pall. in radioactive calcium solutions. *Trudy vses. nauchno-issled. Inst. morsk. rȳb. Khoz. Okeanogr.*, **69**, 49–57.

MOXHAM, R. M. (1964). Some aerial observations on the terrestrial component of the environmental γ-radiation. In J. A. S. Adams and M. Lowder (Eds), *The Natural Radiation Environment*. Chicago University Press, Chicago. pp. 737–746.

MÜNNICH, K. O. and ROETHER, W. (1967). Transfer of bomb ^{14}C and tritium from the atmosphere to the ocean. Internal mixing of the ocean on the basis of tritium and ^{14}C profiles. In *Radioactive Dating and Methods of Low-level Counting*, Proc. Symp. Monaco, March 1967. International Atomic Energy Agency, Vienna. pp. 93–103.

MURRAY, C. N. and FUKAI, R. (1978). Measurement of $^{239+240}Pu$ in the northwestern Mediterranean. *Estuar. coast. mar. Sci.*, **6**, 145–151.

MURRAY, C. N., KAUTSKY, H., HOPPENHEIT, M., and DOMIAN, M. (1978). Actinide activities in water entering the northern North Sea. *Nature, Lond.*, **276**, 225–230.

MURTHY, M. S. S., VENKATARAMAN, G., and DATTA, S. (1973). Revised parameters for use in Loevinger's *β* point-source dose-distribution function. *J. nucl. Med.*, **14**, 846–849.

NEIFAKH, A. A. and DONTSOVA, G. V. (1962). A radiation study of the role of the nucleus in cytochrome oxidase activity in fish embryos. *Biokhimiya*, **27**, 339–348.

NEIFAKH, A. A. and ROTT, N. N. (1958). The ways in which radiation damage appears in early development in fish. *Dokl. Akad. Nauk. SSSR*, **119**, 261–264.

NELSON, D. J. and BLAYLOCK, B. G. (1963). The preliminary investigation of salivary gland chromosomes of *Chironomus tentans* Fabr. from the Clinch River. In V. Schultz and A. W. Klement (Eds), *Radioecology*, Proc. 1st Nat. Symp. on Radioecology, Fort Collins, Colorado, September 1961. Reinhold Publishing Corp., New York. pp. 367–372.

NELSON, D. M. and LOVETT, M. B. (1978). Oxidation state of plutonium in the Irish Sea. *Nature, Lond.*, **276**, 599–601.

NELSON, S. G., ANDERSEN, A. C., MOMENI, M. H., and YEO, R. R. (1976). Attempted sterilization of sexually undifferentiated fry of *Tilapia zilli* by ^{60}Co gamma-ray irradiation. *Progve Fish Cult.*, **38**, 131–134.

NEUSTROEV, G. V. and PODYMAKHIN, V. N. (1966a). Respiration of *Salmo salar* L. spawn in radioactively contaminated waters. *Radiobiologiya*, **6**, 115–116.

NEUSTROEV, G. V. and PODYMAKHIN, V. N. (1966b). On the rate of development of salmon (*Salmo salar*, L) roe under conditions of radioactive pollution of the hydrosphere with ^{90}Sr–^{90}Y. *Radiobiologiya*, **6**, 321–323.

NEWCOMBE, H. B. (1973). 'Benefit' and 'harm' from exposure of vertebrate sperm to low doses of ionizing radiation. *Hlth Phys.*, **25**, 105–107.

NEWCOMBE, H. B. and MCGREGOR, J. F. (1967). Major congenital malformations from irradiations of sperm and eggs. *Mutat. Res.*, **4**, 663–673.

NEWCOMBE, H. B. and MCGREGOR, J. F. (1972). Increased embryo production following low doses of radiation to trout spermatoza. *Radiat. Res.*, **51**, 402–409.

NEWCOMBE, H. B. and MCGREGOR, J. F. (1975). Dose–response relationships for the production of body malformations in trout by exposure of sperm to low doses of radiation. *Radiat. Res.*, **61**, 519–525.

NILOV, V. I. (1974). Accumulation of strontium–90 and yttrium–90 and its effect on the eggs and larvae of *Ctenopharyngodon idella*. In G. P. Andrushaytis, G. G. Polikarpov, R. M. Aleksakhin, D. G. Fleishman, and M. P. Leinert (Eds), *Radioecology of Water Organisms*, Vol. 3, The Effects of Ionizing Radiation on Water Organisms (USAEC) translation from the Russian, **AEC-tr-7592**). pp. 36–40.

NISHIWAKI, Y., KIMURA, Y., and HONDA, Y. (1979). Experimental methods for radiobiological studies with developing fish eggs. In *Methodology for Assessing Impacts of Radioactivity on Aquatic Ecosystems*. (Tech. Rep. Int. atom. Energy Ag., **190**). International Atomic Energy Agency, Vienna. pp. 195–209.

NOSHKIN, V. E. and BOWEN, V. T. (1973). Concentrations and distributions of long-lived fallout radionuclides in open ocean sediments. In *Radioactive Contamination of the Marine Environment*, Proc. Symp. Seattle, July 1972. International Atomic Energy Agency, Vienna. pp. 671–686.

NOSHKIN, V. E., BOWEN, V. T., WONG, K. M., and BURKE, J. C. (1973). Plutonium in north Atlantic Ocean organisms; ecological relationships. In D. J. Nelson (Ed.), *Radionuclides in Ecosystems*, Vol. 2, Proc. Symp. Oak Ridge, Tennessee, May 1971. United States Atomic Energy Commission, Oak Ridge. pp. 681–688.

NOZAKI, Y., THOMSON, J., and TUREKIAN, K. K. (1976). The distribution of ^{210}Pb and ^{210}Po in the surface waters of the Pacific Ocean. *Earth Planet. Sci. Lett.*, **32**, 304–312.

NYDAL, R. and LÖVSETH, K. (1970). Prospective descrease in atmospheric radiocarbon. *J. geophys. Res.*, **75**, 2271–2278.

OECD-NEA (1980). *Review of the Continued Suitability of the Dumping Site for Radioactive Waste in the Northeast Atlantic*. Organization for Economic Co-operation and Development Nuclear Energy Agency, Paris.

OGANESYAN, S. A. (1973). Histogenesis and functioning of the hypophysis and thyroid gland in the larvae of Atlantic salmon exposed to ionizing radiation. In B. P. Sorokin (Ed.), *Effect of Ionizing Radiation on the Organism* (USAEC translation from the Russian, **AEC-tr-7418**). United States Atomic Energy Commission, Oak Ridge, Tennessee. pp. 53–65.

OLSON, P. R. (1973). Accumulation of calcium-45 in developing coho salmon eggs and fry reared in varying concentrations of stable calcium. In D. J. Nelson (Ed.), *Radionuclides in Ecosystems*, Vol. 2, Proc. Symp. Oak Ridge, Tennessee, May 1971. United States Energy Commission, Oak Ridge. 866–874.

OPHEL, I. L., HOPPENHEIT, M., ICHIKAWA, R., KLIMOV, A. G., KOBAYASHI, S., NISHIWAKI, Y., and SAIKI, M. (1976). Effects of ionizing radiation on acquatic organisms. In *Effects of Ionizing Radiation on Aquatic Organisms and Ecosystems* (Tech. Rep. Int. atom. Energy Ag., **172**). International Atomic Energy Agency, Vienna. pp. 57–86.

OSTERBERG, C. L. (1962). Fallout radionuclides in euphausiids. *Science, N. Y.*, **138**, 529–530.

OSTERBERG, C. L., SMALL, L. and HUBBARD, L. (1963). Radioactivity in large marine plankton as a function of surface area. *Nature, Lond.*, **197**, 883–884.

PALMER, H. E. and BEASLEY, T. M. (1967). ^{55}Fe in the marine environment and in people who consume ocean fish. In B. Aberg and F. P. Hungate (Eds), *Radioecological Concentration Processes*. Pergamon Press, London. pp. 259–262.

PARCHEVSKAYA, D. S. (1969). Estimation of absorbed doses of β-radiation in small-sized globular hydrobionts. *Radiobiologiya*, **9**, 281–285.

PARCHEVSKAYA, D. S. (1972). Estimation of absorbed doses of β-radiation in globular hydrobionts of small size. In G. G. Polikarpov (Ed.), *Marine Radioecology* (USAEC translation from the Russian, **AEC-tr-7299**). United States Atomic Energy Commission, Oak Ridge, Tennessee. pp. 165–173.

PARKER, F. L. (1965). United States and Soviet Union waste-disposal standards. *Nucl. Saf.*, **6**, 433–436.

PATIN, S. A., PECHKURENKOV, V. L., and SHEKHANOVA, I. A. (1971). Kinetics and mechanism of accumulation of plutonium by *Misgurnus fossilis* spawn. *Radiobiologiya*, **11**, 742–746.

PECHKURENKOV, V. L. (1970). Appearance of chromosome aberrations in larvae of loach (*Misgurnus fossilus*, L) developing in solutions of strontium–90 and yttrium–90 of various activities. *Sov. Genet.*, **6**, 1323–1332.

PECHKURENKOV, V. L. (1976). Chronic irradiation of embryonic loach (*Misgurnus fossilis*, L) using EDTA and caffeine. *Radiobiologiya*, **16**, 587–592.

PECHKURENKOV, V. L. (1977). Dependence of the aberrant anaphase yield in loach (*Misgurnus fossilis*, L) embryos on the dose rate of chronic irradiation. *Radiobiologiya*, **17**, 907–909.

PECHKURENKOV, V. L., SHEKHANOVA, I. A., DOKHOLYAN, V. K., and KOSTROV, B. P. (1974). The effect of combined radiation with α-, β- and γ-rays on the embryonal development of loach. *Trudy vses. nauchno-issled. Inst. morsk rȳb. Khoz. Okeanogr.*, **100**, 80–83.

PECHKURENKOV, V. L., SHEKHANOVA, I. A., and TELYSHEVA, I. G. (1972a). The effect of chronic exposure to small doses of irradiation on the embryonic development in fish and the validity of various assessment methods. *Vop. Ikhtiol.*, **12**, 84–93.

PECHKURENKOV, V. L., SHEKHANOVA, I. A., and TELYSHEVA, I. G. (1972b). Results of research on the effect of chronic exposure to various concentrations of radionuclides on the embryogenesis of fish. *Trudy vses. nauchno-issled. Inst. morsk. rȳb. Khoz. Okeanogr.*, **85**, 9–26.

PENTREATH, R. J. (1973a). The roles of food and water in the accumulation of radionuclides by marine teleost and elasmobranch fish. In *Radioactive Contamination of the Marine Environment*, Proc. Symp. Seattle, July 1972. International Atomic Energy Agency, Vienna. pp. 421–436.

PENTREATH, R. J. (1973b). The accumulation and retention of ^{65}Zn and ^{54}Mn by the plaice, *Pleuronectes platessa*, L. *J. exp. mar. Biol. Ecol.*, **12**, 1–18.

PENTREATH, R. J. (1976). Some further studies on the accumulation and retention of ^{65}Zn and ^{54}Mn by the plaice *Pleuronectes platessa* L. *J. exp. mar. Biol. Ecol.*, **21**, 179–189

PENTREATH, R. J. (1977a). Radionuclides in marine fish. *Oceanogr. mar. Biol. An. Rev.*, **16**, 365–460.

PENTREATH, R. J. (1977b). The accumulation of ^{110m}Ag by the plaice, *Pleuronectes platessa*, L., and the thornback ray, *Raja clavata*, L. *J. exp. mar. Biol. Ecol.*, **29**, 315–325.

PENTREATH, R. J. (1978a). ^{237}Pu experiments with the plaice, *Pleuronectes platessa. Mar. Biol.*, **48**, 327–335.

PENTREATH, R. J. (1978b). ^{237}Pu experiments with the thornback ray, *Raja clavata. Mar. Biol.*, **48**, 337–342.

PENTREATH, R. J. (1980) Radioactive materials—the aquatic environment. In Sir Kenneth Blaxter (Ed.), *Food Chains in Human Nutrition*. Applied Science, Barking. pp. 385–397.

PENTREATH, R. J. and JEFFERIES, D. F. (1971). The uptake of radionuclides by I-group plaice (*Pleuronectes platessa*) off the Cumberland coast, Irish Sea. *J. mar. biol. Ass. U.K.*, **51**, 963–976.

PENTREATH, R. J., JEFFERIES, D. F., LOVETT, M. B., and NELSON, D. M. (1980a). The behaviour of transuranic and other long-lived radionuclides in the Irish Sea and its relevance to the deep sea disposal of radioactive wastes. In *Marine Radioecology*, Proc. 3rd. NEA Seminar, Tokyo, October 1979. Organization for Economic Cooperation and Development, Paris. pp. 203–221.

PENTREATH, R. J. and LOVETT, M. B. (1978). Transuranic nuclides in plaice (*Pleuronectes platessa*) from the north-eastern Irish Sea. *Mar. Biol.*, **48**, 19–26.

PENTREATH, R. J., WOODHEAD, D. S., HARVEY, B. R., and IBBETT, R. D. (1980b). A preliminary assessment of some naturally-occurring radionuclides in marine organisms (including deep sea fish) and the absorbed dose resulting from them. In *Marine Radioecology*, Proc. 3rd NEA Seminar, Tokyo, October 1979. Organization for Economic Cooperation and Development, Paris. pp. 291–302.

PENTREATH, R. J., WOODHEAD, D. S., and JEFFERIES, D. F. (1973). Radioecology of the plaice (*Pleuronectes platessa*, L) in the northeast Irish Sea. In D. J. Nelson (Ed.), *Radionuclides in Ecosystems*, Vol. 2. Proc. Symp. Oak Ridge, Tennessee, May 1971. United States Atomic Energy Commission, Oak Ridge. pp. 731–737.

PETERSON, K. R. (1970). An empirical model for estimating world-wide deposition from atmospheric nuclear detonations. *Hlth Phys.*, **18**, 357–378.

PILLAI, K. C., SMITH, R. C., and FOLSOM, T. R. (1964). Plutonium in the marine environment. *Nature, Lond.*, **203**, 568–571.

PITKYANEN, G. B. (1971). Results of the incubation of pike eggs (*Esox lucius*, L) in solutions of ^{90}Sr and ^{137}Cs mixture. *Trudy Instituta ekologii rastenij i zhivotnykh (Sverdlovsk)*, **78**, 149–153.

PODYMAKHIN, V. N. (1973a). Some data on the dependence of dose–effect for eggs of the Atlantic salmon. In B. P. Sorokin (Ed.), *Effect of Ionizing Radiation on the Organism* (USAEC translation from the Russian, **AEC-tr-7418**). United States Atomic Energy Commission, Oak Ridge, Tennessee. pp. 31–35.

PODYMAKHIN, V. N. (1973b). Natural radiation loads on the eggs of marine and freshwater organisms. In B. P. Sorokin (Ed.), *Effect of Ionizing Radiation on the Organism* (USAEC translation from the Russian, **AEC-tr-7418**). United States Atomic Energy Commission, Oak Ridge, Tennessee. pp. 142–156.

POLIKARPOV, G. G. (1966). *Radioecology of Aquatic Organisms*, North Holland Publishing Co., Amsterdam.

POLIKARPOV, G. G. (1973). Radioecology of marine plants and animals. In V. M. Klechkovskii, G. G. Polikarpov, and R. M. Aleksakhin (Eds), *Radioecology*. Wiley, New York. pp. 311–322.

POLIKARPOV, G. G. (1979). Experimental methods for radiobiological investigations with developing fish eggs. In *Methodology for Assessing Impacts of Radioactivity on Aquatic Ecosystems* (Tech. Rep. Int. atom. Energy Ag., **190**). International Atomic Energy Agency, Vienna. pp. 173–194.

POLIKARPOV, G. G. and GAMEZO, N. V. (1966). Radiosensitivity of the eggs of the ocean perch and gilthead (effect of ^{90}Sr–^{90}Y). *Gidrobiol. Zh.*, **2**, 66–70.

POLIKARPOV, G. G. and IVANOV, V. N. (1961). The effect of ^{90}Sr–^{90}Y on the developing anchovy eggs. *Vop. Ikhtiol.*, **1**, 583–589.

POLIKARPOV, G. G. and IVANOV, V. N. (1962a). Accumulation of radioactive isotopes of Sr and Y in eggs of marine fishes. *Radiobiologiya*, **2**, 207–210.

POLIKARPOV, G. G. and IVANOV, V. N. (1962b). The harmful effect of ^{90}Sr–^{90}Y on the early development period of *Mullus barbatus*, the hybrid *Crenilabrus tinca* x *C. quinquemaculatus*, *Trachurus trachurus* and *Engraulis encrasicholus*. *Dokl. Akad. Nauk SSSR*, **144**, 219–222.

PRESTON, A. (1966). Site evaluations and the discharge of aqueous radioactive wastes from civil nuclear power stations in England and Wales. In *Disposal of Radioactive Wastes into Seas, Oceans and Surface Waters*, Proc. Symp. Vienna, May 1966. International Atomic Energy Agency, Vienna. pp. 725–737.

PRESTON, A. (1967). The concentration of ^{65}Zn in the flesh of oysters related to the discharge of cooling pond effluent from the CEGB nuclear power station at Bradwell-on-Sea, Essex. In B. Aberg and F. P. Hungate (Eds), *Radioecological Concentration Processes*, Proc. Symp. Stockholm, April 1966. Pergamon Press, London. pp. 995–1004.

PRESTON, A. (1968). The control of radioactive pollution in a North Sea oyster fishery. *Helgoländer wiss. Meeresunters.*, **17**, 269–279.

PRESTON, A. (1969). Aquatic monitoring programmes. In *Environmental Contamination by Radioactive Materials*, Proc. Seminar, Vienna, March 1969. International Atomic Energy Agency, Vienna. pp. 309–324.

PRESTON, A. (1970). Concentration of iron-55 in commercial fish species from the north Atlantic. *Mar. Biol.*, **6**, 345–349.

PRESTON, A. (1971). The United Kingdom approach to the application of ICRP standards to the controlled disposal of radioactive waste resulting from nuclear power programmes. In *Environmental Aspects of Nuclear Power Stations*, Proc. Symp. New York, August 1970. International Atomic Energy Agency, Vienna. pp. 147–157.

PRESTON, A. (1972). Artificial radioactivity in freshwater and estuarine systems. *Proc. R. Soc. Lond. (B)*, **180**, 421–436.

PRESTON, A., DUTTON, J. W. R., and HARVEY, B. R. (1968). Detection, estimation and radiological significance of silver–110m in oysters in the Irish Sea and the Blackwater estuary. *Nature, Lond.*, **218**, 689–690.

PRESTON, A. and JEFFERIES, D. F. (1967). The assessment of the principal public radiation exposure from, and the resulting control of, discharges of aqueous radioactive waste from the United Kingdom Atomic Energy Authority factory at Windscale, Cumberland. *Hlth Phys.*, **13**, 477–485.

PRESTON, A. and JEFFERIES, D. F. (1969a). The ICRP critical group concept in relation to the Windscale sea discharge. *Hlth Phys.*, **16**, 33–46.

PRESTON, A. and JEFFERIES, D. F. (1969b). Aquatic aspects in chronic and acute contamination situations. In *Environmental Contamination by Radioactive Materials*, Proc. Seminar, Vienna, March 1969. International Atomic Energy Agency, Vienna. pp. 183–211.

PRESTON, A., JEFFERIES, D. F., and MITCHELL, N. T. (1971). Experience gained from the controlled introduction of liquid radioactive waste to coastal waters. In *Nuclear Techniques in Environmental Pollution*, Proc. Symp. Salzburg, October 1970. International Atomic Energy Agency, Vienna. pp. 629–644.

PRESTON, A., JEFFERIES, D. F., and MITCHELL, N. T. (1978). The impact of caesium–134 and caesium–137 on the marine environment. In *Radioactive Effluents from Nuclear Fuel Reprocessing Plants*, Proc. Seminar, Karlsruhe, November 1977. Commission of the European Communities, Luxembourg. pp. 401–419.

PRESTON, A. and MITCHELL, N. T. (1973). Evaluation of public radiation exposure from the controlled marine disposal of radioactive waste (with special reference to the United Kingdom). In *Radioactive Contamination of the Marine Environment*, Proc. Symp. Seattle, July 1972. International Atomic Energy Agency, Vienna. pp. 575–593.

PRESTON, A., MITCHELL, N. T., and JEFFERIES, D. F. (1974). Experience gained in applying the ICRP critical group concept to the assessment of public radiation exposure in control of liquid radioactive waste disposal. In *Population Dose Evaluation and Standards for Man and his Environment*, Proc. Seminar Portoroz, May 1974. International Atomic Energy Agency, Vienna. pp. 131–146.

PROKOF'YEVA-BEL'GOVSKAYA, A. A.(1961). Radiation damage to chromosomes at early developmental stages of salmon. *Tsitologiya*, **3**, 437–445.

PURDOM, C. E. (1966). Radiation and mutation in fish. In *Disposal of Radioactive Wastes into Seas, Oceans and Surface Waters*, Proc. Symp. Vienna, May 1966. International Atomic Energy Agency, Vienna. pp. 861–867.

PURDOM, C. E. and WOODHEAD, D. S. (1973). Radiation damage in fish. In J. H. Schröder (Ed.), *Genetics and Mutagenesis of Fish*. Springer-Verlag, Berlin. pp. 67–73.

RAMA, KOIDE, M. and GOLDBERG, E. D. (1961). Lead-210 in natural waters. *Science, N.Y.*, **134**, 98.

RAMSTER, J. W. (1973). The residual circulation of the northern Irish Sea with particular reference to Liverpool Bay. *Fish. Res. Tech. Rep., MAFF Direct. Fish. Res., Lowestoft*, **5**, 1–21.

RAVERA, O. (1967). The effects of X-rays on the demographic characteristics of *Physa acuta* (Gastropoda: Basommatophora). *Malacologia*, **5**, 95–109.

RUGH, R. (1960). General biology: gametes, the developing embryo, and cellular differentiation. In M. Errera and A. Forssberg (Eds), *Mechanisms in Radiobiology*, Vol. 2. Academic Press, New York. pp. 1–94.

RUGH, R. and CLUGSTON, H. (1955). Effects of various levels of X-irradiation on the gametes and early embryos of *Fundulus heteroclitus*. *Biol. Bull. mar. biol. Lab., Woods Hole*, **108**, 318–325.

RUGH, R. and GRUPP, E. (1959). Ionizing radiations and congential anomalies in vertebrate embryos. *Acta Embryol. Morph. exp.*, **2**, 257–268.

RUSSELL, L. B. (1954). The effect of radiation on mammalian pre-natal development. In A. Hollaender (Ed.), *Radiation Biology*, Vol. 2, Part 2. McGraw-Hill, New York. pp. 861–918.

SALMON, L. (1961). Analysis of gamma-ray scintillation spectra by the method of least squares. *Rep. U.K. atom. Energy Auth.*, **AERE-R3640**, United Kingdom Atomic Energy Authority, Harwell. 1–24.

SALMON, L. (1963). Computer analysis of gamma-ray spectra from mixtures of known nuclides by the method of least squares. *Rep. U.K. atom. Energy Auth.*, **AERE-M1140**, United Kingdom Atomic Energy Authority, Harwell. 1–27.

SARMIENTO, J. L., FEELY, H. W., MOORE, W. S., BAINBRIDGE, A. E., and BROECKER, W. S. (1976). The relationship between vertical eddy diffusion and buoyancy gradient in the deep sea. *Earth Planet. Sci. Lett.*, **32**, 357–370.

SCHREIBER, B. (1967). Radionuclides in marine plankton and coastal sediments. In B. Aberg and F. P. Hungate (Eds), *Radioecological Concentration Processes*, Proc. Symp. Stockholm, April 1966. Pergamon Press, London. pp. 753–770.

SCHRÖDER, J. H. (1969a). X-ray mutations in the poecilid fish, *Lebistes reticulatus*, Peters. *Mutat. Res.*, **7**, 75–90.

SCHRÖDER, J. H. (1969). Die Variabilität quantitativer Merkmale bei *Lebistes reticulatus*, Peters, nach ancestraler Röntgenbestrahlung. *Zool. Beitr. (N. F.)*, **15**, 237–265.

SCHRÖDER, J. H. (1969c). Inheritance of radiation-induced spinal curvature in the guppy, *Lebistes reticulatus*. *Can. J. Genet. Cytol.*, **11**, 937–947.

SCHRÖDER, J. H. (1969d). Quantitative changes in breeding groups of *Lebistes* after irradiation. *Can. J. Genet. Cytol.*, **11**, 955–960.

SCHRÖDER, J. H. (1969e). Radiation-induced spermatogonial exchange between the X and Y chromosomes in the guppy. *Can. J. Genet. Cytol.*, **11**, 948–954.

SCHRÖDER, J. H. (1973). Teleosts as a tool in mutation research. In J. H. Schröder (Ed.), *Genetics and Mutagenesis in Fish*. Springer-Verlag, Berlin. pp. 91–99.

SCHRÖDER, J. H. (1979). Methods for screening radiation-induced mutations in fish. In *Methodology for Assessing Impacts of Radioactivity on Aquatic Ecosystems* (Tech. Rep. Int. atom. Energy Ag., **190**). International Atomic Energy Agency, Vienna. pp. 381–402.

SCHRÖDER, J. H. (1980). Radiation-induced mutations in fish. In N. Egami (Ed.), *Radiation Effects in Aquatic Organisms*, Proc. Symp. Kanagawa, Japan, May 1979. Japan Scientific Societies Press, Tokyo/University Park Press, Baltimore. pp. 217–222.

SCHRÖDER, J. H. and HOLZBERG, S. (1972). Population genetics of *Lebistes (Poecilia) reticulatus*, Peters (Poeciliidae, Pisces). I. Effects of irradiation-induced mutations on the segretation ratio in post-irradiation F_1. *Genetics, Princeton*, **70**, 621–630.

SELIGMAN, H. (1955). The discharge of radioactive waste products into the Irish Sea. Part I. First experiments for the study of movement and dilution of released dye in the sea. In *Proceedings of the International Conference on the Peaceful Uses of Atomic Energy*, Vol. 9. Geneva. pp. 701–711.

SEYMOUR, A. H. and LEWIS, G. B. (1964). Radionuclides of Columbia River origin in marine organisms, sediments and water collected from the coastal and offshore waters of Washington and Oregon. *Rep. Lab. radiat. biol., Univ. Wash.*, **UWFL-86** (TID-4500. 39 ed.), 1–78.

SHANNON, L. V. (1973). Marine α-radioactivity off southern Africa. 3. Polonium-210 and lead-210, *Investl. Rep. Div. Sea Fish Un. S. Afr.*, **100**, 34 pp.

SHANNON, L. V. and CHERRY, R. D. (1967). Polonium-210 in marine plankton. *Nature Lond.*, **216**, 352–353.

SHANNON, L. V., CHERRY, R. D., and ORREN, M. J. (1970). Polonium-210 and lead-210 in the marine environment. *Geochim. cosmochim. Acta*, **34**, 701–711.

SHEKHANOVA, I. A. (1970). Accumulation and distribution of water soluble radioisotopes in fish eggs during embryogenesis. *Trudy vses. nauchno-issled. morsk. rӯb. Khoz. Okeanogr.*, **69**, 19–27.

SHEKHANOVA, I. A., BELMAKOV, V. S., LAPIN, V. I., LYASHENKO, A. G. and MILORADOV, G. K. (1970). Effect of radioisotopes dissolved in the water on the fish roe during embryonic development. *Trudy vses. nauchno-issled. morsk. rӯb. Khoz. Okeanogr.*, **69**, 34–48.

SHEKHANOVA, I. A. and PECHKURENKOV, V. L. (1968). Accumulation of strontium-90 and yttrium-90 dissolved in the water and its effect on the embryonic development of the loach. *Vop. Ikhtiol.*, **8**, 689–701.

SHEKHANOVA, I. A. and PECHKURENKOV, V. L. (1969). The effect of concentration of strontium-90 and yttrium-90 dissolved in the water on breeding loach and their progeny. *Vop. Ikhtiol.*, **9**, 338–349.

SHEKHANOVA, I. A., VORONINA, E. A., KLYASHTORIN, L. B., PECHKURENKOV, V. L., and YARZHOMBEK, A. A. (1974). The effect of ionizing radiation on the biological condition of fish. In G. P. Andrushaytis, G. G. Polikarpov, R. M. Aleksakhin, D. G. Fleishman, and M. P. Leinert (Eds), *Radioecology of Water Organisms*, Vol. 3, The Effects of Ionizing Radiation on Water Organisms (USAEC translation from the Russian, **AEC-tr-7592**). United States Atomic Energy Commission, Oak Ridge, Tennessee. p. 2–12.

SHEPHERD, J. G. (1975). The application of the critical group concept. *Tech. Note, MAFF Direct. Fish. Res., Lowestoft*, **RL4/75**, 1–17.

SIEVERT, R. M. (1959). The work of the International Commission on Radiological Protection. In W. G. Morley and K. Z. Morgan (Eds), *Progress in Nuclear Energy*, Series XII, Health Physics, 1, Proc. 2nd Int. Conf. on the Peaceful Uses of Atomic Energy Geneva, 1958. Pergamon Press, London. pp. 3–9.

SOLBERG, A. N. (1938a). The susceptibility of the germ cells of *Oryzias latipes* to X-irradiation and recovery after treatment. *J. exp. Zool.*, **78**, 417–439.

SOLBERG, A. N. (1938b). The susceptibility of *Fundulus heteroclitus* embryos to X-radiation. *J. exp. Zool.*, **78**, 441–469.

SOMAYAJULU, B. L. K. and GOLDBERG, E. D. (1966). Thorium and uranium isotopes in seawater and sediments. *Earth Planet. Sci. Lett.*, **1**, 102–106.

SPIERS, F. W. (1956). Radiation units and theory of ionization dosimetry. In G. J. Hine and G. L. Brownell (Eds), *Radiation Dosimetry*. Academic Press, New York. pp. 1–47.

SQUIRE, R. D. (1970). The effects of acute gamma irradiation on the brine shrimp *Artemia*: II. Female reproductive performance. *Biol. Bull. mar. biol. Lab., Woods Hole*, **139**, 375–385.

SQUIRE, R. D. and GROSCH, D. S. (1970). The effects of acute gamma irradiation on the brine shrimp *Artemia*: I. Life spans and male reproductive performance. *Biol. Bull. mar. biol. Lab., Woods Hole*, **139**, 363–374.

SRIVASTAVA, P. N. and RATHI, S. K. (1967). Radiation-induced castration in the female teleostean fish, *Heteropneustes fossilis*, Bloch. *Experientia*, **23**, 229–230.

STRAND, J. A., FUJIHARA, M. P., BURDETT, R. D., and POSTON, T. M. (1977). Suppression of the primary immune response in rainbow trout, *Salmo gairdnerii*, sublethally exposed to tritiated water during embryogenesis. *J. Fish. Res. Bd. Can.*, **34**, 1293–1304.

STRAND, J. A. FUJIHARA, M. P., TEMPLETON, W. L., and TANGEN, E. G. (1973a). Suppression of *Chondrococcus columnaris* immune response in rainbow trout sublethally exposed to tritiated water during embryogenesis. In *Radioactive Contamination of the Marine Environment*, Proc. Symp. Seattle, July 1972. International Atomic Energy Agency, Vienna. pp. 543–549.

STRAND, J. A., TEMPLETON, W. L., and TANGEN, E. G. (1973b). Accumulation and retention of tritium (tritiated water) in embryonic and larval fish, and radiation effect. In D. J. Nelson (Ed.), *Radionuclides in Ecosystems*, Vol. 1, Proc. Symp. Oak Ridge, Tennessee, May 1971. United States Atomic Energy Commission, Oak Ridge. pp. 445–451.

SZABO, B. J. (1967). Radium content in plankton and seawater in the Bahamas. *Geochim. cosmochim. Acta*, **31**, 1321–1331.

TARKHANOV, I. (1896). Experiments upon the action of Roentgens' X-rays on organisms. *Izv. S-Peterb. biol. Lab.*, **1**, 47–52.

TELYSHEVA, I. G. and SHULYAKOVSKII, Y. A. (1974). Radioresistance of the eggs of some industrial fish. In G. P. Andrushaytis, G. G. Polikarpov, R. M. Aleksakhin, D. G. Fleishman, and M. P. Leimert (Eds), *Radioecology of Water Organisms*, Vol. 3, The Effects of Ionizing Radiation on Water Organisms (USAEC translation from the Russian, **AEC-tr-7592**). United States Atomic Energy Commission, Oak Ridge, Tennessee. pp. 50–56.

TEMPLETON, W. L. (1966). Resistance of fish eggs to acute and chronic irradiation. In *Disposal of Radioactive Wastes into Seas, Oceans and Surface Waters*, Proc. Symp. Vienna, May 1966. International Atomic Energy Agency, Vienna. pp. 847–859.

TEMPLETON, W. L., BERNHARD, M., BLAYLOCK, B. G., FISHER, C., HOLDEN, M. J., KLIMOV, A. G., METALLI, P., MUKHERJEE, R., RAVERA, O., SZTANYIK, L., and VAN HOEK, F. (1976). Effects of ionizing radiation on aquatic populations and ecosystems. In *Effects of Ionizing Radiation on Aquatic Organisms and Ecosystems* (Tech. Rep. Int. atom. Energy Ag., **172**). International Atomic Energy Agency, Vienna. pp. 89–102.

TEMPLETON, W. L., NAKATANI, R. E., and HELD, E. E. (1971). Radiation Effects. In *Radioactivity in the Marine Environment*. National Academy of Sciences, Washington. pp. 223–239.

TEMPLETON, W. L. and PRESTON, A. (1966). Transport and distribution of radioactive effluents in coastal and estuarine waters of the United Kingdom. In *Disposal of Radioactive Wastes into Seas, Oceans and Surface Waters*, Proc. Symp. Vienna May 1966. International Atomic Energy Agency, Vienna. pp. 267–289.

THOMPSON, S. E., BURTON, C. A., QUINN, D. J., and NG, Y. C. (1972). Concentration factors of chemical elements in edible aquatic organisms. **UCRL-50564** (*Rev. 1*), University of California, Livermore. pp. 1–77.

THOMSON, J. and TUREKIAN, K. K. (1976). ^{210}Po and ^{210}Pb distributions in ocean water profiles from the eastern South Pacific. *Earth Planet. Sci. Lett.*, **32**, 297–303.

TILL, J. E. (1976). *The Toxicity of Uranium and Plutonium to the Developing Embryos of Fish*, Ph.D. Thesis Georgia Institute of Technology.

TILL, J. E. (1978). The effect of chronic exposure to ^{238}Pu(IV) citrate on the embryonic development of carp and fathead minnow eggs. *Hlth Phys.*, **34**, 333–343.

TILL, J. E. and FRANK, M. L. (1977). Bioaccumulation, distribution and dose from ^{241}Am, ^{244}Cm and ^{238}Pu in developing fish embryos. In *Proc. 4th. Int. Congr. Int. Radiat. Prot. Ass.*, Vol. 2. Paris, April 1977. pp. 645–648.

TIMOFEEVA, N. A. and AL'SHITS, L. K. (1970). Influence of chronic irradiation on the development of pike eggs (*Esox lucius*). *Trudy Instituta ekologii rastenij i zhivotnykh (Sverdlovsk)*, **74**, 8–11.

TIMOFEEVA, N. A., KULIKOV, N. V., and AL'SHITS, L. K. (1971). Action of ^{90}Sr–^{90}Y upon embryonic development of some representatives of freshwater fish and molluscs. *Trudy Instituta ekologii rastenij i zhivotnykh (Sverdlovsk)*, **78**, 145–148.

TRABALKA, J. R. and ALLEN, C. P. (1977). Aspects of fitness of a mosquito-fish, *Gambusia affinis*, population exposed to chronic low-level environmental radiation. *Radiat. Res.*, **70**, 198–211.

TSYTSUGINA, V. G. (1971). On the cytogenetic action of incorporated radionuclides at early stages of development of Black Sea fishes. *Trudy Instituta ekologii rastenij i zhivotnykh (Sverdlovsk)*, **78**, 154–159.

TSYTSUGINA, V. G. (1972). Effect of incorporated radionuclides in chromosome apparatus of ocean fish. In G. G. Polikarpov (Ed.), *Marine Radioecology* (USAEC translation from the Russian. **AEC-tr-7299**). United States Atomic Energy Commission, Oak Ridge, Tennessee. pp. 157–165.

TSYTSUGINA, V. G. (1973). Effect of incorporated radionuclides on the chromosomal apparatus of marine fish. In B. P. Sorokin (Ed.), *Effect of Ionizing Radiation on the Organism* (USAEC translation from the Russian, **AEC-tr-7418**). United States Atomic Energy Commission, Oak Ridge, Tennessee. pp. 157–165.

TSYTSUGINA, V. G. (1975). Karyology of marine fish and the action of radionuclides on their chromosome apparatus. In G. G. Polikarpov (Ed.), *Artificial and Natural Radionuclides in Marine Life*. Israel Program for Scientific Translations, Jerusalem. pp. 16–39.

UENO, A. M. (1974). Incorporation of tritium from tritiated water into nucleic acids of *Oryzias latipes* eggs. *Radiat. Res.*, **59**, 629–637.

UMANSKII, S. R. (1968). Influence of irradiation on the composition of messenger RNA during early loach embryogenesis. *Radiobiologiya*, **8**, 347–353.

UNSCEAR (1962). *A Report of the United Nations Scientific Committee on the Effects of Atomic Radiation* (General Assembly Official Records: 17th Session; Suppl. No. 16(A/5216), United Nations, New York.

UNSCEAR (1966). *A Report of the United Nations Scientific Committee on the Effects of Atomic Radiation.* (General Assembly Official Records: 21st Session; Suppl No. 14(A/6314), United Nations, New York.

UNSCEAR (1972). *A Report of the United Nations Scientific Committee on the Effect of Atomic Radiation.* 1. Levels, United Nations, New York.

UNSCEAR (1977). *A Report of the United Nations Scientific Committee on the Effects of Atomic Radiation. Sources and Effects of Ionizing Radiation*, United Nations, New York.

VOLCHOK, H. L., BOWEN, V. T., FOLSOM, T. R., BROECKER, W. S., SCHUERT, E. A. and BIEN, G. S. (1971). Oceanic distributions of radionuclides from nuclear explosions. In *Radioactivity in the Marine Environment*. National Academy of Sciences, Washington. pp. 42–89.

WADLEY, G. W. and WELANDER, A. D. (1971). X-rays and temperature: combined effects on mortality and growth of salmon embryos. *Trans. Am. Fish. Soc.*, **100**, 267–275.

WALDEN, S. J. (1973). Effects of tritiated water on the embryonic development of the three-spine stickleback, *Gasterosteus aculeatus*, L. In D. J. Nelson (Ed.), *Radionuclides in Ecosystems*. Vol. 2. Proc. Symp. Oak Ridge, Tennessee, May 1971. United States Atomic Energy Commission, Oak Ridge. pp. 1087–1090.

WARD, E., BEACH, S. A., and DYSON, E. D. (1971). The effect of acute X-irradiation on the development of the plaice, *Pleuronectes platessa*, L. *J. Fish Biol.*, **3**, 251–260.

WASSON, M. M. and MITCHELL, N. T. (1973). Impact of ten years liquid waste management practices on environmental safety at Bradwell power station. *Rep. Central Electricity Generating Board* (**HPM 371**), London. pp. 1–7.

WELANDER, A. D. (1954). Some effects of X-irradiation of different embryonic stages of the trout (*Salmo gairdnerii*). *Growth*, **18**, 227–255.

WELANDER, A. D., DONALDSON, L. R., FOSTER, R. F., BONHAM, K., and SEYMOUR, A. H. (1948). Effects of roentgen rays on embryos and larvae of chinook salmon. *Growth*, **12**, 203–242.

WELANDER, A. D., WADLEY, G. W., and DYSART, D. K. (1971). Growth and fecundity of rainbow trout (*Salmo gairdnerii*) exposed to single sublethal doses of X-rays during the eyed embryo stage. *J. Fish Res. Bd Can.*, **28**, 1181–1184.

WHITE, J. C. (1964). Fractionated doses of X-radiation: a preliminary study of effects on teleost embryos. *Int. J. Radiat. Biol.*, **8**, 85–91.

WHITE, J. C. and ANGELOVIC, J. W. (1966). Tolerances of several marine species to Co-60 irradiation. *Chesapeake Sci.*, **7**, 36–39.

WONG, K. M., BURKE, J. C., and BOWEN, V. T. (1970). Plutonium concentration in organisms of the Atlantic Ocean. *Rep. Woods Hole oceanogr. Instn* (NYO–2174–117), 11 pp.

WOODHEAD, D. S. (1970). The assessment of the radiation dose to developing fish embryos due to the accumulation of radioactivity by the egg. *Radiat. Res.*, **43**, 582–597.

WOODHEAD, D. S. (1973a). Levels of radioactivity in the marine environment and the dose commitment to marine organisms. In *Radioactive Contamination of the Marine Environment*, Proc. Symp. Seattle, July 1972. International Atomic Energy Agency, Vienna. pp. 499–525.

WOODHEAD, D. S. (1973b). The radiation dose received by plaice (*Pleuronectes platessa*) from the waste discharge into the northeast Irish Sea from the fuel reprocessing plant at Windscale. *Hlth Phys.*, **25**, 115–121.

WOODHEAD, D. S. (1974). The estimation of radiation dose rates to fish in contaminated environments, and the assessment of the possible consequences. In *Population Dose Evaluation and Standards for Man and his Environment*, Proc. Seminar, Portoroz, May 1974. International Atomic Energy Agency, Vienna. pp. 555–575.

WOODHEAD, D. S. (1976). Influence of acute irradiation on induction of chromosome aberrations in cultured cells of the fish, *Ameca splendens*. In *Biological and Environmental Effects of Low-Level Radiation*, Vol. 1, Proc. Symp. Chicago, November 1975. International Atomic Energy Agency, Vienna. pp. 67–76.

WOODHEAD, D. S. (1977). The effects of chronic irradiation on the breeding performance of the guppy, *Poecilia reticulata* (Osteichthyes : Teleostei). *Int. J. Radiat. Biol.*, **32**, 1–22.

WOODHEAD, D. S. (1979). Methods of dosimetry for aquatic organisms. In *Methodology for Assessing Impacts of Radioactivity on Aquatic Ecosystems* (Tech. Rep. Int. atom. Energy Ag., **190**). International Atomic Energy Agency, Vienna. pp. 43–96.

WOODHEAD, D. S. and PENTREATH, R. J. (1980). A provisional assessment of radiation regimes in deep ocean environments. In *Proceedings of the 2nd International Ocean Dumping Symposium, Woods Hole, April 1980*. (In press).

WORTHINGTON, L. V. (1976). On the north Atlantic circulation. *John Hopkins oceanogr. Stud.*, **6**, 1–110.

YAMAMOTO, M. and EGAMI, N. (1974). Effects of X-irradiation on the fine structure of hepatocytes and pituitary gonado-trophic cells of the laying medaka, *Oryzias latipes*. *J. Fac. Sci. Tokyo Univ.*, Sect. IV., **13**, 211–218.

YOSHIMURA, N., ETOH, H., EGAMI, N., ASAMI, K. and YAMADA, T. (1969). Notes on the effects of β-rays from ^{90}Sr–^{90}Y on spermatogenesis in the teleost, *Oryzias latipes*. *Annotnes zool. jap.*, **42**, 75–79.

ZESENKO, A. Y. and IVANOV, V. N. (1966). Accumulation of phosphorus-32 by developing eggs of marine fishes. *Vop. Ikhtiol.*, **6**, 575–578.

Marine Ecology Vol. 5, Part 3
Edited by Otto Kinne

3. POLLUTION DUE TO HEAVY METALS AND THEIR COMPOUNDS

G. W. BRYAN

(1) Introduction

Stimulated by factors such as the mercury problem, increased public concern about marine pollution and the development of rapid analytical techniques, there has over the past 10 yr been a spectacular increase in research on metallic contamination in the sea. In parallel with this there has been an upsurge in work on the biogeochemistry of metals and the 2 topics are inextricably linked.

Of the metals we shall be considering, about half are biologically essential and this group comprises Fe, Cu, Zn, Co, Mn, Mo, Se, Cr, Ni, V, As and Sn (DA SILVA, 1978). As metalloproteins or metal–protein complexes they occur in a very large number of enzymes; for example, more than 90 zinc-containing enzymes and proteins have been discovered, and the addition of zinc increases the activity of many other enzymes (VALLEE, 1978). Metals are also constituents of oxygen carriers, the best known being haemoglobin (Fe) and haemocyanin (Cu), and may also possess a structural role. For example, iron has been shown to be responsible for the hardness of the radular teeth of molluscs including the chiton *Cryptochiton stelleri* (TOWE and LOWENSTAM, 1967) and the limpet *Patella vulgata* (RUNHAM and co-authors, 1969); and an analagous role has been suggested for zinc in the jaws of nereid polychaetes and copper in the jaws of glycerid polychaetes (BRYAN and GIBBS, 1980; GIBBS and BRYAN, 1980; Table 3-9). In addition, an antipredatory role has been proposed for the high vanadium levels in some ascidians (STOECKER, 1980) and for the high copper concentration in the gills of the polychaete *Melinna palmata* (GIBBS and co-authors, 1981; Table 3-9).

In excess, the essential metals are toxic and can thus be considered alongside the non-essential metals which include Ag, Al, Be, Cd, Hg, Pb, Sb, Ti, and Tl. DA SILVA (1978) has discussed possible modes of action of metals on living systems and these are: (i) binding at centres where metals are not normally bound; mercury, for example, can bind to ligands under conditions where none of the essential metals could bind; (ii) substitution of essential elements particularly at the active sites of enzymes; (iii) binding which changes the conformation and reactivity of enzymes; (iv) replacement of groups like phosphate by others such as arsenate having similar dimensions; (v) precipitation of, for example, phosphate groups by the formation of very insoluble metal phosphates; (vi) alteration of membrane permeability by combination of metals with various groups in proteins; (vii) replacement of elements with electrochemical roles, such as potassium by, for example, thallium.

Because, the natural concentrations of trace metals in the sea are generally very low, the possibilities of contamination from man's activities are high. The accumulation of trace metals in marine organisms by concentration factors of the order of 10^3 to 10^5 times is commonly found and in contaminated water tissue levels may be reached which are toxic to the organism or its prey. From the public health point of view, the main problem is one of avoiding the excessive intake of metals from contaminated foodstuffs of marine origin. Mercury, cadmium, and lead are considered the most hazardous metals and the first 2 occur on various black lists. From the marine environment point of view, the main problem is that of preventing biological deterioration while at the same time using the enormous capacity of the sea for diluting, dispersing, and inactivating wastes. In this respect, the more abundant metals such as zinc and copper may sometimes be of greater hazard than mercury or cadmium.

Since metals are natural constituents of the marine environment, studies on metal contamination encounter the problem of distinguishing between natural levels and those which are enhanced from anthropogenic sources. For example, there are a number of examples where high residue levels in organisms, formerly thought to be evidence for contamination, appear to be perfectly natural. Even when contamination is patently obvious, and based on experimental evidence deleterious effects would be expected, it has in fact proved very difficult to observe them.

Thus one of the main objects of the review is to try and reconcile the experimental evidence for the toxicity of accumulated metals to marine organisms with the situation observed in the field in contaminated areas. However, before discussing this aspect it is necessary to consider the various processes leading to the contamination of the marine environment with metals and their accumulation by marine species.

The units of concentration used in this paper are ppb (parts per 10^9 or $\mu g\ l^{-1}$) for water and ppm dry weight (parts per 10^6 or $\mu g g^{-1}$) for tissues and sediments unless a wet weight basis is specified.

(2) Analysis of metals

Particularly with oceanic sea water, avoidance of contamination during collection and subsequent manipulation prior to analysis probably poses as many problems as the analysis itself. Thus, the gradual lowering of 'normal' oceanic trace-metal concentrations (Table 3-3) has been achieved by increased care in both sampling and analysis, and good examples include lead (SCHAULE and PATTERSON, 1980), mercury (MUKHERJI and KESTER, 1979) and cadmium, copper, nickel, and zinc (BRULAND and co-authors, 1979). Problems are created in all types of sea water analysis by the salt matrix. Thus, although recent advances in furnace atomic absorption have made it possible to analyse iron, manganese, and zinc directly (STURGEON and co-authors, 1979), for measurement of the lowest levels preconcentration and removal from the salt matrix is still desirable. BRULAND and co-authors (1979) used a dithiocarbamate extraction into chloroform followed by back extraction into nitric acid and carbon furnace atomic absorption to detect low oceanic concentrations (Cd, Cu, Ni, Zn). On the other hand, KINGSTON and co-authors (1978) used a development of the Chelex-100 resin method to preconcentrate prior to graphite furnace atomic absorption (Cd, Co, Cu, Fe, Mn, Ni, Pb, Zu). None of these methods can cope simultaneously with all trace metals and others must be analysed separately. Mercury is generally measured by cold-vapour atomic absorption and a preconcentration step involving amalgamation with gold is sometimes involved (MUKHERJI and KESTER, 1979).

Arsenic and other metals or metalloids (Sb, Se, Sn) forming gaseous hydrides can be liberated from sea water by reaction with sodium borohydride and measured in a suitable detector (ANDREAE, 1977). A technique of this type has recently been used to measure inorganic and organic forms of tin in sea water (BRAMAN and TOMPKINS, 1979). The inorganic and organic tin hydrides are scrubbed from solution, trapped by freezing, separated by warming and detected in a hydrogen–air flame-emission type detector. A method of this type has the added attraction of measuring different metallic species and many recent methods have been developed with this object in mind. This is particularly true of recent developments in the electrochemistry of sea water such as the use of anodic stripping voltammetry to determine the chemical forms of dissolved metals (BATLEY and FLORENCE, 1976). The number of metals possible to analyse by anodic stripping voltammetry is at present limited but, since preconcentration is not necessary and several metals can be analysed almost simultaneously, the method has great attractions: it seems ideal for use in the field and can also be automated (ZIRINO and co-authors, 1978). For example, GILLAIN and co-authors (1979) have described a method for the measurement of Bi, Cd, Cu, Pb, Sb, and Zn in sea water by differential pulse anodic stripping voltammetry, and SIPOS and co-authors (1980) have produced a voltammetric method for mercury.

Other methods for multi-element analysis of sea water which have recently been proposed include the simultaneous determination of 5 metals (Cu, Fe, Mn, Ni, Zn) by inductively coupled plasma atomic emission spectrometry following preconcentration (BERMAN and co-authors, 1980), and analysis of 12 metals (Ag, As, Au, Cd, Co, Cr, Fe, Hg, Mo, Sb, Se, Zn) by neutron activation of freeze-dried sea water followed by electrolytic separation from the salt matrix and gamma spectrometry (JØRSTAD and SALBU, 1980).

With biological tissues and sediments, the preconcentration of trace metals is usually not necessary and for many of the more abundant metals such as iron, zinc and copper flame atomic absorption is an ideal technique. For less abundant metals such as silver and cadmium the sensitivity of atomic absorption can be extended with the use of carbon furnace techniques; however, at the lowest levels, matrix problems in sediments and calcareous tissues cannot always be adequately removed by simultaneous background correction and some form of separation becomes necessary. For example, the great advantage of cold-vapour mercury methods and the hydride methods for arsenic and tin, mentioned earlier, is that they separate the metals from the salt matrix before atomization.

Although not generally available, neutron activation analysis is regarded as being particularly reliable at trace concentrations. For example, GRIMANIS and co-authors (1978; Table 3-17) analysed fish by neutron activation and used both instrumental (Co, Fe, Sb, Zn) and radiochemical techniques (As, Cd, Cu, Hg, Se) to separate the elements. On the other hand, in studies on mussels KARBE and co-authors (1978; Table 3-11) used instrumental neutron activation analysis for all elements (Ag, As, Cd, Co, Cr, Fe, Hg, Ni, Se, Sn, Zn).

Polarographic and voltammetric methods, although less commonly used for biological samples than the preceding methods, are extremely useful for detecting low concentration of cadmium, lead, and copper and the equipment is comparatively inexpensive

(NURNBERG, 1979). A method which has recently received much attention is inductively coupled plasma emission spectrography; this has the attraction of being a multi-element method and has been used for Ag, Al, Ba, Cd, Co, Cr, Cu, Fe, Mn, Mo, Ni, Pb, Sn, Ti, V, and Zn (KEITH and TELLIARD, 1979). Unfortunately no single method has proved ideal for all metals and for really comprehensive coverage several methods are generally required. Thus, a combination of X-ray fluorescence, spark-source mass spectrometry, and atomic absorption spectrometry was used by HAMILTON and co-authors (1979) to obtain comprehensive analyses of sediments from the Severn estuary (Table 3-6).

The methods discussed so far have been suitable for the analysis of pieces of tissue. However, the development of energy-dispersive X-ray microanalysis facilities on electron microscopes has begun to revolutionize research on the intracellular distribution of metals (CHANDLER, 1977). This technique has been used with great success in studies on the absorption, translocation, and storage of metals such as iron and lead in bivalve molluscs (GEORGE and COOMBS, 1977b; see also p. 1319).

(3) Metals in the Marine Environment

(a) Sources and Inputs

Anthropogenic Sources of Metals

Some idea of the scale upon which metals are used by man is obtained by considering the rates at which they are mined. Approximate annual levels of production are: 10^3 tons (Se); 10^4 to 10^5 tons (Ag<V<Hg<Cd<As<Sb<Mo); 2×10^5 to 5×10^5 tons (Sn<Ni); 10^6 to 10^7 tons (Cr<Pb<Zn<Cu<Mn<Al); 4×10^8 tons (Fe) (MINERALS YEARBOOK, 1970). All stages of metal production are potential sources of metallic contamination. Vast quantities of tailings are produced by mining processes and their disposal at sea has sometimes proved convenient: for example, WALDICHUK (1978) has described the disposal of tailings into Rupert Inlet in British Columbia from a copper mine dealing with 38,000 tons of rock daily. In addition, the problems posed by acidic mine drainage waters having parts per million (ppm) concentrations of heavy metals have been stressed by FÖRSTNER and WITTMANN (1979), and this is a source which can remain long after the mines have closed. Although mining declined in southwest England 80 yr ago, the input to the Fal estuary from the Carnon river, which drains many old mines, is still about 260 tons of zinc, 21 tons of copper, and 5·5 tons of arsenic annually, most of it in solution (Table 3-5; p. 1301). Smelting processes have proved to be sources of contamination in many sea areas, and in Sweden, for example, are a significant source of metal input to the Baltic Sea (HELLSTRÖM, 1979). In the USA, CRECELIUS and co-authors (1975) assessed the input of wastes from the Tacoma copper smelter into Puget Sound and showed that the annual input of arsenic included 20 to 70 tons in liquid effluent, 1500 tons in slag dumped into the Sound and a proportion of the 200 tons of arsenic trioxide released as stack dust: another important by-product was antimony. Although dredging for tin in shallow water has been practised for many years, the possibilities of recovering metals from deeper water are being seriously considered (OWEN, 1977). For example, an assessment is at present being carried out to predict the

environmental effects of mining the metalliferous sediments which occur in the deeper parts of the Red Sea (KARBE and co-authors 1980, unpubl.).

Virtually all industrial processes involving water are potential sources of metallic contamination in estuaries and coastal areas. Typical examples include the manifold uses of chromium. cadmium, and zinc, particularly in plating and galvanizing (ASSOCIATION EUROPEENNE OCEANIQUE, 1977), and the use of silver in photography (BARD and co-authors, 1976). However, mercury has provided the most notable examples. Effluent containing methyl mercury from the use of mercury as a catalyst in the production of acetaldehyde caused the tragedy at Minamata Bay in Japan in the 1950s, and also caused a second outbreak of Minamata disease near the mouth of the Agano river in 1965 (TSUBAKI and IRUKAYAMA, 1977). The most significant source of mercury contamination in other industrialized countries has been shown to be chlor-alkali factories which produce chlorine and caustic soda electrolytically with the use of a flowing mercury cathode. Fortunately, these sources and others such as the use of mercury-based slimicides have now been minimized.

Electricity power stations are significant sources of heavy metals. The burning of coal in particular releases metals to the atmosphere (BERTINE and GOLDBERG, 1971) and disposal of the fly-ash at sea is a common occurrence (Table 3-1). In addition, the use of sea water for cooling has through corrosion sometimes led to local pollution problems, particularly with copper (MARTIN and co-authors, 1977). Another source of this metal is effluent from desalination installations (ROMERIL, 1977).

Sewage disposal is a significant source of metals to the sea, either through direct discharge or the dumping of sludge produced by treatment. The metal content of the dumping of sludge produced by treatment. The metal content of the sewage depends on many factors, particularly the degree to which domestic wastes are enhanced by industrial wastes: an analysis of sewage sludge is included in Tablc 3-1 and shows the variability in composition observed in Plymouth.

Contamination from ships is most obvious in docks and harbours where very high concentrations of metals may build up in the sediments: these may subsequently be dredged and dumped at sea (see examples in Table 3-1). Such high concentrations come from the use of copper, tin, and mercury in anti-fouling paints and other metals including lead, zinc, and chromium in preservative paints (BELLINGER and BENHAM, 1978; YOUNG and co-authors, 1979). Accidental release of metallic ores through shipwreck seems unlikely to create the sort of problems posed by oil and some other chemicals. However, concern was certainly felt when the *Cavtat* with a cargo of 325 tons of alkyl-lead petrol additive sank in Adriatic coastal waters. Fortunately, all but 7% was salvaged and the effects were minimal (TIRAVANTI and BOARI, 1979).

Inventories of Metal Inputs in Coastal Waters

Under natural conditions, rivers provide the most important inputs of metals to coastal regions. These metals are derived largely from the mechanical and chemical weathering of rocks, plus components which are washed from the atmosphere in rainfall and may be marine or terrestrial in origin. The proportion of soluble to particulate metals in river water varies widely, but the soluble concentrations are generally low (Table 3-3) and, therefore, easily increased by contamination, particularly at low pH.

Table 3-1

Examples of quantities and composition of materials dumped in United Kingdom coastal waters (compiled from the sources indicated)

Material	Quantity dumped (metric tons yr^{-1})							Source
	Cd	Cr	Cu	Hg	Pb	Zn	Total solids dumped	
Dredged spoil	17	910	860	17	1770	3760	14.1×10^6 (1977)	MURRAY and NORTON (1979)
Sewage sludge	13	210	260	4	196	761	2.53×10^5 (1976)*	
Fly ash	0	14	40	0	20	30	7.5×10^5 (1976)*	
Industrial wastes	0	7	41	0	2	70	3.86×10^4 (1976)*	
	Concentration (ppm dry weight)							
	Cd	Cr	Cu	Hg	Pb	Zn		
Dredged spoil								
Manchester Docks	1.9	126	155	20.7	5080	687		MURRAY and NORTON (1979)
Portsmouth	<0.2	36	10	<0·07	21	53		
Sewage sludge								
Plymouth (1975)	5.6–376	50–1272	123–617	0.3–24	10–470	93–1660		EAGLE and co-authors (1979a)
Coal fly ash								
NE England	<2	18–19	30–56	<1	20–60	36–40		EAGLE and co-authors (1979b)
Colliery waste								
NE England	<2	5	120	—	230	540		
Dumping in relation to other inputs in dumping area off mouth of Humber estuary (metric tons yr^{-1})								
	Cd	Cr	Cu	Hg	Pb	Zn		
Dumping sewage sludge	0.1	10	7.7	0.18	5.1	31		MURRAY and co-authors (1980)
Dumping industrial waste	0	0.4	33	0	0.1	19		
Leaving Humber estuary†								
soluble	14	41	68	0.53	28	401		
particulate	<2.2	158	95	0.84	147	339		
Atmospheric input in rain over 1810 km^2	1.8	2.6	17	0.02	19	50		

*Licensed.

†Apart from river input to this estuary there are very significant inputs of sewage, industrial discharges, and dredged spoil from Humber ports.

With the increasing availability of information, attempts have been made to compare the magnitudes of natural and anthropogenic inputs to the sea and to identify the most significant sources.

From studies on dated sediment cores, GOLDBERG and co-authors (1978a,b) calculated that anthropogenic inputs to Narragansett Bay (USA) exceeded natural inputs by factors of 79 for copper, 56 for lead, and 21 for zinc: in Chesapeake Bay (USA), on the other hand, the 2 types of input were approximately equal. In this latter area, HELZ (1976) calculated the inputs from different sources. About half of the total input of cobalt, iron, manganese, nickel, and zinc could be attributed to river discharge, shore erosion and sea-water advection from the ocean. On the other hand, the principal source of chromium and copper was direct industrial discharge, that for cadmium was municipal waste water, and dust and rain were the principal sources of lead. Similarly, YOUNG and co-authors (1977) compared metal inputs to southern Californian coastal waters from municipal waste water outfalls, storm runoff, power station cooling water and dry aerial fallout. Waste water was the principal source of most metals but about half of the lead input was attributed to aerial fallout. Much of this atmospheric lead emanates from its use in lead alkyl 'antiknock' additives in petrol and its impact on concentrations in coastal waters appears to be very significant. For example, HODGE and co-authors (1978) calculated that leaching from aerosols could account for all lead in the upper 100 m of southern Californian coastal waters. The atmospheric input of other metals can also be quite significant near industrialized areas. Estimates by TOPPING (1974) indicated that atmospheric input in rainfall was possibly the most significant single source of lead, zinc, copper, and cadmium in the Firth of Clyde in Scotland. Similarly, CAMBRAY and co-authors (1979) calculated that the annual inputs of various soluble metals in rainfall were equivalent to 10 to 55% of the totals dissolved in the waters of the North Sea and that the total inputs in rainfall were approximately equal to those from the River Rhine (Table 3-2). Inputs from the River Humber to the North Sea (Table 3-1) are far exceeded by the total rainfall inputs in Table 3-2.

The dumping of wastes in some coastal areas provides a significant source of metal input. Around Britain dumped materials include dredged spoil from harbours and docks, power station fly-ash, sewage sludge, colliery wastes and various other industrial wastes. The magnitude of these operations and the composition of some of the materials are summarized in Table 3-1. Other important materials which are dumped include red mud, a mixture consisting largely of the oxides of iron and aluminium which arises from the production of aluminium from bauxite (BLACKMAN and WILSON, 1973), and liquid wastes such as ferrous sulphate and sulphuric acid resulting from the production of titanium dioxide: for example, 1800 tons daily were dumped at a site in the German Bight (RACHOR and GERLACH, 1978).

Atmospheric Inputs to the Oceans

As we have already seen, the atmospheric input of metals in coastal waters can be of considerable significance. Since the influences of other forms of input tend to be localized near the coast, the relative importance of atmospheric sources of input is likely to increase with distance from the land. Its potential significance is also increased by the discovery that the concentrations of several metals in oceanic waters are lower than was formerly supposed.

Table 3-2

Input of metals to the North Sea in rainwater compared with other inputs (compiled from the sources indicated)

Metals	Amounts in rainwater Total (ppb)	Soluble (ppb)	Annual soluble input in rain (% of soluble content of North Sea)	Total annual input to North Sea in rainwater (metric tons)	Annual input of R. Rhine (metric tons)	Soluble input via Straits of Dover (metric tons)
As	<2	<2	—	<460	1000	9000
Cd	—	<1.7	<25	<390	200	600
Co	0.38	0.17	9.8	88	—	—
Cr	3.2	0.99	—	740	1000	1000
Cu	24	15	35	5600	2000	2000
Fe	450	84	55	105,000	80,000	6000
Hg	0.03	—	2.4*	7	100	100
Mn	17.5	<12	<20	4100	6000	1500
Ni	<8	6.3	18	<1900	2000	1500
Pb	25	17.5	46	5800	2000	300
Sb	0.60	0.35	—	140	—	—
Se	0.54	0.50	—	125	—	—
V	5.4	3.7	—	1250	—	—
Zn	70	55	20	16,000	20,000	7000
			CAMBRAY and co-authors (1979)		WEICHART (1973)	BURTON (1978)

*Total values used.

The main natural sources of metals in the atmosphere have been listed by NRIAGU (1979) as: (i) windblown dust; (ii) forest fires; (iii) volcanic particles; (iv) particles produced by vegetation; (v) sea salt spray; and the main anthropogenic sources as: (i) mining; (ii) primary and secondary non-ferrous metal production; (iii) iron and steel production; (iv) industrial applications; (v)combustion of coal, oil, wood, and waste; (vi) phosphate fertilizer manufacture; (vii) miscellaneous. Apart from the different total quantities of metals which are involved, the volatility of the metal is perhaps the most important factor governing its release to the atmosphere; thus values for the ratio anthropogenic/natural input decrease in the order Pb > Hg > Ag > Mo > b > e > A > Zn > Cd > Cu > Sn > Ni > V > Cr > Co > Mn > Fe (LANTZY and MACKENZIE (1979). Ratios calculated for a global assessment by NRIAGU (1979) are 18 for lead (mainly from petrol), 9 for cadmium, 7 for zinc, 3 for copper (non-ferrous metal production being a major source) and about 2 for nickel (mainly from oil combustion).

Lead is almost certainly the most important heavy metal contaminant in the atmosphere, and PATTERSON (1978) has compared this form of input with other forms under present conditions and 'natural' conditions. Anthropogenic inputs of lead to the oceans are given as: atmospheric, 40,000 tons (37,000 tons from petrol); dissolved in rivers and sewers, 60,000 tons (much of it derived from aerosols); solids, 200,000 tons. The corresponding natural inputs are 1000, 13,000, and 100,000 tons respectively.

Effects of Inputs on Concentrations in Sea Water

A search of the literature for what might be regarded as baseline concentrations in sea water reveals a great deal of variability. There are various reasons for this: (i) the natural input of heavy metals to the sea is not uniform; (ii) the distribution is affected by agencies such as biological transport; (iii) many of the variations result from sample contamination and the use of different methods of analysis. The latter are arguably the greatest sources of variation in the literature values. For example, BRULAND and co-authors (1979) showed that by paying particular attention to the elimination of sample contamination, total concentrations of copper, cadmium, nickel, and zinc in sea water are lower than had previously been realized: moreover the results gave consistent profiles with concentrations, particularly of zinc and cadmium, increasing with depth. Levels of cadmium in the northeast Pacific showed a close relation to levels of phosphate and nitrate (BRULAND and co-authors, 1978), a concentration of 0·0045 ppb being found in nutrient depleted surface waters increasing to 0·125 ppb at the depth of the phosphate and nitrate maxima. These results are consistent with the idea that, in addition to nutrients, heavy metals are absorbed by planktonic organisms in the photic zone and transferred to deeper water when the dead organisms and the faeces and moults of zooplankton sink. Indeed, in the oceans the residence times of many metals (i.e. the average time for which the metal remains in the water before it is deposited) are thought to be primarily controlled by faecal deposition (BEWERS and YEATS, 1977; CHERRY and co-authors, 1978). A comparison is made in Table 3-3 between some of the recent concentrations for sea water and the typical values given earlier by BREWER (1975). It is clear that the concentrations of several metals, such as zinc and lead, formerly regarded as normal were too high. This fact, and the discovery that concentrations of metals including zinc and cadmium are often lowest at the surface, certainly

Table 3-3

Typical metal concentrations in natural waters (ppb) (compiled from the sources indicated in the table footnotes)

Metal	Fresh water[1]	Oceanic sea water		Probable major species in sea water	
		Older[2] values	Some recent values	Normal salinity[2,8,14]	Salinity <10‰[14]
Ag	0·3	0·04	0·01[3]	$AgCl_2^-$	
Al	<30	2	0·85[4]	$Al(OH)_4^-$	
As	2	3·7	2.2–3.3[5]	$HAsO_4^{2-}$, $H_2AsO_4^-$	
Cd	0·07	0·1	0·015–0·118[6]	$CdCl^+$, $CdCl_2^0$ [14]	Cd^{2+}, $CdCl^+$
Co	0·05	0·05	<0·01[7]	$CoCO_3^0$, Co^{2+} [14]	Co^{2+}, $CoCO_3^0$
Cr	0·5	0·3	0·13[8]	CrO_4^{2-}, $Cr(OH)_3$ [8]	
Cu	1·8	0·5	0·092–0·24[6]	$Cu(OH)_2^0$, Cu humic, $CuCO_3^0$	Cu humic, $Cu(OH)_2^0$
Fe	<30	2	0·46[7]	$Fe(OH)_2^+$, $Fe(OH)_4^-$	
Hg	0·01	0·03	0·011–0·033[9] 0·004[15]	$HgCl_4^{2-}$, $HgCl_3^-$ [14]	Hg humic, $HgCl_2^0$
Mn	<5	0·2	0·08[10]	Mn^{2+} [14]	Mn^{2+}
Mo	1	10	10·7[11]	MoO_4^{2-}	
Ni	0·3	1·7	0·228–0·693[6]	$NiCO_3^0$, Ni^{2+} [14]	Ni^{2+}, $NiCO_3^0$
Pb	0·2	0·03	0·001–0·015[12]	$PbCO_3^0$, $Pb(CO_3)_2^{2-}$	
Sb	0·1	0·24	0·37–0·7[5]	$Sb(OH)_6^-$	
Se	0·1	0·2	0·05–0·12[13]	SeO_3^{2-}	
Sn	0·03	0·01	0·0005[16]	$SnO(OH)_3^-$	
Ti		1	—	—	
Tl		0·01	—	—	
V	0·9	2·5	1·19[11]	$H_2VO_4^-$	
Zn	10	4·9	0·007–0·64[6]	Zn^{2+}, $ZnCl^+$ [14]	Zn^{2+}

1. Forstner and Wittmann (1979); 2. Brewer (1975); 3. Burrell (1977); 4. Hydes (1979); 5. Romanov and Ryabinin (1976); 6. Bruland and co-authors (1979); 7. Danielsson (1980); 8. Cranston and Murray (1978); 9. Gardner (1975); 10. Bewers and co-authors (1976); 11. Morris (1975); 12. Schaule and Patterson (1980); 13. Burton and co-authors (1980); 14. Mantuora and co-authors (1978); 15. Mukherji and Kester (1979); 16. Hodge and co-authors (1979).

increases the possibility that concentrations in oceanic surface waters may be changed by atmospheric input.

As we saw at the end of the previous section, the atmospheric input of anthropogenic lead is very significant and, since the concentration of lead in deep ocean waters is extremely low (0·001 ppb), the higher level found in surface waters (0·015 ppb) is almost certainly the result of atmospheric input (SCHAULE and PATTERSON, 1980). Although atmospheric input from anthropogenic sources has been invoked to explain elevated levels of other volatile metals including mercury and copper in both continental shelf surface waters (WINDOM and SMITH, 1979; WINDOM and TAYLOR, 1979) and ocean surface waters (GARDNER, 1975; BOYLE and co-authors, 1977), these conclusions are not unequivocal. GARDNER, for example, found a mean of 0·011 ppb of mercury in waters from the southern hemisphere compared with 0·033 ppb in the more heavily industrialized northern hemisphere. Studies on the atmospheric influx of mercury to the North and South Atlantic by MILLWARD and GRIFFIN (1980) support this picture. However, against this, no recent increases over the natural inputs of mercury, cadmium and vanadium from the atmosphere to the Greenland ice sheet have been observed, although the rates for lead and zinc are 2 to 3 times normal (HERRON and co-authors, 1977; APPELQUIST and co-authors, 1978).

Generally speaking, analyses of oceanic waters are carried out without prior filtration since this eliminates one of the sources of analytical contamination and, in any case, the concentrations of particulate metals are low. For example, BRULAND and co-authors (1979) found that less than 1% of the cadmium, copper, nickel, and zinc in deep water was associated with particles and only a few per cent near the surface. However, in shallow coastal waters and estuaries, where the particular fraction may be very significant, filtration, usually through a 0·45 μm membrane, is an accepted way of arbitrarily separating particulate from 'soluble' forms of metals although this latter category includes colloidal forms also. A summary of concentrations in some coastal areas and enclosed seas is given in Table 3-4. Even in the Baltic Sea (8–12‰S), which since it is surrounded by industrialized areas and receives a high fresh water input might be expected to show really obvious signs of contamination, the concentrations are not very different from those of the Gulf of St Lawrence where contamination is thought to be largely absent. Indeed, YEATS and co-authors (1978) have concluded that the sensitivity of coastal waters to anthropogenic emissions is quite low; so that warnings of changes due to anthropogenic inputs are much more easily obtained by monitoring the inputs rather than coastal sea water. Although there is certainly evidence for the enhancement of 'soluble' metal concentrations in areas such as the Irish Sea (Table 3-4), where levels of mercury, lead, and zinc seem abnormally high, it is generally only in the inner regions of bays, and in harbours and estuaries that really high concentrations have been measured, as, for example, in the Gulf of Fos, Corpus Christi Harbour, and Restronguet Creek (Table 3-4). The reasons why higher concentrations of 'soluble' metals have not been observed in coastal areas lies not only in the capacity of the sea to dilute inputs, but also in the processes leading to the deposition of metals in sediments: these are particularly obvious in estuaries.

Summary—Sources and Inputs

(i) Metals from anthropogenic sources reach the sea via rivers and outfalls, atmospheric fallout, dumping, marine mining and drilling, and from ships.

Table 3-4

Examples of the range of metal concentrations in coastal waters and estuaries (ppb) (compiled from the sources indicated)

Metal	Gulf of Alaska (mean)	Gulf of St Lawrence (mean)*	Baltic Sea (range)*	Irish Sea (mean)	NW Mediterranean (range)	Rhine Estuary (range)
Cd	0·03	0·07	0·03–0·06	0·04	0·06–0·8	0·3–0·6
Cu	0·20	0·61	0·6–1·0	0·66	0·2–21·6	2·5–7·5
Fe	—	3·4	0·3–0·9	0·18	—	—
Hg	0·007	—	—	0–0·44†	—	—
Mn	—	0·98	—	1·95	—	—
Ni	0·65	0·39	0·6–0·9	0·71	—	—
Pb	0·04	—	0·05–0·2	0·11	—	0·6–1·0
Zn	0·3	1·8	1·5–3·5	4·2	1·9–10	43–70
	BURRELL (1977)	YEATS and co-authors (1978)	MAGNUSSON and WESTERLUND (1980)	PRESTON and co-authors (1972a)	FUKAI and HUYNH-NGOC (1976)	DUINKER and KRAMER (1977)

Metal	Raritan Bay (range)	Bristol Channel (range)	Gulf of Fos (range)	Corpus Christi Harbour (range)	Sørfjord (range)	Restronguet Creek (S‰ 34–18)	Derwent estuary (range)
Cd	—	0·4–9·4	0·05–11·5	2–78	0·5–9	—	<0·5–15
Cu	2·9–13·4	0·6–5·4	2·2–17·0	—	1–23	5·8–126	—
Fe	—	1·5–10·5	—	—	—	0·5–3	—
Hg	—	0·009–0·07†	—	—	0·003–0·41‡	—	0·1–16
Mn	—	0·4–5·9	—	—	—	11–520	—
Ni	—	0·2–3·0	—	—	—	—	—
Pb	1·3–7·5	0·35–13·0	0·4–13·7	—	1–92	—	4–16
Zn	—	2·7–44	3·2–400	4–480	8–900	35–2520	6–1500
	WALDHAUER and co-authors (1978)	ABDULLAH and ROYLE (1974)	BENON and co-authors (1978)	HOLMES and co-authors (1974)	MELHUUS and co-authors (1978)		BLOOM and AYLING (1977)

*Unfiltered water. †GARDNER (1978). ‡EIDE and co-authors (1979).

(ii) With the increasing availability of information it has become possible to compare the magnitude of natural and anthropogenic inputs; and in some instances the latter exceed the former by more than an order of magnitude.

(iii) The relative importance of different routes of input has been assessed. Although, as might be expected, the most contaminated sites are in confined areas such as estuaries, fjords, and bays receiving direct inputs from rivers and outfalls, atmospheric input has been shown to be more significant than might be imagined in some coastal areas. The potential importance of this source further from land has been increased by the discovery that some surface-water levels are much lower than was formerly thought.

(b) The Fate of Metals in the Sea

Estuarine Deposition

Since some of the most heavily industrialized areas of the world are sited on the banks of estuaries, these waters and those of other confined areas are particularly at risk from metallic contamination. In the past, the discharge of wastes to rivers and estuaries was based on the assumption that they would be carried to the open sea and dispersed. The truth of the matter is rather different, and TUREKIAN (1977) has emphasized the efficiency of estuaries as traps where heavy metals become deposited in the sediments. Deposition from solution is based on the very low solubilities of some metals such as iron in saline waters, on the tendency for particulate materials to adsorb metals from the water, and on the adsorption and modification of metal species by organisms. On the other hand, metals may be returned to solution by desorption from particles or by remobilization from sediments. Thus, although the trend is towards deposition, what happens in any particular estuary will depend on the dominating processes.

It is now fairly well established that 2 of the most important components in the estuarine deposition process are iron and the humic materials which stabilize it and other colloidal constituents of river water (SHOLKOVITZ, 1978; MOORE and co-authors, 1979). The increase in salinity and sometimes pH during estuarine mixing leads to the flocculation of iron oxides, some humic substances, and other colloidal particles such as clays, which are then deposited together with their adsorbed or coprecipitated trace metals. DAVIES and LECKIE (1978) concluded from experimental studies that the adsorption of metals may be due to binding by colloid particles coated with humic compounds rather than to binding by simple oxide surface sites. Removal of heavy metals from the water and their deposition is assisted by biological processes, such as the production of faeces or pseudofaeces by filter-feeders and the incorporation of metals into tissues. The effect of biological processes on the deposition and distribution of pollutants has been reviewed by LEE and SWARTZ (1980).

Good examples of the deposition of metals are provided by studies on the input of acid metalliferous mine wastes into the sea. The examples in Fig. 3-1 show how the 'soluble' concentrations of iron, copper, zinc, and manganese change during estuarine mixing when the Carnon river enters Restronguet Creek, a branch of the Fal estuary system in southwest England. The initial fall in concentration is the result of dilution with uncontaminated water from another river, but the cross-hatched areas below the theoretical lines for simple mixing of sea water and river water show the degree of flocculation of the different metals; whereas iron is almost completely removed from 'solution' and copper

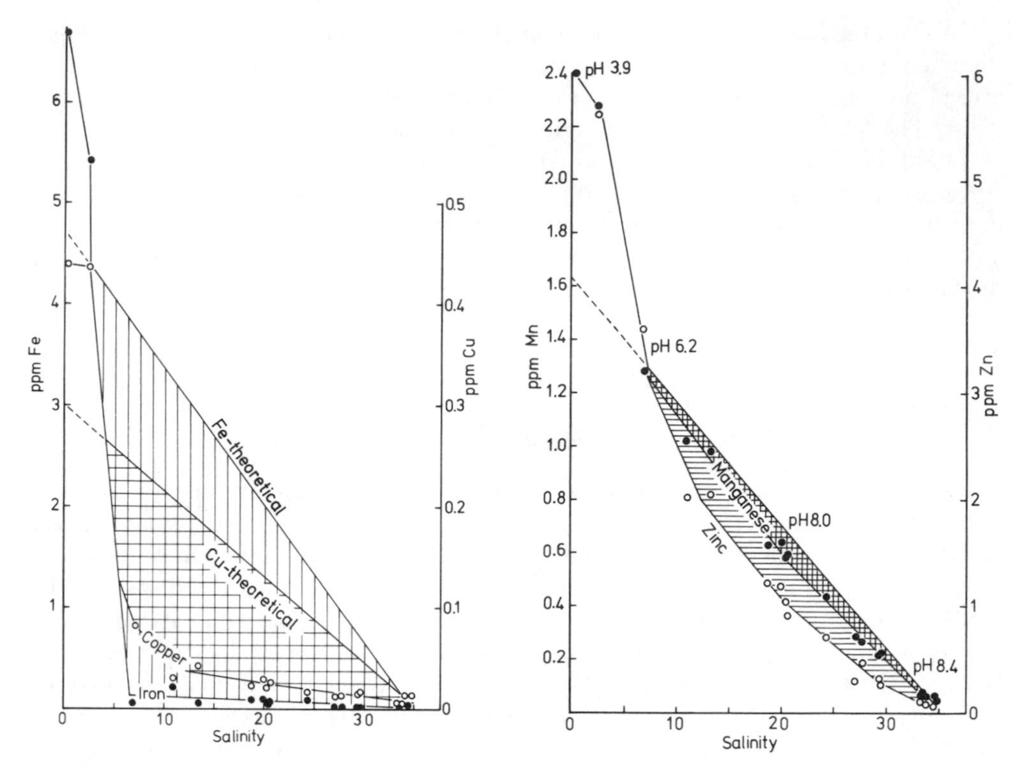

Fig. 3-1: Restronguet Creek. Relations between salinity and 'soluble' concentrations of 4 metals. Shaded areas below theoretical lines for simple dilution indicate degree of removal of metal from solution. (After BRYAN, 1980; reproduced by permission of Biologische Anstalt Helgoland)

largely so, manganese and zinc exhibit far less tendency to be removed (i.e. they show more conservative behaviour). A comparison between concentrations in the inflowing water (largely in solution, pH 3.9) and the creek sediments is shown in Table 3-5. In addition, the degrees of retention of metals by the sediments, assuming that iron is 100% retained, are also shown and compared with values for other areas. The high degree of retention of lead and arsenic may relate to their adsorption by oxides of iron, or, as is more likely for copper, by the associated humic materials. Possibly by virtue of their high concentrations in the river, manganese and zinc show little retention, as also does cadmium which forms strong chloride complexes in sea water and is not readily removed from solution (SHOLKOVITZ, 1978).

Studies by FOSTER and co-authors (1978) on the input of acid mine wastes via the Afon Goch into Dulas Bay in North Wales gave similar results. It was concluded that losses of copper and zinc from solution occurred on to hydrated ferric hydroxide above pH values of 4·5 and 6 respectively and that concentrations of manganese were affected solely by dilution.

Although the behaviour of iron (Fig. 3-1) is commonly found, the degree of removal of other metals (cross-hatched areas) varies considerably between estuaries. Working in the uncontaminated Beaulieu estuary in southern England, HOLLIDAY and LISS (1976) showed that dissolved zinc and manganese behaved conservatively and a similar obser-

Table 3-5

Relation between metal concentrations in Carnon River and Restronguet Creek sediment (After BRYAN, 1980; reproduced by permission of Biologische Anstalt Helgoland)

As	Cd	Cu	Fe	Mn	Pb	Zn
Total concentration in River Carnon (ppb)						
160	22	615	7310	957	40	7585
Total concentration in Restronguet Creek sediment (ppm)						
1080*	3*	3190	60,900	403	379	2400
Percentage retention of river input by sediment assuming Fe = 100%						
81	1·6	62	100	5	114	3·8
Percentage retention by southeastern USA estuaries (WINDOM, 1976)						
—	17	—	100	64	—	—
Percentage retention in Chesapeake Bay (HELZ, 1976)						
—	22	20	89	26	23	26

*THORNTON and co-authors (1975)

vation was made by ELDERFIELD and co-authors (1979) for zinc (max. 50 ppb) in the Conwy estuary in North Wales where some mining contamination occurs. On the other hand, DUINKER and NOLTING (1978) found that the behaviour of zinc in the Rhine estuary was not conservative, and nor was that of copper or cadmium. They concluded that trace-metal concentrations in the Rhine sediments reflect the mixing and removal processes occurring in the water column and that only a limited fraction of the metal input reaches the sea. In other estuaries, peak concentrations of soluble metals have been observed at intermediate salinities; these may be the result of desorption from suspended particles or the remobilization of dissolved metals from the sediments. For example, BEWERS and YEATS (1978) concluded that metals were desorbed from particles of fresh water origin in low-salinity waters of the St Lawrence estuary and ETCHEBER (1979) made comparable observations in the Gironde estuary. On the other hand, increased concentrations of soluble manganese, copper, and zinc in the middle reaches of the Tamar estuary have been attributed to their remobilization from the sediments (MORRIS and co-authors, 1978).

Metals in Sediments—Concentrations and Chemistry

Values from uncontaminated and contaminated coastal and estuarine sediments from the United Kingdom are summarized in Table 3-6 and compared with the values for shale rock. Some of the differences between values result from differences in methodology, since some of the sediments are finer than others and concentrations tend to be higher in the finer fractions. Similarly some methods of digestion are more complete than others. For example, digestion with nitric acid probably dissolves most metals of anthropogenic origin but leaves untouched some of the most tightly bound metals in the mineral lattices. The tin ore cassiterite, for example, is not attacked by nitric acid and

Table 3-6

Concentrations of metals in some United Kingdom sediments (ppm dry weight except Al, Fe, Ti, %). Some of the highest values are underlined (Compiled from the sources indicated)

		Coastal areas			
				Firth of Clyde (Scotland)	
Metal	Shale	Torbay (Devon) clean area (mean)	NE Coastal industrialized (mean)	Control area (mean)	Sewage sludge dump area (max)
Ag	0·07		1·0	<0·2	5 (underlined)
Al (%)	8·0				
As	13			8	24
Ba	580				
Be	3				
Cd	0·3	0·37	0·2	3·4	7
Co	19	6·6	7·1	34	40
Cr	90	9·9	9·9	64	308
Cu	45	4·2	8·0	37	208
Fe (%)	4·72			5·3	6·1
Hg	0·4	0·07	0·10	0·1	2·2
Mn	850	128	242	1118	1000
Mo	2·5			0·3	10
Ni	68	7·2	10	50	70
Pb	20	31	45	86	320
Sb	1·5				
Se	0·6			<0·5	<0·5
Sn	6			19	100
Ti (%)	0·46				
Tl	1·4				
V	130		26	252	400
Zn	95	25	74	165	826
	TUREKIAN and WEDEPOHL (1961)	TAYLOR (1979)		MACKAY and co-authors (1972) HALCROW and co-authors (1973)	

	Estuaries								
Metal	Avon (Devon) clean (typical)	Teign Barytes mining (max)	Restronguet Creek mining wastes (max)	Gannel old lead mines (max)	Poole Harbour industrial wastes (max)	Loughor tin plate manufacture (max)	Severn industrial inc. smelting (mean)	Mersey industrial inc. chlor-alkali (max)	Humber industrial inc. TiO_2 (max)
Ag	0·1	1·0	4·1	2·9	3·8	0·2	0·5		
Al (%)	6·21		7·33				6·52		
As	13	74	2520	233		22	15		
Ba		2000†					386		
Be							1·6		
Cd	0·3	1·8	1·2	3·0	12	1·1	1·9	4·2	
Co	10	18	22	40	16	13	16		30
Cr	37	35	37	29	90	799	145		201
Cu	19	68	2540	217	98	47	94		160
Fe (%)	1·94	2·16	5·76	3·32	3·57	3·81	3·87		9·2
Hg	0·12	0·36	0·22	0·09		0·13	0·15	11·3*	
Mn	417	777	559	1160	180	631	868		1098
Mo							3·8		
Ni	28	30	32	49	49	31	73		63
Pb	39	382	290	2175	145	77	101	349‡	221
Sb							2·2	2·9	
Se							0·05		
Sn	28	176	1730	305		320	91		
Ti (%)							0·4		1·2
Tl							1·1		
V							98		2031
Zn	98	375	3510	1215	386	220	296	800	433
		Bryan and co-authors (1980) †Merefield (1976)					Hamilton and co-authors (1979)	Leatherland and Burton (1974) *Bartlett and co-authors (1978) ‡Head and co-authors (1980)	Jaffe and Walters (1977)

can only be dissolved after fusion with ammonium iodide. Most, if not all, of the sediments in Table 3-6 are inhabited by macroscopic organisms and, as will be seen later, the availability of these sediment-bound metals to the biota is of particular interest. Other examples of sediment contamination are illustrated in Table 3-7 and demonstrate the incredibly high levels which have been reached in some confined areas.

The question of the remobilization of metals in sediments and their return to the overlying water has been studied both in the laboratory and in the field. For example, LU and CHEN (1977) exposed 3 types of contaminated sediment, ranging from sandy silt to silty clay, to sea water under oxidizing, slightly oxidizing, and reducing conditions and followed the concentrations in the water for several months. It was observed that with reducing conditions in the water the release of iron and manganese from the sediment was enhanced, whereas the release of cadmium, nickel, lead, and zinc increased as conditions became more oxidizing. A practical example of changes in the mobility of metals caused by redox changes in the water is given by the work of HOLMES and co-authors (1974). In the long (13 km), narrow harbour at Corpus Christi, Texas, they observed peak concentrations of 11,000 ppm of zinc and 130 ppm of cadmium in the sediments of the middle reaches but much lower levels at either end. During summer when water in the harbour stagnates, levels of zinc and cadmium from industrial effluent increased in the oxidizing surface waters and maxima of 480 and 78 ppb respectively were observed. However, the deeper water in the harbour becomes anoxic and metals from the water above are precipitated with the sulphide ions present. This results in low concentrations of metals in the deeper water and a metal-sulphide-rich sediment. In winter, high winds and lower temperatures increase the circulation of water between the harbour and the bay, so that metal-poor, oxygen-rich water enters along the bottom and leaves along the surface. The desorption of metals from the harbour sediment by this circulation transfers metals to the bay where they can be redeposited on the bay sediments. It was concluded that the harbour sediments may form a source of metals in the bay long after the discharge of industrial effluent has stopped.

Redox conditions in the sediments are certainly very important and in estuarine sediments, where the organic content is often of the order of 5 to 10%, a surface layer a few millimetres or centimetres thick is usually oxidized, the deeper sediments being dominated by sulphides. Since the sulphides of iron and manganese are more soluble than the oxides, their concentrations in the interstitial water of deeper sediments are often of the order of several parts per million (ppm) and a significant fraction of the iron, but not manganese, is organically bound (KROM and SHOLKOVITZ, 1978). Although the sulphides of other metals such as zinc and copper have extremely low solubilities, interstitial water concentrations which exceed those of the overlying water have frequently been found. It is thought that metals such as zinc, copper, cobalt, nickel, and lead are released from association with the oxides of iron and manganese when the latter dissolve and are stabilized at relatively high interstitial water concentrations by dissolved organic matter (ELDERFIELD and HEPWORTH, 1975); although another possibility is that they exist as polysulphide or bisulphide complexes (GARDNER, 1974). In the Conwy estuary sediments, ELDERFIELD and co-authors (1979) observed about 100 ppb of zinc in the interstitial water compared with about 10 ppb (max. 50 ppb) in the overlying water. A partial barrier to the diffusion of zinc from the interstitial to the estuary waters seems to be provided by the more rapid re-oxidation of iron and manganese at the sediment–water interface which scavenges the dissolved zinc. These remobilization and cycling processes can lead to the enrichment of iron, manganese and, to a lesser extent, zinc in the surface sediments on which many species feed. ELDERFIELD and HEPWORTH (1975) have pointed out that the redistribution of metals—especially if it leads to enrichment near the surface—will make it more difficult to assess anthropogenic inputs from sediment analyses. Thus SKEI and PAUS (1979) suggested that the enrichment of manganese in the upper part of a dated core from Ranafjord was partially explained by its redistribution in the sediment, although it was assumed that higher values for other metals were of anthropogenic origin. However, ERLENKEUSER and co-authors (1974) did not observe manganese enrichment in a dated core from the western Baltic Sea and were able to conclude that increases in the concentrations of cadmium, copper, lead, and zinc in the sediment by factors of 7, 2, 4, and 3 over the past 130 yr parallel industrial growth as reflected in the consumption of fossil fuels over this period.

In addition to the influence of factors such as redox conditions and the presence of dissolved organic matter on remobilization, the action of burrowing deposit-feeders on the release of metals is likely to be significant. The work of ALLER (1978) has shown experimentally that such organisms can appreciably increase the flux rates of manganese to the overlying water and this is likely to apply to other metals also. In shallow waters and marshes, angiosperms such as *Spartina* which absorb metals from sediments via their roots may also assist in the remobilization of metals following the death of the leaves and stems (see also p. 1316). In the field, the influence of sediment remobilization on concentrations in the overlying water is not easily separated from other effects. It is most obvious for manganese, but enhanced concentrations of copper and zinc in the waters of the Tamar estuary have also been attributed to this source by MORRIS and co-authors (1978). A study of the loss of mercury from contaminated sediments in Bellingham Bay by BOTHNER and co-authors (1980) revealed a concentration as high as 3·5 ppb in the interstitial water from anoxic sediments, which was 126 times the level in the overlying water, but only 0·01 to 0·06 ppb was found in oxidized surface sediments. Thus, although mercury fluxes could be measured from the anoxic sediments, losses could not be detected from the oxidized sediment. Mercury may be released mainly as organic or polysulphide complexes (LINDBERG and HARRISS, 1974). The sediments are also regarded as the source of methyl mercury in the environment, but its production is not thought to be the principal route by which mercury is lost from sediments (WINDOM and co-authors, 1976). Oxidizing rather than reducing conditions are most favourable for the microbiological methylation of mercury in sediment (FAGERSTRÖM and JERNELÖV, 1973); however, certainly *in vitro*, methylation has been observed under both aerobic and anaerobic conditions (BERDICEVSKY and co-authors, 1979). Low salinity also appears to favour methylation of mercury under experimental conditions (BLUM and BARTHA, 1980). In the field, BARTLETT and co-authors (1978) found that sediments from the Mersey estuary contained a total of 1·3 to 11·3 ppm of mercury of which about 0·5% was in the methyl form, lower percentages in other sediments being attributed to the increased presence of mercury as sulphide. In spite of the fact that sediments contain methyl mercury and are presumed to be the source of this compound in the environment, its measurement in sea water has not met with much success. EGAWA and TAJIMA (1977) reported 0·005 to 0·032 ppb of methyl mercury in waters from the worst part of Minamata Bay, where there are hundreds of ppm of mercury in the sediments, but offshore concentrations were below the detection limit of 0·005 ppb.

Table 3-7

Concentrations of metals in sediments from a range of contaminated areas (ppm dry weight, except Fe %). Highest values are underlined (Compiled from the sources indicated)

Metal	Gulf of Bothnia industrialized (mean)	Baltic proper industrialized (max)	Elbe Estuary industrialized (max)	Sörfjord smelting (max)	Gunnekleivfjord chlor-alkali plant (max)	Minamata Bay acetalaldehyde plant (max)	Saguenay Fjord chlor-alkali plant (max)
Ag				190			
As	15·6						
Cd	4·1	8·1	38	850			
Co	25		30				
Cr			770				
Cu	33	283	1250	12,000			
Fe (%)	7·5						
Hg	2·8	5	35		350	908	218
Mn	6410						
Ni	70	24	136				
Pb	67	400	655	30,500			
Sb				1080			
Sn				1350			
Zn	129	2090	2450	118,000			
	HALLBERG (1979)	OLAUSSON and co-authors (1977)	MÜLLER and FÖRSTNER (1975)	SKEI and co-authors (1972)	SKEI (1978)	TSUBAKI and IRUKAYAMA (1977)	LORING and BEWERS (1978)

Metal	Long Island Sound (max)	Raritan Bay industrialized (max)	Corpus Christi Harbour industrialized (max)		Puget Sound smelter and chlor-alkali (max)	New Bedford Harbour industrialized (max)	Derwent estuary Zn refinery and chlor-alkali (max)
Ag	6·7					40	
As					10,000	45	
Cd	4·2	15	130	21		76	862
Co	14·8						137
Cr	276	260		82		3200	258
Cu	269	1230		120		7250	>400
Fe (%)				1·23		1·4	16·1
Hg	2·2			18	100	3·8	1130
Mn	1218			257		180	8900
Ni	42	50		17		550	42
Pb	210	985		316		560	>1000
Sb					12,500		
Sn							
Zn	291	815	11,000	4055		2300	>10,000
	GREIG and co-authors (1977a)	GREIG and MCGRATH (1977)	HOLMES and co-authors (1974)	NEFF and co-authors (1978)	CRECELIUS and co-authors (1975)	SUMMERHAYES and co-authors (1977)	BLOOM and AYLING (1977)

DAVIES and co-authors (1979) estimated that the minimum level of methyl mercury in the waters of the Firth of Forth was 0·00006 ppb or only 0·3% of the total concentration.

Other metals or metalloids may be methylated in the sediments but the evidence is not unequivocal. ANDREAE (1979) was unable to find evidence for the biomethylation of arsenic in the interstitial waters of oxidized or reduced sediments, although methylated forms of arsenic are common in the environment (see also pp. 1311 and 1326). Some experimental evidence for the methylation of lead has been discovered in fresh water sediments (WONG and co-authors, 1975). Coastal sediments have been suggested as a source for alkyl lead in the atmosphere (HARRISON and LAXEN, 1978); however, THOMPSON and CRERAR (1980) concluded that, at least in anoxic sediments, methylation contributes little to the mobilization of the metal. CHAU and co-authors (1980) were unable to detect tetra-alkyl lead compounds in fresh water sediments or plants but found that they accounted for about 10% of the total lead content in fish. Percentages ranging from 9·5% to 81% have been found in marine tissues, including fish muscle and liver and lobster digestive gland (SIROTA and UTHE, 1977), but the original source of the tetra-alkyl lead is uncertain (see also p. 1388). Organo-tin compounds have been detected in the waters of Narragansett Bay (HODGE and co-authors, 1979) and may be of natural origin or could stem from the use of these compounds as biocides.

Summary—Deposition and Sediments

(i) Interactions between fresh water (or other discharges) and sea water tend to promote the transfer of metals from water to sediments. Thus, rather than being dispersed to the open sea, metal inputs to estuaries and other confined areas may become trapped in highly metallic sediments.

(ii) The metals in sediments are not necessarily inert and, depending on the conditions, remobilization may occur sometimes followed by the return of the dissolved metal to the overlying water. A particularly important facet of this process is the microbiological methylation of mercury (and possibly other metals) in the sediment, which is thought to be the ultimate source of much of this rather toxic compound in the environment.

(c) Bioaccumulation of Metals

Biological Availability from Solution

From experimental and theoretical studies on the complexation of metals by humic materials in natural waters, MANTOURA and co-authors (1978) concluded that in fresh water more than 90% of the copper and mercury may be complexed by humics but less than 11% of cadmium, cobalt, manganese, nickel, and zinc. When fresh water is diluted with sea water, the proportions of various ligands and competing ions change and so also does the speciation of metals. Between salinities of 0‰ and 10‰S, the speciation of copper, for example, is still dominated by humics. However, in full strength sea water inorganic species, principally $Cu(OH)_2^0$ and $CuCO_3^0$, usually become dominant although some organic complexes remain (Table 3-3). The concentration of ionic copper Cu^{2+} in sea water is thought to be quite low, but it is this form which determines the bioavailability of copper to phytoplankton (SUNDA and GUILLARD, 1976) and fish eggs (ENGEL and

SUNDA, 1979). Its actual level depends not only on the total concentration of copper and the extents to which it is complexed by organic or inorganic ligands, but also on pH.

In the case of cadmium, the free ion Cd^{2+} is the dominant form in fresh water but, with increasing salinity, chloride complexes including $Cd\,Cl^{+}$ and $Cd\,Cl^{O}$ become dominant (MANTOURA and co-authors, 1978). Since the free ion appears to be the most readily available form, chloride complexation has been used by ENGEL and FOWLER (1979) to explain the decreased toxicity of cadmium to the shrimp *Palaemonetes pugio* at high salinity. To show that toxicity was a function of cadmium ion concentration rather that salinity, the Cd^{2+} concentration was also changed by varying the levels of total cadmium and of a chelator, nitrilotriacetic acid, at constant salinity. For zinc also, the free ion is the available species to phytoplankton (ANDERSON and co-authors, 1978) although, unlike cadmium, it appears to be the principal species in both fresh water and sea water (MANTOURA and co-authors, 1978; Table 3-3).

The speciation of metals is not simply dependent on the mixing of fresh water and sea water since significant changes can also be effected by biological activity and may occur through the introduction of organic chelators into the water by seaweeds or phytoplankton (GNASSIA-BARELLI and co-authors, 1978). The influence of organic materials can vary considerably. Thus, GEORGE and COOMBS (1977a) found that cadmium absorption by the mussel *Mytilus edulis* was enhanced by ligands such as alginic acid, humate and EDTA, whereas RAINBOW and co-authors (1980) found that uptake by the barnacle *Semibalanus balanoides* was unaffected by alginic acid and humate but inhibited by EDTA. Other changes in the speciation of metals are brought about through their metabolism by organisms. As we saw in the previous section, methyl mercury produced by biomethylation of inorganic mercury in the sediments may diffuse into the overlying water, where, probably by virtue of its lipid solubility, it is much more biologically available than the inorganic form (CORNER and RIGLER, 1958). Although arsenate is the major species of arsenic in sea water, it can be converted to arsenite and methylated species by phytoplankton (ANDREAE, 1979) and also converted back to arsenate by bacteria to complete the cycle (SANDERS, 1979a; 1980). That changes in oxidation state can influence availability was shown by KLUMPP (1980a): 4 times as much arsenate as arsenite was absorbed from solution by the brown seaweed *Fucus spiralis*.

Biological Availability from Sediments

Although soluble species of metals are generally considered to be of greater biological availability than sediment-bound forms, the much higher concentrations in sediments may render them more important as sources of metals in some species. LUOMA and JENNE (1977) studied the absorption of ^{65}Zn, ^{60}Co, and ^{110m}Ag by the deposit-feeding bivalve *Macoma balthica* from labelled artificial sediments including oxides of iron or manganese, precipitated carbonates, powdered clam shells, and organic detritus: they corrected for the uptake of tracer desorbed from the particles by enclosing control animals in dialysis sacs. It was shown that the availability of the metal was dependent on the type of sediment, the degree of availability being inversely related to the strength of sediment binding; for example, ^{65}Zn was more readily absorbed from powdered clam shells than from iron oxide particles, since the latter bind metals more strongly. A comparison was also made between the ability of the bivalve to absorb the metal radionuclides and that of chemical extractants to remove them from the sediments.

Bioaccumulation of ^{65}Zn for example, was most closely related to the amount extracted by 1N ammonium acetate and bioaccumulation of ^{109}Cd to the amount extracted with 70% ethanol (LUOMA and JENNE, 1976).

Under field conditions, the biologically most significant part of the sediment is almost certainly the oxidized surface layer, on which many species feed, rather than the reduced subsurface sediments which are dominated by sulphides. It is thought that the distribution of trace metals between the various sediment components or substrates depends on the strength of metal–substrate binding and on the abundance of the various substrates. As we saw earlier, oxides of iron and humic substances are intimately associated with depositional processes, and not surprisingly they are among the most important metal-binding substrates in surface sediments. Recognition of the various substrates and assessment of their associated trace metals depends mainly on chemical extraction procedures, often inherited from soil chemists. Although these methods are useful, they are rarely if ever specific, so that descriptions of extracted metals as 'adsorbed' or 'organically' bound are operational rather than accurate (LUOMA and BRYAN, 1981). Attempts to find a single extractant with which to predict the biological availability of sediment-bound metals have not, in general, been very successful. NEFF and co-authors (1978) who exposed the bivalve *Rangia cuneata*, the polychaete *Neanthes arenaceodentata* and other species to sediments in experimental tanks were unable to relate the uptake of metals by the organisms to the concentrations removed from the sediments by 5 extractants including ammonium acetate, acetic acid, hydroxylamine hydrochloride, acidic hydrogen peroxide and a total acid digestion. LUOMA and BRYAN (1978, 1979), compared the concentrations of heavy metals in the deposit-feeding bivalve *Scrobicularia plana* with those of different sediment extracts at sites in 20 different estuaries. This enabled them to study not only the influence of different sediment–metal concentrations on those of the organisms but also the influence of variations in the concentrations of metal-binding substrates. For example, it was shown that, although the relationship between concentrations of lead in *S. plana* and those of 1N HCl extracts of surface sediment was poor, the relationship to the Pb/Fe ratios in the extracts was one of almost direct proportionality. A similar result for arsenic in the same species was found by LANGSTON (1980) (Fig. 3-2) and both results are interpreted as indicating that the availability of the metal is controlled by the amount of 'reactive' iron oxide in the sediment, with increased levels of iron tending to inhibit the absorption of the metal. This simple approach could not be used for all metals and in the case of zinc the best relationship, although not one of great predictive value, was found between the concentration in the animal and that extracted with 1N ammonium acetate (LUOMA and BRYAN, 1979). In the case of copper, some influence of iron on its availability could be discerned, but some of the results defied explanation (LUOMA and BRYAN, 1982). On the other hand at the same sites the concentrations of copper in *Nereis diversicolor* were very closely related to the total sediment concentrations and to the amounts of copper extracted with 1N HCl (see also Fig. 3-3). This implies that different sediment fractions are more readily available to different species.

There has been a great deal of work on the availability of metals from dumped wastes. LEE and co-authors (1978) examined the possibility of using chemical extractants to predict the availability of metals to the marsh plants *Spartina alterniflora* and *S. patern* growing at dredged material disposal sites. Of the various extractants considered DPTA (diethylenetriamine penta acetic acid) was the most successful. The availability of

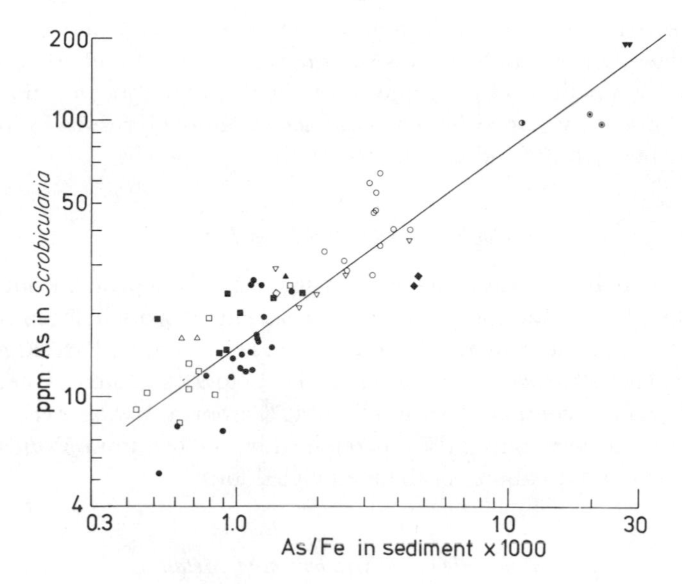

Fig. 3-2: *Scrobicularia plana*. Relationship between the concentration of arsenic in the body on a dry-weight basis and the As/Fe ratio in 1N-HCl extracts of surface sediments from a range of United Kingdom estuaries. (After LANGSTON, 1980; Arsenic in U.K. estuarine sediments and its availability to benthic organisms. *Journal of the Marine Biological Association of the United Kingdom*; reproduced by permission of Cambridge University Press)

metals from power station fly-ash is considered to be low (NORTON and ROLFE, 1978). Fly-ash does, however, contain some soluble metals and CRECELIUS (1980) showed that around 25% of the total arsenic, and selenium was rapidly dissolved in sea water but less than 1% of zinc, cobalt, and iron. In an experimental study, RYTHER and co-authors (1979) exposed different organisms to coal fly-ash over which sea water was flowing. Analyses of *Mya arenaria* and *Nereis virens* after 4 months' exposure showed some enhancement of concentration over controls maintained in sand; however, this enhancement seemed to be explained to some degree by fly-ash contained in the gut.

The results of studies on the release of metals during the dredging of waterways and the dumping of spoil seem to indicate that the release is generally at a sufficiently low level and of such short duration that availability for uptake from the water is extremely limited (PATRICK and co-authors, 1977; SUSTAR and WAKEMAN, 1977; BRANNON, 1978). Material deposited offshore during dumping operations and having abnormally high metal content is likely to retain some biological availability to deposit feeders if the surface of the deposit becomes oxidized. However, in the absence of continuous input and due to leaching by the relatively uncontaminated sea water, material dredged from a contaminated estuary and then dumped, largely in a reduced form, would not be expected, even on oxidation, to possess its original availability. Analyses of organisms from areas of dumping tend to confirm this view. MACKAY and co-authors (1972) observed rather higher concentrations of metals in several species from a sludge dumping area in the Firth of Clyde and some increase in concentrations in scallops *Placopecten magellanicus* appeared to result from the dumping of sewage sludge and acid wastes from a titanium dioxide plant off the US mid-Atlantic coast (PESCH and co-authors, 1977). Studies on other dump sites off the same coast showed that concentrations of several metals increased in the gastropod *Busycon canaliculatum* although results for other species were less convincing (GREIG and co-authors, 1977c).

Biological Availability From Food

Just as the availability of heavy metals from sediments is dependent not only on the total concentration but on the composition of the sediment as a whole, so the uptake of metals from food is influenced by the dietary matrix. For example, PENTREATH (1976d) showed that the plaice *Pleuronectes platessa* retained 80% to 93% of methyl mercury from a diet of *Nereis diversicolor* compared with 4% to 42% from a *Mytilus edulis* diet. When inorganic mercury was used only 11% was retained by the fish from *N. diversicolor*, thus illustrating the greater availability of the methylated form.

Accumulation and Metabolism of Metals

The accumulation of abnormal concentrations of metals by the tissues of marine organisms is the basis for much of our concern about metallic contamination; first, because the accumulation of high concentrations by commercial species might prove harmful to man and other higher vertebrates; and second, because the productivity of marine or estuarine organisms may be affected. In the following account, evidence concerning the absorption, storage, and excretion of metals will be considered in some of the most important marine groups. However, even for some extremely important groups, including marine bacteria (JONES and co-authors, 1976a), marine protozoans, coelenterates, echinoderms, and tunicates there is insufficient information on the processes of accumulation to provide a coherent account.

Marine Plants

Phytoplankton

The adsorption and accumulation of metals by phytoplankton have been reviewed in depth by DAVIES (1978). From studies on the uptake of zinc in *Phaeodactylum tricornutum* and mercury in *Isochrysis galbana* and *Dunaliella tertiolecta*, DAVIES concludes that uptake is a passive process involving rapid adsorption on to the cell membrane followed by diffusion-controlled transport into the cytoplasm at rates proportional to the concentration of surface-bound metal. Binding to protein may control the level of zinc in the cell because during the population growth cycle, the concentration reaches a maximum and then decreases as the amount of protein in the cell declines. A similar pattern has been observed for nickel in the same species (SKAAR and co-authors, 1974). It was also shown that in phosphate-starved cells the capacity for nickel accumulation was low and was enhanced by pretreatment with phosphate, presumably due to the synthesis of new binding sites.

Although the concentration in the cell is influenced by its metabolic state, it is also markedly influenced by the concentration of metal in the medium. For example, SKAAR and co-authors (1974) found that concentrations of nickel in the cells of *Phaeodactylum tricornutum* were proportional to those of the medium up to a level of about 750 ppb. In the same species and in *Cricosphaera elongata*, a similar conclusion was drawn for copper by BENTLEY-MOWAT and REID (1977); and—although this was not observed for copper in *Ditylum brightwellii* by CANTERFORD and co-authors (1979)—concentrations of zinc, lead, and cadmium tended to reflect those of the medium. That concentrations in phytoplankton reflect those of the medium under field conditions also has been shown by EIDE and co-authors (1979) who cultured *Phaeodactylum tricornutum* at sites having different sea water zinc levels.

Since the culture media used in experimental studies on phytoplankton often contain organic chelators, a number of studies have been concerned with the influence of natural and synthetic chelators on metal accumulation by phytoplankton; for example, SUNDA and GUILLARD (1976) showed that in such media the availability of copper is reduced and in *Thalassiosira pseudonana* the concentration in the cells is related to the cupric ion (Cu^{2+}) activity of the water and not to the total copper concentration. The availability of zinc to the same species appears to be dependent upon the availability of Zn^{2+}, and it has been concluded that if complexation of zinc occurred to a sufficient degree in uncontaminated water, growth of this species could easily be limited by deficiency (ANDERSON and co-authors, 1978).

Seaweeds

Experimental work on the uptake of zinc by the green alga *Ulva lactuca* and the red alga *Porphyra umbilicalis* has suggested that passive adsorption or ion-exchange processes are involved (GUTKNECHT, 1961, 1963, 1965). Uptake is promoted by increased pH, and both uptake and loss are enhanced by increased temperature and the presence of light. Light appears to be particularly essential for the absorption of cadmium by *P. umbilicalis* and, unlike zinc, its accumulation seems to depend largely on the synthesis of protein uptake sites (MCLEAN and WILLIAMSON, 1977). Zinc absorption in *Laminaria* sp. is enhanced by these same factors, but is inhibited by the presence of other metals including cadmium, manganese, and copper (GUTKNECHT, 1963; BRYAN, 1969). The overall accumulation of zinc by brown seaweeds appears to be mainly a net uptake process, since losses are slow. Losses do occur, however, and in *Ascophyllum nodosum* field experiments have shown them to be faster in summer than in winter (EIDE and co-authors, 1980). These authors suggest that the uptake and release of zinc and cadmium require an input of metabolic energy whereas the accumulation of lead may depend on ion-exchange processes. It has been concluded by KLUMPP (1980a) that arsenic accumulation by brown seaweeds requires energy derived from respiration. The rate of uptake increases with temperature but is independent of salinity and pH; it also depends on the form of arsenic and arsenate is more readily absorbed by *Fucus spiralis* than arsenite.

In brown seaweeds there are several components capable of binding metals. The affinity of alginate extracted from *Laminaria digitata* for metals is: Pb > Cu > Cd > Ba > Sr > Ca > Co > Ni, Zn, Mn > Mg (HAUG, 1961); the affinity of fucoidan from *Ascophyllum nodosum* is: Pb > Ba > Cd > Sr > Cu > Fe > Co > Zn > Mg > Mn > Cr > Ni > Hg > Ca (PASKINS-HURLBURT and co-authors, 1978); and that of polyphenols from *Fucus vesiculosus* is: Cu > Pb > Ni > Zn > Co > Cd > Mn > Be > Ca > Mg > Sr (RAGAN and co-authors, 1979). Polysaccharides such as alginates and fucoidan in the cell walls or intercellular spaces confer some metal-exchange capacity on the weed, but the relatively irreversible uptake of zinc and possibly other metals is thought to be explained by its binding to material in membrane-bound vacuoles—possibly polyphenol-rich vacuoles called physodes (SKIPNES and co-authors, 1975).

The concentrations of metals reached in seaweeds tend to reflect those dissolved in the surrounding water, although it is possible that some, such as copper and lead, may also be absorbed from contact with particulates (SEELIGER and EDWARDS, 1977). Concentration factors relating levels in the weed to those of the water are summarized in Table 3-8. One of the parameters on which the concentration factor depends is the age of the weed, and in *Ascophyllum nodosum* concentrations of zinc, copper, lead, and mercury increase markedly with age (MYKLESTAD and co-authors, 1978). The same authors showed that when *A. nodosum* was transferred from a contaminated to an uncontaminated site, the concentration in new growth was related to that of the new environment In the older tissues, however, only 20% to 30% of the zinc and mercury was lost in 5 months, but cadmium and lead were lost more readily, suggesting that they are bound differently.

Species such as *Fucus vesiculosus* are sometimes able to tolerate more than a 1000 ppm of copper or zinc in their tissues, suggesting that mechanisms exist for immobilizing metals (cf. Table 3-28). In other forms such as the fresh water green alga *Scenedesmus acutiformis* copper is immobilized in nuclear inclusions (SILVERBERG and co-authors, 1976) and, in higher plants, inclusions of zinc phosphate have been observed; thus the presence of zinc can lead to phosphorus deficiency (FILIPPIS and PALLAGHY, 1975). Arsenic in the form of arsenate is a competitor of phosphate for uptake by seaweeds. It is not immobilized but is converted to organic forms having an apparently lower toxicity (SANDERS, 1979b); analyses of a large number of species showed that, on average, nearly 80% of the arsenic in brown algae and about 50% of that in red and green algae was in an organic form.

Angiosperms

Since they are rooted in the sediments, marine angiosperms provide a potential route by which metals can be returned to the overlying water from the sediment. Concentrations of metals in the tissues tend to reflect levels in the environment. For example, STENNER and NICKLESS (1975) found concentrations of zinc, copper, and lead in eelgrass *Zostera* sp. from an area of heavy contamination to be 1 to 2 orders of magnitude higher than normal (cf. Table 3-29). Similarly AUGIER and co-authors (1978) showed that both *Z. marina* and *Posidonia oceanica* reflect levels of mercury in the environment.

Although metals including zinc and manganese are released to the water during the conversion of eelgrass to detritus (WOLFE and co-authors, 1975) this does not necessarily imply that there is a net transfer of metals from the sediments to the overlying water. Certainly in the case of cadmium, the metal is absorbed by the leaves from solution and translocated to the root rhizome and thus provides a route from the water to the sediment rather than a means of remobilization from the sediment (FARADAY and CHURCHILL, 1979; BRINKHUIS and co-authors, 1980).

In the case of marsh grasses such as *Spartina alterniflora* there is evidence that the formation of detritus from the dead plants provides a means of conveying metals from salt-marsh sediments to other areas and possibly a means of returning sediment-bound metals to the water column. WINDOM (1975) has compared the amounts of metals

Table 3-8

Concentration factors in seaweeds ($\mu g\ g^{-1}$ dry tissue $\mu g^{-1}\ ml^{-1}$ water). All values in thousands (Compiled from the sources indicated)

	Ag	Cd	Cr	Cu	Mn	Ni	Pb	Zn	Source
Phaeophyceae									
Fucus (UK coast)	5	2·7		4·5	23	2·8	2·4	20	PRESTON and co-authors (1972a)
vesiculosus (Restronguet Creek)				27	4·6			11	BRYAN and HUMMERSTONE (1973c)
(Menai Straits)		10·5	11·2	6·4	19	6·8	2·9	10	FOSTER (1976)
(Raritan Bay)				3·6–7·4			13–24		SEELIGER and EDWARDS (1977)
(Bristol Channel)		15		10	2·3–22			18	MORRIS and BALE (1975)
(Sørfjord)		4·2–13		4·8–19			3·2–26	7·1–24	MELHUUS and co-authors (1978)
Ascophyllum nodosum									
(Menai Straits)		9	7	8·6	3·9	4·6	2·4	13	FOSTER (1976)
(Sørfjord)		7·1–11		6–20			1·2–10	12–18	MELHUUS and co-authors (1978)
Laminaria digitata (UK coast)				16·3	1·7			17·5	BRYAN (1969)
Rhodophyceae									
Porphyra umbilicalis (UK coast)	1·9	6·6		6·3	7	1·1	2·0	10	PRESTON and co-authors (1972a)
Chlorophyceae									
Blidingia minima (Raritan Bay)				7·1–18·3			27–82		SEELIGER and EDWARDS (1977)
Enteromorpha (Sørfjord)		3·9–16		4·3–10·2			2–34	8–38	MELHUUS and co-authors (1978)
Enteromorpha linza (Raritan Bay)				5·6–7·7			20–45		SEELIGER and EDWARDS (1977)
Ulva (Raritan Bay)				4·7–8·6			17–49		

accumulated annually by the leaves and shoots of *S. alterniflora* (i.e. the maximum amounts which may be lost as detritus when the plants die) with the amounts gained by sedimentation from the water. Uptake by the plants from the sediments amounted to about 3% of the copper, cadmium, and iron gained by sedimentation and 17% of the mercury. It was also concluded that the release of metals from the plants probably has an insignificant effect on concentrations in the water column. Other evidence for the remobilization of metals from sediments by *Spartina* has been reviewed by LEE and co-authors (1978), who also considered the relations between concentrations in the roots and shoots and those in various sediment extracts. They observed significant relationships between concentrations of cadmium, copper, and zinc in the plants and the amounts extracted from the sediments, which were of dredged material, with diethylenetriamine penta acetic acid.

Marine Animals

Coelenterates

The concentrations of metals in a few species are shown in Table 3-16. However, little is known about the metabolism of metals by coelenterates and most of the available information is to be found in papers on the toxicity of metals which are discussed on p. 1377.

Polychaetes

In many species the uptake of metals from solution probably takes place over the whole body surface, but in the sabellid *Eudistylia vancouveri*, where the abdomen is enclosed in an unirrigated tube, the branchial crown is thought to be the main site for copper absorption (YOUNG and co-authors, 1979a). Although some cadmium is absorbed from solution over the body surface of *Glycera dibranchiata*, significant uptake also occurs via the intestine, the absorbed metal rapidly binding to coelomic proteins including haemoglobin (RICE and CHIEN, 1979). There is no evidence that active transport processes are involved in the absorption of metals from solution. CHIPMAN (1966) concluded that the absorption of chromate ions by *Hermione hystrix* was a passive process, and BRYAN (1976a), working with *Nereis diversicolor*, reached similar conclusions for a range of metals. For manganese the rate of uptake is directly proportional to the external concentration over the range 1 to 10,000 ppb, but the relations for other metals are less exact, the rates being relatively lower at higher concentrations. In the case of zinc, this is because uptake is more closely related to its adsorption on to the surface of the body during the uptake process (BRYAN and HUMMERSTONE, 1973a). However, the rapid initial uptake of methyl mercury in *Glycera dibranchiata* is thought to be facilitated by its solubility in the lipid cell membrane (MEDEIROS and co-authors, 1980).

Experimental studies have shown that in *Nereis diversicolor* the rates of absorption of manganese, zinc, and copper are increased at lower salinities (BRYAN and HUMMERSTONE, 1973a,b; JONES and co-authors, 1976b). One possible explanation, particularly in the case of manganese, may be found in the work of FLETCHER (1970), who showed that the concentration factor for calcium in the coelomic fluid increases with decreasing salinity and was able to explain this, without invoking active transport, in terms of an increase in the potential difference (negative inside) across the body wall. The influence of salinity may also be the the result of changes in the speciation of metals, as has already been discussed from cadmium (p. 1311), and changes brought about by competition from calcium and magnesium for uptake sites. Although changing a parameter such as salinity may alter the rate of metal absorption in an organism, it may affect excretion similarly, with the result that there is no net change in concentration. Field observations on *N. diversicolor* suggest that this is so for zinc, which shows every sign (Fig. 3-3) of being effectively regulated, as also does iron (BRYAN and HUMMERSTONE, 1973a; BRYAN and GIBBS, 1980). For copper also, BRYAN and HUMMERSTONE (1971) could find no unequivocal evidence for the influence of salinity under field conditions. On the other hand, it was concluded that the concentration of readily exchangeable manganese in *N. diversicolor* is dependent on the high level of Mn^{2+} ions ($\lesssim$300 to >1000 ppb) in the interstitial water and its salinity (BRYAN and HUMMERSTONE, 1973b). Although the interstitial water may be a source of other metals including zinc (RENFRO, 1973), their dissolved concentrations are usually so much lower than those of manganese that the importance of uptake following direct surface contact with sediment particles or through ingestion needs to be considered. Experiments by UEDA and co-authors (1976) showed that when *N. japonica* were exposed directly to sediments labelled with ^{115m}Cd, they absorbed 6 times more tracer than animals held in the water overlying the sediment. When the same species, lying in sand, were fed with sewage sludge having enhanced metal levels, there was no marked increase in concentrations of cadmium, chromium, copper, nickel, lead, and zinc in the body although it was calculated that a few per cent of the metals had been absorbed (MORI and KURIHARA, 1979). RAY and co-authors (1980) have con-

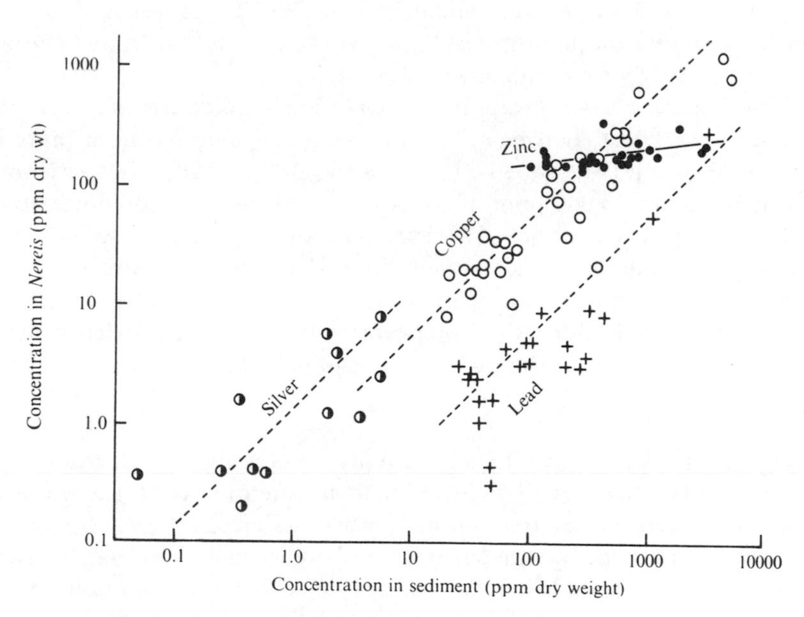

Fig. 3-3: *Nereis diversicolor*. Relationship between concentrations of metals in body and total concentrations in sediments from estuaries in south-west England. (After BRYAN, 1976a; in A. P. M. Lockwood (Ed.), *Effects of Pollutants on Aquatic Organisms*; reproduced by permission of Cambridge University Press)

cluded that in *N. virens* cadmium is absorbed from the interstitial water rather than from the sediment and that the metal is not excreted. However, field observations on *N. diversicolor* have suggested that dissolved and sediment-bound cadmium are both important sources (LOUMA and BRYAN, 1982b). In addition, NEUHOFF (in THEEDE, 1980) found that cadmium-contaminated food (*Mytilus* + 30 to 100 ppm dry) reduced the growth rate in *N. succinea*. Concentrations of other metals in *N. diversicolor*, especially copper, are strongly related to those of the sediments (Fig. 3-3), but there is still uncertainty about what in the sediment constitutes the available fraction.

Only a few studies have considered the influence of metallic form on uptake. For example, CHIPMAN (1966) showed that chromium was available to *Hermione hystrix* in the soluble hexavalent form but not in the relatively insoluble trivalent form and, in *Cirriformia spirabrancha*, the uptake of copper from solution was unaffected by the presence of dissolved yellow organic matter (MILANOVICH and co-authors, 1976).

There is not much information on the detoxification and excretion of metals in polychaetes. In nereid species (Table 3-9) as much as 50% of the body zinc may be incorporated in the jaws (BRYAN and GIBBS, 1980), but this is not regarded as a detoxification mechanism since the jaw concentrations (~15,000 ppm) are unrelated to environmental levels. BRYAN and HUMMERSTONE (1973a) concluded that since the absorption of zinc does not appear to be controlled, its regulation at a relatively constant level in the body (Fig. 3-3) is dependent upon the excretion of the metal. On the other hand, in copper-contaminated *Nereis diversicolor* (Fig. 3-3) the metal is detoxified and stored in membrane-bound vesicles in the epidermal cells (BRYAN, 1976a). It is also found in the nephridia, but there is little evidence for copper excretion when contaminated animals are placed in uncontaminated conditions. Silver seems to be detoxified and stored by a similar mechanism (BRYAN, 1976a), and lead is also stored although whether by the same system is unknown.

As in *Nereis diversicolor*, concentrations of copper in the soft parts of *Nephthys hombergi* and *Glycera convoluta* reflect changes in the environment, whereas concentrations of zinc tend to be more constant (BRYAN, 1976b; GIBBS and BRYAN, 1980). In species of *Glycera*, the jaws contain about 16,000 ppm of copper (Table 3-9) and sometimes account for more than 60% of the body burden. However, as with zinc in nereid species, this is not thought to be part of a detoxification system since the copper concentration of the jaw is unaffected by environmental variations. The same applies to the high gill copper content of *Melinna palmata* (Table 3-9) which appears to be a chemical defence mechanism rendering the exposed parts of the body distasteful to small fish.

Bivalve molluscs—mussels

Although mussels (*Mytilus* sp.) do not normally contain high concentrations of heavy metals (Table 3-11), they have by virtue of their potential as indicators of metallic contamination, received more attention than other molluscs.

The absorption of metals by mussels has been studied in some detail. For example, SCHULZ-BALDES (1974) found that rates of uptake of lead from solution and the concentrations achieved after 40 d of uptake were proportional to the external concentration over the range 5 to 5000 ppb. The rate of loss into clean water turned out to be proportional to the concentration of lead in the body. Thus, on exposure to contaminated sea water, the concentration of lead in the body increases until the rate of excretion equals the rate of intake. Assuming equal inputs of lead from food and water, it was calculated that equilibrium at a concentration factor on a dry weight basis of $3{\cdot}5 \times 10^4$ would take more than 230 d to be achieved. Further studies revealed uptake of lead by the gills into the blood to be the rapid phase of accumulation, followed by a slower process, involving the transport of dissolved lead to the kidney, its uptake by pinocytosis and the formation within the kidney cells of membrane-bound vesicles or lysosomes containing granules (SCHULZ-BALDES, 1978a,b). In addition to lead, the granules contain calcium and phosphorus and excretion occurs when the vesicles are pinched off and eliminated in the urine.

There is evidence too that absorption of some lead from solution occurs by pinocytosis of the gill epithelium (COOMBS and GEORGE, 1978). However, this process does not seem to be involved in the absorption of cadmium from solution, and uptake by facilitated diffusion has been proposed by GEORGE and COOMBS (1977a). There appear to be 2 routes for the storage and ultimate removal of cadmium from the bloodstream: the first, via the kidney, resembles that for lead (GEORGE and PIRIE, 1979); the second involves storage in membrane-bound vesicles in the digestive gland and loss in the faeces (JANSSEN and SCHOLZ, 1979). The chemical form of cadmium within the membrane-bound vesicles of the kidney and digestive gland remains uncertain, but some of it may be in the form of Cd-metallothionein (GEORGE and co-authors, 1979). Proteins of the Cd-metallothionein type have been found in the digestive gland of exposed mussels by SHOLZ (1980), and similar proteins for binding copper in the gills by VIARENGO and co-authors (1980). Another possible route for the removal or detoxification of cadmium (and lead) is by incorporation into the shell (STURESSON, 1978). The various pathways for the uptake and removal of zinc are summarized in Fig. 3-4.

Experimental studies with several metals including zinc, manganese, cadmium, and selenium have indicated that food and other types of particles may be a more significant source of metals than dissolved species (PENTREATH, 1973a; FOWLER and BENAYOUN, 1974, 1976). This is almost certainly true of iron which is very insoluble in sea water and usually occurs as particles of ferric hydroxide. GEORGE and co-authors (1976) have shown that these particles are absorbed not only via the digestive system, where digestion is intracellular, but also by pinocytosis through the gills—a process which seems likely to be energy-dependent (GEORGE and co-authors, 1977). The iron is subsequently transferred to other tissues by circulating amoebocytes and, although some loss occurs via the gut and kidney, a significant amount is lost through the byssus gland, and byssal threads, as also is arsenic (ÜNLÜ and FOWLER, 1979).

Experimental studies suggest that for most metals, the concentration in the mussel will come to reflect that of the environment (Table 3-10). It is on this property that the use of mussels as indicators of metallic contamination is based. Exact proportionality is not always found, since there is sometimes a tendency for the concentration factor (animal/environment) to decrease at higher levels. However, erratic behaviour has been reported for copper in mussels by PHILLIPS (1976), and BRYAN and HUMMERSTONE (1977) expressed doubts about the ability of mussels to reflect changes in the availability of silver.

Accumulation of heavy metals by mussels is influenced by various factors; these may affect uptake or excretion or both processes. Not least important is the form of the metal in the environment. For example, GEORGE and COOMBS (1977a) showed that the rate of uptake of dissolved cadmium and the concentration achieved by the animal was doubled by complexation with EDTA, humic acid, alginic acid or pectin. In the absence

Table 3-9

Metal levels in whole bodies and tissues of some polychaete worms (Compiled from the sources indicated)

Polychaete	Concentration (ppm dry weight)													Source
	Ag	As	Cd	Co	Cr	Cu	Fe	Hg	Mn	Ni	Pb	Sn	Zn	
Nereis (Hediste) diversicolor (estuaries of S.W. Britain)														
minimum	0·05	4	0·03	0·7	0·1	10	181	0·01	4·1	0·6	0·2	0·06	91	Bryan and co-authors (1980)
maximum	24	87	10·1	24·5	10·4	1430	871	2·5	51·9	15·6	1190	3·9	510	
Nereis (Neanthes) virens (Tamar estuary)														
whole body	0·40		0·22			18·3	489		5·3		5·6		110	Bryan and Gibbs (1980)
jaws	<0·1		0·10			14	600		795		8·7		16,110	
% in jaws	<0·05		0·11			0·16	0·26		31·8		0·37		30·9	
Nereis (Neanthes) japonica (experimental study)														
whole body			0·4		8	19				5	4		140	Mori and Kurihara (1979)
Perinereis nuntia (experimental study)														
whole body			0.5		7	24				8	5		110	Mori and Kurihara (1979)

Glycera gigantea											
(English Channel)											
whole body					29	153	6·7			479	GIBBS and BRYAN (1980)
jaws					14,180	270	low			2150	
% in jaws					60	0·2	—			0·6	
Nephthys hombergi											
(Cleddau estuary)											
whole body	0·49	0·18	0·92	0·45	13	588	5·2	2·1	3·4	173	
Melinna palmata											
(English Channel)											
whole body	5·0				1109	285	13·2			97	GIBBS and co-authors (1981)
gills					10,910	598	21			392	
% in gills					40·8	3·7	9·1			13·8	
Cirriformia luxuriosa											
whole body		1·0		4·8	48	235	9·0	13·1	0·4	57	EMERSON and co-authors (1975)
Notomastus tenuis											
(Los Angeles Harbour)											
whole body		0·6		33	46	181	13·5	7·8	1·2	59	

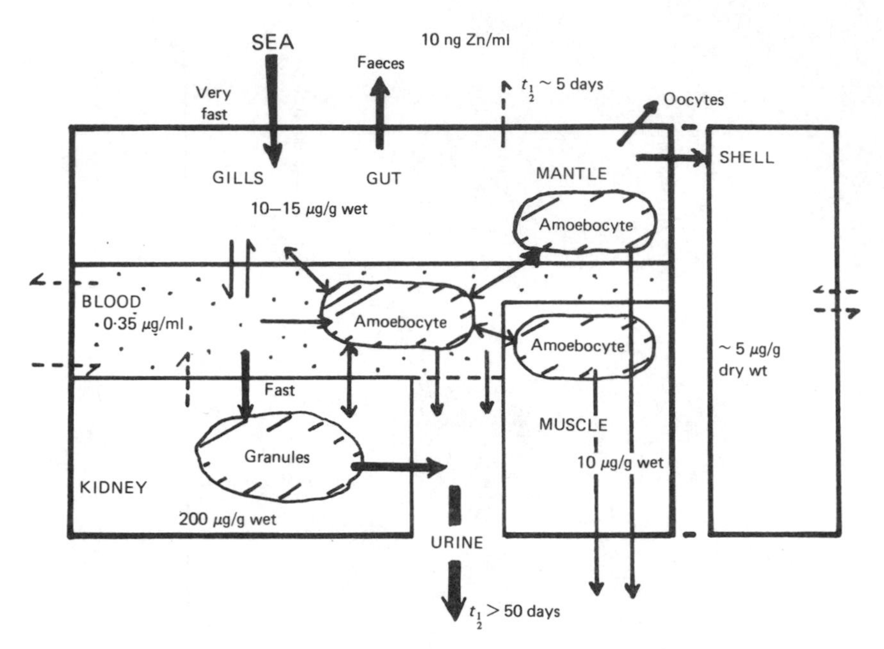

Fig. 3-4: *Mytilus edulis*. A model for zinc dynamics. Solid arrows: known pathways; dashed arrows: possible routes. Size of tissue spaces is proportional to the proportion of total body zinc in that tissue (including its amoebocytes)—except the blood which contains only 0.5% of the total zinc. Concentrations in the tissues are also shown. (After GEORGE and PIRIE, 1980; Metabolism of zinc in the mussel *Mytilus edulis*: a combined ultrastructural and biochemical study *Journal of the Marine Biological Association of the United Kingdom*; reproduced by permission of Cambridge University Press)

of these reagents there was an initial lag in the uptake process, suggesting that ionic cadmium must be complexed before it can be absorbed—an example of facilitated diffusion. Similarly, complexation with EDTA or citrate increased the rate of absorption of ferric hydroxide; this was attributed to the possible effects of these reagents on the process of pinocytosis (GEORGE and COOMBS, 1977b).

Other factors sometimes influencing the uptake process are changes in salinity, temperature, and concentrations of other heavy metals in the water which might compete for binding sites; examples are shown in Table 3-10. Low salinity enhances the absorption of cadmium (cf. p. 1311) but not zinc, and changing the temperature has little effect on the rates of uptake of several metals including zinc and cadmium, thus supporting the idea that they are passively absorbed.

Among factors which influence metal concentrations in mussels are size, sex, reproductive condition, and seasonal variation. BOYDEN (1977) showed that concentrations of iron, manganese, nickel, lead, and zinc were generally higher in small specimens whereas the level of cadmium was independent of size. In *Choromytilus meridionalis*, WATLING and WATLING (1976a) found significant differences between the concentrations of some metals in the 2 sexes (Table 3-11) and LOWE and MOORE (1979a) showed that lysosomes containing zinc granules were more abundant in the kidney of male *Mytilus edulis* than in females. In the latter the storage of zinc granules in oocyte lysosomes appears to represent an additional route for excretion. Seasonal variations in the concentrations of a range of metals have been studied in *M. galloprovincialis* by MAJORI and co-authors (1978b) who found that, generally speaking, concentrations were higher in winter than in summer.

Other bivalve molluscs

Although in other bivalves the same principles may apply to the absorption, detoxification, and excretion of metals as were observed in mussels, the emphasis on these processes varies and leads to appreciable differences between the concentrations of metals in the tissues of different species (Table 3-11).

The absorption of mercury by the oyster *Crassostrea virginica* consists of a rapid initial phase followed by a slower irreversible phase (MASON and co-authors, 1976), a situation similar to that for lead in mussels. However, experiments with inhibitors have suggested that the absorption of mercury over the gills of oysters may be energy dependent (WRENCH, 1978). Although in mussels the rate of cadmium uptake is almost independent of temperature, suggesting passive absorption, an approximate doubling of rate for every 10 °C increase in temperature was observed in *Nucula proxima* and *Mulina lateralis* (JACKIM and co-authors, 1977). On the other hand, the rate of cadmium absorption was increased at low salinity in all 3 species.

A considerable body of evidence suggests that particulates rather than the water are the principal source of metals in bivalves, including oysters (SAYLER and co-authors, 1975), scallops (BRYAN, 1973), and the deposit-feeding clam *Scrobicularia plana* (BRYAN and UYSAL, 1978). In oysters, wandering amoebocytes are particularly important in the absorption, storage, transport, and ejection of foreign materials. Accumulation by these cells is largely responsible for the high concentrations of zinc and copper which are characteristic of oysters and largely explains why the contrast between the concentrations of metals in different oyster tissues is less than in other bivalves (Table 3-11). Working with *Ostrea edulis*, GEORGE and co-authors (1978) showed that the 2 metals are immobilized in membrane-limited vesicles in 2 different types of granular amoebocyte: copper is associated with sulphur and zinc with phosphorus. It was calculated that the amoebocytes may contain as much as 13,000 ppm of copper and 25,000 ppm of zinc on a wet weight basis compared with 8 and 30 ppm respectively in the cell-free body fluid. The accumulation of metals is not simply dependent on the concentrations in the amoebocytes, because RUDDELL and RAINS (1975) found that the level of zinc in species of *Crassostrea* was proportional to the number of basophilic amoebocytes. The same species possesses a possible detoxification mechanism for cadmium in the form of a low molecular weight (7400) binding protein, which, however, differs chemically from the metallothioneins of vertebrates (RIDLINGTON and FOWLER, 1979).

The immobilization of metals in membrane-bound vesicles has been recognized in a range of species. In *Mercenaria mercenaria* exposed to high levels of dissolved mercury, the metal was detoxified in cytosomes of the mantle epithelium (FOWLER and co-authors, 1975). Deposition of metals including cadmium also occurs in mineral concretions occupying the apical vacuoles of the kidney cells of this species (CARMICHAEL and co-authors, 1980); these concretions consist mainly of calcium phosphate but contain several per cent of zinc, iron, and manganese as well (DOYLE and co-authors 1978). High concentrations of manganese, zinc, and other metals in the kidneys of the scallop

Table 3-10

Some experimental studies on the uptake and loss of metals by the mussels (*Mytilus edulis* or *M. galloprovincialis*) (Compiled from the sources indicated)

Metal	Range (ppb)	Uptake period (days)	Proportionality between body level and that of medium	Effects on uptake rate from solution of: Lower salinity	Higher temperature	Presence of other metals	Half-time for slow component of loss (days)	
As (arsenate)	1–100*	13	Relatively less uptake at high concentrations	Increase	Increase	—	32	ÜNLÜ and FOWLER (1979)
Cd (chloride)	0·1–100*	2	Proportional	—	No effect	Zn no effect	307	FOWLER and BENAYOUN (1974)
	0–20	35	Approx. proportional	Increase	Increase at low S‰	Zn, Pb, Cu, no effect	—	PHILLIPS (1976)
	10–100	18	Approx. proportional				14–29	SCHOLZ (1980)
Cu (chloride)	0–40	35	Poor relation	Increase	Increase	Affected by Zn, Pb, Cd	—	PHILLIPS (1976)
Fe (hydroxide)	7·4–500	25	Proportional	—	—	—	—	GEORGE and co-authors (1976)
	400–4000	90–150	Approx. proportional	—	—	—	—	WINTER (1972)

Hg	0·005–0·1*	46	Relatively less uptake at high concs.	—	—	—	—	Majori and co-authors (1978a)
(chloride)	~1*	35	—	—	Slight increase	—	82	Fowler and co-authors (1978)
(methyl)	0·1*	35	—	—	Increase	—	63	
Ni (chloride)	18–107	28	Roughly proportional	—	—	—	—	Friedrich and Filice (1976)
Pb (nitrate)	5–5000	40	Nearly proportional	—	—	—	about 80	Schulz-Baldes (1974)
	0–40	35	Roughly proportional	Decrease	No effect	Zn, Cd, Cu, no effect	—	Phillips (1976)
Sb (potassium antimonyl tartrate)	150–5000	30–186	Roughly proportional	—	Increase	—	17·8	Walz (1979)
Se (selenite)	0·1–100*	21	Nearly proportional	—	Increase	Hg no effect	81	Fowler and Benayoun (1976)
Tl (sulphate)	0–100	40	Nearly proportional	—	—	—	about 1	Zitko and Carson (1975)
V (vanadyl chloride)	2–100*	21	Relatively less uptake at high concentrations	Increase	No effect	—	100	Miramand and co-authors (1980)
Zn (chloride)	0–400	35	Roughly proportional	No effect	No effect	Cd, Cu, Pb, no effect	—	Phillips (1976)

* *Mytilus galloprovincialis*.

Table

Metal levels in some bivalve molluscs

Species	Concentration (ppm dry weight) Ag	Al	As	Cd	Co	Cr	Cu
Mytilus edulis							
Poole, UK	—	—	15*	3·7			7
California	0·18	270		11·2	—	1·7	18·8
Helgoland	0·16	—	11·2	1·7	0·42	1·5	—
Mytilus galloprovincialis (min.)	0·10			0·4	0·5	0·5	2·4
(NW Mediterranean) (max.)	1·89			5·9	7·4	28·8	15·4
Chloromytilus meridionalis (S Africa)							
(female)	0·3			0·9	0·9	1·4	7·7
(male)	0·3			0·9	1·0	1·4	5·9
Chlamys opercularis (English Channel)							
(whole)	10·4	49		5·5	0·33	2·2	15·4
(digestive gland)	77	43		27	1·03	4·7	36·7
(kidney)	35	58		41	15·1	6·6	1285
(striated muscle)	—	—		—	0·03	—	1·3
Placopecten magellanicus	1·8	338		20·9	0·56	3·1	7·3
(E coast USA slight cont.)				21·3		4·3	10·1
Cerastoderma edule (Poole, UK)	0·1		5·1*	1·0	4·3	2·6	3
Pitar morrhuana							
(Rhode Island slight cont.)	1·8			1·4	1·3	7·3	19·6
Mercenaria mercenaria							
(Southampton water)			4·5*	1·4			23·4
Scrobicularia plana (SW Britain)							
(min.)	<0·1		5	0·2	2	0·5	9
(max.)	259		191	42·7	106	23·8	752
% of total metal burden in digestive gland from Tamar Estuary animals	15%		36%	95%	94%	80%	38%
Macoma balthica (SW Britain)							
(min.)	0·3		11	0·1	1·0	0·8	21
(max.)	122		46	9·4	7·4	16·3	224
Crassostrea gigas							
(Knysna estuary, S Africa)	1·9			3·7			32
(Tomales Bay, California)	1·7			9			86
(Redwood City, California)	82			40·9			860
(Poole, UK)				4·6			200
(Tasmania) (min.)				0			47
(max.)				173			860
Crassostrea margaritacea							
(Knysna estuary, S Africa)	2·6			2·5			17
Crassostrea virginica							
(Tomales Bay, California)	6·4			12·2			273
(Redwood City, California)	81·5			39·7			1510
(Deer Island, Mississippi)				2			
(Connecticut estuaries) (min.)				4·6			214
(max.)				107			4304
Visceral mass				13·4			918
Mantle				10·2			1450
Gill				10·1			936
Muscle				9·9			332
Ostrea edulis							
(Knysna estuary, S Africa)	6·4			3·1			38
(Poole, UK)			2·6*	5·9			86
(Restronguet Creek, UK) green oyster	0·4			7·9	0·41	0·34	2610
Ostrea angase (Western Port, Australia)							
(min.)				0·97			39
(max.)				2·64			104

*LEATHERLAND and BURTON (1974); †SMITH and BURTON (1972).

(Compiled from the sources indicated)

Concentration (ppm dry weight)								
Fe	Hg	Mn	Ni	Pb	Sb	Sn	Zn	Source
87	0·42	3	5	7	0·047*		94	BOYDEN (1975)
430		6·6	1·8	0·36			161	STEPHENSON and co-authors (1979)
118	0·33		1·6				106	KARBE and co-authors (1978)
149		3·3	0·9	2·7			97	FOWLER and OREGIONI (1976)
2220		69·8	14·1	117			644	
66		12	5·2	1·8			97	WATLING and WATLING (1976a)
54		6	5·7	2·1			54	
113		158	1·6	12·0			462	BRYAN (1973)
853		30	4·2	10·2			132	
330		17,300	78·2	827			40,800	
10·7		4	0·17	0·55			62	
803		30	4·4	3·6			105	PESCH and co-authors (1977)
—	1·05	—	12·0	8·5			111	PALMER and RAND (1977)
221	0·86*	2·6	35·5	0·3	0·034*		57	BRYAN (unpublished)
272		82	11·2	11·1			203	EISLER and co-authors (1978)
109	0·17*		9·4	7·8	0·013*	0·23†	177	BOYDEN (1977)
248	0·02	10	1·1	5		<0·1	256	BRYAN and co-authors (1980)
9770	1·3	339	22·4	1080		14·0	4920	
10%	65%	73%	83%	85%		58%	87%	
210	0·12	8	1·0	2·3		<0·1	396	
1950	1·3	356	12·7	36·5		1·7	1790	
128		16	1·6				396	WATLING and WATLING (1976b)
368		40					442	STEPHENSON and co-authors (1979)
668		61					3550	
228		18	3	6			1766	BOYDEN (1975)
							1390	RATKOWSKI and co-authors (1974)
							57.600	
57		2	1·6				886	WATLING and WATLING (1976b)
260		34					3570	STEPHENSON and co-authors (1979)
148		13					19,300	
	0·4			5				HARVEY and KNIGHT (1978)
	0·15	4·1					3895	FENG and RUDDY (1975)
	0·79	29·6					15,948	
	0·64	10·2					10,921	
	0·78	10·9					13,646	
	0·88	19·2					9225	
	0·74	6·4					3285	
167		6	1·7				660	WATLING and WATLING (1976b)
394	1·2*	7	3	8	0·01*		1966	BOYDEN (1975)
219		16·7	0·24	2·51			17,100	BRYAN (unpub.)
612		7·8		0·5			878	HARRIS and co-authors (1979)
2585		16·7		1·9			3051	

Chlamys opercularis are attributable to similar excretory concretions (Table 3-11) and this seems to be a feature of all scallop species (CARMICHAEL and co-authors, 1979). In *Pecten maximus* the 5 to 15 μm granules consist chiefly of the mixed phosphates of calcium, manganese, and zinc, but include lower concentrations of magnesium, copper, iron, and cadmium (GEORGE and co-authors, 1980). Concretions also occur in the kidneys of *Tridacna maxima* and *Pinna nobilis:* although they consist primarily of calcium phosphate and oxalate, they also contain metals such as zinc and cobalt (HIGNETTE, 1979).

In most bivalves that have been studied experimentally, tissue concentrations tend to reflect those in the water, although direct proportionality is not necessarily found (e.g. oysters: SHUSTER and PRINGLE, 1969; ZAROOGIAN and co-authors, 1979). In fact, under field conditions, tissue concentrations may be more closely related to those of the sediments. AYLING (1974) concluded that concentrations of zinc, cadmium, and lead (but not copper and chromium) in *Crassostrea gigas* were related to those of the sediments, and in the deposit-feeder *Scrobicularia plana* concentrations of several metals including lead (LUOMA and BRYAN, 1978) and arsenic (fig. 3-2) are related to those of the sediments. In addition to the availability of the metal, size and season exert a considerable influence on bivalve tissue concentrations. The influence of size was studied by BOYDEN (1977) in 6 species, including *Ostrea edulis, Mercenaria mercenaria,* and *Pecten maximus.* He found that concentrations of cadmium, copper, iron, manganese, nickel, lead, and zinc either fell gradually with increasing size or were independent of size: an exception was cadmium in *P. maximus,* since the concentration increased from about 20 ppm in animals of 2 g tissue weight (dry) to 100 ppm in those of 8 g. In this species, 90% of the cadmium lies in the digestive gland compared with less than 2% in the kidneys (BRYAN, 1973), and therefore it may be less readily excreted than metals such as zinc and lead which occur largely in the kidneys. In *S. plana* where the majority of metals are concentrated in the digestive gland (Table 3-11), excretion seems to be rather inefficient and concentrations of cadmium, zinc, and lead, for example, increase markedly with increasing size, particularly in animals from contaminated estuaries (BRYAN and UYSAL, 1978; BRYAN and HUMMERSTONE, 1978). Transferred to an uncontaminated site, contaminated *S. plana* revealed that much of the decrease in concentration of lead, for example, could be attributed to tissue growth, rather than to a reduction in body burden. This is one of the factors responsible for seasonal concentration changes in bivalves; others which may also be concerned include changes in temperature, availability of metals, and metabolic rate (BRYAN, 1973).

Metal levels in oysters tend to vary with season, and FRAZIER (1975 concluded that whereas changes in the body burdens of manganese and iron are associated with shell deposition, changes in levels of copper, zinc, and cadmium correlate in some way with gonad development and spawning. A rapid loss of 33% to 50% of these metals by *Crassostrea virginica* in late summer was partially attributed to loss of gametes, but amoebocytes must be involved as well since GREIG and co-authors (1975a) established that the spawned eggs contain 50 to 100 times lower concentrations of zinc and copper than the whole body. Spawning has also been suggested as a major route for the loss of mercury by CUNNINGHAM and TRIPP (1973), who studied its depuration from contaminated animals. Recent work on the depuration of other metals including zinc, copper, cadmium, and lead from *C. virginica* documents that they are lost fairly readily (SCRUDATO and ESTES, 1976; ZAROOGIAN, 1979; ZAROOGIAN and co-authors, 1979). The latter followed the loss of lead well after the end of the spawning period and observed a biological half-time of 5.5 wks. Probably the activities of amoebocytes are largely responsible for the loss of copper and zinc, but whether this extends to all other metals is less certain.

Gastropod molluscs

An experimental study of the absorption of dissolved copper by *Busycon canaliculatum* showed that the gills are the main site of uptake and the rate of absorption is proportional to the external concentration (BETZER and PILSON, 1975). The metal is transported around the body bound non-specifically to haemocyanin and excess copper is accumulated by the digestive gland. There is good evidence that, in reality, the food is the main source of heavy metals in gastropods. According to YOUNG (1975, 1977) this is so for iron and zinc in the carnivore *Nucella lapillus* and in the herbivore *Littorina obtusata.* In the same species the diet is also the major source of arsenic (KLUMPP, 1980b). A similar conclusion can be drawn from the work of STEWART and SCHULZ-BALDES (1976) on the absorption of lead from algae by *Haliotis.*

It can be inferred from the distribution of most metals in the tissues of *Haliotis tuberculata* (Table 3-12) that they are stored in the digestive gland and excreted via the kidneys; however, these organs seem to have both functions to some extent (CROFTS, 1929). Copper and nickel have distributions rather different from the other metals, copper occurring in haemocyanin in the blood and nickel being concentrated in the mantle. Copper storage granules have been found in *Littorina littorea* by MARTOJA and co-authors (1980), and CROFTS (1929) reported granules containing iron in the digestive gland, crop, and right kidney of *Haliotis tuberculata,* as well as 'chalk granules' in its digestive gland. Granules of this latter type also occur in the digestive gland of *Nucella lapillus;* they contain high levels of phosphate, calcium, zinc, magnesium, and copper (IRELAND, 1979). This observation may explain, in part at least, the ability of *N. lapillus* to tolerate high concentrations of zinc and copper. STENNER and NICKLESS (1974a) found 2900 ppm of zinc, 905 ppm of copper, and 780 ppm of cadmium in individuals from the Bristol Channel compared with 'normal' values of 345 ppm, 70 ppm, and 36 ppm respectively. Cadmium in *N. lapillus* and other species, including *Patella vulgata* and *Littorina littorea*, is detoxified and stored as the metallothionein (NOËL-LAMBOT and co-authors, 1978a, 1980). Some copper in contaminated *P. vulgata* also occurs in this form (HOWARD and NICKLESS, 1977), but not zinc which is presumably stored in granules of the type already referred to in *N. lapillus.*

There is little information about the routes of excretion in gastropods. However, a study by BROWN and BROWN (1965) on the fate of thorium dioxide following its injection into species of *Bullia,* showed that the particles were picked up by phagocytic amoebocytes and eliminated into the mantle cavity via the heart wall, pericardial cavity, renopericardial canal and kidney lumen.

Metal levels in various species are shown in Table 3-12. The concentration is governed to some extent by the size of the organism and, in the majority of cases, values decrease with increasing size or are independent of size (BOYDEN, 1977). Cadmium tends to be an exception and, especially in contaminated areas, its concentration increases with increasing size; in *P. vulgata* from the Bristol Channel, this is directly related to the build-up of Cd-metallothionein which is absent in young limpets (NOËL LAMBOT and co-authors, 1980). UEDA and TAKEDA (1979) observed that total concentrations of mercury and methyl mercury increased with increasing size in both muscle

Table

Metal levels (ppm dry weight) in some gastropod and

	Ag	Al	As	Concentration Cd	Co	Cr	Cu
Herbivorous gastropods							
Haliotis tuberculata							
(English Channel)							
(whole)	2·9	67	—	5·6	0·44	0·88	39
(digestive gland)	9·9	393	—	46	3·6	2·1	17·6
(right kidney)	5·5	—	—	43	1·1	6·4	24·2
(muscle)	0·35	—	—	0·24	0·12	0·38	12·1
Haliotis rufescens (Californian coast)							
(digestive gland)	13·9	—	—	184	—	1·4	11·2
(muscle)	2·8	—	—	0·17	—	—	2·6
Olivella biplicata (Californian coast)							
(whole)	2·1	84		2·3			177
Littorina littorea (English Channel)							
(whole)	2·5	286	14*	2·0	1·6	<2·5	70
Acmaea scabra (Californian coast)							
(min. values) (whole)				6·2			3·3
Patella vulgata (English Channel)							
(whole)	1·7	249	24*	12	1·4	<2	10
Crepidula fornicata							
(English Channel) (whole)	5·5	311	8·7*	1·1		<3·2	132
Carnivorous gastropods							
Buccinum undatum (whole)			11*	6·5	7·3		123
Scaphander lignarius (whole)				6·8	26·4		44·5
Nucella lapillus (English Channel)							
(whole)	2·5	42	38*	23	0·8	<1·4	66
Polinices lewisii (Californian coast)							
(whole)	0·7	51		0·3			115
Cephalopods							
Octopus vulgaris							
(Mediterranean Sea)							
(whole)				1·2			260
(digestive gland)				50			2500
(muscle)				0·08			26
Sepia officinalis (gills)			198	0·11			
Alloteuthis subulata							
(English Channel) (whole)	1·0	72		1·4	0·7	<0·4	146
Sthenoteuthis bartrami (Japan coast)							
(digestive gland)‡					—		2160
(muscle)					0·01		5
Loligo opalescens (S California)							
(digestive gland)	45·9			121			8370
Symplectoteuthis oualaniensis							
(digestive gland)	24·1			782			1720
Ommastrephes bartrami							
(Pacific Ocean) (digestive gland)	12·1			287			195

*Leatherland and Burton (1974); †Smith and Burton (1972); ‡Converted to dry weight basis × 4.

cephalopod molluscs (Compiled from the sources indicated)

Concentration								
Fe	Hg	Mn	Ni	Pb	Sb	Sn	Zn	Source
474	—	3·3	13·6	2·1	—	—	98	BRYAN and co-authors (1977)
4920	—	19·1	1·3	14·6	—	—	656	
1200	—	16·6	6·8	21·1	—	—	298	
30·3		0·49	0·29	0·68	—	—	38	
—	0·12	—	2·6	9	—	—	556	ANDERLINI (1974)
—	0·03	—	0·7	<0·1	—	—	42	
530		13·5	1·8	8·2			127	SCHWIMER (1973)
435	0·47*	26	3·2	9	0·095*		99	BRYAN (1976b)
620		4·5	6·1				32	FLEGAL (1979)
973	0·16*	6	1·1	9	0·07*		107	BRYAN (1976b)
376	0·30*	130	5·7	2·3	0·039*	0·71†	111	BRYAN (1976b)
65	0·38*	6·0	5·6	9·1	0·012*	0·33†	508	BOYDEN (1977)
		7·9	54·0	44·5			984	BOYDEN (1977)
234	0·44*	13	1·3	5	0·011*		351	BRYAN (1976b)
313		6·5		5·0			132	SCHWIMER (1973)
140		5					150	MIRAMAND and GUARY (1980)
700		7					1450	
30		3					70	
	0·90				0·026		99	LEATHERLAND and BURTON (1974)
71		3	<1	0·2			83	BRYAN (1976b)
324		2					312	UEDA and co-authors (1979)
4		0·5					48	
87							449	MARTIN and FLEGAL (1975)
319							513	
399							163	

and midgut from *Buccinum tenuissimum*. About 95% of the mercury in muscle was in the methyl form compared with 65% in the midgut, and it was suggested that demethylation may occur in this organ.

A comparison between concentrations of metals in *Littorina littorea* and in the brown seaweed *Fucus vesiculosus* indicated that whereas concentrations of silver, cadmium, nickel, and lead in the winkle tended to be proportional to those in the weed, the levels of copper and zinc did not increase in proportion—suggesting that they are to some extent regulated (BRYAN, 1979). According to BETZER and PILSON (1975) copper is regulated rather crudely by *Busycon canaliculatum,* and both ANDERLINI (1974) and BRYAN and co-authors (1977) concluded that zinc is probably regulated in muscle tissue in *Haliotis rufescens* and *H. tuberculata*.

Cephalopod molluscs

There is a paucity of information on heavy metals in cephalopods—certainly not sufficient to obtain a coherent picture. *Octopus vulgaris* obtains copper largely from its diet of crustaceans (GHIRETTI and VIOLANTE, 1964) and its branchial hearts appear to be implicated in metal metabolism; they accumulate cobalt very strongly (NAKAHARA and co-authors, 1979). Branchial hearts and the digestive gland of *Sepia officinalis* contain storage bodies (SCHIPP and HEVERT, 1978); a role for the branchial gland in the metabolism of copper and zinc has been proposed by DECLEIR and co-authors (1978). The high concentrations observed in the digestive gland of different species are shown in Table 3-12. There is not much evidence for the removal of metals from cephalopods, but copper and zinc are lost in the urine and rectal fluid of *Octopus dofleini* (POTTS and TODD, 1965).

Crustaceans

There is a considerable body of information on the absorption and loss of metals by crustaceans, although no single species has been exhaustively studied (cf. Table 3-13). In sea water the most readily available forms of metals such as cadmium and copper are probably the divalent ions (ENGEL and FOWLER, 1979; YOUNG and co-authors, 1979b); however, in many cases the food may be a more important source of metal than the water (Table 3-13). There is no definite rule, since in the same species both views may be correct, depending on the metal, the ambient concentrations and the type of food. In the euphausiid *Meganyctiphanes norvegica,* for example, the food is the principal source of zinc, cadmium, and selenium, but not mercury, at natural levels in the sea (FOWLER and co-authors, 1976a).

In decapod crustaceans the absorption of zinc, and probably several other metals, from solution is almost certainly a passive process involving adsorption on to the cuticle of the gills and inward diffusion, probably attached to organic molecules (BRYAN, 1971). Zinc is transported to the tissues bound to blood proteins, mainly haemocyanins, and can be stored in the hepatopancreas or excreted via the antennal gland, gut or gills. The importance of these various routes depends not only on the level of the metal in the body but also on the species; since losses over the gills seem to be important in the crab *Carcinus maenas,* losses in the urine are appreciable in the lobster *Homarus gammarus* and losses in the faeces are virtually the only route in the fresh water crayfish *Austropotamobius pallipes* (BRYAN, 1967). As a result of these processes, excess zinc, whether absorbed from the water or from food, can to a large extent be eliminated. Thus BRYAN (1971) showed that for a 500-fold increase in the level in sea water, the concentration of zinc in the blood of the crab *C. maenas* was only doubled in 32d, and that of the whole body was quadrupled. Although concentrations in the gills and hepatopancreas increased markedly, as might be expected from their functions, no changes were observed in muscle or gonads. This ability to regulate zinc is not an experimental artefact, since crabs exposed to very abnormal concentrations of zinc and copper in Restronguet Creek do not show the degree of change observed in some other phyla (cf. Table 3-28). Copper by virtue of its association with haemocyanin is also probably regulated fairly efficiently by decapods, but this process is superimposed on concentration variations which occur naturally during the moult cycle and are generally greater for copper than zinc (MARTIN, 1975). Metals can be removed from the body by incorporation in the moulted shell, and as much as 50% of the total cadmium was removed in this way from the shrimp *Lysmata seticaudata* (FOWLER and BENAYOUN, 1974). Some experimental evidence for the regulatory ability of decapods towards zinc, cadmium, lead, and mercury is given in Table 3-14; it shows that, unlike zinc, exposure to other metals increases the concentration in muscle tissue—an indication of poorer regulatory ability, although not necessarily reflecting an inability to detoxify the metal.

Mechanisms of detoxification in decapod crustaceans include the induction of heavy-metal-binding proteins: cadmium and zinc metalloproteins have been induced in the hepatopancreas of the crab *Scylla serrata* (OLAFSON and co-authors, 1979), cadmium and copper metallothioneins have been found in the hepatopancreas of wild *Cancer pagurus* (OVERNELL and TREWHELLA, 1979), and cadmium, zinc, and copper-binding proteins have been reported in the hepatopancreas of *Carcinus maenas* (RAINBOW and SCOTT, 1979). Granular storage mechanisms also exist: excess copper is stored as granules in the hepatopancreas of the shrimp *Crangon crangon* (DJANGMAH, 1970) and in the hepatopancreatic caeca of the amphipod *Corophium volutator* (ICELY and NOTT, 1980). The fixation of zinc in the antennal gland of *Crangon crangon* has been described by ELKAIM and CHASSARD-BOUCHAUD (1978); lead is associated with calcium storage granules in the hepatopancreas of *Carcinus maenas* (HOPKIN and NOTT, 1979). An interesting mechanism for the detoxification and removal of cadmium from heavily contaminated shrimp *Penaeus duorarum* appears to involve incorporation in the gills followed by sloughing off the affected parts (NIMMO and co-authors, 1977). The existence of such detoxification mechanisms often leads to the presence of high metal levels in specific organs such as hepatopancreas and gills (Tables 3-14 and 3-15).

In crustaceans, by far the highest concentrations of zinc, copper, and cadmium have been found in species of barnacles, although they are not stored as metallothioneins (RAINBOW and co-authors, 1980; Table 3-15). Copper granules occur in the parenchyma cells of the midgut of contaminated barnacles *Semibalanus balanoides* (WALKER, 1977); the presence of zinc phosphate granules leads to high body concentrations even in uncontaminated animals (WALKER and co-authors, 1975). These metals are not permanently fixed, and are gradually lost from contaminated animals in uncontaminated water (WALKER and FOSTER, 1979)

Echinoderms

The concentrations of metals in a few species are shown in Table 3-16. However, there is very little information about the metabolism of metals in echinoderms. An exception is the work of BUCHANAN and co-authors (1980) on burrowing spatangoids including

Table 3-13

Some experimental studies on uptake and loss of metals in crustaceans (Compiled from the sources indicated)

Metal	Group	Species	Remarks	Source
As	Shrimp	*Lysmata seticaudata*	Absorption via food chain and metabolism of organic and inorganic forms	WRENCH and co-authors (1979)
	Crab	*Carcinus maenas*	Absorption from food in different chemical forms and subsequent excretion	UNLÜ (1979)
	Lobster	*Homarus americanus*	Organoarsenic comes from food and is not synthesized from arsenate. Trimethyl arsonium lactate identified	COONEY and BENSON (1980)
Cd	Shrimp	*Crangon crangon*	Accumulation from solution by tissues at different concentrations and subsequent loss	DETHLEFSEN (1977)
	Shrimp	*Penaeus duorarum*	Accumulation from solution by tissues over a wide range of concentrations; histological effects	NIMMO and co-authors (1977)
		Palaemonetes pugio	Accumulation from solution by tissues; uptake from food may be less important than uptake from water	
		Palaemonetes vulgaris	Accumulation from solution by tissues over a range of concentrations	
	Crab	*Carcinus maenas*	Accumulation from solution by tissue; increased uptake at lowered salinity	WRIGHT (1978); WRIGHT and BREWER (1979)
		Cancer pagurus	Accumulation from food and induction of Cd-metallothionein in hepatopancreas	OVERNELL and TREWHELLA (1979)
	Barnacle	*Semibalanus balanoides*	Accumulation reduced by EDTA in water but not by presence of humate or alginate; no Cd-metallothionein formed	RAINBOW and co-authors (1980)
	Copepod	*Pseudodiaptomus coronatus*	Absorption from water and food; water more important	SICK and BAPTIST (1979)
Co	Shrimp	*Crangon crangon*	Absorption of cobalt-60 from water and food	WEERS (1975)

Cu	Shrimp	*Crangon crangon*	Variation in concentrations of tissues with moult; storage of excess copper as granules in hepatopancreas	DJANGMAH (1970)
	Barnacle	*Semibalanus balanoides*	Accumulation in field for over 2 yr and seasonal variation; loss from contaminated barnacles	WALKER and FOSTER (1979)
Fe	Crab	*Carcinus maenas*	Uptake of soluble and particulate iron-59	MARTIN (1977)
	Barnacle	*Semibalanus balanoides*	Absorption mainly from food	YOUNG (1974)
Hg	Lobster	*Homarus americanus*	Accumulation from low concentration in solution by tissues	THURBERG and co-authors (1977)
	Shrimp	*Lysmata seticaudata*	Methyl mercury much less readily excreted than inorganic form	FOWLER and co-authors (1978)
		Palaemon debilis	Mercury absorbed from water rather than sediment under field conditions	LUOMA (1977a)
	Euphausiid	*Meganyctiphanes norvegica*	Flux of mercury through organism; importance of absorption from water	FOWLER and co-authors (1976a)
Mn	Lobster	*Homarus gammarus*	Absorption by tissues from food and water and loss by different routes; importance of uptake from food	BRYAN and WARD (1965)
Pb	Crab	*Macropodia rostrata*	Uptake by tissues from low concentrations in water	CHAISEMARTIN and co-authors (1978)
Se	Shrimp	*Lysmata seticaudata*	Absorption by tissues from water and food and subsequent loss	FOWLER and BENAYOUN (1976, 1977)
Zn	Crab	*Carcinus maenas*	Absorption by tissues from food and water and loss by different routes; Zn regulation	BRYAN (1966)
		Carcinus maenas	Absorption from food and water suggests that the food pathway may not be of such overriding importance as sometimes assumed, especially in shrimp	RENFRO and co-authors (1975)
	Shrimp	*Lysmata seticaudata*		
	Barnacle	*Semibalanus balanoides*	Accumulation in field for over 2 yr and seasonal variation; loss from contaminated animals	WALKER and FOSTER (1979)
	Barnacle	*Semibalanus balanoides*	Absorption mainly from food	YOUNG (1974)
	Euphausiid	*Meganyctiphanes norvegica*	Flux of zinc through organism; importance of absorption from food	SMALL and co-authors (1973)

Table 3-14

Influence of low sea-water metal concentrations (ppm) on tissue metal levels of 4 decapod crustaceans (Compiled from the sources indicated)

Species	Concentration in water (ppb)	Hepato-pancreas	Gills	Muscle	Exoskeleton	Blood	Whole animal	Source
			Cadmium† (13-d exposure)					
Crangon crangon	Control	2·97	0·92	0·14	0·1	—	—	Dethlefsen (1977)
	20	14·6	13·1	0·23	0·74	—	—	
			Mercury* (30-d exposure)					
Homarus americanus	Control	0·12	0·14	0·23				Thurberg and co-authors (1977)
	3	4·12	40·9	0·54				
	6	15·2	85·3	1·0				
			Lead* (7-d exposure)					
Macropodia rostrata	8	0·003	0·0086	0·0026	0·00017	0·0078	0·009	Chaisemartin and co-authors (1978)
	102	0·037	0·112	0·031	0·00116	0·093	0·128	
			Zinc* (mean 28-d exposure)					
Homarus gammarus	Control	36·6	14·9	15·1	4·6	7·3		Bryan (1964)
	100	47·8	28·3	13·7	8·4	8·3		

* Wet tissue. † Dry tissue.

Brissopsis lyrifera and *Echinocardium cordatum*. The connective tissue layer of the large intestine contains an extracellular deposit of ferric phosphate amounting to 30% of the tissue weight in large specimens of *Brissopsis*. It is thought that the source of iron, which accumulates with age, is that in the interstitial water of sediment ingested under reducing conditions. Comparatively high concentrations of other metals including lead are also found in the large intestine (Table 3-16).

Tunicates

Several species including *Ascidia nigra* are notable for accumulating high concentrations of vanadium in their blood cells or vanadocytes (Table 3-16), while in other species such as *Molgula manhattensis* the blood cells accumulate iron (MACARA and co-authors, 1979). From field and experimental work on a wide range of species, STOECKER (1980) has concluded that high vanadium contents are involved in chemical defence. Some of the earlier evidence that different species of tunicates may accumulate high concentrations of other metals including niobium and tantalum has been reviewed by BRYAN (1976b), but, apart from tissue analyses (Table 3-16) there is a dearth of information on the metabolism of metallic contaminants by tunicates.

Fish

PENTREATH (1973b) has suggested that the absorption of trace metals by the gills of plaice *Pleuronectes platessa* is a passive process, since a favourable concentration gradient may be achieved by the adsorption of metals on to mucus covering the gills. This may be so, since COOMBS and co-authors (1972) found levels of copper and zinc in mucus from this species to be 100 times and 20 times, respectively, greater than those of sea water. However, experimental work on plaice has shown that the food is probably the main source of iron, cobalt, zinc, manganese, methyl mercury, cadmium, and silver (PENTREATH, 1973b, c, 1976,a, b, c, d, 1977a, b). Estimates of the retention of metals from food (*Nereis diversicolor*) gave values including 89% for methyl mercury, 11% for inorganic mercury and less than 5% for silver and cadmium. Even at these low levels of retention, it was calculated that at fairly natural levels in the environment, only for inorganic mercury (PENTREATH, 1976d) does the contribution of the water approach that of the food. RADOUX and BOUQUEGNEAU (1979) have concluded that uptake of inorganic mercury from the water is the major route in *Serranus cabrilla* and FUJIKI and co-authors (1978) have drawn a similar conclusion for methyl mercury in *Chrysophrys major*. In the ray *Raja clavata* the importance of food as a source of trace metals, including mercury, is greater than in the plaice (PENTREATH, 1973d, 1976e); this may be because, unlike teleosts, elasmobranchs do not drink sea water. (Volume I: HOLLIDAY, 1971). Although it might be expected that uptake of metals from the water would be more important in young fish, PENTREATH (1976a) concluded that, even in the earliest stages, food is the major pathway for the uptake of zinc and manganese in plaice.

By comparison with levels in most invertebrate phyla, concentrations of metals in fish tend to be on the low side (BRYAN, 1976b; Table 3-17). Mercury is the principal exception to this rule, since in some species very high concentrations have been observed, a point that will be discussed further when the biomagnification of metals along food chains is considered (cf. p. 1353). In many fish, most of the mercury in the body occurs as the more toxic methyl mercury; it is presumably absorbed in this form since it does not appear to be methylated by the fish (PENTREATH, 1976b; HUCKABEE

Table

Metal concentrations (ppm dry weight) in a range of

Species	Ag	As	Cd	Co	Cr	Cu	Fe	Hg
Acartia clausi (copepod)	—	2·9	0·61	0·28	3·3	55	738	0·29
Balanus amphitrite (barnacle, body)					3·9	54		1·14
Semibalanus balanoides			10–28			232		
(barnacle, body)			60			3232		
*Squilla mantis** (stomatopod)			1·8	1·2		63	66	0·6
Corophium volutator (amphipod)						77	494	
Meganyctiphanes						259	325	
norvegica (euphausiid)			1·3			66	140	
*Crangon crangon** (shrimp)			1·2		2·8	92		0·6
Pandalus borealis (shrimp)			0·9			95		0·4
Carcinus maenas (crab)			2·4		0	33	1345	
Homarus gammarus (lobster)				0·08		94	20	
hepatopancreas			12	0·36		1074	40	
muscle			0·3	0·012		22	3·7	
Decapod crustaceans (geometric mean of literature values)	0·4	30	1	0·2	0·3	70	160	0·4

* Wet weight values × 4.
† See p. 1302.

and co-authors, 1978). However, RUDD and co-authors (1980) recently reported that methylation by the intestinal contents had been observed in 6 species of fresh water fish. In addition, some species including *Fundulus heteroclitus* and *Salmo gairdneri* are capable of slow demethylation (RENFRO and co-authors, 1974; OLSON and co-authors, 1978). The ability to demethylate methyl mercury seems to be well developed in some predatory species, particularly marlin, where very high total concentrations are found (Table 3-17). This was suggested by RIVERS and co-authors (1972) to account for the low proportion of organic mercury observed in both liver and muscle tissue from the Pacific blue marlin *Makaira nigricans* (ampla). In marine mammals, selenium appears to be involved in the detoxification of mercury (cf. p. 1345); in *Makaira nigricans* and the black marlin *Makaira indica*, concentrations of both elements appear to be closely related in kidney, liver, and muscle (MACKAY and co-authors, 1975; SHULTZ and ITO, 1979). While experimental work with *F. heteroclitus* has shown that treatment with selenium changes the tissue distribution of methyl mercury (SHELINE and SCHMIDT-NIELSEN, 1977), how they are associated in fish remains unknown—although it might be as a protein.

In sea-water-adapted eels *Anguilla anguilla*, the induction of the protein metallothionein has been shown to detoxify inorganic mercury in the gills (BOUQUEGNEAU, 1979). This low-molecular-weight protein seems to be particularly concerned with the detoxification of cadmium in fish and has, following exposure, been found in the liver of *Fundulus heteroclitus* (PRUELL and ENGELHARDT, 1980), gills and liver of eels (NOËL-LAMBOT and co-authors, 1978b), and gills, liver, and kidney of the coho salmon *Oncorhynchus kisutch*

3-15

crustacean species (Compiled from the sources indicated)

Mn	Ni	Pb	Sb	Se	Sn	V	Zn	Locality	Source
9·3	—	—	0·31	1·9	—	—	1270	Elefsis Bay	ZAFIROPOULOS and GRIMANIS (1977)
		10·2						Grado Lagoon	BARBARO and co-authors (1978)
							27,840	Southend	RAINBOW and co-authors (1980)
							113,250	Dulas Bay†	
2·1	2·9	1·5					59	Gulf of Trieste	MAJORI and co-authors (1978c)
32		60					104	Menai Strait	ICELY and NOTT (1980)
33		60					109	Dulas Bay†	
2·0							85	Mediterranean Sea	FOWLER and co-authors (1976b)
		12·8					88	England and Wales coast	MURRAY (1979)
		2·7					103	Oslofjord	NEELAKANTAN (1976)
673	15·6	38					76	Ireland	WILSON (1980)
128							66	Scotland	BRYAN (1976b)
12		<0·8					83		
3·1		<0·8					59		
40	1	1	0·02	4	2	2	80		BRYAN (1976b)

(REICHERT and co-authors, 1979). Zinc and copper metallothionein occurs in the liver of normal eels and it is suggested that this protein has a role in the metabolism of these essential metals (NOEL-LAMBOT and co-authors, 1978b); lead, however, does not appear to associate with specific detoxification proteins in fish (REICHERT and co-authors, 1979). Detoxification of arsenic in the brown trout *Salmo trutta,* by conversion to less toxic organic forms, is probably carried out by the intestinal flora (PENROSE, 1975).

Heavy metals are probably lost from fish in the urine, via the liver, bile duct and gut, via the gills and also over the body surface. However, there is little information about the relative importance of different routes. VARANASI and MARKEY (1978) injected lead and cadmium into sea-water-adapted salmon *Oncorhynchus kisutch* and observed rapid uptake by the skin followed by a gradual release. They suggest that the skin and scales act as a temporary detoxification site, especially for lead—suggesting a relation to calcium metabolism—and that mucus turnover is an important process for lead and cadmium excretion in fish. In the same species there is also a possibility that protein-bound cadmium may be excreted with the gill mucus (REICHERT and co-authors, 1979). Losses of arsenic from the liver to the gut in the bile were observed by PENROSE (1975) following its injection into the trout *Salmo trutta.*

Several metals, including copper, zinc, molybdenum, and cadmium but not methyl mercury, can be regulated by some fresh water fish (see reviews by BRYAN, 1976b, 1979). In marine teleosts there is evidence for the regulation of copper and zinc, but not mercury, in *Pleuronectes platessa* (SAWARD and co-authors, 1975; PENTREATH, 1976a, b; MILNER, 1979; Table 3-18). Manganese and iron may also be regulated (CROSS and

Table

Metal concentrations (ppm dry weight) in some coelenterates,

Species	Ag	As	Cd	Co	Cr	Cu	Fe
Coelenterates							
Tealia felina	0·05	72*	0·07*	<0·15	0·37	57	730
Alcyonium digitatum	<0·03		4·1	0·57	<0·41	9·7	250
Echinoderms							
Ophioderma longicauda	0·07			0·2	0·46		130
Echinaster sepositus	0·25			0·38	0·83		170
Marthasterias glacialis	7·3	5·8*		0·09	1·6		110
Henricia sanguinolenta	0·6		<3·5	0·19	0·46	8·2	1800
Pisaster brevispinus (ray)	2·4		0·7			15·8	102
Asterias rubens		10*	1·6*	<0.3		4·3	37
Sphaerechinus granularis	0·32			0·66	6·4		1100
Paracentrotus lividus	0·32			0·28	4·8		600
Arbacia lixula	0·30			0·32	13		620
Strongylocentrus droebachiensis		3·9	1·5			2·9	495
Brissopsis lyrifera (test)			0·2			3	867
(gonads)			0·4			7	184
(large intestine)			—			7	17·3%
Holothuria tubulosa	0·05			0·11	0·8		74
Tunicates							
Ascidia nigra (tunic)						0	160
(body)						100	230
(blood cells)						30	270
Ciona intestinalis	0·021			0·52	5·5		880
Phallusia sp.		33	0·43			7·8	
Halocynthia sp.		27	0·28			6·1	
Microcosmus sulcatus	0·031			1·9	6·6		840
Styela clava		4·8					
Botrylus schlosseri		6·6	2·7				

* LEATHERLAND and BURTON (1974); ‡SMITH (1970); †DANSKIN (1978).

co-authors, 1973), possibly also cadmium (WESTERNHAGEN and co-authors, 1980; Table 3-18) but not lead or chromium.

Regulation is never perfect: for example, flounder *Platichthys flesus* from the Severn estuary having a diet of *Macoma balthica* (non-regulator) contained higher levels of zinc than flounder having a diet of crustaceans and small fish which tend to regulate zinc (HARDISTY and co-authors, 1974). Mercury shows little sign of being regulated and a linear relation between concentrations in sea water and those in the flesh of teleosts has been observed in the field (GARDNER, 1978; Table 3-18). Concentration factors (wet

3-16

echinoderms, and tunicates (Compiled from the sources indicated)

Hg	Mn	Ni	Pb	Sb	Se	Sn	V	Zn	Source
0·86*	9·3	3·3	2·6	0·022*				280	RILEY and SEGAR (1970)
	3·7	17	24					46	
				0·01	1·9			59	PAPADOPOULOU and co-authors (1976)
				0·020	4·4	0·14‡		120	
0·92*				0·020	3·0	0·26‡		65	
	34	3·7	<0·5					240	RILEY and SEGAR (1970)
		14	31					53	SCHWIMER (1974)
0·22*	6·5	1·5	2·3	0·010*				220*	RILEY and SEGAR (1970)
				0·13	0·80			100	PAPADOPOULOU and co-authors (1976)
				0·07	1·1	0·15‡		54	
				5·5	2·4			95	
								34	BOHN (1979)
	592			1				3	BUCHANAN and
	41			4·4				239	co-authors (1980)
	842			721				259	
				0·05	3·4			36	PAPADOPOULOU and co-authors (1976)
							320		MACARA and co-authors (1979)
							2120		
							27,760	100	PAPADOPOULOU and
				0·16	1·2		92†		KANIAS (1977)
0·18	148				5·51			88	KOSTA and co-authors (1978)
0·29	94				2·51	2·3‡	175†	67·2	
				0·1	5·1			180	PAPADOPOULOU and KANIAS (1977)
0·13				0·15				—	LEATHERLAND and BURTON (1974)
0·57				0·26				135	

basis) ranged from $2{\cdot}9 \times 10^3$ in the least contaminated to $10{\cdot}6 \times 10^3$ in the most contaminated inshore waters, possibly reflecting the tendency for concentration–size regressions in fish to steepen in contaminated areas (RENZONI, 1976). This is because the ability of fish to excrete methyl mercury, the most common form in fish, is rather limited (PENTREATH, 1976c) and in many species equilibrium with the environment is never achieved. Thus concentrations of mercury in fish muscle frequently increase with size (Fig. 3-5) and the slope of the graph tends to increase with increasing levels of contamination. Size-concentration relationships for other metals usually show variable patterns.

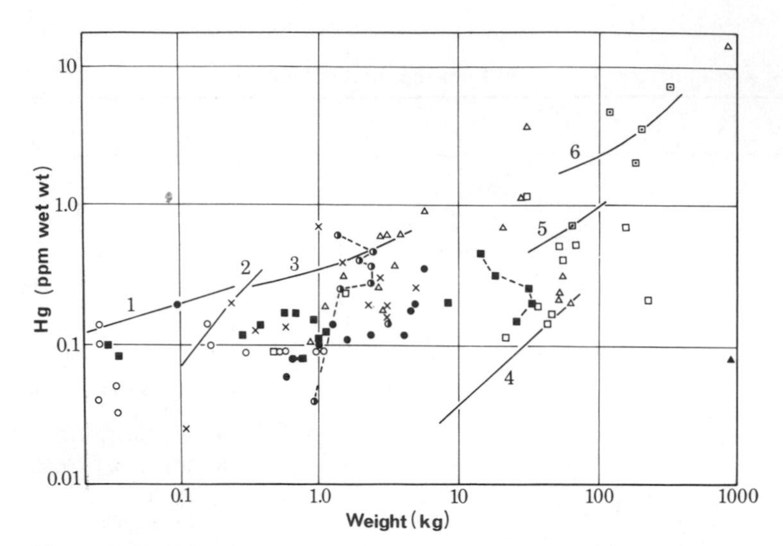

Fig. 3-5: Relationship between approximate weight of fish and muscle mercury concentration with the use of mean values from literature. ○ clupeids; ● gadoids; ■ flatfish; ◑ sablefish; □ mackerel, tuna, swordfish; ⊡ marlin; △ sharks; ▲ basking shark; × others.* Continuous lines: concentration-weight relation for (1) *Merlangus merlangus* (CLERK and co-authors, 1974); (2) *Moxocephalus quadricornis* (NUORTEVA and HÄSÄNEN, 1975); (3) *Pomatomus saltatrix* (CROSS and co-authors, 1973); (4) *Neothunnus albacora* (MENASVETA and SIRIYONG, 1977); (5) *Tetrapturus audax*; (6) *Makaira nigricans* (SHOMURA and CRAIG, 1974). Broken lines: geographic variation in sablefish *Anaplopoma fimbria* and halibut *Hippoglossus stenolepis* (HALL and co-authors, 1976a, b). (Original)*

For example, concentrations of arsenic clearly increased with increasing size in the muscle and liver of the sculpin *Myoxocephalus scorpius*, whereas trends were less obvious for cadmium, copper, lead, and zinc (BOHN and FALLIS, 1978). In white muscle from blue fish *Pomatomus saltatrix* and a bathy-demersal fish *Antimora rostrata*, CROSS and co-authors (1973) showed that concentrations of manganese, iron, copper, and zinc, either fell with increasing size or remained constant, whereas that of mercury increased. Similarly in the dogfish *Squalus mitsukurii* concentrations of cadmium, copper, iron, and zinc tend to be higher in embryos than the mother, whereas mercury levels are highest in the mother (TAGUCHI and co-authors, 1979; Table 3-17).

Marine mammals

A summary is given in Table 3-19 of some of the more recent values for heavy metal concentrations in seals, whales, sea otter, and dugong. The outstanding results are the high concentrations of cadmium in the kidneys of some species and the high concentrations of mercury and selenium in some livers.

*Other data from BECKETT and FREEMAN (1974); CHILDS and GAFFKE (1973); ESTABLIER (1972); FREEMAN and co-authors (1974); GILMARTIN and REVELANTE (1975); GREIG and co-authors (1975, 1977b); KNAUER and MARTIN (1972); PETERSON and co-authors (1973); RIVERS and co-authors (1972); ROBERTSON and co-authors (1975); SHULTZ and co-authors (1976); WALKER (1976).

For cadmium, the overall picture seems to resemble that in humans: the concentrations are very low in the tissues of young animals and tend to increase with age, particularly in kidney and liver (cf. ROBERTS and co-authors, 1976; MARTIN, 1979; Table 3-19). Similarly, cadmium is stored in these tissues in the form of Cd-metallothionein (OLAFSON and THOMPSON, 1974; LEE and co-authors, 1977).

Concentrations of mercury are usually much higher in the liver than in the kidney and in both tissues the concentrations increase with age; this trend is much less obvious in tissues such as muscle and brain where levels tend to remain low (Table 3-19). In the sea lion *Zalophus californianus,* virtually all the mercury in muscle exists in the methyl form, compared with about 50% in brain, 17% in kidney, and only 2% in liver (BUHLER and co-authors, 1975). Since the fish on which these animals feed contain a high proportion of methyl mercury, the large amount of inorganic mercury in the liver and kidney suggests that these tissues are the site of a demethylation system. KOEMAN and co-authors (1973) have shown that the high mercury levels in the livers of marine mammals are usually equalled on an atomic basis by those of selenium, and several examples are given in Table 3-19. In addition MARTIN and co-authors (1976) observed that in sea lions with a high liver mercury level there was a 1 : 1 : 1 relation for mercury, selenium, and bromine, suggesting that the latter elements are in some way involved in the demethylation and detoxification of mercury. Experiments in which seals were fed with methyl mercury for up to 24 wk showed that concentrations of organic and inorganic mercury increased in the tissues and that in the liver and kidney, but no other tissue, the level of selenium also increased (VEN and co-authors, 1979). However, it was not possible to identify the role of selenium in the demethylation mechanism. To a small degree the formation of Hg-metallothionein seems to be involved, since it accounted for about 20% of the mercury in sea-lion kidney but only 2 to 3% in the liver (LEE and co-authors, 1977). One explanation of the mercury–selenium relationship has been given by MARTOJA and VIALE (1977) who observed that storage granules in the liver of the whale *Ziphius cavirostris* consisted of mercuric selenide.

Other tissues where mercury is stored, at least in seals, are the fur and claws and their concentrations tend to be indicative of those of the animal as a whole (FREEMAN and HORNE, 1973). For example, in a large male grey seal *Halichoerus grypus* the levels of total mercury on a wet basis were; fur 12 ppm, claw 9.8 ppm, liver 30 ppm, kidney 5.7 ppm, muscle 1.6 ppm, and blubber 0.086 ppm.

In addition to age, the diet of marine mammals appears to have an important effect on their body mercury concentrations. For example, SERGEANT and ARMSTRONG (1973) found more mercury in the grey seal *Halichoerus grypus,* which has a diet of large fish and cephalopods, than in the harp seal *Pagophilus groenlandicus,* which feeds on small fish and crustaceans. Similarly, NAGAKURA and co-authors (1974) observed only 0.01 to 0.03 ppm (wet) of mercury in muscle from the fin whale *Balaenoptera physalis,* which feeds on krill, but more than 1 ppm in muscle from the sperm whale *Physeter catodon* which feeds on fish and squid (Table 3-19).

Although high mercury levels have been found in the tissues of marine mammals in areas where contamination would not be expected, there is evidence that contamination may sometimes be involved. For example, ROBERTS and co-authors (1976) concluded that, if age is taken into account, concentrations of mercury in the livers of seals *Phoca* sp. increase in areas of increasing industrialization: Canadian Arctic and Atlantic coast < West Scotland < East Anglia < Netherlands coast.

Table

Metal concentrations (ppm dry weight) in a range of

Species	Ag	As	Be	Cd	Co	Cr	Cu	Fe	Hg	Mn
Morone saxatilis (striped bass)										
Liver*	0·32	2·8	<0·004	1·2	—	—	8·4	—	2·2	—
Muscle*	0·012	1·0	<0·004	0·12	—	—	1·4	—	1·4	—
Myoxocephalus scorpius (shorthorn sculpin)										
Liver		81		41			7·6			
Muscle		40		1·4			4·1			
Pseudopleuronectes americanus (flounder)										
Liver*	0·56	—	—	0·56	—	<2·4	26	—	0·28	4·8
Muscle*	<0·4	—	—	<0·4	—	<1·2	4·4	—	0·24	1·2
Sargus annularis (sabaris)										
Liver	—	30	—	<0·3	0·41		31	420	1·1	
Muscle	—	2·5	—	<0·3	0·035		1·7	28	0·86	
Makaira indica (black marlin)										
Liver*		4		36·8			18·4		41·6	
Muscle*		2·4		3·6			1·6		29·2	
Squalus mitsukurii (dogfish)										
Adult muscle*				0·024			1·12		4·8	
Embryo muscle*				0·028			1·8	13·6	0·13	
Geometric mean literature values										
Whole	0·1	10		0·2	0·1	0·5	3	50	0·4	10

* Wet weight values × 4.

Seabirds

Some examples of metal concentrations in the tissues of seabirds are shown in Table 3-20. The distribution of cadmium between different tissues resembles the pattern in mammals, low levels being found in muscle and very high levels in kidney. Similarly age is important and in the great skua *Catharacta skua,* which is a top predator, concentrations of cadmium in the kidney and liver increase with age (FURNESS and HUTTON, 1979). In addition to the analyses of the puffin *Fratercula arctica* given in Table 3-20, OSBORN and co-authors (1979) found very little cadmium in the brain, skin, and feathers but 16 ppm in the gonads. As in other phyla, metallothioneins appear to be involved in the detoxification of cadmium, and bind copper and zinc also (BROWN and co-authors, 1977).

Fairly high concentrations of mercury occur in the liver and kidneys of birds (Table 3-20) and the proportion of methyl mercury in the tissues is generally high (GARDNER and co-authors, 1978; OSBORN and co-authors, 1979). There is no definite evidence for the demethylation of methyl mercury in the liver or kidney and the evidence for a

3-17

fish species (Compiled from the sources indicated)

Ni	Pb	Sb	Se	Sn	Te	Tl	V	Zn	Locality	Source
8	3·2	<0·04	2·64	1·32	0·56	≤0·02	0·16	144	Chesapeake	HEIT (1979)
4	2·0	<0·04	1·2	1·2	0·4	0·024	0·12	15·2	Bay	
	0·3							100	Strathcona	BOHN and FALLIS
	—							43	Sound	(1978)
3·2	<2·4							168	New York	GREIG and WENZLOFF
2	<1·2							20·8	Bight	(1977)
		<0·02	6·6					83	Aegean Sea	GRIMANIS and
		0·03	1·4					71		co-authors (1978)
	2·8		21·6					190	NE Australia	MACKAY and
	2·4		8·8					34·4		co-authors (1975)
								10	Japan	TAGUCHI and
								14·4		co-authors (1979)
1	3	0·02	1	—	—	—	1	80	Worldwide	BRYAN (1976b)

mercury–selenium relationship in these tissues is much weaker than it is in marine mammals, there being no signs of a 1 : 1 atom ratio (KOEMAN and co-authors, 1975; STONEBURNER and co-authors, 1980). In the great skua *Catharacta skua* concentrations of selenium exceed those of mercury in the liver and kidneys; however, the levels are related and increase markedly with age (FURNESS and HUTTON, 1979; Table 3-20). In addition, concentrations of cadmium and selenium are closely related in the kidney and it is suggested that selenium might play a role in the detoxification of both mercury and cadmium. The results in Table 3-20 are not inconsistent with the idea that levels of mercury and cadmium in the liver and kidney are dependent on the diet, the highest being found in fish-eating and bird-eating birds.

Routes for the removal of metals include excretion in the faeces and incorporation in the feathers or eggs. Studies on the loss of methyl mercury from contaminated mallards *Anas platyrhynchos* revealed that incorporation in the growing feathers is particularly important (STICKEL and co-authors, 1977). Analysis of feathers provides a means of monitoring mercury levels in birds without killing them and concentrations in the

Table 3-18

Influence of sea-water metal concentrations (ppm) on tissue metal levels of fish (Compiled from the sources indicated)

Species	Concentration in: Water (ppb)	Liver	Pancreas	Spleen	Gills	Kidneys	Muscle	Backbone	Skin	Source
				Cadmium* (96-d exposure)						
Limanda limanda (dab)	Control	0·43	—	—	0·22	—	0·07	0·054	0·20	WESTERNHAGEN and co-authors (1980)
	5	1·69	—	—	2·96	—	0·18	0·126	0·29	
	50	6·83	—	—	5·73	—	0·14	0·362	0·80	
				Copper† (100-d exposure)						
Pleuronectes platessa (plaice)	4·9	30‡	—	—		—	15–30	—		SAWARD and co-authors (1975)
	8·8	71‡	—	—		—	29·5	—		
	17·8	147‡	—	—		—	42	—		
	47·6	567‡	—	—		—	19·1	—		
				Chromium(VI)† (44-d exposure)						
Citharichthys stigmaeus (sand dab)	Control	ND	—	—		—	ND	—	1·4	SHERWOOD and WRIGHT (1976)
	16	ND	—	—		—	ND	—	3·2	
	95	0·2	—	—		—	0·1	—	3·5	
	550	1·9	—	—		—	0·6		11	
	3000	13	—	—		—	3·2	—	32	

				Mercury* (from field)						
Teleost fish	0·018						0·16	—		GARDNER (1978)
	0·034						0·26	—		
	0·058						0·64	—		
				Mercury† (57-d exposure)						
Mugil auratus	Control	0·67	—	—	0·32	—	0·47	—	—	ESTABLIER and
(golden grey mullet)	100	344	—	—	32	—	10·3	—	—	co-authors (1978)
				Methyl mercury† (45-d exposure)						
	Control	0·62	—	1·61	0·08	3·94	0·36	—	—	
	8	111	—	77·8	24·4	78·9	11·8	—	—	
				Lead† (36-d exposure)						
Gillichthys mirabilis	Control	0·48	—	11·7	8·77		0·33	2·69	2·28	SOMERO and
(goby)	2650 (100% SW)	9·04	—	197	69·1		2·48	26·8	29·7	co-authors (1977a)
	2650 (25% SW)	14·6	—	365	119		4·71	45·9	66·1	
				Zinc‡ (25-d exposure)						
Scyliorhinus canicula	Control	47	230	87	—	120	57	—	—	FLOS and co-authors
(dogfish)	15000	242	620	147	—	93	68	—	—	(1979)

* Wet tissue. † Dry tissue. ‡ Viscera. ND, Not detectable.

Tab

Metal concentrations (ppm wet weight) in mari

Location, species	Age, length or weight	Cd			Cu			Hg†		
		Muscle	Liver	Kidney	Muscle	Liver	Kidney	Muscle	Liver	Kidne
E. Wadden Sea										
Phoca vitulina	Increase									
(common seal)	with age		0·01–0·2	0·06–0·38		2·6–17	2·3–4·0			
United Kingdom	Increase									
Phoca vitulina	with age	0·13	0·2–0·8	0·1–0·6				1·4	1·5–106	0·2–4·
Finland										
Phoca hispida (ringed seal)	Adult							0·47–1·6	14–300	2·8–5·
Canadian Arctic										
Phoca hispida	10·2 yr							0·91	16·1	
									(0·89)	
Okhotsk Sea										
Histriophoca fasciata	Adult	0·17	2·57							
(ribbon seal)										
California	Pup		—	—		146*	28*		9·6*	4·6*
Zalophus californianus	11.5 yr		15·1*	115*		86*	22*		747*	28·4*
(sea lion)	Adult	0·08	1·6	7·2				0·84	96	5·4
	Adult							(0·84)	(1·5)	(0·7
Bay of Fundy										
Phocoena phocoena	0·5 yr							0·4	0·75	0·48
(harbour porpoise)	8 yr							1·52	17·2	4·3
								(89%)	(17%)	(42%
Cumberland Island										
Globocephala macrorhyncha	3·68 m ♀		19·8	27·1					454	55·7
(short finned pilot whale)	3·85 m ♂		11·3	27·8					56·9	4·7
Japan										
Globocephala melaena	3·4 m							4·16		
(pilot whale)										
Antarctic Ocean										
Balaenoptera physalis	17·4–22·2 m							0·01–0·03		
(fin whale)								(<0·008)		
Antarctic Ocean and N Pacific										
Physeter catodon	9·3–15·5 m							1·32		
(sperm whale)								(0·93)		
California										
Enhydra lutris	1·2 kg	ND*	1·0*	ND*	13·7*	171*	41*	0·65*	1·16*	0·59
(sea otter)	31·5 kg	5·0*	22·3*	350*	4·2*	133*	31*	3·67*	7·67*	1·76
Australia										
Dugong dugon	Cd, Zn increase	<0·1–	<0·1–	0·2–309*	0·4–2·9*	9·1–608*	2·7–16·6*			
(dugong)	with age;	<0·2*	58·8*							
	Cu decreases									

ND, Not detected. * Dry weight. WW/DW ~ 3·5. † Figures in parentheses are methyl mercury.

3-19

mammals (Compiled from the sources indicated)

Se			Hg/Se atom ratio	Pb			Zn			
Muscle	Liver	Kidney	Liver	Muscle	Liver	Kidney	Muscle	Liver	Kidney	Source
					0·1–0·57	0·14–0·55		27–56	16·3–32·5	Drescher and co-authors (1977)
				1·2	2·3	1·2				Roberts and co-authors (1976)
0·44–0·92	6·1–110	2·5–3·3	0·98							Kari and Kauranen (1978)
	9·4		0·67							Smith and Armstrong (1978)
							23	52		Hamanaka and co-authors (1977)
	4·1*	6·1*	1·21*					505*	117*	Martin and co-authors (1976)
	260*	22*	0·88*					220*	173*	
										Buhler and co-authors (1975)
				1·1*	1·3*	2·2*				Braham (1973)
										Gaskin and co-authors (1979)
	59·4	9·55	3·0							Stoneburner (1978)
	22·8	2·98	0·98							
0·87										Arima and Nagakura (1979)
										Nagakura and co-authors (1974)
										Nagakura and co-authors (1974)
							109*	282*	123*	Martin (1979)
							191*	270*	251*	
				<0·3–<0·5*	<0·1–<0·3*	<0·1–<0·3*	32–113*	219–4183*	74·4–278*	Denton and co-authors (1980)

Table

Concentrations of metals in the tissues of some

Species, remarks	Ag Muscle	Ag Liver	Ag Kidney	Cd Muscle	Cd Liver	Cd Kidney	Cu Muscle	Cu Liver	Cu Kidney
	Concentration (ppm dry weight)								
Catharacta skua (great skua)‡ (Shetland; diet—fish and birds)					1–31	13–336			
Fulmarus glacialis (fulmar) (St Kilda; diet—fish)				3·4	49	228			
Fratercula arctica (puffin) (St Kilda; diet—fish)				<1·9	20	114			
Puffinus puffinus (manx shearwater) (St Kilda; diet—fish)				2·7	16	94			
Sterna fuscata (sooty tern)† (Florida; diet—juvenile fish)				4·4	15	94			
Melanitta perspicillata (surf scoter)† (British Columbia; diet—mussels)		0·12–0·56						42–44	
Aythya marila (great scaup)† (British Columbia; diet—plant forage)		0·16–1·28						69–78	
Somateria mollissima (eider) (Norway; diet—invertebrates)	2	44	7	2	13	25	13	367	43
Larus fuscus fuscus (lesser black-backed gull) (Norway; diet—fish, carrion)	2	3	1	1	4	10	14	17	14
Larus argentatus (Herring gull) (Britain; diet—scavenger)				ND	<1·9	14			

ND, Not detectable. * Largely methyl mercury. † Wet weight × 4. ‡ Concentrations increase with age.

feathers and liver were related in the great skua (FURNESS and HUTTON, 1979). It is thought that high levels of mercury in the eggs reflect elevated blood levels produced by recent dietary intake and are not necessarily related to high tissue levels (FIMREITE and co-authors, 1980). Thus HEINZ (1974) observed 6 to 9 ppm of mercury (wet) in eggs from mallards having a diet containing 3 ppm (wet) of methyl mercury, whereas normal eggs contained less than 0·05 ppm. Eggs in fish-eating species contain more than this: FIMREITE and co-authors (1980) reported 0·8 ppm (wet) in the eggs of uncontaminated gannets *Morus bassarus*, FURNESS and HUTTON (1979) found a mean of 0·47 ppm (wet) in eggs of the great skua *Catharacta skua* and STONEBURNER and co-authors (1980) found 8 ppm (wet) in eggs from the sooty tern *Sterna fuscata*, although there was no evidence that this was of anthropogenic origin

3-20

sea birds (Compiled from the sources indicated)

Concentration (ppm dry weight)									
Hg			Se			Zn			
Muscle	Liver	Kidney	Muscle	Liver	Kidney	Muscle	Liver	Kidney	Source
	3·2–30·4	3·4–19·5		6·7–34·6	13·3–89·1				Furness and Hutton (1979)
1·6	25*	13·4*				58	364	310	Osborn and co-authors (1979)
<1·4	4·5*	5·0*				47	118	164	
0·9	10·2*	4·7*				46	141	176	
2·6	6·1	6·6	27	91	165				Stoneburner and co-authors (1980)
	3·7–8·5						124–144		Vermeer and Peakall (1979)
	1·0						161–162		
0·8†	2·4					33	204	117	Lande (1977)
1·2†						55	89	116	
<0·7	2·2	<2·0				59	67	93	Nicholson (1981)

Biomagnification of Metals Along Food Chains

Preston and co-authors (1972b) have pointed out that it is the initial accumulation of metals from sea water by phytoplankton which provides much of the momentum for their transfer along food chains. Of 18 metals considered by Bryan (1976b), in various groups of marine organisms, mercury was one of the few where mean levels in fish exceed those in phytoplankton or seaweed on a dry weight basis and, as we have already seen, it is the least well regulated of trace metals.

Since mercury is one of the most important metallic contaminants, some of the evidence for and against its biomagnification along food chains will be considered. There is not much evidence for amplification of mercury levels in the sequence invertebrates to small fish (cf. Knauer and Martin, 1972), but there is more evidence when larger fish

Table 3-21

Mean inorganic and methyl mercury concentrations (ppm dry weight) in organisms from a contaminated salt marsh (Based on GARDNER and co-authors, 1978)

Material	Inorganic mercury	Methyl mercury
Sediments (0–5 cm, 9 sites)	0·63	<0·001
Spartina (leaf, 10 sites)	0·05	<0·001–0·002
*Annelids (3 species)	0·87	0·13
*Bivalves (4 species)	1·45	0·15
(1 species, 2 sites)	1·49	0·26
Gastropod (1 species, 10 sites)	3·25	0·25
Crustacean (1 species, 5 sites)	0·24	0·28
*Echinoderm (1 species)	0·10	0·01
*Fish (11 species, muscle)	0·32	1·04
(6 species, liver)	0·96	1·57
Terrestrial mammals (4 species, muscle)	2·7	2·2
(4 species, liver)	5·6	4·3
Birds (12 species, muscle)	0·8	3·0
(12 species, liver)	5·5	8·2

* From river running through marsh.

are considered. Much of the mercury in fish is found in muscle, and Fig. 3-5 compares concentrations in a wide range of species with the weight of the whole fish. Generally speaking, there is an increase in trophic level and mercury concentration ranging from plankton-feeding clupeids on the left to large predacious species on the right; this supports the conclusion of RATKOWSKY and co-authors (1975) that mercury concentration is related to the position in the food chain. However, leaving aside the problems of analytical and geographical variability, it is also clear from Fig. 3-5 that in some individual species concentrations of mercury increase markedly with size or age. Thus, CROSS and co-authors (1973) proposed that the higher concentrations in large predacious fish may be as much a function of time as of trophic level, if it is assumed that the larger predacious fish are on average longer lived. Both factors are probably important since, although the high levels in some fish can be explained by old age, the very high levels in marlin and some sharks tend to support the trophic-level idea. So also does the contrast in levels between the very large (although possibly quite young) plankton-feeding basking shark *Cetorhinus maximus* and the other sharks listed in Fig. 3-5. In marine mammals also, both factors seem to be important: NAGAKURA and co-authors (1974) found only 0·01 to 0·03 ppm (wet) of mercury in the muscle of baleen whales, which feed on krill, but 0·54 to 1·57 ppm, increasing with size, in muscle from the sperm whale which feeds on squid and fish.

Any tendency for total mercury levels to be amplified along food chains should be much more obvious for methyl mercury, since it generally accounts for an increasing proportion of the total in going from some invertebrates with as little as a few per cent to almost 100% in some predacious fish. In addition, the apparent ability of some species of marlin (Fig. 3-5) and marine mammals to demethylate methyl mercury (p. 1345) suggests that this mechanism has evolved in response to biomagnification. Biomagnification of methyl mercury is also strongly suggested by the results presented in Table 3-21, although there is likely to be some age effect as well.

A certain amount of mystery surrounds the ultimate source of methyl mercury: its production in sediments by micro-organisms has been recognized, but its detection in sea water has so far proved almost impossible (p. 1307). It is possible that the formation of methyl mercury may occur at other points in the food chain, perhaps due to intestinal microflora; indeed, methylation by the intestinal contents of fresh water fish has been reported (RUDD and co-authors, 1980).

Thus, although for a number of metals examples can be found where concentrations in individual predators exceed those of their prey, when the overall situation is considered, the only heavy metal for which there is evidence of general bioamplification is mercury or, more specifically, methyl mercury.

Bioaccumulation of Metals—Summary

Biological availability

Recent evidence indicates that free ions are biologically the most available inorganic species of trace metals in sea water. The concentration of ions is controlled not only by the total metal concentration but also by inorganic and organic complexation; the former is largely dependent on salinity and pH, the latter on the type of organic ligand. The influence of organic complexes varies from almost complete prevention of metal uptake to no effect, or even enhancement if complexation removes competing metals.

In oxidized surface sediment the availability of metals to deposit feeders appears to be controlled by the relative concentrations of various metal-binding substrates such as oxides of iron or organics. The availability of metals from food is clearly dependent on the chemical form of the metal and the composition of the food matrix.

Water versus food as source of metals

The diet is the principal source of metals in marine mammals and birds and, at the other end of the spectrum, water is the principal source in bacteria, phytoplankton, and seaweed. Between these extremes there is a good deal of uncertainty. However, since mercury is most readily adsorbed from solution (see Table 3-24) it is the metal for which the water route might be expected to be most important. This is generally the case: the majority of examples suggest the water route to be important or at least significant for mercury, whereas for most other metals the food route is more important.

Uptake of metals

Although there is evidence in some micro-organisms for the active transport of trace heavy metals (BRYAN, 1976a), evidence from most groups suggests that uptake from water at least is a passive process. Rates of uptake in *Nereis diversicolor* decrease in the order $\mathrm{Hg} > \mathrm{Cu} > \mathrm{Ag} > \mathrm{Cr} = \mathrm{Pb} > \mathrm{Zn} > \mathrm{Cd} > \mathrm{Ni}$, Co, Se, and this seems to be fairly typical. However, in molluscs, pinocytosis has been shown to be involved in the uptake and transfer of metals and, since endocytosis is a common phenomenon (see review by SILVERSTEIN and co-authors, 1977), it is likely to be important in the transfer of metals in other groups also.

Table 3-22

Some properties of biological indicators of heavy-metal contamination (Compiled from the sources indicated in the table footnotes; after BRYAN, 1980; reproduced by permission of Biologische Anstalt Helgoland)

Use as indicator:
+ Appears to act as indicator
++ Particularly good accumulator
( ) Months for transplanted organisms to equilibrate with new environment
? Doubt about use
R Definite regulatory ability

Species	Feeding type	Substrate	Estuarine tolerance (SPOONER and MOORE, 1940) Upstream limit in Tamar estuary (km from mouth)	Ag	Cd	Cu	Cr	Hg	Ni	Pb	Zn	Ref.
Ascophyllum nodosum	—	Rock	11		+	+		+		+	+	1, 2, 3
				(New growth relates to new environment but old tissue may never equilibriate)								
Fucus vesiculosus	—	Rock	21	+	+	+	+	+	+	+	+	2, 4, 5, 6, 7, 8
Mytilus edulis	Filter	Rock	15	?[6]	+	?[11]	+	+	+	+	+	9, 10, 12, 13
						(>5 m?)		(~3 m)		(>4 m)	(>5 m)	10, 12, 14, 15
Ostrea edulis	Filter	Sediment/ stones	10		+	++					++	9, 10
					(>5 m)	(>5 m?)					(<5 m)	10
Cerastoderma edule	Filter	Sediment	13	+	+	+			++	?[6]	+	6, 9
Scrobicularia plana	Deposit	Sediment	18	+	+	?	+	+	+	++	++	6
				(8 m)	(>12 m)	(8 m)			(>12 m)	(>12 m)	(>12 m)	17
Macoma balthica	Deposit	Sediment	15	+	+	?	+	+	+	+	++	6
Nereis diversicolor	Deposit/ Omnivore	Sediment	23	+	+	+	+	+	+	+	R	6, 19
Littorina littorea	Herbivore	Rock/ Sediment	11	+	+	++R	?		+	+	+R	6, 21, 22
Littorina obtusata	Herbivore	Rock/weed/ sediment	11	+	++	++			+	+	+R	6, 23
Patella vulgata	Herbivore	Rock	8	+	++	+				+	+	6, 22, 24
					(3 m?)	(>3 m)					(>3 m)	20
Nucella lapillus	Carnivore	Rock	5	+	++	++					++	6, 22
					(>4 m)	(>4 m)					(3 m)	20

Species	Feeding type	Substrate	Estuarine tolerance (SPOONER and MOORE, 1940) Upstream limit in Tamar estuary (km from mouth)	Influence of increasing size or age. Concentration rises (+) falls (−) remains constant (0): Ag	Cd	Cu	Cr	Hg	Ni	Pb	Zn	Ref.
Ascophyllum nodosum	—	Rock			+	+		+		+	+	3
				Seaweed values higher in older tissues								
Fucus vesiculosus	—	Rock	21			+				+	+	4
Mytilus edulis	Filter	Rock	15		0	−		+[12]	−	−	−	10
Ostrea edulis	Filter	Sediment/ stones	10		0	0			−	−	0	10
Cerastoderma edule	Filter	Sediment	13		−	−			0	−	−	16
Scrobicularia plana	Deposit	Sediment	18	+/−	+	+/−	+	+	+	+	+	6, 7, 18
Macoma balthica	Deposit	Sediment	15	0	+	0	0		+	0	+	8
Nereis diversicolor	Deposit/ Omnivore	Sediment	23	Variable but not usually marked								
Littorina littorea	Herbivore	Rock/ Sediment	11		0	0				−	−	10
Littorina obtusata	Herbivore	Rock/weed/ sediment	11									
Patella vulgata	Herbivore	Rock	8		+	−				−	−	10
Nucella lapillus	Carnivore	Rock	5								−	8

1. HAUG and co-authors (1974); 2. MELHUUS and co-authors (1978); 3. MYKLESTAD and co-authors (1978); 4. BRYAN and HUMMERSTONE (1973c); 5. MORRIS and BALE (1975); 6. BRYAN and HUMMERSTONE (1977); 7. SEELIGER and EDWARDS (1977); 8. Unpublished; 9. BOYDEN (1975); 10. BOYDEN (1977); 11. PHILLIPS (1977); 12. DAVIES and PIRIE (1978); 13. YOUNG, D. R. and co-authors (1979); 14. SIMPSON (1979); 15. MAJORI and co-authors (1978a); 16. BOYDEN (1974); 17. BRYAN and HUMMERSTONE (1978); 18. BRYAN and UYSAL (1978); 19. BRYAN (1974); 20. STENNER and NICKLESS (1974a); 21. BRYAN (1979); 22. NICKLESS and co-authors (1972); 23. YOUNG (1975); 24. PRESTON and co-authors (1972a).

Methods of detoxification

These include (i) binding to non-specific high-molecular-weight proteins or polysaccharides (all groups); (ii) binding to specific low-molecular-weight proteins of the metallothionein type (most animals but less certain in plants); (iii) immobilization in intracellular inclusions of different types; a high calcium level in some granules suggests a link between metal detoxification and calcium metabolism (most animals and some plants); (iv) immobilization by incorporation in shell, skeleton, fur, feathers, etc.; (v) demethylation of methyl mercury in marine mammals, some fish and possibly other groups such as birds and gastropod molluscs; (vi) detoxification of arsenic by conversation to less toxic organic forms in marine plants and possibly some animals, although in the latter the intestinal flora may be involved.

Methods of removal from the body

These include (i) loss over body surface by diffusion and excretion—comprising losses in secretions and granules and by diapedesis of amoebocytes; (ii) urinary excretion as fluid or granules; (iii) loss via alimentary tract; (iv) loss by moulting of exoskeleton, fur or feathers; (v) incorporation in eggs.

Relations between levels in biota and environment

There is little evidence that organisms actively prevent the absorption of metals and, therefore, the relations between body concentrations and those in the environment may largely be determined by mechanisms for immobilization and excretion.

Three types of relation can be recognized:

(i) The organism excretes the metal at a rate proportional to the body burden. If intake increases, the body level rises until intake is balanced by excretion. Thus the concentration in the body tends to be proportional to environmental availability and usually remains constant or falls with increasing size or age (e.g. Pb in *Mytilus edulis*).

(ii) The organism has limited powers of excretion and tends to detoxify and store metals. In this case, the concentration in the organism may still be roughly proportional to environmental availability, but—unless the organism grows fast enough to dilute the metal—the level in the body tends to increase with age and no equilibrium level is reached (e.g. Hg in marlin).

(iii) In response to increased absorption, the organism increases the efficiency of excretion and, therefore, the concentration in the body increases relatively little for a considerable increase in environmental availability (e.g. Zn in crabs).

Generally speaking, the more highly evolved forms including fish and decapod crustaceans are the best regulators and the essential metals, such as zinc and copper, are better regulated than the non-essential, such as lead, cadmium, and especially mercury.

Biomagnification of metals

Although it is possible to find many examples where predators contain higher heavy-metal concentrations than their prey, when the overall situation is considered only mercury, or more specifically methyl mercury, shows appreciable signs of being biologically magnified as a result of food-chain transfer.

(4) Biological Monitoring of Metallic Contamination

While the analysis of sea water certainly has a place in the detection of metallic contamination, it is at present far from ideal as a method of routine monitoring. The possibilities of inadvertent sample contamination are high, the number of metals that can be readily analysed is limited and, particularly in stratified estuaries, considerable sampling effort is required to obtain an integrated picture. By comparison, analyses of sediments are much easier and have provided valuable information not only about recent metallic inputs (cf. FÖRSTNER and WITTMANN, 1979) but also, from dated cores, about the history of contamination (cf. ERLENKEUSER and co-authors, 1974; GOLDBERG and co-authors, 1978b).

At present, analyses of water and sediments are rarely carried out with regard to biological availability, although attempts are being made to rectify this (pp. 1310 and 1315). Since biological availability is one of the prerequisites for pollution, there is a strong argument for the analysis of biological indicators. Such indicator organisms should be good accumulators of metals, and should reflect changes in environmental availability. For this reason, organisms having an ability to regulate metals are unsuitable. Desirable properties are that the organism should be widely distributed, common, accessible, easily recognized, relatively stationary, available at all times of year and, for estuarine purposes, sufficiently tolerant of low salinities and high suspended solids to penetrate a reasonable distance upstream (PHILLIPS, 1977, 1980). The properties of intertidal benthic organisms which have proved useful as indicators in European waters are summarized in Table 3-22, and an indication of the relative merits of some of these species as accumulators is given in Table 3-23. Because different species have different distributions ranging from rocky shores to muddy estuaries and absorb metals from different sources, there is no universal indicator organism. Thus, any reasonable monitoring programme should involve analyses of several species such as a seaweed, a filter feeder, and a deposit feeder to try and assess contamination in different forms. It was found in the Looe estuary, for example, that particulate silver was readily available to the deposit-feeding bivalves *Scrobicularia plana* and *Macoma balthica* but not to *Mytilus edulis* and *Fucus vesiculosus*, 2 commonly used indicators (Table 3-23). Because of the influence of factors such as size (Table 3-22) and condition on tissue concentrations, the use of indicator organisms requires uniformity of sampling. With *S. plana* , for example, it was necessary to analyse animals of 4 cm shell length collected from mid-tide level in either spring or autumn to avoid the breeding season (Fig. 3-6; BRYAN and co-authors, 1980).

Because of their world-wide distribution and potential as indicators, species of *Mytilus* have become the subject of various monitoring programmes of the 'Mussel Watch' type (GOLDBERG and co-authors, 1978b; DAVIES and PIRIE, 1980). Partly because it has been the subject of so much study, *Mytilus* spp. appear to have more than their share of problems as indicators, particularly with regard to obtaining organisms of comparable size, condition, and so on at different sites. Various ideas have been put forward to avoid the influence of changes in condition which occur particularly during the breeding season. COSSA and co-authors (1979) suggested that only immature individuals should be used, and other authors tried to correct by expressing their results as μg(standard ind.)$^{-1}$ rather than as a concentration (e.g. SIMPSON, 1979). Some of these problems may also be avoided by analysis of a single tissue, such as the digestive gland, rather

Table 3-23

Comparison between metal indicator species from a single site in a silver–lead contaminated estuary. Highest concentrations (ppm dry tissue) in organisms are in boxes (After BRYAN and co-authors, 1980; reproduced by permission of the authors)

Species (group), sediment	Mean dry weight soft parts (g)	Metal												
		Ag	Cd	Co	Cr	Cu	Fe	Mn	Ni	Pb	Zn	As	Hg	Sn
		East Looe estuary (2 km from mouth)												
Scrobicularia plana (bivalve)	0·148	56·4	1·2	**11·5**	**3·3**	300	603	25	14·5	**189**	**1120**	**94**	0·47	**0·87**
Macoma balthica (bivalve)	0·037	**81·9**	0·2	3·7	3·3	**338**	788	24	6·9	61	1010	46	**0·97**	—
Nereis diversicolor (ragworm)	0·019	4·2	0·4	3·9	0·3	60	415	8	3·5	25	258	23	0·11	0·18
Mytilus edulis (mussel)	0·240	0·2	2·3	0·6	2·5	9	401	6	2·6	45	113	21	0·39	0·81
Cerastoderma edule (cockle)	0·278	2·4	0·7	1·6	1·9	10	565	6	**54**	5·3	54	21	0·26	0·44
Littorina littorea (winkle)	0·180	30·0	1·6	1·1	0·5	154	361	25	2·8	17	232	37	0·21	0·13
Patella vulgata (limpet)	0·530	5·6	**5·6**	0·6	0·5	18	**1160**	6	2·3	30	145	33	0·26	—
Fucus vesiculosus (seaweed)	—	0·7	1·1	4·7	1·9	12	946	**315**	11·3	17	190	38	0·09	0·63
Surface sediment (HNO_3 digest; fusion for Sn)	—	1·3	0·2	9·8	37	43	2·61%	382	35	104	133	8·7	0·16	63

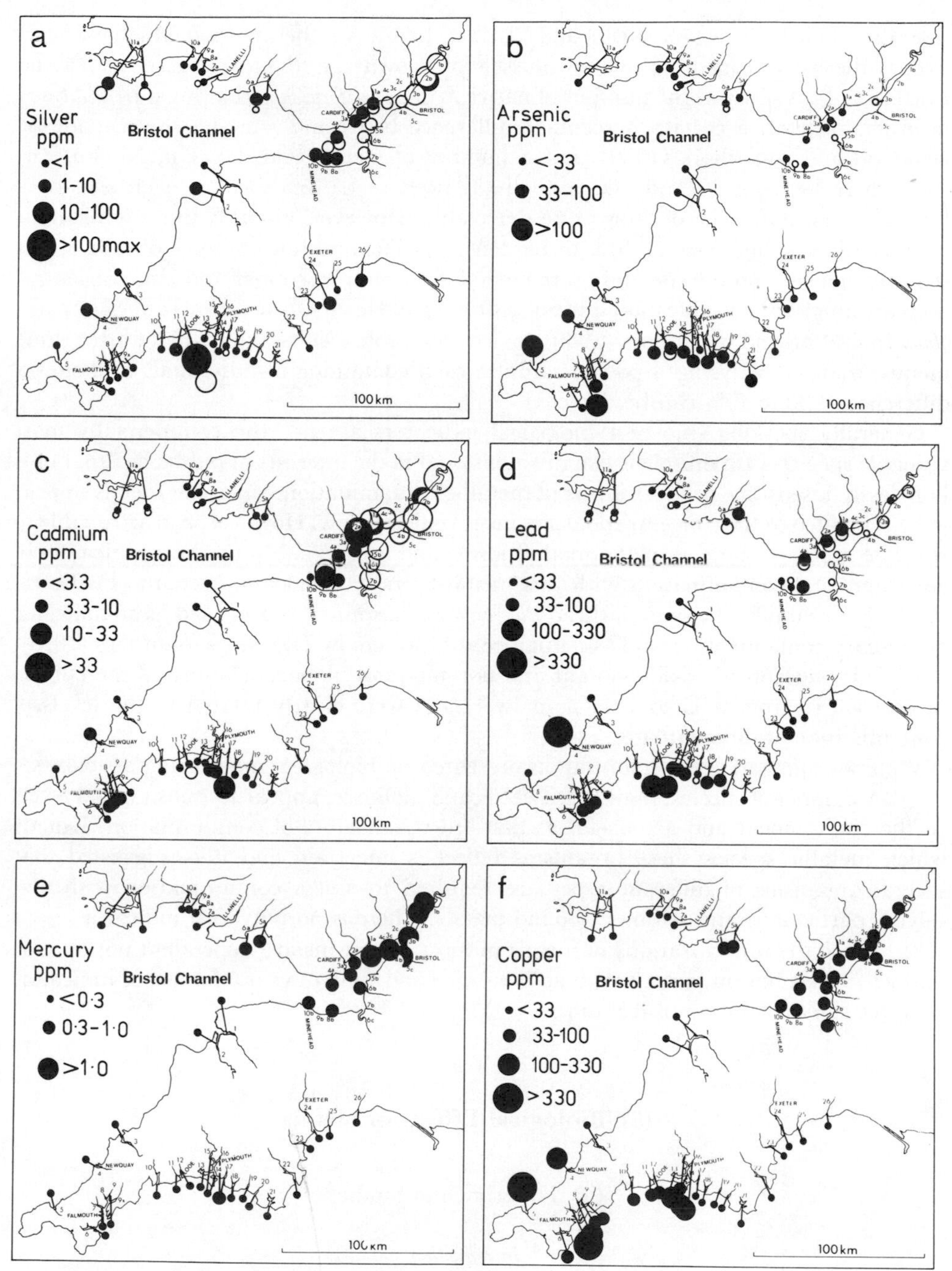

Fig. 3-6: Southwest Britain. Geographical distribution of 4 metals in whole soft parts of *Scrobicularia plana* (a, b, c, d); open circles are equivalent levels in *Macoma balthica*, concentration ranges being divided by 0·59 for silver, 6·2 for cadmium and 3·6 for lead. (c, f) Geographical distribution of 2 metals in tissues of *Nereis diversicolor*. Results are for individual sites in Severn estuary (1a–4a, 1b–10b, 1c–6c); in estuaries 1–26 and 5a–11a, are the ranges within which most of the values for different sites occur. Restronguet Creek is No. 7. (After BRYAN and co-authors, 1980.) (Original)

than the whole body (ALEXANDER and YOUNG, 1976). Another method which is attractive, is the use of caged, cultivated mussels of known age as indicators (DAVIES and PIRIE, 1978). A study of the number of native *Mytilus californianus* that must be analysed in order to detect a certain percentage difference between 2 sites has been made by GORDON and co-authors (1980). For a number of metals (Cd, Cr, Cu, Ni, Pb, Zn) somewhere between 20 and 100 individuals need to be analysed at each site for a concentration difference of 20% to be detectable. However, about 3 times fewer individuals allow a difference of 40% to be detected. The absolute number of individuals required depends on the site and on the metal; for example, copper and zinc concentrations are much less variable than those of cadmium and lead. Observations on *Scrobicularia plana* by BRYAN and co-authors (1980) gave comparable results, indicating, for the same metals, that by analysing 3 pooled samples each containing 6 individuals at 2 sites a difference of 30 to 40% can be detected.

Generally speaking, the best biological indicators are not the commercially most valuable species. Fish muscle is usually monitored in the interests of public health, but is hopelessly insensitive to most forms of metallic contamination, since the metals appear to be regulated or the concentrations are inconveniently low. However, fish are unable to regulate mercury (largely in the methyl form) and are arguably the best indicators of environmental contamination with this metal if size is taken into account (PHILLIPS, 1977). For example, the sand flathead *Platycephalus bassensis* was proposed as an indicator of mercury contamination in Tasmanian coastal waters by DIX and co-authors (1975), since it is widespread, easily caught and non-migratory; concentrations of mercury in muscle varied from 0·03 to 1·06 ppm (wet) and were clearly related to the levels of contamination at different sites.

We may summarize the information presented on biological monitoring as follows:

(i) A number of species, mainly seaweeds and molluscs, appear to reflect metal levels in the environment and are useful as first-order indicators of contamination. Exactly which metallic species these organisms reflect is uncertain and it is suggested that several organisms of different types are required to assess contamination with dissolved, particulate, and sediment-bound metals. There is no universal indicator.

(ii) Since the concentrations of metals in most organisms are dependent not only on ambient levels but on factors such as body size and time of year, the use of indicators requires standardization of technique.

(5) Biological Effects of Metals

(a) Experimental Studies

Introduction

There is no doubt that the LC_{50} (median lethal concentration) approach to metal toxicity in marine species has provided a wealth of information about the effects of different metals on different species and has revealed the many factors—chemical, biological, and environmental—which modify toxicity (cf. BRYAN, 1976b). However, even after long exposure, LC_{50} concentrations rarely fall within the range of concentrations observed in the most contaminated waters (cf. Table 3-4). Experiments with some larval forms, which may accumulate metals rapidly, provide exceptions and will be considered in subsequent sections. In many of the larger species, absorption of the metal (and hence an effective dose) is often so slow that sometimes only extremely high concentrations are effective in toxicity tests having death as the end point. Even experiments of long duration are not entirely satisfactory in animals, since the absorption of metals from food is rarely considered although it is the primary source in many species.

In the search for effects at realistic environmental levels, increasing numbers of studies have considered the sublethal effects of metals and have progressively increased in complexity from simple experiments on growth to multiparameter physiological, biochemical, and histopathological studies on exposed test organisms.

The following discussion is largely concerned with recent experimental evidence for the influence of *realistic* metal concentrations on different marine groups and is roughly divided into the evidence for effects on individual organisms, the evidence from cultured populations and the evidence from experimental ecosystems.

Effects on Individual Organisms

Seaweeds

The majority of studies have been concerned with the influence of metals on the growth of seaweeds. In *Laminaria digitata* an approximately 50% reduction in growth resulted from exposure to 25 ppb of copper, 50 ppb of silver, 100 ppb of zinc, and 500 ppb of lead (BRYAN, 1976b). Growth in *Ectocarpus siliculosus* was retarded by only 10 ppb of copper (MORRIS and RUSSELL, 1973) and a slightly higher level affected *Pelvetia canaliculata* (STRÖMGREN, 1980a). Studies by the same author (1980b) on the effects of mercury, lead, and cadmium on 5 species including *Pelvetia canaliculata, Fucus vesiculosus,* and *Ascophyllum nodosum* showed that apical growth was retarded by about 10 ppb of mercury but was insensitive to hundreds of ppb of lead and cadmium: in fact growth was accelerated in *Pelvetia* by 45 to 2600 ppb of lead or 1.5 to 1040 ppb of cadmium. However, in *Laminaria saccharina* growth was significantly reduced by 190 ppb of cadmium (MARKHAM and co-authors, 1980). Arsenic does not appear to be particularly toxic to seaweeds and the metabolism of *Sargassum* was unaffected by 20 to 200 ppb (BLAKE and JOHNSON, 1976). The number of studies on the small stages of seaweeds is very limited. However, BONEY (1971) used the sporelings of intertidal red algae to demonstrate the much greater toxicity of organic than inorganic mercury compounds. The same author has considered the influence of iron ore dust on the eggs of *Fucus serratus* and shown that adhesion of particles may remove them from levels in the water where spermatozoid release occurs and may reduce the attractive power of the egg for the spermatozoid (BONEY, 1980).

It is possible to conclude from the results above that the general order of heavy-metal toxicity to seaweeds is: Hg (organic) > Hg (inorganic) > Cu > Ag > Zn > Cd > Pb. By virtue of its abundance copper is arguably the metal most likely to affect algae in the field and one manifestation of this effect is found in the development of copper tolerance in some species. This was first observed in the brown fouling alga *Ectocarpus siliculosus* by RUSSELL and MORRIS (1970) and is thought to be genetically based. Recent studies on *E. siliculosus* point to an exclusion mechanism as the basis for copper tolerance, coupled

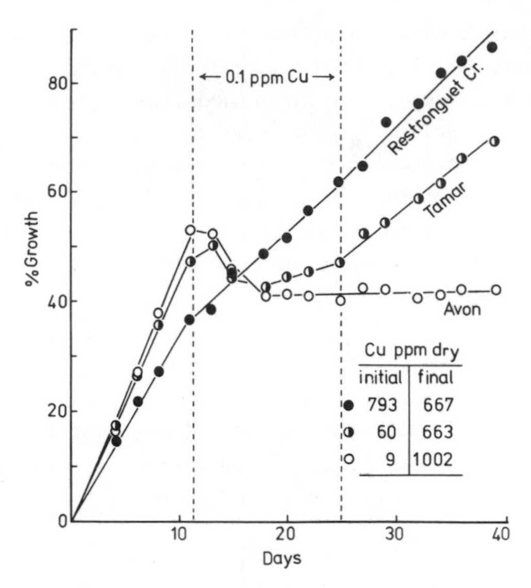

Fig. 3-7: *Fucus vesiculosus*. Effect of Cu on growth and concentrations in plants from 3 contrasting estuaries. Each line represents 3 plants of about 3 cm initial length (13 °C, 17·5‰S, continuous light, no added nutrients). (After BRYAN, 1980; reproduced by permission of Biologische Anstalt Helgoland.)

with an ability to function with a high internal concentration (HALL and co-authors, 1979): co-tolerance to zinc and cobalt may be explained by the same exclusion mechanism (HALL, 1980). A similar explanation may also apply to the tolerance of *Fucus vesiculosus* to copper (Fig. 3-7). Although having a lower initial growth rate, the heavily contaminated weed from Restronguet Creek is able to continue growing in water containing 0·1 ppm of copper which prevents growth in the weed from other estuaries. Analyses of the weed before and after exposure suggest that the tolerant weed is probably less permeable to copper and that this coupled with growth dilution helps to limit the internal concentration. In some species, the production of extracellular products having the capacity to bind metals can ameliorate their effects.

Polychaetes

The evidence available suggests that polychaetes are fairly resistant to heavy metals (Tables 3-24, 3-27). In the short term at least, mercury, copper, and silver are the most toxic; aluminium, chromium, zinc, and lead are less toxic; and the least toxic are cadmium, nickel, cobalt, and selenium. In *Nereis diversicolor* the acute form of toxicity is very dependent on the rate of uptake of the metal, since this determines the speed with which the lethal dose is built up (Table 3-24). Although the dose to an organism is very important it is difficult to define. It is not simply the total level in the body since this will include metals in immobilized forms. The rate of intake is important because this determines whether the organism's detoxification mechanisms can cope or not. For

Table 3-24

Nereis diversicolor. Relationship between rate of metal uptake and toxicity (Original)

Metal	Rate of absorption at 0·1 ppm in 50% SW at 13 °C (ppm dry weight d^{-1})	192-h LC_{50} (ppm)	Concentration in body, lethal in about 192 h (ppm dry weight)
Hg (chloride)	52	0·1	—
Cu (sulphate)	16	0·27	289
Ag (nitrate)	9	0·5	430
Cr (dichromate)	5*	10	1250
Pb (nitrate)	5*	>5	>1000
Zn (sulphate)	1·4	30	2930
Cd (sulphate)	0·4	100	1300
Ni (sulphate)	0·2*	130	2030
Co (chloride)	0·2*	170	2040
Se (selenite)	0·2	100	—

* Estimated by extrapolation from net uptake rates at higher concentrations.

example, McLUSKY and PHILLIPS (1975) observed that the lethal body burden of copper to *Phyllodoce maculata* was about 270 ppm at 100 ppb in solution, but twice this dose absorbed more slowly at 50 ppb was tolerated with little effect. Similarly, a comparison between the lethal level in *N. diversicolor* (Table 3-24) and the maximum concentration observed in the field (Table 3-9) shows that for copper, the apparently toxic dose can in reality be far exceeded—by tolerant forms at least.

Arguably, the metals most likely to affect polychaetes in the field are copper, chromium, and zinc. Curiously, the threshold for lethal copper toxicity lies in the region of 70 to 100 ppb for several species, including *Nereis diversicolor* BRYAN and HUMMERSTONE, 1971), *Neanthes arenaceodentata* (PESCH, 1979), *Phyllodoce maculata* (McLUSKY and PHILLIPS, 1975) and *Cirriformia spirabrancha* (MILANOVICH and co-authors, 1976). In *Hermione hystrix,* CHIPMAN (1966) noted that this animal was affected by long exposure to chromate concentrations of 100 to 500 ppb. These experiments were all carried out under fairly ideal conditions, whereas the work of JONES and co-authors (1976b) showed that the toxicity of copper to *N. diversicolor* is minimal at intermediate salinities to which it is normally acclimatized and increases markedly at lower or higher salinities. Another factor is the development of tolerance to metals which has been observed in *N. diversicolor* and *Nephthys hombergi* from estuaries in southwest England where heavy contamination of sediments and animals have been found (Table 3-9; Fig. 3-6). Tolerance to zinc, copper, silver, and possibly lead has been observed in *N. diversicolor.* The tolerances to zinc, copper, and perhaps lead seem to have developed separately, but that to silver depends on the presence of tolerance to copper or lead (BRYAN, 1976a). Circumstantial evidence suggests that tolerance is genetically determined since it is not removed by growing tolerant organisms in uncontaminated conditions (BRYAN, 1979). The basis of tolerance appears to be a lower permeability to metals, coupled possibly with a more efficient detoxification system (cf. p. 1320).

The majority of studies on the sublethal effects of metals on polychaetes concern their influence on reproduction in cultures and are considered later (p. 1379). However, MÜLLER (1979), who measured the influence of metals on oxygen consumption in *Nereis diversicolor,* gives the effective concentrations as 10 ppb, 50 ppb, and 100 ppb of cadmium, copper, and lead respectively, and these concentrations of cadmium and copper are realistic in relation to those observed in contaminated areas (Table 3-4). Another process that might be expected to be sensitive to metals is the direct uptake of organic nutrients from the water by some polychaetes. However, studies on the uptake of glycine by several species have failed to demonstrate effects at realistic metal levels (RICE and CHIEN, 1977; SIEBERS and EHLERS, 1979).

Bivalve molluscs

There seems little doubt that the embryonic and larval stages of bivalve molluscs are the most vulnerable to heavy metals. Although mercury is the most toxic metal, others including copper, cadmium, and zinc seem most likely to pose problems in the field. For example, 15 ppb of copper was found to produce deformed embryos in *Crassostrea virginica* and 33 ppb proved lethal to the larvae (Table 3-25). The adults, on the other hand, can withstand long exposure to such levels although, through the immobilization of copper, they become green and unpalatable: for example ORTON (1923) reported a value of about 16,000 ppm in a very green specimen of *Ostrea edulis,* from Restronguet Creek (p. 1384). Very high concentrations of zinc also have been measured in contaminated oysters (Table 3-29). There is no evidence that such levels are toxic but they suggest the animals had been exposed to the equivalent of several hundred ppb of zinc, which, as Table 3-25 shows, would be toxic to the young stages.

Exposure to 100 ppb of cadmium for about 15 wk induced poor condition and mortalities in adult *Crassostrea virginica* at a body concentration of around 700 ppm (SHUSTER and PRINGLE, 1969); however no effects were reported by ZAROOGIAN (1980) who found over 290 ppm of cadmium in the tissues after exposure to 15 ppb for 40 wk. About 20 ppb of cadmium in solution was observed to increase mortality and reduce growth in the larvae of *C. gigas* (WATLING, 1978) and such a concentration is of the same order as those observed in the most contaminated sea areas (Table 3-4). In considering these results it should be remembered that in reality metallic contaminants rarely occur singly, and at least in oyster embryos the effects of metals have been shown to be additive (MACINNES and CALABRESE, 1978). In addition, most of the work reported in Table 3-25 refers to the effects of metals on organisms not subjected to other stresses. Thus, from experiments in which salinity, temperature, and copper concentrations were varied, MACINNES and CALABRESE (1979) concluded that copper may produce intolerable effects on the recruitment of oyster embryos during periods of low salinities and low or high temperatures.

Since oysters are fairly tough organisms, greater sensitivity would be expected in other species. A thorough study of the combined effects of temperature, cadmium, and salinity on the development of the larvae of *Mytilus edulis* was made by LEHNBERG and THEEDE (1979), who reported significant effects at 50 ppb of cadmium. Similar work on embryonic development in *Mytilus galloprovincialis* by HRS-BRENKO and co-authors (1977) showed that it was retarded by 100 ppb of lead at high temperature coupled with low salinity. However, mussels seem to be most sensitive to copper. ADEMA and co-authors (1972) observed a high mortality in adult mussels at 25 ppb of copper and some mortality even at 17 ppb, a concentration of 80 ppm being reached in the animal after 30 d. MARTIN (1979) has calculated the lethal level in mussels to be about 60 ppm and reports that 20 ppb in solution is toxic. Mussels are more sensitive to copper during the reproductive period when, due to more rapid accumulation, a critical body level is reached more rapidly (DELHAYE and CORNET, 1975). Other examples of experimental studies on bivalve toxicity are included in Table 3-25. They also point to the significance of copper at concentrations well within the range observed in contaminated waters (Table 3-4). For example, PESCH and co-authors (1979) documented that juvenile scallops *Argopecten irradians* contained 42, 320, and 510 ppm of copper following 42 d of exposure to 1, 5, and 11 ppb respectively. At the highest concentration, mortality was appreciable and even at 5 ppb growth was reduced and mortality increased.

Our own experience suggests that, in the field, metal-contaminated sediments can exert a toxic effect on burrowing bivalves (p. 1384) and, from laboratory experiments, MCGREER (1979) concluded that *Macoma balthica* burrowed more slowly into contaminated sediments and would, if possible, avoid them.

Gastropod molluscs

Most of the information available suggests that adult gastropod molluscs are rather tolerant to heavy-metal toxicity. A partial explanation of this may lie in the protection afforded by the shell in some species and the fact that toxicity studies have considered metals in solution although the principal route of uptake is probably the food. In *Bullia digitalis* the 96-h LC_{50} concentration for cadmium was 900 ppb and after 24-d exposure no sublethal effects could be detected below 100 ppb (CUTHBERT and co-authors, 1976). However, in *Haliotis rufescens* the 96-h LC_{50} concentration for copper was only 65 ppb, and lower concentrations produced histopathological abnormalities of the gills. Curiously, the larvae were rather more tolerant, the LC_{50} concentration being 114 ppb (MARTIN and co-authors 1977). *Busycon canaliculatum* is much more resistant and adults survived copper levels of 200 to 500 ppb for 54 to 77 d (BETZER and YEVICH, 1975). The lowest effective concentrations have been observed in embryos. For example REINHART and MYERS (1975) showed that the number of eye and tentacle abnormalities in embryos of *Urosalpinx cinerea* was increased by exposure to 10 ppb of mercury.

Crustaceans

Although a few recent studies have been concerned with the long-term effects of low dissolved metal concentrations on adult decapod crustaceans (THURBERG and co-authors, 1977; PESCH and STEWART, 1980), the bulk of experimental work was concerned with effects on planktonic species, including larval decapods and copepods (Table 3-26). Several groups of workers have studied the interaction between metallic contamination and the stresses produced by suboptimal combinations of temperature and salinity such as may be found in estuaries (cadmium: VERNBERG and co-authors, 1974; zinc: MCKENNEY and NEFF, 1979). Not surprisingly, low metal concentrations are more toxic in combination with other stresses and, in part at least, this is dependent on changing the absorption rate of the metal. Thus ENGEL and FOWLER (1979) showed that the greater toxicity of cadmium to *Palaemonetes pugio* at low salinity is related to the higher concentration of ionic cadmium, in the water, resulting from reduced chloride complexation (see also p. 1311).

Table

Toxicity of metals to some bivalve molluscs. Concentration

Species, parameter studied	Result	Metals Ag	As (arsenite)	Cd	Cr	Cu
Crassostrea virginica						
Mortality, embryos	48 h LC_{50}	9	7500	3800	10,300	128
Deformity, embryos	48 h EC_{50}	24	—	—	—	15
Mortality, larvae	12 d LC_{50}	25	—	—	—	33
Reduced shell growth, juveniles	33% in 47 d	—	—	—	—	—
Reduced shell growth, adults	Inhibition in 20 wk	—	—	<100	>100	>50
Condition, adults	Poor in 7 wk	—	—	—	—	—
Crassostrea gigas						
Mortality of 5 d larvae	Increased			20		
5 d larvae	96 h LC_{50}			50		
16 d larvae	96 h LC_{50}			200		
25 d spat	96 h LC_{50}			1000		
3 month spat	96 h LC_{50}			2000		
Settlement, larvae	Delay	—	—	—	—	—
Ostrea edulis						
Mortality, larvae	48 h LC_{50}					
Mortality, adults	48 h LC_{50}					
Mercenaria mercenaria						
Mortality, embryos	48 h LC_{50}	21				
Mortality, larvae	8-10 d LC_{50}	32				16
Argopecten irradians						
Mortality, 20–30 mm juveniles	96 h LC_{50}	33	3490	1480		310*
Growth, 11–17 mm juveniles	42 d EC_{50}					5·8
Mytilus edulis						
Behaviour, adults	Shell closure					21
Venerupis decussata						
Burrowing activity	60 d EC_{50}					10
Protothaca staminea						
Mortality, adults	30 d LC_{50}					~30
Effect on gills	cytotoxic response?					7

Experiments on other groups, including copepods, mysids, and brine shrimp, have shown that at low metal levels the fecundity of some species is reduced (e.g. cadmium: PAFFENHÖFER and KNOWLES, 1978; copper: REEVE and co-authors, 1977; Table 3-26), development is delayed and mortality increased. It can be concluded from Table 3-26 that concentrations of copper, cadmium, and zinc, lying withing the ranges observed in contaminated waters (Table 3-4) are likely to inhibit reproduction and development in some species of crustaceans, especially when conditions are not optimal. However, in

3-25

in water (ppb) (Compiled from the sources indicated)

Metals					
Hg	Ni	Pb	Zn	Conditions	Source
6	1190	2460	340	Static, 26 °C, 25‰ S	CALABRESE and co-authors (1973)
11	—	—	206	Static, 20 °C, 26‰ S	MACINNES and CALABRESE (1978)
12	1200	—	—	Static, 25 °C, 24‰ S	CALABRESE and co-authors (1977)
10	—	—	—		CUNNINGHAM (1976)
—	—	—	>200	Flow, 20 °C, 31‰ S	SHUSTER and PRINGLE (1969)
—	—	100	—	Flow, 20 °C, 31‰ S	PRINGLE and co-authors (1968)
				Static, 26 °C, 34‰ S	WATLING (1978)
—	—	—	125	Static, 21 °C, 29‰ S	BOYDEN and co-authors (1975)
1–3·3 4200				Static, 15 °C, 34‰ S	CONNOR (1972)
4·8	310	780	166	Static, 26 °C, 25‰ S	CALABRESE and NELSON (1974)
15	5700	—	195	Static, 25 °C, 24‰ S	CALABRESE and co-authors (1977)
89				Static, 20 °C, 25‰ S	NELSON and co-authors (1976)
				Flow, 9·5–15 °C, 27·4–31·5‰ S	*PESCH and co-authors (1979)
				Flow, 12–15 °C, ~34‰ S	MANLEY and DAVENPORT (1979)
				Flow, 15 °C, 35‰ S	STEPHENSON and TAYLOR (1975)
				Flow, 12·3 °C, 31·2‰ S	ROESIJADI (1980)

making this prediction some uncertainty remains, because, in some species at least, resistance to contamination can be increased by the development of metal tolerance. The induction of tolerance by pre-exposure to sublethal levels of copper has been reported in *Artemia salina* by SALIBA and KRZYZ (1976), but no change in tolerance to mercury was found when this principle was applied to post-larval *Penaeus setiferus* by GREEN and co-authors (1976). Increased tolerance to metals in populations from contaminated areas has certainly been observed; examples include tolerance to zinc in

Table 3-26

Effects of low metal concentrations (ppb) on crustaceans (Compiled from the sources indicated)

Metal	Animal	Species	Temperature (°C)	Salinity (‰)	Concentration (ppb)	Effect	Source
Ag	Crab larvae	*Carcinus maenas*	15	—	1	Photopositive behaviour depressed by prolonged exposure	AMIARD (1976)
Cd	Shrimp	*Palaemonetes pugio*	6·5–14	27–31·5	200–300	Incipient LC_{50}	PESCH and STEWART (1980)
	Hermit crab	*Pagurus longicarpus*	2·5–7	27–31·5	70	Incipient LC_{50}	PESCH and STEWART (1980)
	Shrimp	*Palaemonetes pugio*	15–25	5–30	50	Increased mortality as salinity decreased	VERNBERG and co-authors (1977)
	Larval crab	*Rhithropanopeus harrisii*	20–35	10–30	50	Hatch to megolopa development time increased under non-optimal conditions	ROSENBERG and COSTLOW (1976)
	Copepod	*Tigriopus japonicus*	20–22	—	44	Time to reach F_2 generation more than doubled	D'AGOSTINO and FINNEY (1974)
	Larval crab	*Eurypanopeus depressus*	25	30	10	Development time from megalopa to juvenile crab extended	MIRKES and co-authors (1978)
	Mysid	*Mysidopsis bahia*	20–28	15–23	6·4	Number of young produced by ♀ halved: 10·6 ppb gave 90% mortality at 23 d	NIMMO and co-authors (1978)
	Copepod	*Pseudodiaptomus coronatus*	20	30	5	50% reduction in reproduction rate but not other effects	PAFFENHÖFER and KNOWLES (1978)
	Lobster	*Homarus americanus*	18–20	24–26	3	Elevated gill tissue O_2 consumption loss of Mg sensitivity in heart muscle transaminase after 30 d	THURBERG and co-authors (1977)
			18–23	27	6	Increased glycolysis.	GOULD (1980)
	Larval crab	*Uca pugilator*	13 combinations		1	Increased mortality under suboptimal conditions	VERNBERG and co-authors (1974)

Cu	Larval shrimp	*Pandalus danae*	8·7–10·3	29·8–30·6	5	Delay of 5 d in transformation from Stage 4 to Stage 5 zoea	Young, J. S. and co-authors (1979)
	Copepod	*Tigriopus japonicus*	20–22	—	64	Time to reach F_2 generation more than doubled	D'Agostino and Finney (1974)
	Copepod	*Tigriopus japonicus*	20–22	—	6·4 Cu + 4·4 Cd	Time to reach F_2 generation more than doubled	D'Agostino and Finney (1974)
	Copepod	*Acartia tonsa*	21	—	10	Reduced fecundity	Reeve and co-authors (1977)
	Larval copepod	*Euchaeta japonica*	8·5 or 10	—	6–7	Reduced survival beyond second naupliar stage	Lewis and co-authors (1972)
	Copepod	*Acartia clausi*	14	—	1	Reduced fecundity	Moraitou-Apostolopoulou and Verriopoulos (1979)
	Barnacle nauplii	*Balanus improvisus*	20	15, 30	50	Reduced swimming speed	Lang and co-authors (1980)
Hg	Larval shrimp	*Palaemonetes vulgaris*	26·5–28	33–34	5·6	30% mortality of unfed larvae in 48-h	Shealy and Sandifer (1975)
	Larval crab	*Eurypanopeus depressus*	25	30	1·8	Depressed swimming rate of early zoeal stage	Mirkes and co-authors (1978)
	Post-larval shrimp	*Penaeus setiferus*	21–24	25	1	No effect on growth, respiration or moulting rate: 96 h LC_{50} = 17 ppb	Green and co-authors (1976)
	Brine shrimp	*Artemia* sp.	20	—	1 (methyl)	Reduced survival of nauplii from treated parents	Cunningham and Grosch (1978)
Pb	Larval crab	*Rhithropanopeus harrisii*	23·5	20	50	Hatch to megalopa development time increased	Benijts-Claus and Benijts (1975)
Zn	Larval shrimp	*Palaemonetes pugio*	20–35	3–31	250	Little or no survival of larvae at high temperature/low salinity	McKenney and Neff (1979)
	Larval crab	*Rhithropanopeus harrisii*	23·5	20	25	Hatch to megalopa development time increased	Benijts-Claus and Benijts (1975)

Carcinus maenas (Table 3-28), tolerance to copper in *Arcartia clausi* (MORAITOU-APOSTOLOPOULOU and VERRIOPOULOS, 1979), tolerance to copper and lead in the fresh water isopod *Asellus meridianus* (BROWN, 1978), and acclimation to cadmium in the copepod *Tisbe holothuriae* (HOPPENHEIT, 1977).

Echinoderms

Early work showing that egg and larval development in echinoderms was sensitive to metals (see also Volume III: KINNE, 1977) has been confirmed by more recent studies. For example, HESLINGA (1976) showed that larval skeletal development was inhibited by 20 ppb of copper in *Echinometra mathaei,* and BOUGIS and co-authors (1979) used the sensitivity of the larvae of *Paracentrotus lividus* to copper and its modification by different waters as a means of assessing sea water quality. Similarly, KOBAYASHI (1977) has shown that the formation of the pluteus larva in *Peronella japonica* is sensitive to metals. The percentage of normal pluteus larvae present 24 h after insemination was reduced by about 10 ppb of mercury, 8ppb of copper, 14 ppb of zinc, 600 ppb of cadmium, and 420 ppb of chromium(VI)

Fish

Studies on the influence of low metal concentrations of fish are largely concerned with the long-term effects of soluble forms. CALABRESE and co-authors (1975) showed that exposure of the flounder *Pseudopleuronectes americanus* to 5 ppb of cadmium for 60 d caused oxygen consumption to be depressed in the excised gills. No effects were observed on various haematological parameters such as plasma protein level and plasma osmolality, nor were any histopathological effects found. However, the synthesis of zinc metallo-enzymes was stimulated so that, although under attack from cadmium, their functions could continue at a near-normal rate (GOULD, 1977). Adaptation to cadmium exposure was also observed by DAWSON and co-authors (1977); oxygen consumption of excised gill tissue from the striped bass *Morone saxatilis* was depressed after 30-d exposure to 0·5 to 5 ppb of cadmium; it returned to near-normal after 90-d. At 1 ppb of mercury, oxygen consumption was unaffected in the same species, but changes observed at 5 ppb suggested an ability to adapt to the metal.

Growth was depressed in juvenile plaice *Pleuronectes platessa* by 5 ppb of cadmium, but this concentration did not affect the dab *Limanda limanda* during 96-d of exposure (WESTERNHAGEN and co-authors, 1980). On the other hand, 50 ppb of cadmium caused fin erosion and mortality in both species; however, since some fish recovered from fin erosion, it was concluded that the effects of cadmium might not be caused by poisoning *per se* but by the action of pathogenic bacteria exacerbated by cadmium stress.

Although the osmoregulatory system of fish might be expected to be sensitive to metallic inhibitors, this has not been found to be the case for cadmium (MACINNES and co-authors, 1977) or methyl mercury (SCHMIDT-NIELSEN and co-authors, 1977). The latter observed no effect from injected methyl mercury on electrolyte transport or Na–K-ATPase activity in the gills of *Pseudopleuronectes americanus* and concluded that this lack of effect might be attributed to enzyme adaptation or to the synthesis of metallothionein. Induction of mercury metallothionein in the gills of sea-water adapted eels has been observed to give protection against the inorganic form of the metal (BOUQUEGNEAU, 1979).

In addition to mercury and cadmium, copper seems to be quite toxic to fish. SAWARD and co-authors (1975) observed decreased growth in *Pleuronectes platessa* exposed to 10 to 100 ppb of copper for 100-d, but this was not lethal (Table 3.18). Marine fish appear to be fairly insensitive to long exposure to ppm concentrations of lead, chromium, and zinc although these metals are accumulated by some of the tissues (SHERWOOD and WRIGHT, 1976; SOMERO and co-authors, 1977b; FLOS and co-authors, 1979; Table 3–18). The ability of fish to adapt and tolerate metallic contamination as described above is obviously of great ecological advantage, however, as GOULD (1977) has pointed out, adaptive processes involving chemical changes will be a drain on metabolic energy.

Since they might be expected to be least capable of coping with metals, fish eggs and larval stages have been the subject of a great deal of recent work. Copper appears to be one of the most toxic metals, the ionic form Cu^{2+} being largely responsible for its effect (ENGEL and SUNDA, 1979). Indeed, it has been suggested that the cupric ion activity which suppresses hatching of the eggs in spot *Leiostomus xanthurus* is so low that natural levels of copper might in the absence of organic ligands inhibit hatching.

Effects on fertilization and embryonic development in Baltic spring spawning herring have been detected at 10 ppb of copper and 5 ppb of zinc or cadmium in water with a salinity of about 5·7‰S (OJAVEER and co-authors, 1980). In the Atlantic herring *Clupea harengus* egg mortalities were observed at 30 ppb of copper by BLAXTER (1977), and in the Pacific herring *C. harengus pallasi* significant mortality of embryos was observed at 35 ppb (RICE and HARRISON, 1978); on the other hand, embryos of the anchovy *Engraulis mordax* were 6 times less sensitive (RICE and HARRISON, 1979). Pacific herring eggs are less sensitive to cadmium than copper: ALDERDICE and co-authors (1979a,b) showed that the eggs were more susceptible to mechanical damage at a combination of ≥1000 ppb of cadmium and ≤20‰S and the volume of the egg was reduced by 100 ppb of cadmium. It was concluded that the ultimate effect of cadmium could be the production of weaker eggs having a modified buoyancy. The results for other estuarine species—including *Pseudopleuronectes americanus, Menidia menidia*, and *Fundulus heteroclitus*—suggest that their eggs and larvae are relatively insensitive to cadmium levels less than 100 ppb, even at unfavourable temperatures and salinities (MIDDAUGH and DEAN, 1977; VOYER and co-authors, 1977, 1979). Development in *F. heteroclitus* was influenced by 30 ppb of inorganic or organic mercury but was very insensitive to either cadmium or lead (WEIS and WEIS, 1977a,b). Organic forms of lead may be more toxic since MARCHETTI (1978) found that for larvae of *Morone labrox* the 48-h LC_{50} concentration for tetraethyl lead was 65 ppb.

Marine mammals

Experimental studies in which harp seals *Pagophilus groenlandicus* were fed with methyl mercury chloride revealed that doses of 25 mg Kg^{-1} body weight d^{-1} were lethal in 20 to 26 d (RONALD and co-authors, 1977). Exposure to 1% of this dose resulted in high levels of methyl mercury in the tissues after 90-d, but no obvious effects were observed. For example, the concentration in muscle was 40 ppm (wet) after 90-d compared with only 0·5 ppm in control animals, suggesting that seals are extremely resistant to this form of contamination.

The possibility that metals might be involved with sickness has been studied in sea lions *Zalophus californianus* by BUHLER and co-authors (1975), but no firm conclusions could be drawn. Similarly, the possibility that the elemental composition of the tissues might be involved in a high incidence of premature births in the same species was studied by MARTIN and co-authors (1976). There was certainly evidence for an imbalance of mercury, selenium, and bromine in premature pups, but no evidence that this had anything to do with metallic contamination.

Heavy metals have also been implicated in the stranding of whales (see also Volume II: KINNE, 1975, p. 844): STONEBURNER (1978) observed Hg/Se atom ratios of about 3 in the livers of 2 non-gravid female pilot whales compared with normal values of about 1 in 2 other stranded animals and suggested that the mercury–selenium detoxification system may have broken down (Table 3-19; see also p. 1345). In the same vein, VIALE (1978) considers that there may be a relation between the beaching of whales in the Mediterranean and pollution, particularly the dumping of wastes from titanium dioxide production (p. 1295).

Seabirds

Heavy metal residues have on various occasions been cited as a possible contributary factor in seabird kills, perhaps as a result of being remobilized in starving birds, as appears to happen with pesticides. For example, a recent report in *Marine Pollution Bulletin* (**11**, No. 1) names organic lead as the prime suspect in a recent bird-kill in the Mersey estuary (see also p. 1388). In addition, ingested lead shotgun pellets have proved toxic to some species. There is also evidence for the toxic effects of mercury to fish-eating birds, particularly to the eggs (VERMEER and PEAKALL, 1977). Unlike marine mammals, birds may be unable to demethylate mercury and it is certainly transferred to the eggs. Mallards *Anas platyrhynchos* on a diet having 3 ppm (wet) of methyl mercury produced eggs containing 6 to 9 ppm (wet) compared with $< 0{\cdot}05$ ppm in controls (HEINZ, 1974). The birds were not visibly affected by 12 months on this diet, but egg laying was reduced and embryonic and duckling mortalities were increased. Eggs from fish-eating birds may be more tolerant than those of mallards, since 2 to 16 ppm (wet) has been reported as having no apparent effect on hatching and fledging in herring gulls (VERMEER and PEAKALL, 1977) and 8 ppm (wet) seems to be a natural level in eggs of the sooty tern *Sterna fusca* (STONEBURNER and co-authors, 1980).

Effects on Experimental Populations

Experimental work with trace metals in cultured organisms ranges from that on microscopic species such as bacteria having very short generation times, to that on much larger organisms such as polychaete worms. One of the possible problems with this type of work is that the most easily cultured organisms are frequently the most physiologically resistant and, therefore, may not be entirely representative of the majority of species (for details consult Volume III).

Bacteria

Some marine bacteria appear to be quite resistant to metals, since studies on strains of deep-sea bacteria by YANG and EHRLICH (1976) showed that growth was inhibited by a minimum of 1000 ppb of copper and nickel and 5000 ppb of cobalt. On the other hand, other work has revealed effects on glucose assimilation at about 1 ppb of mercury and 2 ppb of copper (AZAM and co-authors, 1977; GILLESPIE and VACCARO, 1978). The effective form of copper is the cupric ion Cu^{2+}, the level of which depends on the total amount of copper and the degree of inorganic and organic complexation (SUNDA and GILLESPIE, 1979).

An extremely important property of bacteria is their ability to develop metal-resistant strains—a useful account of this, including the role of extrachromosomal factors or plasmids which carry the genes conferring resistance, was given by SILVER and co-authors (1976). One of the most important facets of tolerance concerns the ability of some tolerant species to volatilize or methylate mercury in sediments (p. 1307). OLSON and co-authors (1979) who worked on mercury-resistant strains from Chesapeake Bay have shown that volatilization of mercury by reduction of Hg^{2+} or organic mercury to metallic mercury Hg° is plasmid mediated; a loss of resistance and the capability to volatilize mercury turned out to be associated with the loss of plasmids in 2 strains. Two of the resistant strains were able to methylate mercury, but this seems to be mediated by a genetic system different from that involved in volatilization. On exposure of the mercury-resistant bacteria to other metals (Co, Cd, Zn, Ni, Cr, Pb) it was found that each type was also resistant to at least one other metal, although the resistance mechanism, in the case of cadmium at least, is thought to involve an exclusion process. Probably through a linkage via plasmids, bacterial resistance to metals also appears to be related to antibiotic resistance (DEVANAS and co-authors, 1980).

Phytoplankton

Virtually all the information available before 1977 on metal toxicity to marine phytoplankton has been reviewed by DAVIES (1978). He considered the relevance of phytoplankton culture experiments to the situation in the field and stressed the fact that in this type of system the toxicity of metals is strongly dependent on the composition of the culture medium, the density of the culture and metal losses through adsorption, precipitation or volatilization. Thus, the concentration of metal added at the start gives little indication of the amount incorporated in the cells on which the inhibition of growth actually depends. That this is so has been demonstrated for mercury in *Isochrysis galbana* by DAVIES (1974) and for copper in *Thalassiosira pseudonana* by SUNDA and GUILLARD (1976). Although it seems inevitable that many studies with phytoplankton cultures tend to underestimate metal toxicity, there are a number of examples where extremely low initial concentrations have affected sensitive populations.

Copper appears to be extremely toxic to some species, and NIELSEN and ANDERSEN (1970) found that additions of 1 to 2 ppb produced measurable inhibition of growth in *Nitzschia palea*; 10 ppb produced complete inhibition. Similarly, growth of the dinoflagellate *Scrippsiella faeroense* was inhibited at 1 ppb, and 20 ppb were lethal (SAIFULLAH, 1978). The toxic form of copper is the cupric ion (SUNDA and GUILLARD, 1976; JACKSON and MORGAN, 1978); it has been suggested by a number of workers that variations in its concentration may control some species of phytoplankton even at natural levels of total copper (NIELSEN and ANDERSEN, 1970; ANDERSON and MOREL, 1978). The latter suggested that normally growth of the dinoflagellate *Gonyaulax tamarensis* is inhibited by natural cupric ion concentrations which do not affect other species, but that if the organic chelating capacity of the water increases and binds the copper a dinoflagellate bloom results.

Resistance of metals varies widely among species and whereas *Thalassiosira pseudonana* and *Skeletonema costatum* were affected by 10 ppb of copper, *Phaeodactylum tricornutum* was almost unaffected by 250 ppb (JENSEN and co-authors, 1976). Similarly, a few ppb of mercury inhibited the growth of *Isochrysis galbana*), but hundreds of ppb had no effect on *Dunaliella tertiolecta* (DAVIES, 1976). Because of the difficulty of extrapolating the results of laboratory culture experiments to the field there has been a recent trend towards the use of cultures in the field and the use of natural populations. The former is illustrated by the work of EIDE and co-authors (1979) who cultured laboratory species in dialysis sacs immersed in the zinc-contaminated waters of Norwegian fjords. In Orkdalsfjord (Zn 7 to 54 ppb; Cu 1·3 to 9·4 ppb) *S. costatum* would not grow, but growth in *T. pseudonana* responded to variations in contamination; in Sørfjord where the minimum concentration of zinc was about 100 ppb neither this species nor *S. costatum* could survive, but *P. tricornutum* survived at a reduced level of growth in waters containing around 500 ppb of zinc. An important point here is that in laboratory culture, *P. tricornutum* was unaffected by 4000 ppb of zinc plus 250 ppb of copper (BRAEK and co-authors, 1976).

From experiments on the effect of zinc on the growth of multispecies cultures of phytoplankton, KAYSER (1977) concluded that mixtures of species are more sensitive than single species and that this might also apply in the field. Certainly, studies on natural populations of phytoplankton have shown them to be quite sensitive: thus, DAVIES and SLEEP (1979/1980) found that the carbon fixation rate was depressed by about 15 ppb of zinc or 1 to 2·5 ppb of copper—levels often exceeded in contaminated areas. In a recent study by THOMAS and co-authors (1980) the toxicity of a mixture of 10 metals (As, Cd, Cr, Cu, Hg, Ni, Pb, Sb, Se, Zn) was studied in natural populations. The composition of this mixture approximated the concentrations found in moderately contaminated waters (e.g. 0·15 ppb Hg, 0·75 ppb Cd, 3 ppb Cu) but it had no effect on phytoplankton growth. However, a 5-fold increase in concentration did inhibit growth, and deletion of metals from the mixture showed that copper and mercury were the cause of toxicity. It was also observed that the dominance of certain species in natural phytoplankton was changed by metal contamination.

Most of the evidence points to mercury and copper as being particularly toxic to phytoplankton. However, the effects of zinc cannot be ignored since it is such a common contaminant, and there is certainly some evidence for the influence of low levels of lead and cadmium, although generally speaking their toxicity is low (DAVIES, 1978). For example, RIVKIN (1979) observed inhibition of growth in *Skeletonema costatum* at less than 10 ppb of lead. BERLAND and co-authors (1976) established inhibition of growth in *Cylindrotheca closterium* at 5 ppb of cadmium, and KAYSER and SPERLING (1980) recorded inhibition in *Prorocentrum micans* at about 1 ppb of cadmium.

The mechanisms by which species of phytoplankton handle metals are only gradually becoming understood. In the case of arsenate, which competes with phosphate and can inhibit primary productivity in *Skeletonema costatum* at very low levels, conversion to an organic form occurs, one of the products being the relatively non-toxic dimethylarsine (ANDREAE and KLUMPP, 1979; SANDERS, 1979c). The high tolerance of *Dunaliella tertiolecta* to mercury is thought to be partially dependent on a slow rate of uptake, but largely dependent on the precipitation of the metal within the cell as sulphide, since the species produces hydrogen sulphide (DAVIES, 1976). In *S. costatum* no evidence could be found for the detoxification of mercury by metallothionein, although low-molecular-weight proteins may detoxify copper and zinc (CLOUTIER-MANTHA and BROWN, 1980). The adaptation of a population to metal contamination has been studied by LI (1979) who observed physiological adaptations in *Thalassiosira weissflogii* following prolonged exposure to sublethal levels of cadmium. In addition, the development of tolerant strains sometimes occurs; thus, a strain of *S. costatum* isolated from Sørfjord, where hundreds of ppb of zinc are found in the water, was more tolerant to the metal than an isolate from Trondheimsfjord (BRAEK and co-authors, 1980). Rather more is known about mechanisms of resistance in fresh water species. In *Chlorella vulgaris* it was observed that copper resistance in a tolerant strain was due to the exclusion of the metal, since copper accumulated within the cell was equally toxic to both normal and resistant strains (FOSTER, 1977). The development of zinc tolerance in the same species also depends on an exclusion mechanism but mercury resistance is primarily associated with an increase in the capacity of the cells to volatilize it out of solution by enzymic means (FILIPPIS and PALLAGHY, 1976). Another possible mechanism by which species may detoxify their surroundings is by the production and extrusion of organic chelators. For example, GNASSIA-BARELLI and co-authors (1978) observed that the toxicity of copper to *Cricosphaera elongata* was reduced by extracellular products.

Protozoans

The few species of marine ciliates which have been studied appear to be relatively metal tolerant. GRAY and VENTILLA (1973) found that in *Cristigera* spp., a sediment-dwelling form, growth of the population was reduced by about 10% in the presence of 25 ppb of $HgCl_2$, 300 ppb of $Pb(NO_3)_2$, and 250 ppb of $ZnSO_4$, the combined effect of all 3 being rather more than additive. *Euplotes vannus* is an even more tolerant ciliate: 100 ppb of mercury had no effect on population growth and 100 ppb of copper produced only 50% inhibition (PERSOONE and UYTTERSPROT, 1975). Total inhibition of growth required 1000 ppb of mercury, 10,000 ppb of copper and 100,000 ppb of cadmium, lead, and zinc. As little as 1 ppb of $HgCl_2$ reduced the growth of *Uronema marinum* by 10%; the 24-h LC_{50} was 6 ppb. However, the equivalent values for $PbCl_2$ were 5000 to 10,000 ppb and 60,000 ppb; for $ZnCl_2$, 20,000 ppb and 400,000 ppb (PARKER, 1979).

The basis of tolerance, for some metals at least, may depend on their immobilization in membrane-bound vesicles such as lysosomes (NILSSON, 1979). In *Eutreptia* sp., a *Euglena*-like organism found in water containing 30,000 ppb of zinc, high concentrations were observed in the chloroplasts (SIMS, 1975). Studies by ALBERGONI and co-authors (1980) on *Euglena gracilis* revealed that a low-molecular-weight glycoprotein plays a protective role by chelating copper in the cell and also, following excretion, by limiting copper uptake from the medium. The protein also chelates zinc but not cadmium. Protection from cadmium within the cell appears to be provided by a high-molecular-weight protein which is not excreted, and thus the metal is accumulated.

Although PARKER (1979) was unable to induce increased tolerance by exposing *Uronema marinum* to sublethal concentrations of mercury, lead, and zinc for 18 d, BERK and co-authors (1978) found that tolerance to mercury in *U. nigricans* was increased within a single generation by feeding with mercury-laden bacteria.

Coelenterates

Our knowledge of the effects of metals on coelenterates is largely confined to marine hydroids. In *Eirene viridula* morphological abnormalities were produced by as little as 1 to 3·3 ppb of mercury, 30 to 60 ppb of copper, 100 to 300 ppb of cadmium, 1000 to 3000 ppb of lead and 1500 to 3000 ppb of zinc (KARBE, 1972). Consequently, KARBE proposed *E. viridula* as a possible bioassay organism. This theme has been continued by

STEBBING (1976) who showed that the colonial growth rate of *Campanularia flexuosa* was sensitive to as little as 1·6 to 1·7 ppb of mercury, 10 to 13 ppb of copper and 110 to 280 ppb of cadmium. The effect of these metals involves increased lysosomal hydrolase activity, and by employing a cytochemical staining reaction the effective threshold was reduced to 0·17 ppb of mercury, 1·2 to 1·9 ppb of copper and 40 to 75 ppb of cadmium (MOORE and STEBBING, 1976). Pretreatment of the colonies with 10 ppb of copper failed to induce additional tolerance, although it was suggested that sequestration of copper occurs in endodermal cell lysosomes.

The effect of cadmium on the hydroid *Clava multicornis* at the various temperatures and salinities showed that after a 6-w exposure the sublethal threshold concentration of 200 ppb was unaffected by combinations of 5 to 20 °C and 10 to 25‰S (FISCHER, 1978). Another hydroid, *Laomedea loveni*, was more sensitive to cadmium. Concentrations producing irreversible retraction of 50% of the hydranths after 7-d exposure varied from 3 ppb at 17 °C and 10‰S to 80 ppb at 7·5 °C and 25‰S (THEEDE and co-authors, 1979a). Uptake studies demonstrated that accumulation of the metal occurred more rapidly at higher temperatures, but was insensitive to a salinity change from 10 to 35‰S—although, in theory, reduced chloride complexation should make the metal more readily available at low salinity (p. 1311).

Table 3-27

Effects of metals on survival and reproduction of cultured polychaete worms (After REISH, 1978; PETRICH and REISH, 1979; modified; reproduced by permission of the Revue Internationale d'Océanographie Médicale, Springer-Verlag)

Test conditions and effects	Concentration in water (ppm) Hg	Cu	Al	Cr	Zn	Pb	Ni	Cd
				Ctenodrilus serratus				
96 h LC_{50}	0·09	0·3	0·48	4·3	7·1	14	17	>20
Conc. significantly suppressing reproduction	0·05	0·1	<0·5	0·05	0·5	1·0	0·5	2·5
				Neanthes arenaceodentata				
96 h LC_{50}	0·02	0·3	>2	3·2	1·8	7·7	49	12·1
Conc. significantly suppressing reproduction	ND	ND	ND	0·05	0·32	3·1	ND	1·0
				Capitella capitata				
96 h LC_{50}	<0·1	0·2	2	5·0	10·7	11·4	>50	5·8
Conc. significantly suppressing reproduction	ND	ND	ND	0·1	0·56	0·2	ND	0·56
				Ophryotrocha diadema				
96 h LC_{50}	0·09	0·16	ND	>5·0	2·7	14·0	ND	4·2
Conc. significantly suppressing reproduction	>0·05	>0·1	ND	>0·5	0·5	>1·0	ND	1·0

ND, Not determined.

Polychaetes

Studies on the effects of metals on the reproduction of polychaete worms in culture have shown that effective concentrations are generally much lower than the LC_{50} values (Table 3-27). This is particularly true for chromium and probably reflects its slow uptake by the worm—and hence the slower attainment of a toxic dose. In *Ctenodrilus serratus*, for example, reproduction is affected by equal concentrations of chromium and mercury, although the LC_{50} concentrations differ by a factor of almost 50—a classical example of the difficulty of extrapolating the results of acute toxicity tests.

A study of cadmium accumulation by fed *Ophryotrocha diadema* in culture by KLÖCKNER (1979) showed that after 64 d exposure to 10, 100, and 1000 ppb, concentrations of 79, 315, and 1708 ppm were reached by the worms. The levels in 62 to 116-day-old worms remained approximately the same when 3 generations were subjected to continuous contamination, although very little cadmium was transmitted in the eggs.

RØED (1980) studied the effects of 200 and 400 ppb of cadmium on *Ophryotrocha labronica* at 3 different salinities (20, 25, 30‰S) over 3 generations. Low salinity and cadmium caused a reduction in growth rate, prolonged the time to reach sexual maturity, and reduced size at maturity. At the lowest salinity and 400 ppb of cadmium, the first generation was unable to produce viable larvae: at the higher salinities, there was some evidence for an increase in tolerance to cadmium between the second and third generations. In comparable experiments with the archiannelid *Dinophilus gyrociliatus*, increased tolerance to cadmium and mercury acting together was observed at 30‰S but at 25‰S the adaptation mechanism seemed to be neutralised (RØED, 1979).

These results and those in Table 3-27 tend to confirm the impression given by field observations that many polychaetes are quite tolerant to heavy metals.

Effects in Experimental Ecosystems

By enclosing natural planktonic populations in plastic bags or tubes it has become possible to study defined self-sustaining populations over a long period of time. The scale of the enclosures varies enormously and only the very largest are able to sustain as many as 4 trophic levels.

KUIPER (1977) used enclosures having a volume of 1·4 m^3 to study the influence of mercury on a North Sea coastal plankton community. When 5 ppb were added to this system, the mercury disappeared rapidly from the water some being deposited in the sediment. However, mercury reduced the growth of phytoplankton as long as the level in the water exceeded 1·5 ppb. In addition, the second phytoplankton bloom was delayed—probably due to a lag in the mineralization of organic material in the sediment—and methylation of mercury in the sediment was also detected. Studies in the same enclosures on the effect of mercury on periphyton, showed that, although the addition of 0·5 ppb had no effect on the biomass, the species composition was changed (GROLLE and KUIPPER, 1980). When the influence of 1·3 ppb of cadmium on phytoplankton was studied in the 68 m^3 Saanich Inlet enclosures (CEPEX) no effects were observed (KREMLING and co-authors, 1978).

Studies on the large enclosures at Loch Ewe (95 m^3) and Saanich Inlet (1300 m^3) have been reviewed by GRICE and MENZEL (1978), DAVIES and GAMBLE (1979), and STEELE (1979). Following the addition of 1 ppb of inorganic mercury to the Loch Ewe system there was, by comparison with the control, a transitory effect on the phytoplankton but no overall effect on the structure or size of the phytoplankton or zooplankton populations (DAVIES and GAMBLE, 1979). Mercury was, however, lost rapidly from the water, the reactive mercury level falling to about 0·2 ppb in only 4 d. On the other hand, the addition of 10 ppb caused a 10-fold decrease in the biomass of zooplankton and other obvious effects. Essentially similar results over 72 d were observed in the Saanich Inlet enclosures, with marked transitory effects on bacteria and phytoplankton at 1 ppb but fairly obvious effects at 5 ppb (GRICE and MENZEL, 1978). However, reduced growth of chum salmon at the 5 ppb level was attributed to the reduced availability of zooplankton for food rather than to the direct effect of mercury.

Although some effects have been observed with the addition of 1ppb of mercury (and 10 ppb of copper) to these systems, it has proved difficult in some cases to separate the effects of metals from the added stresses of enclosure. In fact, STEELE (1979) points out that large-tank experiments involving a simple phytoplankton—*Tellina tenuis*—0-group plaice food chain have proved to be more sensitive to mercury and copper than the more natural enclosed ecosystems. Effects on the carbon fixation rate of phytoplankton in the tanks were observed at a copper concentration about 4 ppb higher than the control (SAWARD and co-authors, 1975) and at 0·1 ppb of mercury various effects were observed including a decrease in the growth of plaice (STEELE, 1979). However, unlike the enclosures, the water in the tank experiments was gradually replaced and the metal levels maintained by additions

Experimental Toxicity Studies—Summary

(i) The relative toxicity of metals may be exemplified as follows:

(a) Natural phytoplankton (growth); Hg >> Cu > Pb > As(V) > Zn = Cd > Ni, Cr(VI), Sb, Se, As(III) (HOLLIBAUGH and co-authors, 1980).

(b) Phytoplankton *Ditylum brightwellii* (growth); Hg > Ag > Cu > Pb > Cd > Zn > Tl (CANTERFORD and CANTERFORD, 1980).

(c) Seaweeds (growth); Hg (org.) > Hg (inorg.) > Cu > Ag > Zn > Cd > Pb (p. 3–75).

(d) Hydroid *Eirene viridula* (morphological abnormalities); Hg > Cu > Cd > Pb > Zn (KARBE, 1972).

(e) Polychaete *Ctenodrilus serratus* (reproduction); Hg = Cr(VI) > Cu > Al > Zn > Ni > Pb > Cd (REISH, 1978; PETRICH and REISH, 1979).

(f) Mollusc embryos *Crassostrea virginica* (48 h LC_{50}); Hg > Ag > Cu > Zn > Ni > Pb > Cd > As(III) > Cr(III) > Mn (CALABRESE and co-authors, 1973).

(g) Echinoderm larvae *Peronella japonica* (development); Hg = Cu > Zn > Cr(VI) > Cd (KOBAYASHI, 1977).

(h) Copepod *Nitocra spinipes* (96 h LC_{50}); Hg > Zn > Cd = Cu > Co > Ni > Al > Fe > Mn (BENGTSSON, 1978).

Generally speaking the most rapidly absorbed metals—mercury, copper, and silver—are the most toxic, but for others—including cadmium, lead, zinc, and chromium(VI)—the toxicities are extremely variable depending not only on the permeability of the particular species to the metal but also on its capacity to detoxify the metal. The length of toxicity experiments is quite important since a metal such as cadmium or chromium which is absorbed slowly might ultimately prove more toxic relative to say mercury than would be predicted from short-term experiments.

(ii) Factors affecting the toxicity of a metal in a particular species are:

(a) Form of the metal in the environment; this governs its availability for bioaccumulation.

(b) Presence of other metals or poisons; their influence may range from a more-than-additive increase in toxicity on the one hand to antagonistic effects on the other.

(c) Environmental factors (temperature, salinity, pH, dissolved oxygen, light); these may influence toxicity both by changing the speciation of the metal and by changing the physiological stress on the organism, to which that of the metal is often additive.

(d) Condition of the organism (phase of life cycle, age, size, sex, nutritional state, etc.) is extremely important; for example the embryonic and larval stages of bivalve molluscs are far more vulnerable than the adults.

(e) Ability to avoid the worst conditions by, for example, shell closure.

(f) Ability to adapt to metals: improvements in the ability to detoxify metals may be induced in individuals of some species by exposure to sublethal levels and in populations from contaminated areas metal tolerance sometimes has a genetic basis. Reduced metal permeability is a fairly common genetic adaptation.

(iii) Life functions most sensitive to metals:

There are no clear conclusions to be drawn, since the numbers and types of study which have been carried out vary so much between different phyla. However, growth of individuals or of populations has generally proved to be particularly sensitive to metals. Even more important, embryonic and larval development has been shown to be vulnerable to low metal levels in various phyla, and so also has fecundity, although the number of studies on this topic is fairly limited. Oxygen utilization and the functions of various enzymes also number among the most metal-sensitive processes, although it is not always clear how significant some of these observations are in terms of the survival of organisms and populations. For instance, there is evidence that such metal-induced changes can be compensated for, at least in some species, by adaptive processes, although this presumably involves the utilization of additional energy.

(iv) Experimental toxicity in relation to real environmental concentrations:

Information from the foregoing pages on the toxicity of 5 metals is compared in Fig. 3-8 with the sea water concentrations from different areas also reported in this review. The figure does not include the vast proportion of toxicity data which concerns completely unrealistic concentrations, and is also biased in the sense that negative results are excluded. In addition, the water concentrations are far from comprehensive but are thought to cover fairly accurately the full range of values observed in the field.

Mercury is undoubtedly the most toxic metal but there is a paucity of evidence for its toxicity at concentrations below 1 ppb. However, the tendency for mercury to disappear from experimental solutions probably means that in many experiments actual exposure levels were perhaps half the nominal concentrations or even less. Even so, sea water concentrations exceeding 0·5 ppb have only rarely been found and levels exceeding 1 ppb in Fig. 3-8 are confined to Minamata Bay (Japan) and a high value from the Derwent estuary (Tasmania) (Table 3-4).

Although generally less toxic than mercury, copper possesses considerable toxicity to some species of phytoplankton, embryonic bivalve molluscs, and crustaceans. These toxic concentrations are well within the range of values observed in various contaminated areas. However, the toxicity of copper appears to be dependent upon the concentration of free cupric ions and this is very dependent on the presence of dissolved organic

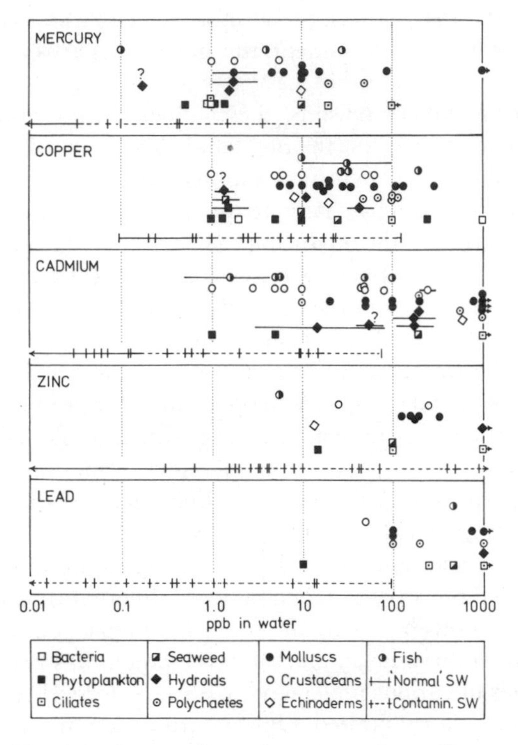

Fig. 3-8: Comparison of concentrations affecting organisms under experimental conditions with those actually found in 'normal' and contaminated sea areas. Only values given in this review are used. (Original)

ligands. Thus the potential effects of a relatively high dissolved copper level may be ameliorated by, for example, the presence of humic acids. The toxic level of cupric ions to some species may be extremely low and several workers have suggested that in the absence of chelators copper may be toxic to some species at natural concentrations.

There is certainly evidence from some species of phytoplankton, crustaceans, and fish that effects of cadmium may be expected at concentrations such as have been observed in the field. On the other hand, there is a lot of evidence supporting the view that cadmium has little effect on the majority of species.

Although zinc is not regarded as being particularly toxic and is the best regulated of all metals, it is a very common contaminant and potentially toxic concentrations have been observed in a number of different sea areas.

The threat from inorganic lead to marine organisms seems to be very small, since there is little evidence that environmental concentrations ever approach toxic levels. However, it may be somewhat premature to ignore lead completely in view of our general ignorance about the environmental cycling of the relatively toxic alkyl lead compounds (pp. 1310 and 1374).

There is little evidence to suggest that any of the other metals or metalloids in solution is likely to pose problems except under rather exceptional circumstances. Having said this, it is perhaps worth pointing out that practically all the toxicity data relate to concentrations in solution when we know that in many animals food and particulates are the major sources for accumulation. For example, some of our recent work on the toxicity of contaminated sediments to bivalve molluscs indicates that in the field the toxicity of sediment-bound copper is more important than that of the water with which it may be in equilibrium (BRYAN and GIBBS, 1983).

(v) Predictions from experimental toxicity data:

If concentrations of dissolved metals in coastal areas are assumed to be Hg, 0·005 to 0·05 ppb; Cu, 0·2 to 1 ppb; Cd, 0·03 to 0·07 ppb; Zn, 0·5 to 2 ppb; Pb, 0·01 to 0·1 ppb, then an increase in concentrations by an order of magnitude would be expected to produce some effects in the cases of copper and possibly mercury. However, an increase of nearer to 2 orders of magnitude would be required for cadmium or zinc to produce significant effects and arguably an increase of 3 orders of magnitude would be required before the toxicity of lead became apparent(Fig. 3-8).

Sea areas where potentially toxic concentrations of dissolved metals have been found are largely confined to estuaries, fjords, and enclosed bays (Table 3-4). Estuarine organisms tend to be physiologically tolerant species and might be expected to be rather insensitive to metals. Under optimal environmental conditions this may be so, but recent experimental work on the toxicities of metals under different temperature/salinity regimes has shown that under suboptimal conditions to which such species are sometimes exposed, the toxicity of a particular metal may be increased by as much as an order of magnitude. Thus there is little doubt that these confined areas are at greatest risk and the conditions in some of them will be considered next.

(b) Effects of Metals in Contaminated Areas

In this section, the question of the environmental impact of metals will be considered by reference to evidence from some of the most heavily contaminated areas of the world.

Minamata Bay—Mercury

The most important example has been the contamination of Minamata Bay in Japan with mercury and has been the subject of a book by TSUBAKI and IRUKAYAMA (1977). The metal was used as a catalyst in the production of acetaldehyde and it is estimated that over 80 tons were discharged to the bay between 1932 and 1968. Even before the recognition of Minamata disease in local inhabitants in 1956, there were signs in the bay that all was not well; the disappearance of oysters from the hulls of ships near the factory outfall was noticed and dead fish and octopus were found floating near the shore. In addition, birds were affected and some domestic animals, especially cats. By 1959 to 1960 it had been recognized that Minamata disease was similar to organo-mercury poisoning and the source was tracked down to sea food. It was thought at first that the source of mercury to the organisms was the sediments where maximum values of the order of 6 to 900 ppm were found. However, it was eventually shown that the main source of the problem was methyl mercury in the factory effluent rather than any that was produced in the sediments. Analyses of various fish species gave values ranging over 7 to 23 ppm (wet) in 1961 but, following the closure of the acetaldehyde factory in 1968 concentrations fell to 0·1 to 1·1 ppm of which 0·01 to 0·23 ppm was methyl mercury. In

1959 the rock-dwelling bivalve *Hormomya mutibalis* contained about 100 ppm (dry) of mercury of which about 30% may have been in the methyl form; however, concentrations had fallen to about 10 ppm by the early 1960s. Even in the 1960s the sediment-dwelling *Venus japonica* maintained values in the range 8 to 84 ppm (dry) and it was only following the final closure of the factory that levels fell sharply over a few months so that total mercury levels from 1969 to 1972 were, with a few exceptions, less than 5 ppm of which less than 0·2 ppm was in the methyl form. Recent studies in Minamata Bay indicate that mercury levels in the water have fallen by about an order of magnitude since the 1960s when maximum values of 1·5 and 3·6 ppb were observed (HOSOHARA and co-authors, 1961; MATIDA and KUMADA, 1969; KUMAGAI and NISHIMURA, 1978). Sediment concentrations have also fallen but values as high as 100 ppm are still found at some sites (KUMAGAI and NISHIMURA, 1978) and there are plans to remove or cover the contaminated sediments by 1983 (TSUBAKI and IRUKAYAMA, 1977; see also p. 1398).

Restronguet Creek—Copper, Zinc, and Arsenic

This estuary (Fig. 3-6) is a branch of the Fal estuary system (UK) and is heavily contaminated with metals from metalliferous mining (Fig. 3.1; Tables 3-4, 3-5). The creek is 4 km long, almost dries out at low tide and has been contaminated for over 200 yr. Examples of copper and zinc concentrations in organisms approaching their upstream limits of distribution are given in Table 3-28 and ratios showing their enhancement above normal values are also given. Enhancement for copper varies from about 2 orders of magnitude in seaweed and polychaete worms to low levels in the crab and flounder where regulation of the metal almost certainly occurs. Regulation by more advanced forms probably explains the variability in zinc enhancement both directly and because some predators may feed on organisms which already regulate.

Enhanced arsenic concentrations are also found in seaweeds and invertebrates from the creek (KLUMPP and PETERSON, 1979; LANGSTON, 1980). Although copper is one of the most toxic metals and the influence of zinc and arsenic might be expected to be additive, the variety of species in the creek clearly results from their ability to handle metals. For example, this is inherent in oysters where the immobilization of copper (and zinc) in a granular form in the amoebocytes leaves the animals green in colour and inedible but otherwise unaffected (Table 3-28). Other species appear to have adapted following exposure to metals. *Carcinus maenas* from the upper creek were particularly tolerant to zinc, most of the larger creek animals in one experiment being unaffected by 10 ppm over a period of 38 d whereas 50% of the normal crabs died within 6 d. The more tolerant animals were generally less permeable to zinc and perhaps better equipped to excrete it; however, the induction of Zn-metallothionein may also be involved, having been observed in other species of crabs by OLAFSON and co-authors (1979). In the polychaete *Nereis diversicolor* from the creek, tolerance to both copper and zinc may be genetically based (BRYAN, 1976a) and this may also be true of copper tolerance in *Fucus vesiculosus* (Fig. 3-7).

Since there seems to be no non-metallic contaminants of any significance in the creek, any effects on the distribution of the biota should be attributable to metals. Owing to the development of tolerance in various species, the effects are certainly not as clear-cut as might be expected. Bivalve molluscs appear to be most vulnerable; for example,

Scrobicularia plana is absent from large areas of the intertidal muds where, under normal conditions, it would account for an appreciable fraction of the total biomass (BRYAN and GIBBS, 1983).

Sørfjord—Zinc, Cadmium, Lead

This branch of Hardangerfjord in Norway is about 40 km long and has a zinc smelter at its upper end. Concentrations of zinc, cadmium, lead, and other metals in the sediments and waters are extremely high (Tables 3-7, 3-4) and there is appreciable enhancement of concentrations in seaweeds and invertebrates but not in fish (Table 3-28). No thorough study of the effects of metals on the area has been published but the most obvious are at the upper end of the fjord where, for example, seaweeds and mussels are scarce or absent (STENNER and NICKLESS, 1974a). Again, there is evidence for the development of tolerance, and a zinc-adapted strain of the diatom *Skeletonema costatum* has been isolated from the fjord (BRAEK and co-authors, 1980)

Derwent Estuary—Zinc, Cadmium, Mercury

This Tasmanian estuary is contaminated with a range of metals from a zinc refinery and with mercury from a chlor-alkali plant and from the earlier use of mercury slimicides in a paper-making plant (BLOOM and AYLING, 1977). Extremely high concentrations have been reported in the sediments and waters (Tables 3-7, 3-4) and in some species of invertebrates (BLOOM and AYLING, 1977). Very high concentrations of zinc and cadmium have been found in oysters, one value for zinc being perhaps the highest ever found (Table 3-29). On the other hand, with the exception of mercury (Table 3-29) the influence of contamination on fish-muscle concentrations has been shown to be small (EUSTACE, 1974). To the reviewer's knowledge, there is no evidence concerning the effects of metals on the biota of the estuary.

Severn Estuary and Bristol Channel—Cadmium, Silver, etc.

As a result of industrial processes such as smelting, the waters of the Severn estuary (Fig. 3-6) are contaminated with cadmium (1·4 to 9·4 ppb) and other metals (ABDULLAH and ROYLE, 1974). Detectable cadmium contamination extends for some 200 km and Table 3-28 illustrates the enhancement of both cadmium and silver levels in species collected towards their upstream limits of distribution. Enhancement of cadmium in all species is consistently high, reflecting the inability of most organisms to regulate cadmium levels. The apparent indifference of these organisms to such high cadmium levels resides in most cases in the induction of metallothioneins (Table 3-28). In addition to cadmium, there is appreciable enhancement of silver, particularly in burrowing species and seaweeds, but not in herbivorous gastropods.

There is no unequivocal evidence that metals or indeed other contaminants have affected the distribution of species in this area. Evidence collected by CLARK (1971) suggested that the variety of fish in the area has been reduced; but a study of faunistic records covering the last 30 yr by METTAM (1979) indicates that the total number of species in the estuary has remained relatively constant and that there is no evidence for long-term environmental changes. As with many other areas, however, contamination of the Severn and Bristol Channel is of long standing (CLIFTON and HAMILTON, 1979)

Table 3-28

High metal concentrations (dry basis) in biota from 3 contaminated areas (After BRYAN, 1980**; modified; reproduced by permission of Biologische Anstalt Helgoland)

		Restronguet Creek (UK)				
		Copper		Zinc		
Site (km from mouth)	Species	ppm	Restr. Cr. / 'normal'	ppm	Restr. Cr. / 'normal'	Remarks
2	*Fucus vesiculosus*	1612	160	2040	20	Tolerance to Cu partly the result of exclusion mechanism (Fig. 3-7)
2	*Nereis diversiclor*	1170	58	290	2	Cu and Zn tolerance based on lower permeability and Cu storage[1]
2	*Nephtys hombergi*	2116	118	483	2	Possibly same as *N. diversicolor*
2	*Scrobicularia plana*	150	4	2580	7	Slight tolerance to Cu.[1] May avoid worst by shell closure
0	*Ostrea edulis*	3870	45	14,900	9	Storage of Cu and Zn in amoebocytes—green oysters[2]
−1	*Chlamys varia*	54	4	2070	4	Storage of Zn in kidney granules (Zn = 7·6% of dry kidney)
0·5	*Littorina obtusata*	1300	13	828	7	Cu probably immobilized in granules[10]
2	*Carcinus maenas*	191	2–3	149	2	Zn tolerance based partly on lower permeability and possibly metallothionein
4	*Platichthys flesus* (liver)	118	2	203	1	May avoid worst conditions and tends to regulate both metals

		Severn estuary (UK)				
		Cadmium		Silver		
			Severn		Severn	
Site	Species	ppm	'normal'	ppm	'normal'	Remarks
Sharpness	*Fucus vesiculosus*	58	58	6·6	33	Unknown how metals handled
Sharpness	*Nereis diversicolor*	10	50	24	60	Ag may be handled like Cu[1]
Sharpness	*Macoma balthica*	9·4	47	100	100	Unknown how metals handled
Fontygary B.	*Mytilus edulis*[3]	60	60	—	—	Cd metallothionein induced[4]
Portishead	*Patella vulgata*	289	29	12	7	Cd metallothionein induced[6]
Sheperdine	*Littorina obtusata*	199	40	9	3	Cd metallothionein possibly involved[9]
Flatholm	*Nucella lapillus*[3]	725	72	—	—	Cd metallothionein induced[9]
Clevedon	*Carcinus maenas*[5]	~100	100	—	—	Cd metallothionein induced[7]
Oldbury	*Platichthys flesus*[8]	~5	—	—	—	Cd metallothionein possibly involved

		Sørfjord (Norway)					
		Cadmium		Lead		Zinc	
			Sørfjord		Sørfjord		Sørfjord
Site (km from mouth)	Species	ppm	'normal'	ppm	'normal'	ppm	'normal'
25	*Fucus vesiculosus*[11]	13	13	250	125	4700	47
25	*Ascophyllum nodosum*	11	11	105	50	3480	35
25	*Mytilus edulis*[12]	140	74	3100	1550	2370	23
25	*Asterias rubens*	18	46*	460	200*	1500	7*
0	*Gadus morhua*[12] (muscle)	0·35	3·5	2	2	33	3
	(gill)	1·4	1	12	2·4	105	2
25	*Pleuronectes limanda* (muscle)	0·7	—	14	—	45	—
	(gill)	2·5	—	50	—	240	—

*'Normal' values from Table 3-16

** Compiled from 1. BRYAN (1976a, 1979); 2. GEORGE and co-authors (1978); 3. NICKLESS and co-authors (1972); 4. NOËL-LAMBOT (1976); 5. PEDEN and co-authors (1973); 6. HOWARD and NICKLESS (1977); 7. JENNINGS and co-authors (1979); 8. HARDISTY and co-authors (1974); 9. NOËL-LAMBOT and co-authors (1978a); 10. MARTOJA and co-authors (1980); 11. MELHUUS and co-authors (1978); 12. STENNER and NICKLESS (1974b)

and there is no certain baseline with which to compare. Obviously, more sensitive methods of assessing the presence or absence of deleterious effects are required and some additional evidence for this estuary is considered on p. 1389.

Other Areas

Examples from other areas where, in most cases, appreciable levels of contamination have been observed in the biota, are shown in Table 3-29. There is only limited evidence in these papers for the effects of heavy metals, based largely on observations of the distribution of organisms. For example, RENZONI and co-authors (1973) observed a scarcity of organisms near an Italian chlor-alkali factory, but it was uncertain to what extent mercury contamination was responsible. One of the very few studies linking faunal differences with residue levels was carried out by LANDE (1977) who showed that the bottom fauna of Orkdalsfjorden, part of Trondheimsfjord in Norway, differed from that of other areas and attributed this to the influence of mining wastes.

In the autumn of 1979 more than 2000 birds were killed in the Mersey estuary (UK): these included waders (1336 dunlin *Calidris alpina)*, gulls (368 black-headed gulls *Larus ridibundus* and wildfowl. It has been concluded by HEAD and co-authors (1980) and BULL) and co-authors (1983) that lead is an important factor in the mortalities. Sources of input to the estuary include lead refining, paint manufacture, alkyl lead production, and sewage; also the sediments contain moderately high concentrations of lead (Table 3-6). Analyses of affected dunlin revealed 11 to 34 ppm (wet) of lead in the livers compared with less than 1 ppm in the controls; most important, 27 to 78% of the metal was in the tri- or dialkyl form. Trialkyl lead also accounted for 31 to 67% of the totals in *Macoma balthica* (2 to 4 ppm wet) and *Nereis diversicolor* 2·7 to 5·2 ppm wet) and it is thought that the birds died from alkyl lead poisoning through eating contaminated prey.

Monitoring the Effects of Metallic Contamination

Based on the foregoing evidence, it is obvious that even at high levels of metallic contamination some species are still able to thrive. This ability is based on the presence in some organisms of remarkably efficient detoxification or regulatory mechanisms and the development in other species of tolerant strains. Most of the evidence, such as it is, for the influence of metals in the field comes from studies on the presence or absence of species in the contaminated areas. However, it is quite difficult, even in an area like Restronguet Creek, where one is dealing with 'pure' metal contamination, to state unequivocally that the absence of a certain species is caused by metallic contamination. Fortunately, in this same area there are similar creeks with which comparison can be made, but there is no baseline for the creek itself BRYAN and GIBBS (1983). This type of biological study although frequently very useful has certain disadvantages since it is very time consuming, the response is not usually specific to any particular contaminant and it is not very sensitive since it must be judged against a background of natural variations.

A recent review (ICES, 1978b) has considered the feasibility of monitoring effects before they are obviously detectable in biological surveys. The authors considered the various contaminant-related effects that have been observed in laboratory experiments under headings such as biochemical, morphological, physiological, and behavioural and considered that assessment of effects of the first 3 types may be useful under field conditions. An approach to effects monitoring was suggested which would include: (i) attempts to study the well-being of organisms in terms of the incidence of, for example, morphological deformities and tumours; (ii) experimentation to link effects to contaminant levels in organisms and in the environment; (iii) the use of bioassays to identify areas of poor water quality.

One of the most intractable problems in effects monitoring is that of relating effects to specific pollutants, in our case metals. Since it demonstrates exposure, and also presumably the utilization of energy by organisms to detoxify metals, tissue-metal analysis is an obvious preliminary to any attempt to solve this problem. Even this approach requires care to make it specific: as a result of the regulation of some metals by fish muscle, analysis of this tissue, rather than perhaps the gills or viscera, is unlikely to show whether the fish has been exposed to contamination.

In Restronguet Creek (p. 1384) it was reasonable to assume that ecological effects were related to metals, but it was impossible from analyses of copper and zinc in the biota and simple toxicity tests to say with any certainty which metal was having the greater impact on different species. However, Fig. 3-7 demonstrates the development of a very marked tolerance to copper in *Fucus vesiculosus* from the creek. Similar experiments have shown that the weed is not abnormally tolerant to zinc, thus indicating that copper has the more important effect on this species. LUOMA (1977b) has advocated the study of differences in tolerance between normal and polluted populations towards specific toxins as a means of identifying which contaminants are having most impact. Even this method may not be completely specific since inducible and genetically based tolerance mechanisms are not necessarily specific to single metals (pp. 1365 and 1375).

Various studies following the reciprocal transfer of organisms between uncontaminated and contaminated areas have proved useful in assessing effects. For example, changes in metal residue levels and condition in transferred *Scrobicularia plana* were used to assess what at first sight had appeared to be a deleterious effect of lead and other metals. However, the occurrence of healthier animals having even higher metal residues at another site suggested that metals were not primarily responsible (BRYAN and HUMMERSTONE, 1978). Concentration gradients in the field often provide suitable conditions for studies relating effects to environmental levels. For example, SHORE and co-authors (1975) showed that there appeared to be a correlation between increasing levels of cadmium in the limpet *Patella vulgata* along the Severn estuary–Bristol Channel and reduced ability to utilise glucose (Table 3-28).

The use of bioassay techniques to monitor water quality is normally non-specific. However, the contributions of metals to the effect of low quality water from the Bristol Channel on the hydroid *Campanularia flexuosa* was measured by the improvement observed when they were removed from the water with Chelex resin (STEBBING, 1979, 1980). A system for culturing phytoplankton within dialysis bags exposed to the waters of metal-contaminated fjords has been described by EIDE and co-authors (1979). Since the effects on growth observed at different sites need not have been due to metals, that in the main they were was strongly suggested by the relationship between the effects and the metal residues accumulated at the same time by the cells.

Although, as we have seen above, there are ways of improving the specificity of effects monitoring towards metals or groups of metals, many effects which can be detected in generally polluted areas cannot be attributed with any certainty to particular contamin-

Table 3-29

Some studies in metal-contaminated areas (Compiled from the sources indicated)

Country	Sea area	Source	Main contaminants and other metals	Examples of contaminant levels	Source
Canada	Powell River	Pulp mills	Zn (Cd, Cu)	19,400 ppm Zn in *Crassostrea gigas*, compared with 1–2000 ppm normally	Nelson and Goyette (1976)
Canada	Howe Sound	Mining	Cu (As, Fe, Mn, Pb, Se, Zn)	1500 ppm of Cu in viscera of *Mytilus edulis*, compared with low value of 18 ppm	Stump and co-authors (1979)
Canada	Belledure Harbour	Smelting	Cd Zn	Over 100 ppm of Cd in *M. edulis*	Uthe and Zitko (1980)
Newfoundland	Moreton's Harbour	Mining	As	2600 ppm maximum in sediments, but very limited dispersion	Penrose and co-authors (1975)
Germany	German Bight	Acid-iron wastes	Fe	No certainty that dumping 1800 tons d^{-1} of acid-iron wastes for 6 yr had any effect on fauna in middle of dump area	Rachor and Gerlach (1978)
Germany	Kiel Fjord	Industry	Cd	34 ppm in *Mytilus edulis*, compared with low value of 1·1 ppm	Theede and co-authors (1979b)
Ireland	Avoca estuary	Mining	Cu, Zn (Al, Cd, Cr, Fe, Mn, Ni, Pb)	221 and 474 ppm of Cu and Zn in *Mytilus edulis*, compared with low values of 8 and 85 ppm	Wilson (1980)
Italy	Tyrrhenian Sea coast	Chlor-alkali plant	Hg	4·5 ppm (wet) in crab *Pachygrapsus marmoratus*, compared with low value of 0·3 ppm; some possible effects	Renzoni and co-authors (1973)

Norway	Trondheimfjord	Mining	Cu, Zn (Ag, Cd, Cr, Fe, Hg, Ni)	88 ppm in *Mytilus edulis*, compared with low value of 5 ppm; faunal effects	LANDE (1977)
Spain	Rio Tinto estuary	Mining	Cu, Pb, Zn (Cd)	1350 ppm Cu, 1800 ppm Pb and 1480 ppm Zn in *Zostera* sp., compared with low values of 9, 6, 100 ppm	STENNER and NICKLESS (1975)
Tasmania	Derwent estuary	Zn refinery	Cd, Cu, Zn	173 and 57,600 ppm of Cd and Zn in *Crassostrea gigas*, compared with low values of 0 and 1350	RATKOWSKY and co-authors (1974)
		Chlor-alkali plant	Hg	1·06 ppm (wet) in muscle of fish *Platycephalus bassensis*, compared with low value of 0·03	DIX and co-authors (1975)
United Kingdom	Poole Harbour	Industry	Cd (Cu, Fe, Mn, Ni, Pb, Zn)	54 ppm of Cd in *Ostrea edulis*, compared with low value of 5·9 ppm; possible effects in oyster hatchery	BOYDEN (1975)
	Looe estuary	Mining and industry?	Ag, Pb (Cd, Co, Cr, Fe, Mn, Ni, Pb, Zn)	184 and 473 ppm of Ag and Pb in *Scrobicularia plana*, compared with low values of <1 and <25 ppm; no clear effects	BRYAN and HUMMERSTONE (1977)
	Dulas Bay	Mining	Cu, Zn	Very high levels of Cu and Zn in barnacles *Semibalanus balanoides* (cf. Table 3-15); very limited fauna	FOSTER and co-authors (1978)
	Estuaries of SW Britain	Different types	13 metals	Evidence for very high concentrations in some benthic species; no effects data	BRYAN and co-authors (1980)
United States	Turtle River	Chlor-alkali plant	Hg	9·4 ppm in *Littorina irrorata* and fairly high in many other species	GARDNER and co-authors (1978)
	Bellingham Bay	Chlor-alkali plant	Hg	Organisms contain about 10 times normal concentrations	RASMUSSEN and WILLIAMS (1975)
	South San Francisco Bay	Industrial/Urban	Ag, Cu, Zn	220 and 440 ppm of Ag and Cu in *Macoma baltica*, compared with low values of 7 and 74 ppm	LUOMA and CAIN (1979)

ants; and it may be that the contaminants act jointly. For example, although tests in a rotary flow apparatus showed that the physical performance of the cod *Gadus morhua* from a polluted area was lower than normal it was impossible to ascertain the specific cause (OLOFSSON and LINDAHL, 1979). Similarly, studies in the Bristol Channnel have demonstrated deleterious effects in mussels as inferred by various physiological, biochemical, and cytological indices of condition (BAYNE and co-authors, 1979) and the high incidence of diseased specimens (LOWE and MOORE, 1979b). Attempts are being made to refine these methods and perhaps arrive at indices that might measure the effects of single classes of pollutant. For example, HODSON and co-authors (1977) showed that in the rainbow trout *Salmo gairdneri* the activity of erythrocyte D-amino lavulinic acid dehydratase was depressed by as little as 10 ppb of lead in fresh water. The effect was specific to lead since the enzyme was not inhibited by near-toxic levels of mercury, copper, cadmium, and zinc. In addition, the effect could be observed within 2 wk and was, therefore, proposed as a short-term indicator of the potentially long-term effect of lead. A comparable result was obtained in the mullet *Mugil auratus* following exposure to 500 ppb of lead in sea water for 2 wk and coincided with a fall in haemoglobin concentration (KRAJNOVIC-OZRETIC and OZRETIC, 1980). The effects remained constant for up to 82 d, although the level of lead in the blood increased continuously. Whether the enzyme is specifically inhibited by lead *in vivo* is uncertain since exposure to cadmium, copper, and mercury *in vitro* resulted in inhibition. PEQUEGNAT and WASTLER (1980) regard the inhibition of catalase activity as being a specific response to heavy metals. They observed inhibition of liver catalase in the killifish *Fundulus heteroclitus* following exposure for about 2 months at sewage and mud dump sites: the fish were retained in novel mesh containers (Pelagic and Benthic Biotal Monitors) near the sea surface and near the sea bed.

The evidence for relationships between the diseases of fish and shellfish and pollution has been reviewed by SINDERMANN (1979). He concluded that for some diseases, such as fin erosion (p. 1372) and ulcers in fish, there is fairly good evidence for a relation with pollution and some evidence that this conclusion might extend to certain tumours and skeletal abnormalities. However, he also observed that there is at present no evidence for linkages between certain diseases and specific contaminants.

Effects of Metals in Contaminated Areas—Summary

(i) In some heavily contaminated estuaries enhancement of metal concentrations by as much as 3 orders of magnitude has been observed in some species. The extent of enhancement depends both on the species and on the metal and for the majority of metals studied is most obvious in plants and invertebrate species which do not regulate metals. Enhancement is least obvious in fish, where the ability to regulate metals is perhaps most highly developed, and also depends on whether prey organisms regulate or not. Mercury is exceptional in that it does not appear to be regulated in any species and in contaminated areas is the only metal for which enhanced concentrations are usually most obvious in fish rather than invertebrates.

(ii) Effects of metals on species distribution have been observed in a few heavily contaminated areas. However, in spite of the discovery of high residue levels in many areas, there is a paucity of evidence on effects which can unequivocally be attributed to metals. There are several reasons for this including: (a) the relative insensitivity and nonspecificity of monitoring methods based on changes in populations and species distributions, etc.; (b) the present lack of specificity to particular groups of pollutants in most of the more sensitive methods of effects monitoring based on various morphological, physiological, biochemical, and cytological indices of reduced performance and increased stress; (c) the ability of many species to detoxify metals and in some cases to adapt to them.

(iii) The importance of combining effects monitoring with residue analysis is stressed and some of the ways in which effects monitoring might be made more specific towards metals are discussed

(c) Effects of Metals from Marine Sources on Man

Occupational hazards involving metals are well known, but it is evident that much lower levels of metals in the diet may be implicated in human disease (HAMILTON, 1979). Some diseases result from the deficiency of essential metals such as iron, zinc, copper, and cobalt (MOYNAHAN, 1979), while others are attributable to the toxicity of essential or non-essential metals. Toxicity is, however, not simply a function of dietary concentrations but rather of imbalances between concentrations of different metals. Thus STOCKS and DAVIES (1964) who established a relationship between the incidence of stomach cancer and the Zn/Cu ratio in soils, have suggested that an imbalance of metals in the diet can eventually lead to disease. UNDERWOOD (1979) has stressed the importance of interactions between dietary trace metals and cites, as an example, the interaction between mercury and selenium (p. 1345). He suggests that tolerable levels of mercury in diets are likely to be higher where the intake of selenium is also high and lower in areas of low selenium status. Thus, in fact, there is no single tolerable level for a toxic metal, since this will depend not only on its chemical form but also on its ratio to metals with which it interacts.

Generally speaking, the essential metals such as zinc, copper, and iron, for which efficient homoeostatic mechanisms are found in man, are regarded as having low toxicities whereas the non-essential metals including mercury, cadmium, and lead are thought to have a much lower margin of safety.

Mercury

With few exceptions, fish are the most significant source of methyl mercury in the diet and, in Japan, Minamata disease, which has the symptoms of methyl mercury poisoning, occurred in persons such as fishermen who had consumed large quantities of fish containing 7 to 23 ppm (wet) of mercury (TSUBAKI and IRUKAYAMA, 1977). There is a relation between the intake of mercury and concentrations in human hair and blood which can be used for monitoring. A study on mercury (WORLD HEALTH ORGANIZATION, 1976) has indicated that the levels likely to produce effects in sensitive individuals are of the order of 0·2 to 0·5 ppm in blood and 50 to 125 ppm in hair, corresponding to an intake of several thousand μg of mercury per week. Apart from Japanese examples, blood and hair values in this range have been found in individuals from fish-eating populations in both Sweden and Canada who ingested fish containing several ppm of mercury. No definite symptoms of poisoning have been found in Sweden, but SKERFVING and co-authors (1974) reported chromosome damage in some individuals. Studies on native Canadians (Indians and Eskimos) have not conclusively demonstrated symptoms of Minamata disease, although there are indications of some effects (CHARLEBOIS, 1978). These problems in Sweden and Canada have stemmed more from the contamination of fresh water fish with mercury than marine fish, although the latter have been involved. Even in the absence of contamination, high levels of mercury are found in the tissues of some marine fish and mammals (Tables 3-17, 3-19), and studies have been carried out on Canadian native populations where seals, for example, are an important source of food. However, although enhanced levels of methyl mercury have been found in Eskimos, there is no evidence that this diet has any effect and it is suggested that such populations are likely to be adapted to their diet (SMITH and ARMSTRONG, 1975).

Studies on fish-eating populations from other areas of the world have been reviewed by HAXTON and co-authors (1979). They examined mercury (and selenium) intake and blood and hair levels in individuals from fishing communities on the northeast Irish Sea, an area affected by the industrial discharge of mercury. Although the weekly intake of mercury by some individuals exceeded the FAO/WHO (1972) provisional tolerable weekly intake of 300 μg per person (200 μg methyl mercury), the maximum value for blood was 0·026 ppm and for hair was 11·3 ppm. This is about an order of magnitude lower than the levels thought likely to approach the toxic threshold in humans and an order of magnitude higher than values for individuals having an average or lower than average intake of uncontaminated fish.

A point that emerges from the vast amount of monitoring data on fish muscle is that, although concentrations in a specific species are enhanced by contamination, some of the highest values, observed mainly in large fish, appear to be perfectly natural (Fig. 3-5). Thus the dietary limits such as 0·5 or 1 ppm set for saleable fish in several countries have led to the curtailment of some fisheries where there is no evidence for contamination. It might be possible to relax such limits if it could be shown that the potential toxicity to man of the mercury in these species (often in the methyl form) is ameliorated by, for example, selenium in the same tissue. For this reason, BERNHARD (1978) suggested that limits should not apply to pelagic species such as tuna and swordfish, where high mercury concentrations in the tissues are likely to be accompanied by high selenium concentrations. Similarly, SUZUKI and co-authors (1980) proposed that the lack of evidence for effects of methyl mercury on inhabitants of the Tokara Islands having hair concentrations of around 50 ppm might be explained by the high Se/Hg atom ratios (0·3 to 88) in most dietary fish.

Cadmium

Since cadmium is a cumulative poison with a half-life in man of 16 to 33 yr (FAO/WHO, 1972), its intake should be kept as low as possible. While in the United Kingdom, for example, the weekly intake is 105 to 210 μg ind.$^{-1}$ (Ministry of Agriculture, Fisheries and Food, 1973), the FAO/WHO (1972) provisional tolerable weekly intake is 400 to 500 μg ind.$^{-1}$. In fish muscle, cadmium levels are normally low (Table 3-17) and are relatively unaffected by contamination (Table 3-18). However, high concentrations have been found in some invertebrates from both contaminated and uncontaminated areas (Table 3-28) and any problems with cadmium would be expected to come from this source since the tolerable intake could easily be exceeded. There is no evidence that people have been permanently affected by eating contaminated shellfish, although high concentrations of cadmium and zinc in some Tasmanian oysters appear to have caused nausea and vomiting (RATKOWSKY and co-authors, 1974). Copper in oysters is known to have a similar effect, although its presence is normally revealed by the green colour of the tissues.

Lead

In the United Kingdom, for example, the average weekly intake of lead in food is about 1400 μg ind.$^{-1}$ (Ministry of Agriculture, Fisheries and Food, 1972) compared with the FAO/WHO (1972) provisional weekly intake of 3000 μg ind.$^{-1}$ for adults. As with cadmium, levels of lead in fish muscle are generally low and not markedly influenced by contamination. The highest levels have been found in invertebrates from heavily contaminated sites (Table 3-28) but such areas are usually very localized. There is no evidence to suggest that lead in foodstuffs of marine origin poses any special problems to man although, in the absence of much information about the abundance of organic and possibly more toxic forms of lead in the marine environment (pp. 1310 and 1374), it may be too soon to ignore lead completely. Indeed, the recent introduction of limits for fish (2 ppm wet) and shellfish (10 ppm wet) in England and Wales is evidence of some anxiety.

Other Metals and Metalloids

Relatively high concentrations of arsenic are found in most marine tissues and exceed the limits of a few ppm usually laid down for foodstuffs (Tables 3-15, 3-17). Recent work on the clearance of arsenic from man following its absorption from fish (FREEMAN and co-authors, 1979) has confirmed earlier studies showing that arsenic in marine tissues exists in an organic form of low toxicity having a short residence time in man.

High concentrations of selenium have been found in some marine tissues, generally in association with mercury which it appears to detoxify (Tables 3-19, 3-20). Interest in selenium has also been generated by the claims that it has anti-carcinogenic properties in man (UNDERWOOD, 1979).

Other potentially toxic metals including silver, chromium, nickel, antimony, tin, and vanadium have rarely if ever been found at significant concentrations in fish muscle. Enhanced levels have been found in invertebrates from some sites but as far as I am aware there is no evidence that these metals are likely to pose problems in marine foodstuffs.

Effects on Man—Summary

(i) Persons at risk from metal contamination of marine origin are obviously those who eat large amounts of fish or shellfish from estuarine or coastal areas which are associated with industry. Others who may also be at risk are individuals having diets containing a large proportion of species which are naturally high in various metals. Species of fish, mammals, and birds are most likely to contain particularly high concentrations of mercury and selenium in their edible tissues, whereas high concentrations of other metals are more usually associated with shellfish, particularly molluscs.

(ii) With the exception of Minamata disease in Japan there is at present little unequivocal evidence that foodstuffs of marine origin have caused any permanent forms of metal poisoning. However, the provisional tolerable weekly intake rates for mercury, cadmium, and lead are so low that persons eating contaminated sea food regularly can easily exceed them.

(iii) Studies on communities living in highly mineralized areas have indicated that diseases relating to metal contamination may only become apparent after extremely long exposure and that metals should not be considered in isolation since the interactions between different elements are important in the induction of some diseases. The problems of obtaining unequivocal evidence from epidemiological studies are considerable and, therefore, it is not surprising that it is difficult to be more specific about the influence of foodstuffs of marine origin on man. The most obvious risk comes from high levels of mercury in marine tissues and there is reason to suppose that high levels of cadmium and lead are also undesirable. However, the threat from other metals in marine tissues appears to be small, although based on the present evidence it would seem unwise to assume that there are no risks.

(6) Assessment of the Present state of Knowledge About Metal Contamination of Oceans and Coastal Waters

Contamination of the marine environment with metals is extremely common, particularly in the industrialized northern hemisphere. The highest levels of contamination occur in confined areas such as estuaries, fjords, and bays where not only is dilution limited but the interactions between inputs and sea water tend to promote the deposition of metals in the sediments. This trapping effect certainly decreases the input of dissolved metals to coastal waters although, augmented by dumping and atmospheric inputs, contamination is detectable in many areas. Atmospheric fallout over coastal waters is a significant source of input for some of the more volatile metals such as lead, copper and zinc and, for lead at least, the influence of atmospheric contamination extends to oceanic surface waters. The chemical changes undergone by metals in the water and sediments and their cycling in the environment are at present the subject of a great deal of research, since the pollution potential of a metal is largely based on its form and hence its biological availability. Precisely which forms are most readily available is gradually being revealed. Demonstrably important dissolved species are the free ions of metals such as copper, cadmium and zinc, and also methyl mercury which appears to be remobilized from the sediments. At natural levels the measurement of these dissolved species poses many problems and, in fact, the much higher concentrations in particulates and food organisms often render them more important as sources for marine animals. However, so little is known about the relevance of different forms of metals in the environment that the analysis of biological indicator species is arguably the best method at present available for assessing levels of contamination.

Studies on the bioaccumulation and metabolism of metals have concentrated on the major groups, particularly phytoplankton, molluscs, and crustaceans. Evidence from most phyla suggests that uptake from solution is generally a passive process, although in molluscs pinocytosis is implicated in the uptake and transfer of metals and in some micro-organisms active transport appears to be involved. In many animals, however, uptake from solution is far less important than the dietary route. From studies on metal metabolism, it is increasingly evident that in many species even the non-essential metals are handled more effectively than was previously suspected. Thus, although there is little evidence for processes which actively prevent the intake of metals, most groups are equipped with mechanisms for the detoxification, storage, and elimination of metals. The relation between the concentration of a particular metal in an organism and environmental availability is a function of these mechanisms: it ranges from the situation where metals are regulated against environmental variations to that where the metal is accumulated with increasing age and no equilibrium level is ever reached. Generally speaking, the more highly evolved organisms are the best regulators and the essential metals such as zinc and copper are better regulated than the non-essential—especially mercury. Whether or not biomagnification occurs with increasing trophic level along food chains is in part related to the ability of organisms at different levels to control their body concentrations. At present, the only convincing evidence for biomagnification is for mercury although even here the picture is clouded by the fact that in fish, for example, concentrations also increase with age or size. As a result of these processes it seems likely that the high mercury concentrations observed in some large predatory fish and marine mammals are unrelated to contamination.

Our knowledge of the effects of metals is largely based on experimental observations and much of this work has involved unrealistically high concentrations of dissolved metals. Mercury is the most toxic metal in virtually all taxonomic groups examined and, although the relative toxicity of other metals varies, that of copper is consistently high. A remarkable number of factors have been shown to influence metal toxicity. They include the various environmental factors, particularly temperature and salinity, the condition of the organism and the ability of some species to adapt to metallic contamination. The most sensitive life functions include embryonic and larval development, growth and fecundity. Oxygen utilization and the functions of various enzymes are demonstrably sensitive, although there is evidence that some effects can be compensated for by adaptive processes, possibly with the expenditure of additional energy. Predictions based on experimental evidence suggest that increasing dissolved concentrations of copper and mercury in coastal waters by an order of magnitude would be expected to produce measurable effects; equivalent increases for cadmium, zinc, and lead would need to be greater. However, such predictions are based on total soluble concentrations when we know that the effective fraction is generally much smaller and, through complexation, not necessarily directly related to the total concentration. In addition, the biological availability of metals from other sources such as sediments is ignored by these predictions.

When real situations are considered, the fact is that only in the most heavily contaminated areas have effects been observed which can confidently be attributed to metallic contamination. Reasons for this lack of evidence include problems caused by the coexistence of metals with other contaminants, the lack of sensitivity in monitoring methods based on population and species distribution—particularly in the absence of baseline information, the lack of metal specificity in more sensitive monitoring methods based on indices of performance and stress, and the adaptation of some populations to metallic contamination. It is also a fact that such knowledge as we have about the situation in the field stems mainly from work in temperate areas of the world, and it seems reasonable to suppose that in tropical areas any effects might be enhanced.

With the exception of Minamata disease in Japan, there is no unequivocal evidence that metal-contaminated foodstuffs of marine origin have caused any permanent forms of illness in man. In several areas of the world, however, diets of mercury-contaminated fish have resulted in human tissue levels approaching the supposed threshold for Minamata disease. A question mark remains with regard to the desirability of eating some species of fish having naturally high concentrations of methyl mercury until they can be shown to be harmless—possibly through the coexistence of selenium and mercury in the same tissues. Any problems with cadmium, lead or other metals would be expected to stem from contaminated shellfish rather than fish, since concentrations in the latter are less sensitive to contamination. Thus the accumulation of copper and zinc by oysters has caused problems in many areas. However, as with the effects of metals on the marine biota in the field, only the obvious effects of metals on man have been recognized: certainly persons eating contaminated seafood regularly can easily exceed the intake rates for metals at present regarded as tolerable. It can be concluded that serious problems with marine foodstuffs seem most likely to result from contamination with mercury or possibly cadmium, whereas effects on the marine environment seem at least as likely to stem from contamination with abundant metals such as copper and zinc as from contamination with mercury or cadmium

(7) Management of Metal Pollution

(a) Abatement of Contamination

Studies on the abatement of contamination have so far largely been concerned with the reduction of mercury inputs following the realization that it was an urgent pollution problem.

At Minamata Bay (see also p. 1383) closure of the offending acetaldehyde factory in 1968 lead to a sharp drop in mercury concentrations in indicator species over a period of a few months. Mean concentrations from samples of *Venus japonica* taken at 4 sites on several occasions fell from 17 ppm (3 to 60 ppm dry) in 1967 to 3·7 ppm (0·3 to 16 ppm) in 1969 and 5·4 ppm (0·3 to 16 ppm) in 1970, a reduction of about 70% in mercury residues (FUJIKI, 1973). Although this reduction was rapidly achieved, the presence of sediments—sometimes containing more than 100 ppm of mercury—delayed further improvements. The results of monitoring several species of fish in recent years has shown that in some species the flesh contains higher concentrations than the Japanese limit of 0·4 ppm (wet) and this is thought to be the result of persistent contamination (HIROTA and co-authors, 1979). Fish confined in pens are being used to monitor any increase in mercury levels which may result from disturbance during the removal of the heavily contaminated sediments from the bay by dredging. The comprehensive plans for the clean-up of the Bay have been described by ISHIKAWA and IKEGAKI (1980).

Following a reduction of 95 to 98% in mercury output from an Italian chlor-alkali factory it was observed that in 28 to 29 months the concentration in the crab *Pachygrapsus marmoratus* fell by 78% and the fall in other species ranged from 20 to 31% (RENZONI, 1976). It was noted that complete recovery would probably be delayed by the large amounts of mercury built up in the sediments. A fall of more than 50% in the mercury level of fish muscle was observed a year after the introduction of a waste treatment plant at a caustic soda factory in Thailand (SUCKCHAROEN and LODENIUS, 1980). Similarly, mercury levels in eelpout *Zoarces viviparus* appear to have fallen following the reduction of inputs to the Ems estuary (ESSINK, 1980). From studies on Saguenay Fjord in Canada, following the regulation of mercury input from the chlor-alkali industry, LORING and BEWERS (1978) concluded that although concentrations in the water should return to near-natural levels in 2 yr, decontamination of mercury-rich sediments by flushing and burial will take decades. These results from different areas all point to the possibility of making rapid reductions in mercury concentrations in the biota by removing the soluble input, but also indicate a much longer time scale for the removal of the influence of sediment-bound mercury.

It is of course not always easy to stop the input of wastes. The input of metals in runoff from areas where the soils have been contaminated by mining or aerial fallout from smelting or the use of lead in petrol cannot be switched off. For example, examination of a dated core from the Tamar estuary by CLIFTON and HAMILTON (1979) showed that levels of copper, zinc, and lead rose sharply following a peak of metalliferous mining in the mid-nineteenth century and have remained constant for the past 100 yr, although mining largely ceased 80 yr ago. This has been caused by the weathering of spoil heaps and, although the estuary is regarded as unpolluted, concentrations in the biota are abnormally high (BRYAN and HUMMERSTONE, 1971, 1973c; BRYAN and UYSAL, 1978).

(b) Monitoring

Although there is a great deal of information available on metal residues in sea water, sediments, and biota in different parts of the world, studies have often tended to be unconnected and the results difficult to compare because of the use of different analytical methods. Over the past few years attempts have been made at both national and international levels to introduce an element of cooperation. In the United Kingdom, for example, the coordinating body for marine monitoring by various agencies, including government departments, water authorities, industry, and universities, is the Marine Pollution Monitoring Management Group. They have formulated policies (DOE, 1979) to extend marine monitoring beyond the analysis of residues in commercial fish and shellfish, which was at one time of primary concern (MURRAY, 1979), to the monitoring of inputs, the monitoring of trends with biological indicators and the study of the marine chemistry of pollutants and its relation to biological availability and effects, if any.

International cooperation is becoming increasingly important since some forms of pollution are not confined by national boundaries. Under the auspices of the International Council for the Exploration of the Sea (ICES) the monitoring of baseline metal concentrations in fish and shellfish has covered the Baltic Sea (ICES, 1977a), the North Sea (ICES, 1977b) and the North Atlantic (ICES, 1977c, 1980) and in connection with this work, several analytical intercalibration exercises have been carried out (ICES, 1978c). The availability of information on the inputs of metals to the sea from rivers and pipelines, the atmosphere and dumping, lags far behind that on biological monitoring to which it is ultimately related. However, some figures for inputs of metals from 10 northwest European nations are available (ICES, 1978a), although there are many gaps in the information.

The present evidence for metallic contamination of the Mediterranean Sea has been reviewed by BERNHARD (1978). Cooperation between the Mediterranean countries has been established under the auspices of the United Nations Environmental Programme (OSTERBERG and KECKES, 1977; JUDA, 1979) and includes programmes for monitoring heavy metals and their possible effects.

(c) Practical Application of Results

The principles of the application of the results of laboratory and field experiments to the derivation of standards for the discharge of contaminants to the sea, while at the same time protecting the environment have been described by PRESTON (1979). This involves the recognition of the critical target for the metal which may be man or some section of the marine biosphere, the setting of exposure standards which may contain substantial margins of safety and the application of the standards to individual situations. PRESTON concludes:

> 'The way ahead is reasonably clear: develop standards, if necessary on a purely empirical basis, with the use of a hypothesis such as that of linear dose–response, unless threshold phenomena can be clearly shown to exist; apply the standards to individual situations based on the application of critical path techniques; define acceptable rates of introduction and associated environmental concentrations; monitor introduction and, where appropriate, the environment by using d.w.ls (environmental quality standards) as the field criteria demonstrating compliance with standards' (PRESTON, 1979, p. 622).

A present bone of contention is whether the emission standards for pollutant discharges should be uniformly applied over a whole country or countries or whether, as advocated above, emissions should be controlled in the light of the capacity of each sea area, estuary and so on to accept the contaminant, taking account of all environmental factors. The arguments, environmental, practical and economic, for and against these 2 views have been discussed by HOLDGATE (1979) who points out that the uniform emission view tends to reject the use of the disposal capacity of the sea as a resource. Certainly in the case of metals, our own experience of over 50 estuaries of different types would support the contention that if there are metallic emissions they should be tailored to individual areas. The importance of the form of input and the local environmental conditions on the biological availability of metals may be illustrated by the following example. In Restronguet Creek (UK), which receives acid mine wastes containing as much as 1 ppm of dissolved copper, and where the sediments contain 3000 ppm of copper, the deposit-feeding bivalve *Scrobicularia plana* contains 150 ppm. On the other hand, in the Erme estuary (UK), where there is no comparable copper input and the sediments contain only 30 ppm, *S. plana* have been found containing around 700 ppm—due apparently to some subtlety in the chemical composition of the sediments which are very reducing in character.

(8) Conclusions and Future Prospects

Ten years ago there was very little information on metal contamination in the sea and even now after the publication of thousands of papers there are still question marks against many aspects of the problem. For example, there is a shortage of reliable information about metallic inputs to the sea, although in some areas it has been possible to draw up inventories. Even less is known about the relevance of different inputs in terms of their ability to cause pollution since, with few exceptions, we do not know which chemical forms in the field are most readily available or toxic to marine organisms.

Minamata disease, for example, would probably not have been caused if the effluent responsible had contained only inorganic mercury rather than a proportion of the more readily available methyl form. However, in the cases where the relevant chemical forms are known it has proved difficult to measure them except at high concentrations. A great deal of research is at present devoted to studying the speciation of metals in sea water by electrochemical methods since these appear to have the greatest potential for monitoring biologically significant dissolved metallic species in the field either directly or with a minimum of sample treatment. Biologically available metals are not confined to those in the water, and in the sediments, for example, there is great uncertainty about the available fractions although they are known to include metals in the interstitial water and sediment-bound forms. In addition, considerable doubt still exists concerning the extent to which metals are returned to the overlying water following remobilization in the sediments including the role of the sediments as the ultimate source of methyl mercury and some other organo-metal compounds in the sea.

Although the literature on the concentrations of metals in marine organisms is formidable and we talk about some organisms as biological indicators of contamination, which metallic forms these species are indicating is largely a matter of speculation—particularly in animals—where several pathways lead from metal input to bioaccumulation. Even in mussels where biochemical and physiological studies in the laboratory have given a clear insight into the absorption, metabolism, and excretion of a range of metals, the importance of different sources of metals in the field is largely unknown. Indeed, the whole question of biological availability and the real behaviour of metals in marine organisms is still remarkably little understood. At present we know most about bivalve molluscs and crustaceans, but far less about fish, and almost nothing about groups such as coelenterates, echinoderms, and tunicates with little commercial value. The ability of most species to detoxify metals to some extent is now well established; and thus studies on the metabolism of metals are relevant not only to problems of bioaccumulation but to the effects of metals also.

Our present knowledge about the toxicity of metals is almost totally based on laboratory experimental work. Much of this has been carried out at concentrations which are so unrealistic that it is questionable whether the results are relevant. However, there is now a reasonable amount of experimental evidence suggesting that in some species processes such as fecundity, embryonic development, and growth are inhibited at dissolved concentrations of the same order as levels observed in some contaminated areas. It is also clear that other species are not affected perceptibly and can be continuously cultured in the presence of remarkably high metal concentrations. Predictions based on the experimental evidence would lead one to expect that in moderately contaminated areas a careful study should reveal effects on sensitive species and that in heavily contaminated areas the effects should be really obvious. In the latter, some effects undoubtedly occur, although the ability of many species to adapt and the emergence of tolerant strains ameliorates the overall effect. It is in moderately contaminated areas that real uncertainty arises about the effects of metals, since at present there is very little unequivocal evidence that they have much effect. This does not mean that there are no deleterious effects caused by metals, but that at present we are largely unable to detect them against a background of biological variation, the effect of natural environmental stressors or other groups of contaminants, and the absence in many contaminated areas of a natural baseline with which to compare. Sensitive biochemical and physiological indices of environmental stress have been and are being developed in various species. These indices are not usually contaminant specific. However, coupled with residue analyses and used along contaminant gradients or at sites receiving different suites of contaminants, such indices should enable the detection of the most important stressors. The use of field conditions for testing such methods is of particular importance, since the conditions encountered in vulnerable areas such as estuaries are not reproducible in the laboratory. There is evidence that ultimately it might be possible to identify the effects of individual metals by, for example, their inhibition of specific enzymes. However, it should be borne in mind that such an effect may not necessarily be deleterious to the organism or the population as a whole: that is, it may not be an indication of pollution. There is an obvious need for considerably more field-orientated research in many metal-contaminated areas to ascertain whether they can also be described as polluted. This information, coupled with that on inputs and metal speciation should provide a sounder basis than exists at present for the setting of environmental standards.

Literature Cited (Chapter 3)

ABDULLAH, M. I. and ROYLE, L. G. (1974). A study of dissolved and particulate trace elements in the Bristol Channel. *J. mar. biol. Ass. U.K.*, **54,** 581–597.

ADEMA, D. M. M. and SWAAF-MOOY, S. I. de, and BAIS, P. (1972). Laboratoriumonderzoek over de invloed van koper op mosselen (*Mytilus edulis*). *T.N.O. Nieuws.*, **27**, 482–487.

ALBERGONI, V., PICCINNI, E., and COPPELLOTTI, O. (1980). Response to heavy metals in organisms. I. Excretion and accumulation of physiological and non-physiological metals in *Euglena gracilis*. *Comp. Biochem. Physiol.*, **67c**, 121–127.

ALDERDICE, D. F., ROSENTHAL, H., and VELSEN, F. P. J. (1979a). Influence of salinity and cadmium on capsule strength in Pacific herring eggs. *Helgoländer. Meeresunters.*, **32**, 149–162.

ALDERDICE, D. F., ROSENTHAL, H., and VELSEN, F. P. J. (1979b). Influence of salinity and cadmium on the volume of Pacific herring eggs. *Helgoländer wiss. Meeresunters.*, **32**, 163–178.

ALEXANDER, G. V. and YOUNG, D. R. (1976). Trace metals in southern Californian mussels. *Mar. Pollut. Bull., N.S.*, **7**, 7–9.

ALLER, R. C. (1978). Experimental studies of changes produced by deposit feeders on pore water, sediment and overlying water chemistry. *Am. J. Sci.*, **278**, 1185–1234.

AMIARD, J. -C. (1976). Les variations de la phototaxie des larves de crustacés sous l'action de divers polluants métalliques : mise au point d'un test de toxicité subléthale. *Mar. Biol.*, **34,** 239–245.

ANDERLINI, V. C. (1974). The distribution of heavy metals in the red abalone, *Haliotis rufescens*, on the California coast. *Archs environ. Contam. Toxicol.*, **2**, 253–265.

ANDERSON, D. M. and MOREL, F. M. M. (1978). Copper sensitivity of *Gonyaulax tamarensis*. *Limnol. Oceanogr.*, **23**, 283–295.

ANDERSON, M. A., MOREL, F. M. M., and GUILLARD, R. R. L. (1978). Growth limitation of a coastal diatom by low zinc ion activity. *Nature, Lond.*, **276**, 70–71.

ANDREAE, M. O. (1977). Determination of arsenic species in natural waters. *Analyt. Chem.*, **49,** 820–823.

ANDREAE, M. O. (1979). Arsenic speciation in seawater and interstitial waters: the influence of biological-chemical interactions on the chemistry of a trace element. *Limnol. Oceanogr.*, **24,** 440–452.

ANDREAE, M. O. and KLUMPP, D. W. (1979). Biosynthesis and release of organo-arsenic compounds by marine algae. *Environ. Sci. Technol.*, **13**, 738–741.

APPELQUIST, H., JENSEN, K. O., and SEVEL, T. (1978). Mercury in the Greenland ice sheet. *Nature, Lond.*, **273**, 657–659.

ARIMA, S. and NAGAKURA, K. (1979). Mercury and selenium content of Odontoceti. *Bull. Jap. Soc. scient. Fish.*, **45**, 623–626.

ASSOCIATION EUROPEENE OCEANIQUE (1977). Metallic effluents of industrial origin in the marine environment—Association Européene Océanique. *A report for the Directorate-General for industrial and Technological Affairs for the Environment and Consumer Protection Service of the European Communities.* Graham and Trotman, London. pp. 1-216.

AUGIER, H., GILLES, G., and RAMONDA, G. (1978). Recherche sur la pollution mercurielle dans le Golf de Fos (Méditerranée, France) : degré de contamination par le mercure des phanérogames marines *Posidonia oceanica* Delile, *Zostera noltü* Horneman et *Zostera marina* (L.). *Revue int. Océanogr. Méd.*, 51-52, 55-69.

AYLING, G. M. (1974). Uptake of cadmium, zinc, copper, lead and chromium in the Pacific oyster, *Crassostrea gigas* grown in the Tamar river, Tasmania. *Wat. Res.*, **8**, 729–738.

AZAM, F., VACCARO, R. F., GILLESPIE, P. A., MOUSSALLI, E. I., and HODSON, R. E. (1977). Controlled ecosystem pollution experiment: effect of mercury on enclosed water columns. II. Marine bacterioplankton. *Mar. Sci. Communs*, **3**, 313–329.

BARBARO, A., FRANCESCON, A., POLO, B., and BILIO, M. (1978). *Balanus amphitrite* (Cirripedia : Thoracica)—a potential indicator of fluoride, copper, lead, chromium and mercury in north Adriatic lagoons. *Mar. Biol.*, **46**, 247–257.

BARD, C. C., MURPHY, J. J., STONE, D. L., and TERHAAR, C. J. (1976). Silver in photoprocessing effluents. *J. Wat. Pollut. Control Fed.*, **48**, 389–394.

BARTLETT, P. D., CRAIG, P. J., and MORTON, S. F. (1978). Total mercury and methyl mercury levels in British estuarine and marine sediments. *Sci. Total Environ.*, **10**, 245–251.

BATLEY, G. E. and FLORENCE, T. M. (1976). Determination of the chemical forms of dissolved cadmium, lead and copper in seawater. *Mar. Chem.*, **4**, 347–363.

BAYNE, B. L., MOORE, M. N., WIDDOWS, J., LIVINGSTONE, D. R., and SALKELD, P. (1979). Measurement of the responses of individuals to environmental stress and pollution : studies with bivalve molluscs. *Phil. Trans. R. Soc. B.*, **286**, 563–581.

BECKETT, J. S. and FREEMAN, H. C. (1974). Mercury in swordfish and other pelagic species from the western Atlantic Ocean. *Spec. scient. Rep. natn oceanic atmos. Adm. U.S. (Fisheries)*, No. 675, 154-159.

BELLINGER, E. G. and BENHAM, B. R. (1978). The levels of metals in dock-yard sediments with particular reference to the contributions from ship-bottom paints. *Environ. Pollut.*, **15**, 71–81.

BENGTSSON, B. -E. (1978). Use of a harpacticoid copepod in toxicity tests. *Mar. Pollut. Bull., N.S.*, **9**, 238–241.

BENIJTS-CLAUS, C. and BENIJTS, F. (1975). The effect of low lead and zinc concentrations on the larval development of the mud-crab *Rhithropanopeus harrisii* Gould. In J. H. Koeman, and J. J. T. W. A. Strik, (Eds), *Sublethal Effects of Toxic Chemicals on Aquatic Animals*, Proc. Swed. Neth. symp., Wageningen, 2–5 Sept. 1975. Elsevier, Oxford. pp. 43-52.

BENON, P., BLANC, F., BOURGADE, B., DAVID, P., KANTIN, R., LEVEAU, M., ROMANO, J.-C., and SAUTRIOT, D. (1978). Distribution of some heavy metals in the Gulf of Fos. *Mar. Pollut. Bull., N.S.*, **9**, 71–75.

BENTLEY-MOWAT, J. A. and REID, S. M. (1977). Survival of marine phytoplankton in high concentrations of heavy metals, and uptake of copper. *J. exp. mar. Biol. Ecol.*, **26**, 249–264

BERDICEVSKY, I., SHOYERMAN, H., and YANNAI, S. (1979). Formation of methyl mercury in the marine sediment under *in vitro* conditions. *Environ Res.*, **20**, 325–334.

BERK, S. G., HENDRICKS, D. L., and COLWELL, R. R. (1978). Effects of ingesting mercury-containing bacteria on mercury tolerance and growth rates of ciliates. *Microb. Ecol.*, **4**, 319–330.

BERLAND, B. R., BONIN, D. J., KAPKOV, V. I., MAESTRINI, S. Y., and ARLHAC, D. P. (1976). Action toxique de quatre métaux lourds sur la croissance d'algues unicellulaires marines. *C.r. hebd. Séanc. Acad. Sci., Paris*, **282**, 633–636.

BERMAN, S. S., MCLAREN, J. W., and WILLIE, S. N. (1980). Simultaneous determination of five trace metals in seawater by inductively coupled plasma atomic emission spectrometry with ultrasonic nebulization. *Analyt. Chem.*, **52**, 488–492.

BERNHARD, M. (1978). Heavy metals and chlorinated hydrocarbons in the Mediterranean. *Ocean Mgmt*, **3**, 253–313.

BERTINE, K. K. and GOLDBERG, E. D. (1971). Fossil fuel combustion and the major sedimentary cycle. *Science, N.Y.*, **173**, 233–235.

BETZER, S. B. and PILSON, M. E. Q. (1975). Copper uptake and excretion by *Busycon canaliculatum* L. *Biol. Bull. mar. biol. Lab., Woods Hole,* **148**, 1–15.

BETZER, S. B. and YEVICH, P. P. (1975). Copper toxicity in *Busycon canaliculatum* L. *Biol. Bull, mar. biol. Lab., Woods Hole,* **148**, 16–25.

BEWERS, J. M., SUNDBY, B., and YEATS, P. A. (1976). The distribution of trace metals in the western North Atlantic off Nova Scotia. *Geochim. cosmochim. Acta,* **40**, 687–696.

BEWERS, J. M. and YEATS, P. A. (1977). Oceanic residence times of trace metals. *Nature, Lond.,* **268**, 595–598.

BEWERS, J. M. and YEATS, P. A. (1978). Trace metals in the waters of a partially mixed estuary. *Estuar. coast. mar. Sci,* **7**, 147–162.

BLACKMAN, R. A. A. and WILSON, K. W. (1973). Effects of red mud on marine animals. *Mar. Pollut. Bull., N.S.,* **4**, 169–171.

BLAKE, N. J. and JOHNSON, D. L. (1976). Oxygen production–consumption of the pelagic *Sargassum* community in a flow-through system with arsenic additions. *Deep Sea Res.,* **23**, 773–778.

BLAXTER, J. H. S. (1977). The effect of copper on the eggs and larvae of plaice and herring. *J. mar. biol. Ass. U.K.,* **57**, 849–858.

BLOOM, H. and AYLING, G. M. (1977). Heavy metals in the Derwent estuary. *Environ. Geol.,* **2,** 3–22.

BLUM, J. E. and BARTHA, R. (1980). Effect of salinity on methylation of mercury. *Bull. environ. Contam. Toxicol.,* **25**, 404–408.

BOHN, A. (1979). Trace metals in fucoid algae and purple sea urchins near a high Arctic lead/zinc ore deposit. *Mar. Pollut. Bull., N. S.,* **10**, 325–327.

BOHN, A. and FALLIS, B. W. (1978). Metal concentrations (As, Cd, Cu, Pb, and Zn) in shorthorn sculpins, *Myxocephalus scorpius* (Linnaeus) and Artic char, *Salvelinus alpinus* (Linnaeus), from the vicinity of Strathcona Sound, Northwest Territories. *Wat. Res.,* **12**, 659–663.

BONEY, A. D. (1980). Effects of seawater suspensions of iron ore dust on *Fucus* oospheres. *Mar. Pollut. Bull., N.S., 11,* 41-43.

BONEY, A. D. (1971). Sub-lethal effects of mercury on marine algae. *Mar. Pollut. Bull., N.S.,* **2**, 69–71.

BOTHNER, M. H., JAHNKE, R. A., PETERSON, M. L., and CARPENTER, R. (1980). Rate of mercury loss from contaminated estuarine sediments. *Geochim. Cosmochim. Acta,* **44**, 273–285.

BOUGIS, P., CORRE, M. C., and ÉTIENNE, M. (1979). Sea urchin larvae as a tool for assessment of the quality of sea water. *Annls Inst. Océanogr., Paris,* **55**, 21–26.

BOUQUEGNEAU, J. M. (1979). Evidence for the protective effect of metallothionein against inorganic mercury injuries to fish. *Bull. environ. Contam. Toxicol.,* **23**, 218–219.

BOYDEN, C. R. (1974). Trace element content and body size in molluscs. *Nature, Lond.,* **251,** 311–314.

BOYDEN, C. R. (1975). Distribution of some trace metals in Poole Harbour, Dorset. *Mar. Pollut. Bull., N.S.,* **6**, 180–186.

BOYDEN, C. R. (1977). Effect of size upon metal content of shellfish. *J. mar. biol. Ass. U.K.,* **57,** 675-714.

BOYDEN, C. R., WATLING, H., and THORNTON, I. (1975). Effect of zinc on the settlement of the oyster *Crassostrea gigas. Mar. Biol.,* **31.** 227–234.

BOYLE, E. A., SCLATER, F. R., and EDMOND, J. M. (1977). The distribution of dissolved copper in the Pacific. *Earth Planet. Sci. Lett.,* **37**, 38–54.

BRAEK, G. S., JENSEN, A., and MOHUS, A. (1976). Heavy metal tolerance of marine phytoplankton. III. Combined effects of copper and zinc ions on cultures of four common species. *J. exp. mar. Biol. Ecol.,* **25**, 37–50.

BRAEK, G. S., MALNES, D., and JENSEN, A. (1980). Heavy metal tolerance of marine phytoplankton IV. Combined effect of zinc and cadmium on growth and uptake in some marine diatoms. *J. exp. mar. Biol. Ecol.,* **42**, 39–54.

BRAHAM, H. W. (1973). Lead in the California sea lion *Zalophus californianus. Environ. Pollut.* **5,** 253–258.

BRAMAN, R. S. and TOMPKINS, M. A. (1979). Separation and determination of nanogram amounts of inorganic tin and methyltin compounds in the environment. *Analyt. Chem.,* **51,** 12–19.

BRANNON, J. M. (1978). Evaluation of dredged material for pollution potential. *Dredged Material Research Program Report Series,* DS-78-6. U.S. Army, Washington, D.C. pp. 1–39.

BREWER, P. G. (1975). Minor elements in sea water. In J. P. Riley and J. Skirrow (Eds), *Chemical Oceanography,* Vol I, (2nd ed.), Academic Press, London and New York. pp. 415-496.

BRINKHUIS, B. H., PENELLO, W. F. and CHURCHILL, A. C. (1980). Cadmium and manganese flux in the eelgrass *Zostera marina. Mar. Biol.,* **58**, 187–196.

BROWN, A. C. and BROWN, R. J. (1965). The fate of Thorium dioxide injected into the pedal sinus of *Bullia* (Gastropoda : Prosobranchiata). *J. exp. Biol.,* **42**, 509–19.

BROWN, B. E. (1978). Lead detoxification by a copper-tolerant isopod. *Nature, Lond.,* **276**, 388–390.

BROWN, D. A., BAWDEN, C.A., CHATEL, K. W., and PARSONS, T. R. (1977). The wildlife community of Iona Island Jetty, Vancouver, B.C., and heavy-metal pollution effects. *Environ. Conserv.,* **4**, 213–216.

BRULAND, K. W., KNAUER, G. A., and MARTIN, J. H. (1978). Cadmium in northeast Pacific waters. *Limnol. Oceanogr.,* **23**, 618–625.

BRULAND, K. W., FRANKS, R. P., KNAUER, G. A., and MARTIN, J. H. (1979). Sampling and analytical methods for the determination of copper, cadmium, zinc, and nickel at the nanogram per liter level in sea water. *Analytica chim. Acta,* **105**, 233–245.

BRYAN, G. W. (1964). Zinc regulation in the lobster *Homarus vulgaris.* Tissue zinc and copper concentrations. *J. mar. biol. Ass. U.K.,* **44**, 549–563.

BRYAN, G. W. (1966). The metabolism of Zn and ^{65}Zn in crabs, lobsters and freshwater crayfish. In B. Aberg and F. P. Hungate (Eds), *Radioecological Concentration Processes.* Pergamon Press, Oxford and New York. pp. 1005–1016.

BRYAN, G. W. (1967). Zinc regulation in the freshwater crayfish (including some comparative copper analyses). *J. exp. Biol.,* **46**, 281–296.

BRYAN, G. W. (1969). The absorption of zinc and other metals by the brown seaweed *Laminaria digitata. J. mar. biol. Ass. U.K.,* **49**, 225–243.

BRYAN, G. W. (1971). The effects of heavy metals (other than mercury) on marine and estuarine organisms. *Proc. R. Soc. B,* **177**, 389–410.

BRYAN, G. W. (1973). The occurrence and seasonal variation of trace metals in the scallops *Pecten maximus* (L.) and *Chlamys opercularis* (L.) *J. mar. biol. Ass. U.K.,* **53**, 145–166.

BRYAN, G. W. (1974). Adaptation of an estuarine polychaete to sediments containing high concentrations of heavy metals. In F. W. Vernberg and W. B. Vernberg (Eds), *Pollution and Physiology of Marine Organisms.* Academic Press, London and New York. pp. 123-135.

BRYAN, G. W. (1976a). Some aspects of heavy metal tolerance in aquatic organisms. In A. P. M. Lockwood (Ed.), *Effects of Pollutants on Aquatic Organisms.* Cambridge University Press. pp. 7-34.

BRYAN, G. W. (1976b). Heavy metal contamination in the sea. In R. Johnston (Ed.), *Marine Pollution* Academic Press, London and New York. pp. 185-302.

BRYAN, G. W. (1979). Bioaccumulation of marine pollutants. *Phil. Trans. R. Soc. B,* **286**, 483–505.

BRYAN, G. W. (1980). Recent trends in research on heavy-metal contamination in the sea. *Helgoländer Meeresunters.,* **33**, 6–25.

BRYAN, G. W. and GIBBS, P. E. (1980). Metals in nereid polychaetes : the contribution of metals in the jaws to the total body burden. *J. mar. biol. Ass. U.K.,* **60**, 641–654.

BRYAN, G. W. and GIBBS, P. E. (1983). Heavy metals in the Fal Estuary (Cornwall): a study of long-term contamination by mining waste and its effects on estuarine organisms. *Mar. biol. Ass. U.K.* (occ. publ. 2), 1–112.

BRYAN, G. W. and HUMMERSTONE, L. G. (1971). Adaptation of the polychaëte *Nereis diversicolor* to estuarine sediments containing high concentrations of heavy metals. I. General observations and adaption to copper. *J. mar. biol. Ass. U,K,,* **51**, 845–863.

BRYAN, G. W. and HUMMERSTONE, L. G. (1973a). Adaptation of the polychaete *Nereis diversicolor* to estuarine sediments containing high concentrations of zinc and cadmium. *J. mar. biol. Ass. U.K.,* **53**, 839–857.

BRYAN, G. W. and HUMMERSTONE, L. G. (1973b). Adaptation of the polychaete *Nereis diversicolor* to manganese in estuarine sediments. *J. mar. biol. Ass. U.K.,* **53**, 859–872.

BRYAN, G. W. and HUMMERSTONE, L. G. (1973c). Brown seaweed as an indicator of heavy metals in estuaries in south-west England. *J. mar. biol. Ass. U.K.,* **53**, 705–720.

BRYAN, G. W. and HUMMERSTONE, L. G. (1977). Indicators of heavy-metal contamination in the Looe Estuary (Cornwall) with particular regard to silver and lead. *J. mar. biol. Ass. U.K.,* **57**, 75–92.

BRYAN, G. W. and HUMMERSTONE, L. G. (1978). Heavy metals in the burrowing bivalve *Scrobicularia plana* from contaminated and uncontaminated estuaries. *J. mar. biol. Ass. U.K.*, **58**, 401–419.

BRYAN, G. W., LANGSTON, W. J., and HUMMERSTONE, L. G. (1980). The use of biological indicators of heavy metal contamination in estuaries with special reference to an assessment of the biological availability of metals in estuarine sediments from south-west Britain. *Mar. biol. Ass. U.K.*, (occ. publ. 1), 1–73.

BRYAN, G. W., POTTS, G. W. and FORSTER, G. R. (1977). Heavy metals in the gastropod mollusc *Haliotis tuberculata* (L.). *J. mar. biol. Ass. U.K.*, **57**, 379–390.

BRYAN, G. W. and UYSAL, H. (1978). Heavy metals in the burrowing bivalve *Scrobicularia plana* from the Tamar estuary in relation to environmental levels. *J. mar. biol. Ass. U.K.*, **58**, 89–108.

BRYAN, G. W. and WARD, E. (1965). The absorption and loss of radioactive and non-radioactive manganese by the lobster *Homarus vulgaris*. *J. mar. biol. Ass. U.K.*, **45**, 65–95.

BUCHANAN, J. B., BROWN, B. E., COOMBS, T. L., PIRIE, B. J. S., and ALLEN, J. A. (1980). The accumulation of ferric iron in the guts of some spatangoid echinoderms. *J. mar. biol. Ass. U.K.*, **60**, 631–640.

BUHLER, D. R., CLAEYS, R. R., and MATE, B. R. (1975). Heavy metal and chlorinated hydrocarbon residues in California sea lions (*Zalophus californianus californianus*). *J. Fish. Res. Bd Can.*, **32**, 2391–2397.

BULL, K. R., EVERY, W. J., FREESTONE, P., and OSBORNE, D. (1983). Alkyl lead pollution and bird mortalities on the Mersey Estuary, U.K. 1979–1981. *Environ. Pollut.* (A), 31, 239–259.

BURRELL, D. C. (1977). Natural distribution of trace heavy metals and environmental background in Alaskan shelf and estuarine areas. In *US NOAA Outer Continental Shelf Environmental Assessment Program*, Vol. XIII, Contaminant Baselines. US NOAA, Boulder, Colorado, pp. 291–485.

BURTON, J. D. (1978). Chemical processes in estuarine and coastal waters : environmental and analytical aspects. *J. Instn Wat. Engrs*, **32**, 31–44.

BURTON, J. D., MAHER, W. A., and STATHAM, P. J. (1980). Aspects of the distribution and chemical form of arsenic and selenium in ocean waters and marine organisms. *VI International Symposium, Chemistry of the Mediterranean*, Institute Ruder Bošković, 5–10 May 1980. Rovinj, Zagreb. pp. 39-41.

CALABRESE, A., COLLIER, R. S., NELSON, D. A., and MACINNES, J. R. (1973). The toxicity of heavy metals to embryos of the American oyster *Crassostrea virginica*. *Mar. Biol.*, **18**, 162–166.

CALABRESE, A., MACINNES, J. R., NELSON, D. A., and MILLER, J. E. (1977). Survival and growth of bivalve larvae under heavy metal stress. *Mar. Biol.*, **41**, 179–184.

CALABRESE, A. and NELSON, D. A. (1974). Inhibition of embryonic development of the hard clam, *Mercenaria mercenaria*, by heavy metals. *Bull. environ. Contam. Toxicol.*, **11**, 92–97.

CALABRESE, A., THURBERG, F. P., DAWSON, M. A., and WENZLOFF, D. R. (1975). Sublethal physiological stress induced by cadmium and mercury in the winter flounder, *Pseudopleuronectes americanus*. In J. H. Koeman and J. J. T. W. A. Strik (Eds), *Sublethal Effects of Toxic Chemicals on Aquatic Animals*. Elsevier. pp. 15–21.

CAMBRAY, R. S., JEFFERIES, D. F., and TOPPING, G. (1979). The atmospheric input of trace elements to the North Sea. *Mar. Sci. Commun*, **5**, 175–194.

CANTERFORD, G. S., BUCHANAN, A. S., and DUCKER, S. C. (1979). Accumulation of heavy metals by the marine diatom *Ditylum brightwellii* (West) Grunow. *Aust. J. mar. Freshwat. Res.*, **29**, 613–622.

CANTERFORD, G. S. and CANTERFORD, D. R. (1980). Toxicity of heavy metals to the marine diatom *Ditylum brightwellii* (West) Grunow: correlation between toxicity and metal speciation. *J. mar. biol. Ass. U.K.*, **60**, 227–242.

CARMICHAEL, N. G., SQUIBB, K. S., ENGEL, D. W., and FOWLER, B. A. (1980). Metals in the molluscan kidney: uptake and subcellular distribution of ^{109}Cd, ^{54}Mn and ^{65}Zn by the clam *Mercenaria mercenaria*. *Comp. Biochem. Physiol.*, **65A**, 203–206.

CARMICHAEL, N. G., SQUIBB, K. S., and FOWLER, B. A. (1979). Metals in the molluscan kidney: a comparison of two closely related bivalve species *(Argopecten)*, using X-ray microanalysis and atomic absorption spectroscopy. *J. Fish. Res. Bd Can.*, **36**, 1149–1155.

CHAISEMARTIN, C., CHAISEMARTIN, R. -A., and BRETON, J. -C. (1978). Aspect de lat détresse métabolique chez Macropodia: bioconcentration du plomb et activité de l'aspartate aminotransférase. *C.r. Séanc. Soc. Biol.*, **172**, 1180–1187.

CHANDLER, J. A. (1977). *X-ray Microanalysis in the Electron Microscope*, North-Holland, Amsterdam.

CHARLEBOIS, C. T. (1978). High mercury levels in Indians and Inuits (Eskimos) in Canada. *Ambio*, **7**, 204–210.

CHAU, Y. K., WONG, P. T. S., KRAMAR, O., BENGERT, G. A., CRUZ, R. B., KINRADE, J. O., LYE, J., and Van LOON, J. (1980). Occurrence of tetraalkyl lead compounds in the aquatic environment. *Bull. environ. Contam. Toxicol.*, **24**, 265–269.

CHERRY, R. D., HIGGO, J. J. W., and FOWLER, S. W. (1978). Zooplankton faecal pellets and element residence times in the ocean. *Nature, Lond.*, **274**, 246–248.

CHILDS, E. A. and GAFFKE, J. N. (1973). Mercury content of Oregon groundfish. *Fish. Bull., U.S.*, **71**, 713–717.

CHIPMAN, W. A. (1966). Some aspects of the accumulation of ^{51}Cr by marine organisms. In B. Aberg and F. P. Hungate (Eds), *Radioecological Concentration Processes*. Pergamon Press, Oxford. pp. 931-941.

CLARK, R. B. (1971). Changing success of coastal sport fishing. *Mar. Pollut. Bull., N.S.*, **2**, 153–156.

CLERK, R. dE, VANDERSTAPPEN, R., and VYNCKE, W. (1974). Mercury content of fish and shrimps caught off the Belgian coast. *Ocean Mgmt*, **2**, 117–126.

CLIFTON, R. J. and HAMILTON, E. I. (1979). Lead-210 chronology in relation to levels of elements in dated sediment core profiles. *Estuar. coast. mar. Sci.*, **8**, 259–269.

CLOUTIER-MANTHA, L. and BROWN, D. A. (1980). The effects of mercury exposure on intracellular distribution of mercury, copper and zinc in *Skeletonema costatum* (Grev.) Cleve. *Botanica mar.*, **23**, 53–58.

CONNOR, P. M. (1972). Acute toxicity of heavy metals to some marine larvae. *Mar. Pollut. Bull., N.S.*, **3**, 190–192.

COOMBS, T. L., FLETCHER, T. C., and WHITE, A. (1972). Interaction of metal ions with mucus from plaice (*Pleuronectes platessa* L.). *Biochem. J.*, **128**, 128–129.

COOMBS, T. L. and GEORGE, S. G. (1978). Mechanisms of immobilization and detoxication of metals in marine organisms. In D. J. McLusky and A. J. Berry (Eds), *Physiology and Behaviour of Marine Organisms*. Pergamon Press, Oxford and New York. pp. 179-187.

COONEY, R. V. and BENSON, A. A. (1980). Arsenic metabolism in *Homarus americanus*. *Chemosphere*, **9**, 335–341.

CORNER, E. D. S. and RIGLER, F. H. (1958). The modes of action of toxic agents. III. Mercuric chloride and N-amylmercuric chloride on crustaceans. *J. mar. biol. Ass. U.K.*, **37**, 85–96.

COSSA, D., BOURGET, E., and PIUZE, J. (1979). Sexual maturation as a source of variation in the relationship between cadmium concentration and body weight in *Mytilus edulis* L. *Mar. Pollut. Bull., N.S.*, **10**, 174–176.

CRANSTON, R. E. and MURRAY, J. W. (1978). The determination of chromium species in natural waters. *Analytica Chim. Acta*, **99**, 275–282.

CRECELIUS, E. A. (1980). The solubility of coal fly ash and marine aerosols in sea water. *Mar. Chem.*, **8**, 245–250.

CRECELIUS, E. A., BOTHNER, M. H., and CARPENTER, R. (1975). Geochemistries of arsenic, antimony, mercury and related elements in sediments of Puget Sound. *Environ. Sci. Technol.*, **9**, 325–333.

CROFTS, D. (1929). Haliotis. *L.M.B.C. Mem. typ. Br. mar. Pl. Anim.*, **29**, 1–174.

CROSS, F. A., HARDY, L. H., JONES, N. Y., and BARBER, R. (1973). Relation between total body weight and concentrations of manganese, iron, copper, zinc, and mercury in white muscle of bluefish (*Pomatomus saltatrix*) and a bathyal-demersal fish *Antimora rostrata*. *J. Fish. Res. Bd Can.*, **30**, 1287–1291.

CUNNINGHAM, P. A. (1976). Inhibition of shell growth in the presence of mercury and subsequent recovery of juvenile oysters. *Proc. natn. Shellfish. Ass.*, **66**, 1–5.

CUNNINGHAM, P. A. and GROSCH, D. S. (1978). A comparative study of the effects of mercuric chloride and methyl mercury chloride on reproductive performance in the brine shrimp, *Artemia salina*. *Environ. Pollut.*, **15**, 83–89.

CUNNINGHAM, P. A. and TRIPP, M. R. (1973). Accumulation and depuration of mercury in the American oyster *Crassostrea virginica*. *Mar. Biol.*, **20**, 14–19.

CUTHBERT, K. C., BROWN, A. C., and ORREN, M. J. (1976). Toxicity of cadmium to *Bullia digitalis* (Prosobranchiata: Nassaridae). *Trans. R. Soc. S. Afr.*, **42**, 203–208.

D'AGOSTINO, A. and FINNEY, C. (1974). The effect of copper and cadmium on the development of *Tigriopus japonicus*. In F. J. Vernberg and W. B. Vernbery (Eds), *Pollution and Physiology of Marine Organisms*. Academic Press, New York. pp. 445–463.

DANIELSSON, L. G. (1980). Cadmium, cobalt, copper, iron, iron, lead, nickel and zinc in Indian Ocean water. *Mar. Chem.*, **8**, 199–215.

DANSKIN, G. P. (1978). Accumulation of heavy metals by some solitary tunicates. *Can. J. Zool.*, **56**, 547–551.

DA SILVA, J. J. R. F. (1978). Interaction of the chemical elements with biological systems. In R. J. P. Williams and J. J. R. F. Da Silva (Eds), *New trends in bioinorganic chemistry*. Academic Press, London. pp. 449–484.

DAVIES, A. G. (1974). The growth kinetics of *Isochrysis galbana* in cultures containing sub-lethal concentrations of mercuric chloride. *J. mar. biol. Ass. U.K.*, **54**, 157–169.

DAVIES, A. G. (1976). An assessment of the basis of mercury tolerance in *Dunaliella tertiolecta*. *J. mar. biol. Ass. U.K.*, **56**, 39–57.

DAVIES, A. G. (1978). Pollution studies with marine plankton. Part II. Heavy metals. *Adv. mar. Biol.*, **15**, 381–508.

DAVIES, A. G. and SLEEP, J. A. (1979). Photosynthesis in some British coastal waters may be inhibited by zinc pollution. *Nature, Lond.*, **277**, 292–293.

DAVIES, A. G. and SLEEP, J. A. (1980). Copper inhibition of carbon fixation in coastal phytoplankton assemblages. *J. mar. biol. Ass. U.K.*, **60**(4), 841–850.

DAVIES, I. M., GRAHAM, W. C., and PIRIE, J. M. (1979). A tentative determination of methyl mercury in seawater. *Mar. Chem.*, **7**, 111–116.

DAVIES, I. M. and PIRIE, J. M. (1978). The mussel *Mytilus edulis* as a bio-assay organism for mercury in seawater. *Mar. Pollut. Bult., N.S.*, **9**, 128–132.

DAVIES, I. M. and PIRIE, J. M. (1980). Evaluation of a 'mussel watch' project for heavy metals in Scottish coastal waters. *Mar. Biol.*, **57**, 87–93.

DAVIES, J. A. and LECKIE, J. O. (1978). Effect of adsorbed complexing ligands on trace metal uptake by hydrous oxides. *Environ. Sci. Technol.*, **12**, 1309–1315.

DAVIES, J. M. and GAMBLE, J. C. (1979). Experiments with large enclosed ecosystems. *Phil. Trans. R. Soc.* B, **286**, 523–544.

DAWSON, M. A., GOULD, E., THURBERG, F. P., and CALABRESE, A. (1977). Physiological response of juvenile striped bass, *Morone saxatilis*, to low levels of cadmium and mercury. *Chesapeake Sci.*, **18**, 353–359.

DECLEIR, W., VLAEMINCK, A., GELADI, P., and GRIEKEN, R. van. (1978). Determination of protein-bound copper and zinc in some organs of the cuttlefish *Sepia officinalis* L. *Comp. Biochem. Physiol.*, **60B**, 347–350.

DELHAYE, W. and CORNET, D. (1975). Contribution to the study of the effect of copper on *Mytilus edulis* during reproductive period. *Comp. Biochem. Physiol.*, **50A**, 511–518.

DENTON, G. R. W., MATSH, H., HEINSOHN, G. E., and BURDON-JONES, C. (1980). The unusual metal status of the dugong *Dugong dugon*. *Mar. Biol.*, **57**, 201–219.

DEPARTMENT OF THE ENVIRONMENT (DOE) (1979). Monitoring the marine environment 1978–83: the way ahead. The second report of the Marine Pollution Monitoring Management Group 1977–1978. *Pollution Report.*, **6**, 1–23.

DETHLEFSEN, V. (1977). Uptake, retention and loss of cadmium by brown shrimp (*Crangon crangon*). *Ber.dt.wiss.Kommn Meeresforsch.*, **26**, 137–152.

DEVANAS, M. A., LICHFIELD, C. D., MCCLEAN, C., and GIANNI, J. (1980). Coincidence of cadmium and antibiotic resistance in New York Bight apex benthic microorganisms. *Mar. Pollut. Bull., N.S.*, **11**, 264–269.

DIX, T. G., MARTIN, A., AYLING, G. M., WILSON, K. C., and RATKOWSKY, D. A. (1975). Sand flathead (*Platycephalus bassensis*), an indicator species for mercury pollution in Tasmanian waters. *Mar. Pollut. Bull., N.S.*, **6**, 112–144.

DJANGMAH, J.S. (1970). The effects of feeding and starvation on copper in the blood and hepatopancreas, and on blood proteins of *Crangon vulgaris* (Fabricius). *Comp. Biochem. Physiol.*, **32**, 709–731.

DOYLE, L. J., BLAKE, N. J., WOO, C. C., and YEVITCH, P. (1978). Recent biogenic phosphorite: concretions in mollusc kidneys. *Science, N.Y.*, **199**, 1431–1433.

DRESCHER, H. E., HARMS, U., and HUSCHENBETH, E. (1977). Organochlorines and heavy metals in the harbour Seal *Phoca vitulina* from the German North Sea coast. *Mar. Biol.*, **41**, 99–106.

DUINKER, J. C. and KRAMER, C. J. M. (1977). An experimental study on the speciation of dissolved zinc, cadmium, lead and copper in River Rhine and North Sea water, by differential pulsed anodic stripping voltammetry. *Mar. Chem.*, **5**, 207–228.

DUINKER, J. C. and NOLTING, R. F. (1978). Mixing, removal and mobilization of trace metals in the Rhine Estuary. *Neth. J. Sea Res.*, **12**, 205–223.

EAGLE, R. A., HARDIMAN, P. A., NORTON, M. G., and NUNNY, R. S. (1979a). The field assessment of effects of dumping wastes at sea: 4. A survey of the sewage sludge disposal area off Plymouth. *Fish. Res. tech. Rep., Lowestoft*, No. 50. 1–24.

EAGLE, R. A., HARDIMAN, P. A., NORTON, M. G., NUNNY, R. S., and ROLFE, M. S. (1979b). The field assessment of effects of dumping wastes at sea: 5. The disposal of solid wastes off the north-east coast of England. *Fish. Res. tech. Rep., Lowstoft*, No. 51, 1–34.

EGAWA, H., and TAJIMA, S. (1977). Determination of trace amounts of methyl-mercury in sea water. In S. A. Peterson and K. K. Randolph (Eds), *Management of Bottom Sediments Containing Toxic Substances* (Ecological Research Series EPA-600/3-77-083). US Environmental Protection Agency, Office of Research and Development, Corvallis, Oregon. pp. 96–106.

EIDE, I., JENSEN, A., and MELSOM, S. (1979). Application of *in situ* cage cultures of phytoplankton for monitoring heavy metal pollution in two Norwegian fjords. *J. exp. mar. Biol. Ecol.*, **37**, 271–286.

EIDE, I., MYKLESTAD, S., and MELSOM, S. (1980). Long-term uptake and release of heavy metals by *Ascophyllum nodosum* (L.) Le Jol. *Environ. Pollut. (Series A).*, **23**, 19–28.

EISLER, R., BARRY, M. M., LAPAN, R. L., Jr., TELEK, G., DAVEY, E. W., and SOPER, A. E. (1978). Metal survey of the marine clam *Pitar morrhuana* collected near a Rhode Island (USA) electroplating plant. *Mar. Biol.*, **45**, 311–317.

ELDERFIELD, H. and HEPWORTH, A. (1975). Diagenesis, metals and pollution in estuaries. *Mar. Pollut. Bull., N.S.*, **6**, 85–87.

ELDERFIELD, H., HEPWORTH, A., EDWARDS, P. N., and HOLLIDAY, L. M. (1979). Zinc in the Conwy river and estuary. *Estuar. coast. mar. Sci.*, **9**, 403–422.

ELKAIM, B. and CHASSARD-BOUCHAUD, C. (1978). Concentrations minérales détéctees par spectrographie des rayons X chez *Crangon crangon (Crustacé décapode) au cours du cycle d'intermue. Relations avec le milieu. Cah. Biol. mar.*, **19**, 459–471.

EMERSON, R. R., SOULE, D. F., OGURI, M., CHEN, K. Y., and LU, J. (1975). Heavy metal concentrations in marine organisms and sediments collected near an industrial waste outfall. *Proc. Int. Conf. Environmental Sensing & Assessment* Vol. 1, Section 6-7. Las Vegas, Nevada, pp. 1–5.

ENGEL, D. W. and FOWLER, B. A. (1979). Factors influencing cadmium accumulation and its toxicity to marine organisms. *Environ. Hlth. Perspect.*, **28**, 81–88.

ENGEL, D. W. and SUNDA, W. G. (1979). Toxicity of cupric ion to eggs of the spot *Leiostomus xanthurus* and the Atlantic silverside *Menidia menidia*. *Mar. Biol.*, **50**, 121–126.

ERLENKEUSER, H., SUESS, E., and WILLKOMM, H. (1974). Industrialization affects heavy metal and carbon isotope concentrations in recent Baltic Sea sediments. *Geochim. Cosmochim. Acta.* **38**, 823–842.

ESSINK, K. (1980). Mercury pollution in the Ems estuary. *Helgoländer Meeresunters.*, **33**, 111–121.

ESTABLIER, R. (1972). Concentración de mercurio en los tejidos de algunos peces, moluscos y crustáceos del golfo de Cádiz y caladeros del noroeste africano. *Investigación pesq.*, **36**, 355–364.

ESTABLIER, R., GUTIÉRREZ, M., and ARIAS, A. (1978). Acumulación y efectos histopathlógicos del mercurio inorgánico y organico en la lisa (*Mugil auratus Risso*). *Investigación pesq.*, **42**, 65–80.

ETCHEBER, H. (1979). Répartition et comportement du Zn, Pb, Cu, et Ni dans l'estuaire de la Gironde. *Bull. Inst. Géol. Bassin d'Aquitaine*, **25**, 121–147.

EUSTACE, I. J. (1974). Zinc, cadmium, copper and manganese in species of finfish and shellfish caught in the Derwent Estuary, Tasmania. *Aust. J. mar. Freshwat. Res.*, **25**, 209–220.

FAGERSTRÖM, T. and JERNELÖV, A. (1973). Formation of methyl mercury from pure mercuric sulphide in aerobic organic sediment. *Wat. Res.*, **5**, 121–122.

FAO/WHO JOINT EXPERT COMMITTEE ON FOOD ADDITIVES (1972). *Tech. Rep. Ser. Wld Hlth Org.*, **505**, 11–16.

FARADAY, W. E., and CHURCHILL, A. C. (1979). Uptake of cadmium by the eelgrass *Zostera marina*. *Mar. Biol.*, **53**, 293–298.

FENG, S. Y. and RUDDY, G. M. (1975). Concentrations of zinc, copper, cadmium, manganese and mercury in oysters (*Crassostrea virginica*) along the Connecticut coast. In *3rd Int. Ocean Devel. conf., Tokyo, Japan, 5-8 August 1975,* Vol. 4. Organizing Committee, Int. Ocean Development Conf., Tokyo, Japan. pp. 109–130.

FILIPPIS, L. F. DE and PALLAGHY, C. K. (1975). Localization of zinc and mercury in plant cells. *Micron*, **6**, 111–120.

FILIPPIS, L. F. DE and PALLAGHY, C. K. (1976). The effects of sub-lethal concentrations of mercury and zinc on *Chlorella.* III. Development and possible mechanisms of resistance to metals. *Z. Pflphysiol.,* **79**, 323–335.

FIMREITE, N., KVESETH, N., and BREVIK, E. M. (1980). Mercury, DDE and PCBs in eggs from a Norwegian gannet colony. *Bull. environ. Contam. Toxicol.,* **24**, 142–144.

FISCHER, H. (1978). Hydroids in biotest: *Clava multicornis* exposed to cadmium. *Kieler Meeresforsch.* (Suppl. 4), 327–334.

FLEGAL, A. R. (1979). Trace element concentrations of the rough limpet, *Acmaea scabra,* in California. *Bull. environ. Contam. Toxicol.,* **20**, 834–839.

FLETCHER, C. R. (1970). The regulation of calcium and magnesium in the brackish water polychaete *Nereis diversicolor* (O.F.M.) *J. exp. Biol.,* **53**, 425--443.

FLOS, R., CARITAT, A., and BALASCH, J. (1979). Zinc content in organs of dogfish (*Scyliorhinus canicula* L.) subject to sublethal experimental aquatic zinc pollution. *Comp. Biochem. Physiol.,* **64C**, 77–81.

FÖRSTNER, U. and WITTMANN, G. T. W. (1979). *Metal Pollution in the Aquatic Environment*, Springer-Verlag, Berlin, Heidelberg, New York.

FOSTER, P. (1976). Concentrations and concentration factors of heavy metals in brown algae. *Environ. Pollut.,* **10**, 45–53.

FOSTER, P., HUNT, D. T. E., and MORRIS, A. W. (1978). Metals in an acid mine stream and estuary. *Sci. Total Environ.,* **9**, 75–86.

FOSTER, P. L. (1977). Copper exclusion as a mechanism of heavy metal tolerance in a green alga. *Nature, Lond.,* **269**, 322–323.

FOWLER, B. A., WOLFE, D. A., and HETTLER, W. F. (1975). Mercury and iron uptake by cytosomes in mantle epithelial cells of quahog clams (*Mercenaria mercenaria*) exposed to mercury. *J. Fish. Res. Bd Can.,* **32**, 1767–1775.

FOWLER, S. W. and BENAYOUN, G. (1974). Experimental studies on cadmium flux through marine biota. *Radioactivity in the Sea,* **44,** 1-19 (IAEA, Vienna).

FOWLER, S. W. and BENAYOUN, G. (1976). Influence of environmental factors on selenium flux in two marine invertebrates. *Mar. Biol.,* **37**, 59–68.

FOWLER, S. W. and BENAYOUN, G. (1977). Accumulation and distribution of selenium in mussel and shrimp tissues. *Radioactivity in the Sea,* **61**, 1–8 (I.A.E.A., Vienna).

FOWLER, S. W., HEYRAUD, M., and LA ROSA, J. (1976a). Heavy metal cycling studies: mercury kinetics in marine zooplankton. *Report of the International Laboratory of Marine Radioactivity, Monaco,* 1976 (IAEA-187), 20-33.

FOWLER, S. W., HEYRAUD, M., and LA ROSA, J. (1978). Factors affecting methyl and inorganic mercury dynamics in mussels and shrimp. *Mar. Biol.,* **46**, 267–276.

FOWLER, S. W. and OREGIONI, B. (1976). Trace metals in mussels from the N.W. Mediterranean. *Mar. Pollut. Bull., N.S.,* **7**, 26–29.

FOWLER, S. W., OREGIONI, B., and LA ROSA, J. (1976b). Intercalibration measurements for non-nuclear pollutants: trace metals in pelagic organisms from the Mediterranean Sea. *Report of the International Laboratory of Marine Radioactivity, Monaco,* 1976 (IAEA-187), 110-122.

FRAZIER, J. M. (1975). The dynamics of metals in the American oyster, *Crassostrea virginica.* I. Seasonal effects. *Chesapeake Sci.,* **16**, 162–171.

FREEMAN, H. C. and HORNE, D. A. (1973). Mercury in Canadian seals. *Bull. environ. Contam. Toxicol.,* **10**, 172–180.

FREEMAN, H. C., HORNE, D. A., MCTAGUE, B., and MCMENEMY, M. (1974). Mercury in some Canadian Atlantic coast fish and shellfish. *J. Fish. Res. Bd Can.,* **31**, 369–372.

FREEMAN, H. C., UTHE, J. F., FLEMING, R. B., ODENSE, P. H., ACKMAN, R. G., LANDRY, G., and MUSIAL, C. (1979). Clearance of arsenic ingested by man from arsenic contaminated fish. *Bull. environ. Contam. Toxicol.,* **22**, 224–229.

FRIEDRICH, A. R. and FILICE, F. P. (1976). Uptake and accumulation of the nickel ion by *Mytilus edulis*. *Bull. environ. Contam. Toxicol.*, **16**, 750–755.

FUJIKI, M. (1973). The transitional condition of Minamata Bay and the neighbouring sea polluted by factory waste water containing mercury. In S. H. Jenkins (Ed.), *Advances in Water Pollution Research,* Vol. 6. Permagon Press, Oxford. pp. 905-920.

FUJIKI, M., YAMAGUCHI, S., HIROTA, R., TAJIMA, S., SHIMOJO, N., and SANO, K. (1978). Accumulation of methyl mercury in the red sea bream (*Chrysophrys major*) via the food chain. In S. A. Peterson and K. K. Randolph (Eds), *Management of Bottom Sediments Containing Toxic Substances* (EPA-600/3-78-084). Corvallis Environmental Research Laboratory, Corvallis, Oregon pp. 87-94.

FUKAI, R., and HUYNH-NGOC, L. (1976). Copper, zinc and cadmium in coastal waters of the N.W. Mediterranean. *Mar. Pollut. Bull., N.S.*, **7**, 9–13.

FURNESS, R., and HUTTON, M. (1979). Pollutant levels in the great skua *Catharacta skua*. *Environ. Pollut.*, **19**, 261–268.

GARDNER, D. (1975). Observations on the distribution of dissolved mercury in the ocean. *Mar. Pollut. Bull., N.S.*, **6**, 43–46.

GARDNER, D. (1978). Mercury in fish and waters of the Irish Sea and other United Kingdom fishing grounds. *Nature, Lond.*, **272**, , 49–51.

GARDNER, L. R. (1974). Organic versus inorganic trace metal complexes in sulfidic marine waters—some speculative calculations based on available stability constants. *Geochim. Cosmochim. Acta*, **38**, 1297–1302.

GARDNER, W. S., KENDALL, D. R., ODOM, R. R., WINDOM, H. L., and STEPHENS, J. A. (1978). The distribution of methyl mercury in a contaminated salt marsh ecosystem. *Environ. Pollut.*, **15**, 243–251.

GASKIN, D. E., STONEFIELD, K. I., SUDA, P., and FRANK, R. (1979). Changes in mercury levels in harbor porpoises from the Bay of Fundy, Canada, and adjacent waters during 1969-1977. *Archs environ. Contam. Toxicol.*, **8**, 733–762.

GEORGE, S. G., CARPENE, E., COOMBS, T. L., OVERNELL, J., and YOUNGERSON, A. (1979). Characterisation of cadmium-binding proteins from mussels *Mytilus edulis* (L), exposed to cadmium. *Biochim. biophys. Acta*, **580**, 225–233.

GEORGE, S. G. and COOMBS, T. L. (1977a). The effects of chelating agents on the uptake and accumulation of cadmium by *Mytilus edulis*. *Mar. Biol.*, **39**, 261–268.

GEORGE, S. G. and COOMBS, T. L. (1977b). Effects of high stability iron-complexes on the kinetics of iron accumulation and excretion in *Mytilus edulis* (L.). *J. exp. mar. Biol. Ecol.*, **28**, 133–140.

GEORGE, S. G. and PIRIE, B. J. S. (1979). The occurrence of cadmium in sub-cellular particles in the kidney of the marine mussel, *Mytilus edulis*, exposed to cadmium. *Biochim. biophys. Acta*, **580**, 234–244.

GEORGE, S. G. and PIRIE, B. J. S. (1980). Metabolism of zinc in the mussel, *Mytilus edulis* (L.), : a combined ultrastructural and biochemical study. *J. mar. biol. Ass. U.K.*, **60**, 575–590.

GEORGE, S. G., PIRIE, B. J. S., CHEYNE, A. R., COOMBS, T. L., and GRANT, P. T. (1978). Detoxication of metals by marine bivalves: an ultrastructural study of the compartmentation of copper and zinc in the oyster *Ostrea edulis*. *Mar. Biol.*, **45**, 147–156.

GEORGE, S. G., PIRIE, B. J. S., and COOMBS, T. L. (1976). The kinetics of accumulation and excretion of ferric hydroxide in *Mytilus edulis* (L.) and its distribution in the tissues. *J. exp. mar. Biol. Ecol.*, **23**, 71–84.

GEORGE, S. G., PIRIE, B. J. S. and COOMBS, T. L. (1977). Metabolic characteristics of endocytosis of ferritin by gills of a marine bivalve mollusc. *Trans. biochem. Soc.*, **5**, 136–137.

GEORGE, S. G., PIRIE, B. J. S., and COOMBS, T. L. (1980). Isolation and elemental analysis of metal-rich granules from the kidney of the scallop, *Pecten maximus* (L). *J. exp. mar. Biol. Ecol.*, **42**, 143–156.

GHIRETTI, F. and VIOLANTE, U. (1964). Richerche sul metabolismo del rame in *Octopus vulgaris*. *Boll. zool.*, **31**, 1081–1092.

GIBBS, P. E. and BRYAN, G. W. (1980). Copper—the major metal component of glycerid polychaete jaws. *J. mar. biol. Ass. U.K.*, **60**, 205–214.

GIBBS, P. E., BRYAN, G. W., and RYAN, K. P. (1981). Copper accumulation by the polychaete *Melinna palmata* : an antipredation mechanism? *J. mar. biol. Ass. U.K.*, **61**, 707–722.

GILLAIN, G., DUYCKAERTS, G., and DISTECHE, A. (1979). Direct and simultaneous determinations of Zn, Cd, Pb, Cu, Sb and Bi dissolved in sea water by differential pulse anodic stripping voltammetry with a hanging mercury drop electrode. *Analytica Chim. Acta,* **106**, 23–37.

GILLESPIE, P. A. and VACCARO, R. F. (1978). A bacterial bioassay for measuring the copper-chelation capacity of seawater. *Limnol. Oceanogr.,* **23**, 543–548.

GILMARTIN, M. and REVELANTE, N. (1975). The concentration of mercury, copper, nickel, silver, cadmium, and lead in the northern Adriatic anchovy, *Engraulis encrasicholus,* and sardine, *Sardina pilchardus. Fish. Bull. natn oceanic atmos. Adm. U.S.,* **73**, 193–201.

GNASSIA-BARELLI, M., ROMEO, M., LAUMOND, F., and PESANDO, D. (1978). Experimental studies on the relationship between natural copper complexes and their toxicity to phytoplankton. *Mar. Biol.,* **47**, 15–19.

GOLDBERG, E. D., BOWEN, V. T., FARRINGTON, J. W., HARVEY, G., MARTIN, J. H., PARKER, P. L., RISEBOROUGH, R. W., ROBERTSON, W., SCHNEIDER, E., and GAMBLE, E. (1978a). Mussel watch. *Environ, Conserv.,* **5**, 101–125.

GOLDBERG, E. D., HODGE, V., KOIDE, M., GRIFFIN, J., GAMBLE, E., BRICKER, O. P., MATISOFF, G., HOLDREN, G. R., and BRAUN, R. (1978b). A pollution history of Chesapeake Bay. *Geochim. Cosmochim. Acta,* **42**, 1413–1425.

GORDON, M., KNAUER, G. A., and MARTIN, J. H. (1980). *Mytilus californianus* as a bioindicator of trace metal pollution: variability and statistical considerations. *Mar. Pollut. Bull., N.S.,* **11**, 195–198.

GOULD, E. (1977). Alteration of enzymes in winter flounder, *Pseudopleuronectes americanus,* exposed to sublethal amounts of cadmium chloride. In F. J. Vernberg, A. Calabrese, F. B. Thurberg, and W. B. Vernberg (Eds), *Physiological Response of Marine Biota to Pollutants.* Academic Press, New York. pp. 209-224.

GOULD, E. (1980). Low-salinity stress in the American lobster *Homarus americanus,* after chronic sublethal exposure to cadmium: biochemical effects. *Helgoländer Meeresunters,* **33**, 36–46.

GRAY, J. S. and VENTILLA, R. J. (1973). Growth rates of sediment living marine protozoa as a toxicity indicator for heavy metals. *Ambio,* **2**, 118–121.

GREEN, F. A., Jr, ANDERSON, J. W., PETROCELLI, S. R., PRESLEY, B. J., and SIMS, R. (1976). Effect of mercury on the survival, respiration and growth of postlarval white shrimp, *Penaeus setiferus. Mar. Biol.,* **37**, 75–81.

GREIG, R. A. and MCGRATH, R. A. (1977). Trace metals in sediments of Raritan Bay. *Mar. Pollut. Bull., N.S.,* **8**, 188–192.

GREIG, R. A., NELSON, B. A., and NELSON, D. A. (1975). Trace metal content in the American oyster. *Mar. Pollut. Bull., N.S.,* **6**, 72–73.

GREIG, R. A., REIDS, R. N., and WENZLOFF, D. R. (1977a). Trace metal concentrations in sediments from Long Island Sound. *Mar. Pollut. Bull., N.S.,* **8**, 183–188.

GREIG, R. A. and WENZLOFF, D. R. (1977). Trace metals in finfish from the New York Bight and Long Island Sound. *Mar. Pollut. Bull., N.S.,* **8,** 198-200.

GREIG, R. A., WENZLOFF, D. R., and SHELPUK, C. (1975). Mercury concentrations in fish, North Atlantic offshore waters—1971. *Pestic. Monit. J.,* **9**, 15–20.

GREIG, R. A., WENZLOFF, D. R., SHELPUK, C., and ADAMS, A. (1977b). Mercury concentrations in three species of fish from North Atlantic offshore waters. *Archs environ. Contam. Toxicol.,* **5**, 315–323.

GREIG, R. A., WENZLOFF, D. R., ADAMS, A., NELSON, B., and SHELPUK, C. (1977c). Trace metals in organisms from ocean disposal sites of the Middle Eastern United States. *Archs environ. Contam. Toxicol.,* **6**, 395–409.

GRICE, G. D. and MENZEL, D. W. (1978). Controlled ecosystem pollution experiment : effect of mercury on enclosed water columns. VIII. Summary of results. *Mar. Sci. Communs,* **4**, 23–31.

GRIMANIS, A. P., ZAFIROPOULOS, D., and VASSILAKI-GRIMANI, M. (1978). Trace elements in the flesh and liver of two fish species from polluted and unpolluted areas of the Aegean Sea. *Environ. Sci. Technol.,* **12**, 723–726.

GROLLE, T. and KUIPER, J (1980). Development of marine periphyton under mercury stress in a controlled ecosystem experiment. *Bull. environ. Contam. Toxicol.,* **24**, 858–865.

GUTKNECHT, J. (1961). Mechanism of radioactive zinc uptake in *Ulva lactuca. Limnol Oceanogr.,* **6**, 426–431.

GUTKNECHT, J. (1963). ^{65}Zn uptake by benthic marine algae. *Limnol. Oceanogr.*, **8**, 31–38.

GUTKNECHT, J. (1965). Uptake and retention of cesium 137 and zinc 65 by seaweeds. *Limnol. Oceanogr.*, **10**, 58–66.

HALCROW, W., MACKAY, D. W., and THORNTON, I. (1973). The distribution of trace metals and fauna in the Firth of Clyde in relation to the disposal of sewage sludge. *J. mar. biol. Ass. U.K.*, **53**, 721–739.

HALL, A. (1980). Heavy metal co-tolerance in a copper-tolerant population of the marine fouling alga, *Ectocarpus siliculosus* (Dillw.) Lyngbye. *New Phytol.*, **85**, 73–78.

HALL, A., FIELDING, A. H., and BUTLER, M. (1979). Mechanisms of copper tolerance in the marine fouling alga *Ectocarpus siliculosus*—evidence for an exclusion mechanism. *Mar. Biol.*, **54**, 195–199.

HALL, A. S., TEENY, F. M., and GAUGLITZ, E. J., Jr. (1976a). Mercury in fish and shellfish of the northeast Pacific. II. sablefish, *Anoplopoma fimbria. Fish. Bull. natn oceanic atmos. Adm. U.S.*, **74**, 791–797.

HALL, A. S., TEENY, F. M., LEWIS, L. G., HARDMAN, W. H., and GAUGLITZ, E. J., Jr. (1976b). Mercury in fish and shellfish of the northeast Pacific. I. Pacific halibut, *Hippoglossus stenolepis. Fish. Bull. natn oceanic atmos. Adm. U.S.*, **74**, 783–789.

HALLBERG, R. O. (1979). Heavy metals in the sediments of the Gulf of Bothnia. *Ambio*, **8**, 265–269.

HAMANAKA, T., KATO, H., and TSUJITA, T. (1977). Cadmium and zinc in ribbon seal, *Histriophoca fasciata*, in the Okhotsk Sea. *Spec. Vol. Res. Inst. N. Pacific Fish.*, 547-561.

HAMILTON, E. I. (1979). *The Chemical Elements and Man*, Charles C. Thomas, Springfield, Illinois.

HAMILTON, E. I., WATSON, P. G., CLEARY, J. J., and CLIFTON, R. J. (1979). The geochemistry of recent sediments of the Bristol Channel–Severn Estuary System. *Mar. Geol.*, **31**, 139–182.

HARDISTY, M. W., KARTAR, S., and SAINSBURY, M. (1974). Dietary habits and heavy metal concentrations in fish from the Severn Estuary and Bristol Channel. *Mar. Pollut. Bull., N.S.*, **5**, 61–63.

HARRIS, J. E., FABRIS, G. J., STATHAM, P. J., and TAWFIK, F. (1979). Biogeochemistry of selected heavy metals in Western Port, Victoria, and use of invertebrates as indicators with emphasis on *Mytilus edulis planulatus. Aust. J. mar. Freshwat. Res.*, **30**, 159–178.

HARRISON, R. M. and LAXEN, D. P. H. (1978). Natural source of tetraethyl lead in air. *Nature, Lond.*, **275**, 738–740.

HARVEY, E. J. and KNIGHT, L. A., Jr. (1978). Concentration of three toxic metals in oysters (*Crassostrea virginica*) of Biloxi and Pascagoula, Mississippi estuaries. *Wat. Air Soil Pollut.*, **9**, 255–261.

HAUG, A. (1961). The affinity of divalent metals to different types of alginate. *Acta chem. scand.*, **15**, 1794–1795.

HAUG, A., MELSOM, S., and OMANG, S. (1974). Estimation of heavy metal pollution in two Norwegian fjord areas by analysis of the brown alga *Ascophyllum nodosum. Environ. Pollut.*, **7**, 179–192.

HAXTON, J., LINDSAY, D. G., HISLOP, J. S., SALMON, L., DIXON, E. J., EVANS, W. H., REID, J. R., HEWETT, C. J., and JEFFERIES, D. F. (1979). Duplicate diet study on fishing communities in the United Kingdom: mercury exposure in a 'critical group'. *Environ. Res.*, **18**, 351–368.

HEAD, P. C., D'ARCY, B. J. and OSBALDESTON, P. J. (1980). The Mersey estuary bird mortality, autumn–winter 1979 summary report. *North West Water Scientific Report*, **DSS-EST-80-2.**

HEINZ, G. (1974). Effects of low dietary levels of methyl mercury on Mallard reproduction. *Bull. environ. contam. Toxicol.*, **11**, 386–392.

HEIT, M. (1979). Variability of the concentrations of seventeen trace elements in the muscle and liver of a single striped bass, *Morone saxatilis. Bull. environ. Contam. Toxicol.*, **23**, 1–5.

HELLSTRÖM, T. (1979). *Flow in the Sea of Heavy Metals from a Smelter Industry*, Swedish Water and Air Pollution Research Institute, Stockholm.

HELZ, G. R. (1976). Trace element inventory for the northern Chesapeake Bay with emphasis on the influence of man. *Geochim. Cosmochim. Acta*, **40**, 573–580.

HERRON, M. M., LANGWAY, C. C., WEISS, H. V., and CRAGIN, J. H. (1977). Atmospheric trace metals and sulphate in the Greenland ice sheet. *Geochim. Cosmochim. Acta*, **41**, 915–920.

HESLINGA, G. A. (1976). Effects of copper on the coral-reef echinoid *Echinometra mathaei. Mar. Biol.*, **35**, 155–160.

HIGNETTE, M. (1979). Composition des concrétions minerales contenues dans les reins de 2 mollusques Lamellibranches : *Pinna nobilis* L. et *Tridacna maxima* (Röding). *C.r. hebd. Séanc. Acad. Sci., Paris, D,* **289**, 1069–1072.

HIROTA, R., FUJIKI, M., IKEGAKI, Y., and TAJIMA, S. (1979). Management of bottom sediments containing toxic substances. *Proceedings of the 4th U.S.—Japan Experts Meeting, October 1978, Tokyo.* U.S. Environmental Protection Agency, EPA-600/3-79-102.

HODGE, V., JOHNSON, S. R., and GOLDBERG, E. D. (1978). Influence of atmospherically transported aerosols on surface ocean water composition. *Geochem. J.,* **12**, 7–20.

HODGE, V., SEIDEL, S. L., and GOLDBERG, E. D. (1979). Determination of tin(IV) and organotin compounds in natural waters, coastal sediments and macro algae by atomic absorption spectrometry. *Analyt. Chem.,* **51**, 1256–1259.

HODSON, P. V., BLUNT, B. R., SPRY, D. J., and ANSTEN, K. (1977). Evaluation of erythrocyte delta-amino levulinic acid dehydratase as a short-term indicator in fish of a harmful exposure to lead. *J. Fish. Res. Bd Can.,* **34**, 501–508.

HOLDGATE, M. W. (1979). *A Perspective of Environmental Pollution,* Cambridge University Press.

HOLIDAY, F. G. T. (1971). Salinity: animals. Fishes. In O. Kinne (ED.), *Marine Ecology,* Vol. I, Environmental Factors, Part 2. Wiley, London. pp. 997-1083.

HOLLIBAUGH, J. T., SEIBERT, D. L. R., and THOMAS, W. H. (1980). A comparison of the acute toxicities of 10 heavy metals to phytoplankton from Saanich Inlet, B.C., Canada. *Estuar. coast. mar. Sci,* **10**, 93–105.

HOLLIDAY, L. M. and LISS, P. S. (1976). The behaviour of dissolved iron, manganese and zinc in the Beaulieu Estuary, S. England. *Estuar. coast. mar. Sci.,* **4**, 349–353.

HOLMES, C. W., SLADE, E. A., and MCLERRAN, C. J. (1974). Migration and redistribution of zinc and cadmium in a marine estuarine system. *Environ. Sci. Technol.,* **8**, 255–259.

HOPKIN, A. P. and NOTT, J. A. (1979). Some observations on concentrically structured, intracellular granules in the hepatopancreas of the shore crab *Carcinus maenas* (L.). *J. mar. biol. Ass. U.K.,* **59**, 867–877.

HOPPENHEIT, M. (1977). On the dynamics of exploited populations of *Tisbe holothuriae* (Copepoda, Harpacticoida). V. The toxicity of cadmium : response to sub-lethal exposure. *Helgoländer wiss. Meeresunters,* **29**, 503–523.

HOSOHARA, K., VEZUMA, H., KAWASAKI, K., and TSURUTA, T. (1961). Studies on the total amount of mercury in sea waters. *J. chem. Soc. Japan,* **81**, 1479–1480.

HOWARD, A. G. and NICKLESS, G. (1977). Heavy metals complexation in polluted molluscs. I. Limpets (*Patella vulgata* and *Patella intermedia*). *Chemico-Biol. Interactions,* **16**, 107–114.

HRS-BRENKO, H., CLAUS, C., and BUBIC, S. (1977). Synergistic effects of lead, salinity and temperature on embryonic development of the mussel *Mytilus galloprovincialis. Mar. Biol.,* **44**, 109–115.

HUCKABEE, J. W., JANZEN, S. A., BLAYLOCK, B. G., TALMI, Y., and BEAUCHAMP, J. J. (1978). Methylated mercury in Brook Trout (*Salvelinus fontinalis*) : absence of an *in vivo* methylating process. *Trans. Am. Fish. Soc.,* **107**, 848–852.

HYDES, D. J. (1979). Aluminum in seawater: control by inorganic processes. *Science, N.Y.,* **205**, 1260–1262.

ICELY, J. D. and NOTT, J. A. (1980). Accumulation of copper within the 'Hepatopancreatic' caeca of *Corophium volutator* (Crustacea: Amphipoda). *Mar. Biol.,* **57**, 193–199.

ICES (INTERNATIONAL COUNCIL FOR THE EXPLORATION OF THE SEA) (1977a). Studies of the pollution of the Baltic Sea by the ICES/SCOR Working Group on the study of pollution of the Baltic. *Cooperative Research Report,* No. 63, 1-97.

ICES (INTERNATIONAL COUNCIL FOR THE EXPLORATION OF THE SEA) (1977b). The ICES coordinated monitoring programmes 1975 and 1976. *Cooperative Research Report,* No. 72, 1–26.

ICES (INTERNATIONAL COUNCIL FOR THE EXPLORATION OF THE SEA) (1977c). A baseline study of the level of contaminating substances in living resources of the North Atlantic. *Cooperative Research Report,* No. 69, 1-82.

ICES (INTERNATIONAL COUNCIL FOR THE EXPLORATION OF THE SEA) (1978a). Input of pollutants to the Oslo Commission area. *Cooperative Research Report,* No. 77, 1–57.

ICES (INTERNATIONAL COUNCIL FOR THE EXPLORATION OF THE SEA) (1978b). On the feasibility of effects monitoring. *Cooperative Research Report,* No. 75, 1–42.

ICES (INTERNATIONAL COUNCIL FOR THE EXPLORATION OF THE SEA) (1978c). Report on intercalibration analyses in ICES North Sea and North Atlantic baseline studies. *Cooperative Research Report,* No. 80, 1-52.

ICES (INTERNATIONAL COUNCIL FOR THE EXPLORATION OF THE SEA) (1980). Extension to the baseline study of contaminant levels in living resources of the North Atlantic. *Cooperative Research Report,* No. 95, 1-57.

IRELAND, M. P. (1979). Distribution of metals in the digestive gland–gonad complex of the marine gastropod *Nucella lapillus. J. mollusc. Stud.,* **45**, 322–327.

ISHIKAWA, T. and IKEGAKI, Y. (1980). Control of mercury pollution in Japan and the Minamata Bay clean-up. *J. Wat. Pollut. Control Fed.,* **52**, 1013–1018.

JACKIM, E., MORRISON, G., and STEELE, R. (1977). Effects of environmental factors on radiocadmium uptake by four species of marine bivalves. *Mar. Biol.,* **40**, 303–308.

JACKSON, G. A. and MORGAN, J. J. (1978). Trace metal–chelator interactions and phytoplankton growth in seawater media : theoretical analysis and comparison with reported observations. *Limnol. Oceanogr.,* **23**, 268–282.

JAFFE, D. and WALTERS, J. K. (1977). Intertidal trace metal concentrations in some sediments from the Humber Estuary. *Sci. Total Environ.,* **7**, 1–15.

JANSSEN, H. H. and SCHOLZ, N. (1979). Uptake and cellular distribution of cadmium in *Mytilus edulis. Mar. Biol.,* **55**, 133–141.

JENNINGS, J. R., RAINBOW, P. S., and SCOTT, A. G. (1979). Studies on the uptake of cadmium by the crab *Carcinus maenas* in the laboratory. II. Preliminary investigation of cadmium-binding proteins. *Mar. Biol.,* **50**, 141–149.

JENSEN, A. and RYSTAD, B., and MELSOM, B. (1976). Heavy metal tolerance of marine phytoplankton. 2. Copper tolerance of three species in dialysis and batch cultures. *J. exp. mar. Biol. Ecol.,* **22**, 249–256.

JONES, G. E., ROYLE, L. G., and MURRAY, L. (1976a). The assimilation of trace metal ions by the marine bacteria, *Arthrobacter marinus* and *Pseudomonas cuprodurans*. In J. M. Sharpley and A. M. Kaplan (Eds), *Proceedings of the Third International Biodegradation Symposium*. Applied Science Publishers, Barking, England. pp. 889–898.

JONES, L. H., JONES, N. V., and RADLETT, A. J. (1976b). Some effects of salinity on the toxicity of copper to the polychaete *Nereis diversicolor. Estuar. coast. mar. Sci.,* **4**, 107–111.

JØRSTAD, K. and SALBU, B. (1980).Determination of trace elements in seawater by neutron activation analysis and electrochemical separation. *Analyt. Chem.,* **52**, 672–676.

JUDA, L. (1979). The regional effort to control pollution in the Mediterranean Sea. *Ocean Mgmt,* **5**, 125–150.

KARBE, L. (1972). Marine Hydroiden als Test organismen zur Prüfung der Toxizität von Abwasserstoffen. Die Wirkung von Schwermetallen auf Kolonien von *Eirene viridula. Mar. Biol.,* **12**, 316–328.

KARBE, L., SCHNIER, C., and NIEDERGESÄSS, A. (1978). Trace elements in mussels (*Mytilus edulis*) from German coastal waters. Evaluation of multi-element patterns with respect to their use for monitoring programmes. *Gesellschaft für Kernenergieverwertung in Schiffbau und Schiffahrt mbH,* **GKSS 78/E/51**.

KARI, T. and KAURANEN, P. (1978). Mercury and selenium contents of seals from fresh and brackish waters in Finland. *Bull. environ. Contam. Toxicol.,* **19**, 273–280.

KAYSER, H. (1977). Effect of zinc sulphate on the growth of mono- and multispecies cultures of some marine plankton algae. *Helgoländer Meersunters.,* **30**, 682–696.

KAYSER, H. and SPERLING, K.-R. (1980). Cadmium effects and accumulation in cultures of *Prorocentrum micans* (Dinophyta). *Helgoländer Meeresunters.,* **33**, 89–102.

KEITH, L. H. and TELLIARD, W. A. (1979). Priority pollutants. I. A perspective view. *Environ. Sci. Technol.,* **13**, 416–423.

KINGSTON, H. M., BARNES, I. L., RAINS, T. C., and CHAMP, M. A. (1978). Separation of eight transition elements from alkali and alkaline earth elements in estuarine and seawater with chelating resin and their determination by graphite furnace atomic absorption spectrometry. *Analyt. Chem.,* **50**, 2064–2070.

KINNE, O. (1975). Orientation in space: animals. Mammals. In O. Kinne (Ed.), *Marine Ecology,* Vol. II, Physiological Mechanisms, Part 2. Wiley, London. pp. 709-916.

KINNE, O. (1977). Cultivation of animals. Research cultivation. In O. Kinne (Ed.), *Marine Ecology,* Vol. III, Cultivation, Part 2. Wiley, Chichester. pp. 936–967.

KLÖCKNER, K. (1979). Uptake and accumulation of cadmium by *Ophryotrocha diadema* (Polychaeta). *Mar. Ecol. Prog. Ser.*, **1**, 71–76.

KLUMPP, D. W. (1980a). Characteristics of arsenic accumulation by seaweeds *Fucus spiralis* and *Ascophyllum nodosum*. *Mar. Biol.*, **58**, 257–264.

KLUMPP, D. W. (1980b). Accumulation of arsenic from water and food by *Littorina littoralis* and *Nucella lapillus*. *Mar. Biol.*, **58**, 265–274.

KLUMPP, D. W. and PETERSON, P. J. (1979). Arsenic and other trace elements in the waters and organisms of an estuary in SW England. *Environ. Pollut.* **19**, 11–20.

KNAUER, G. A. and MARTIN, J. H. (1972). Mercury in a marine pelagic food chain. *Limnol. Oceanogr.*, **17**, 868–876.

KOBAYASHI, N. (1977). Preliminary experiments with sea urchin pluteus and metamorphosis in marine pollution bioassay. *Publs Seto mar. biol. Lab.*, **24**, 9–21.

KOEMAN, J. H., PEETERS, W. H. M., KOUDSTAAL-HOL, C. H. M., TJIOE, P. S., and DE GOEIJ, J. J. M. (1973). Mercury–selenium correlations in marine mammals. *Nature, Lond.*, **245**, 385–386.

KOEMAN, J. H., VEN, W. S. M. VAN DE, GOEIJ, J. J. M. DE, TJIOE, P. S., and HAAFTEN, J. L. VAN. (1975). Mercury and selenium in marine mammals and birds. *Sci. Total Environ.*, **3**, 279–287.

KOSTA, L., RAVNIK, V., BYRNE, A. R., ŠTIRN, J., DERMELJ, M., and STEGNAR, P. (1978). Some trace elements in the waters, marine organisms and sediments of the Adriatic by neutron activation analysis. *J. radioanalyt. Chem.*, **44**, 317–332.

KRAJNOVIC-OZRETIC, M. and OZRETIC, B. (1980). The ALA-D activity test in lead-exposed grey mullet *Mugil auratus*. *Mar. Ecol. Prog. Ser.*, **3**, 187–191.

KREMLING, K., PIUZE, J., BRÖCKEL, K. VON, and WONG, C. S. (1978). Studies on the pathways and effects of cadmium in controlled ecosysystem enclosures. *Mar. Biol.*, **48**, 1–10.

KROM, M. D. and SHOLKOVITZ, E. R. (1978). On the association of iron and manganese with organic matter in anoxic marine pore waters. *Geochim. Cosmochim, Acta*, **42**, 607–611.

KUIPER, J. (1977). An experimental approach in studying the influence of mercury on a North Sea coastal plankton community. *Helgoländer. Meersunters.*, **30**, 652–665.

KUMAGAI, M. and NISHIMURA, H. (1978). Mercury distribution in seawater in Minimata Bay and the origin of particulate mercury. *J. oceanogr. Soc. Japan*, **34**, 50–56.

LANDE, E. (1977). Heavy metal pollution in Trondheimsfjorden, Norway, and the recorded effects on the fauna and flora. *Environ. Pollut.*, **12**, 187–198.

LANG, W. H., FORWARD, R. B. Jr, MILLER, D. C. and MARCY, M. (1980). Acute toxicity and sublethal behavioural effects of copper on barnacle nauplii (*Balanus improvisus*). *Mar. Biol.*, **58**, 139–145.

LANGSTON, W. J. (1980). Arsenic in U.K. estuarine sediments and its availability to benthic organisms. *J. mar. biol. Ass. U.K.*, **60**, (4), 869–881.

LANTZY, R. J. and MACKENZIE, F. T. (1979). Atmospheric trace metals: global cycles and assessment of man's impact. *Geochim. Cosmochim. Acta*, **43**, 511–525.

LEATHERLAND, T. M. and BURTON, J. D. (1974). The occurrence of some trace metals in coastal organisms with particular reference to the Solent region. *J. mar. biol. Ass. U.K.*, **54**, 457–468.

LEE, C. R., SMART, R. M., STURGIS, T. C., GORDON, R. N., and LANDIN, M. C. (1978). *Prediction of Heavy Metal Uptake by Marsh Plants Based on Chemical Extraction of Heavy Metals From Dredged Material*, US Army Engineer Waterways Experiment Station, Vicksburg (Tech. Rep. D-78-6).

LEE, H. and SWARTZ, R. C. (1980). Biological processes affecting the distribution of pollutants in marine sediments. Part II. Biodeposition and bioturbation. In R. A. Baker (Ed.), *Contaminants and Sediments*, Vol. 2. Ann Arbor Science, Ann Arbor, Mich. pp. 555-606.

LEE, S.S., MATE, B. R., VON DER TRENCK, K. T., RIMERMAN, R. A., and BUHLER, D. R. (1977). Metallothionein and the subcellular localization of mercury and cadmium in the California sea lion. *Comp. Biochem. Physiol.*, **57**(C), 45–53.

LEHNBERG, W. and THEEDE, H. (1979). Kombinierte Wirkungen von Temperatur, Salzgehalt und Cadmium auf Entwicklung, Wachstum and Mortalität der Larven von *Mytilus edulis* aus der westlichen Ostsee. *Helgoländer Meeresunters.*, **32**, 179–199.

LEWIS, A. G., WHITFIELD, P. H., and RAMNARINE, A. (1972). Some particulate and soluble agents affecting the relationship between metal toxicity and organism survival in the calanoid copepod *Euchaeta japonica*. *Mar. Biol.*, **17**, 215–221.

LI, W. K. W. (1979). Cellular composition and physiological characteristics of the diatom *Thalassiosira weissflogii* adapted to cadmium stress. *Mar. Biol.*, **55**, 171–180.

LINDBERG, S. E. and HARRISS, R. C. (1974). Mercury–organic matter associations in estuarine sediments and interstitial water. *Environ. Sci. Technol.*, **8**, 459–462.

LORING, D. H. and BEWERS, J. M. (1978). Geochemical mass balances for mercury in a Canadian fjord. *Chem. Geol.*, **22**, 309–330.

LOWE, D. M. and MOORE, M. N. (1979a). The cytochemical distribution of zinc (Zn II) and iron (Fe III) in the common mussel, *Mytilus edulis*, and their relationship with lysosomes. *J. mar. biol. U.K.*, **59**, 851–858.

LOWE, D. M. and MOORE, M. N. (1979b). The cytology and occurrence of granulocytomas in mussels. *Mar. Pollut. Bull., N.S.*, **10**, 137–141.

LU, J. C. S. and CHEN, K. Y. (1977). Migration of trace metals in the interfaces of seawater and polluted surficial sediments. *Environ. Sci. Technol.*, **11**, 174–182.

LUOMA, S. N. (1977a). The dynamics of biologically available mercury in a small estuary. *Estuar. coast. mar. Sci.* **5**, 643–652.

LUOMA, S. N. (1977b). Detection of trace contaminant effects in aquatic ecosystems. *J. Fish. Res. Bd Can.*, **34**, 436–439.

LUOMA, S. N. and BRYAN, G. W. (1978). Factors controlling the availability of sediment-bound lead to the estuarine bivalve *Scrobicularia plana*. *J. mar. biol. Ass. U.K.*, **58**, 793–802.

LUOMA, S. N. and BRYAN, G. W. (1979). Trace metal bioavailability : modeling chemical and biological interactions of sediment-bound zinc. In E. A. Jenne (Ed.), *Chemical Modeling in Aqueous Systems*. American Chemical Society, Washington, D.C. pp. 577-609.

LUOMA, S. N. and BRYAN, G. W. (1981). A statistical assessment of the form of trace metals in oxidized surface sediments employing chemical extractants. *Sci. Total Environ.*, **17**, 165–196.

LUOMA, S. N. and BRYAN, G. W. (1982). A statistical assessment of environmental factors controlling the concentrations of heavy metals in the burrowing bivalve *Scrobicularia plana* and the polychaete *Nereis diversicolor*. *Estuar. coast. Shelf Sci.*, **15**, 95–108.

LUOMA, S. N. and CAIN, D. J. (1979). Fluctuations of copper, zinc and silver in tellenid clams as related to freshwater discharge—South San Francisco Bay. In: *San Fransisco Bay—the Urbanized Estuary*. California Academy of Sciences, San Francisco. pp. 231-246.

LUOMA, S. N. and JENNE, E. A. (1976). Estimating bioavailability of sediment-bound trace metals with chemical extractants. In D. D. Hemphill (Ed.), *Trace Substances in Environmental Health*. University of Missouri, Columbia. pp. 343-351.

LUOMA, S. N. and JENNE, E. A. (1977). The availability of sediment-bound cobalt, silver and zinc to a deposit-feeding clam. In H. Drucker and R. E. Wildung (Eds), *Biological Implications of Metals in the Environment*, U.S. N.T.I.S., Conf - 750929. pp. 213-231.

MACARA, I. G., MCCLEOD, G. C., and KUSTIN, K. (1979). Tunichromes and metal ion accumulation in tunicate blood cells. *Comp. Biochem. Physiol.*, **63B**, 299–302.

MCGREER, E. R. (1979). Sublethal effects of heavy metal contaminated sediments on the bivalve *Macoma balthica*. *Mar. Pollut. Bull., N.S.*, **10**, 259–262.

MACINNES, J. R. and CALABRESE, A. (1978). Response of embryos of the American oyster, *Crassostrea virginica*, to heavy metals at different temperatures. In D. S. McLusky and A. J. Berry (Eds), *Physiology and Behaviour of Marine Organisms*. Pergamon Press, Oxford and New York. pp. 195-202.

MACINNES, J. R. and CALABRESE, A. (1979). Combined effects of salinity, temperature and copper on embryos and early larvae of the American oyster, *Crassostrea virginica*. *Archs environ. Contam. Toxicol.*, **8**, 553–562.

MACINNES, J. R., THURBERG, F. P., GREIG, R. A., and GOULD, E. (1977). Long-term cadmium stress in the cunner, *Tautogolabrus adspersus*. *Fishery Bull. natn oceanic atmos. Adm. U.S.*, **75**, 199–203.

MACKAY, D. W., HALCROW, W., and THORNTON, I. (1972). Sludge dumping in the Firth of Clyde. *Mar. Pollut. Bull., N.S.*, **3**, 7–11.

MACKAY, N. J., KAZACOS, M. N., WILLIAMS, R. J., and LEEDOW, M. I. (1975). Selenium and heavy metals in black marlin. *Mar. Pollut. Bull., N.S.*, **6**, 57–60.

MCKENNEY, C. L. and NEFF, J. M. (1979). Individual effects and interactions of salinity, temperature and zinc on larval development of the grass shrimp *Palaemonetes pugio*. 1. Survival and developmental duration through metamorphosis. *Mar. Biol.*, **52**, 177–188.

McLean, M. W. and Williamson, F. B. (1977). Cadmium accumulation by the marine red alga *Porphyra umbilicalis. Physiologia Pl.*, **41**, 268–272.
McLusky, D. S. and Phillips, C. N. K. (1975). Some effects of copper on the polychaete *Phyllodoce maculata. Estuar. coast. mar. Sci.*, **3**, 103–108.
Magnusson, B. and Westerlund, S. (1980). The determination of Cd, Cu, Fe, Ni, Pb and Zn in Baltic Sea water. *Mar. Chem.*, **8**, 231–244.
Majori, L., Nedoclan, G., Modonutti, G. B., and Daris, F. (1978a). Methodological researches on the phenomenon of metal accumulation in the *Mytilus galloprovincialis* and on the possibility of using biological indicators as test-organisms of marine metal pollution. *Revue int. Océanogr. méd.*, **49**, 81–87.
Majori, L., Nedoclan, G., Modonutti, G. B., and Daris, F. (1978b). Study of the seasonal variations of some trace elements in the tissues of *Mytilus galloprovincialis* taken in the Gulf of Trieste. *Revue int. Océanogr. méd.*, **49**, 37–40.
Majori, L., Nedoclan, G., Modonutti, G. B., and Daris, F. (1978c). Metal content in some species of fish in the northern Adriatic Sea. Comparison of 2 sample areas. *Revue int. Océanogr. méd.*, **49**, 41–43.
Manley, A. R. and Davenport, J. (1979). Behavioural responses of some marine bivalves to heightened seawater copper concentrations. *Bull. environ. Contam. Toxicol.*, **22**, 739–744.
Mantuora, R. F. C., Dickson, A., and Riley, J. P. (1978). The complexation of metals with humic materials in natural waters. *Estuar. coast. mar. Sci.*, **6**, 387–408.
Marchetti, R. (1978). Acute toxicity of alkyl leads to some marine organisms. *Mar. Pollut. Bull., N.S.*, **9**, 206–207.
Markham, J. W., Kremer, B. P., and Sperling, K.-R. (1980). Effects of cadmium on *Laminaria saccharina* in culture. *Mar. Ecol. Prog. Ser.*, **3**, 31–39.
Martin, J. H. (1979). *Bioaccumulation of Heavy Metals by Littoral and Pelagic Marine Organisms,* U.S. Environmental Protection Agency, Environmental Research Laboratory, Narragansett (EPA-600/3-79-038).
Martin, J. H., Elliott, P. D., Anderlini, V. C., Girvin, D., Jacobs, S. A., Risebrough, R. W., DeLong, R. L., and Gilmartin, W. G. (1976). Mercury–selenium–bromine imbalance in premature parturient California sea lions. *Mar. Biol.*, **35**, 91–104.
Martin, J. H. and Flegal, A. R. (1975). High copper concentrations in squid livers in association with elevated levels of silver, cadmium and zinc. *Mar. Biol.*, **30**, 51–55.
Martin, J. L. M. (1975). Le cuivre et le zinc chez *Cancer irroratus* (Crustace: Decapode): metabolisme compare au cours du cycle d'intermue. *Comp. Biochem. Physiol.*, **51A**, 777–784.
Martin, J. L. M. (1977). Etude de la contamination de *Carcinus maenas* (L.) par le fer 59. *Revue int. Océanogr. méd.*, **45-46,** 3-16.
Martin, J. L. M. (1979). Schema of lethal action of copper on mussels. *Bull. environ. Contam. Toxicol.*, **21**, 808–814.
Martin, M., Stephenson, M. D., and Martin, J. H. (1977). Copper toxicity experiments in relation to abalone deaths observed in a power plant's cooling waters. *Calif. Fish Game*, **63**, 95–100.
Martoja, M., Tue, V. T., and Elkaim, B. (1980). Bioaccumulation du cuivre chez *Littorina littorea* (L.) (Gastérpode prosobranche): signification physiologique et écologique. *J. exp. mar. Biol. Ecol.*, **43**, 251–270.
Martoja, R. and Viale, D. (1977). Accumulation de granules de séléniure mercurique dans le foie d'Odontocètes (Mammiféres, Cétacés): un mécanisme possible de détoxication du méthyl-mercure par le sélénium. *C. R. hebd. Séanc. Acad. Sci. Paris (Séries D)*, **285**, 109–112.
Mason, J. W., Cho, J. H., and Anderson, A. C. (1976). Uptake and loss of inorganic mercury in the eastern oyster (*Crassostrea virginica*). *Archs environ. Contam. Toxicol.*, **4**, 361–376.
Matida, Y. and Kumada, H. (1969). Distribution of mercury in water, bottom mud and aquatic organisms of Minamata Bay, the River Agano and other water bodies in Japan. *Bull. Freshwat. Fish. Res. Lab.* Tokyo, **19**, 73–93.
Medeiros, D. M., Cacwell, L. L., and Preston, R. L. (1980). A possible physiological uptake mechanism of methylmercury by the marine bloodworm (*Glycera dibranchiata*). *Bull. environ. Contam. Toxicol.*, **24**, 97–101.
Melhuus, A., Seip, K. L., Seip, H. H., and Myklestad, S. (1978). A preliminary study of the

use of benthic algae as biological indicators of heavy metal pollution in Sørfjorden, Norway. *Environ. Pollut.*, **15**, 101–107.

MENASVETA, P. and SIRIYONG, R. (1977). Mercury content of several predacious fish in the Andaman Sea. *Mar. Pollut. Bull., N.S.*, **8**, 200–204.

MEREFIELD, J. R. (1976). Barium build-up in the Teign Estuary. *Mar. Pollut. Bull., N.S.*, **7**, 214–216.

METTAM, C. (1979). Faunal changes in the Severn Estuary over several decades. *Mar. Pollut. Bull., N.S.*, **10**, 133–136.

MIDDAUGH, D. P. and DEAN, J. M. (1977). Comparative sensitivity of eggs, larvae and adults of the estuarine teleosts, *Fundulus heteroclitus* and *Menidia menidia* to cadmium. *Bull. environ. Contam. Toxicol.*, **17**, 645–652.

MILANOVICH, F. P., SPIES, R., GURAM, M. S., and SYKES, E. E. (1976). Uptake of copper by the polychaete *Cirriformia spirabrancha* in the presence of dissolved yellow organic matter of natural origin. *Estuar. coast. mar. Sci.*, **4**, 585–588.

MILLWARD, G. E. and GRIFFIN, J. H. (1980). Concentrations of particulate mercury in the Atlantic marine atmosphere. *Sci. Total Environ.*, **16**, 239–248.

MILNER, N. J. (1979). Zinc concentrations in juvenile flatfish. *J. mar. biol. Ass. U.K.*, **59**, 761–775.

MINERALS YEAR BOOK (1970). *U.S. Bureau of Mines*, US Department of Interior, Washington, D.C.

MINISTRY OF AGRICULTURE, FISHERIES AND FOOD (MAFF) (1972). *Survey of Lead in Food. Working Party on the Monitoring of Foodstuffs for Heavy Metals* (2nd rep.), H.M. Stationery Office, London.

MINISTRY OF AGRICULTURE, FISHERIES AND FOOD (MAFF) (1973). *Survey of Cadmium in Food. Working Party on the Monitoring of Foodstuffs for Heavy Metals* (4th rep.), H.M. Stationery Office, London.

MIRAMAND, P. and GUARY, J. C. (1980). High concentration of some heavy metals in tissues of the Mediterranean octopus. *Bull. environ. Contam. Toxicol.*, **24**, 783–788.

MIRAMAND, P., GUARY, J. C., and FOWLER, S. W. (1980). Vanadium transfer in the mussel *Mytilus galloprovincialis*. *Mar. Biol.*, **56**, 281–293.

MIRKES, D. Z., VERNBERG, W. B., and DECOURSEY, P. J. (1978). Effects of cadmium and mercury on the behavioural responses and development of *Eurypanopeus depressus* larvae. *Mar. Biol.*, **47**, 143–147.

MOORE, M. N. and STEBBING, A. R. D. (1976). The quantitative cytochemical effects of three metal ions on a lysosomal hydrolase of a hydroid. *J. mar. biol. Ass. U.K.*, **56**, 995–1005.

MOORE, R. M., BURTON, J. D., WILLIAMS, P. J. LeB., and YOUNG, M. L. (1979). The behaviour of dissolved organic material, iron and manganese in estuarine mixing. *Geochim. Cosmochim. Acta*, **43**, 919–926.

MORAITOU-APOSTOLOPOULOU, M. and VERRIOPOULOS, G. (1979). Some effects of sub-lethal concentrations of copper on a marine copepod. *Mar. Pollut. Bull., N.S.*, **10**, 88–92.

MORI, T. and KURIHARA, Y. (1979). Accumulation of heavy metals in polychaetes *Perinereis nuntia* var. *Vallata* and *Neanthes japonica*. *Sci. Rep. Tohôku Univ. (d. Biol.)* S. 4, **37**, 299–303.

MORRIS, A. W. (1975). Dissolved molybdenum and vanadium in the north-east Atlantic Ocean. *Deep Sea Res.*, **22**, 49–54.

MORRIS, A. W. and BALE, A. J. (1975). The accumulation of cadmium, copper, manganese and zinc by *Fucus vesiculosus* in the Bristol Channel. *Estuar. coast. mar. Sci.*, **3**, 153–163.

MORRIS, A. W., MANTOURA, R. F. C., BALE, A. J., and HOWLAND, R. J. M. (1978). Very low salinity regions of estuaries : important sites for chemical and biological reactions. *Nature, Lond.*, **274**, 678–680.

MORRIS, O. P. and RUSSELL, G. (1973). Effect of chelation on the toxicity of copper. *Mar. Pollut. Bull., N.S.*, **4**, 159–160.

MOYNAHAN, E. J. (1979). Trace elements in man. *Phil. Trans. R. Soc. B*, **288**, 65–79.

MUKHERJI, P. and KESTER, D. R. (1979). Mercury distribution in the Gulf Stream. *Science, N.Y.*, **204**, 64-66.

MÜLLER, D. (1979). Subletale und letale schädigungen von vertretern der lebensgemeinschaft der auBenelewatten durch die schwermetalle kupfer, cadmium und blei. *Arch. Hydrobiol. (Suppl.)*, **43**, 289–346.

MÜLLER, G. and FÖRSTNER, U. (1975). Heavy metals in sediments of the Rhine and Elbe estuaries: mobilization or mixing effect? *Environ. Geol.*, **1**, 33–39.

MURRAY, A. J. (1979). Metals, organochlorine pesticides and PCB residue levels in fish and shellfish landed in England and Wales during 1974 (MAFF Direct. Fish. Res., Lowestoft). *Aquat. Environ. Monit. Rep.*, **2**, 1–11.

MURRAY, L. A. and NORTON, M. G. (1979). The composition of dredged spoils dumped at sea from England and Wales. *Fish. Res. Tech. Rep., Lowestoft,* **52**, 1–10.

MURRAY, L. A., NORTON, M. G., NUNNY, R. S., and ROLFE, M. S. (1980). The field assessment of the effects of dumping wastes at sea. 6. The disposal of sewage sludge and industrial waste off the River Humber (MAFF Direct. Fish. Res., Lowestoft). *Fish. Res. Tech. Rep.*, **55**, 1–35.

MYKLESTAD, S., EIDE, I., and MELSOM, S. (1978). Exchange of heavy metals in *Ascophyllum nodosum* (L.) Le Jol. *in situ* by means of transplanting experiments. *Environ. Pollut.*, **16**, 277–284.

NAGAKURA, K., ARIMA, S., KURIHARA, M., KOGA, T., and FUJITA, T. (1974). Mercury content of whales. *Bull. Tokai reg. Fish. Res. Lab.*, **78**, 41–46.

NAKAHARA, M., KOYANAGI, T., UEDA, T., and SHIMIZU, C. (1979). Peculiar accumulation of cobalt-60 by the branchial heart of octopus. *Bull. Jap. Soc. scient. Fish.*, **45**, 539.

NEELAKANTAN, B. (1976). Distribution of heavy metals in the northern shrimp *Pandalus borealis* from the Oslofjord. *Fish. Technol.*, **13**, 20–25.

NEFF, J. W., FOSTER, R. S., and SLOWEY, J. F. (1978). *Availability of Sediment-assorbed Heavy Metals to Benthos with Particular Emphasis on Deposit-feeding Infauna,* Environmental Laboratory, US Army Engineer Waterways Experiment Station, Vicksburg, Mississippi (Tech. Rep. D.78-42).

NELSON, D. A., CALABRESE, A., NELSON, B., MACINNES, J. R., and WENZLOFF, D. R. (1976). Biological effects of heavy metals on juvenile bay scallops, *Argopecten irradians,* in short-term experiments. *Bull. environ. Contam. Toxicol.*, **16**, 275–282.

NELSON, H. and GOYETTE, D. (1976). *Heavy Metal Contamination in Shellfish with Emphasis on Zinc Contamination of the Pacific Oyster, Crassostrea gigas,* Environmental Protection Service, Environment Canada, West Vancouver, Canada (Report No. EPS 5-PR-76-2).

NICHOLSON, J. K. (1981). The comparative distribution of zinc, cadmium and mercury in selected tissues of the herring gull, *Larus argentatus. Comp. Biochem. Physiol.*, **68C**, 91–94.

NICKLESS, G., STENNER, R., and TERRILLE, N. (1972). Distribution of cadmium, lead and zinc in the Bristol Channel. *Mar. Pollut. Bull., N.S.*, **3**, 188–190.

NIELSEN, E. S. and ANDERSEN, S. W. (1970). Copper ions as a poison in sea and freshwater. *Mar. Biol.,* **6,** 93-97.

NILSSON, J. R. (1979). Intracellular distribution of lead in *Tetrahymena* during continuous exposure to the metal. *J. Cell. Sci.,* **39**, 383–396.

NIMMO, D. W. R., LIGHTNER, D. V., and BAHNER, L. H. (1977). Effects of cadmium on the shrimps, *Penaeus duorarum, Palaemonetes pugio,* and *Palaemonetes vulgaris.* In F. J. Vernberg, A. Calabrese, F. D. Thurberg, and W. B. Vernberg (Eds), *Physiological Responses of Marine Biota to Pollutants.* Academic Press, New York. pp. 131-183.

NIMMO, D. R., RIGBY, R. A., BAHNER, L. M., and SHEPPARD, J. M. (1978). The acute and chronic effects of cadmium on the estuarine mysid, *Mysidopsis bahia. Bull. environ. Contam. Toxicol.,* **19**, 80–85.

NOËL-LAMBOT, F. (1976). Distribution of cadmium, zinc and copper in the mussel *Mytilus edulis.* Existence of cadmium-binding proteins similar to metallothioneins. *Experientia,* **32**, 324–326.

NOËL-LAMBOT, F., BOUQUEGNEAU, J. M., FRANKENNE, F., and DISTECHE, A. (1978a). Le rôle des métallothioneines dans le stockage des métaux lourds chez les animaux marins. *Revue int. Océanogr. méd.,* **49**, 13–20.

NOËL-LAMBOT, F., BOUQUEGNEAU, J. M., FRANKENNE, F., and DISTECHE, A. (1980). Cadmium, zinc and copper accumulation in limpets (*Patella vulgata*) from the Bristol Channel with special reference to Metallothioneins. *Mar. Ecol. Prog. Ser.,* **2**, 81–89.

NOËL-LAMBOT, F., GERDAY, C., and DISTECHE, A. (1978b). Distribution of Cd, Zn and Cu, in liver and gills of the eel, *Anguilla anguilla,* with special reference to metallothioneins. *Comp. Biochem. Physiol.,* **61C**, 177–187.

NORTON, M. G. and ROLFE, M. S. (1978). The field assessment of effects of dumping wastes at sea: 1. An introduction. *Fish. Res. Tech. Rep., Lowestoft,* **45**, 1–9.

NRIAGU, J. O. (1979). Global inventory of natural and anthropogenic emissions of trace metals to the atmosphere. *Nature, Lond.*, **279**, 409–411.

NUORTEVA, P., and HÄSÄNEN, E. (1975). Bioaccumulation of mercury in *Myxocephalus quadricornis* (L.) (Teleostei, Cottidae) in an unpolluted area of the Baltic. *Annls zool. fenn.*, **12**, 247–254.

NÜRNBERG, H. W. (1979). Polarography and voltammetry in studies of toxic metals in man and his environment. *Sci Total Environ.*, **12**, 35–60.

OJAVEER, E., ANNIST, J., JANKOWSKI, H., PALM, T., and RAID, T. (1980). On effect of copper, cadmium and zinc on the embryonic development of Baltic spring spawning herring. *Finn. mar. Res.*, **247**, 135–140.

OLAFSON, R. W., KEARNS, A., and SIM, R. G. (1979). Heavy metal induction of metallothionein synthesis in the hepatopancreas of the crab *Scylla serrata. Comp. Biochem. Physiol.*, **62B**, 417–424.

OLAFSON, R. W. and THOMPSON, J.A.J.(1974). Isolation of heavy metal binding proteins from marine vertebrates. *Mar. Biol.*, **28**, 83–86.

OLAUSSON, E., GUSTAFSSON, O., MELLIN, T., and SVENSSON, R. (1977). *The Current Level of Heavy Metal Pollution and Eutrophication in the Baltic Proper*, Goteborg Maringeologiska Laboratoriet, Göteborg (Meddelande fran Maringeologiska Laboratoriet No. 9).

OLOFSSON, S. and LINDAHL, P. E. (1979). Decreased fitness of cod (*Gadus morhua* L.) from polluted waters. *Mar. environ. Res.*, **2**, 33–45.

OLSON, B. H., BARKAY, T., and COLWELL, R. R. (1979). Role of plasmids in mercury transformation by bacteria isolated from the aquatic environment. *Appl. environ. Microbiol.*, **38**, 478–485.

OLSON, K. R., SQUIBB, K. S., and COUSINS, R. J. (1978). Tissue uptake subcellular distribution and metabolism of $^{14}CH_3HgCl$ and $CH_3{}^{203}$ Hg Cl by rainbow trout, *Salmo gairdneri. J. Fish. Res. Bd Can.*, **35**, 381–390.

ORTON, J. H. (1923). An account of investigations into the cause or causes of the unusual mortality among oysters in English oyster beds during 1920 and 1921. *Fishery Invest., Lond.* (Series 2), **6** (3), 1–199.

OSBORN, D., HARRIS, M. P., and NICHOLSON, J. K. (1979). Comparative tissue distribution of mercury, cadmium and zinc in three species of pelagic seabirds. *Comp. Biochem. Physiol.*, **64C**, 61–67.

OSTERBERG, C. and KECKES, S. (1977). The state of pollution in the Mediterranean Sea. *Ambio*, **6**, 321–326.

OVERNELL, J. and TREWHELLA, E. (1979). Evidence for the natural occurrence of (cadmium, copper)–metallothionein in the crab *Cancer pagurus. Comp. Biochem. Physiol.*, **64C**, 69–76.

OWEN, R. M. (1977). An assessment of the environmental impact of mining on the Continental Shelf. *Mar. Mining*, **1**, 85–102.

PAFFENHÖFER, G.-A. and KNOWLES, S. C. (1978). Laboratory experiments on feeding, growth and fecundity of and effects of cadmium on *Pseudodiaptomus coronatus. Bull. mar. Sci.*, **28**, 574–580.

PALMER, J. B. and RAND, G. M. (1977). Trace metal concentrations in two shellfish species of commerical importance. *Bull. environ. Contam. Toxicol.*, **18**, 512–520.

PAPADOPOULOU, C. and KANIAS, G. D. (1977). Tunicate species as marine pollution indicators. *Mar. Pollut. Bull., N.S.*, **8**, 229–231.

PAPADOPOULOU, C., KANIAS, G. D., and MORAITOPOULOU-KASSIMATI, E. (1976). Stable elements of radioecological importance in certain echinoderm species. *Mar. Pollut. Bull., N.S.*, **7**, 143–144.

PARKER, J. G. (1979). Toxic effects of heavy metals upon cultures of *Uronema marinum* (Ciliophora : Uronematidae). *Mar. Biol.*, **54**, 17–24.

PASKINS-HURLBURT, A. J., SKORYNA, S. C., TANAKA, Y., MOORE, W., Jr, and STARA, J. F. (1978). Fucoidan: it binding of lead and other metals. *Botanica mar.*, **21**, 13–22.

PATRICK, W. H., GAMBRELL, R. P., and KHALID, R. A. (1977). Physicochemical factors regulating solubility and bioavailability of toxic heavy metals in contaminated dredged sediment. *J. environ. Sci. Hlth*, **A12**, 457–492.

PATTERSON, C. C. (1978). Transport of industrial lead to the seas by aerosols. In *Workshop on the Tropospheric Transport of Pollutants to the Ocean*. U.S. National Academy of Sciences, Washington, D.C. pp. 45–54.

PEDEN, J. D., CROTHERS, J. H., WATERFALL, C. E., and BEASLEY, J. (1973). Heavy metals in Somerset marine organisms. *Mar. Pollut. Bull., N.S.*, **4**, 7–9.

PENROSE, W. R. (1975). Biosynthesis of organic arsenic compounds in brown trout (*Salmo trutta*). *J. Fish. Res. Bd Can.*, **32**, 2385–2390.

PENROSE, W. R., BLACK, R., and HAYWARD, M. J. (1975). Limited arsenic dispersion in sea water, sediments, and biota near a continuous source. *J. Fish. Res. Bd Can.*, **32**, 1275–1281.

PENTREATH, R. J. (1973a). The accumulation from water of ^{65}Zn, ^{54}Mn, ^{58}Co and ^{59}Fe by the mussel, *Mytilus edulis*. *J. mar. bio., Ass. U.K.*, **53**, 127–143.

PENTREATH, R. J. (1973b). The accumulation and retention of ^{65}Zn and ^{54}Mn by the plaice, *Pleuronectes platessa* L. *J. exp. mar. Biol. Ecol.*, **12**, 1–18.

PENTREATH, R. J. (1973c). The accumulation and retention of ^{59}Fe and ^{58}Co by the plaice, *Pleuronectes platessa* L. *J. exp. mar. Biol. Ecol.*, **12**, 315–326.

PENTREATH, R. J. (1973d). The accumulation from sea water of ^{65}Zn, ^{54}Mn, ^{58}Co and ^{59}Fe by the thornback ray, *Raja clavata* L. *J. exp. mar. Biol. Ecol.*, **12**, 327–334.

PENTREATH, R. J. (1976a). Some further studies on the accumulation and retention of ^{65}Zn and ^{54}Mn by the plaice, *Pleuronectes platessa* L. *J. exp. mar. Biol. Ecol.*, **21**, 179–189.

PENTREATH, R. J. (1976b). The accumulation of inorganic mercury from sea water by the plaice, *Pleuronectes platessa* L. *J. exp. mar. Biol. Ecol.*, **24**, 103–119.

PENTREATH, R. J. (1976c). The accumulation of organic mercury from sea water by the plaice, *Pleuronectes platessa* L. *J. exp. mar. Biol. Ecol.*, **24**, 121–132.

PENTREATH, R. J. (1976d). The accumulation of mercury from food by the plaice, *Pleuronectes platessa* L. *J. exp. mar. Biol. Ecol.*, **25**, 51–65.

PENTREATH, R. J. (1976e). The accumulation of mercury by the thornback ray, *Raja clavata* L. *J. exp. mar. Biol. Ecol.*, **25**, 131–140.

PENTREATH, R. J. (1977a). The accumulation of ^{110m}Ag by the plaice, *Pleuronectes platessa* L., and the thornback ray, *Raja clavata* L. *J. exp. mar. Biol. Ecol.*, **29**, 315–325.

PENTREATH, R. J. (1977b). The accumulation of cadmium by the plaice, *Pleuronectes platessa* L. and the thornback ray, *Raja clavata* L. *J. exp. mar. Biol. Ecol.*, **30**, 223–232.

PEQUEGNAT, W. E. and WASTLER, T. A. (1980). Field bioassays for early detection of chronic impacts of chemical wastes upon marine organisms. *Helgoländer Meeresunters.*, **33**, 531–545.

PERSOONE, G. and UYTTERSPROT, G. (1975). The influence of inorganic and organic pollutants on the rate of reproduction of a marine hypotrichous ciliate: *Euplotes vannus* Muller. *Revue int. Océanogr. méd.*, **37–38**, 125–151.

PESCH, C. E. (1979). Influence of three sediment types on copper toxicity to the polychaete *Neanthes arenaceodentata*. *Mar. Biol.*, **52**, 237–245.

PESCH, G., REYNOLDS, B., and ROGERSON, P. (1977). Trace metals in scallops from within and around 2 ocean disposal sites. *Mar. Pollut. Bull. N.S.*, **8**, 224–228.

PESCH, G. and STEWART, N. (1980). Cadmium toxicity to three species of estuarine invertebrates. *Mar. environ. Res.*, **3**, 145–156.

PESCH, G., STEWART, N., and PESCH, C. E. (1979). Copper toxicity to the bay scallop (*Argopecten irradians*). *Bull. environ. Contam. Toxicol.*, **23**, 759–765.

PETERSON, C. L., KLAWE, W. L., and SHARP, G. D. (1973). Mercury in tunas: a review. *Fishery Bull. natn. oceanic atmos. Adm.*, **71**, 603–613.

PETRICH, S. M. and REISH, D. J. (1979). Effects of aluminium and nickel on survival and reproduction in polychaetous annelids. *Bull. environ. Contam. Toxicol.*, **23**, 698–702.

PHILLIPS, D. J. H. (1976). The common mussel *Mytilus edulis* as an indicator of pollution by zinc, cadmium, lead and copper. I. Effects of environmental variables on uptake of metals. *Mar. Biol.*, **38**, 59–69.

PHILLIPS, D. J. H. (1977). The use of biological indicator organisms to monitor trace metal pollution in marine and estuarine environments—a review. *Environ. Pollut.*, **13**, 281–317.

PHILLIPS, D. J. H. (1980). Quantitative Aquatic Biological Indicators, Applied Science Publishers, Barking.

POTTS, W. T. W. and TODD, M. (1965). Kidney function in the octopus. *Comp. Biochem. Physiol.*, **16**, 479–489.

PRESTON, A. (1979). Standards and environmental criteria: the practical application of the results of laboratory experiments and field trials to pollution control. *Phil. Trans. R. Soc. B*, **286**, 611–624.

PRESTON, A., JEFFERIES, D. F., DUTTON, J. W. R., HARVEY, B. R., and STEELE, A. K. (1972a). British Isles coastal waters: the concentrations of selected heavy metals in sea water, suspended matter and biological indicators—a pilot survey. *Environ. Pollut.*, **3**, 69–82.

PRESTON, A., JEFFERIES, D. F., and PENTREATH, R. J. (1972b). The possible contributions of radioecology to marine productivity studies. *Symp. zool. Soc. Lond.*, **29**, 271–284.

PRINGLE, B. H., HISSONG, D. E., KATZ, E. L., and MULAWKA, S. T. (1968). Trace metal accumulation by estuarine molluscs. *J. sanit. Engng Div. Am. Soc. civ. Engrs*, **94**, 455–475.

PRUELL, R. J. and ENGELHARDT, F. R. (1980). Liver cadmium uptake, catalase inhibition and cadmium thionein production in the killifish (*Fundulus heteroclitus*) induced by experimental cadmium exposure. *Mar. environ. Res.*, **3**, 101–111.

RACHOR, E. and GERLACH, S. A. (1978). Changes of macrobenthos in a sublittoral sand area of the German Bight, 1967–1975. *Rapp. P.-v. Réun. Cons. int. Explor. Mer*, **172**, 418–431.

RADOUX, D. and BOUQUEGNEAU, J. M. (1979). Uptake of mercuric chloride from sea water by *Serranus cabrilla*. *Bull. environ. Contam. Toxicol.*, **22**, 771–778.

RAGAN, M. A., SMIDSRØD, O., and LARSEN, B. (1979). Chelation of divalent metal ions by brown algal polyphenols. *Mar. Chem.*, **7**, 265–271.

RAINBOW, P. S. and SCOTT, A. G. (1979). Two heavy metal binding proteins in the midgut gland of the crab *Carcinus maenas*. *Mar. Biol.*, **55**, 143–150.

RAINBOW, P. S., SCOTT, A. G., WIGGINS, E. A., and JACKSON, R. W. (1980). Effect of chelating agents on the accumulation of cadmium by the barnacle *Semibalanus balanoides*, and complexation of soluble Cd, Zn and Cu. *Mar. Ecol. Prog. Ser.*, **2**, 143–152.

RASMUSSEN, L. F. and WILLIAMS, D. C. (1975). The occurrence and distribution of mercury in marine organisms in Bellingham Bay. *NW. Sci.*, **49**, 87–94.

RATKOWSKY, D. A., DIX, T. G., and WILSON, K. C. (1975). Mercury in fish in the Derwent Estuary, Tasmania, and its relation to the position of the fish in the food chain. *Aust. J. mar. Freshwat. Res.*, **26**, 223–231.

RATKOWSKY, D. A., THROWER, S. J., EUSTACE, I. J., and OLLEY, J. (1974). A numerical study of the concentration of some heavy metals in Tasmanian oysters. *J. Fish. Res. Bd Can.*, **31**, 1165–1171.

RAY, S., MCLEESE, D., and PEZZACK, D. (1980). Accumulation of cadmium by *Nereis virens*. *Archs. environ. Contam. Toxicol.*, **9**, 1–8.

REEVE, M. R., WALTER, M. A., DARCY, K., and IKEDA, T. (1977). Evaluation of potential indicators of sub-lethal toxic stress on marine zooplankton (feeding, fecundity, respiration, and excretion): controlled ecosystem pollution experiment. *Bull. mar. Sci.*, **27**, 105–113.

REICHERT, W. L., FEDERIGHI, D. A., and MALINS, D. C. (1979). Uptake and metabolism of lead and cadmium in Coho salmon (*Oncorhynchus kisutch*). *Comp. Biochem. Physiol.*, **63C**, 229–234.

REINHART, K. and MYERS, T. D. (1975). Eye and tentacle abnormalities in embryos of the Atlantic oyster drill, *Urosalpinx cinerea*. *Chesapeake Sci.*, **16**, 286–288.

REISH, D. J. (1978). The effects of heavy metals on polychaetous annelids. *Revue int. Océanogr. Méd.*, **49**, 99–104.

RENFRO, J. L., SCHMIDT-NIELSEN, B., MILLER, D., BENOS, D., and ALLEN, J. (1974). Methyl mercury and inorganic mercury: uptake, distribution and effect on osmoregulatory mechanism in fishes. In F. J. Vernberg and W. B. Vernberg (Eds), *Pollution and Physiology of Marine Organisms*. Academic Press, London, New York. pp. 101–122.

RENFRO, W. C. (1973). Transfer of ^{65}Zn from sediments by marine polychaete worms. *Mar. Biol.*, **21**, 305–316.

RENFRO, W. C., FOWLER, S. W., HEYRAUD, M., and LA ROSA, J. (1975). Relative importance of food and water pathways in the bio-accumulation of zinc. *J. Fish. Res. Bd Can.*, **32**, 1339–1345.

RENZONI, A. (1976). A case of mercury abatement along the Tuscan coast. *25th Congress and Plenary Assembly of ICSEM Workshop on Marine Pollution (22–23 October 1976)*. Split, Yogoslavia.

RENZONI, A., BACCI, E., and FALCIAI, L. (1973). Mercury concentration in the water, sediments and fauna of an area of the Tyrrhenian coast. *Revue int. Océanogr. méd.*, **31–32**, 17–45.

RICE, D. W., Jr. and HARRISON, F. L. (1978). Copper sensitivity of Pacific herring, *Clupea harengus pallasi*, during its early life history. *Fishery Bull. natn. oceanic atmos. Adm., U.S.*, **76**, 347–356.

RICE, D. W., Jr. and HARRISON, F. L. (1979). *Copper Sensitivity of the Northern Anchovy, Engraulis mordax, During its Early Life History*, US Nuclear Regulatory Commission, Washington D.C. (NUREG/CR–0748).

RICE, M. A. and CHIEN, P. K. (1977). The effects of divalent cadmium on the uptake kinetics of glycine by the polychaete, *Neanthes virens. Wasmann J. Biol.*, **35**, 137–143.

RICE, M. A. and CHIEN, P. K. (1979). Uptake, binding and clearance of divalent cadmium in *Glycera dibranchiata* (Annelida : Polychaeta). *Mar. Biol.* **53**, 33–39.

RIDLINGTON, J. W. and FOWLER, B. A. (1979). Isolation and partial characterization of a cadmium binding protein from the American oyster (*Crassostrea virginica*). *Chemico-Biol. Interactions*, **25**, 127–138.

RILEY, J. P. and SEGAR, D. A. (1970). The distribution of the major and some minor elements in marine animals. I. Echinoderms and coelenterates. *J. mar. biol. Ass. U.K.*, **50**, 721–730.

RIVERS, J. B., PEARSON, J. E., and SHULTZ, C. D. (1972). Total and organic mercury in marine fish. *Bull. environ. Contam. Toxicol.*, **8**, 257–266.

RIVKIN, R. B. (1979). Effects of lead on growth of the marine diatom *Skeletonema costatum. Mar. Biol.*, **50**, 239–247.

ROBERTS, T. M., HEPPLESTON, P. B., and ROBERTS, R. D. (1976). Distribution of heavy metals in tissues of the common seal. *Mar. Pollut. Bull. N.S.*, **7**, 194–196.

ROBERTSON, T., WAUGH, G. D., and MOL, J. C. M. (1975). Mercury levels in New Zealand snapper *Chrysophrys auratus. N.Z. J. mar. Freshwat. Res.*, **9**, 265–272.

RØED, K. H. (1979). The effects of interacting salinity, cadmium, and mercury on population growth of an archiannelid, *Dinophilus gyrociliatus. Sarsia*, **64**, 245–252.

RØED, K. H. (1980). Effects of salinity and cadmium interaction on reproduction and growth during three successive generations of *Ophryotrocha labronica* (Polychaeta). *Helgoländer Meeresunters.*, **33**, 47–58.

ROESIJADI, G. (1980). Influence of copper on the clam *Protothaca staminea*: effects on gills and occurrence of copper-binding proteins. *Biol. Bull. mar. biol. Lab., Woods Hole*, **158**, 233–247.

ROMANOV, A. S. and RYABININ, A. I. (1976). The investigation of arsenic and stibium in the Atlantic Ocean waters. *Morsk. Gidrofiz. Issled. Sevastopol*, **3**, 162–172.

ROMERIL, M. G. (1977). Heavy metal accumulation in the vicinity of a desalination plant. *Mar. Pollut. Bull., N.S.*, **8**, 84–87.

RONALD, K., TESSARO, S. V., UTHE, J. F., FREEMAN, H. C., and FRANK, R. (1977). Methylmercury poisoning in the harp seal *Pagophilus groenlandicus. Sci. Total Environ.*, **8**, 1–11.

ROSENBERG, R. and COSTLOW, J. D., Jr. (1976). Synergistic effects of cadmium and salinity combined with constant and cycling temperatures on the larval development of two estuarine crab species. *Mar. Biol.*, **38**, 291–303.

RUDD, J. W. M., FURUTANI, A., and TURNER, M. A. (1980). Mercury methylation by fish intestinal contents. *Appl. environ. Microbiol.*, **40**, 777–782.

RUDDELL, C. L. and RAINS, D. W. (1975). The relationship between zinc, copper and the basophils of two crassostreid oysters, *C. gigas* and *C. virginica. Comp. Biochem. Physiol.*, **51A**, 585–591.

RUNHAM, N. W., THORNTON, P. R., SHAW, D. A., and WAYTE, R. C. (1969). The mineralization and hardness of the radular teeth of the limpet *Patella vulgata* L. *Z. Zellforch. mikroski. Anat.*, **99**, 608–626.

RUSSELL, G. and MORRIS, O. P. (1970). Copper tolerance in the marine fouling alga. *Ectocarpus siliculosus. Nature, Lond.*, **228**, 288–289.

RYTHER, J., LOSORDO, T. M., FURR, A. K., PARKINSON, T. F., GUTENMAN, W. H., PAKKALA, I. S., and LISK, D. J. (1979). Concentration of elements in marine organisms cultured in seawater flowing through coal-fly ash. *Bull. environ. Contam. Toxicol.*, **23**, 207–210.

SAIFULLAH, S. M. (1978). Inhibitory effects of copper on marine dinoflagellates. *Mar. Biol.*, **44**, 299–308.

SALIBA, L. J. and KRZYZ, R. M. (1976). Acclimation and tolerance of *Artemia salina* to copper salts. *Mar. Biol.*, **38**, 231–238.

SANDERS, J. G. (1979a). Microbial role in the demethylation and oxidation of methylated arsenicals in seawater. *Chemosphere*, **8**, 135–137.

SANDERS, J. G. (1979b). The concentration and speciation of arsenic in marine macro-algae. *Estuar. coast mar. Sci.*, **9**, 95–99.

SANDERS, J. G. (1979c). Effects of arsenic speciation and phosphate concentration on arsenic inhibition of *Skeletonema costatum* (Bacillariophyceae). *J. Phycol.*, **15**, 424–428.

SANDERS, J. G. (1980). Arsenic cycling in marine systems. *Mar. environ. Res.*, **3**, 257–266.

SAWARD, D., STIRLING, A., and TOPPING, G. (1975). Experimental studies on the effects of copper on a marine food chain. *Mar. Biol.*, **29**, 351–361.

SAYLER, G. S., NELSON, J. D., Jr, and COLWELL, R. R. (1975). Role of bacteria in bioaccumulation of mercury in the oyster *Crassostrea virginica*. *Appl. Microbiol.*, **30**, 91–96.

SCHAULE, B. and PATTERSON, C. (1980). The occurrence of lead in the northeast Pacific and the effects of anthropogenic inputs. In M. Branica and Z. Konrad (eds), *Proc. Int. Experts Discussion on Lead : Occurrence, Fate and Pollution in the Marine Environment*. Pergamon Press, Oxford. pp. 31–43.

SCHIPP, R. and HEVERT, F. (1978). Distribution of copper and iron in some central organs of *Sepia officinalis* (Cephalopoda). A comparative study by flameless atomic absorption and electron microscopy. *Mar. Biol.*, **47**, 391–399.

SCHMIDT-NIELSEN, B., SHELINE, J., MILLER, D. S., and DELDONNO, M. (1977). Effect of methyl mercury upon osmoregulation, cellular volume, and ion regulation in winter flounder, *Pseudopleuronectes americanus*. In F. J. Vernberg, A. Calabrese, F. P. Thurberg, and W. B. Vernberg (Eds), *Physiological Responses of Marine Biota to Pollutants*. Academic Press, New York. pp. 105–117.

SCHOLZ, N. (1980). Accumulation, loss and molecular distribution of cadmium in *Mytilus edulis*. *Helgöländer Meeresunters.*, **33**, 68–78.

SCHULZ-BALDES, M. (1974). Lead uptake from sea water and food, and lead loss in the common mussel *Mytilus edulis*. *Mar. Biol.*, **25**, 177–193.

SCHULZ-BALDES, M. (1978a). Lead transport in the common mussel *Mytilus edulis*. In D. S. McLusky and A. J. Berry (Eds), *Physiology and Behaviour of Marine Organisms*. Pergamon Press, Oxford and New York. pp. 211–218.

SCHULZ-BALDES, M. (1978b). Transport and deposition of lead within *Mytilus edulis*, with respect to the use of mussels as monitoring organisms. *Coun. Meet. int. Coun. Explor. Sea (C.M.I.C.E.S.)*, **1978/E : 13**, 1–10.

SCHWIMER, S. R. (1973). Trace metal levels in three subtidal invertebrates *Veliger*, **16**, 95–102.

SCRUDATO, R. J. and ESTES, E. L. (1976). Depuration of copper and zinc by *Crassostrea virginica* (American oyster). *Tex. J. Sci.*, **27**, 437–441.

SEELIGER, U. and EDWARDS, P. (1977). Correlation coefficients and concentration factors of copper and lead in seawater and benthic algae. *Mar. Pollut. Bull., N.S.*, **8**, 16–19.

SERGEANT, D. E. and ARMSTRONG, F. A. J. (1973). Mercury in seals from eastern Canada. *J. Fish. Res. Bd Can.* **30**, 843–846.

SHEALY, M. H., Jr and SANDIFER, P. A. (1975). Effects of mercury on survival and development of the larval grass shrimp *Palaemonetes vulgaris*. *Mar. Biol.*, **33**, 7–16.

SHELINE, J. and SCHMIDT-NIELSEN, B. (1977). Methyl mercury–selenium : interaction in the killifish, *Fundulus heteroclitus*. In F. J. Vernberg, A. Calabrese, F. P. Thurberg, and W. B. Vernberg (Ed), *Physiological Responses of Marine Biota to Pollutants*. Academic Press, New York. pp. 119–130.

SHERWOOD, M. J. and WRIGHT, J. L. (1976). Uptake and effects of chromium on marine fish. In *Southern California Coastal Water Research Project Report for year ended 30 June 1976*. El Segundo, California. pp. 123–128.

SHOLKOVITZ, E. R. (1978). The flocculation of dissolved Fe, Mn, Al, Cu, Ni, Co and Cd during estuarine mixing. *Earth Planet. Sci. Lett.*, **41**, 77–86.

SHOMURA, R. S. and CRAIG, W. L. (1974). Mercury in several species of billfishes taken off Hawaii and southern California. *Spec. scient. Rep. natn. oceanic atmos. Adm. U.S.* (Fisheries), **675**, 160–163.

SHORE, R., CARNEY, G., and STYGALL, T. (1975). Cadmium levels and carbohydrate metabolism in limpets. *Mar. Pollut. Bull., N.S.*, **6**, 187–189.

SHULTZ, C. D., CREAR, D., PEARSON, J. E., RIVERS, J. B., and HYLIN, J. W. (1976). Total and organic mercury in the Pacific blue marlin. *Bull. environ. Contam. Toxicol.*, **15**, 230–234.

SHULTZ, C. D. and ITO, B. M. (1979). Mercury and selenium in blue marlin, *Makaira nigricans*, from the Hawaiian islands. *Fish. Bull. Fish Wildl. Serv. U.S.*, **76**, 872–879.

SHUSTER, C. N. and PRINGLE, B. H. (1969). Trace metal accumulation by the American oyster, *Crassostrea virginica. Proc. natn. Shellfish. Ass.*, **59**, 91–103.

SICK, L. V. and BAPTIST, G. J. (1979). Cadmium incorporation by the marine copepod *Pseudodiaptomus coronatus. Limnol. Oceanogr.*, **24**, 453–462.

SIEBERS, D. and EHLERS, U. (1979). Heavy metal action on transintegumentary absorption of glycine in two annelid species. *Mar. Biol.*, **50**, 175–179.

SILVER, S., SCHOTTEL, J., and WEISS, A. (1976). Bacterial resistance to toxic metals determined by extrachromosomal R factors. In J. M. Sharpley and A. M. Kaplan (Eds), *Proceedings of the Third International Biodegradation Symposium*. Applied Science Publishers, Barking, England. pp. 899–917.

SILVERBERG, B. A., STOKES, P. M., and FERSTENBERG, L. B. (1976). Intranuclear complexes in a copper-tolerant green alga. *J. Cell Biol.*, **69**, 210–214.

SILVERSTEIN, S. C., STEINMAN, R. M., and COHN, Z. A. (1977). Endocytosis. *A. Rev. Biochem.*, **46**, 669–722.

SIMPSON, R. D. (1979). Uptake and loss of zinc and lead by mussels (*Mytilus edulis*) and relationships with body weight and reproductive cycle. *Mar. Pollut. Bull., N.S.*, **10**, 74–78.

SIMS, M. A. (1975). An electron microscope autoradiographic investigation of the accumulation of zinc-65 by a species of *Eutreptia. Experientia*, **31**, 426–427.

SINDERMANN, C. J. (1979). Pollution-associated diseases and abnormalities of fish and shellfish: a review. *Fish. Bull. U.S.*, **76**, 717–749.

SIPOS, L., NÜRNBERG, H. W., VALENTA, P., and BRANICA, M. (1980). The reliable determination of mercury traces in sea water by subtractive differential pulse voltammetry at the twin gold electrode. *Analytica Chim. Acta*, **115**, 25–42.

SIROTA, G. R. and UTHE, J. F. (1977). Determination of tetra-alkyllead compounds in biological materials. *Analyt. Chem.*, **49**, 823–825.

SKAAR, H., RYSTAD, B., and JENSEN, A. (1974). The uptake of ^{63}Ni by the diatom *Phaeodactylum tricornutum. Physiologia Pl.*, **32**, 353–358.

SKEI, J. M. (1978). Serious mercury contamination of sediments in a Norwegian semi-enclosed bay. *Mar. Pollut. Bull., N.S.*, **9**, 191–193.

SKEI, J. and PAUS, P. E. (1979). Surface metal enrichment and partitioning of metals in a dated sediment core from a Norwegian fjord. *Geochim. Cosmochim. Acta*, **43**, 239–246.

SKEI, J. M., PRICE, N. B., and CALVERT, S. E. (1972). The distribution of heavy metals in sediments of Sörfjord, West Norway. *Wat. Air Soil Pollut.*, **1**, 452–461.

SKERFVING, S., HANSSON, K., MANGS, C., LINDSTEN, J., and RYMAN, N. (1974). Methylmercury-induced chromosome damage in man. *Environ. Res.*, **7**, 83–98.

SKIPNES, P., ROALD, T., and HAUG, A. (1975). Uptake of zinc and strontium by brown algae. *Physiologia Pl.*, **34**, 314–20.

SMALL, L. F., FOWLER, S. W., and KĔCKĔS, S. (1973). Flux of zinc through a macroplanktonic crustacean. *Radioactivity in the Sea*, **34** (IAEA, Vienna).

SMITH, J. D. (1970). Tin in organisms and water in the Gulf of Naples. *Nature, Lond.*, **225**, 103–104.

SMITH, J. D. and BURTON, J. D. (1972). The occurrence and distribution of tin with particular reference to marine environments. *Geochim. Cosmochim. Acta*, **36**, 621–629.

SMITH, T. G. and ARMSTRONG, F. A. J. (1975). Mercury in seals, terrestrial carnivores, and principal food items of the Inuit, from Holman, N.W.T. *J. Fish. Res. Bd. Can.*, **32**, 795–801.

SMITH, T. G. and ARMSTRONG, F. A. J. (1978). Mercury and selenium in ringed and bearded seal tissues from Arctic Canada. *Arctic*, **31**, 75–84.

SOMERO, G. N., CHOW, T. J., YANCEY, P. H., and SNYDER, C. B. (1977a). Lead accumulation rates in tissues of the estuarine teleost fish, *Gillichthys mirabilis*: salinity and temperature effects. *Arch environ. Contam. Toxicol.*, **6**, 337–348.

SOMERO, G. N., YANCEY, P. H., CHOW, T. J., and SYNDER, C. B. (1977b). Lead effects on tissue and whole organism respiration of the estuarine teleost fish, *Gillichthys mirabilis. Archs environ. Contam. Toxicol*, **6**, 349–354.

SPOONER, G. M. and MOORE, H. B. (1940). The ecology of the Tamar Estuary. Part VI. An account of the macrofauna of the intertidal muds. *J. mar. biol. Ass. U.K.*, **24**, 283–330.

STEBBING, A. R. D. (1976). The effects of low metal levels on a clonal hydroid. *J. mar. biol. Ass. U.K.*, **56**, 977–994.

STEBBING, A. R. D. (1979). An experimental approach to the determinants of biological water quality. *Phil. Trans. R. Soc. B*, **286**, 465–481.

STEBBING, A. R. D. (1980). The biological measurement of water quality. *Rapp. P.-v. Réun. Cons. int. Explor. Mer*, **179**, 310–314.

STEELE, J. H. (1979). The uses of experimental ecosystems. *Phil. Trans. R. Soc. B*, **286**, 583–595.

STENNER, R. D. and NICKLESS, G. (1974a). Absorption of cadmium, copper and zinc by dog whelks in the Bristol Channel. *Nature, Lond.*, **247**, 198–199.

STENNER, R. D. and NICKLESS, G. (1974b). Distribution of some heavy metals in organisms in Hardangerfjord and Skjerstadfjord, Norway. *Wat. Air Soil Pollut.*, **3**, 279–291.

STENNER, R. D. and NICKLESS, G. (1975). Heavy metals in organisms of the Atlantic coast of S.W. Spain and Portugal. *Mar. Pollut. Bull., N.S.*, **6**, 89–92.

STEPHENSON, M. D., GORDON, R. M., and MARTIN, J. H. (1979). Biological monitoring of trace metals in the marine environment with transplanted oysters and mussels. In *Bio-accumulation of Heavy Metals by Littoral and Pelagic Marine Organisms* (Ecological Research Series EPA-600/3-79-038). US Environmental Protection Agency, Office of Research and Development, Narragansett, Rhode Island. pp. 12–50.

STEPHENSON, R. R. and TAYLOR, D. (1975). The influence of EDTA on the mortality and burrowing activity of the clam (*Venerupis decussata*) exposed to sub-lethal concentrations of copper. *Bull environ. Contam. Toxicol.*, **14**, 304–318.

STEWART, J. and SCHULZ-BALDES, M. (1976). Long-term lead accumulation in abalone (*Haliotis* spp.) fed on lead-treated brown algae (*Egregia laevigata*). *Mar. Biol.*, **36**, 19–24.

STICKEL, L. F., STICKEL, W. H., MCLANE, M. A. R., and BRUNS, M. (1977). Prolonged retention of methyl mercury by mallard drakes. *Bull. environ. Contam. Toxical.*, **18**, 393–400.

STOCKS, P. and DAVIES, R. I. (1964). Zinc and copper contents of soils associated with the incidence of cancer of the stomach and other organs. *Br. J. Cancer*, **18**, 14–24.

STOECKER, D. (1980). Relationships between chemical defense and ecology in benthic ascidians. *Mar. Ecol. Prog. Ser.*, **3**, 257–265.

STONEBURNER, D. L. (1978). Heavy metals in tissues of stranded short-finned pilot whales. *Sci. Total Environ.*, **9**, 293–297.

STONEBURNER, D. L., PATTY, P. C., and ROBERTSON, W. B., Jr (1980). Eivdence of heavy metal accumulations in sooty terns. *Sci. Total Environ.*, **14**, 147–152.

STRÖMGREN, T. (1980a). The effect of dissolved copper on the increase in length of 4 species of intertidal fucoid algae. *Mar. envir. Res.*, **3**, 5–13.

STRÖMGREN, T. (1980b). The effect of lead, cadmium and mercury on the increase in length of 5 intertidal fucales. *J. exp. mar. Biol. Ecol.*, **43**, 107–119.

STUMP, I. G., KEARNEY, J., D'AURIA, J. M., and POPHAM, J. D. (1979). Monitoring trace elements in the mussel, *Mytilus edulis*, using X-ray energy spectroscopy. *Mar. Pollut. Bull., N.S.*, **10**, 270–276.

STURESSON, U. (1978). Cadmium enrichment in shells of *Mytilus edulis*. *Ambio*, **7**, 122–125.

STURGEON, R. E., BERMAN, S. S., DESAULNIERS, A., and RUSSELL, D. S. (1979). Determination of iron, manganese, and zinc in seawater by graphite furnace atomic absorption spectrometry. *Analyt. Chem.*, **51**, 2364–2369.

SUCKCHAROEN, S. and LODENIUS, M. (1980). Reduction of mercury pollution in the vicinity of a caustic soda plant in Thailand. *Wat. Air Soil Pollut.*, **13**, 221–227.

SUMMERHAYES, C. P., ELLIS, J. P., STOFFERS, P., BRIGGS, S. R., and FITZGERALD, M. G. (1977). Fine grained sediment and industrial waste distribution and dispersal in New Bedford Harbor and Western Buzzards Bay, Massachusetts. *Tech. Rep. Woods. Hole oceanogr. Inst, WHO1–76–115*, pp. 1–110.

SUNDA, W. G. and GILLESPIE, P. A. (1979). The response of a marine bacterium to cupric ion and its use to estimate cupric ion activity in seawater. *J. mar. Res.*, **37**, 761–777.

SUNDA, W. and GUILLARD, R. R. L. (1976). The relationship between cupric ion activity and the toxicity of copper to phytoplankton. *J. mar. Res.*, **34**, 511–529.

SUSTAR, J. F. and WAKEMAN, T. H. (1977). Dredging conditions influencing the uptake of heavy metals by organisms. In *Management of Bottom Sediments Containing Toxic Substances* (Ecological Research Series EPA–6–3–77–083). US Environmental Protection Agency, Office of Research and Development, Corvallis, Oregon. pp. 246–252.

SUZUKI, T., SATOH, H., YAMAMOTO, R., and KASHIWAZAKI, H. (1980). Selenium and mercury in foodstuffs from a locality with elevated intake of methyl mercury. *Bull. environ. Contam. Toxicol.*, **24**, 805–812.

TAGUCHI, M., YASUDA, K., TODA, S., and SHIMIZU, M. (1979). Study of metal contents of elasmobranch fishes. Part 1. Metal concentration in the muscle tissues of a dogfish, *Squalus mitsukurii*. *Mar. environ. Res.*, **2**, 239–249.

TAYLOR, D. (1979). The effect of discharge from 3 industrialised estuaries on the distribution of heavy metals in coastal sediments of the North Sea. *Estuar. coast. mar. Sci.*, **8**, 387–393.

THEEDE, H. (1980). Physiological responses of estuarine animals to cadmium pollution. *Helgoländer Meeresunters.*, **33**, 26–35.

THEEDE, H., SCHOLZ, N., and FISCHER, H. (1979a). Temperature and salinity effects on the acute toxicity of cadmium to *Laomedea loveni* (Hydrozoa) *Mar. Ecol. Prog. Ser.*, **1**, 13–19.

THEEDE, H., ANDERSSON, I., and LEHNBERG, W. (1979b). Cadmium in *Mytilus edulis* from German coastal waters. *Ber. dt. wiss. kommn Meeresforsch.*, **27**, 147–155.

THOMAS, W. H., HOLLIBAUGH, J. T., SEIBERT, D. L. R., and WALLACE, G. T., Jr (1980). Toxicity of a mixture of ten metals to phytoplankton. *Mar. Ecol. Prog. Ser.*, **2**, 213–220.

THOMPSON, J. A. J., and CRERAR, J. A. (1980). Methylation of lead in marine sediments. *Mar. Pollut. Bull., N.S.*, **11**, 251–253.

THORNTON, I., WATLING, H., and DARRACOTT, A. (1975). Geochemical studies in several rivers and estuaries used for oyster rearing. *Sci. total Environ.*, **4**, 325–345.

THURBERG, F. P., CALABRESE, A., GOULD, E., GREIG, R. A., DAWSON, M. A., and TUCKER, F. K. (1977). Response of the lobster, *Homarus americanus*, to sublethal levels of cadmium and mercury. In F. J. Vernberg, A. Calabrese, F. P. Thurberg, and W. B. Vernberg (Eds), *Physiological Responses of Marine Biota to Pollutants*. Academic Press, New York. pp. 185–197.

TIRAVANTI, G. and BOARI, G. (1979). Potential pollution of a marine environment by lead alkyls: the Cavtat incident. *Environ. Sic. Techol.*, **13**, 849–854.

TOPPING, G. (1974). The atmospheric input of some heavy metals to the Firth of Clyde and its relation to other inputs. *Coun. Meet. int. Coun. Explor. Sea (C.M.-I.C.E.S.)*, **1974/E : 32**.

TOWE, K. M. and LOWENSTAM, H. A. (1967). Ultrastructure and development of iron mineralization in the radular teeth of *Cryptochiton stelleri* (Mollusca). *J. Ultrastruct. Res.*, **17**, 1–13.

TSUBAKI, T. and IRUKAYAMA, K. (1977). *Minamata Disease*, Kodansha, Tokyo and Elsevier, Amsterdam.

TUREKIAN, K. K. (1977). The fate of metals in the oceans. *Geochim. Cosmochim. Acta*, **41**, 1139–1144.

TUREKIAN, K. K. and WEDEPOHL, K. H. (1961). Distribution of the elements in some major units of the earth's crust. *Bull. geol. Soc. Am.*, **72**, 175–192.

UEDA, T., NAKAHARA, M., ISHII, T. SUZUKI, Y., and SUKUKI, H. (1979). Amounts of trace metals in marine cephalopods. *J. Radiat. Res.*, **20**, 338–342.

UEDA, T., NAKAMURA, R., and SUZUKI, Y. (1976). Comparision of ^{115m}Cd accumulation from sediments and sea water by polychaete worms. *Bull. Jap. Soc. Scient. Fish.*, **42**, 299–306.

UEDA, T. and TAKEDA, M. (1979). Total and methylmercury levels in three species of whelks. *Bull. Jap. Soc. scient. Fish.*, **45**, 763–769.

UNDERWOOD, E. J. 1979). Trace elements and health: an overview. *Phil. Trans. R. Soc. B*, **288**, 5–14.

ÜNLÜ, M. Y. (1979). Chemical transformation and flux of different forms of arsenic in the crab *Carcinus maenas*. *Chemosphere*, **8**, 269–275.

ÜNLÜ, M. Y. and FOWLER, S. W. (1979). Factors affecting the flux of arsenic through the mussel *Mytilus galloprovincialis*. *Mar. Biol.*, **51**, 209–219.

UTHE, J. F. and ZITKO, V. (1980). Cadmium pollution of Belledure Harbour, New Brunswick, Canada. *Can. Tech. Rep. Fish Aquat.* Sci., No. 963, 1–107

VALLEE, B. L. (1978). Zinc biochemistry and physiology and their derangements. In R. J. P. Williams and J. R. R. F. Da Silva (Eds), *New Trends in Bio-inorganic Chemistry*, Academic Press, London. pp. 11–57.

VARANASI, U. and MARKEY, D. (1978). Uptake and release of lead and cadmium in skin and mucus of coho salmon (*Oncorhynchus kisutch*). *Comp. Biochem. Physiol.*, **60C**, 187–191.

VEN, W. S. M. VAN DE, KOEMAN, J. H. and SVENSON, A. (1979). Mercury and selenium in wild and experimental seals. *Chemosphere*, **7**, 539–555.

VERMEER, K. and PEAKALL, D. B. (1977). Toxic chemicals in Canadian fish-eating birds. *Mar. Pollut. Bull., N.S.*, **8**, 205–210.

VERMEER, K. and PEAKALL, D. B. (1979). Trace metals in seaducks of the Fraser River delta intertidal area, British Columbia. *Mar. Pollut. Bull., N.S.*, **10**, 189–193.

VERNBERG, W. B., DECOURSEY, P. J., KELLY, M., and JOHNS, D. M. (1977). Effects of sublethal concentrations of cadmium on adult *Palaemonestes pugio* under static and flow-through conditions. *Bull. environ. Contam. Toxicol.*, **17**, 16–24.

VERNBERG, W. B., DECOURSEY, P. J., and O'HARA, J. (1974). Multiple environmental factor effects on physiology and behaviour of the fiddler crab, *Uca pugilator*. In F. J. Vernberg and W. B. Vernberg (Eds), Pollution and *Physiology of Marine Organisms*. Academic Press, New York. pp. 381–425.

VIALE, D. (1978). Evidence of metal pollution in cetacea of the western Mediterranean. *Annls Inst. océanogr., Monaco*, **54**, 5–16.

VIARENGO, A., PERTICA, M., MANCINELLI, G., ZANICCHI, G., and ORUNESU, M. (1980). Rapid induction of copper-binding proteins in the gills of metal exposed mussels. *Comp. Biochem. Physiol.*, **67C**, 215–218.

VOYER, R. A., HELTSCHE, J. F., and KRAUS, R. A. (1979). Hatching success and larval mortality in an estuarine teleost, *Menidia menidia* (Linnaeus), exposed to cadmium in constant and fluctuating salinity regimes. *Bull. environ. Contam. Toxicol.*, **23**, 475–481.

VOYER, R. A., WENTWORTH, C. E., Jr, BARRY, E. P., and HENNEKEY, R. J. (1977). Viability of embryos of the winter flounder *Pseudopleuronectes americanus* exposed to combinations of cadmium and salinity at selected temperatures. *Mar. Biol.*, **44**, 117–124.

WALDHAUER, R., MATTE, A., and TUCKER, R. E. (1978). Lead and copper in the waters of Raritan and Lower New York Bays. *Mar. Poll. Bull. N.S.*, **9**, 38–42.

WALDICHUK, M. (1978). Disposal of mine wastes into the sea. *Mar. Pollut. Bull., N.S.*, **9**, 141–143.

WALKER, G. (1977). 'Copper' granules in the barnacle *Balanus balanoides*. *Mar. Biol.*, **39**, 343–349.

WALKER, G. and FOSTER, P. (1979). Seasonal variation of zinc in the barnacle *Balanus balanoides* (L.) maintained on a raft in the Menai Strait. *Mar. environ. Res.*, **2**, 209–221.

WALKER, G., RAINBOW, P. S., FOSTER, P., and HOLLAND, D. L. (1975). Zinc phosphate granules in tissue surrounding the midgut of the barnacle *Balanus balanoides*. *Mar. Biol.*, **33**, 161–166.

WALKER, T. I. (1976). Effects of species, sex, length and locality on the mercury content of school shark *Galeorhinus australis* (Macleay) and gummy shark *Mustelus antarcticus* Guenther from south-eastern Australian waters. *Aust. J. Mar. Freshwat. Res.*, **27**, 603–616.

WALZ, F. (1979). Uptake and elimination of antimony in the mussel, *Mytilus edulis*. *Veroff. Inst. Meeresforsch. Bremerh.*, **18**, 203–215.

WATLING, H. R. (1978). Effect of cadmium on larvae and spat of the oyster *Crassostrea gigas* (Thunberg). *Trans. R. Soc. S. Afr.*, **43**, 125–134.

WATLING, H. R. and WATLING, R. J. (1976a). Trace metals in *Choromytilus meriodionalis*. *Mar. Pollut. Bull.*, **7**, 91–94.

WATLING, H. R. and WATLING, R. J. (1976b). Trace metals in oysters from Knysna estuary. *Mar. Pollut. Bull.*, **7**, 45–48.

WEERS, A. W. van (1975). Uptake of cobalt-60 from sea water and from labelled food by the common shrimp *Crangon crangon*. In *Impacts of nuclear releases into the aquatic environment*. Vienna, IAEA, pp. 349–361.

WEICHART, G. (1973). Pollution of the North Sea. *Ambio*, **2**, 99–106.

WEIS, J. S. and WEIS, P. (1977a). Effects of heavy metals on development of the killifish, *Fundulus hereroclitus*. *J. Fish Biol.*, **11**, 49–54.

WEIS, P. and WEIS, J. S. (1977b). Methyl mercury teratogenesis in the killifish, *Fundulus heteroclitus*. *Teratology*, **16**, 317–326.

WESTERNHAGEN, H. von, DETHLEFSEN, V., and ROSENTHAL, H. (1980). Correlation between cadmium concentration in the water and tissue residue levels in dab, *Limanda limanda* L., and plaice, *Pleuronectes platessa* L. *J. mar. biol. Ass. U.K.*, **60**, 45–58.

WILSON, J. G. (1980). Heavy metals in the estuarine macrofauna of the east coast of Ireland. *J. Life Sci. R. Dubl. Soc., 1980*, 183–189.

WINDOM, H. L. (1975). Heavy metal fluxes through salt-marsh estuaries. In L. E. Cronin (Ed.), *Estuarine Research*, Vol. 1. Academic Press, London. pp. 137–152.

WINDOM, H. L. (1976). *Geochemical Interactions of Heavy Metals in Southeastern Salt Marsh Environments* (EPA–600/3–76–023), U.S. Environmental Protection Agency, Skidaway Institute of Oceanography, Savannah, Georgia.

WINDOM, H., GARDNER, W., STEPHENS, J., and TAYLOR, F. (1976). The role of methyl mercury production in the transfer of mercury in a salt marsh ecosystem. *Estuar. coast. mar. Sci.*, **4**, 579–583.

WINDOM, H. L. and SMITH, R. G. (1979). Copper concentrations in surface waters off the south-eastern Atlantic coast, U.S.A. *Mar. Chem.*, **7**, 157–163.

WINDOM, H. L. and TAYLOR, F. E. (1979). The flux of mercury in the South Atlantic Bight. *Deep Sea Res.*, **26A**, 283–292.

WINTER, J. E. (1972). Long-term laboratory experiments on the influence of ferric hydroxide flakes on the filter-feeding behaviour, growth, iron content and mortality in *Mytilus*. In M. Ruivo (Ed.), *Marine Pollution and Sea Life*, Fishing News (Books) Ltd, London. pp. 392–396.

WOLFE, D. A., THAYER, G. W., and ADAMS, S. M. (1975). Manganese, iron, copper and zinc in an eelgrass (*Zostera marina*) community. In C. E. Cushing (Ed.), *Radioecology and energy Resources*, Procs of 4th Nat. Symp. on Radio-ecology 12–14 May 1975. Oregon State Univ., Corvallis, Oregon (The Ecological Soc. of America, Spec. Publ. No. 1).

WONG, P. T. S., CHAU, Y. K., and LUXON, P. L. (1975). Methylation of lead in the environment. *Nature, Lond.*, **253**, 263–264.

WORLD HEALTH ORGANIZATION, (1976). *Environmental Health Criteria*. 1. Mercury. Geneva.

WORLD HEALTH ORGANIZATION, (1977). *Environmental Health Criteria*. 3. Lead. Geneva.

WRENCH, J. J. (1978). Biochemical correlates of dissolved mercury uptake by the oyster *Ostrea edulis*. *Mar. Biol.*, **47**, 79–86.

WRENCH, J., FOWLER, S. W., and ÜNLÜ, M. Y. (1979). Arsenic metabolism in a marine food chain. *Mar. Pollut. Bull., N.S.*, **10**, 18–20.

WRIGHT, D. A. (1978). Heavy metal accumulation by aquatic invertebrates. *Appl. Biol.*, **3**, 331–394.

WRIGHT, D. A. and BREWER, C. C. (1979). Cadmium turnover in the shore crab *Carcinus maenas*. *Mar. Biol.*, **50**, 151–156.

YANG, S. H. and EHRLICH, H. L. (1976). Effect of 4 heavy metals (Mn, Ni, Cu and Co) on some bacteria from the deep sea. In J. M. Sharpley and A. M. Kaplan (Eds), *Proceedings of the Third International Biodegradation Symposium*. Applied Science Publishers, Barking, England. pp. 867–874.

YEATS, P. A., BEWERS, J. M., and WALTON, A. (1978). Sensitivity of coastal waters to anthropogenic trace metal emissions. *Mar. Pollut. Bull., N.S.*, **9**, 264–268.

YOUNG, D. R., ALEXANDER, G. V., and MCDERMOTT-EHRLICH, D. (1979). Vessel-related contamination of southern California harbours by copper and other metals. *Mar. Pollut Bull., N.S.*, **10**, 50–56.

YOUNG, D. R., JAN, T.-K., and MOORE, M. D. (1977). Metals in power plant cooling water discharges. *Southern California Coastal Water Research Project Annual Report*, El Segundo, California. pp. 25–38.

YOUNG, J. S., BUSCHBOM, R. L., GURTISEN, J. M., and JOYCE, S. P. (1979a). Effects of copper on the sabellid Polychaete, *Eudistylia vancouveri* I. Concentration limits for copper accumulation. *Arch. environ. Contam. Toxicol.*, **8**, 97–106.

YOUNG, J. S., GURTISEN, J. M., APTS, C. W., and CRECELIUS, E. A. (1979b). The relationship between the copper complexing capacity of sea water and copper toxicity in shrimp zoea. *Mar. environ. Res.*, **2**, 265–273.

YOUNG, M. L. (1974). *The Transfer of ^{65}Zn and ^{59}Fe along 2 Marine Food Chains*, PhD Thesis, University of East Anglia.

YOUNG, M. L. (1975). The transfer of ^{65}Zn and ^{59}Fe along a *Fucus serratus* (L.) – *Littorina obtusata* (L.) food chain. *J. mar. biol. Ass. U.K.*, **55**, 583–610.

YOUNG, M. L. (1977). The roles of food and direct uptake from water in the accumulation of zinc and iron in the tissues of the dogwhelk, *Nucella lapillus* (L.). *J. exp. mar. Biol. Ecol.*, **30**, 315–325.

ZAFIROPOULOS, D. and GRIMANIS, A. P. (1977). Trace elements in *Acartia clausi* from Elefsis Bay of the Upper Saronikos Gulf, Greece. *Mar. Pollut. Bull., N.S.*, **8**, 79–81.

ZAROOGIAN, G. E. (1979). Studies on the depuration of cadmium and copper by American oyster *Crassostrea virginica*. *Bull. environ. Contam. Toxicol.*, **23**, 117–122.

ZAROOGIAN, G. E. (1980). *Crassostrea virginica* as an indicator of cadmium pollution. *Mar. Biol.*, **58**, 275–284.

ZAROOGIAN, G. E., MORRISON, G., and HELTSCHE, J. F. (1979). *Crassostrea virginica* as an indicator of lead pollution. *Mar. Biol.*, **52**, 189–196.

ZIRINO, A., LIEBERMAN, S. H., and CLAVELL, C. (1978). Measurement of Cu and Zn in San Diego Bay by automated anodic stripping voltammetry. *Environ. Sci. Technol.*, **12**, 73–79.

ZITKO, V. and CARSON, W. V. (1975). Accumulation of thallium in clams and mussels. *Bull. environ. Contam. Toxicol.*, **14**, 530–533.

Marine Ecology Vol. 5, Part 3
Edited by Otto Kinne

4. OIL POLLUTION AND ITS MANAGEMENT

R. JOHNSTON

(1) Introduction

(a) Man's Developing Need for Oil

The first observations of oil affecting land and fresh water were made before biblical times and before the records of people worshipping the mysterious fire that burned forever at seeps from fractured rock strata. Inquisitive man was intrigued by the unusual physical properties of oil. He soon found it burned, and later that it would yield liquid oil and greasy, waxy or tarry residues. Oil in these forms could be used by early man for lighting, for waterproofing, for decorating pottery and so on. In the Middle Ages, apothecaries prized various oil preparations as intended remedies for many of the ailments which abounded in those times. In the twentieth century, crude oil, which is now one of the resources most abundantly taken from the earth, provides the raw materials from which many synthetic pharmaceuticals are made to treat diseases in plants, animals, and man. Paradoxically, a wide range of toxic substances are also made. Hydrocarbons in gaseous, liquid, and solid form can be refined, reformed and converted into derivatives on a very large scale and hence industrial fuels, chemicals and plastics in the form of fibres, films, die-casting materials are familiar in every realm of living. The prime use of oil and oil products is as a source of energy; fuel is burned and the heat released is used either directly, indirectly or by conversion into work. For how long we continue to burn up this finite resource with little heed to future generations is an urgent issue that man today must face. Meanwhile, oil is a great and glorious provider of mobility, warmth, comfort and luxury for the lucky ones.

Oil pollution as it is popularly understood arises from the exploration, extraction, stabilization, transport, storage, and refining of crude oil (Fig. 4-1) and also thereafter in the subsequent manufacture and handling of products. Losses can arise in the operations and transfers at every stage. Atmospheric hydrocarbon pollution is popularly recognized only in relation to smog and where products are exposed for example at garages and filling stations. Quantitatively, much greater atmospheric hydrocarbon pollution arises from the generation of power in electricity generating stations, refineries, large furnaces, fixed plant and heavy traction units.

The term 'pollution' is defined here according to WHO/UN: Any kind of man-made reduction in the quality of the marine environment. The term 'contamination' is used in relation to a reduction in quality of food or material for man or animal. There is no intention that pollution consistently implies damage on a scale that would merit popular publicity, or that contamination necessarily implies noxious or disgusting quality.

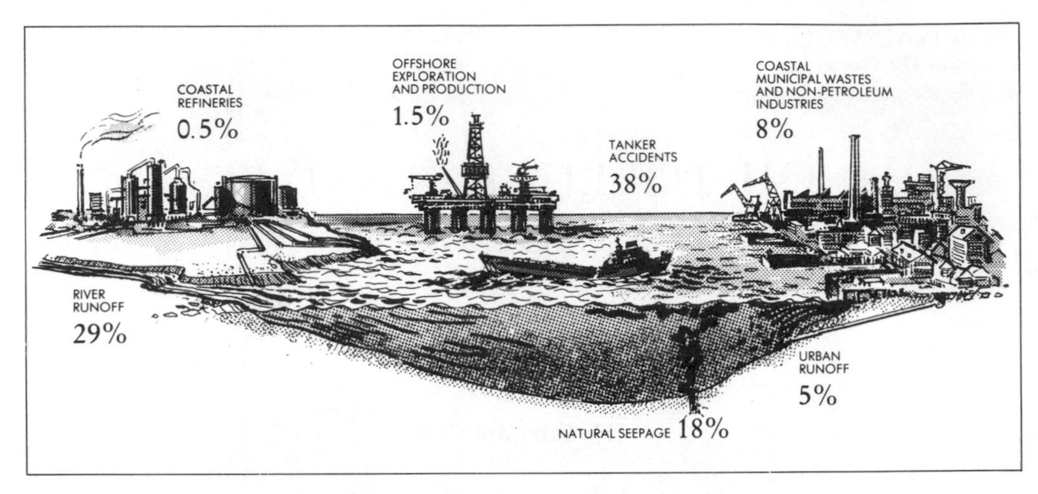

Fig. 4-1: Petroleum in the marine environment. (After RAY, 1981; redrawn; reproduced by permission of the Association of Européene Océanique)

Because the literature on oil pollution is so extensive and in order to convey other interpretations a literature appendix (p. 1580 to 1582) has been provided for guidance. Views on the importance of oil pollution may be gleaned from the GESAMP (1977) review and the many workshops and studies on inputs and effects. Detailed accounts of major oil spill incidents are listed and a few important references to the principal specialist topics. Legal aspects are of interest in the wider scene. A compilation of the more accessible and comprehensive bibliographies is included for those seeking detailed literature references on specific topics.

(b) Cause for Concern

Concern about oil pollution gets its most immediate boost from major oil spills—particularly in recent years from errant oil tankers—but also from blowouts, oil pipelines and storage tanks as a result of operational failure and error, or deliberate acts of sabotage or war.

The US Academy of Sciences is currently reviewing its well publicized estimate of a total 6·113 million tons of oil reaching the seas of the world (NAS, 1975) but there is little expectation of a major revision. COWELL (1978, unpubl.) provides an interim revised and reduced estimate of 4·951 million tons based on improvements in oil handling and oil technology over the intervening period. Subsequently, however, there have been recent disastrous oil spill incidents, like that of the grounding of the *Amoco Cadiz* and the Ixtoc I blowout, which can only increase the latest figures.

Oil exploration and exploitation—earlier restricted to sources on land or in shallow seas in an oil belt girdling the world—are now rapidly extending to land areas all over the world and into territorial seas everywhere at ever increasing depths and into ever more adverse environments.

Regardless of increasing oil costs, the demand for oil continues at a high rate and the frenzy to explore and exploit escalates in the face of mounting difficulties. There is evident reason for environmentalists to clamour more and more for restraint in the use of a scarce resource, for proper caution in new oil developments and for the diminution of the present level of oil pollution.

The menace of oil pollution is driven home in television and newspaper coverages of successive disasters, and equally forcefully by the first-hand witness of tragic seabird mortalities caused by smaller incidents. The number of seabird casualities and the magnitude and frequency of oil spills are contentious statistics, but no one would deny that there is reason for concern at least on humanitarian grounds.

Equally forcibly, residents and holiday-makers alike are acutely aware of filthy oil slicks and extensive oil films in busy ports and seaways. Who has not experienced the disgust of soiled clothing or grease-besmirched skin on what used to be clean holiday beaches and attractive remote coasts. Oil on or near the sea and seashore has already become an offensive menace to an intolerable degree, and until much greater guilt perturbs those at fault or responsible for mishandling oil in any form one can only predict more and more widespread loss of amenity and mounting criticism.

To complete this narrow humanistic view of the miseries of oil pollution, one should add the nuisance of living in the hydrocarbon-laden atmosphere of the neighbourhood of oil industry and the smog of city streets. These all detract forcefully from the great benefits everyone derives from oil as a raw material and an energy source.

The impact of oil on sea and seashore brings in its trail a wide range of threats of varying severity to marine micro-organisms, plants, and animals. These are at risk not only as individuals or populations but also as components of an intricate network of interdependent assemblages that we can recognize in the vast and complex marine ecosystem. Present knowledge does not extend to a fully proven case establishing every likely impact of oil pollution but it is possible to unravel many strands of this network. Each strand must also assess the effects of many forms of crude oil and related components and compounds and of their degradation products. It is beginning to be possible to make approximate estimates of the overall threat to the seas and marine resources.

Holistic ecological appraisals of oil pollution must take into account many facets in addition to the evaluation of the impacts on marine species and on the food web. They extend into concepts of how to quantify oil effects, how to add together inputs from the many direct and indirect sources, how to evaluate oil spills and necessary emissions, and how to make generalizations, rationalizations, and simplifications of the total problem. It is hard to contest the claim by the petroleum industry that oil is one of the world's greatest boons to mankind giving manifold major benefits and causing only at worst trivial dis-benefits, but cost–benefit analysis gives a myopic view of the global problem.

Like other all-pervading problems—such as dirt, contamination of food, corrosion of metals, rotting of timber, damage to stone and, of course, the impact of disease—oil pollution is encountered every day by virtually everybody. It is so common that it is unnoticed—or if noticed, promptly forgotten. Somewhere, somehow, someone or some form of life encounters the impact of oil in a chronic, hurtful, damaging or fatal form every day or many times a day. It is only by adopting a total ecological approach that one becomes aware not only of the visible, possibly dramatic incidents and their immediate repercussions but also of the much wider virtually continuous events of undramatic damage and the constant all-pervading cloud, unseen and unnoticed, of uncertain portent. This ecological assessment attempts to integrate these levels of effect and their numerous interactions.

The costs to man and life on earth of global oil pollution have rarely been assessed.

The cost of marine oil pollution is only a little less complicated, but various attempts have been made to express the losses of marine resources and disturbance of the fishing industry in financial terms to be added to the costs of clean-up. Costings are necessary for reclaiming damages from those spilling oil. Surely the true costs of continuous on-going marine oil pollution as it affects animals and plants are also worth calculating even if the reckoning is incomplete.

(c) Oil Pollution and its Management

Once it reaches the sea there is no way in which oil can be totally recovered. Remedial measures for oil spills are difficult, inefficient, and restricted in application. They are also expensive and can be damaging to the environment. Scope exists for good management of clean-up operations but most would agree that such efforts are unrewarding. Prevention is better than cure, no more so than for oil pollution which has no cure.

The tools for management are sound knowledge, identification of responsibility, and means of effective control. At best, the science of oil pollution is complex and incomplete. The responsibilities for global oil pollution lie not only with the oil industry but also with every industry and everyone. There are serious deficiencies in existing controls, standards and methodologies and overall total absence of harmonization.

Above all is the fact that the management of oil pollution does not fall on any single group or agency and consequently deficiencies are not recognized.

Management in the oil and shipping industries and on the part of their trade associations can justly claim to be striving to fulfil their responsibilities, and their scientists show lively awareness. In most other aspects of oil pollution, experts and administrators seem content to perpetuate attitudes and practices that have progressed little since the beginning of the century.

Only better awareness and an improved partnership between expert and administrator can bring about reduction in global oil pollution and, more important, work towards optimum utilization of oil on a long-term plan. It is the duty of scientists to assist management in these objectives and to direct management towards more meaningful control of oil pollution and better integrated monitoring programmes for the total environment and ecosystem.

The role of the scientist is to recognize oil pollution for what it is, examine comprehensively what damage it does to the sea and declare the action that is necessary and possible to remedy these effects. Inevitably this means dedicated research into the problems of fate of oil and the effects of oil on organisms and communities.

It is logical to begin by looking at the pathways by which oil reaches the sea and by quantifying these having regard to the validity of the available data. The global picture only becomes clear if one looks at air, land, and fresh water, as well as at the sea which is the centre of interest.

(2) Oil Inputs to the Sea and their Characteristics

It is clear from estimates of the mean annual quantities of oil reaching the sea (Table 4-1) that marine transportation—including tanker accidents—is an important component, only equalled or perhaps exceeded by the much more elusive input from rivers. Inputs related to oil transportation are initiated by direct deposit of oil into the sea. In ecological terms, the initial impact for some unfortunate marine creature is 100% oil.

Table 4-1
Estimate of oil reaching the seas of the world, 1978 (Based on recent world budgets: in millions of tons; updated by COWELL: British Petroleum Company; for a European Parliamentary Hearing, Paris, 4 July, 1978)

Offshore oil production		0·06
Coastal oil refineries		0·06
Industrial waste		0·15
Municipal waste		0·30
Urban runoff		0·40
River runoff		1·40
Natural seeps		0·60
Atmospheric rainout		0·60
Marine transportation		
bilges and bunkering	0·12	
load on top	0·11	
non-load on top	0·50	
dry docking	0·25	
tanker accidents	0·30	
dry cargo accidents	0·10	
terminal loading	0·001	
		1·38
Total		4·95

Some river-borne municipal and industrial inputs may well begin upstream in the same way, but by the time a river enters an estuary most of the oil will evaporate, dissolve or emulsify and much of the remainder is associated with particulate matter. Hence its overall ecological impact is less acute but much more disperse, affecting much greater volumes of water.

(a) Marine Transportation

The figure for oil in most of the transporation sub-heads is a value derived from physical measurement of amounts found in bilges, tanks and pipework, tank bottoms, spills at terminals and losses from damaged tankers and can be taken as realistic for total oil. This oil is as diverse as the many crude-oil sources it incorporates together with 'tank bottoms', the waxy heavy fractions that separate in cargo tanks *en voyage* and also a wide range of products ranging from light liquid fractions and petrol (gasoline) to heavy fuel oils kept liquid in heated tanks. The toxicological score of each 'oil' must vary widely and the probable target organisms also depend on the hydrocarbons emitted and their manner of emission. Oil derivatives may be included as a minor component but oil breakdown products are essentially absent on entry.

(b) Land-Derived Inputs

Oil in river outflows, municipal and industrial wastes and urban runoff will rarely be identifiable as coming from a single source. It will be modified by interaction with water, weathering, passage through soil, sewage works and other biodegradative routes. Its probable make-up would include traffic-contaminated runoff from highways (which themselves are often based on tarred stone chips) and escapes or discharges of used lubricating oils and greases from sundry domestic, service and industrial sources and, of course, land-based spills of petrol, heating and engineering oils. One suspects that much of the more toxic lighter fractions will have escaped into the air or have been metabolized. Of the more persistent fractions it has been suggested that polycyclic aromatic hydrocarbons which may include carcinogens are among the important groups.

Because about 90% of all petroleum is used primarily as a source of energy (Table 4-2) in static or mobile machines for combustion, these operations constitute a major point of origin for oil pollution of air, land, and fresh water. It is hard to descry the exclusive man-made inputs to the air and to the terrestrial environment, and little is known about the dynamics of the exchanges into water. In consequence, the approach usually adopted is to regard hydrocarbons used as fuel as giving rise initially to atmospheric pollution within which subsequent damage to the terrestrial environment is included. Virtually nothing is known about the inputs of air-borne hydrocarbons into rivers; the usual analytical methods are so designed that any contribution would largely be excluded. Our uncertain knowledge of hydrocarbons in fresh water extends almost exclusively to the remaining 10% of oil used as lubricants and non-fuels.

The main non-fuel products include naphtha, essentially a feedstock for manufacturing chemical products; bitumen, mainly used as a construction material for roads, roofs, etc.; lubricating oils, white spirits, widely used as solvents; and waxes, used in solid form and blended into greases. Of these, in terms of marine pollution potential, naphthas are relatively toxic; bitumen is a source of polycyclic aromatic hydrocarbons; lubricants are chiefly discarded as waste oils; white spirit has an indirect air-borne input and waxes reach the sea only as greases included under waste lubricants. Table 4-3 summarizes the general features of the normal disposal of these non-fuel petroleum fractions.

From Table 4-3 it is clear that lubricants give rise to the major disposal problem in effluents. Lubricants of numerous types are in use, many of them in widespread small applications. Since lubricants have usually a short life, there is considerable turnover generating substantial amounts of waste oils and greases. About one-sixth is held in permanent use or is recycled, about one-third is burned during use or on recovery leaving about one-half which is either dumped in a controlled manner or finds its way into soil, water, or sewers. Chemical additives are widely used to improve the performance of hydrocarbons as lubricants, in amounts varying with application but on average about 4% (see KORTE, 1977 for details). Some of these chemical additives are considerably more toxic than oil and most are at least as resistant to degradation. Scope exists for approaching complete recycling or controlled destruction of waste oils from static engines but the elimination of uncontrolled disposal of waste lubricants from mobile units is more difficult where there is no focal point for waste oil collection. It is equally necessary to provide an efficient routine collection of waste oils from interceptor traps, sumps and other oil separation units at engineering sites; otherwise considerable quantities of waste lubricants may find their way via surface runoff and sewers into domestic water treatment plants which have limited potential for coping with oil.

Characterization of river-borne, municipal, and industrial inputs is complex because of the extremely diverse nature of the component inputs—including appreciable levels of pesticides, PCBs, and other petrochemical derivatives. Quantification is generally made in terms of 'persistent oils and greases' by a solvent extraction stage followed by weigh-

Table 4-2

Material balance of crude oil and its products and losses, 1974
(After KORTE, 1977; reproduced by permission of the author)

Millions of tons yr^{-1}	Loss	Product
Crude oil produced		2862
Direct crude burning	29	
Operational discharges	1·4	
Accidental spills	0·2	
Evaporation losses	1·2	
Crude oil into refineries		2830
Operational discharges	0·3	
Evaporation losses	8·1	
Combustion in refineries	67·7	
Products from refineries		2754
Fuels		2418
Non-fuels		336
Evaporation losses	8·3	
Operational discharges } Accidental spills	Negligible	
Fuels		2418
Residual fuel oil		839
Gas/Diesel oil		637
Motor gasoline		598
Liquified Petroleum Gas		121
Aviation fuels		111
Refinery gas (not liquified)		52
Kerosenes		52
Evaporation losses	1·1	
Non-fuels		336
Naphthas		130
Bitumen		85
Petroleum coke		46
Lubricants		36
White spirits, SBP		33
Waxes		3
Sulphur		3
Disposal of non-fuels	336	
Fixation: Transfer, Reuse	264	
Emission: Burning	28	
Emission: Evaporation	26	
Dumping: Effluent discharge	19	

ing. It is virtually impossible to calibrate these results in terms of petroleum hydrocarbons, and in toxicological terms these values are meaningless. Persistent oils are unlikely to engender mortalities to aquatic plants and animals as they are finally diluted and dispersed in rivers, and it is unlikely that they have any short-term effects in the marine environment. Concern has been expressed about the carcinogens they may contain (see p. 1550 to 1552).

Table 4-3

Disposal of non-fuel products (10^6 tons yr^{-1}) (After KORTE, 1977; reproduced by permission of the author)

Product	Total quantity	Dumping, effluent discharge, etc.	Evaporation	Burning	Fixation, transfer reuse
Naphtha	130	—	—	—	130
Bitumen	86	1	—	8	77
Petr. coke	46	—	—	—	46
Lubes	36	17	—	13	6
White spirits	33	1	26	5	1
Waxes	3	—	—	2	1
Sulphur	3	—	—	—	3
Total	337	19	26	28	264

(c) Natural Seeps

Natural seeps and biogenic hydrocarbons form the next largest input. Petroleum seepages into the Gulf of Venezuela, on the Venezuelan island of Cabagua, off Puerto de la Cruz, and near Nueva Zamora (Maracaibo) have been known for over 400 yr (COWELL, 1976) and no doubt many other minor emissions of crude oils exist. More recently, underwater seeps in ocean areas have been alleged and some coastal seeps have been intensively studied (HARVEY and co-authors, 1979; SPIES and co-authors, 1980). Oil-drilling activities can weaken rock structure leading to blow-outs (VAN DEN HOEK, 1977; STRAUGHAN, 1977).

Environmental studies in relation to these seeps have made an invaluable contribution to our knowledge of long-term effects of exposure to oil. Quantification of seeps in shallow water is based on a direct estimate of the mean rate of flow of the seep. Many seeps display variable and intermittent flow. The toxicology of oil seeps is discussed on p. 1456 to 1458.

(d) Biogenic Hydrocarbon Inputs

It is generally acknowledged that petroleum hydrocarbons have been formed from primordial aquatic plants and animals under conditions of heat (up to 80 °C) and high pressure in the absence of air. Because of the prolonged period during which fossil hydrocarbons have been out of contact with water and the atmosphere the natural stable isotope ^{13}C is much depleted relative to the stable non-radioactive ^{12}C during maturation and radioactive ^{14}C decreases. Marine biogenic hydrocarbons are largely of recent origin and are derived from a multitude of sources on land, in fresh water, and in the sea. These recent hydrocarbons possess ^{13}C and ^{14}C relative to ^{12}C in proportions similar to, or only slightly depleted from, the ratio in the carbon dioxide of the lower atmosphere and surface waters showing recent parentage associated with photosynthesis and photosynthetic products.

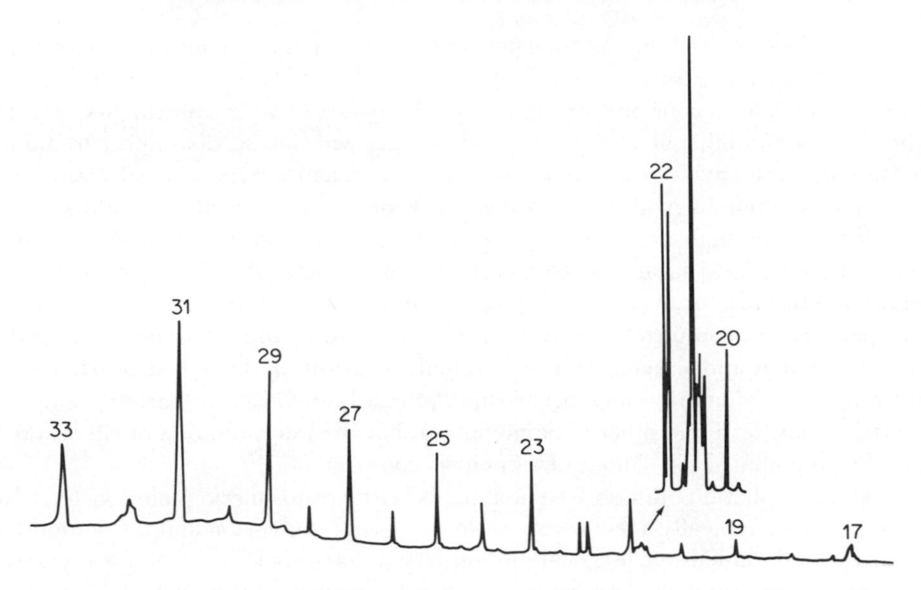

Fig. 4-2: Gas chromatogram of aliphatic hydrocarbons from clean sediment from Dales Voe, Shetland Isles. (Based on information provided by JOHNSTON 1980)

In clean areas the natural hydrocarbons of near-surface sediments show a characteristic array of n-alkanes with a strongly emphasized preference for odd numbered carbon chains (Fig. 4-2). The exact mechanism giving rise to this predominance is not known; it may be related to microbial action and maturation of plant and animal debris. There is no odd–even predominance in the n-alkanes from petroleum.

The most abundant natural hydrocarbons are the simplest gaseous alkanes, methane predominantly with much smaller proportions of ethane, propane, and butane. These gaseous fractions are omitted from the global hydrocarbon estimates as they are generally regarded as rapidly biodegraded.

There is a complex interchange of gaseous hydrocarbons between soil, air, and water. Biogenic production of methane may be at least 10^9 tons yr^{-1}. There is great uncertainty about the annual production of other marine biogenic hydrocarbons. BUNT (1975) estimates 10^{17} tons yr^{-1} but much higher estimates are currently favoured. WENT (1960) estimated that coniferous forests alone release 200 million tons of terpene hydrocarbons annually as microparticles to the atmosphere. It may well be that atmospheric biogenic hydrocarbons carried to the sea in rainout may greatly exceed the amounts of fossil hydrocarbons entering by the same route.

(e) Atmospheric Inputs

Damaging atmospheric hydrocarbon pollution is essentially a local and often transient condition. Vapours from the handling of petrol and other low-boiling-point hydrocarbons can reach lethal concentrations following mishaps in work situations and the general public can be exposed to unpleasant and distressing concentrations near processing, storage, and distribution plants. Hydrocarbon combustion products can also be obnoxious and, especially in some areas, give rise to frequent and persistent photochemical smog. This results from complicated chemical reactions of nitrogen oxides with partly oxidized hydrocarbons.

The very high local concentration of crude oil vapours created over the first few days of the *Amoco Cadiz* oil spill was reported as causing widespread discomfort to animals and man and also gave rise to immediate and subsequent effects on local plant life.

Regional pollution by hydrocarbons affecting wider areas up to 200 km radius is much rarer, because the high degree of natural dispersion precludes such problems in the absence of acute local inputs, almost certainly man made. Possible problems in this category are the dispersion of benzo [α] pyrene and related compounds in urban air and the dispersion of incompletely combusted fuel oil residues which become adsorbed on particulate matter and as aerosols. It is frequently difficult to isolate hydrocarbon pollution from other pollutants which occur simultaneously such as soot(carbon), sulphuric and nitric acids, lead and other toxic metals so that the interpretation of effects on the biota of contaminated air, fallout or rainout is complex.

Global atmospheric pollution is synonymous with stratospheric pollution for which mixing times are typically a few weeks in the east–west dimension and a few months in the north–south dimension. Vertical motion has characteristic times of a few years. In this stratum are found the sources of the major problems relating to excess carbon dioxide, threatened ozone deficiency, enhanced concentrations of sulphur dioxide and oxides of nitrogen. Methane and some of its oxidized derivatives are also involved with these in intricate photochemical reactions and equilibria. The disturbance of chemistry in the stratosphere is widely held to constitute man's most serious infringement on the natural balance of cycles and equilibria.

The quantification of these atmospheric hydrocarbon inputs is highly uncertain. Of prime importance is the combustion in one way or another of gas, oil, oil fractions, oil derivatives, oily wastes, and the many kinds of disposable plastic products. Since use of fuels amounts to at least 88% of crude oil production (natural and produced gas to be added), inefficiency at this stage can generate pollution of considerable, perhaps overbearing magnitude. Values in the percentage range apply to internal and jet combustion engines in certain modes of operation, and the inefficiency of furnaces rises when output above a certain optimum is demanded. Fuel efficiency is greatly sought after in recent engine and furnace design but the proportion of modern improvements may be outweighted by wear and malfunction in the vastly greater resource of old and very old equipment. Typical average efficiencies from large and small units are given by KORTE (1977, Table 4-4), but it is hard to tell how representative these values may now be. An impression of the magnitude of global emissions from oil fuels and gas firing may also be gleaned from Table 4-4. A larger plant is more fuel efficient than smaller units, but owing to higher furnace temperatures forms nitrogen oxides. The particulate matter is generally soot accompanied by some metal oxides, etc.

To add to the uncertainty in the best statistics for emissions at source, there is much still to be known of the fate of hydrocarbons in the atmosphere and of the dynamics of hydrocarbon interchanges between the atmosphere, the land and its living mantle, and the aquatic regime.

The general approach to monitoring hydrocarbons at low concentrations in the atmosphere is to adsorb them quantitatively at ambient temperature on to an appropriate porous polymeric substrate. A solvent extraction or other elution procedure follows, and the constituent hydrocarbons are measured by gas chromatography or gc mass

Table 4-4

Emissions from combustion of fuels, 1974 (After KORTE, 1977, Tables VII and X reproduced by permission of the author)

	Large units*	Small units**	Global oil† ($\times 10^6$ tons yr^{-1})	Global gas‡ ($\times 10^6$ tons yr^{-1})
Aldehydes	0·071	0·240	0·3	0·02
Benz[α]pyrene	$0{\cdot}13 \times 10^{-6}$	$10{\cdot}59 \times 10^{-6}$	10^{-4}	3×10^{-6}
Carbon monoxide	0·004	0·240	0·2	0·007
Hydrocarbons	0·38	0·24	0·6	negl.
Nitrogen oxides as NO_2	12·38	8·57	20	6
Sulphur dioxide	18·68	18·68	60	0·01
Sulphur trioxide	0·28	0·24	—	—
Particulates	0·95	1·43	2	0·4

* kg pollutant m^{-3} fuel, over 1000 h.p.
** kg pollutant m^{-3} fuel, less than 1000 h.p.
† Based on 1736×10^6 tons yr^{-1} fuel oil (excluding large industrial furnaces).
‡ Based on 1549×10^6 tons yr^{-1} oil equivalent to gas.

spectrometry. For total hydrocarbons, adsorption is followed by pyrolysis and the determination of carbon dioxide by infra-red absorptiometry or chemical oxidation and the determination of the amount of residual oxidant.

(f) Inputs from the Oil Industry

Deliberate discharge of treated oily waters arises in relation to the offshore oil extraction industry and the onshore refining and subsequent processing of oil fractions. Broadly speaking the amounts of these oily waters are related to the amounts of oil produced but can vary substantially, being much higher for oil fields approaching the end of their useful lives and for ageing petroleum refineries and petrochemical plants. The quoted values of 0·12 to 0·28 million tons yr^{-1} (Table 4-1) are associated with world production in the region of 3000 million tons yr^{-1} of crude oils.

The discharges from oil production platforms can be regarded as the soluble fraction from fresh crude oil. Discharges from onshore processing can include similar but less well specified inputs together with a highly variable blend of products from the processing. Due account is required of the other chemicals that may accompany hydrocarbons in these outputs and relevant associated operations.

(3) Effluents from Modern Oil Operations

Experience in North Sea offshore oil operations and shore-based industry affords a very useful source of data employing many of the latest developments in oily water treatment, oil handling and of the new as well as long-established refinery and petrochemical installations. North Sea production is about 5% of world production.

(a) Production Water

Recent estimates for the volume and trend of production water arising in the Norwegian and British sectors are summarized in Table 4-5. In future years, as the oil fields become depleted (Fig. 4-3), these estimates will at least double and a joint total in the region of 5000 tons yr^{-1} of dispersed oil would not be unexpected for the end of the 1980s. Oil-field depletion leads to the greater breakthrough of formation (and perhaps injection) water into the oil entering the arrays of boreholes conducting the oil to the production platforms. The higher the extraction rate of oil from the sedimentary bedrock, the greater the risk of water breakthrough; the higher the rate of oily water flow into the separation and treatment units housed in the platforms, the less efficient they are. Because of severe restrictions on space for such plant on production platforms there are penalties on size, and the efficiency of oil removal for any given design capacity is inversely related to flow rate. The problems encountered in the treatment of oily waters on offshore platforms have been discussed in some detail by READ (1980).

Production water (initially formation water) has been in contact with crude oil under conditions of high pressure and temperatures not far short of 100 °C. At this stage the crude oils contain gaseous hydrocarbons and all the light liquid fractions including benzene and toluene. As a result the water is saturated with a wide range of oil components and a considerable concentration of endemic organic compounds. On reaching the platform, excess gas and some of the lighter liquid fractions are released at successive stages of pressure reduction causing some of the less soluble heavy-oil fractions to form emulsions or oil droplets in the aqueous phase. In the separators a complex particle

Table 4-5

North Sea offshore oily water discharges. (a) Norwegian Shelf (Based on SCHREINER, 1980); (b) UK Shelf (Based on READ, 1980)

(a) Norwegian Shelf	Maximum volume (m^3 d^{-1})	Concentration max (mg l^{-1})	Maximum as oil (tons yr^{-1})
Ekofisk	4000	25	37
Cod	400	30	4
Edda	700	30	8
Eldfisk	3000	30	33
Albuskjell	2100	30	23
West Ekofisk	nil	—	—
Frigg (Nor.)	600	30	7
Statfjord A	60,000	25	548
Tar	900	30	10

(b) UK Shelf	Oil in predicted discharges 1980*	1985*
Displacement water	45	55
Production water	765	2430
Operations	10	10
Total oil content	820 tons yr^{-1}	2495 tons yr^{-1}

* The original paper (READ, 1980) should be consulted for information on the basis of these predictions.

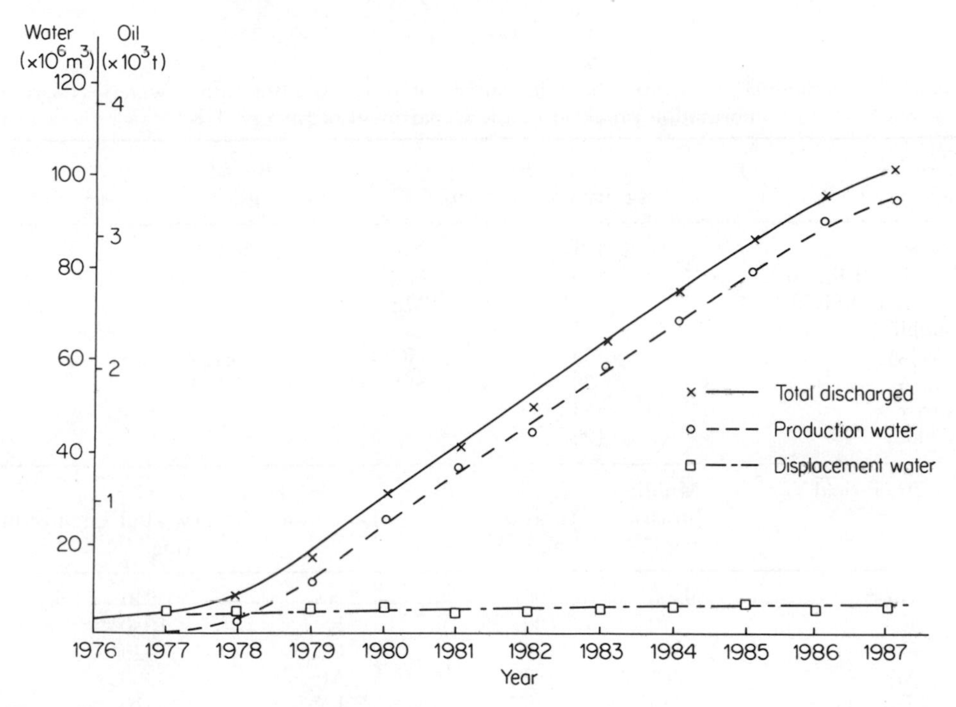

Fig. 4-3: Predicted oily water discharges from UK shelf operations. (After READ, 1980; reproduced by permission of the author)

fractionation occurs between soluble, dispersed, and liquid oil phases. In general, all hydrocarbons are much more soluble in oil than in water, but the theoretical partition coefficients may be modified by other organic components of the formation water. For North Sea installations, the concentration of oil in the treated water is in the range 15 to 40 mg l^{-1}. As there is no absolute method capable of measuring total oil, these figures relate to oil as measured by the IMCO method (p. 1491) together with the endemic organic compounds arising from the formation water. The toxicology and possible ecological impact of these discharges is discussed on pp. 1507–1509.

Although geochemists have studied formation water in considerable detail, mainly with regard to variations in composition in terms of mineral elements, the organic composition is very complex. LYSYJ and co-authors (1981) describe the hydrocarbon composition of production/formation water from 3 platforms in Cook Inlet, Alaska. Total organic carbon was generally in the range 400 to 500 mg C l^{-1} of which a variable amount was as suspended oil which could be reduced by treatment to about 36 mg l^{-1} with aromatic hydrocarbons proportionally higher in the treated water. Much of the dissolved non-volatile organic matter (140 to 400 mg l^{-1}) could not be accounted for.

Some unpublished UK results for production/formation water are of considerable interest. They show a wide range in brine strengths and composition and quite large departures from the mineral composition of sea water. Most brines precipitate iron oxides and hydroxides for a considerable period after samples are drawn. Some of these brines also form copious precipitates of alkaline earth carbonates and hydroxides along with some of the less soluble hydrocarbons when mixed with sea water. Not much is

Table 4-6

Some data submitted regarding the salt content of production/formation waters (Based on information provided by the Department of Energy, UK)

(a)	Total solids (g dry wt l^{-1})	(b)	Buchan (g l^{-1})
Forties	101	Na	58
Thistle (21.9.78)	<5	Ca	16
Thistle (29.11.78)	27·5	Mg	1.03
Dunlin	7	Cl	122
Beryl A	35	HCO_3	0·122
Auk A	78	SO_4	<0·001
Argyll	43		
Buchan	196		

(c) Brent field	Middle Jurassic (mg l^{-1})	Lower Jurassic (mg l^{-1})	(d)	58° 41′N 1° 17′E (Date from Matthew Hall Engineering) (mg l^{-1})
Na	9600	5100	Na	21,240
K	170	180	K	1030
Ca	65	55	Ca	420
Mg	260	480	Mg	50
Total Fe	7·3	6	Fe	10
Dissolved Fe	0·2	2·3	Ba	21
Ba	39	39	St	6
Cl	14,400	13,472	Cl	31,620
CO_3	Nil	Nil	HCO_3	5030
HCO_3	1460	433	SO_4	26
SO_4	33	21	CO_3	Nil
Diss. solids	26,710	23,910	OH	Nil
Susp. solids	140	92	NO_3	16
Specific conductivity at 25 °C = 0·0236 ohm metres pH 6·6 No bacteria were detected			Specific conductivity at 15·6 °C = 0·158 ohm metres pH 6·7 Saturated with CO_2 and methane at 20 p.s.i. and 65 °C	

(a) Total solids in North Sea formation waters.
(b) Salts from Buchan Field.
(c) Formation water from Brent Field (Data: courtesy of Department of Energy).
(d) Formation water from 58° 41′N 1° 17′E (Data: courtesy of Matthew Hall Engineering Ltd, London, and Shell UK Exploration and Production, Aberdeen, Scotland).

known about the identity of the non-hydrocarbon organics and the whole topic needs much more detailed study (Tables 4-6, 4-7). The organic matter in sea water can considerably influence the solubility of oil or of specific hydrocarbons; its effect on the toxicity of oily waters has not been determined.

(b) Displacement Water

In the process of transferring oil emerging from the array of boreholes to either a pipeline to shore or a buoy system for loading tankers at sea, temporary storage is

Table 4-7

Tentative analysis of production water. Sample of Forties Field production water, 11.12.79. (Data: courtesy of BP Grangemouth Ltd, UK)

Total solids by evaporation 87·5 g l^{-1}
'Salinity' by chloride titration 87·5 g l^{-1}
Suspended solids 0·38 g l^{-1} as received, the amount increases with time due to precipitation of Fe, etc.
pH 7·0 (as received)
Soluble organic carbon by UV irradiation and CO_2 by IR spectrometry:

Unfiltered	459 mg Cl^{-1}
Filtered	465 mg Cl^{-1}

	Filtered	Unfiltered
Iron	33 mg l^{-1}	65·5 mg l^{-1}
Manganese	3·9 mg l^{-1}	4·0 mg l^{-1}
Cobalt	50 μg l^{-1}	110 μg l^{-1}
Nickel	60 μg l^{-1}	60 μg l^{-1}
Zinc	18 μg l^{-1}	—
Lead	<3 μg l^{-1}	—
Copper	3·3 μg l^{-1}	—
Cadmium	2·7 μg l^{-1}	—
Mercury	—	<2 μg l^{-1}

No detectable radioactivity
Nitrogen

Total *ca.* 70 mg N l^{-1}	Ammonium—N	38·7 mg l^{-1}
	Sol. organic—N	28·2 mg l^{-1}
	particulate—N	13·0 mg l^{-1}
	nitrate + nitrite—N	0·062 mg l^{-1}

Phosphorus
Soluble reactive phosphate 0·87 mg l^{-1} (filtered)
Total phosphate—no result, interferences
Total hydrocarbons about 10 mg l^{-1} (gravimetric)
Total hydrocarbons (IR method) 50–60 mg l^{-1}
Other components suspected: aliphatic and aromatic carboxylic acids; no esters found.

necessary. There must be a break in the system to allow for pressure balancing in the pipeline, and in the case of tanker loading from a buoy system interim storage allowing oil to accumulate in time for the next load. Since it is unsafe to store crude oil in open tanks it is usual to adopt storage over sea water. This contact between oil and water gives rise to displacement water. There are 2 general methods namely: (i) enclosed storage of oil above a permanent volume of water that is displaced into a reserve tank as oil accumulates and water is pumped back to displace oil as needed; (ii) open-ended storage with more or less free access of sea water beneath the oil tank. Where there is storage of oil above a flexible water bed the loss of oil in displacement water is appreciable (SCHREINER, 1980). Compared to the oil spill situation, however, the condition of the oil/water interface in storage tanks is essentially tranquil and the rate of loss of oil is low.

(c) Contaminated Rainwater, etc.

On most production platforms separate arrangements have been installed to cope with (i) oil-contaminated rain and wash water, and (ii) the occasional small amounts of oil lost from leaks and necessarily released during repair and maintenance of pipework, valves, pumps, etc. Usually platform rainwater drainage is conducted to an intermediate tank from which the settled water layer is passed to the oily water separators; at sites where routine small oil 'spills' are expected there are drains connected to oil interceptors from which the trapped oil is recovered.

(d) Oily Ballast Water

On shore, deliberate discharges of treated oily ballast are associated with those loading terminals which provide the reception facilities for oily ballast water. In addition some refineries have the capacity to provide oily ballast reception treatment.

In principle, onshore processing of oily ballast water appears to be similar to that for production of water offshore; however, the operations differ in the scope of treatment available on shore, and in the continuity of the differing activities which generate the oily water. Tankers in ballast calling at a terminal to load oil cargoes discharge their oily ballast, which usually amounts to 1/3 of their cargo capacity, simultaneously with cargo loading. Both rates are adjusted, and sector tank filling and emptying are balanced to a rigorous schedule, so that the stability of the tanker and safety of its crew are ensured.

The normal stages in oily ballast treatment begin with initial reception in large enclosed tanks where settlement (usually for 15 to 24 h) permits the first gross stage of watery oil and oily water separation to take place. The oily water is later drained off and passes to the treatment plant. The watery oil is pumped to storage and further settlement; then the emulsion layer is pumped to the special units where the emulsion is broken down by heat treatment aided by chemical additions if necessary. The oily water separator plant usually consists of some kind of plate separator together with one or more accessory stages such as air flotation, chemical flocculation or sand filtration; for sensitive environments a final biotreatment stage with clarifier may be incorporated. Successful operation depends on spreading the load of ballast water to match the plant capacity (tanker turnover planning) and successfully coping with problems associated with certain crude oils or uncooperative mixtures of crude oils. In general a standard of 25 mg l^{-1} treated ballast water should be readily attained and maintained routinely; in practice 10 to 15 mg l^{-1} is more usual. Full treatment can reliably meet a standard of 5 mg l^{-1} with operational values running close to 3 mg l^{-1}.

(e) Segregated Ballast Water

Since it would be difficult to envisage a scheme by means of which oil production platforms can accept and process oily ballast water in addition to their own oily waters, at all North Sea platforms where oil is loaded at special mooring buoys, only tankers carrying permanent ballast or segregated ballast are used.

Segregated ballast tankers loading at offshore loading buoys or shore terminals arrange to pump out the ballast water in step with the filling of cargo compartments. In theory this ballast water should be oil free but in practice—due to flexing movements of the ships' structural members when under way and possibly also due to minor cracks in these compartments—oil in varying degree enters and contaminates this water. The water taken aboard will differ in salinity and chemical composition and in biological characteristics, including bacteria and parasites from the surrounding water at the point of discharge. If the reception area has restricted water exchange and there are sensitive or vulnerable biota, then even this apparent improvement over oily ballast water may have its own particular disadvantages. Oily ballast water is unlikely to transmit viable invasive marine species, except bacteria and viruses.

(f) Refinery and Petrochemical Effluents

Oil refineries have acquired an unflattering reputation with the general public as a source of oil pollution and are a special target for environmentally-minded extremists. In a doctrinal sense, refineries occupy a key role in man's exploitation of one natural resource, vital for the promotion and sustenance of today's high-technology culture. Amenity-wise, refineries have limited visual, aural, and nasal appeal. In terms of aquatic pollution they cannot honestly be accused of creating undue waste problems in relation to the bulk of material processed and the value of the work done, the products manufactured and the money earned on behalf of the whole community. Compared to many other industries their proven record for reduction in waste effluents and in conservation of energy and water stands high.

At the beginning of 1976 there were 256 refineries operating in the United States capable of processing 15×10^6 bbl d^{-1} (RESCORLA, 1977) and in Western Europe 166 refineries capable of processing $20{\cdot}2 \times 10^6$ bbl d^{-1}.

> 'In the United States, the construction of 12 grass root refineries has been proposed since 1970 for the east coast but has been delayed in a majority of cases as a result of environmental objections. Another factor in the delay was the soaring construction costs and, as a result, it is not anticipated that any new refineries will be constructed before 1985' (RESCORLA, 1977, p. 14).

Petroleum refinery statistics for world areas (RESCORLA, 1977) are given in Table 4-8.

In early times an oil refinery was, as the name describes, a plant where crude oil was refined by the very simple basic process of washing out residual brine with water and subjecting the salt-free oil to distillation at atmospheric pressure in a conventional water-cooled still. Modern refineries may bypass the desalting stage, which is essentially a step to reduce corrosion rather than necessary chemistry, and employ highly sophisticated multi-stage fractionating columns backed up by vacuum distillation units for special products such as lubricant feedstock. Few refineries are, however, totally restricted to refining only, and many have integral petrochemical processes to convert the various oil fractions into finished products such as feedstocks, gasoline, diesel oils, aviation fuels. Whereas a simple refinery would be expected to generate a typical oily water effluent, the introduction of catalytic and cracking processes (which are in principle dry reactions) promotes the possibility of more complex wastes generally from stripping processes and the removal of unwanted elements such as sulphur. In addition there are the usual oily waters from site drainage, storage tank bottoms, production water washings, and water treatment regeneration.

Major progress in reducing aquatic oil pollution has followed from the evolving trend from once-through water cooling to recycled enclosed systems and to air cooling.

The quality of waste water from refineries varies according to day-to-day demands for product ranges and batch-wise special production runs. Detailed information is given by

Table 4-8

Petroleum refineries and product consumption statistics (After RESCORLA, 1977; reproduced by permission of the UNEP, Paris)

	Latin America	Middle East	North America	Communist areas	Western Europe	Africa	Asia & Pacific	World Total
Number of refineries (1976)	90	34	294	100	166	38	109	837
Refinery capacity (1976) ($\times 10^3$ bbl d^{-1})	7690	3288	17,119	12,406	20,238	1320	9778	71,839
Product consumption 1975 (estimate) ($\times 10^3$ bbl d^{-1})	3784	1669	18,085	10,199	13,030	960	7855	55,582

JONES (1973) along with the US. Environmental Protection Agency recommendations regarding technological devices for effluent treatment and control and corresponding environmental standards to be met.

Because of the diversity of contaminants in refinery discharges the usual routine analyses are the familiar non-specific determinants such as suspended solids, BOD, pH and extractable oils and greases together with specific items such as phenols, sulphides, and ammonia.

According to LUND (1971), in the early seventies it was the practice at one European refinery to collect the spent caustic, phenols, and thiophenols in a closed system for tank storage. Periodically the contents of the tank were pumped into an oil tanker in ballast for discharge into the mid-Atlantic Ocean. This report is in conflict with the general trend of refinery waste management now and the dramatic evidence of improvement reported by CONCAWE (1980) for European refineries (Fig. 4-4).

In 1969 the oil content of refinery discharges in terms of refining capacity was 0·015% (150 tons oil loss for every 1×10^6 tons throughput) and this has fallen to 0·004% for 1974. Rather less headway was made in reducing BOD and phenols, (see inset on Fig. 4-4). From these figures the annual European refinery oil loss was 54,860 tons in 1969 and 39,942 tons in 1974 which puts current annual oil loss at about 240 tons per refinery.

Usually refineries are located either inland or on flat inner estuarine sites. As a result, geographical constraints often accentuate their amenity impact and promote the unfortunate close conjunction of refineries with urban centres. With regard to impact on the estuarine environment, refinery locations may suffer from less than ideal conditions for effluent dilution and dispersal and may be adjacent to sensitive ecological communities.

(g) Chemicals Used in the Oil Industry

Although this chapter is devoted to oil pollution, it would be wrong not to mention the use of oil-based drilling muds and a broad range of toxic chemicals. In general all these substances are used only at certain specific stages of development and often they come only into limited contact with the sea. Typical applications are the use of bactericides to combat sulphate-reducing bacteria, the treatment of water for borehole injection and the protection of underwater pipelines and other pipework.

In every phase of oil production a constant battle is waged against sulphate-reducing bacteria. These anaerobes are dependent on low-molecular-weight organic substrates as carbon source; to replace dissolved oxygen, any available sources of bound oxygen in nitrate, sulphate, oxides of iron, etc., are reduced. As a result, ammonia, hydrogen sulphide, and finely reduced iron are produced causing accelerated corrosion of pipes and tanks since the protective oxide layer is continuously eroded. The bacterial slime and other particulate matter can cause operational difficulties and all the while valuable oil is being consumed. Activity of this type can become prevalent, sometimes equivalent to a general epidemic, in pipework, process units and storage tanks all along the pathways from borehole to refinery and is also a serious problem in the handling of specialized oil products such as aviation fuel which for safety reasons must meet high standards for freedom from particulate matter and corrosion entities.

Oil in the oil-bearing strata is essentially sterile and has attained static equilibrium over geological epochs with the water and rocks with which it has contact. On breaching these strata there are risks of allowing the entry of sulphate-reducing bacteria and along

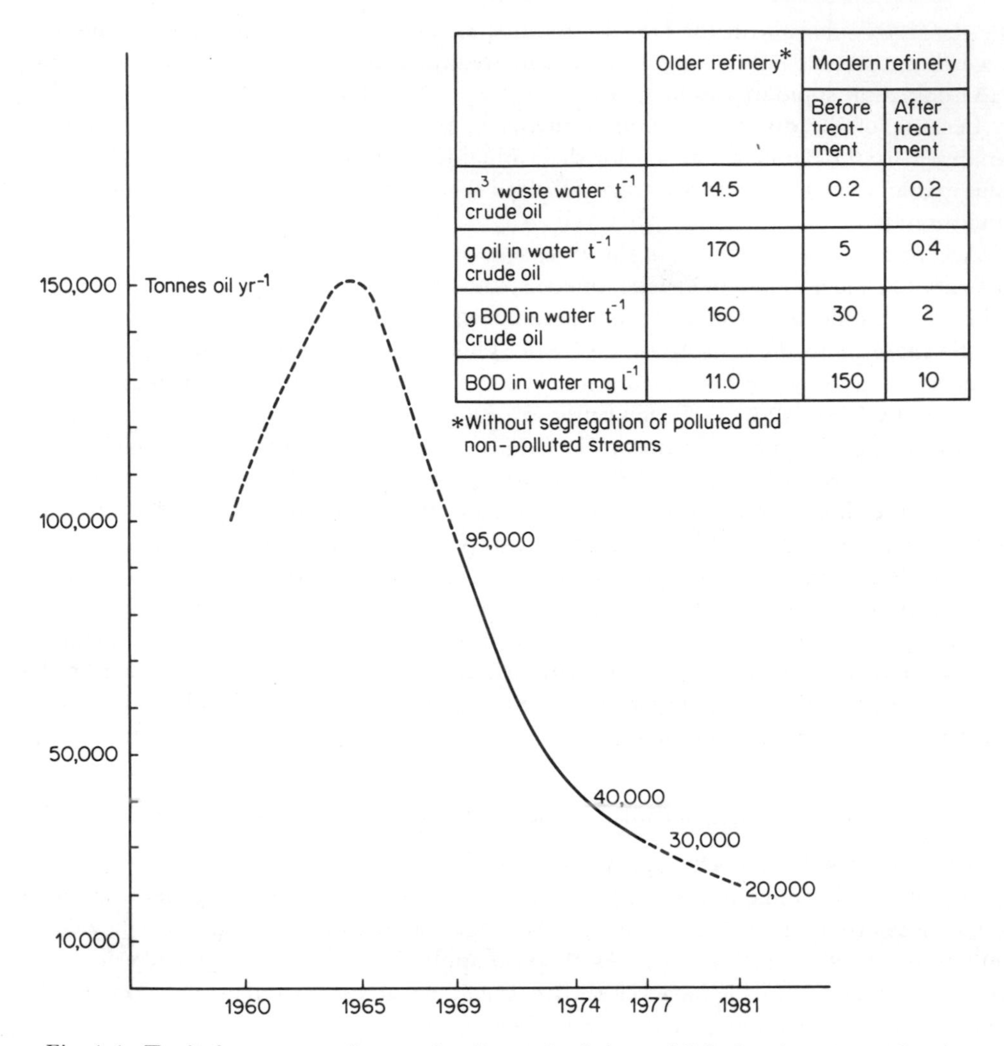

	Older refinery*	Modern refinery	
		Before treat-ment	After treat-ment
m^3 waste water t^{-1} crude oil	14.5	0.2	0.2
g oil in water t^{-1} crude oil	170	5	0.4
g BOD in water t^{-1} crude oil	160	30	2
BOD in water mg l^{-1}	11.0	150	10

*Without segregation of polluted and non-polluted streams

Fig. 4-4: Typical wastewater flows and pollutant loads in established and modern refineries in OECD Europe. (After CONCAWE, 1980; modified; reproduced by permission of CONCAWE)

with them drilling mud or injection water, thus setting up conditions for the growth and abundant development of these bacteria. It is essential then at all stages of borehole drilling and installation to eliminate or counteract any entry of bacteria. This is achieved by the use of biocides, variously termed bactericides, slimicides, mud inhibitors, etc. To fulfill these functions, powerful and persistent toxic chemicals must be used, compatible with the technological properties of drilling mud or injection water, and such chemicals will be toxic to marine life. For these chemicals to function they must be water soluble since bacterial growth prevails in trapped water and water/oil interfaces and cannot exist in oil alone.

The release of these oil chemicals to the environment only occurs when used mud is jettisoned or is carried overboard on rock cuttings or when injection water reaches the oily water separators and is discharged. Because drilling muds are recycled until exhausted, and since drilling is a fairly limited and enormously expensive aspect of oil extraction, there are quite discrete limits to their usage. Recently, GETTLESTON (1980) has provided a very useful and detailed account of the effects of oil and gas drilling on the marine environment. He also lists the many components of water-based drilling muds and quotes a typical composition and usage rates. Among the biocides he lists aldehydes, quaternary amines, diamine salts and chlorinated phenols. Of these the chlorinated phenols have the greatest potential for accumulation and persistence along with LC_{50} toxicities in the range 0·2 to 2 mg l^{-1}, similar to some of the complex quaternary amines which are less persistent. The use of chlorinated phenols has been discouraged in North Sea applications for fear of causing taint to fish, and similar US reaction is quoted by GETTLESTON (1980). He concludes that there may be some local physical impact on the benthos from drilling muds and cuttings but in general any chemical impact is very local and no contamination of biota attributable to mud chemicals can be traced. MONAGHAN and co-authors (1980) provide further data on mud use and the toxicity (LC_{50}) of component mud chemicals; neither report deals with oil-based drilling muds (see p. 1454).

The purpose of injection water is to create a rising water table to displace the oil from interstices of rock formations, and its controlled entry below the lens of oil is most carefully engineered to avoid—as far as technical expertise can ensure—the penetration of the introduced water through the oil thus defeating the objective of the exercise. In consequence—in theory at least and by every practical stratagem—injection water should not be returned to the platform, at worst in minimal proportion. Injection water must be most carefully filtered, primarily so that no extra particulate matter or slimy material is introduced which would possibly increase the problems of sustaining flow in the bedrock and in the interfaces between bedrock and well pipes. In addition, filtration helps to reduce bacterial numbers, lower the oxygen content (prior to deoxygenation) and remove potential nutrient sources. Bacteria are further depressed by treatment with chlorine (or similar halogen) or a bactericide. Special chemicals may also be added to promote the flow of oil and keep the interstices open.

Hydrogen sulphide gas is poisonous, and in the ambience of oil platform working its normally obnoxious smell may not be noticed or heeded and several cases of workmen being overcome and killed by this gas when entering process chambers and related restricted spaces have occurred. Biocides and other means of controlling sulphate-reducing bacteria are, therefore, in regular and in some instances continuous use on production platforms in quantities appropriate to the size of the units treated. These chemicals—especially when offshore costs are considered—are very expensive and their use is carefully monitored by the operators.

In addition to biocides, mud chemicals and chlorine (which is extensively used to dose injection water), oxygen scavengers and corrosion inhibitors are used in preparing pipelines prior to commissioning and if need be in association with periodic pipeline cleaning operations. Only the initial chemicals in sea water or fresh water medium used in preparing the pipelines prior to commissioning need be discharged in controlled fashion to the sea. The plugs or segments of cleaning solutions in routine 'pigging' of the pipelines are usually dispatched landwards and disposed of ashore. Chemicals are employed to a minor extent in the processing of natural gas at platforms and at shore-based reception plants with very few of these chemicals finding an entry into the sea.

It is appropriate to add a cautionary remark concerning the use of chlorine and other halogens as antiseptics. The practice has been in widespread common use for many years in relation to drinking water and over the past decade has extended rapidly in some countries into the disinfection of processed sewage and industrial wastes, especially food wastes. Chlorination gives rise to a wide range of halogenated residues normally in low concentration but amounting to many tons per year. Chloroform, carbon tetrachloride, chlorinated aromatic hydrocarbons and phenols, etc., are undesirable contaminants now occurring widely. Data on the amounts of halogenated compounds, their persistence and bioaccumulation, and their direct and indirect health effects are a relatively recent topic for study (e.g. JOLLEY and co-authors, 1978). The total global inputs of these substances to the sea may far exceed those of pesticides and polychlorinated biphenyls but their environmental importance is as yet unknown. That 'the practice of chlorinating waste waters containing aromatic hydrocarbons can no longer be regarded favourably' (SMITH and CARPENTER, 1977) seems a sensible guideline. To date no reports have been forthcoming on the possible environmental repercussions of the use of chlorine in the offshore oil industry.

The industrial practice causing most concern to marine environmental scientists is the use of oil-based drilling muds. These have ordinary mud components but water is replaced by a mineral oil as suspension medium.

(h) Oil-Based Drilling Muds

Oil-based muds are employed where particularly hard rocks are encountered or where it is desired to drill strongly deviating holes. The hydrocarbon base is usually diesel or gas oil. Special provisions are introduced to minimize the oil pollution from the contaminated cuttings. The rock cuttings passing back up the drill stem are separated on special screens and the oil-based mud is recycled. At the end of the drilling phase the oil-based mud is normally brought ashore for disposal. Rock cuttings are processed by high pressure washing or by burning or by various patent cleansing processes, but so far national authorities and oil companies alike are aware that the clean-up methods seldom fulfil the designer's expectations when operated under field conditions. The partly cleaned cuttings pass down a water-filled shaft and are ejected into the sea some tens of metres above bottom. Any oil separated from the cuttings in the water-filled shaft is recovered by pumping the upper part of the water column to the oily water separators. Because the remaining oily mud adheres tenaciously to the rock cuttings, dispersal of oil from the mounds of drill cuttings is slow, and it is claimed (SCHREINER, 1980) that even the present imperfect arrangements protect the environment against pollution.

To illustrate the scale of these operations involving oil-based muds GRAHL-NIELSON and co-authors (1980) have described recent experience at Statfjord A in the Norwegian sector of the North Sea where 4 wells have been drilled out of a programme requiring 32 wells. At the primary screens where the mud is separated for recycling, drill cuttings for clean-up contained 128 g hydrocarbons kg^{-1} of wet cuttings. One option for cleansing oil-contaminated drill cuttings is to wash them with diesel oil, but as in the UK sector this was not successful. Confirming this, GRAHL-NIELSON reports an increase in hydrocarbon content to 177 g kg^{-1}. On the basis of assurances from the operators and their mud experts, the Norwegian government had set a target standard of not more than 50 l hydrocarbons m^{-3} of cuttings as discharged. In practice the amount of oil on a volume/

volume basis varied between 74 l m^{-3} and 480 l m^{-3} with a mean of 298 l m^{-3}. In the process of drilling the first 4 wells, 1700 m^3 cuttings were generated. Of this an estimated 450 m^3 cuttings had been exposed to a burning process for clean-up, together with a quantity of used mud, and the remaining 1250 m^3 cuttings had been disposed of as described initially. Burning has also been tried in the UK sector but is of limited success since complete combustion is not achieved.

At this platform the chute discharged the washed cuttings 65 m above the sea floor forming a mound which was 15 m high. Passage through the water-filled discharge trunk was found to remove on average some 20 % of the attached oil. On this basis an estimate of 300 m^3 (250 tons) of hydrocarbon resembling diesel oil was deemed to be associated with this mound of dumped cuttings.

Under the prevailing tidal currents, which did not exceed 43 cm s^{-1}, most of the oil remained bound in the mud attached to the small pieces of cut rock and the small amounts of soluble components generated an elliptical pattern of enhanced hydrocarbons in bottom water within an area of some 4 km^2. The pattern of hydrocarbon concentration found from field surveys was in general agreement with that predicted from continuous records of current speed and direction 1 m above bottom.

Continued efforts are being made at all North Sea platforms, where oil-based muds must of necessity be employed, to develop oil-recovery or removal processes that are more effective. The present use of oil-based muds is confined to 14 out of 73 wells currently being drilled for production or water injection in the UK sector. Should the programme of 32 wells at Statfjord be completed on the lines of the experience described above, this would entail the release of a further 2100 m^3 diesel-type oil. Norwegian scientists are planning to continue their studies and to include examination of the benthos of the area. Further discussion of oil concentrations in the vicinity of production platforms is given on p. 1465 to 1471 and 1474, 1475.

(4) Lessons from Oil Seeps

(a) Background Papers

Natural submarine oil seeps have been described for Santa Barbara Channel, California; Alaska Peninsula, western Gulf of Alaska; Gulf of Mexico and Caribbean Sea. It seems highly unlikely that seeps are confined solely to the mid and north American continent but these are the only ones for which marine studies appear to have been reported (see for example the bibliographies listed in the literature appendix and the article by GEYER and GIAMMONA, 1980). Most of the published observations on oil seeps relate to the Santa Barbara blowout and the natural chronic escape of crude oil from the weak strata off Coal Oil Point, southern California (USA).

Detailed and comprehensive observations were made on the impact of the blowout on virtually every component of the coastal and shallow water ecosystems (STRAUGHAN, 1971). The beached dead seabirds from January to March 1969 totalled 3600. They comprised mainly swimming species such as loons, grebes, and cormorants—with few gulls, terns, and shore birds though many were present in the area. Marine mammals apparently sufferred no ill effects from the spill. Fish populations and diversity of fish species changed little and there were no reports of tainted commercial fish species, crustaceans or cephalopods. Damage occurred to shore species such as barnacles, marine grass and attached algae. Subsequently there was no reduction in the breeding of intertidal sessile barnacles (2 spp.) but the lower intertidal stalked barnacle and the mussel *Mytilus californianus* showed reduced breeding rates. The latter species is here at the warm limit of its geographical spread. Within a year substantial recuperation occurred in virtually all of the affected plant and animal species. Damage was much less than predicted when the blowout commenced; the main factors limiting the damage being : (i) the relatively low toxicity of the oil by the time it covered the 10.6 km from site to shore; (ii) the modifying effect of other natural seeps in the area over a long period; and (iii) abnormally heavy rainfall and subsequent flooding causing above-normal inshore water renewal.

According to SPIES and co-authors (1980) no less than 2000 individual seeps have been located along the southern California borderland with some corresponding 22 faults and cross-faults. Some seeps emit only gas. Off Coal Oil Point in the Santa Barbara Channel at depths of 20 to 35 m, tarry oil escapes into the sea—usually at a rate of 50 to 70 bbl d^{-1} (about 10 m^3 d^{-1}) but on occasion up to 100 bbl d^{-1} (nearly 20 m^3 d^{-1}) according to STRAUGHAN (1977) or 15 to 400 bbl d^{-1} as revised by FISCHER (1978). This seep is the best known and largest submarine oil intrusion.

Sediments from sites chosen as controls had background levels of CCl_4-extractable material generally below 100 mg l^{-1}, never in excess of 200 mg l^{-1} as determined by an analysis using infra-red (units are as quoted in the original, STRAUGHAN, 1977, which presents no analytical details). Macrobenthic infauna (2 mm mesh size), regarded as sensitive to petroleum hydrocarbon were found in sediments of 1000 mg l^{-1}, and even 10,000 mg l^{-1}, extractable organics. Extractable organics above 100 or 200 mg l^{-1} are interpreted by STRAUGHAN as weathered petroleum residues. Petroleum hydrocarbons as determined by the gas chromatographic methods employed were absent from the edible parts of large molluscs such as abalone and mussel in samples (6 to 30 m depth) from control sites and from Coal Oil Point but were detected in the viscera of 4 out of 21 molluscs at the latter site. Abalone grew at virtually the same rate and had the same body weight for shell length at all the sites and the reproductive cycles were similar. No malformed or histologically abnormal molluscan gonads were found among 2200 samples examined. Earlier analyses had found 30 to 210 μg hydrocarbon g^{-1} in 'whole animal samples' of *Mytilus californianus* from Coal Oil Point. Other samples gave an average of 130 $\mu g\ g^{-1}$ compared to 34 $\mu g\ g^{-1}$ near the Los Angeles sewage outfall and 7 $\mu g\ g^{-1}$ at Pismo Beach. No petroleum hydrocarbons were found in mussel samples from the legs of two oil production platforms.

STRAUGHAN (1977, p. 566) emphasizes that

> 'these data must be extrapolated with care because this (i.e. that marine invertebrates can live and breed in areas of chronic exposure to crude oil) may not be true in areas where completely different types of petroleum are being produced'.

He goes on to reject the argument that animals living near natural seepages have evolved a higher tolerance of hydrocarbons on the grounds that these species have pelagic larvae and that in consequence recruitment is not specifically limited to the areas near seeps. There may be a gradual increase in tolerance in the animals after settlement.

BARSDATE (1972) describes effects of natural oil seeps at Cape Simpson, Alaska. In ponds where the water was in contact with old tars and asphalts, phytoplankton productivity and abundance were high, accompanied by large populations of bacteria. Productivity was substantially less in oil-free ponds and in ponds containing fresh, low-viscosity oil. It was thought that phytotoxicity might limit the development of phytoplankton in waters in contact with fresh oil and that, at much lower levels of soluble hydrocarbon stress, productivity was high possibly because of reduced grazing pressure.

(b) Recent Research

SPIES and co-authors (1980) have produced a more comprehensive ecological study of the 1000 m^2 Isla Vista seep area which includes many 0·25 to 2·0 m^2 patches of active seepage where droplets leave the generally fine, sandy sediments. Gas bubbles containing methane with small amounts of C_2-C_5 alkanes are common in the active patches and have petrogenic characteristics. The benthos in seep and comparison areas were evaluated using the Shannon–Weiner diversity index, evenness value, dominance–diversity curves, and Spearman's ranking correlation. In addition, patterns of population fluctuation for the 10 most abundant species were compared for the sets of 15 bimonthly surveys between 1975 and 1978. The first benthic analyses showed that the main difference in faunal composition was the occurrence of a large number of oligochaetes at the seep station, some 10 times higher than at the comparison site. In general, the patterns of population fluctuation, numbers of species and community densities were in accord with higher densities at the seep station revealing peak differences from the comparison station at times when densities were highest. Also the fluctuation of some species populations were out of phase with community abundance, especially at the seep station which together with the above trends suggests that community stability differs at the 2 sites. Tests of persistence stability supported previous indications of larger fluctuations at the seep station and relatively greater community stability at the comparison station. Deliberate disturbance of the seabed in the seep area induced a sharp fall in the number of individuals and dramatic redistribution of just about the usual number of species. Something approaching complete recovery was achieved in 8 wk.

The relationship between benthic community and oil seeps giving support to enhanced numbers is of great interest. Does the enhancement (i) arise from utilization of the added organic-rich substrate, (ii) simply reflect adaptation of some opportunistic organisms, or (iii) merely demonstrate that the seep is not a toxic environment? The third dubious option is soon dismissed since the seep oil is shown to be as toxic as Prudhoe Bay crude oil; hence there is no quibble that seep oil might be innocuous. To test the organic enrichment hypothesis the microbial biomass was determined by measuring ATP at the centre of an intrusion, around its margin, and at the regular sampling sites. Microbial ATP ratings in sediment samples were higher at the centre than around the periphery which in turn were higher than the regular sites. Macrofauna (excluding nematodes) were less abundant (4 individuals of 3 species) at the centre of intrusion than at the margin (33 individuals of 11 species) which in turn was less rich than the regular stations (100 individuals of 50 species) as sampled by corer. Hence microbial abundance is not reflected in benthic richness. For nematodes, there were few at the centre but many more at the margin than at the regular sites. Other studies on ^{13}C abundance in 2 species of deposit-feeding polychaetes supported the hypothesis that seep-oil carbon, deficient in ^{13}C, made a significant extra contribution to their carbon uptake. Hence the evidence in support of the suggestion that the supplementary organic input enhances the ecosystem does not provide a simple yes/no interpretation.

A bioassay using early-stage larvae of the locally common starfish *Patiria miniata* showed no support for the adaptation hypothesis, just as STRAUGHAN (1976) had shown earlier using *Strongylocentrotus purpuratus* for Santa Barbara crude oil. Furthermore, examination of *P. miniata* gonad tissue failed to reveal any induction of aryl hydrocarbon hydroxylase, though there is literature support for induction in polychaetes by No. 2 Fuel Oil. Induction of the hydroxylase was demonstrated, however, for a bottom-feeding fish. Hence the evidence once again does not produce a clear-cut answer applicable to all levels of the ecosystem.

Probably the most striking feature of these seep studies is the lack of evidence of widespread mortalities or even of an explicitly hostile environment in the neighbourhood of these hydrocarbon intrusions. The ecological disturbance could be compared to other forms of organic-rich introductions—such as sewage, food remains, etc.—when scale is taken into account. Many seeps are very slow, much slower than the 100 l min^{-1} off Coal Oil Point which is regarded as the most active intrusion. In terms of oil input this 2 tons d^{-1} (max.) is comparable with many single oil industry treated discharges but probably does not match these in terms of oil availability since much of the seep oil is admixed with bottom sediment and remains as a slowly weathering tarry mound. All the work on seeps relates to fairly shallow warm waters with generally good open circulation. As SPIES and co-authors (1980) comment, it remains at present an open question whether the effects they have seen in a broad study of the seep community are representative of chronic oil pollution world-wide.

(5) Hydrocarbons in Sea and Sediment

(a) Oceanographic Surveys

As stark numbers, (Table 4-9), the total hydrocarbon concentrations in sea and ocean give very little idea of the ecological picture behind them. There is much to be learned from the make-up of the complex hydrocarbon mixture, how concentration and quality vary with depth and how seasonal factors are demonstrable in hydrocarbon amount and composition. Although it would be hard to demonstrate any likely effects of the general overall hydrocarbon concentrations in oceans and open seas, all reports, involving even modest numbers of samples, contain occasional unexpectedly high values here and there at the surface, in the water column, and in bottom sediments. These high values are more frequent along tanker traffic and shipping routes, in coastal waters, and near oil-related operations. Because of the variety of methods of approximating for total hydrocarbons or specific groups of hydrocarbons it is impossible to resolve this rather ragged and unsystematic information into a neat set of charts of distribution as for temperature or salinity.

The main purpose of this section is to provide the reader with the observed range and frequency of values against which ecological observations and interpretations may be placed. Biogenic hydrocarbons and other components of the natural organic matter in surface film, water column, and upper seabed sediment have important bearing on oil concentrations and on the microbial populations. The realm of organic matter in the sea

Table 4-9

Hydrocarbons in sea water (After BOEHM* and co-authors, 1978; reproduced by permission of the authors)

Location	Concentration ($\mu g\ l^{-1}$)	Comments	Reference
Georges Bank region	0·2–98	Gas chromatography (GC)	This study
South Texas OCS	0·1–2·0	Paraffins only	BERRY and BRAMMER (1977)
Alaska OCS		GC	SHAW and co-authors (1977)
Gulf of Mexico Loop Current	0–75	GC	ILIFFE and CALDER (1974)
West African coast	10–95	GC	BARBIER and co-authors (1973)
French coast	46–137	GC	BARBIER and co-authors (1973)
Open ocean (Atlantic)	1–50	IR	BROWN and co-authors (1973)
	<6	Fluorescence	GORDON and KEIZER (1974)
	20	1–3 mm Fluorescence	GORDON and KEIZER (1974)
Mediterranean Sea	2–200	Surface (IR)	BROWN and co-authors (1975)
	2–8	Subsurface (IR)	BROWN and co-authors (1975)
Atlantic	0·5–6		BROWN and co-authors (1975)
Baltic Sea	50–60	Non-aromatics	ZSOLNAY (1972)
Gulf of Mexico (coastal)	1–0·6	n-alkanes only	PARKER and co-authors (1972)
Galveston Bay area	8		BROWN and co-authors (1973)
New York Bight	1–21		BROWN and co-authors (1973)
Gulf of Venezuela	50		BROWN and co-authors (1973)
Bedford Basin, Nova Scotia	1–60		KEIZER and GORDON (1973)
Gulf of St Lawrence	1–15		LEVY and WALTON (1973)
Narragansett Bay	8·5	GC	DUCE and co-authors (1972)
	5–15	GC	BOEHM (1977)
Woods Hole harbour	11	GC	STEGEMAN and TEAL (1973)

* Consult this paper for full details.

is vast and only a few aspects immediately associated with oil pollution can be mentioned.

The evidence of tar balls, tar specks, oil patches, oil slicks, and oil sheen is well documented for the surface of the sea and no voyager can make a cruise without encountering some of this evidence. This multiple macrosampling from visual reports, together with aerial patrols and surveys, provides wider substantiation of the tiny spot samples taken by scientists at infrequent and widely spaced intervals. Virtually all tar is of petrogenic origin but occasionally some oil films are formed from zooplankton breakdown and quite often from factory ships processing fish and offal.

Extensive data on surface and subsurface petroleum hydrocarbons and extractable organic matter are given by MYERS and GUNNERSON (1976) for a broad coverage of the Pacific Ocean and mainly coastwise around the Atlantic Ocean and the east coast of Africa incorporating many of the routes used by crude oil tanker traffic. The overall hydrocarbon composition of the samples and of some of the relevant crude oils is shown in Fig. 4-5. The authors describe how surface samples were collected using a clean bucket; subsurface samples were collected on a number of cruises at fixed depths (*ca.* 3 m and 10 m) by means of pipeline supplies through the vessels' hull. Hydrocarbon concentrations ranged over several orders of magnitude; surface volatile hydrocarbons from 0·01 $\mu g\ l^{-1}$ (lower detection limit) to 4·0 $\mu g\ l^{-1}$; samples from 3 m and 10 m, 0·01 to 2·0 $\mu g\ l^{-1}$; with median values near 0·10 $\mu g\ l^{-1}$. These volatile fractions contained C_4 to C_8 hydrocarbons of which benzene, toluene, and xylenes made up 75% of the total.

The non-volatile hydrocarbons showed a wider range; 0·01 to 56 $\mu g\ l^{-1}$ at the surface and 0·01 to 32 $\mu g\ l^{-1}$ at 10 m. Cycloparaffins were predominant followed by iso-paraffins and aromatics indicating a pervasive distribution of petrogenic and biogenic hydrocarbons on and in ocean waters. Coastal routes and tanker routes gave higher and more variable values. In deeper oceanic waters the values were mostly below 1 $\mu g\ l^{-1}$. A trend for higher values in the thermocline was observed.

A positive relationship was established between extractable organic matter and non-volatile hydrocarbons for samples from different water depths.

Microlayer surface sampling of the Sargasso Sea gave 14 to *ca.* 59 $\mu g\ l^{-1}$ 'total hydrocarbons'; subsurface values (0·2 to 0·3 m) ranged from 0·13 to 239 $\mu g\ l^{-1}$ (WADE and QUINN, 1975). Off the west coast of Africa dissolved hydrocarbons ranged from 10 to 43 $\mu g\ l^{-1}$ for samples over water-column depths from 50 to 4500 m with little change in detailed hydrocarbon components (gc–ms). Surface samples at Brest and Roscoff contained 137 and 46 $\mu g\ l^{-1}$ respectively (BARBIER and co-authors, 1973).

Much higher ranges were found for enclosed sea areas. Total hydrocarbons (by an IR method) in the surface microlayer of the Mediterranean Sea varied between 0·26 and 2·97 mg l^{-1}, compared to 0·0 to 0·1 mg l^{-1} at 1 m depth. Petroleum-derived hydrocarbons were 0·10 to 2·05 mg l^{-1} in the microlayer and averaged only 0·02 ± 0·001 mg l^{-1} at 1 m. Higher values were recorded in inshore waters (MIKHAYLOV, 1979). Significant correlations were established between petroleum hydrocarbons and chlorinated organic pesticides, and also their metabolites in the surface film.

SIMONOV and co-authors (1980) illustrate the spread of oil pollution into the Arctic Sea water and ice. No quantitative results are given in the English version available, but total hydrocarbon determinations were made using infrared method. Distribution of n-alkanes (gc) between ice, sub-ice water, and sub-zero bottom water is discussed in relation to partition by low-temperature processes. The n-alkanes dominated (83% on

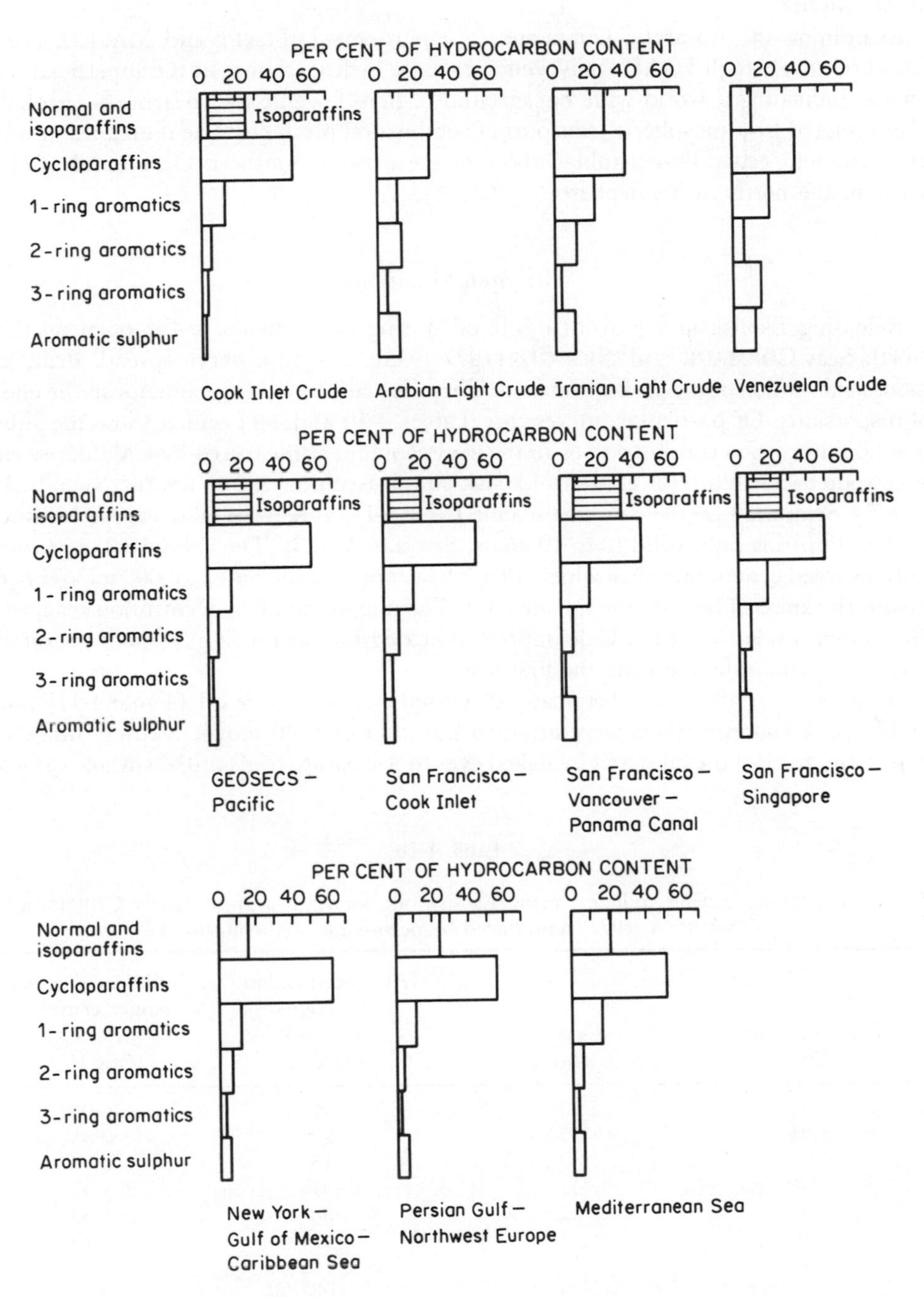

Fig. 4-5: Group composition of hydrocarbons extracted from ocean waters and in reference crude oils. (After MYERS and GUNNERSON, 1976)

average) the fractions extracted and had the gc characteristics of petrogenic (man-made) inputs.

Examining the Antarctic fauna and its environment, PLATT and MACKIE (1980) found relatively high PCAH (polycyclic aromatic hydrocarbons) in the superficial sediments, indicating a world-wide background of non-biogenic hydrocarbons—probably disseminated by atmospheric transport of combustion products. The n-alkanes in water and sediment generally resembled those of the pristine Southern Ocean and of clean areas of the northern hemisphere

(b) Spill Situations

Releasing Ekofisk and Kuwait crude oil during experiments in the open northern North Sea, CORMACK and NICHOLS (1977) were able to observe spread, drift, and mousse formation of the slicks produced. They also conducted experiments on the effects of dispersants. Of particular interest are Tables 4-10 and 4-11 which show the subsequent hydrocarbon concentrations in the water column as measured by UV fluorescence spectrometry. Table 4-10 relates to Ekofisk oil released during fine weather (winds: 1 to 3 knots, Sea state 1 on the Beaufort scale); Table 4-11, to Kuwait oil released chemically dispersed during light wind (8 to 10 knots, Sea state 2 to 3). The calm weather ensured initial spread of an intact slick which after 3 h occupied a calculated 20,000 m^2 and had a 27-μm thickness of largely unemulsified oil. The maximum oil concentration reached at 30 cm depth was 18 mg l^{-1}. Calculations indicated that as much as 40% of the oil was lost to the atmosphere during the first 8 h.

In open-sea trials using chemically dispersed Kuwait crude oil (Table 4-11) much higher peak concentrations were attained but after only 30 min it became difficult to detect any trace of the oil patch by naked eye. In 3 separate trials initial surface values of

Table 4-10

Oil concentrations during spill test using Ekofisk oil; winds: 1–3 knots (After CORMACK and NICHOLS, 1977; reproduced by permission of the authors)

Time after spill (h)	Depth (m)	Oil concentration under edge of main slick (mg l^{-1})	Oil concentration under centre of main slick (mg l^{-1})
$\frac{1}{2}$	2	2·49	2·03
$1\frac{1}{2}$	2	2·22	0·83
3	2	1·15	0·79
4	2	0·94	3·95
8	2	1·88	1·63
8	5	0·17	0·19
8	10	0·10	0·07
8	15	0·08	0·07
11	5	0·02	0·04
11	10	0·02	0·02
11	15	0·02	0·03
21	2	0·59	1·49

Table 4-11

Oil concentrations at 1 m during spill test using chemically dispersed Kuwait oil; winds: 8–10 knots (After CORMACK and NICHOLS, 1977; reproduced by permission of the authors)

Time after spill (min)	Concentration of Kuwait crude oil in mg l^{-1} in the upper metre of water		
	Run 1	Run 2	Run 3
0	34·4	24·2	0·85
1		15·8	
2	47·8		8·7
2·5		12·2	
5		9·4	
7	17·8		3·5
10		5·2	
15			1·7
18	1·9		
25		4·2	
40	8·8		1·35
50		1·9	
80			1·5
100	2·2	0·8	

34,400, 24,200, and 850 μg l^{-1} were found which illustrates the problem of adequate sampling. The maximum possible concentration would be much higher (theory, 1200 mg l^{-1}). Subsequent samplings over frequent brief intervals were made over a period of 100 min. The general level of dispersed oil after 30 min was about 1 mg l^{-1}, which is about the 48 h LD_{50} value for zooplankton and fish larvae.

Sea trial spills of 14 tons gas oil (b.p. 200 to 360 °C) under conditions of moderate winds (15—20 knots, Sea state 4 to 5) gave rise to initial concentrations of 1500 μg l^{-1} (at 1·5 m), after 90 min 460 μg l^{-1} and over the entire patch to 17·5 m the average value was 290 μg l^{-1} equivalent to 18% of the oil introduced. A similar test using leaded gasoline under moderately rough weather proved difficult to study. Within 1 h the slick could not be identified and in only one of the subsurface samples taken was any gasoline identified. The gasoline spread and evaporated rapidly and could not be adequately studied.

BOEHM and co-authors (1979) determined hydrocarbon levels in the Georges Bank region over repeated surveys during 1977 subsequent to the *Argo Merchant* oil spill. The regional mean values are summarized in Table 4-12. The table shows particulate petrogenic hydrocarbons as distinct from total hydrocarbons using a correction factor based on pristane as index of biogenic hydrocarbons. Spill oil, rich in higher aromatics—including the *Argo Merchant* No. 6 Cargo Oil in winter 1977—influenced the dissolved aromatic hydrocarbon distributions throughout the year. Land-derived hydrocarbons from plant and degraded petroleum sources were also present.

Particulate petrogenic hydrocarbons were prevalent following the spill. Paraffinic and degraded tar were widely distributed throughout the 30,000 square miles surveyed and were not sharply focused on shipping lanes. Whereas most open-ocean particulate hydrocarbon gc profiles are characteristic of pelagic tar, in this region the profiles showed

Table 4-12

Hydrocarbon concentrations ($\mu g\ l^{-1}$) near Georges Bank and Nantucket Shoals after the *Argo Merchant* oil spill (After BOEHM and co-authors, 1979; reproduced by permission of the authors)

	Particulate				Dissolved		Surface microlayer
	Total $(f_1 + f_2)_2$		Aliphatic, pristane free (f_1[PF])		Total $(f_1 + f_2)$		
Season	Near surface	Near bottom	Near surface	Near bottom	Near surface	Near bottom	Total $(f_1 + f_2)$
Winter (Feb. 1977)	0·57 ± 0·35	0·90 ± 1·21	0·28 ± 0·29	0·46 ± 0·73	29·6 ± 15·3	35·3 ± 24·8	48·1 ± 24·5
Spring (May 1977)	1·86 ± 1·62	0·92 ± 0·71	0·39 ± 0·52	0·23 ± 0·14	12·3 ± 12·7	10·0 − 9·8	18·4 ± 7·9
Summer (Aug.1977)	0·75 ± 0·74	0·26 ± 0·32	0·35 ± 0·53	0·09 ± 0·08	1·0 ± 1·0	1·9 ± 2·7	8·4 ± 4·4
Fall (Nov. 1977)	0·27 ± 0·34	0·18 ± 0·11	0·06 ± 0·10	0·04 ± 0·04	0·2 ± 0·2	0·8 ± 1·5	$45·8_3$
Winter (Feb. 1978)	0·10 ± 0·08	—	0·05 ± 0·07	—	0·4 ± 0·6	—	—

Averaged over maximal 12 stations.
Aliphatic hydrocarbons = f_1; olefinic and aromatic hydrocarbons = f_2.
Single analysis.

considerable variation in weathering characteristics perhaps associated with oceanographic mixing factors.

Surface microlayer hydrocarbons averaged between 2 and 9 times the concentrations at 20 to 30 cm depth (1 sample 92 times). Surface hydrocarbons showed a predominance of pelagic tar (0·1 to 1000-μm diameter particles).

The most significant single aliphatic hydrocarbon in the zooplankton, accounting for more than 98 % of the aliphatics and 60 % of the total hydrocarbons was pristane. It approached 500 $\mu g\ g^{-1}$ in winter and summer, and 2000 $\mu g\ g^{-1}$ during spring bloom. This pristane seems to have been biosynthesized by the zooplankton from phytoplankton precursors, including phytol. Any petrogenic hydrocarbons present are thought to be masked or diluted by the greatly predominating biogenics.

Although occasional oil slicks and, more commonly, oil sheens cover a proportion of the ocean surface there is no evidence that such films have any significant effect on gaseous transfers between air and sea (KINSEY, 1973).

(c) Surveys of Oil and Bacteria

A joint German, American, and Norwegian team undertook a 3 yr programme to investigate the distribution of hydrocarbons in the North Sea together with studies on various bacterial groups, including hydrocarbon degraders. Surface films, water column, and sediments were sampled at 138 stations during 1975, 1976, and 1977 (Interim reports: GUNKEL and GASSMANN, 1977; CROW and co-authors, 1977; OPPENHEIMER and co-authors, 1977). The latest interim paper (presented by GUNKEL and co-authors at the Sixth National Congress of Microbiology, 6 to 9 July, 1977; Santiago de Compostela, Spain) was published in 1980. The impressive research already undertaken is not yet complete in aspects such as taxonomic identification of yeasts, oil-degrading micro-organisms, phytoplankton, and zooplankton. During the study, annual oil production grew from 10 million tons in 1975 to 60 million tons in 1977; production in 1980 was about 150 million tons.

Neuston samples were examined for tar balls. The weight of the tar balls varied from 0 to 7·1 mg m^{-2} in 1975 and 0 to 6·6 mg m^{-2} in 1976, with the mean changing from 0·32 mg m^{-2} to 0·40 mg m^{-2}—statistically not a significant increase. Close examination showed that tar balls in the far north were unlike those prevailing around the oil fields and to the south, and may have come from the North Atlantic Ocean. Sampling of neuston and tar balls was greatly affected by the sea state. The water-borne hydrocarbons at 10 m depth were determined by infra-red absorption spectrometry following carbon tetrachloride extraction and Florisil treatment in 1975, and by the IGOSS ultraviolet emission fluorescence method in 1975 and 1976. Optical measurements were converted to total alkanes and total aromatics, respectively, by calibration using Ekofisk crude oil.

Hydrocarbons at 10 m depth as measured by the IR method, ranged from 17 to 652 $\mu g\ l^{-1}$ (mean 146 $\mu g\ l^{-1}$); by the UV method, from 0 to 2·06 $\mu g\ l^{-1}$ (mean 0·35 $\mu g\ l^{-1}$) in 1975, and from 0·39 to 88·0 $\mu g\ l^{-1}$ (mean 5·89 $\mu g\ l^{-1}$) in 1976. Total hydrocarbon by the IR method measures biogenic and fossil hydrocarbons; the UV method is more specifically for the aromatic component of fossil hydrocarbons. In general the distribution of tar balls in the surface layer (Figs 4-6, 4-7) and of hydrocarbons at 10 m (Fig 4-8) documented the strong influence of the oil fields. Simultaneous determinations of oil-degrading bacteria followed a similar pattern (Fig. 4-9) and so did the proportion of

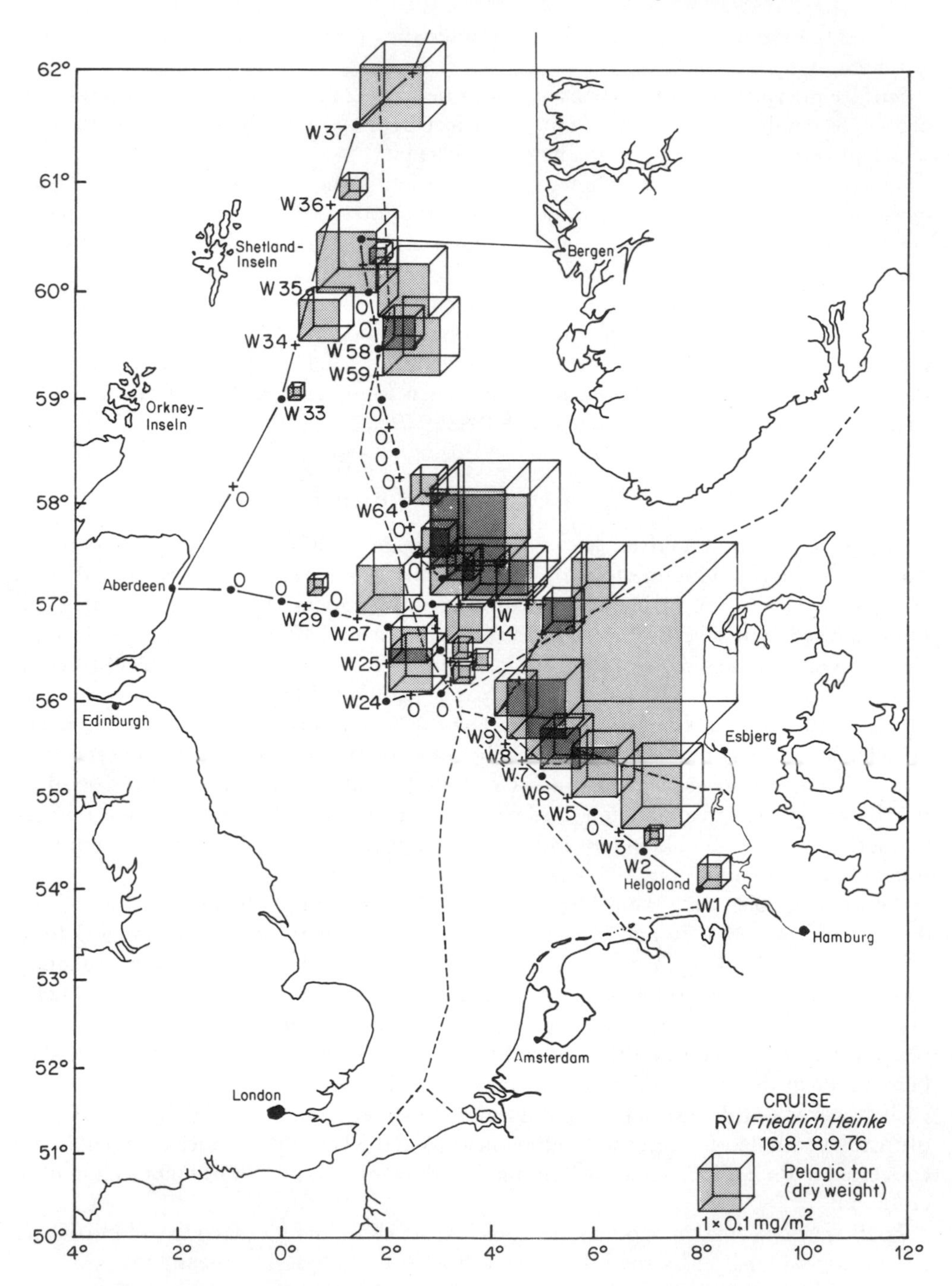

Fig. 4-6: Distribution of pelagic tar in the North Sea, 1976. (After GUNKEL and co-authors, 1980; reproduced by permission of the authors)

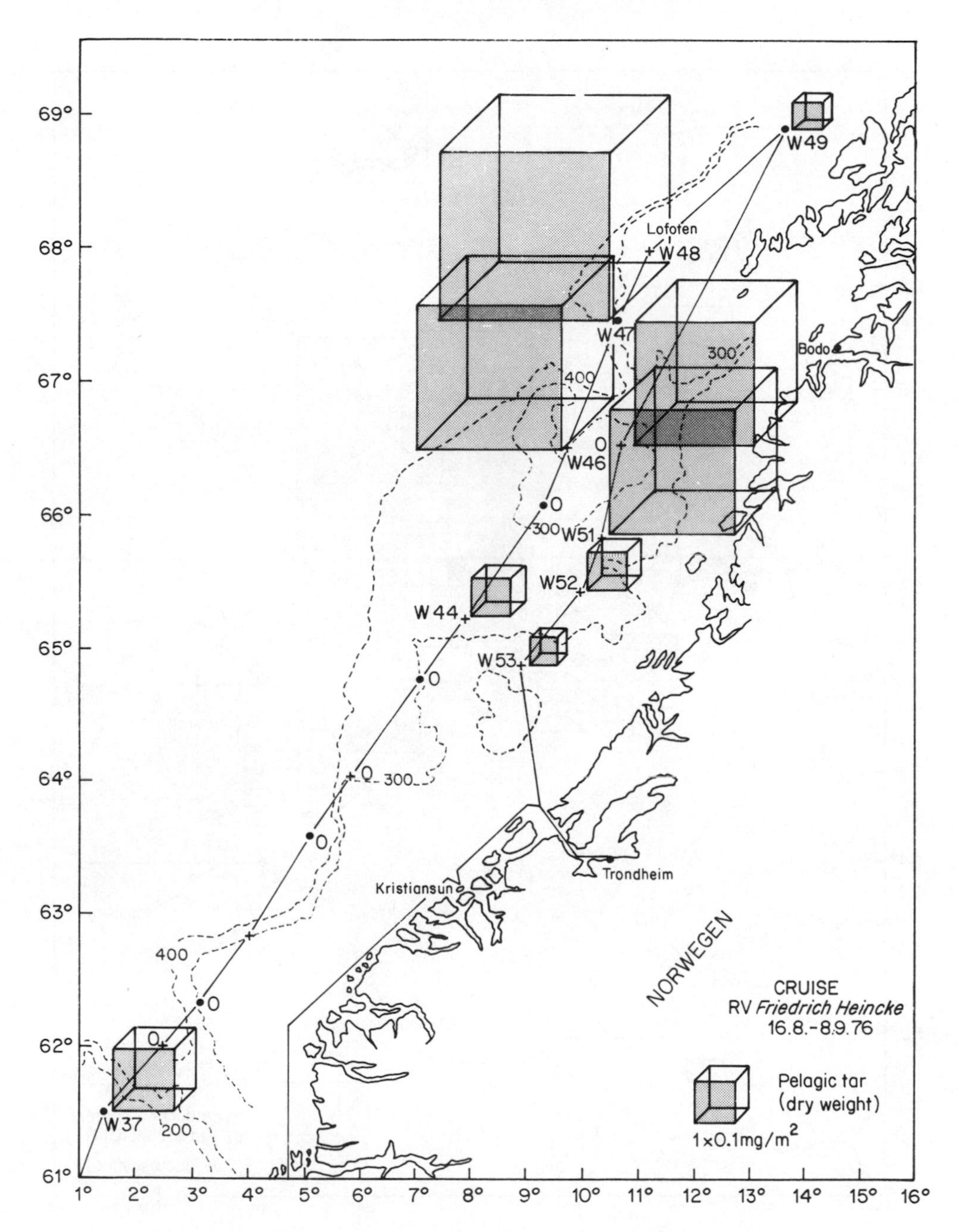

Fig. 4-7: Distribution of pelagic tar in the Norwegian Sea, 1976. (After GUNKEL and co-authors, 1980; reproduced by permission of the authors)

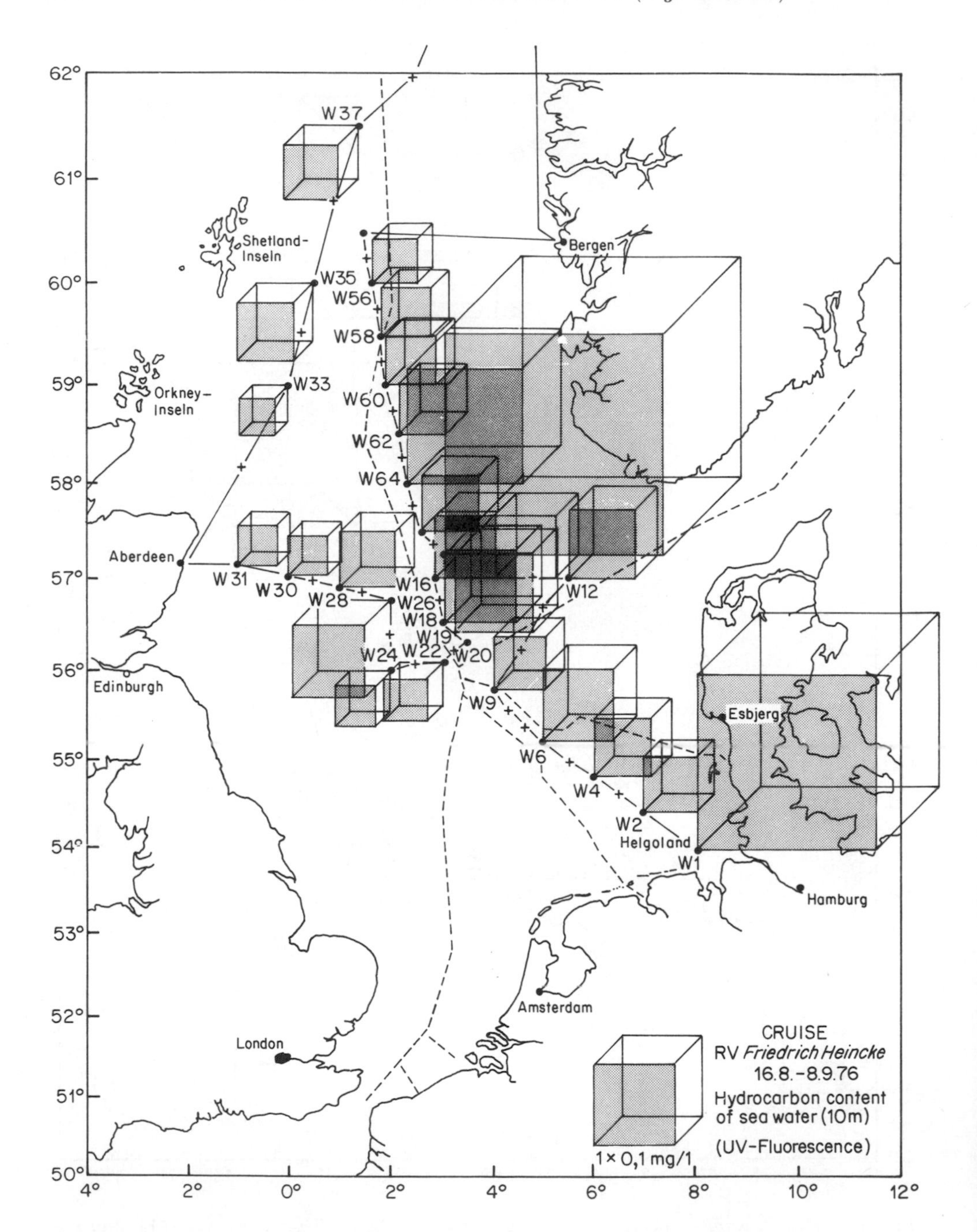

Fig. 4-8: Hydrocarbon concentration in the North Sea at 10 m, 1976; measured by UV fluorescence; scale basis unit cube: 4·4 μg hydrocarbons l^{-1}. (After GUNKEL and co-authors, 1980; reproduced by permission of the authors)

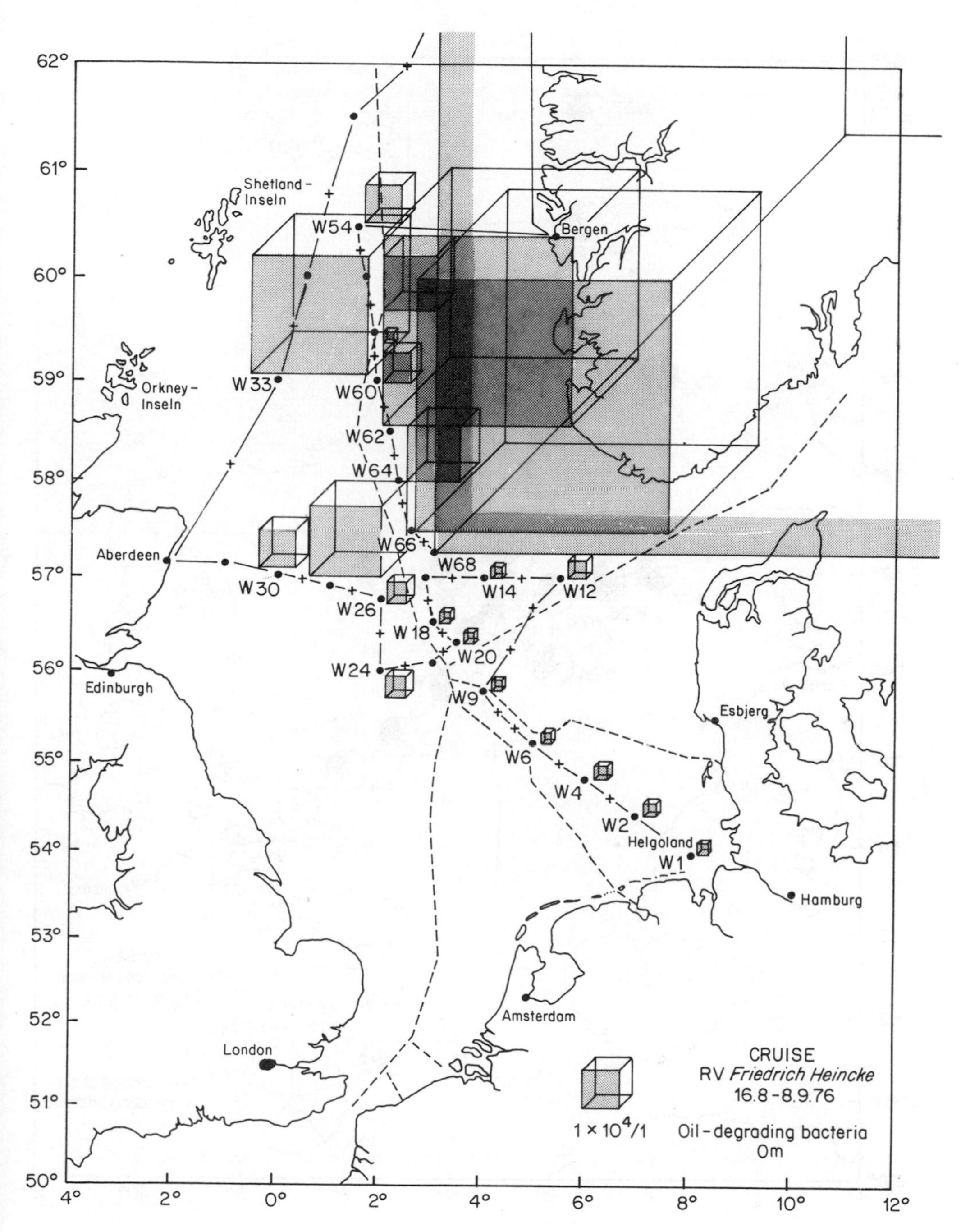

Fig. 4-9: Distribution of hydrocarbon-degrading bacteria in the uppermost 5 cm of water in the North Sea, 1976. (After GUNKEL and co-authors, 1980; reproduced by permission of the authors)

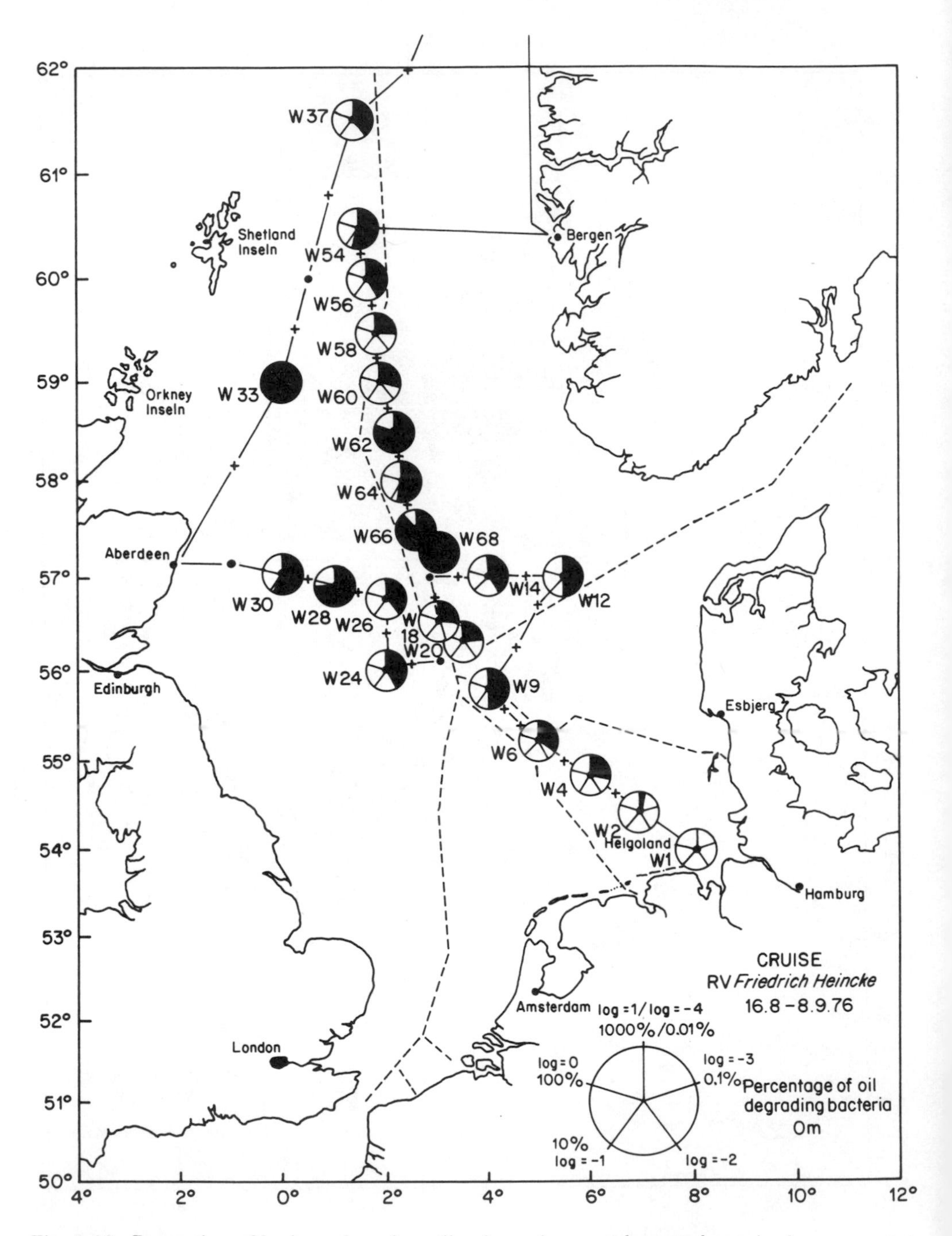

Fig. 4-10: Proportion of hydrocarbon-degrading bacteria to total saprophytes in the uppermost 5 cm of water in the North Sea, 1976; log scale. (After GUNKEL and co-authors, 1980; reproduced by permission of the authors)

oil-degrading to saprophytic bacteria (Fig. 4-10). All the distributions were highly variable in detail, but the strong influence of oil production areas and industrialized coastal areas was evident—to a lesser extent the influence of Baltic, Atlantic, Norwegian, and English Channel waters.

The oil-spill situation allied to continuing oil-related activity is found at the new North Sea oil reception plant and export terminal at Sullom Voe in the Shetland Islands. As yet the proportions of oil-degrading bacteria to total saprophytes are very much lower than have been measured close to North Sea production platforms and near the outfall of the River Elbe. Nevertheless, rates of mineralization of naphthalene are 10 to 200 times background levels and benz [α] pyrene mineralization may be 2 to 3 times increased (DAVIES and co-authors, 1981a).

Because of the influence of land, rivers, and nutrient-and organic-rich effluents, bacteria populations increase towards the coast. The data of GUNKEL (1973) clearly show an inverse relationship between total heterotrophic bacteria, some 50 miles west-northwest of the River Elbe, and salinity, though presence of fresh water may not be the sole causal factor at the sampling point. As in other waters, oil-oxidizing marine bacteria, on average, accounted for about 10% of the total marine heterotrophs during the year. On the section from the Cable Buoy to River Elbe there is a fairly regular increase in total marine heterotrophs, from 2×10^5 cells l^{-1} to 8×10^6 cells l^{-1}; despite the decrease in salinity from 32 to 1‰ S the marine oil-degraders increase from 5×10^3 to 2×10^5 at the mouth of the Elbe. On average, water from the sea surface yielded almost twice as many oil-oxidizing bacteria as at 1 m depth.

Although bacterial numbers from sediments at the mouth of the Elbe were about 100 times higher than 5 miles seawards, the numbers thereafter remained steady to 50 miles Of the average 4×10^9 marine heterotrophs l^{-1}, about $1{\cdot}6 \times 10^8$ were oil oxidizers. There was no obvious correlation between sediment particle size (silt to coarse sand and small stones) and hydrocarbon content or bacterial numbers.

Away from the main Elbe influence the numbers of oil-oxidizing bacteria in intertidal sand samples collected on beaches at Helgoland ranged from 5×10^7 to 1×10^9 with an overall average of almost 20% oil degraders among the heterotrophs compared to 5% for the offshore sediment population. Even higher proportions were found near high-tide mark. These findings confirm that where oil tends to be most prevalent at the sea surface, at the seabed, and at the highwater level in or on beach sand, bacterial oil-degraders are also more numerous than in the adjacent water column.

(d) Detailed Hydrocarbon Surveys

To no lesser extent than in the oceans and shelf seas already described, the concentrations of hydrocarbons—whether in terms of gross measures or group determinations—exhibit a wide degree of variability with time and space, especially in the surface microlayer.

This is evident from the survey around Britain (mainly 1972 to 1976 results) described by HARDY and co-authors (1977b) and summarized briefly in Tables 4-13, 4-14. Progressively more detailed hydrocarbon chemistry has been introduced in the biennial inshore–offshore surveys since 1972. Table 4-15 shows the 1978 results (MACKIE, 1980; unpubl.). These surveys start from an inshore area characterized by refinery, petrochemical, and oil-terminal activities set amid shipping, the refuelling of cargo and

Table 4-13

n-Alkanes (C_{15}–C_{33}) in surface film; North Sea, 1972–1976 (After HARDY and co-authors, 1977a; reproduced by permission of the authors)

Site category	No. of samples	Range of amounts ($\mu g\ m^{-2}$)	Mean ($\mu g\ m^{-2}$)	Standard deviation ($\mu g\ m^{-2}$)
All	62	4·3–158	33·6	27·8
North Sea (Oil fields)	10	13–75	32·8	23·4
Open sea (Celtic Sea)	7	4·3–7·7	5·7	1·4
Urban	8	25–158	62·9	45·4
Refinery	5	23–148	64·2	50·3
Industrial	10	13–90	38·1	22·4

Table 4-14

n-Alkanes (C_{15}–C_{33}) in North Sea, 1 m depth, 1972–1976 (After HARDY and co-authors, 1977b; reproduced by permission of the authors)

Site category	No. of samples	Range of amounts ($\mu g\ l^{-1}$)	Mean ($\mu g\ l^{-1}$)	Standard deviation ($\mu g\ l^{-1}$)
All	85	0·2–24	2·0	2·0
North Sea (Oil fields)	7	2·4–7·1	4·5	1·6
Open sea (Celtic Sea)	6	0·2–0·8	0·57	0·16
Urban	6	0·4–2·5	0·92	0·85
Refinery	5	0·8–2·1	1·1	0·52
Industrial	10	0·4–2·0	0·82	0·55

naval vessels, heavy and light industry of many kinds and important urban concentrations. The gradient along the axis of the Firth of Forth is sampled thence through relatively clean waters to the neighbourhood of the Forties oil field. Such a progression is evident from Table 4-15; it shows higher ranges of all the components in the upper estuary dropping off seawards. The Bell Rock site reveals unusually high levels in this set of samples with A and B more representative of Scottish coastal water. Values for many components increase near the Forties oilfield, but the high polycyclic aromatics—notable in the upper Forth estuary—are not reproduced. This tends to confirm many other indications that these aromatics arise mainly from outfalls, rivers, land sources, and urban air.

(e) Sediment Surveys

Hydrocarbons in sediments from the open North Sea, measured by an IR method (Florisil treatment proved effective) showed less variability than levels in surface film

Table 4-15

Concentration (10^{-9} g l^{-1}) of certain PCAH in water (10 m) on a survey between Kincardine on Forth and the Forties Field, 1978. Kinc: Kincardine; Long: Longannet; Gran: Grangemouth; Queens: Queensferry; Hound Pt: Hound Point; Bass: Bass Rock; Bell: Bell Rock; Forties: Forties Field (After MACKIE, unpublished, reproduced by permission of the MAFF Torry Research Station, Aberdeen)

	Kinc	Long	Gran	Queens	Hound Pt	Bass	Bell	oo° 04′W	00° 42′E	Forties
Naphthalene	1·80	0·13	8·76	1·27	1·25	0·14	4·49	0·09	0·07	0·18
Methyl naphthalene	12·00	ND	11·28	4·84	1·02	0·34	4·96	0·03	0·03	1·80
Dimethyl substituted	18·90	ND	34·04	19·29	4·0	7·80	14·93	0·05	0·05	12·84
Dibenzthiophene	ND	ND	2·18	1·75	0·78	1·17	0·89	0·49	0·83	0·81
Phenanthrene	0·90	2·59	2·06	1·36	1·87	0·85	0·21	0·72	0·49	0·41
Methyl phenanthrenes and anthracenes	1·50	ND	3·13	3·44	1·52	2·79	1·70	0·64	0·59	1·39
Fluoranthene	3·20	0·13	2·18	1·06	1·07	0·76	0·60	0·40	0·19	0·29
Pyrene	1·60	0·20	2·96	1·51	1·04	0·44	0·28	0·30	0·34	0·23
Benzanthracenes and benzphenanthrenes	ND	ND	1·17	1·17	0·66	0·76	0·41	1·04	0·84	0·38
Benz[α]pyrene	6·50	0·34	0·37	0·43	0·16	0·13	0·11	0·49	0·78	0·07

ND, Not determined.

Table 4-16

n-Alkanes (C_{15}–C_{33}) in North Sea sediments, 1972–1976 (After HARDY and co-authors, 1977a; reproduced by permission of the authors)

Site category	No. of samples	Range of amounts (dry sediment) ($\mu g\ g^{-1}$)	Mean ($\mu g\ g^{-1}$)	Standard deviation ($\mu g\ g^{-1}$)
All	60	0·1–42·5	2·69	6·44
North Sea (Oil fields)	14	0·1–4·8	1·13	1·15
Urban	7	0·2–1·5	0·77	0·54
Refinery	5	0·2–2·1	1·04	0·76
Industrial	9	0·6–22·7	6·00	7·13

and water column. The range found by GUNKEL and co-authors (1980) was 3·9 to 45·4 mg kg^{-1} calculated on a dry weight sediment with a mean of 21·0 mg kg^{-1}.

Other general results for hydrocarbons in offshore and inshore sediments are summarized in Table 4-16. The range in concentration and the scatter of values for replicates taken within small areas are higher than for the surface microlayer or the water column but much of the variation comes from a small number of industrial sites marked by high concentration. Difference in concentration between sediment and water is also marked by differences in the component hydrocarbons, e.g. the n-alkanes (p. 1441). No consistent associations could be made between the quantities or detailed n-alkane profiles of the hydrocarbons from sediment, water column, and surface film. These analyses did not attempt to discriminate between biogenic and petroleum hydrocarbons.

Similar ranges for total saturated hydrocarbons (pentane extracts) were reported by SLEETER and co-authors (1978), namely 10 to 60 mg kg^{-1} inside the Bermudan reef, 3 to 10 mg kg^{-1} outside the reef, and neglible concentrations (about 0·3 mg kg^{-1}) in waters 15 km offshore. Further studies (SLEETER and co-authors, 1979) gave 1·47 ± 1·42 mg kg^{-1} inside and 0·57 ± 0·51 mg kg^{-1} outside the reef. Using the unresolved complex mixture as an index of petroleum, the biogenic hydrocarbons from surface sediment almost exactly doubled the concentration of petroleum for sediments inside the reef. Outside the reef about equal quantities were found in each category. Concentrations of total aliphatics were only 0·27 ± 0·17 mg kg^{-1} inside and 0·08 ± 0·05 mg kg^{-1} outside at 10 to 13 cm sediment depth, containing about equal amounts of biogenic and petrogenic components.

Occasional surveys have also been made around UK offshore installations selected on the basis of the type of drilling in operation or the amount of oily water discharge. The Forties oilfield in production since 1974 has been established largely without the use of oil-based drilling muds; to date (1981) its treated oily water discharges do not include a large amount of production water. The Beryl oilfield, like the Forties, is fairly isolated but because of the drilling conditions has used oil-based muds almost from the beginning. The Brent field is one of a number of oilfields in the east Shetland basin, an area of expanding development and capable of producing a major fraction of North Sea oil and gas.

The influence of these 3 different oilfields on their surroundings is illustrated in Fig. 4-11(a)-(c). It shows the concentration of oil in fluorescence equivalents. The relationship between Beryl field sediments and those along the axis of the Firth of Forth are clearly seen in Table 4-17a, b (MASSIE and co-authors, 1981). The outright dominance of the oil base of the drilling mud is demonstrated only in the aromatics profile for sediment 0.5 mile south of Beryl; the others at this distance are markedly oil-enhanced but there are also indications of crude oil and combustion residues (see Table 4-17b). At 1 or 2 miles distance the profiles appear intermediate between the oil-enhanced areas and the open North Sea characterized principally by air-borne residues from combustion. The sediment samples from the inner Firth of Forth show marked enhancement from oil, air-borne and especially land-derived petrogenic inputs.

Studies on the microbial mineralization of ^{14}C substrates demonstrate much higher rates near Beryl than near Forties or Brent. These findings clearly consolidate the field work of GUNKEL and co-authors (1980).

In the heavily contaminated sediments of New York Bight the total saturated hydrocarbons were consistently 10 to 40 mg kg^{-1} in the outer area, rising to 2290 mg kg^{-1} in the dumping area. The corresponding aromatic hydrocarbons were 14 to 54 mg kg^{-1} with a top value of 4240 mg kg^{-1}, coinciding with the previous peak, but in general the ratio saturated/aromatic hydrocarbons showed no consistent trend with other known measures of sediment contamination. Most of the dump-site sediment hydrocarbons revealed prominent profiles of unresolved complex mixtures typical of petroleum. The outer sediments gave more pronounced n-alkane peaks in the range C_{19} to C_{28} (KOONS and THOMAS, 1979), showing evidence of less marked petroleum contamination.

Although some account must be taken of the various methodologies employed, these reports serve well enough to illustrate the wide ranges in hydrocarbons and oil-oxidizing bacteria between open sea environments and those affected by oil or associated with the dense urban and industrial centres. The sporadic nature of oil pollution is also evident.

(6) Chemical Aspects of Oil Pollution

(a) Oil Analysis and Analytical Objectives

The hard truth about the chemistry of oil and oil products is that all crude oils and many oil products have never been analysed completely into single hydrocarbon components either qualitatively or quantitively. From partial analysis it is predicted that a crude oil has many thousand or even a million individual components. It could take a skilled analyst a year to extract, separate, identify, and quantify just 100 components in one oil. Examine for example the gas chromatograms of a sample of *Argo Merchant* No. 6 Fuel Oil in Fig. 4-12 which shows the saturated hydrocarbons with numbered n-alkane peaks and Fig. 4-13 which shows the aromatic hydrocarbons. Most crude oils are equally complex over these boiling-point ranges and contain in addition many lighter and a very great number of heavier components. It is not surprising, therefore, that in routine field investigations research chemists limit their precise identifications and quantifications to only a very minor proportion of the n-alkanes and aromatics (Fig. 4-14), and only extend the intricacy of their analyses for more detailed work on a very restricted number of samples. Total hydrocarbon estimates are made from quantifying known peaks and relating these to some standard oil (CLARK and BROWN, 1979;

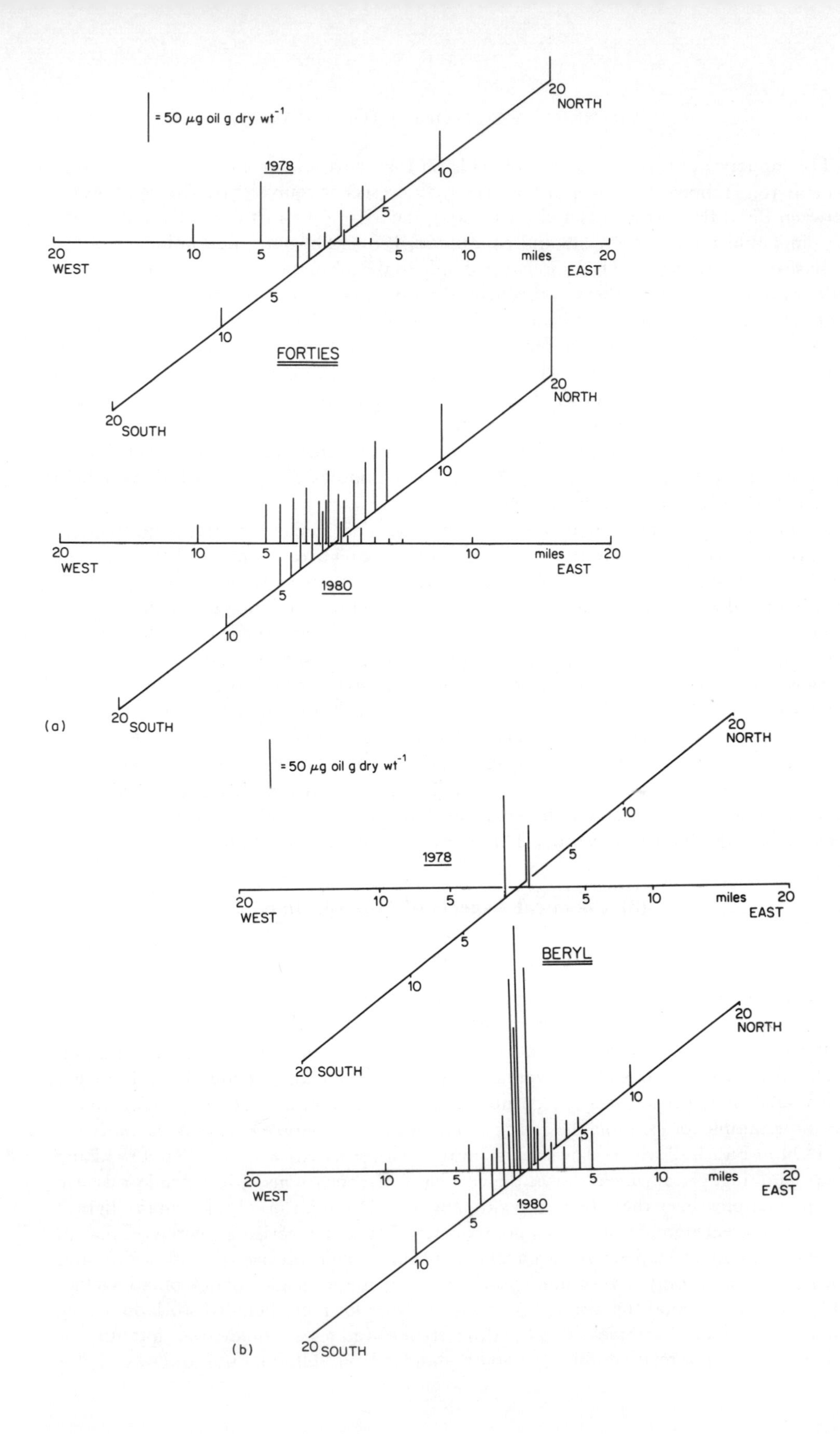

= 50 μg oil g dry wt⁻¹
1978
20
WEST
10
5
5
10
miles
20
EAST
20
NORTH
10
5
5
10
20
SOUTH
FORTIES
20
NORTH
10
20
WEST
10
5
10
miles
20
EAST
5
1980
10
20
SOUTH
(a)
= 50 μg oil g dry wt⁻¹
20
NORTH
10
5
1978
20
WEST
10
5
5
10
miles
20
EAST
5
10
BERYL
20 SOUTH
20
NORTH
10
5
20
WEST
10
5
5
10
miles
20
EAST
1980
5
10
(b)
20
SOUTH

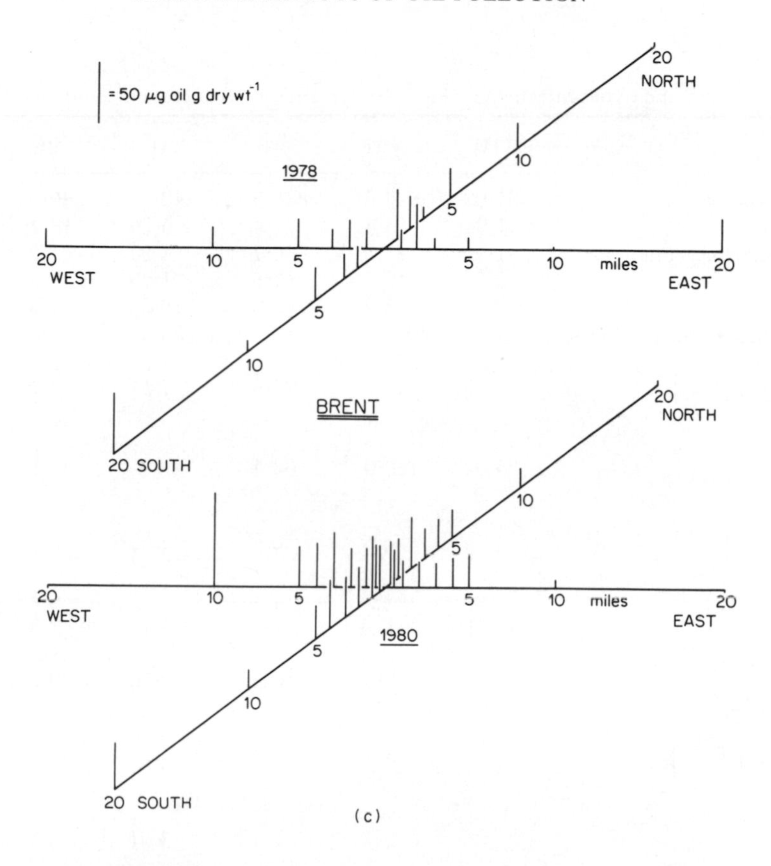

Fig. 4-11: Total oil by IGOSS fluorescence measurement (mg oil kg^{-1} sediment) in bottom samples collected around oil fields in the North Sea, 1978, 1980. (a) Forties field; (b) Beryl field; (c) Brent field. (After MASSIE and co-authors, 1981; reproduced by permission of the DAFS, Marine Laboratory, Aberdeen, Scotland)

unpubl.). It is not surprising then that in the oil industry the analysis of oil is directed towards parameters that aid handling, selection and processing of crudes and the best yield of products that can satisfy trade specifications. Table 4-18 illustrates the industrially important parameters of representative crude oils from sources world-wide. The main groups (paraffins, naphthalenes, other aromatics, non-volatile residue: boiling above 370 °C) reveal considerable differences between crude oils of different origin and also in crude oils from neighbouring fields.

Unfortunately for the chemist, the complexities do not stop here. Crude oil seldom comes into contact with marine life in fully fresh condition. On exposure to the sea several changes set in immediately: (i) loss of volatile components to the atmosphere, at least temporarily; (ii) a complex array of compounds dissolve in sea water and are diluted, transported, and dispersed according to hydrographic conditions; (iii) the surface slick emulsifies absorbing progressively larger amounts of sea water; (iv) depending on the sea conditions, oil becomes dispersed as globules from a drop to colloidal size.

Table

Aromatic hydrocarbons ($\mu g\ kg^{-1}$ dry weight) in sediments around Beryl field (After

	*1W	½ W	½E	1E	2E	3E
Naphthalenes	31·03	622·6	834·5	48·97	40·5	22·60
Biphenyls	1·94	5·8	71·4	5·86	13·2	6·50
Acenaphthenes + Fluorenes	1·91	47·1	36·7	4·82	5·2	1·94
**DBF's	0·76	12·1	11·6	1·99	2·4	0·98
***DBT's	8·32	87·9	218·3	10·83	4·8	0·82
Anthracenes + Phenanthrenes	64·24	2615·3	869·7	55·01	56·5	18·00
3 ⬡ + 1 ⬠	5·35	46·1	22·6	7·11	8·1	2·42
4 ⬡	35·02	129·0	62·4	49·25	45·1	13·08
4 ⬡ + 1 ⬠	5·13	6·9	9·8	3·24	16·2	7·02
Benzo(e)pyrene	1·80	4·2	4·1	1·02	6·8	2·73
Benzo(a)pyrene	0·86	1·1	1·7	0·45	3·5	1·12
Other 5 ⬡	0·26	2·5	10·3	0·56	20·7	8·37
5 ⬡ + 1 ⬠		4·4	10·5	2·37	42·2	15·97
6 ⬡		2·7	7·3	3·91	26·5	11·37
Total ($mg\ kg^{-1}$) aromatics	0·157	3·588	2·171	0·195	0·292	0·113
Total oil by IGOSS Fluor ($mg\ kg^{-1}$)	40·0	150·0	97·0	42·0	29·0	23·0

* 1W: 1 mile west of Beryl 'A' platform.
**DBF's: Dibenzfurans
***DBT's: Dibenzthiophenes

As time goes on, chemical change becomes allied to physical change through microbial degradation and photochemical oxidation. These and many other minor biological and chemical processes generate a great many partially degraded oil derivatives to add to any remaining crude oil hydrocarbons.

Exact figures vary according to oil and sea conditions, but Fig. 4-15 shows the trends for a Venezuelan crude oil (Guanipa) at 2 °C in the laboratory with mechanical agitation, aeration, UV-rich radiation; all at constant level over a period of over 500 h (MacGregor and McLean, 1977)

It is not possible to demonstrate such changes more realistically for an oil spill situation because usually the oil is not confined and water renewal goes on continuously. Evaporative loss of about 50% of dissolved natural gas and volatile components occurred in a few hours in relation to the 'Ekofisk' blowout, more rapid mousse formation has

4-17a

MASSIE and co-authors, 1981; modified; reproduced by permission of the authors)

10E	5N	1N	½N	½S	1S	2S	3S	4S	5S
29·40	20·28	8·24	7170	14,797	596·3	28·80	24·90	6·38	0·96
7·93	3·69	0·46	1032	1111	149·9	7·40	9·00	2·33	0·61
2·37	2·23	1·64	678	183	82·6	2·80	3·40	0·84	0·26
1·02	1·16	0·31	189	45	18·7	1·20	1·50	0·39	0·09
0·52	1·06	5·24	2346	3560	788·1	7·14	4·65	0·80	0·30
17·51	15·69	195·21	12,484	4158	4112·4	47·80	35·50	10·43	5·30
4·21	2·38	19·19	187	24	72·3	3·30	3·90	2·30	1·49
36·42	11·01	57·33	678	248	382·3	12·40	15·00	17·85	12·33
21·56	1·76	31·76	74	1	23·1	3·06	4·85	13·20	5·76
7·04	0·53	11·39	25	0·46	11·0	1·10	2·20	4·96	2·21
3·74	0·26	4·53	7	0·15	3·9	0·64	0·83	2·15	1·07
22·16	0·09	13·90	59		24·4	2·20	5·20	14·60	7·84
44·16	37·81		47		28·6	6·70	12·90	31·14	16·39
25·00		33·14	33		19·4	4·20	8·10	18·47	10·21
0·223	0·060	0·420	25·009	24·128	6·314	0·129	0·132	0·126	0·065
71·0	13·0	36·0	203·0	257·0	191·0	69·0	42·0	33·0	22·0

been reported for other crudes in other waters or none at all in relation to distilled refinery products, and change in specific gravity leading to the sinking of the oil is widely variable in oil spill reports several finding no visible oil after 2 or 3 d (e.g. 'Ekofisk').

Progress to the ultimate breakdown into carbon dioxide and water proceeds quite rapidly: weeks or months for lighter and some middle hydrocarbon ranges; many months to many years for the heavy fractions. In addition to natural organic matter in the sea, generally regarded as up to 200 yr old for the most resistant fractions, there is an incredible variety of petroleum chemicals at all stages of remineralization.

How then should oil be measured if the objectives are not primarily (i) those of industry, (ii) dictated by the ultimate in chemical artistry, (iii) directed towards specific compounds of immediate narrow concern? Should attention be directed at particulate hydrocarbons or soluble fractions? Do available techniques exist that meet ecological demands? These are such demanding questions that no immediate attempt will be made

Table

Aromatic hydrocarbons ($\mu g\ kg^{-1}$ dry weight) in sediments from Inshore, Firth of Forth, modified; reproduced by

	Bell Rock	Bass Rock	Hound Point	Bridges
Naphthalenes	3·03	132·9	581·9	2213·0
Diphenyls	1·23	57·2	279·5	1576·0
Acenaphthene + Fluorenes	0·46	18·1	4·8	679·0
Dibenzfurans	0·21	5·0	24·9	180·5
Dibenzthiophenes	0·30	14·9	106·9	427·4
Anthracenes + Phenanthrenes	5·24	521·5	2845·7	13055·0
3 ⬡ + 1 ⬠	1·14	85·2	564·7	3061·5
4 ⬡	4·56	623·2	2899·3	8238·4
4 ⬡ + 1 ⬠	1·11	72·1	341·0	937·2
Benzo(e)pyrene	0·55	31·5	140·0	396·3
Benzo(a)pyrene	0·24	26·2	110·9	459·7
Other 5 ⬡	1·13	69·2	265·1	1253·3
5 ⬡ + 1 ⬠	1·41	40·1	131·8	547·3
6 ⬡	1·20	60·3	134·3	496·8
Total ($mg\ kg^{-1}$) aromatics	0·022	1·76	8·43	33·5
Total oil by IGOSS Fluor ($mg\ kg^{-1}$)	19·7	196·0	834·0	350·0
	Inshore	Firth of Forth		

to give a ready answer. As the value of the various types of analysis currently practised becomes clearer with respect to the biological issues, perhaps some helpful clues may emerge for further evaluation by some suitable expert group.

Clearly for fresh oil and short-term studies, to know the type of oil or product and how it was used, goes a long way to provide the basic information for a repeat or extension of the work. Similarly, near an oil spill, the identification of the oil, its fate in the water, and its impact over some preliminary period will be capable of unequivocal chemical description. In the case of a steady input of an unaltering type of oily water, it may be possible to relate input to long-term ecological effect but if inputs change greatly over the years and quality varies as might happen in the evolution of a refinery discharge for

4-17b

Control stations, and in reference materials (After MASSIE and co-authors, 1981; permission of the authors)

60° 20′ N 01° 32′ E	59° 59′ N 01° 23′ E	58° 31′ N 01° 05′ E	Beryl A crude, %	Oil-based mud, %	Combustion coal + wood, %
1·10	4·92	7·83	51·51	62·36	0·78
0·14	1·03	2·05	3·70	4·94	0·08
0·46	0·87	3·58	4·98	3·45	0·54
0·27	0·33	0·79	0·91	0·77	0·10
0·77	1·27	3·24	3·69	4·84	0·43
15·90	26·34	146·68	32·99	23·77	42·63
10·11	5·63	43·16	0·26	0·16	15·29
30·13	15·34	133·09	1·74	0·15	32·12
39·27	10·34	106·55	0·03	0	3·50
15·20	4·10	34·46	0·04	0	0·81
5·66	1·66	15·80	0·006	0	0·91
16·57	5·21	44·08	0·12	0	1·86
65·19	13·06	98·36	0·006	ND	0·50
50·75	12·78	78·11	0·014	ND	0·53
0·261	0·103	0·718			
16·0	8·8	53·0			
Control stations					

instance, it then becomes difficult, if not impossible, to correlate input and biotic response.

When it comes to diffuse inputs to the sea from air, land, and water, virtually nothing is known except for the most durable hydrocarbons. Methods of isolation, fractionation, identification, and quantification of polar oil residues remain to be developed for the marine environment. At present methods do exist for a few of these in microbes and in animal and plant tissues. The general field problems in chemical analysis on anything resembling a complete or broad front are virtually insoluble. However, numerous strategies exist for tackling a considerable range of ecologically interesting chemical studies on some of the key issues. These emerge in later sections.

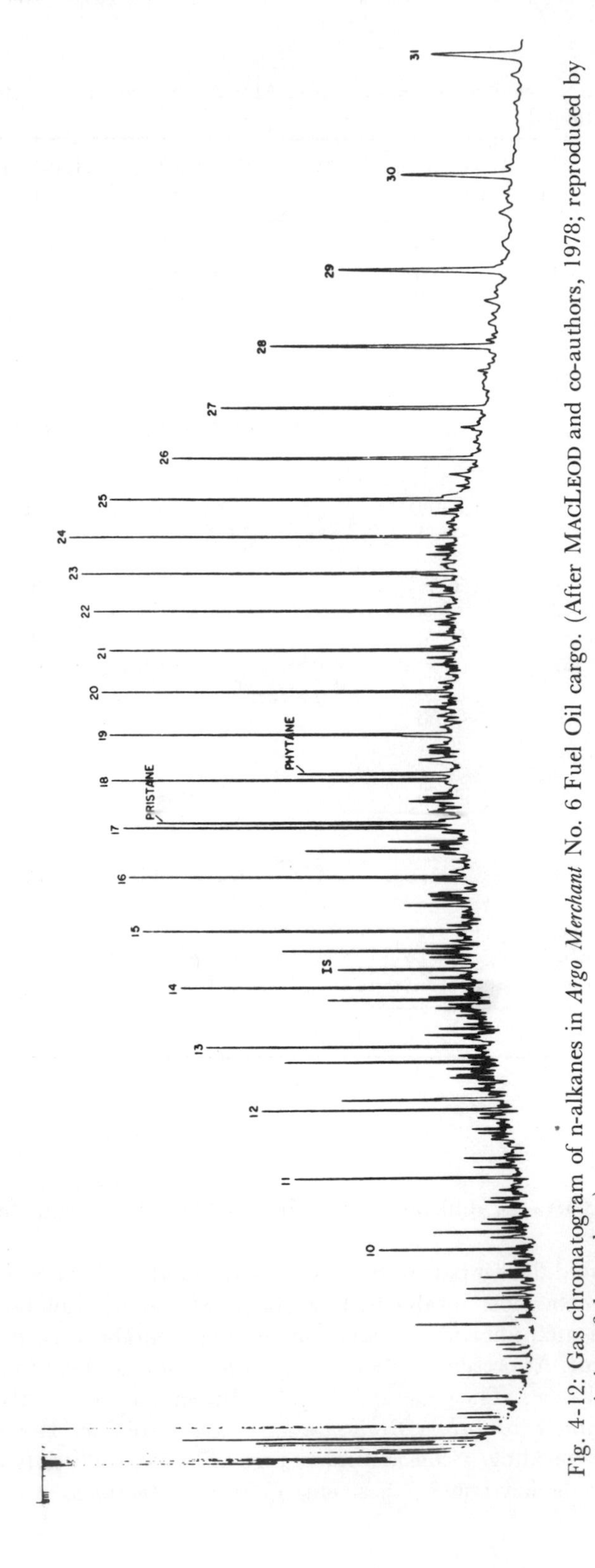

Fig. 4-12: Gas chromatogram of n-alkanes in *Argo Merchant* No. 6 Fuel Oil cargo. (After MACLEOD and co-authors, 1978; reproduced by permission of the authors)

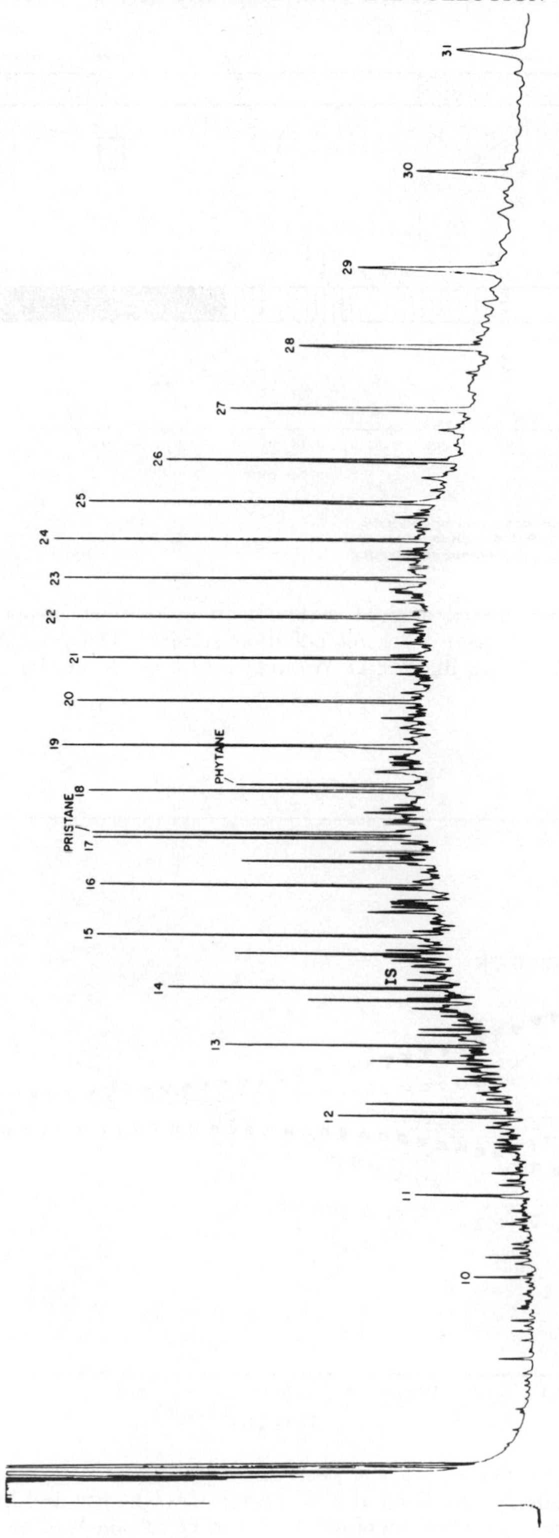

Fig. 4-13: Gas chromatogram of aromatic hydrocarbons in *Argo Merchant* No. 6 Fuel Oil cargo. (After MACLEOD and co-authors, 1978; reproduced by permission of the authors)

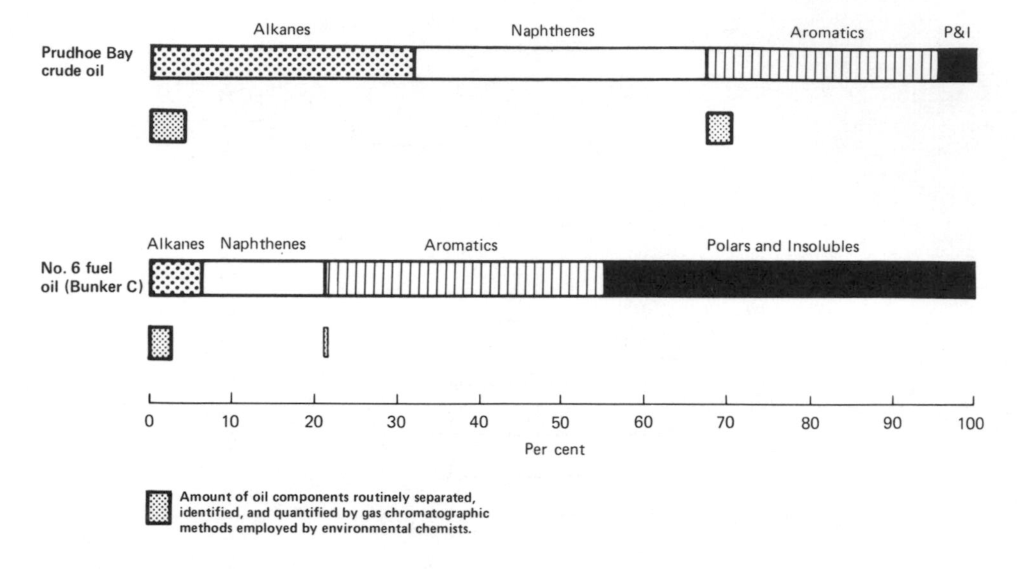

Fig. 4-14: The limited amount of hydrocarbon routinely reported by environmental chemists for oil pollution studies. (Data: courtesy of CLARKE, R. C., Jr., BROWN, D. W. and MACLEOD, W. D., Jr)

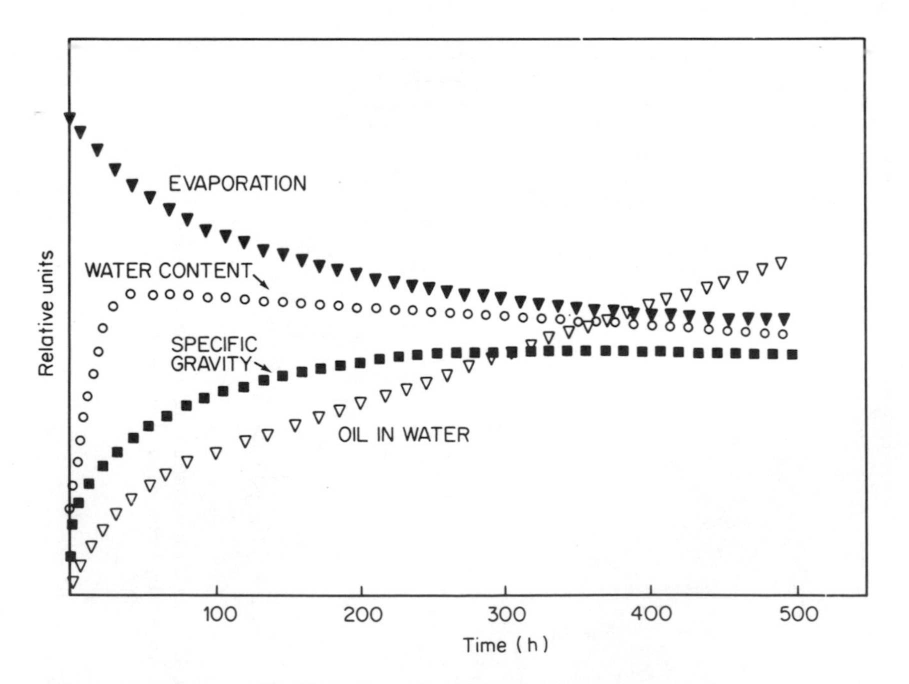

Fig. 4-15: Change with time of key physical parameters measured during a simulated spill of Guanipa crude oil at 2 °C. (After MACGREGOR and MACLEAN, 1977; reproduced by permission of the American Petroleum Institute)

(b) Role of Chemical Analysis

Especially in the last few years great advances have been made in the detailed analysis of crude oil, petroleum products of diverse application and complexity, and compounds derived from these by combustion, mechanical stresses and deliberate or unavoidable biological action. By combinations of fractionation techniques such as distillation, derivatization, partition, and high–pressure liquid chromatography it is possible to resolve the complicated mixture of substances which each fraction constitutes into groups of specific chemical character and narrow range of molecular weight. By application of liquid or gas chromatography each group can be further analysed and subsequently often characterized by reference to appropriate standards; where reference hydrocarbons exist, some or all of the components may be quantified individually. Gas chromatography, infra-red spectrometry, and fluorescence spectrometry employed in various degrees of elaboration are capable of generating spectra of closely related hydrocarbons from which detailed chemical characteristics of oils and oil derivatives can be determined. The highest resolution is attained for straight—chain and branched-chain hydrocarbons by use of high performance glass capillary gas chromatography and for aromatic hydrocarbons using gas chromatography coupled with mass spectrometry. On this basis, many hundreds of compounds have been identified and approximately quantified in crude oils.

Examples of high resolution gas chromatograms of *Amoco Cadiz* mousse are shown in Figs 4-16 and 4-17. The 65 identified and numbered components shown in Fig. 4-17 are characterized in Table 4-19. Component groups can be quantified from peak heights using standards (compare Table 4-16).

In forensic cases, for example as evidence in law for attributing a source to oil from an illegal oil spill, the US Coast Guard Research and Development Center has developed a 'fingerprinting' that is 'scientifically and legally sound' (BENTZ and SMITH, 1979). The major component of the 'fingerprinting' comprises 4 mutually reinforcing analytical approaches, namely infra-red spectroscopy, fluorescence spectroscopy, gas chromatography (glc) and thin-layer chromatography. Computerized data resources enable weathering effects on the oil samples to be taken into account. Essential documentation of all aspects of the spill, source oil, and sampling is vital evidence to support the 'fingerprinting'.

All this illustrates how unique oil samples can be and how difficult it is to generalize about oils in a chemical sense.

When it comes to scientific field surveys the same analytical methods are applicable but there must be prior stages requiring often very meticulous care in sample collection, extraction, clean-up, and group fractionation prior to analysis.

Crude oils contain many alkanes, many cyclic compounds, and a range of hydrocarbons containing hetero-compounds involving atoms such as sulphur, nitrogen, and oxygen. Being of natural origin hydrocarbons contain a wide range of metallic and non-metallic compounds and in spite of careful purification traces of brine salts, especially sodium chloride, are usually present.

The dominant sulphur compounds are mercaptans, organic sulphides, C_1-, C_2-, and C_3-dibenzothiophene isomers and many others in minor amount. Mercaptans give rise to the characteristic odour of crude oils and also of many products. Among the nitrogen compounds are pyridine, quinoline and related hydroquinolines, porphyrines, largely

Table 4-18

Typical industrial data on crude oils (Based on BERGHOFF, 1968; RUMPF, 1969; and ALUND, 1976; in: *Umweltprobleme der Nordsee*, 1980; p. 217)

	Source		Near and Middle East				
	Trade description		Arab light (Berry)	Qatar Dukhan	Iraq Kirkuk	Iranian light	Kuwait
Specification	Density $d_{15°}$	g cm^{-3}	0·831	0·821	0·845	0·856	0·869
	API gravity	API	39	40·9	35·9	33·5	31·3
	Classification	—	PG	G	PG	G	G
	Total sulphur	% S	1·10	1·29	1·95	1·4	2·5
	Kinematic viscosity/38 °C	cST	5·65	~30	4·61	6·41	9·6
	Flow point	°C	−34·4	−20	−36	−6·7	−4
	Vanadium/nickel	ppm/ppm	12/7		20/1	35/13	31/9·6
	C_4 and lighter	vol %		7·9		1·9	2·52
Products on fractional distillation	Benzin fraction (C_5–150 °C)	vol %	10·5	31·3	12·5	8·1	16·65
	Paraffins	%	87·4	58·0	80	50·0	
	Naphthas	%	10·7	20·0	18	33·0	
	Aromatics	%	1·9	22·0	2	17·0	
	Middle distillate (150–370 °C)	vol %	48·9	34·3	53·1	51·3	35·15
	Flow point	°C	−12·2	−15		−37	
	Paraffins	%	66·3		69	54·0	
	Naphthas	%	20·0		21	30·0	
	Aromatics	%	13·7		10	16·0	
	Residual oil (above 370 °C)	vol %	38·0	26·5	34·4	45·4	45·75
	Flow point	°C	24	38	30·0	23·9	
	Total sulphur	% S	2·04	2·34	4·0	2·4	4·16
	Vacuum distillable	%	7·4				19·85
	Vacuum (370–525 °C)						
	Residue (above 525 °C)	%		6·2	18·6		25·90

P = Paraffinic.
N = Naphthenic.
G = Mixed.

associated with naphtha and heavy oil fractions. Oxygen, the least abundant of these extra atoms, gives rise to small amounts of phenols, aromatic aldehydes, and ketones. During refining and as a result of catalytic treatment and hydrogenation, appreciable amounts of new aromatics and olefinic hydrocarbons are produced. Experts are interested in these sulphur and nitrogen compounds and partially oxidized or hydrolysed derivatives as key compounds in some biological effects of oil.

To return to the more general problem of measuring total oil, demand exists outwith and within the oil industry for the routine surveillance of oil in air, municipal and industrial wastes, rivers, estuaries, inshore waters, open seas and oceans, and in many specific applications—for example in oily waters on board tankers, at oily water treat-

Africa			North Sea		Venezuela	
Libya Brega	Nigerian Bonny light	Algerian Hassi M	Norwegian Ekofisk	GB Forties field	Boscan	Bachaquera Zulia
0·824	0·837	0·803	0·843	0·835	0·998	0·954
37·6	37·6	44·7	36·3	38	10·3	16·8
G	G	P	G	G	N	N
0·14	0·13	0·13	0·21	0·29	5·5	2·40
9·0	36·0	1989	42·5	4·2	90,000	1362
+24	−15	−52	20	0	15·6	−23·3
0·5	<0·5/4·0	<0·5/0·7	0·76/1·9	10/7	1200/150	437/75
2·3	2·2		1·0	4·0		
17·3	28·4	11·3	31·0	18·75	4·0	8·5
	34	80·8	56·5	81		27·6
	55	15·5	29·5	17		58·5
	11	3·7	14·0	2		13·9
37·4	38·6	55	29·2	39·80	17·0	20·5
				47·5		19·2
		56·5		38·5		54·8
		32·9	13·1	14		26·0
		10·6				
43·0	7·7	28·18	38·8	36·0	~75	~70
		18	29·4			15·6
0·15	0·39	0·31	0·39	0·65		3·0
22·9				20·40		
				15·60		
21·1						

ment plants, refineries and so on. The concentration range encountered is enormous and methods must take this into account.

In formulating a global picture, problems would be simpler if there were one universal method. This might be possible if there were only one kind of oil, but there are light crudes and the so-called ‘waxy’ or paraffinic crudes, light petroleum fractions, gasoline, fuel oils, heavy fuel oils and a range of asphalts and tars. Above all perhaps, the problem is not so immediately a technical one but a logistic battle involving many bodies in the oil industry, other industry, air and water quality control, river authorities, municipal sewage authorities, port and terminal operators, government bodies and fisheries boards. Each has different sampling requirements and reasons for conducting an analysis. Each also puts a cash value on the analysis with respect to its own needs and commitments. It is, therefore, no surprise that, as discussed in relation to Table 4-1 in

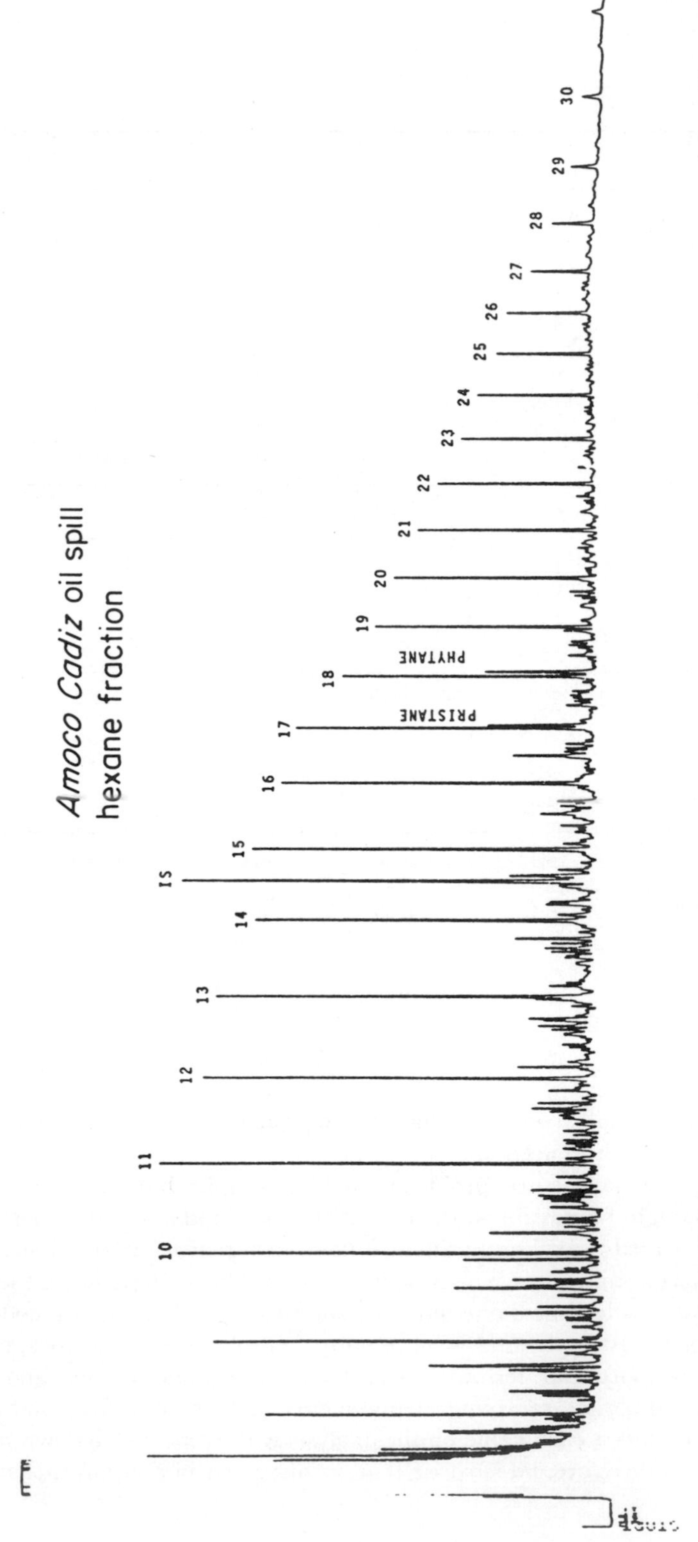

Fig. 4-16: High resolution gas chromatogram of n-alkanes in *Amoco Cadiz* reference crude oil mousse. (After NOAA-EPA, 1978; reproduced by permission of NOAA-EPA)

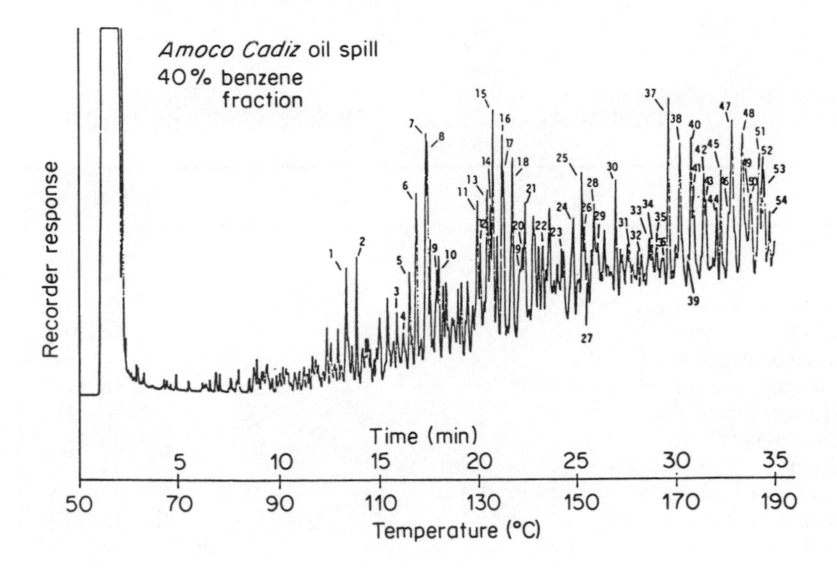

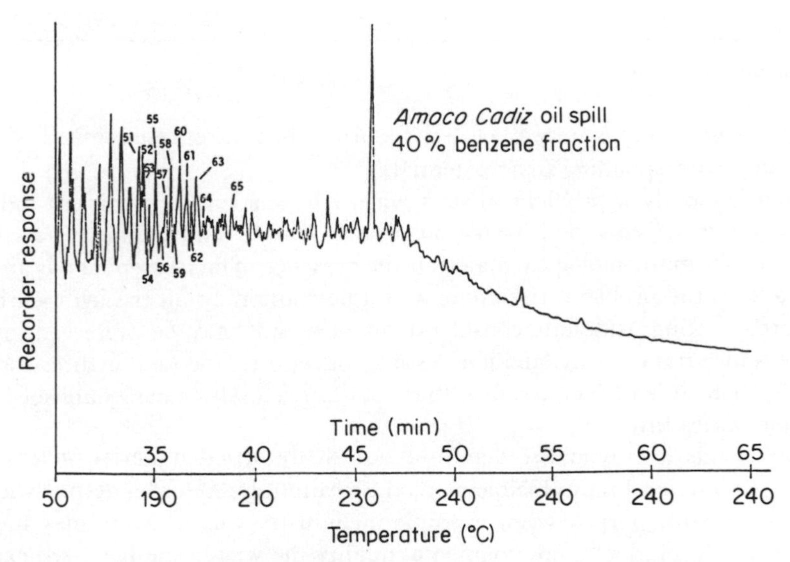

Fig. 4-17: High resolution gas chromatogram of aromatic hydrocarbons in *Amoco Cadiz* reference crude oil mousse. (After NOAA-EPA, 1978; reproduced by permission of NOAA-EPA)

the introduction, very little common ground exists in the measurement of oil in the various categories, and even within these categories there are considerable variations in methods. How many people take pains to formulate a concept of the'oil' referred to in their routine results and consider whether this 'oil' has any real identity capable of being intercalibrated with all the other oils? As a result, these much publicised numbers gleaned from many sources afford only the vaguest impression of the ecological impact of

Table 4-19

Distribution of 65 component aromatic hydrocarbons in *Amoco Cadiz* reference mousse as numbered in Fig. 4-17 (After CALDER and co-authors, 1978; modified; reproduced by permission of NOAA-EPA)

	Peak No. 5
Benzenes C_4-alkyl	3, 19
Biphenyl	4
C_1 naphthalene isomers	1, 2
C_2 naphthalene isomers	5 to 10
C_3 naphthalene isomers	11 to 18
C_4 naphthalene isomers	20 to 26
C_1 fluorene isomers	27 to 29
C_2 fluorene isomers	32 to 36
Phenanthrene	31
C_1 phenanthrene isomers	39, 41 to 43
C_2 phenanthrene isomers	46, 51, (52?)
C_3 phenanthrene isomers	(59)*, 62, (63), 63, 65
Dobenzothiophene	30
C_1 dibenzothiophene isomers	37, 38, 40
C_2 dibenzothiophene isomers	44, 45, 47 to 50
C_3 dibenzothiophene isomers	(52), 58, (59), 60, 61, (63)

* Shared peaks in parenthesis.

oil pollution since there is no way of deriving from them a representative hydrocarbon inventory and corresponding toxic potential.

Sampling is clearly a problem to start with. Air, suspended particles, and rainout demand separate systems. Soil, water, suspended matter, and sediment have different problems and for many biological materials the presence of fats, essential oils and waxes, etc., can add to the problem. Assuming a sample containing all the hydrocarbons has been secured, setting aside any considerations of how it may be preserved intact, the next stage is to extract, purify, and if necessary concentrate the total hydrocarbons prior to quantification (it is further assumed that qualitative analysis has established that the composition varies little).

For inert solids a gravimetric determination of the dried material under standard conditions is a sure and reproducible method. It cannot be so for oil despite widespread adoption for control purposes (for example in industry this measure may meet some routine need). In relation to environmental quality the weight method used exclusively with no better back-up is totally unsuitable. Two distinct methods have been proposed to replace or supplement the weight method, each involving measurement of a specific function of a class of compounds present in the sample, hopefully exclusive to hydrocarbons (LAW, 1980).

Infra-red spectrometry measures the absorption by hydrocarbons and other compounds of light energy of known wavelength either at preset wavelengths or by scanning appropriate bands. Chemical bonds in organic molecules have one or more characteristic absorption wavelengths. Absorption arises from the stretch between C-H represented by CH, CH_2, and CH_2 groups. For technical reasons the CH_2 frequency 3·4 μm has advantages and has been adopted widely. By choosing a solvent such as trichlorotrifluoroethane or carbon tetrachloride interference from CH bonds in the solvent is avoided. In crude oil most of the strongest absorbing groups are in the aliphatic fraction. Synthetic alkanes or benzene prove to be satisfactory standards. The method was adopted by IMCO (1978) as standard method for the calibration of continuous oil-content meters for shipboard or terminal use and is not usually subject to major error in this application. The meter is recalibrated by the method for each different oil. Equally, the method is useful for offshore and shore-based monitoring of treated oily water. In the sea, however, the petroleum is of mixed origin and biogenic hydrocarbons (p. 1492) are also present and will be included in IR measurements.

In 'fingerprinting' an unknown oil by IR, the absorption spectrum between 8 and 17 μm is scanned and absorptivity determined at 21 wavelengths for comparison with standards.

It is also possible to measure absorbance of ultra-violet light (UV). In this method it is the aromatic hydrocarbons that absorb UV light that gives rise to the signal measured and the unknown total oil is determined by relating the absorbance to a reference oil standard. Absorption occurs in a band round about 3650 Å, the exact wavelength varying with molecular structure; simultaneously excited molecules emit light at a longer wavelength between 4000 and 6500 Å. By using a dual scanning device which contours also the emitted light signal in wavelength and fluorescence as the EV source excitor wavelength is progressively varied, elaborate patterns depicting the fluorescent components of the oil are produced and may be regarded as a secondary 'fingerprint'.

Although plant pigments and animal products also produce UV absorbance and UV fluorescence it is not difficult to arrange the excitation wavelength and fluorimetry in such a way that serious interference is eliminated, at least using research instrumentation. Some cross-interference may arise using simple filter fluorometers. Thus it is possible to measure petroleum hydrocarbons, in effect the fraction containing most of the toxic components, using sensitive UV methods with little interference from biogenic materials. The main disadvantages of the UV method relate to the presence of variable amounts of highly fluorescent oil components in some unknown samples—some high, some low relative to the major fractions—so that choice of an appropriate reference material may be critical. Highly purified solvents free of fluorescent contaminants must be used in the extraction stages.

Chromatography is essentially a process of separation which can generate completely pure single compounds in many instances. However, when applied to oils it almost invariably shows dominant single compounds or several closely related compounds superimposed on a background of unresolved clutter which is derived from much smaller amounts of multitudinous overlapping minor hydrocarbons also present.

While relatively inexpensive gas chromatographs will adequately show the general make-up of crude oils, fuels, lubricants, etc., high resolution equipment using support-coated or wall-coated open tubular capillary columns are essential for the determination of individual oil compounds. High performance liquid chromatography can be used in a similar fashion. For quantification of suites of similar hydrocarbons known quantities of pure hydrocarbons are used as internal standards. The height or area of these known peaks is used to derive the amounts represented by the peaks of like substances. Any quantification of complex hydrocarbons is fully reliable only after positive identification of the individual oil compounds by the use of mass spectrometry or other independent means.

As a rough guide to the costs involved—and these can vary appreciably, depending on the routine analytical programmes for which the laboratory is streamed—they range presently from $10 to 50 per gravimetric oil determination, $50 to 100 per IR analysis, $50 to 100 per UV absorbance or single scan spectrofluorescence, about $300 for quantitative SCOT or WCOT gas chromatography, and at least $5000 for gc coupled to mass spectrometry. These costs relate to water or effluent samples. For some sediments and for all biological materials, solvent extracts are much more complex and clean-up processes must be undertaken before the hydrocarbon fraction can be subjected to analysis. These stages greatly increase the cost. One need scarcely mention the importance of analytical costs in relation to the control of pollutant inputs and the assessment of environmental quality. As with most things in life you get what you pay for; for $10 you get an unreliable measure of an anonymous oil, but it costs $5000 to $10,000 for a fully evaluated 'fingerprint' which has legal or ecological validity. Money spent on chemical analysis is an investment. For optimum benefit, a judicious choice (not entirely certain but certainly not a matter of good luck) predisposes the likelihood of the results obtained giving both adequate coverage and sufficient detailed information.

None of the routine methods discussed will discriminate between petrogenic and biogenic hydrocarbons and even some of the research methods give only a debatable indication of their respective contributions (Fig. 4-14). Great care is needed in the choice of methods for environmental monitoring and much of the published information needs careful scrutiny on the basis of the exact methods of sampling and analysis employed.

At this time no one would claim that present knowledge of global hydrocarbons represents a wise investment or a judicious programmed approach. Almost certainly more ecologically-useful knowledge could be obtained with better management without increasing the total cost to the populace.

(c) Biogenic Hydrocarbons

Many species, from microbes to mammals, produce their own brand of hydrocarbons and these contribute directly to the sea or indirectly from land and fresh water. As for petrogenic hydrocarbons, these products subsequently undergo many changes due to physical, chemical, and microbiological processes.

The original hydrocarbons which are minor constituents of phytoplankton (less than 1% by weight) are predominantly alkanes with carbon numbers C_{15}, C_{17}, C_{19}, and C_{21} in addition to carotenes. Seaweeds such as *Laminaria* contribute a single alkane, pentadecane (C_{15}). Some calanoid copepods contain much higher levels of a restricted range of alkanes greatly dominated by C_{17} (pristane), C_{19}, and C_{21} olefins.

River water and land plants contribute a variety of hydrocarbons including leaf waxes showing n-alkanes C_{27}, C_{29}, C_{31}, C_{33}, and unsaturated cyclic and acyclic alkanes.

The evidence on aromatic hydrocarbons of biogenic origin is poor and controversial (CLARK and BROWN, 1977) and on balance it seems unlikely that algae and higher organisms can synthesize them. However, PCAH in large quantities are formed from the burning of wood and in forest fires, and thus contribute indirectly very substantial quantities to the marine environment and are found in sediment samples from clean areas around the British Isles and elsewhere. They are not detectable in clean sea water by existing methods (JOHNSTON, 1980).

As an example of what hydrocarbons may be involved in a clean area, in this instance, the mountain streams flowing into Marmot Basin in the Rocky Mountains, Alberta, Canada, the average level of n-alkanes in these waters was 0·02 to 0·06 $\mu g\ l^{-1}$, predominantly C_{15}–C_{31} with C_{27}, C_{25}, C_{21}, C_{29}, C_{17} in order of decreasing abundance. Odd carbon number n-alkanes were in excess of the intervening even number alkanes by factors of 1·2 to 2·0 (crude oil = 1·0). The isoprenoid hydrocarbons pristane and phytane amounted to only 0·7 ng l^{-1} and total aromatics such as steroids and terpenoids less than 1 ng l^{-1} (TELANG and co-authors, 1980).

(d) Biogenic/Petrogenic Differences

The distinction between biogenic and petroleum hydrocarbons poses no great problem to the analyst in the spill situation. In the absence of such evident clues the detailed gc pattern of alkanes was regarded as fairly reliable (HARDY and co-authors, 1977a) since biogenic material showed disproportionate peak heights with dominant C_{15} and C_{17} peaks while petroleum gave a more regular set of peaks superimposed on a profile of unresolved complex mixture (URCM). Integration of the gc area below this URCM profile has been widely used as a measure of petroleum hydrocarbon presence but usually by adoption of some arbitrary reference standard. The interpretation of the URCM profile is contentious (see MCINTYRE and WHITTLE, 1977, pp. 94 to 96), arguments attributing the complex to petroleum origin and to the natural maturation of recent biogenic products both attracting supporting evidence. There may indeed be an unbroken geochemical relationship spanning geological epochs between recent weathered and matured biogenic hydrocarbons and petroleum. Studies by DASTILLUNG and ALBRECHT (1976) highlight the use of certain pentacyclic triterpanes occurring as diastereomeric isomers as a differentiating factor. Only 1 of the 2 isomers is found in recent marine sediments while fossil fuels contain 2. Using the more representative and more abundant isoprenoid stereoisomers, GASSMANN (1981) showed that in recently biosynthesized 2, 6, 10-trimethyltetradecane, norpristane, pristane, and phytane there was only one discernible peak in the high–performance glass capillary gas chromatograph. In the chromatographs of fossil fuel twin peaks were present in each, attributed to the diagenesis of the biogenic precursors. In 2 North Sea water samples the ratios found for biogenic/petrogenic were 260 : 1 and 3·7 : 1, illustrative of the importance of accurately differentiating the biogenic component in 'oil pollution'.

Fortuitously, QUIRK and co-authors (1980) working in the contrasting realms of fresh water, a blue-green alga, and peat describe their independent discovery of anomalous stereoisomeric ratios as a possible means of attributing isoprenoid alkanes to their particular sources. In river mud the diastereomers of pristane pointed to a mixed biological and petrogenic source. The fresh water unicellular alga *Microcystis aeruginosa* produced hop-22 (29)-ene which was also present in lake water and subsequently found as 17β (H)-hopane in the sediment. In peat water, the 17α (H)-homohopane occurred as the dominant triterpane contrasting with only the merest trace of mixed isomers in the growing moss which could be due to pollutant contamination. The peat isomer could be induced by bacterial action on moss during long storage and decay in dark aerobic conditions.

These approaches exploit earlier work on the geological maturation products of sterols. Work is in progress to further explore these and other chemical indicators of the time scale of the formation of petroleum-like products.

In the hydrocarbon chemistry of sea water the progressive increase in sensitivity and in knowledge of detailed composition, afforded by the most recent analytical developments, marks the beginning of a revolution in the interpretation of environmental chemistry in relation to ecology. A wider insight into marine organic chemistry in general must be expected to unfold exciting new understanding of the role of chemistry in ecology.

(e) Microbial Degradation of Oil

If petroleum concentrations are low enough to exclude lethal and inhibitory effects, many microbes and the tissues of many animals respond to contamination by gradual adaptation enabling hydrocarbons to be broken down or converted into products that do not lead to excessive accumulation in lipid-rich tissues. It will be familiar knowledge that in bio-engineering, microbial species and strains can be effectively used to convert natural gas and some oil components into a wide range of commercial chemical feedstocks, and in effluent treatment works colonies can be established on the inert support media of biotreatment units which are effective in virtually eliminating hydrocarbon solvents and other products from industrial waste waters. However, even these specially selected clones are not able to withstand shock exposure to sudden high concentrations of petroleum hydrocarbons or any undue flux in physical or chemical conditions or access of other contaminants. Indeed it has often been observed that highly selected clones capable of withstanding one form of polluting stress to an unusually high degree, are frequently vulnerable when conditions relating to other stresses are altered by what might be a degree of change tolerated by a less developed wild community (LINDEN and co-authors, 1980).

In a recent symposium the microbial degradation of pollutants in marine environments was discussed in considerable detail, including a section on oil (BOURQUIN and PRITCHARD, 1979). Together with supplementary material, these discussions provide a useful outline of the susceptibility of component oil groups to microbial attack and of the commoner pathways of biochemical transformation of individual hydrocarbons or families of related compounds.

The aliphatic hydrocarbons present in most crude oils form a regular series, primarily and virtually exclusively, of n-alkanes. The microbes considered are chiefly bacteria, yeasts, and fungi but other unicellular organisms such as flagellates, heterotrophic algae, and some protozoans have some capacity for hydrocarbon degradation but are normally less capable and less opportunistic. A large contribution to current knowledge of microbial degradation (of hydrocarbons and pollutants generally) relates to species and communities isolated from soil, fresh water and wastewater treatment plants. It may be helpful to regard the operation of marine microbial degradation as comparable but this extrapolation has not been consistently confirmed for estuary and sea, or adequately investigated over a wide enough range of applications. Since there is, however, a broad resemblance in the general biochemistry and physiology of marine, fresh water, and terrestrial forms of life, this process of generalization is not completely without support.

The n-alkanes range from the gas methane which is the simplest and most uniformly abundant hydrocarbon with petroleum and non-petroleum associations, to liquid n-alkanes including several common to petrogenic (C_6 upwards) and biogenic origins (usually C_{15} to C_{31}) and solid paraffins (above C_{31}). Solubility in sea water, which will be shown to be important with regard to toxicology, is equally relevant to biodegradability since no organisms can survive, let alone metabolize, inside liquid or solid hydrocarbons in the absence of contact water. Two biochemical mechanisms are common: (i) enzyme catalysed reaction with water or hydroxylation which is the general mechanism favoured in the initial attack on n-alkanes; (ii) enzyme catalysed reaction with dioxygen or catabolic oxidation process leading to the cleavage of unsaturated ring systems.

Hence $CH_4 \rightarrow CH_3OH$ is one of the commonest transformations for methane and there is parallel change in at least the lower and some of the middle n-alkane range

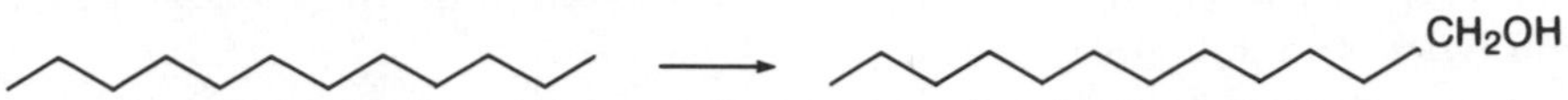

CH_2OH

Such initial hydroxylation can be accompanied by other changes to a lesser degree and a variety of further transformations are known; the change being influenced by the prevailing conditions and by the organism. Alternatively, the transformation may involve change in the C-chain length, e.g. the formation of a mono- and a di-carboxylic acid by *Brevibacterium* sp.

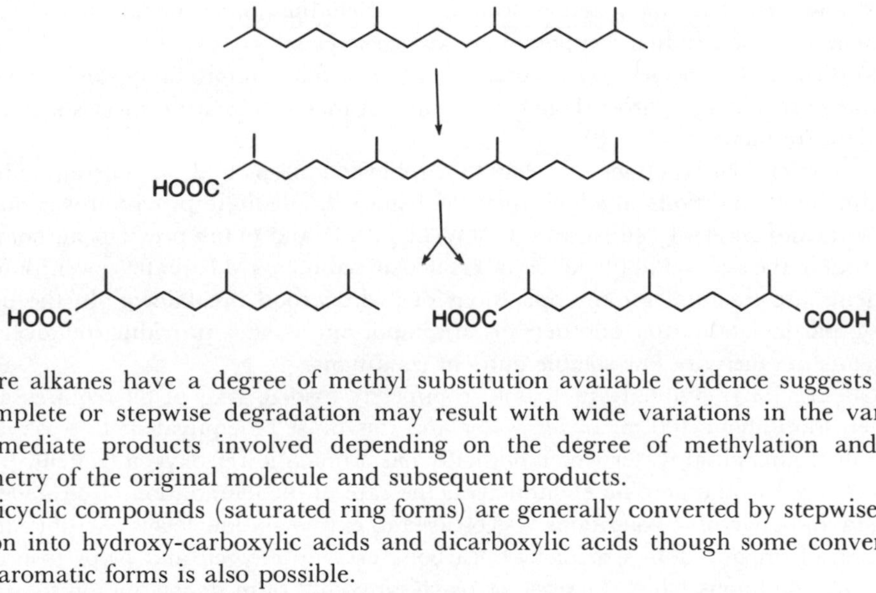

HOOC

HOOC HOOC COOH

Where alkanes have a degree of methyl substitution available evidence suggests that incomplete or stepwise degradation may result with wide variations in the various intermediate products involved depending on the degree of methylation and the geometry of the original molecule and subsequent products.

Alicyclic compounds (saturated ring forms) are generally converted by stepwise oxidation into hydroxy-carboxylic acids and dicarboxylic acids though some conversion into aromatic forms is also possible.

One of the most distinctive features of the reaction sequences brought about by aerobic microbes in the catabolism of benzenoid molecules is the incorporation of the oxygen molecule as a co-substrate in the cleavage of such rings.

aromatic hydrocarbon $\xrightarrow{\text{dioxygen}}$ diphenol $\longrightarrow$ dicarboxylic acid

Where additional functional groups are present, these provide alternative sites for attack. For example, side chains may be subject to oxidation.

With regard to oil components containing sulphur, oxygen or nitrogen, the heteroelement has a strong influence on the susceptibility of the molecule to attack and on the site of attack. Sulphur-containing rings are less amenable to microbial degradative processes than other rings in complex heterocyclic compounds and as a rule oxidation or hydroxylation begins in these other rings.

Complex hydrocarbons containing oxygen substituted rings are more vulnerable to attack and breakdown starts with the oxygen-containing ring.

For saturated ring systems containing nitrogen, hydrolysis of the cyclic imine formed by dehydrogenation of the parent compound is the most usual cleavage mechanism. For unsaturated heterocyclic molecules, such as pyridine carboxylic acids, nicotinic and picolinic acids, hydroxylation at ring position 6 is the initial step followed by one of several types of oxidation.

If hydrocarbons are halogenated, or become chlorinated or brominated by some process for bacterial control at sewage works or at production platforms, the products are much less vulnerable to microbial attack, their toxicity is enhanced and there is increased likelihood of biomagnification. Many studies have been made of organohalogen pesticides and industrial chemicals, and the processes giving rise to breakdown products are broadly well established. It is less well known that there are at least 300 naturally occurring organohalogen compounds, many of marine origin, for which a range of enzymic reactions leading to the elimination of the halogen atom from the main organic residual compound are known to exist.

More than 200 species (60 genera) of bacteria, filamentous fungi, and yeasts are capable of oxidizing hydrocarbons. How many of these are marine species is not clear but they are numerous.

These microbial reactions all relate to aerobic conditions and, in general, under the experimental conditions in which they were studied, adequate provision was made of nutrients and accessory substrates. COLWELL (1974) and many previous authors indicate that in the sea and in the upper layers of oxic sediments, adequate concentrations of nutrients are essential for the operations of hydrocarbon breakdown. In the pelagic phase, particulate matter and detritus are important as sites providing relatively rich nutrients in otherwise low soluble nutrient conditions.

ZOBELL (1964) pointed out that to completely oxidize 1 kg of oil requires all the oxygen from about 400 m^3 of sea water and this must be equivalent to a very great volume of interstitial water which normally has a much lower oxygen content.

Compared to the aerobic environment, the rate of biodegradation of oil under low oxygen and anaerobic conditions is decreased to extremely low levels. Aerobic decomposition of organic debris or of hydrocarbons uses up oxygen and helps to promote reducing conditions which increase or reach an equilibrium depending on the rate at which oxygen is replenished. Any reduction products formed are swept away or re-oxidized when water turnover occurs. At the sea floor, oxygen transfer into pore water falls off radidly especially in fine sediments over depths of a few centimetres or millimetres.

When oxygen becomes depleted, the microbes seek sources of combined oxygen, first from nitrates then from sulphates and the higher oxidized forms of metals, and at greater depth methane-producing bacteria get rid of excess hydrogen by methane formation. In the process, organic substrates such as acetate are cleaved to carbon dioxide which is subsequently reduced to methane. MERTENS (1976) established that about 90% of the sulphate present must be reduced by sulphate-reducing bacteria before anaerobic conditions are sufficiently extreme to support methanogenic species. Since hydrocarbons are already hydrogen-rich compounds they are not suitable as organic substrates for methane-producing bacteria. In contrast, sulphate-reducing bacteria can utilize a wide range of carboxylic acids and other organic substrates and in the process sulphate is reduced to hydrogen sulphide. Under favourable conditions, for example in oil storage tanks and on oil platforms, the hydrogen sulphide thus formed can present a major hazard by greatly accelerating corrosion of iron and concrete and by forming pockets of hydrogen sulphide which is rapidly lethal to man if inhaled.

Few normal benthic animals survive in anaerobic conditions but some forms can live partly immersed in anoxic sediment.

In the dynamics of the oil degradation process perhaps the only unrestricting parameter is the capacity of the hydrocarbon degraders since these can multiply virtually indefinitely at a high rate given optimum substrate conditions. Everywhere in the sea these conditions are limiting, severely limiting or adverse. The most adverse condition is probably dilution in the water column which continuously reduces and disadvantages the bacterial population in relation to the substrate. Presence of a solid substrate favours bacterial proliferation and, simultaneously, access to oil and sometimes nutrients. The amount of particulate matter in offshore water is limiting. Compared to oil as a carbon source in an oil spill situation, the supply of nutrients in forms available to bacteria is severely limiting, very severe when nutrients are seasonally depleted. This shortage of nutrients ensures that the oxygen demand for the biological combustion or partial oxidation of oil does not exceed the rate of supply. Where oil globules occur the associated microbial biomass can sometimes grow to 20 to 60% of the weight of oil.

Higher ratios can be achieved with pure bacterial cultures and pure hydrocarbon substrates but usually pure cultures attack only a narrow spectrum of hydrocarbons. For seeding of oil spills polyvalent or mixed bacterial cultures would be more effective, but (as has just been discussed) this parameter is not usually limiting under natural conditions. The promotion of natural biodegraders might be achieved more readily in suitable sediments than for open sea spills by the introduction of a rich oxygen source such as peroxide in association with oxidized forms of nutrients (GASSMANN, 1978).

(f) Selected Aspects of the Toxicology of Oil

General Considerations

Toxicological results for crude oils, oil products, and single hydrocarbons exist for a wide range of bacteria, yeasts, moulds, fungi, unicellular algae, zooplankton, benthos, marine grasses, intertidal plants and animals, fishes, invertebrates and vertebrates, mammals, birds, and to some extent man. The list is probably incomplete. Marine life has been exposed to hydrocarbons in pure, bacteria-free cultures and all sorts of intermediate groups up to field communities. Bibliographies alone comprise several volumes (see literature appendix). Toxicological studies can be roughly categorized into short-term mortality tests, extended laboratory tests, and model-ecosystem experiments. In addition, oils have been tested along with dispersants in many different ways. A selection from this vast pool of information has been made to illustrate general principles where possible, indications of the more actively toxic oil constituents, and features deemed to be ecologically or commercially important. In seeking useful generalizations OTTWAY (1971) found that the toxicity of 20 crude oils ranked fairly consistently with the proportion of low boiling point fractions, especially the aromatics contained in the distillate below 150 °C.

Phytoplankton Species as Cellular Models

The simplest example is the toxic response of unicellular algae to individual pure hydrocarbon in single culture.

The most comprehensive study which coordinates toxic responses with the physico-chemical properties of individual hydrocarbons has been conducted by HUTCHINSON and co-authors (1979) using several fresh water unicellular algae. The keynote of these studies is the use of the $^{14}CO_2$ uptake method to measure photosynthetic activity under highly reproducible laboratory conditions. Inhibition of photosynthesis in response to added hydrocarbon is measured over a 3-h period. Leakage of ^{14}C-labelled photosynthetic products can also be measured, and loss of potassium and manganese from the cell.

Photosynthetic inhibition may be extrapolated to represent reduction of primary productivity in the photic zone; losses of labelled products, potassium and manganese reflect various aspects of the health of the cell and of its nutritional value in the food chain. These losses also demonstrate effects on the ability of the cell membrane to maintain its integrity and normal function. The physico-chemical measurements on the 38 hydrocarbons, covering a wide range of oil components and organochlorines, provide independent correlation with their lipophilic properties including partition between water and solvent.

The main findings of HUTCHINSON and co-authors are:

(i) Hydrocarbon concentration, expressed as a percentage of fresh water saturation, is statistically related ($p < 0{\cdot}01$) to toxicity as indicated by 50% inhibition of *Chlorella vulgaris* and *Chlamydomonas angulosa*. Hence the toxicity of all hydrocarbons, irrespective of structure, can be predicted from solubility alone. On a molar basis the least soluble hydrocarbons are the most toxic.

(ii) If vapour pressure (mmHg at 25 °C) is used in place of percentage solubility a valid statistical relationship is found with toxicity but not as good as for solubility. Partition coefficient, as determined by a GC headspace technique, is highly correlated to toxicity to algae.

(iii) Differences in response to hydrocarbon toxicity is neither marked nor consistent between the 2 algal species.

(iv) The amount of hydrocarbon inducing the loss of 20% labelled ^{14}C photosynthetic product is significantly correlated ($p < 0{\cdot}001$) with the solubility of the hydrocarbon. This correlation is seen as evidence for partitioning of the hydrocarbon with lipid in the plasmolemma, thus interfering with membrane function.

(v) Even 15-min exposure to hydrocarbon causes complete loss of potassium, indicating a rapid response to change in membrane permeability. Potassium loss is also statistically, highly significantly related to hydrocarbon solubility and toxicity.

(vi) Manganese loss is somewhat more highly correlated than potassium loss with hydrocarbon solubility and toxicity. No attempt was made at detailed interpretation of Mn loss; it it slower than K loss and, again, associated with a membrane effect.

(vii) Chlorinated hydrocarbons behave exactly as other hydrocarbons.

These studies are considered in the wider context of literature reports on other pollutants, such as organochlorines and organophosphorus pesticides. Measures of solubility and partition, such as the partition coefficient for n-octanol and water, were assessed along with other cell properties—such as bioconcentration factors and osmotic balance. HUTCHINSON and co-authors further argue association of these findings based on algal inhibition with toxicity to rainbow trout and a model aquatic ecosystem in which biodegradability was also measured. The common factor in all these interactions is the response of the cell membrane and perhaps cellular disruption. It is claimed that these results have value in the prediction of effects of oil spills on a variety of aquatic organisms.

(7) Effects of Oil and Dispersants

(a) General Aspects

The toxicology of crude oil and its products supports few accurate generalizations and it is hard to summarize the effects on the enormously variable life forms in the sea. In the first few hours or 1 or 2 d of impact it may be adequate to represent the toxicology of petroleum introduced into the sea in terms of oil/organism interactions. In periods of several days to several weeks biodegradation is beginning to be effective, and in the open sea—away from continuous inputs—the microbial population and the overall background level of hydrocarbon contamination tend to sustain a quasi steady-state situation for both. Although it was earlier held that petroleum biodegradation was complete, that carbon dioxide and water were only identifiable products, more sophisticated chemical analyses not possible 10 yr ago are showing stepwise hydroxylation and oxidation of hydrocarbons within microbial cells and, in organs of marine animals, partial excretion of the more polar products formed. The initial transformation is the rate-determining step except perhaps for complex compounds since most of the polar products are rapidly decomposed.

Zooplankton and the Food Web

For fundamental reasons the problems of studying the effects of oil on zooplankton are much greater than is the case with phytoplankton. For example, zooplankton comprises a very wide range of species, covering a broad range of sizes and representing many modes of feeding activity and reproduction; it contains a particularly complex array of metamorphosing and many temporary elements such as eggs and larval stages of fish and benthic invertebrates (to be discussed later), and even migratory benthic species. Studies in the laboratory are also severely limited since many zooplankters are difficult or impossible to cultivate (Volume III: KINNE, 1977). They may take up oil (i) from water, (ii) as oil droplets, (iii) with food or attached to food. Practical manipulations to examine these alternative or simultaneous pathways are exceedingly delicate. It is widely recorded that zooplankton collected from the sea may already contain petroleum in addition to appreciable concentrations of biogenic hydrocarbons. At very high levels of oil, feeding virtually ceases, the production of faecal pellets by a number of different species is substantially reduced, and narcosis is observed (SPOONER and CORKETT, 1974). Over a range of graded concentrations of heating oil or naphthalene, BERDUGO and co-authors (1977) found feeding activity and ingestion altered or completely suppressed, while low levels had no effect (< 50 mg naphthalene l^{-1}).

Ingestion of oil droplets has been widely reported. LC_{50} tests on zooplankters (mainly copepods) reveal short-term survival in concentrations from 1 mg l^{-1} upwards of various oils and oil products (as their water-soluble fractions). These concentrations are many times higher than the general background in sea water (<1 to 140 μg l^{-1}), but these studies concentrate on the cutaneous absorption of hydrocarbons by copepods (mainly) that may have a special resistance to this type of intake. CORNER and co-authors (1976a) established that for the copepod *Calanus helgolandicus* the dietary route was more important for naphthalene. HARRIS and co-authors (1977) came to the same conclusion for *C.helgolandicus* and extended this finding to 6 other species—estuarine, neritic, and oceanic.

In view of the overlap in particle size between dispersed oil and unicellular algae in phytoplankton, HEBER and POULET (1980) examined ingestion of oil droplets with food. Mechanical and natural suspensions of oil in sea water both resolve after mixing into a similar particle size spectrum. When introduced into sea water, some of the oil particles aggregate with food or non-food particles, resulting in contaminated and uncontaminated non-food, and contaminated and uncontaminated good-food which are not differentially ingested. Particle feeders such as euphausiids and copepods are capable of altering their feeding pressure to correspond with the most abundant particles within their range. Some animals thereafter respond to contamination by active avoidance (e.g. cessation of feeding, changes and inhibition of phototropism through sensory mechanisms). On exposure to a suitable size range of particles of Venezuelan crude oil, both feeding and behaviour were modified in *Meganyctiphanes norvegica* and *Calanus hyperborea*, and growth and survival of *Acartia hudsonica* were rapidly and significantly reduced. Groups kept under starvation showed similar decreases in growth and survival. During recovery in clean water, treated and untreated sets showed similar growth rates. HEBER and POULET also considered various aspects of the function of copepod chemoreceptors which clearly trigger a feeding response but do not discriminate against oil, plastic beads or red mud particles. They dismiss the possibility that the chemoreceptors are insensitive to oil and favour an explanation based on the coating of the chemoreceptors with an oil film or on the uncertain or delayed response of chemoreceptors to oil which permits some feeding before the receptors are coated.

Our knowledge of oil effects on zooplankton is dominated by short-term observations and by a very limited coverage of this most variable component of the ecosystem. There is evidence from oil spill studies that species intolerant of laboratory culture are also those specially sensitive to oil.

(b) Enclosed Ecosystems

Two types of pilot-scale laboratory systems have been used to bridge the gap between bench experimentation and field surveys. These are tank systems which have been used for many years to study the dynamics of the food chain in mixed pelagic and bottom communities and large plastic bags (e.g. LEE and TAKAHASHI, 1977) used to enclose for study a column of water and its biota, usually cut off from the seabed. Both types of enclosed ecosystems can be set up so that the stresses of nutrients or pollutants can be evaluated not just for a single species in isolation but for the fairly natural balance of species that survives and develops under the constraints of these systems.

Because the experimental units are much larger than bench units there is scope for more realistic presentation of oil as a pollutant and simulation of chronic or spill exposures. The fate of oil and its effects can be followed over considerable periods of time, generally a period limited only by the stability of the control systems. Enclosed ecosystems are finely poised and after some time communities initially set up in identical fashion can gradually diverge into not only altered communities but also ones of quite different characteristics. The sea bottom is such a potent factor for modifying the environment within large plastic enclosures that these enclosures are normally terminated in the water column. To complete the plastic bag ecosystem model, particular matter reaching the bottom of the enclosure may be transferred to separate chambers on the seabed in order to study related changes in the benthos.

WADE and QUINN (1980) used tanks furnished with water and sediments transported from nearby Narragansett Bay in an attempt to identify the processes that acted on No. 2 Fuel Oil in its transfer from water column to sediment. The fuel oil could only be effectively introduced to the tanks as an emulsion which was applied to give about 180 μg l^{-1} in the water phase. Over the 168 d of the experiment, 78% of the oil introduced was found to associate with particles larger than 45 μm, 22% with particles down to 0·3 μm, and virtually none truly dissolved. The flocculent material which settled over the sediment contained 279 μg g^{-1} equivalent to 30% of the oil attached to the large particles, and 496 μg g^{-1} equivalent to 70% of the oil attached to particles smaller than 45 μm. Clearly marked changes had occurred on sedimentation and diagenesis. Microbial degradation caused large preferential losses of n-alkanes and minor losses of isoprenoids. Over the prolonged experiment the oil added to the sediment amounted to only 9% of that introduced and this added oil was restricted to the upper 2 to 3 cm of sediment. It was hoped that these results would enable the authors to interpret their earlier work on petrogenic and biogenic hydrocarbons in deep undisturbed cores from the bay (WADE and QUINN, 1979).

Much of the research using large plastic bags has been aimed at evaluating food chain relationships among pelagic communities and assessing the production and fate of the related organic matter. Consequently disturbance of these relationships by the introduction of a pollutant such as oil generates many effects on the pelagic community and its operation and, simultaneously subjects the oil to processes of degradation and transfer to the bottom on detritus and faecal pellets. Customarily in order to sustain productivity in these enclosed ecosystems, nutrients are incorporated regularly to maintain algal abundance within a predetermined range (TAKAHASHI and co-authors, 1975).

The large plastic enclosures in Saanich Inlet used by LEE and TAKAHASHI (1977) were principally directed at describing the effects of the water soluble extract of No. 2 Fuel Oil on phytoplankton and determining the changes with time in the capacity of bacterial populations to degrade selected labelled hydrocarbon substrates. Introduction of water-soluble fraction of No. 2 Fuel Oil to give about 20 μg l^{-1} in the plastic bag caused a rapid and marked decline in the diatom population which subsequently did not recover; there was a slower decline in the control enclosure. At 10 μg l^{-1} there was no major difference in diatom survival between treated and untreated bags. Before dosing, the microbial degradation rates for the various ^{14}C-labelled substrates (Table 4–20) were low and variable; after dosing, the rates were more reproducible. Of the substrates

Table 4-20

Microbial degradation rates for ^{14}C-hydrocarbon substrates using water from an ecosystem enclosure before and following addition of No. 2 Fuel Oil (After LEE and TAKAHASHI, 1977; reproduced by permission of the International Council for the Exploration of the Sea)

Hydrocarbon	Concentration ($\mu g\ l^{-1}$)	Collection depth (m)	Incubation time (h)	Time after oil addition (d)	Degradation rate ($\mu g\ l^{-1}\ d^{-1}$) $\times 10^2$*	Turnover time (d)
Benzpyrene	16	5–10	48	0	0	—
Benzpyrene	16	5–10	24	3	1 ± 0·7	1400
Fluorene	30	5–10	48	0	0	—
Fluorene	30	5–10	48	3	0	—
Heptadecane	30	0–5	24	0	7 ± 4	400
Heptadecane	30	0–5	16	3	50 ± 3	60
Methyl naphthalene	50	0–5	24	0	10 ± 6	500
Methyl naphthalene	50	0–5	24	3	26 ± 4	200
Naphthalene	50	0–5	24	0	10 ± 3	500
Naphthalene	50	0–5	16	3	250 ± 7	22
Naphthalene	50	0–5	10	4	100 ± 5	57
Naphthalene	50	5–10	10	4	500 ± 11	10
Octadecane	30	0–5	24	0	16 ± 7	200

* ± Standard deviation.

used to outline the biodegradative processes for the added fuel oil, labelled naphthalene showed the most dramatically increased rate of degradation from low initial values to 5 $\mu g\ l^{-1}\ d^{-1}$ at Day 4. The rates for heptadecane, benzene, and toluene also increased markedly.

It is more difficult to manipulate oil slicks in plastic bags but the results may nevertheless be more meaningful than bench experiments. Using small plastic enclosures LACAZE (1974) studied slicks of Kuwait crude oil on water enclosed from the Rance estuary over a 27 d period covering September and October. Zooplankton were removed by microstraining. As might be expected phytoplankton $^{14}C-$ productivity progressively halved over the period in the control enclosures. In the treated enclosures there was a rapid fall in $^{14}C-$ productivity to about 50% but after 5 d and evaporation of much of the lighter fractions there was some brief recovery followed by collapse to very low levels. The collapse was attributed by LACAZE to the progressive solubilization of oil from the slick and the walls of the enclosure.

The continuing studies using enclosed ecosystems at Loch Ewe on the west coast of Scotland are directly aimed at elucidating oil spill and oily water effluent situations in relation to the ecosystem of the North Sea and where possible North Sea fisheries. As a result, much of the effort has been devoted to observing: (i) the fate of oil; (ii) the effects on phytoplankton and productivity; (iii) the effects on zooplankton including fish eggs and larvae; and (iv) the associated transfer of oil to the bottom and to benthic organisms. These studies are also related to ongoing studies of pelagic and near-bottom food relationships in an effort to construct a realistic quantitative model or simulation of the complete open sea food web (DAVIES and GAMBLE, 1979).

The loch water used receives little fresh water and there are only a few small scattered villages and no industry in this remote area. The control and baseline hydrocarbon concentrations were routinely 1 to 2 $\mu g\ l^{-1}$.

The first approach, avoiding the complications of oil film or particulate oil, was to use the water-soluble fraction of Forties North Sea crude oil (DAVIES and co-authors, 1981a). It was found that only about 10 $\mu g\ l^{-1}$ (max) of oil (estimated by UV fluorescence) could be attained in the enclosure. No biological response was expected nor found over the short period (1 wk) of exposure to $<10\ \mu g\ l^{-1}$. Subsequently a fine oil and water dispersion, prepared by using mechanical emulsification, was used in a single application to bring the oil concentration in the treated ecosystem up to a mean of 100 $\mu g\ l^{-1}$, representing the upper limit of oil in water for the open sea. Repeated sampling in the initial period showed oil concentrations ranging from 40 to 240 $\mu g\ l^{-1}$ in the dosed water declining during the subsequent 4 to 5 wk to about 25 $\mu g\ l^{-1}$. There was clear evidence of rapid transfer of oil to the bottom of the enclosure.

^{14}C-labelled hydrocarbon substrates were employed to measure rates of mineralization of hexadecane, naphthalene, and benzo [α] pyrene. To study how nitrification was also affected a wide range of ^{14}C-labelled amino acids was used. The pattern of hydrocarbon mineralization resembled that reported by LEE and TAKAHASHI (1977). **Hexadecane** mineralization increased from 0·07 $\mu g\ l^{-1}\ d^{-1}$ to 0·16 $\mu g\ l^{-1}\ d^{-1}$ but also in the control, possibly a response to bag material. **Benzo [α] pyrene** stayed at a very low rate averaging 0·005 $\mu g\ l^{-1}\ d^{-1}$. **Naphthalene**, initial rate 0·05 $\mu g\ l^{-1}\ d^{-1}$ rose to a peak (Day 8 to Day 10) of 5·0 $\mu g\ l^{-1}\ d^{-1}$ in samples from the oil-treated enclosure only decreasing gradually to the initial value. The microbial degradation of naphthalene equalled overall or exceeded at times the rate of loss by settlement or biological uptake.

In the upper layers of the control enclosure phytoplankton growth showed a well marked pattern of increasing growth over the first 25 d (reckoned from the date of introducing the oil to the treated ecosystem) followed by a marked decline towards the end of October (Day 30 or so) when the seasonal die-back set in. Some initial inhibition of primary production in the upper 0 to 2·5 m of treated ecosystems was found when the oil concentration was high. After 3 wk when the oil concentration had fallen below 50 $\mu g\ l^{-1}$ there was little between the treated and untreated rate of phytoplankton production. Diatoms survived North Sea crude oil well unlike their rapid decline in the presence of No. 2 Fuel Oil (LEE and TAKAHASHI, 1977).

The most interesting and novel results of this work probably relate to the studies on zooplankton. The dominant species in the untreated enclosure were copepods, mainly the calanoids *Acartia clausi* initially, with *Temora longicornis* and *Pseudocalanus elongatus* becoming gradually more abundant. Cyclopoid copepods were much less numerous mostly *Oithona nana* and some *O. similis*. Various planktonic predators were present, mainly the rapidly reproducing *Lizzia blondina* and *Sarsia gemminifera* and smaller numbers of ctenophores and chaetognaths. The zooplankton species all followed somewhat different but usually parallel trends in abundance and over the period some underwent one or more metamorphoses. In the zooplankton exposed to oil, calanoid copepod numbers fell when oil was introduced but after a few days both treated and untreated populations were liberating about equal numbers of eggs. In the control population 8 to 10 d later, nauplii appeared and thereafter 15 to 35 d later increased numbers of copepodites emerged and adults beginning Day 50. In the oil-treated population no additional nauplii or copepodites emerged and the whole population gradually declined to a very low number.

In the control enclosure the number of predators increased in step with the rising numbers of prey organisms but in the presence of oil the numbers declined after the first few days.

With regard to minor components, no trends were discernible for the less numerous *Oithona* spp. and appendicularians; the larvae of bottom invertebrates were slowly lost from the water column (as expected) but declined abruptly in the oil-treated ecosystem when the oil was introduced.

Of the 30 g oil initially administered as an emulsion, 14 g were recovered from settlement material at the bottom of the enclosure, 7 g were calculated to be in solution or fine suspension or associated with living matter and 9 g could not be accounted for, this fraction was attributed to losses by evaporation and microbial mineralization. Very little oil was adhering to the walls of the enclosure but microbial degradation reaching 1·5 g d^{-1} for naphthalene at peak could readily account for the non-recovery of this fraction.

There is much debate about effects of oil on phytoplankton and zooplankton which are so difficult to study under actual sea conditions. Most short-term tests indicate 50% mortalities in the 0·1 to 5·0 mg l^{-1} range but enclosed ecosystem results are indicating reduced productivity and declining populations in the range 0·01 to 0·05 mg l^{-1} or less for sensitive zooplankters. Hence the total effects of oil spills and chronic pollution from continuous inputs are almost certainly greater than might be demonstrable from *in situ* field sampling.

In terms of the significance of such losses to the wider marine ecosystem most phytoplankton species have a generation period of a few days usually and most zooplankton a few months. Loss of a local population can have no permanent effect on the survival of such species or a very drastic effect on the year's production. However planktonic eggs and larvae of longer lived zooplankters, fish, and some benthos may represent once a year precursors of new stocks and cannot be so readily replaced. Another possible impact on these valuable stocks might be the local diminution or contamination of short-lived plankton as food for these replacement larvae or juveniles at a critical period of their development.

Ecosystem enclosures have very recently been used to study the fate of polycyclic aromatic hydrocarbons by using radiolabelled benz [α] anthracene and dimethylbenz [α] anthracene, (LEE and co-authors, 1982). Adsorption on particulate matter followed by sedimentation removed half the benz [α] anthracene in about 24 h. Photochemical oxidation led to the removal of 50% of the dimethylbenz [α] anthracene in less than 12 h, and once the oxidized products reached the bottom sediment microbial degradation followed.

Oil and Fish

The pathways for oil transport into, within, and out of pelagic and demersal fish are important in relation to the toxicity of hydrocarbons but also in relation to the tainting of fish and the identification of the tissues most likely to incorporate high concentrations.

DIXIT and ANDERSON (1977) exposed *Fundulus similus* to ^{14}C-labelled naphthalene dissolved in cod liver oil by direct introduction into the gut and also by exposing the fish to the water-soluble fraction of No. 2 Fuel Oil at a concentration of 7000 μg l^{-1} of which about 2000 μg l^{-1} were naphthalene and alkylnaphthalenes. By the end of 8 h nearly all the naphthalene placed in the gut had been metabolized by the liver and excreted via the bile and kidney. From samples at 2, 4, and 8 h the gut naphthalene concentration fell off rapidly as did that of the liver. Most of the radioactivity built up in the gall bladder and a small amount in the heart and brain. Very little of the small amounts reaching the gills, head, midsection and tail section musculature remained after 8 h.

When oil fractions were introduced by exposure in solution the sites of accumulation in the fish were similar. After 1 h there was accumulation in the liver, gall bladder, heart, and brain. The gall bladder contained at that time 66,704 $\mu g\ g^{-1}$ of oil and metabolized oil. In very short exposures the heart tissue reached similar concentrations to gills and gut. This might be a clue to a circulation pattern showing entry by the gills and gut and transport of the hydrocarbons in the bloodstream to the liver where detoxification occurs and to the gall bladder for storage. These findings are in agreement with those of ROUBAL and co-authors (1977) for young Coho salmon.

WHITTLE and co-authors (1977) describe similar experiments on herring. Herring is a particularly fatty fish and ^{14}C-labelled hexadecane fed in food was found 45 h after feeding mainly in the muscle (54·3% of the activity in the fish) mesenteric fat (7·2%), liver (6·9%), and brain 0·3%), all of which are lipid-rich tissues. The stomach, pyloric caecum and intestine accounted for 17·5% and the non-lipid muscle residue contained 5·7%. The bile contained only 0·04% of the activity of the fish.

A similar experiment using ^{14}C-labelled benzo [α] pyrene showed that 98·4% of the hydrocarbon remained in the stomach (81·6%), pyloric caecum, and intestine. Some 0·2% was found in the aqueous part of the bile and only 0·07% in the lipid activity of the fish. A relatively large proportion remained with the residual fractions of the tissue extracts. Some 40% of the benzo [α] pyrene was believed to be bound to the stomach wall in a more polar form.

In cod (CORNER and co-authors, 1976b), using the same labelled hydrocarbons, activity was determined in the various tissues 96 h after feeding. There was little or no tendency for deposition in the muscle tissue and the hexadecane and benzo [α] pyrene had been excreted or were in the process of excretion. Cod is a non-fatty fish.

In fish with a herring-type metabolism some of the components of petroleum are quickly assimilated in the muscle while in non-fatty fish like cod these hydrocarbons do not assimilate in the muscle but pass to the liver. Turnover of hydrocarbons in the herring muscle may be more rapid than expected and depuration in cod liver less so (WHITTLE and co-authors, 1974).

On transfer to clean water fish show an excellent capacity to recover with simultaneous progressive loss of hydrocarbon from brain and nervous tissue (DIXIT and ANDERSON, 1977). LEE and co-authors (1972) found benzo [α] pyrene to be readily transferred across the gill membrane into marine fish.

The response of gonadal tissues to hydrocarbon stress has not yet been adequately investigated. The value of these biochemical studies on marine animals is that they establish significant powers of tolerance and recovery. Also, along with the microbial population, animals assist in the transformation and eventual mineralization of hydrocarbons provided that excessive concentrations are not encountered.

(c) Oil, Dispersants, and Reproduction

Before considering the effect of oil or water-soluble fraction of oil on eggs and larvae, it would be opportune to look at effects generally on reproductive processes.

Interferences by hydrocarbons in the pre-reproductive process are puzzling. If the shore crab *Pachygrapsus crassipes* is exposed to naphthalene in solution at concentrations about 10^{6} g l^{-1} for 24 h, the attraction of male to female, a pherome response, is completely blocked and in the absence of naphthalene the crab recovers only after 3 d. At these concentrations damage to chemoreceptors is favoured as the likely explanation (KITTREDGE, 1973; TAKAHASHI and KITTREDGE, 1973; KITTREDGE and co-authors, 1974).

The sperm attractant of the eggs of the brown alga *Dictyota dichotoma* is a conjugated alkene, n-butyl-cyclohepta-2,5-diene. Similarly ectocarpen (a monocyclic unconjugated C_{11} olefin) is found in 2 species of *Ectocarpus*; multifeden (a complex monocyclic conjugated olefin) in *Cutleria*, and fucoserratin (a polyunsaturated C_8 olefin) in 2 species of *Fucus* (BOLAND and co-authors, 1981). No exact quantitative data are given but 0·1 µl droplets of *Dictyota* sperm attractant in a non-active immiscible fluorocarbon solvent were extremely attractive to spermatozoids in a Petri dish. One might speculate the crude oils do not mask the possible hydrocarbon sperm attractants of some eggs but perhaps very small amounts of oil can also be attractive. It would be interesting to find out how general hydrocarbon sperm attractants are in the marine biota and what the exact role of extraneous hydrocarbons might be. Are pheromones also hydrocarbons?

The fertilization of the egg is a sensitive process involving chemotaxis and sperm mobility. Surprisingly, crude oils generally have very little effect on fertilization as reported by ALLEN (1971) for a number of crude and refined oils and sea urchin eggs, LÖNNING (1977) for sea urchin eggs, WINTERS and PARKER (1977) for Kuwait crude and No. 2 Fuel Oils and sand dollar eggs, BAXTER (unpubl.) for up to 250 µg l^{-1} Kuwait oil on herring eggs, RENZONI (1973) for 3 crude oils and the gametes of oysters and mussels. Hence the sexual attraction of males to females is a more vulnerable process than that of the sperm to the egg.

When oil dispersants are involved (LÖNNING, 1977 and earlier papers) effects are much more drastic generally resulting in poor fertilization and impaired embryonic development of the successfully fertilized eggs. If only eggs or sperm are exposed to dispersant solution there is marginally improved fertilization but subsequent embryonic disturbances are found. Dispersants introduce a series of fundamental interferences first with the haploid nucleus of the gamete, and in early development when rapid cleavage should occur this rate is slowed down and differentiation impeded leading to abnormalities in the intestinal and skeletal development. Effects arising from oil plus dispersant are usually somewhat worse than oil alone.

The effect of mineral oils on the development of eggs and larvae of marine species has been reviewed in depth by KÜHNHOLD (1977). In the absence of dispersants the 24-h LC_{50} for crude oil on cod eggs was 12 mg l^{-1}. Equal mortality resulted from 0·1 mg l^{-1} present throughout the development period. KÜHNHOLD (1977) and, in more detail, BAXTER (unpubl.) indicate that the stage of herring embryogenesis most susceptible to crude oil is the early period of development, certainly prior to the formation of the rudimentary backbone or the appearance of the eye spot.

Dispersants enhanced the toxicity of crude oil for eggs and larvae of cod, herring, and plaice. Oils of different origin differ in their toxicity. LINDEN (1975) found that dispersants increased the toxicity of Venezuelan crude oil to herring several hundredfold for modern preparations and a thousandfold for old types. The toxicity of physically dispersed crude oil varied with time and decreased considerably in 24 h and 72 h.

WILSON (1972, 1976) also found high mortalities of the larvae of herring, pilchard, plaice, sole, lemon sole, and haddock exposed to dispersants and refined oil products. Clearly the supervening factor is not just the presence of oil but to an important extent the contact between oil and egg which is promoted by extreme physical mixing and by the enhanced contact and penetration when dispersants are introduced. The relative ineffectiveness of a surface oil film (230 mg l^{-1}) in reducing the survival of juvenile fishes even when confined under aquarium conditions was demonstrated by MIRONOV (1973).

In the absence of more recent work, Table 4-21 from KÜHNHOLD (1977) gives a useful summary of the best data. KÜHNHOLD attributes much of the variation in toxicity of different crude oils to the quantity of soluble aromatic hydrocarbons present, among which the C_{10} to C_{17} components probably contribute the main toxic effect. MOORE and co-authors (1973, 1974) have rationalized short-term toxicity (<24 h) of soluble aromatic hydrocarbons for a wide range of marine classes; longer period toxic effects could be anticipated at 1/10 or less of these concentrations (Table 4–22). (The soluble aromatic fraction is possibly reflected in the percentage of C_5–150 °C 'benzene fraction' shown in Table 4–18 for a range of 12 crude oils, with expected solubilities in the range 500 to 5000 mg l^{-1} for total soluble aromatics.)

We do not have much detail about field effects and safe dilutions but there is probably enough to prescribe great caution in the use of dispersants in situations where spawning of valuable fish and shellfish is under way in the upper water layer or in a shallow turbulent water column. This precaution is particularly necessary since fertilization and spawning frequently occur at times when the animals aggregate in large numbers, the critical season may occur only once a year and may be quite short so that hasty action could have serious effects locally. Also these observations do not take ecological processes into account.

In recent Loch Ewe experiments, using enclosed ecosystems (DAVIES and co-authors, 1981a), fertilized spring-spawned herring eggs were exposed to dilutions to treated oily water effluent (mainly production water) from the Auk North Sea production platform equivalent to 25, 100, and (to be sure) 500 μg l^{-1} hydrocarbons as measured by UV fluorescence against Auk crude oil. In the first trial the most notable effects observed were a prolongation of the hatching period (compare *Argo Merchant* results) but hatching of some of the eggs was significantly earlier than controls. In another set of exposures to these concentrations, mortality of newly hatched herring eggs was significantly increased even at the lowest dosage (25 μg oil l^{-1}). Variation in response is attributed to differences in the condition of the fish and spawning products.

Subsequently, more protracted ecosystem studies were made on the entire self-sustaining food web including larval herring as the top predator. These experiments have not yet been fully worked out but even in the concentration range 5 to 15 μg oil l^{-1} the treated oily effluent gave rise to measurable effects at each trophic level and on the whole ecosystem. The enclosed ecosystem was maintained with periodic addition of nutrients for a period of 100 d. Surprisingly, the surviving larval herring appeared to be less affected by the oily water than the copepod fraction of the zooplankton, the larvae of the bottom invertebrates, and the naupliar stages of the various copepods. The survival of these nauplii is crucial as a suitable food supply for the developing larval herring.

In all these laboratory and ecosystem experiments, purely for practical reasons, the herring eggs are attached to clean etched glass plates which are suspended vertically in

Table 4-21

Toxicity of various oils for fish eggs and larvae (After KÜHNHOLD, 1977; reproduced by permission of the International Council for the Exploration of the Sea)

Fish			Type of test	Type of ore	Manner of application	Volume of test medium	Analysis of medium or oil	Source
Rhombus maeoticus	Blacksea turbot	E	Survival	Russian crude				MIRONOV (1967)
		L	Hatching	Light	OWD	—	None	
Engraulis enrasicholus	anchovy	E	Behaviour	Heavy fuel				
Scorpaena porcus		E	Mortality	Heavy fuel	OWD	—	None	MIRONOV (1969)
Crenilabrus tinca		E						
Gadus morhua	Atlantic cod	E	Mortality	Heavy Bunker C Light Bunker C	SWE of 0·1% (static & flow through)	1, 6 l	None	JAMES (1925)
Gadus morhua		E	Survival time Embryogeneses Hatching	Venezuelan crude Iranian crude	SWE of 0·1, 0·01%	20 l, 1 l, 0·1 l	Total hc (by wt), refinery data of oils, relative aromatic content	KÜHNHOLD, (1973)
		yolk sac larvae	Behaviour recovery LC_{50}	Libyan crude				
Clupea harengus	herring	E L	Survival time Hatching success Morphogenesis behaviour	Venezuelan crude Iranian crude Libyan crude fractions of Iranian crude	SWE of 1 (2)%, dispersions	1 l	None Refinery data for oils	KÜHNHOLD (1969a,b)

E = egg; L = larvae; OWD = Oil in water dispersion. SWE = Sea-water extract.

Table 4-22

Summary of short-term toxicity data for marine flora and fauna (After KÜHNHOLD, 1977; MOORE and co-authors, 1974; reproduced by permission of the International Council for the Exploration of the Sea)

Class of organisms	Estimated concentration (mg l^{-1}) of soluble aromatics causing toxicity
Flora	10–100
Finfish	5–50
Larvae (all species)	0·1–1·0
Pelagic crustaceans	1–10
Gastropods (snails, etc.)	1–100
Bivalves (oysters, clams, etc.)	5–50
Benthic crustaceans (lobster, crabs, etc.)	1–10
Other benthic invertebrates (worms, etc.)	1–10

water. This avoids deposits of oil absorbed on particles giving rise to locally very high oil contamination. In the sea, the eggs are deposited on gravel at the sea floor as a continuous layer or in large patches which would be vulnerable to any oil that might settle to the bottom and one might reasonably expect enhanced effects. On emergence from the egg the larval herring become dispersed in the water column and eventually seek the upper layer, often close to the surface at times.

Juvenile and adult pelagic fish are only exposed to oil pollution at sea in acute concentration in the immediate ambience of freshly spilled oil, especially if dispersants are used.

Most fish have an evident slimy mucus on the external surface which resists adherence of oil. Dispersants tend to destroy the effectiveness of this layer. Fish mortalities from spills of crude oil and oil products have been extensive in some fresh water situations and in marine situations where both oil and fish are largely confined, but for the open sea no severe kills have been recorded even for major disasters (GESAMP, 1977, CROSS and co-authors, 1978). Several reports note a transient decrease in catches of fish in the neighbourhood of recent spills and fish landings decline temporarily because fishermen are reluctant to expose their gear to heavy oil pollution.

Benthos and Sediments

Obvious factors, such as popular appeal, familiar biology, access to specimens for aquaria, ease of sampling and observation, manageable model ecosystems, and extreme exposure to oil, have created a situation in which the study of intertidal benthos has almost eclipsed all others, especially in relation to oil pollution.

A large number of species of molluscs, crustaceans, echinoderms, polychaetes, coelenterates, hydroids and others have been studied as adult and often larval and intermediate juvenile forms. The most extreme impact of oil pollution on these shore communities is reported by CROSS and co-authors (1978), CABIOCH and co-authors (1980) for the disastrous *Amoco Cadiz* oil spill in spring 1978:

> 'On April 2 several million dead molluscs and urchins were present along $2\frac{1}{2}$ km of beach near the village of St Efflam . . . at Rulosquet marsh near Ile de Grand on the same day thousands of dead polychaetes and large numbers of dead crabs covered the marsh surface' (CROSS and co-authors, 1978, p. 212).

> 'Mariculture operations for oysters were severely affected in the Aber Benoit and Aber Wrac'h estuaries and the Bay of Morlaix. Large numbers of oysters were either killed or contaminated by the spill. The holding pens of the commerical lobster operation at Roscoff were heavily oiled and will probably be out of operation for a year', CROSS and co-authors, 1978, p. 213.

CABIOCH and co-authors (1980) describe the long-term effects of the spill and identify the situations where repopulation will be difficult. After an initial phase of sharp and selective mortality lasting no more than a few weeks the effects evolved with time. Fine sand communities (with moderate exposure) were affected to a greater degree than those on mixed sediments (with greater exposure). Recruitment of the residue of the macrobenthic population gradually became normal with an added proliferation of polychaetes in suitable habitats. This recovery took place in sediments retaining high levels (>100 mg kg^{-1}) of hydrocarbons which by that time had lost much of their soluble aromatic components.

The use of old type oil dispersants in such a situation could only have exacerbated these effects, for example compare the reports of the *Torrey Canyon* oil spill (SMITH, 1970) which involved less than half the amount of beached oil.

From hundreds of studies on non-commerical benthos a selected few may be identified as of special interest since they exemplify modes of action of hydrocarbons.

The spill at Long Cove, Maine, introduced a mixture of No. 2 Fuel Oil and jet fuel JP5 into the exposed sediments (DOW, 1978). With time weathering and leaching processes established a maximum level of this mixture at 15 to 25 cm which persisted in substantial quantities for at least 6 yr. Clams (*Mya arenaria*) attempted to recolonize the area after 6 yr, burrowed down through the redistributed clean sediments and on reaching the oil suffered mortalities. This case emphasizes the hazard of thin oils of high toxicity that can penetrate into layers beyond mechanical redistribution by wave action. These layers may be or become anoxic and consequently degradation is extremely slow. Burrowing organisms are particularly at risk in this situation.

The sinking of the barge *Florida* releasing only 700 tons of No. 2 Fuel Oil in September 1969 has also had a long aftermath after initial severe pollution (KREBS and BURNS, 1977). Populations of the small fiddler crab *Uca pugnax* were studied in West Falmouth Bay from the time of the spill to just prior to the report. The spill initially reduced the crab population in the heavily oiled areas and there were subsequent heavy overwinter mortalities. The ratio of females to males considerably declined and all the crabs accumulated oil in their tissues. Ability to move about was impaired and the crabs constructed burrows of abnormal shape. Behavioural disorders displayed by the crabs several years after the spill were attributed by the authors to accumulation of hydrocarbons (possibly substituted aromatics and their metabolites) in the central nervous system of the crabs.

More recently, LAUGHLIN and NEFF (1979) studied the exposure of juvenile mud crabs to a range of phenanthrene concentrations in water under laboratory conditions in relation to gradients of salinity and temperature. The crabs *Rhithropanopeus harrisii* adapted their respiration according to temperature and were tolerant to the wide range of salinities after acclimation. They also withstood osmotic shock following change of salinity. Exposure to phenanthrene elevated respiration rates under stressful salinity and temperature regimes and disturbed the homoeostatic osmoregulatory mechanisms possibly by disrupting the gill membranes and/or upsetting ion transport enzymes in them. Significant effects were found at the lowest dose-rate used, 75 μg l^{-1}. Larval crabs also gave responses of this type but were slightly more sensitive (LAUGHLIN and NEFF, 1979).

The bivalve *Macoma balthica* burrows into the sediment but feeds on the superficial deposits and is, therefore, at risk to stranded oil or oil depositing attached to particulate matter. SHAW and co-authors (1977) undertook extensive statistically designed tests exposing groups of bivalves for 30 d to 500 mg l^{-1} Prudhoe Bay crude oil in the overlying water. This corresponded approximately to 1·22 μl cm^{-2} in the sediment. There were significantly more mortalities in the test exposures but tissue weights, shell length and condition factor did not vary compared to control; therefore, they argue, mortalities were caused not by stress-induced impairment of the animals' normal functions but by metabolic poisoning.

Neanthes arenaceodentata is a polychaete worm normally living in silty mud which feeds by passing mud and detritus through its gut and ejecting the processed material in the form of faecal pellets. When grown for 28 d in water containing only added amounts of the water-soluble fraction of No. 2 Fuel Oil giving 400 μg l^{-1} of total hydrocarbons growth was slowed so that the worms were 30% shorter than controls at the end of the test period. Less food was consumed. The water contained the equivalent of 100 μg l^{-1} of mainly naphthenic hydrocarbons.

ROSSI (1977) planned further experiments to test the uptake of hydrocarbons (mainly napththalene and derivatives) by (i) exposure to treated water, (ii) exposure to artificially hydrocarbon-enriched sediment, and (iii) feeding on a diet dosed with ^{14}C-2-methyl naphthalene.

(i) In the tests for water uptake worms were exposed anaerobically to 150 μg l^{-1} ^{14}C-naphthalene in sea water for 24 h; this concentration is close to the lethal limit. Uptake was rapid and maximal giving 6 μg g^{-1} tissue after only 3 h. *Neanthes arenaceodentata* did not metabolize measurable amounts of the naphthalene over 24 h. When placed in clean sea water with access of air the concentration of labelled naphthalene gradually fell to less than 0·05 μg g^{-1} over 300 h. In a parallel run, an antibiotic was added to suppress microbial activity; this affected neither the uptake nor the depuration rates. Metabolism of naphthalene only occurred under aerobic conditions.

(ii) The sediment used was silty mud (3 parts) blended with the water-soluble fraction of No. 2 Fuel to which was incorporated powdered alfalfa (1 part) as food. The sediment was constantly bathed in flowing clean sea water. Over 28 d the mean total naphthalene content fell for 9 μg g^{-1} wet sediment to 3 μg g^{-1} but none of the 20 replicate samples of *Neanthes arenaceodentata* buried in the mud accumulated detectable amounts of naphthalenes.

(iii) In the experiments in which *N. arenaceodentata* was fed dosed alfalfa, the worms were daily offered small amounts of powdered alfalfa treated with 10 to 15 μg g^{-1} of ^{14}C-2-methyl naphthalene. All the alfalfa was consumed on 16 successive days but no activity was transferred to the worms after accounting for gut contents. The radioactivity was present in the food residues in the faecal pellets.

ROSSI considered that his measurements of the kinetics of uptake from water matched reported values for fish, mussels, and crustaceans. Depuration of unmetabolized naphthalene was essentially complete in 14 d. ROSSI mentions the reported transfer of polycyclic hydrocarbons, resident in natural sediments, to infaunal bivalves in field situations but does not comment on the important point of aerobic or anaerobic conditions in force. The failure of *N. arenaceodentata* to acquire hydrocarbon from treated food is contrary to many other reports relating to fish, crustaceans, copepods, and benthic bivalves. Rossi interprets the lack of hexane-soluble labelled hydrocarbon in *N. arenaceodentata* which would indicate uptake as arising from the strong binding of the polar metabolites in the tissues. Polar radioactivity was also found in the sea water.

The mussel *Modiolus demissus* and the oyster *Crassostrea virginica* were exposed to successive small dosings with No. 2 Fuel Oil in a marsh at the mouth of Chesapeake Bay (LAKE and HERSHNER, 1977). Two weeks after the cessation of 20 wk of dosing, the levels of saturated cyclic and paraffinic hydrocarbons and of mainly naphthenic aromatics were determined. Depuration was measured over the following 4 wk and a further 9 wk. Mussel and oyster both depurated the aromatics at accelerating rates and achieved >90% reduction in content; both had virtually linear depuration rates for the paraffins and retained 30% and 20% at the end of 15 wk. Sulphur compounds were as effectively depurated as the naphthalenes. Depuration in these experiments related to a recovering marsh environment. A summary of the uptake and depuration phases of these and other experiments is given in Table 4-23. Most of the observations show that depuration of the alkylated naphthalenes is much slower than naphthalene. Depuration under field conditions is very much slower than on transfer to clean water.

Biological Accumulation of Hydrocarbons

As our knowledge of hydrocarbons in the marine ecosystem becomes more complete with regard to formation of biogenic hydrocarbons and their subsequent fate within the food web and in the inanimate environment, it becomes more and more difficult to draw hard and fast distinctions between the origins of hydrocarbons found in marine plants and animals. As analytical methods increase in sensitivity a situation approaches when all biota will be deemed to have traces of petrogenic hydrocarbons present on the basis of some criteria adopted and where petrogenic hydrocarbons are abundant one must accept that biogenic hydrocarbons are also there even if it is difficult to demonstrate what may be the common features. It is unlikely that more than a very small refractory residue of today's biogenic products is capable of giving rise to the petroleum of some remote tomorrow but there are strong hints that this type of maturation can be demonstrated to occur for a number of specific triterpanes and isoprenoid alkanes.

In considering the hydrocarbon content of the biota it is also helpful to avoid the assumption that chemical analysis of the necessarily dead tissue describes a permanent state. Almost invariably part, sometimes the major part, of the hydrocarbons found are involved in a continuous dynamic turnover. In the light of much field and experimental observation it is now established that analysis of a specimen reveals perhaps the legacy of exposure to some light fractions persisting a few days, or exposure to fractions progressively more difficult to metabolize surviving weeks, months or years; in addition, this elimination programme is overlaid by successive inputs of oil from the microenvironment of the animal. There are no grounds for assuming that all organisms have equal capacity to eliminate hydrocarbons, either as representative or different species or developmental states of one species or even to some degree as individuals.

The results of MACKIE and co-authors (1974) enable some tentative accumulation factors to be derived for an estuarine environment. These factors are in no way proof that there is a direct pathway involved. Likely sources of petroleum pollution in the Firth of Clyde are crude oil spills and related oily water discharges from a range of Middle East crudes and oil products together with more abundant waste oil residues from urban sources and biogenic hydrocarbons from air, land, and sea, all at some stage of degradation. Individual and mean accumulation factors have been calculated, e.g. Mean accumulation factor for plankton =

$$\frac{\xi\text{Carbon fraction n-alkanes in plankton}}{\xi\text{Carbon fraction n-alkanes in middle depth sea water}}$$

and also the overall accumulation factors for total n-alkanes. For plankton (mainly zooplankton especially copepods) the accumulation factors (see Table 4-24, Column 2) are very high especially for C_{18}, C_{19} and C_{27}, C_{28}, C_{29} with 153,800 for total n-alkanes. Next, in order is 55,800 for herring; 27,700 for hake; 13,100 for dogfish; 5300 for plaice; and 1100 for cod. These species show about average accumulation factors for the C_{24} to C_{32} range of n-alkane fractions. Only plankton and herring show above average factors for C_{18} to C_{20} fractions which perhaps emphasizes the close association between zooplankton and herring as its direct predator. However, these factors are not proof of biochemical pathways. The decrease in accumulation factor and changes in the profile of accumulation with C-number perhaps express progressively less direct dependance on primary herbivores for the fish species hake, dogfish, plaice, and cod. Other related factors may be the normal swimming depth, dependence on other food organisms such as carnivorous plankton, smaller fish, crustaceans, and benthic organisms each more remote from primary production, or as an alternative explanation some increasing capacity to excrete or metabolize hydrocarbons in one or more tissues.

It may also be relevant to derive a similar set of accumulation factors for n-alkanes in sediment in relation to bottom-living organisms and bottom-living fish (see Table 4-25). Starfish (a benthic predator) and whelk (its possible prey) both share a wide carnivorous diet and display a somewhat similar pattern of C_n peaks. There is no evidence of overall hydrocarbon accumulation. *Nephrops* which lives in sticky mud as distinct from whelk (hard bottom) and starfish (universal) shows a different C_n pattern. In sediment and in organisms to some degree there is a pronounced bias for higher accumulation factors for even C numbers; this is in direct contrast to the enrichment of odd numbered alkanes in fish livers (MACKIE and co-authors, 1974).

The results indicate a general and significant accumulation of hydrocarbons, natural or pollution-derived, in the zooplankton but, thereafter, there is only progressive decrease in oil content per unit weight of tissue at each stage along the food chain. Hydrocarbon accumulation is also linked to the richness of lipids in the fish tissues (WHITTLE and co-authors, 1974). Also under normal circumstances accumulation from oil in bottom sediment to benthic animals is at a relatively low level.

Tainting

The creation of off-flavours in marketable fish due to exposure to petroleum hydrocarbons has been most frequently reported for fresh water fish in relation to inland oil

Table 4-23

Hydrocarbon exposure, peak tissue concentration and depuration. M: source reports metabolites formed; T: source reports thiophenes depurated; C: after transfer to clean conditions; F: algal food supplied (Original; compiled from the sources indicated)

Organism	Substrate	Concentration duration	Tissue peak	Depuration time		Source
Spot shrimp						
Pandalus platyceros	^{14}C-naphthalene	8 to 12 $\mu g\ l^{-1}$ up to 36 h	820 $ng\ g^{-1}$	100% 12 h	MC	Varanasi and Malins (1978)
Stages I & V						
Pandalus platyceros	^{14}C + serum albumin	equivalent		100% 12 h	MC	Varanasi and Malins (1978)
Dungeness crabs						
Cancer magister larvae	^{14}C-naphthalene	8 to 12 $\mu g\ l^{-1}$ up to 36 h	220 $ng\ g^{-1}$	100% 12 h	MC	Varanasi and Malins (1978)
Cancer magister larvae	^{14}C + serum albumin	equivalent		100% 12 h	MC	Varanasi and Malins (1978)
Shrimp						
Pandulus hypsinotus	^{3}H-naphthalene	6 $\mu g\ l^{-1}$ 144 h	198 $ng\ g^{-1}$	75% 6 h	MC	Varanasi and Malins (1978)
Polychaete worm						
Neanthes arenaceodentata	^{14}C-naphthalene	150 $\mu g\ l^{-1}$ 24 h	6000 $ng\ g^{-1}$	100% 168 to 192 h	M	Rossi (1977)
Mussel						
Modiolus demissus	No. 2 Fuel Oil	Salt marsh conditions 20 wk	—	35% saturates 88% aromatics 4 wk	T	Lake and Hershner (1977)
Modiolus demissus	No. 2 Fuel Oil	Salt marsh conditions 20 wk	—	68% saturates 98% aromatics 9 wk	T	Lake and Hershner (1977)

Oyster					
Crassostrea virginica	No. 2 Fuel Oil	Salt marsh conditions 20 wk	—	60% saturates 67% aromatics 4 wk	T	LAKE and HERSHNER (1977)
Crassostrea virginica	No. 2 Fuel oil	Salt marsh conditions 20 wk	—	80% saturates 95% aromatics 9 wk	T	LAKE and HERSHNER (1977)
Copepods						
Calanus plumchrus	WSF No. 2 Fuel Oil + labelled tracers	5 $\mu g\ l^{-1}$ 1 to 3 d	—	99% 8 d 99% 28 d	FC	LEE (1975)
Clam						
Rangeata cuneata	Baskets exposed over sediments near platform; Naphthalenes	2 to 4 μg naph g^{-1} sediment 96 d	6300 ng g^{-1}	100% naphthalenes 47 d	C	FUCIK and co-authors (1977)
Rangeata cuneata	Baskets exposed over sediments near platform; Naphthalenes	12 to 22 $\mu g\ g^{-1}$ 104 d	8400 ng g^{-1}	85% naphthalenes 47 d	C	FUCIK and co-authors (1977)
Rangeata cuneata	Baskets exposed over sediments near platform; Naphthalenes	1 to 2 $\mu g\ g^{-1}$ 96 d	Nil	No accumulation		FUCIK and co-authors (1977)
Rangeata cuneata	Baskets exposed over sediments near platform; Naphthalenes	0·5 to 5 $\mu g\ g^{-1}$ 104 d	Nil	No accumulation		FUCIK and co-authors (1977)

Table 4-24

Accumulation factors relative to sea water for plankton, herring, hake, dogfish, plaice, and cod from the Firth of Clyde, Scotland (Original; based on MACKIE and co-authors, 1974)

$$\text{Accumulation factor for plankton} = \frac{\text{n-alkane in plankton}}{\text{n-alkane in sea water (mid-depth)}}$$

Carbon number of n-alkane	Plankton ($\times 10^3$)	Herring ($\times 10^3$)	Hake ($\times 10^3$)	Dogfish ($\times 10^3$)	Plaice ($\times 10^3$)	Cod ($\times 10^3$)
18	296	206	1·3	1·4	6·0	0·1
19	258	194	1·5	2·0	2·5	0·3
20	168	64	1·1	2·1	1·9	0·6
21	138	30	4·8	4·2	3·6	0·8
22	103	28	15	8·6	4·5	1·0
23	115	31	30	10	6·3	1·1
24	135	42	45	14	7·0	1·3
25	132	40	45	14	6·3	1·3
26	147	48	45	16	6·5	1·2
27	160	51	41	18	6·6	1·4
28	158	51	31	18	5·4	1·1
29	168	53	31	18	5·2	2·0
30	121	37	25	15	3·7	1·4
31	113	39	18	14	2·9	1·3
*32	*191	*47	*16	*18	*5·0	*ND
*33	*670	*159	*22	*55	*2·0	*ND
	mean	mean	mean	mean	mean	mean
C_{18}–C_{33}	154×10^3	56×10^3	28×10^3	13×10^3	$5{\cdot}3 \times 10^3$	$1{\cdot}1 \times 10^3$

* Values distorted by small amounts present.
ND, Not detected.

spills. The amount of oil needed to impart a detectable smell or taste is a function of the oil or oil product and this also controls the length of exposure needed, and the time since exposure needed for depuration. The manner of exposure whether by dispersion in water, skin contact, or uptake in food also affects the distribution of oil in edible tissues (WHITTLE, 1978).

Among the commonest means by which seafoods acquire hydrocarbon taint are: (i) physical contact with oil slick or accumulations at the surface or on the seabed; (ii) exposure to high levels of oil in water, with or without added dispersant; (iii) contact with oiled fishing gear, deck, fish boxes, processing plant, etc.; (iv) uptake in food of the organism.

Short-term exposure to direct contact (i.e. (ii) or (iii)) may be discounted if the product is first processed to a fillet, for example, or is shelled before cooking or consumption in the case of certain shellfish.

Fish are also known to acquire off-flavours from 'oily' components of certain items of food occurring in exceptional abundance at certain times of year and positive evidence of hydrocarbons as taint compounds in seafood is rare (HOWGATE and co-authors, 1977).

Table 4-25

Accumulation factors relative to bottom sediment for plaice, whelk, starfish, and Norway lobster from the Firth of Clyde, Scotland (Original; based on MACKIE, and co-authors, 1974)

$$\text{Accumulation factor for plaice} = \frac{\text{n-alkane in plaice}}{\text{n-alkane in superficial sediment}}$$

Carbon number of n-alkane	Plaice (× 1)	Whelk (× 1)	Starfish (× 1)	Norway lobster (× 1)
18	2·9	2·9	1·8	0·8
19	1·1	1·4	0·6	1·2
20	1·3	3·3	0·4	2·1
21	1·8	3·3	0·2	3·7
22	3·5	5·9	0·8	9·7
23	3·2	3·8	0·9	10·2
24	3·0	2·4	0·6	12·7
25	2·3	1·6	0·9	12·0
26	3·3	0·5	0·6	20·6
27	1·8	0·6	0·7	12·1
28	3·2	ND	1·2	26·1
29	0·8	ND	0·5	6·1
30	2·7	ND	1·1	21·9
31	0·4	ND	0·3	3·2
32	1·9	ND	0·5	10·2
33	0·8	ND	0·3	2·2
C_{18}–C_{33}	mean 1·8	mean 1·05	mean 0·66	mean 9·8

ND, Not detected.

Among marine food species the mullet, a coastal fish commonly found swimming near the surface in inshore waters and harbours, is the commonest to be reported to have a kerosene taint. The mullet feeds on algae, plankton, and other small items prone to high levels of oil contamination. Mussels near oil installations are reported to take up as much as 1% flesh weight as 'oil'. Oysters exposed to 0·01 mg South Louisiana crude oil l^{-1} sea water become tainted; SIDHU and co-authors (1972) report 0·01 to 0·02 mg crude oil l^{-1} as the taint threshold for Japanese mackerel, and mullet were tainted in the presence of 5 mg kerosene l^{-1}. DESHIMARU (1971) found tainting in yellow tail fish exposed to 50 mg l^{-1} crude oil for 5 d in sea water. The lower limit for the tainting of river fishes is about 10 to 50 mg of added crude oil or oil product kg^{-1} fresh weight fish.

If it is assumed that 3 mg n-alkanes kg^{-1} fresh weight (equivalent to 10 mg oil) is the threshold level for the tainting of sea fish, then using the accumulation factors in Table 4-23 and the average value of 0·2 μg total n-alkanes l^{-1} of Clyde sea water the calculated amounts of n-alkanes needed to cause taint for the various species is:

Herring 0·264 $\mu g\ l^{-1}$
Hake 0·316 $\mu g\ l^{-1}$
Dogfish 0·439 $\mu g\ l^{-1}$
Plaice 0·776 $\mu g\ l^{-1}$
Cod 2·937 $\mu g\ l^{-1}$

It is by no means proved that taint-giving compounds will behave like the n-alkanes but there are reasonable indications that oily or fatty fish like herring, sprats, mackerel, and mullet (which feeds at a low trophic level) are especially prone to petroleum pollution.

Hydrocarbons and Mussel Watch

There is probably no more comprehensively studied marine invertebrate than the mussel *Mytilus edulis*. To date studies regarding the responses of *M. edulis* have focused on hydrocarbons, organochlorines, metals, and radionuclides. Much is also known about responses to temperature, salinity, wave exposure, substrate types, tides and tidal immersion. The extent of information on life history, feeding, growth, physiology, biochemistry, disease, parasites, behaviour, and other characteristics of individuals and populations is impressive. This fact gave support to and later benefited greatly from the concept of 'Mussel Watch' (GOLDBERG and co-authors, 1978) in an attempt to use the mussel as a recording entity, a logical tool for global marine monitoring.

Compared to the other pollutant groups, oil in its many forms has least permanence in the environment and probably also in mussel tissues. The native mussel, therefore, carries a burden of oil which results from overlapping processes of uptake and depuration of hydrocarbons which the tissues have widely differing capacity to metabolize.

Laboratory studies by ANDERSON and co-authors (1979) have demonstrated (Fig. 4-18) the uptake and induction of depuration for the benthic amphipod *Anonyx laticoxae* in the presence of an extract of Prudhoe Bay oil. During the 120 h exposure the tissues reached an interim peak concentration for naphthalene in 8 h and for methyl naphthalenes in 24 h but the peak for dimethyl naphthalenes was not reached in 120 h. The course of depuration discussed elsewhere demonstrates the persistence of the dimethyl naphthalenes and polynuclear aromatics in bivalves and other animals.

It is possible to make more rigorous *in situ* assays of oil in the environment of a known discharge by exposing cages of uniform clean mussels for an appropriate period, probably not exceeding a few weeks. By this technique it may be possible to determine the uptake and availability of hydrocarbons to the mussel in relation to the amount of oil discharged.

The most widespread and fundamental finding of the Mussel Watch programme was that oil content, as measured by a uniformly applied technique, corresponded broadly with the known history of oil input. Programmes which included more closely sampled sub-areas sometimes disclosed hitherto unsuspected inputs and very detailed sampling revealed variation due to extremely local, often transient, oil incidence of small magnitude. There was some evidence of individual variation in the capacity of mussels in a cluster to manage oil (NRC, 1980).

There are indications that there are 'saturation' levels for individual hydrocarbons or hydrocarbon groups accumulating in mussel tissue believed to be controlled by some factor of lipid metabolism (NRC, 1980). BURNS and SMITH (1977, 1978) set this saturation level at about 30 mg total hydrocarbons g^{-1} body lipid. The upper limit in Mussel Watch samples was about 12 mg total hydrocarbons g^{-1} dry weight body tissues.

An important point emphasized by JENSEN (1981) is that exposure to general industrial and urban wastes is just as effective a source of high levels of n-alkanes (C_{12} to C_{36}) in *Mytilus edulis* as proximity to discharges from the oil industry. The survey relates to

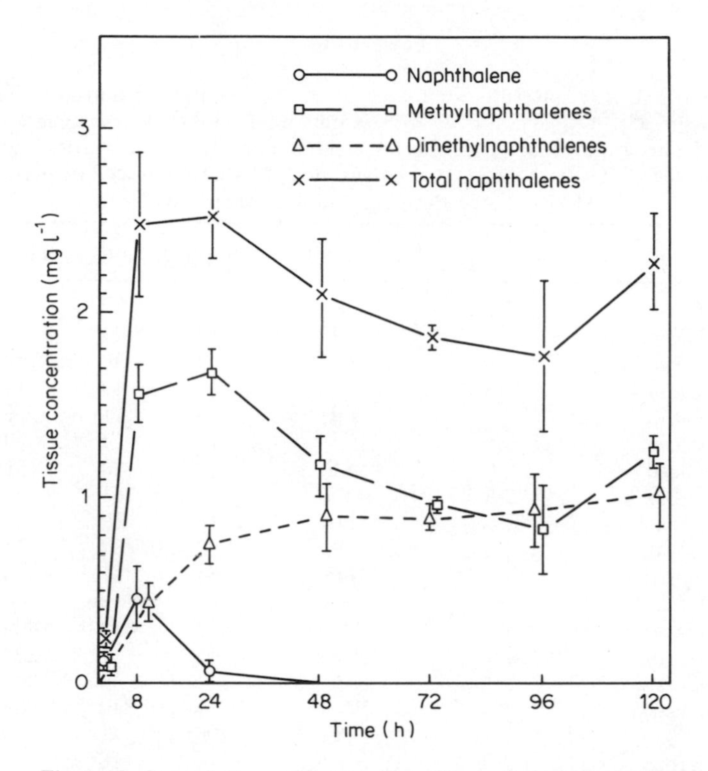

Fig. 4-18: *Anonyx laticoxae*. Uptake of specific naphthalenes during 120 h constant exposure to an extract of Prudoe Bay crude oil containing 506 $\mu g\ l^{-1}$ total hydrocarbons and 31 $\mu g\ l^{-1}$ naphthalenes. Bars: standard errors of the means of 3 samples of 2 individuals each. (After ANDERSON, 1977; modified; reproduced by permission of the American Petroleum Institute)

mussels collected in a variety of Danish coastal environments. The problems of interpretation increase greatly when high resolution gc and gc–ms analyses are employed giving full details of the n-alkanes and aromatics. To eliminate analytical variation due to losses during handling, alkanes below C_{12} and most benzenes are omitted. The results from the Scottish Mussel Watch (1977) are given in a simplified form in Table 4-26; the full analytical data are listed in MACKIE and co-authors (1980). Some idea of the complexity of the individual component variability (aromatics only) has already been shown in Table 4-16. Only pronounced differences were pointed out, many others await interpretation.

In seeking to refine the evaluation of Mussel Watch hydrocarbon data, basic guidelines must be established and useful conclusions already made must be consolidated. The following leads have been identified:

(i) *Quantity*: Oil and individual hydrocarbon analyses demonstrate wide variability in water, sediment, and tissues; consequently, while the general concentration in a group of samples may identify a degree of contamination any one result should not be overemphasized.

Table 4-26

Scottish Mussel Watch, 1977. Results for n-alkanes in the green gland and gut, and aromatics in whole wet tissue. C_{15}–C_{33} alkanes: $\mu g\ g^{-1}$; Naph: total naphthalenes, ng g^{-1}; Anth: total anthracenes and phenanthrenes, ng g^{-1}; HPA: higher polynuclear aromatics, ng g^{-1}; DBT: dibenzothiophene, ng g^{-1} (After MACKIE and co-authors, (1980), reproduced by permission of the MAFF Torry Research Station, Aberdeen)

Station	Type	C_{15}–C_{33}	Alkane source	Naph	Anth	HPA	DBT
1	Clean	4·7	P	64	36	7	1
2	Clean	5·1	O	19	24	91	3
3	Clean	1·4	N/K	9	58	52	3
4	Clean	5·5	N/K	44	49	43	2
5	Clean	2·0	OP	46	11	11	4
6	Clean			18	24	29	3
7	Clean	2·6	SP	8	30	1131	2
8	Urban	2·3	PS	96	61	955	3
9	Urban	4·5	PO	18	12	24	1
10	Urban	5·5*	P	58	63	74	3
11	Industrial	12·6	OS	34	361	1508	24
12	Urban	1·5	PS	10	18	36	3
13	Industrial	14·1	ON/K	122	818	1863	42
14	Clean	7·1	PN/K	13	51	222	2
15	Urban	2·1	S	14	40	306	2
16	Industrial	4·4	PN/K	9	340	888	1
17	Urban	5·7	PO	290	71	39	28
18	Urban	2·0	SN/K	4	16	85	1
19	Urban	1·3*	N/K	2	55	136	1
20	Industrial	5·4	O	206	124	30	16
21	Urban	1·8	SPN/K	21	26	58	2
22	Industrial	4·3	OP	93	273	344	8
23	Urban	1·0	N/K	4	22	103	1
24	Urban	2·7	PN/KS	5	15	471	1
25	Urban	0·4	PSN/K	6	10	96	0·5
26	Industrial	3·7	OSN/K	102	48	99	7
27	Clean	1·5	N/K	67	36	63	8

Alkane sources: P, Phytoplankton; O, Oil; S, Sediment; N/K, Not known.
* Whole animal.

(ii) *n-Alkanes*: the biogenic profile in animals or sediments (e.g. Fig. 4-2) is highly characteristic as is the profile of crude oil (e.g. Fig. 4-16) or oil product (e.g. Fig. 4-13; No. 6 Fuel Oil) hence preponderance of such profiles may be gauged in some mixtures.

(iii) *Weathered n-alkanes* are indicated by the progressive loss of lower alkane peaks: a much weathered sample would show little below C_{25}.

(iv) *Sediment alkanes* show a predominance of odd numbered peaks in the range C_{20} to C_{35}.

(v) *Aromatics* are not produced *de novo* by living marine organisms (HASE and HITES, 1976) and their presence signifies petrogenic contamination. The gc profile may reveal or identify crude oil (e.g. Fig. 4-17) or product (e.g. Fig. 4-14).

(vi) *Weathered aromatics*: bearing in mind the distortions created by selective uptake and depuration, they are marked by the progressive loss of benzenes (if any) naphthalene,

Table 4-27

Mytilus californianus. Light hydrocarbon concentrations in samples from the upper and lower intertidal zone near oil seeps (After STRAUGHAN, undated; reproduced by permission of the author)

	Intertidal zone*		Tissue hydrocarbons†	
Location	Upper	Lower	Upper	Lower
Goleta Point	3·8	16·9	20	20
Carpinteria	0·4	4·6	15	<5
Reef Point	0·2	0·0	<5	<5

* Weight of oil tar (g) collected from sediment samples.
† Lighter hydrocarbon in $\mu g\ g^{-1}$ wet weight tissue.

methyl and dimethyl naphthalenes relative to higher PNAH and DBT. Fine sediments retain PNAH, especially if anaerobic.

(vii) *Identifiable aromatic profiles* have been shown for air-borne hydrocarbon residues (Table 4-16); others are known for asphaltic oils and road tar and it is always possible to characterize specific industrial effluents.

MACKIE and co-authors 1980 have interpreted the Mussel Watch n-alkane profiles as P: phytoplankton; O: oil; S: sediment dominated; N/K if no clear interpretation can be made (Table 4-26, Column 4).

The interpretations of the n-alkanes and on close inspection the aromatics are broadly in agreement with the type of environment, are reasonably harmonious and do not violate local experience. The further indication of dibenzothiophenes (DBT) picks out the 'industrial' stations very well.

It is unfortunate that a special effort has not been made to exploit the advantages of the areas around natural seeps in relation to Mussel Watch. Some data do exist for *Mytilus californianus* but the methods may not be those of Mussel Watch. The main focus of the experiment was to compare beach sediment and mussel tissue from a series of graded oil seep exposures (STRAUGHAN, undated). There is broad agreement between levels of tar in sediment and light hydrocarbon fractions in mussel flesh (Table 4-27).

Hydrocarbon concentrations in the mussel *Mytilus edulis* have also been used to monitor the natural cleansing of the environment subsequent to an oil spill. BLACKMAN and LAW (1981) studied the course of change in mussels and in the ambient water for some 2 yr subsequent to the spill of a large amount of heavy fuel oil from the *Eleni V* (Table 4-28). There is supporting evidence of a further invasion of oil during winter 1979 causing a delay in the depuration of the mussel population.

The composition of hydrocarbons taken up is related to the manner of introduction into the mussel. If the oil fraction is soluble, uptake through gills and other surfaces does not alter the qualitative pattern of the gas chromatograph. Oil in particulate form is taken up preferentially; the gas chromatographs show a relative enrichment of the lower molecular weight saturated and aromatic hydrocarbons in the tissues (RISEBROUGH and co-authors, 1979). The overall hydrocarbon bioconcentration factor from field observations, assuming that the mussels were in equilibrium with the water, is defined by the concentration of hydrocarbons in mussel tussue divided by the concentration in the sur-

Table 4-28

Mytilus edulis. Hydrocarbon concentrations in samples from Corton, UK ($\mu g\ g^{-1}$ wet flesh weight fuel oil equivalents) and ambient water concentrations at time of sampling ($\mu g\ l^{-1}$ fuel oil equivalents) (After BLACKMAN and LAW, 1981; reproduced by permission of the authors)

Date	Days since spill	Mussel flesh	Ambient water
18 May 1978	12	—	19, 24
7 June 1978	32	—	370
31 July 1978	86	—	180
16 October 1978	163	250	150
12 December 1978	210	156	150
25 January 1979	254	265	110
22 February 1979	292	226	260
28 March 1979	326	106	300
4 May 1979	363	80	63
21 November 1979	560	87	86
13 May 1980	740	10	24

rounding water. Values ranged from 10^5 to 10^7 for n-alkanes in the range C_{14} to C_{34} (BURNS and SMITH, 1978). RISEBROUGH and co-authors (1979) report 300,000 to 70 million (Table 4-29). Less consistent relationships have been found for bioaccumulation between sediment and mussel since this comparison introduces much more complex problems of discriminative hydrocarbon uptake by the organism and the manner in which individual hydrocarbons are held on different kind of particles and in micro-organisms. From comparisons of lipid-type extracts of sediment and mussel, the hy-

Table 4-29

Mytilus californianus. Relationships between tissue total saturated hydrocarbons ($ng\ g^{-1}$) and concentrations in 'dissolved' phase and on particulate matter ($ng\ l^{-1}$) (After NRC, 1980; modified; reproduced by permission of the National Research Council, USA)

$$\text{Bioconcentration factor ('dissolved')} = \frac{\text{mussel concentration (ng kg}^{-1})}{\text{'dissolved' concentration (ng l}^{-1})}$$

	C_{14}–C_{34} alkanes concentration	Bioconcentration factor for *Mytilus*	
Palos Verdes			
'dissolved'	4·1 ± 1·3	3×10^7	$n = 2$
particulate	83	2×10^6	
Mytilus	140,000	—	
Goleta Point			
'dissolved'	5·6 ± 1·2	7×10^7	$n = 2$
particulate	970	4×10^5	
Mytilus	390,000	—	
San Nicholas Is.			
'dissolved'	12 ± 2	3×10^5	$n = 2$
particulate	13	3×10^5	
Mytilus	3900	—	

drocarbon content in the former is invariably greater by a factor ranging from tens to thousands. Oil concentrations in surface sediments were higher in the finer samples than in coarse ones.

Perhaps of more importance for man is not so much how mussels tend to record oil exposure in general but how they respond to the more toxic polycyclic aromatic hydrocarbons, especially benzo[α]pyrene. Mussels reveal clearly any exposure to unusually rich sources of polycyclic aromatic hydrocarbons such as runoff from tarred roads (MACKIE and co-authors (1980), oil storage areas and the presence of creosoted harbour timbers. Benzo [α] pyrene concentration in bivalves from the coast of Oregon (MIX and co-authors, 1977) ranged from the limit of detection about 0·1 ng g^{-1} dry weight in clean areas to a peak of 30 ng g^{-1} for industrialized bays; similar values were found by DUNN and STICH (1975) for the Vancouver coasts. EATON and ZITKO (1979) found much higher values for several individual polycyclic aromatic hydrocarbons within 100 m of creosoted harbour timbers.

Birds

It is estimated that annually between 150,000 and 450,000 seabirds die in the North Atlantic Ocean and North Sea as a direct result of oil pollution (GUNKEL and GASSMANN, 1980). The number of seabirds found dead on the seashore is easy to assess and on average between 10 and 25% of them are evidently contaminated with oil to a degree affecting their viability. The normal incidence of seabird mortality due to oil is clearly allied to oil tanker traffic and oil handling activity and may rise to locally dramatic impact when even quite modest oil spills encounter dense aggregations of vulnerable species.

Of what significance to man and to the marine ecosystem are seabirds and threats to their welfare? Birds in general, as well as seabirds, serve as conspicuous indicators of the state of the 'natural' environment and of damage to it. Bird protection is an action call for the defence of nature in ways obvious to and easily understood by the general public.

Inland birds encounter many terrestrial competitors for food and many predators. Birds over the sea have the air almost exclusively to themselves. Uncertain weather and variable food supply have led them to develop strategies for tenacious endurance and their lives tend to be very safe and long, and under favourable conditions their numbers become large and their dispersal vast. Seabirds have not devised means of reproduction on water (though the emperor penguin manages on ice) and they must return to a solid base not too remote from good feeding places to incubate their eggs and rear their young. Seabirds tend to have long, complex life cycles, some with long periods of immaturity, others spend a large part of the year in breeding and can only nest successfully in alternate years. Most pelagic species lay only 1 egg in a season and some do not replace it if it is lost though coastal birds may lay 2 or 3 eggs and replace them readily. The period of moult increases the vulnerability of divers, grebes, sea-duck, auks, and diving-petrels which go out to sea and become flightless while they complete their moult. Oil contamination of eggs reduces the hatchability, and spraying eggs with oil has been used to control numbers in troublesome colonies of gulls and cormorants.

Because oil floats on water it is a constant and concentrated hazard to seabirds wherever it exists. The waterproof plumage does not shake off oil which tends to soak into the feathers reducing their buoyancy and insulation. Increased activity becomes necessary to keep afloat and to maintain body temperature. In severe cases the birds die rapidly from exhaustion and exposure. Many birds are sensitive to smell, some acutely so, and exposure to vapour of crude oil can cause damage to their lung tissue. Birds may also swallow oil after preening and this causes damage to liver, pancreas, kidneys, adrenals, and other organs. Oiled seabirds may sink or drift ashore or simply disintegrate at sea, hence beach surveys underestimate casualties at sea (see BOURNE, 1976 for a comprehensive discussion).

Records of bird deaths due to oil pollution have greatly improved but where prevailing winds drive floating objects consistently offshore there may be little evidence of mortalities. Localities such as estuaries, enclosed bays, narrow channels, and semi-enclosed seas like the Baltic Sea, give the most representative counts; the larger seas and open oceans can conceal major bird catastrophies. Much of the evidence on oil-pollution and its effects on seabirds has been reviewed by BOURNE (1976) who has also attempted to assess the extent of casualties caused by some of the major oil spills world-wide. His studies reveal that seabirds are particularly vulnerable during certain parts of their daily, annual, and life cycles—especially during breeding, moulting, roosting on the water, and feeding. The consequences of oil pollution can be so conspicuously dreadful that a great deal of effort has gone into devising ways to mitigate them including efforts to retrieve and rehabilitate oiled birds.

Some success has been achieved but the available resources fall far short of being able to mitigate the problem on a meaningful scale and the financial and manpower costs are very high.

Damage to local stocks of some seabird species can attain levels where reduction is ecologically significant in terms of the survival of that colony. Small colonies have been wiped out and their niche filled by some competing species. Because of increased vulnerability, susceptibility, and low reproductive capacity, a number of particular seabird species in any region of the world ocean become special targets of this kind of pollution, and with repeated oil spills it is typically these species that suffer most and are least able to make up the losses. The vulnerable species tend to be those that spend a large part of their lives on, rather than above the sea, have small wings, little reserve capacity for prolonged flight, and a low breeding rate.

In ecological terms it is only possible to gauge changes in natural populations by observing live colonies and not from counting dead oily birds. On this basis oil has been established as a definite contributory cause of the marked reduction in numbers of long tailed ducks and velvet scoters migrating through the Baltic Sea, in breeding colonies of puffins and guillemots in southwest England, and in Brittany—possibly also in colonies of puffins and razorbills in Newfoundland (COWELL, 1976). Deaths of several species of gulls in oil spills have not prevented the continued rise in these populations around the British coasts.

Very few attempts have been made at even approximate quantification of the oiling of seabirds in terms of oil (g) per bird at sea in the course of an oil spill. Most casualty figures relate to shore counts of beached birds oiled, disabled or dead. POWERS and RUMAGE (1978) made sightings at sea near the wrecked *Argo Merchant* from the day of stranding to Day 10 and subsequently Days 20 to 30 and 41 to 53. Initially, 1120 birds of 13 species were recorded in a series of 146 valid 10-min counts (where no interfering events occurred and visibility was adequate). Gulls made up 92% of the sightings; of the herring gulls 59% were visibly oiled, and 41% of the great black-backed gulls. Only 12% of the gannets and 9% of the kittiwakes were oiled.

From Days 6 to 40 regular searches of the beaches of Nantucket Island and of Martha's Vineyard yielded 69 live ailing birds and 112 dead birds of 16 species; alcids (49%), gulls (27%), and loons (19%) formed a large part of these casualties. Some 15 specimens representing 5 species showed these common conditions on necropsy: (i) all were underweight as determined by exposure of the keel; (ii) all lacked a layer of body fat; (iii) none had food in the digestive tract.

Away from the tanker the percentage of oiled gulls declined. During the January surveys more birds were present outside the affected area but a larger proportion of the birds within the area were visibly oiled. The authors were concerned that the shore counts of beach birds (i) did not represent the species composition affected offshore and (ii) were not an accurate index of the actual mortalities. In terms of the wider bird populations of the New England coast the dead and damaged birds represented only a minimal impact.

The *Argo Merchant* spill involved heavy fuel oil and occurred in winter when birds were not densely aggregated. The *Amoco Cadiz* spill involved light Arabian crude oil and occurred between mid-March and mid-May when seabirds were on migration to summer resting grounds (CROSS and co-authors, 1978). After some 26 to 28 d, the 2 teams of beach observers counted about 3200 birds killed comprising 33 species. Of these the razorbill *Alca torda*, the guillemot *Uria aalge*, the puffin *Fratercula arctica* and the shag or cormorant represented 80% or 90% of the mortalities. Of these species the puffin mortalities were of most concern since the resting colonies on the nearby islands of Rouzie and Melban were the most southerly in Europe, and, at the same time of the year in 1967, were exposed to some of the *Torrey Canyon* oil. In 1968 the 2000 resting pairs prior to this disaster were reduced dramatically and since then had levelled out at about 800 pairs. In the same way the numbers of breeding pairs of razorbills had fallen from 250 to 90 and of guillemots from 400 to 150. The impact on these three species was reckoned to have endangered the survival of these species in France.

On beaches only lightly affected by oil, gulls were seen to be feeding on the freshly killed intertidal organisms. At heavily oiled beaches killed invertebrates remained uneaten and shorebirds and gulls were conspicuously absent. Gannets, though numerous, did not appear to be acutely affected and accounted for only 2% of beached seabirds.

Because of the large amounts of dead and contaminated intertidal and subtidal food species concern was expressed for chronic hydrocarbon ingestion. A further danger to be looked for was incorporation of oiled seaweeds and other materials in nesting sites. In the southern hemisphere the seabirds most affected in the *Metula* oil spill in the Strait of Magellan were several species of cormorant and penguin; the flocks of gulls, terns, ducks, albatross, and other species escaped lightly (HANN, 1977).

Mammals and Others

Apart from seabirds a considerable number of other animals have become victims of oil pollution. Since all mammals must breathe air, marine mammals could encounter floating oil briefly in oil spill situations and more commonly in thin oil film. Such oil, however, has little effect as it soon sloughs off on diving. It is oil that is washed up on rocky shelves where the mammals haul themselves out of the water that is a greater menace. Aged oil is believed to give rise to ulcerations on the underparts of several types of seals (VAUK, 1973), but conceivably other mammals adopting this behaviour would be vulnerable. Shortly after the *Amoco Cadiz* disaster 3 dead grey seals were found on the northwest coast of Brittany.

Sea otters have been reported to be oiled (BOURNE, 1980), and swans and waders that feed at the shoreline have on occasion been seriously affected (BOURNE, 1976).

Coral reefs are also of animal origin and apart from constituting one of the biologically most productive of all natural communities have created great natural geographical features which act as a living battlement protecting tropical lands and islands against ocean storms. There is, therefore, justifiable concern for the survival of coral reefs which preserve some 400 atolls, many low islands and island chains and thousands of miles of coastline in the warm seas and oceans.

In the normal course of events coral reefs undergo, over a protracted time scale, cycles of degeneration and renewal with a considerable measure of mystery about the events leading to coral death. When coral dies the reef structure gradually weakens and parts may be swept away by storms. Equally, renewal is unpredictable and may or may not serve to restore damaged areas.

As with other small animals, the response to oil pollution is dictated by the manner of encounter with oil, prevailing oil concentrations, and the time scale of exposure. SPOONER (1969) found no damage to coral-reef communities in Taurut Bay in Saudi Arabia in an area of chronic low-level oil pollution. LOPEZ (1978) concluded from studies on oil spills in the vicinity of coral reefs in Puerto Rico that the dilution and dispersion of oil did not penetrate to affect these reefs under the prevailing high energy hydrographic conditions. JOHANNES (1972) has reviewed the early field studies and experimental work much of which relates to unrealistic hydrocarbon concentrations and direct application of oil.

In experiments using sublethal concentrations of the water-soluble part of Iranian crude oil LOYA and RINKEVICH (1979) found that introduction of dilute solutions stimulated the Red Sea coral *Stylophora pistalla* immediately to open its mouthparts followed by premature extrusion of the planula larvae. Premature release of underdeveloped larvae decreases their viability and their chances of successful settlement. The authors observe that almost no new colonization of reef areas is occurring at sites within the coral nature reserve at Eilat where chronic oil pollution prevails, yet at areas free of oil pollution active colonization is taking place.

(d) Sublethal Effects and Organismic Response

General Aspects

It is fundamental to environmental monitoring, field survey or bioassay of primarily sublethal or baseline concentrations of oil that due account is taken of the biochemistry and the physiological pathways of oil in the organism(s) selected as targets. Chronic exposure often leads to and is recognized by the malfunction of a vital organ. This is the final step of a derangement which may be traceable back to a subcellular biochemical disturbance. In studying sublethal effects within the organism 2 approaches may be distinguished: (i) general responses to stressors reflected as broad changes in weight, growth rate, rate of maturation and so on; (ii) more specific responses reflected in changes of certain enzyme levels or loss or gain of enzyme functions. It is usually possible to gauge the exposure thresholds which mark the lowest concentration inducing change yet do not exceed the organism's capacity to recover full function. Stressors can also induce behavioural changes and may be implicated in disease.

Cellular and Physiological Mechanisms

In plants the general approach to identifying the mechanisms by which hydrocarbons take effect has concentrated on cell-membrane integrity and function (e.g. p. 1498). In animals the mechanisms are more complex and vary greatly between species. There are several mechanisms by which oil enters the animal. Thereafter various mechanisms —including complex enzyme reactions—govern storage, transport between organs, and excretion.

Oil, being hydrophobic and lipophilic, tends to move away from aqueous media and to associate with lipids at the surface of the sea and on cell surfaces. Oil also enters animals in food materials and tends to follow the lipid pathway between organs. Transfer into the aqueous phase within an organism may require the presence of emulsifiers or detergents or might follow conversion into more polar derivatives primarily by oxidation or hydroxylation. Oil droplets can pass through the gut of some animals at least, with minimal absorption. But if oil is assimilated, it may be partly or extensively metabolized before excretion. Generally, assimilated oil is enriched in the fatty organs such as the liver relative to muscle tissues and is found in higher concentration in fatty fish, e.g. the herring, than in the less fatty ones (cod, haddock, whiting, etc.).

One example illustrates these steps in more detail. The distribution of hydrocarbons in the various tissues is closely related to the lipid pools in the mussel *Mytilus edulis* and in bivalves generally. During metabolic cycling, hydrocarbons are transported with blood lipid into the liver or other equivalent organ where they are concentrated into lipid vacuoles within the cells. Cytochrome P450-linked reactions progressively convert the component hydrocarbons into arene oxides and hence into transdihydrodiols by epoxide hydrases or into glutathione conjugates, either by glutathiones-s-epoxide transferase or by spontaneous reactions with glutathione. Since at least some of these arene oxide intermediates are recognized as more toxic and carcinogenic than their parent hydrocarbons, these enzyme reactions amount to bioactivation. However, there follow further enzyme oxidations which result in the formation of less toxic products which progress into hydrophilic and readily excreted forms. When very high hydrocarbon levels are accumulated within the lysosomes, damage may ensue as these are spontaneously broken down within the digestive gland.

The relative abundance of mixed function oxidases in bivalves is rather low, so low that their presence was earlier disputed. A limited measure of enhanced enzyme activity can be induced by exposure to some hydrocarbons.

The excretion rate of hydrocarbons in bivalves is high relative to the pathway via bioactivation and deactivation so that the half-life of petroleum is only 16 d (DUNN and STICH, 1976). However, more research is needed into the turnover time of PCAH and other complex hydrocarbons and oxidized intermediates which are known to have protracted lifetimes in mammalian tissues.

Effects on Plants

For unicellular plants and large plants the effects of oil (GESAMP, 1977) have been assessed experimentally in terms of altered photosynthesis, oxygen generation and uptake, and rate of nitrogen, phosphorus, and silicon uptake. Consequently in the sea the overall effect of oil is to modify and usually to reduce primary plant productivity (see Volumes I to IV, *Marine Ecology*). Other approaches might be to look for effects on pigments, cell size and division rate in unicellular algae and also on reproduction in macroalgae.

Effects on Animals

Feeding

For animals, feeding is an essential biological function but there are many problems in the secure interpretation of feeding experiments. Feeding rates have been determined for a large variety of marine animals (see Volumes I to IV, *Marine Ecology*). Almost invariably feeding is reduced and sometimes irreversibly inhibited (e.g. PERKINS, 1976, p.549) where the only food available is affected by crude oil or processed oil; others report a reduction in food detectability, e.g. HYLAND and MILLER (1979) found that 1500 $\mu g\ l^{-1}$ No. 2 Fuel Oil completely blocked food detection by the mud snail *Nassarius obsoletus* and as little as 15 $\mu g\ l^{-1}$ impaired food detection. However, BERDUGO and co-authors (1977) report the exposure of the copepod *Eurytemora affinis* to 10 and 50 mg l^{-1} of ^{14}C-naphthalene for a period of 10 d produced no significant effect on feeding despite accumulation of high concentrations in the tissues. Some researchers observed differences in feeding behaviour where both clean and contaminated foods were offered. Clearly, many factors influence feeding in animals when oil is present and it may be misleading to quote concentrations without full disclosure of at least details of presentation of food, administration of oil, and choice of foods available.

Associated with bioassays based on feeding other important elements are appetite, ability to feed, and, overall, the animal's efficiency in food conversion (Volume II: PANDIAN, 1975). It is not hard to appreciate the relevance of feeding factors in relation to oil pollution of the sea.

Respiration

Measurements of respiration or of gill activity have often been used to determine physiological stress in molluscs, crustaceans, and fishes. Changes in dissolved oxygen are convenient to measure and gill movements can be directly recorded. Frequently, however, it is difficult to evaluate the data because of modifying environmental changes and individual variability. Respiratory activity is a particularly useful criterion for assessing the initial onset of a stressor but it is less useful for deriving long-term effects.

Excretion

There is another important function that in general has some monitoring potential and is particularly relevant to the metabolism of oil. Excretion is normally observed by measuring the release of waste products among which are ammonia, urea, amino acids, purines, and sterols (Volume II: PANDIAN, 1975). Some metabolically produced hydrocarbon derivatives can be studied directly but the easier technique is to study labelled hydrocarbon. By these means the ability and capacity of the organism to cope with oil and individual hydrocarbons can be defined (e.g. CORNER and co-authors, 1973; LEE and RYAN, 1976).

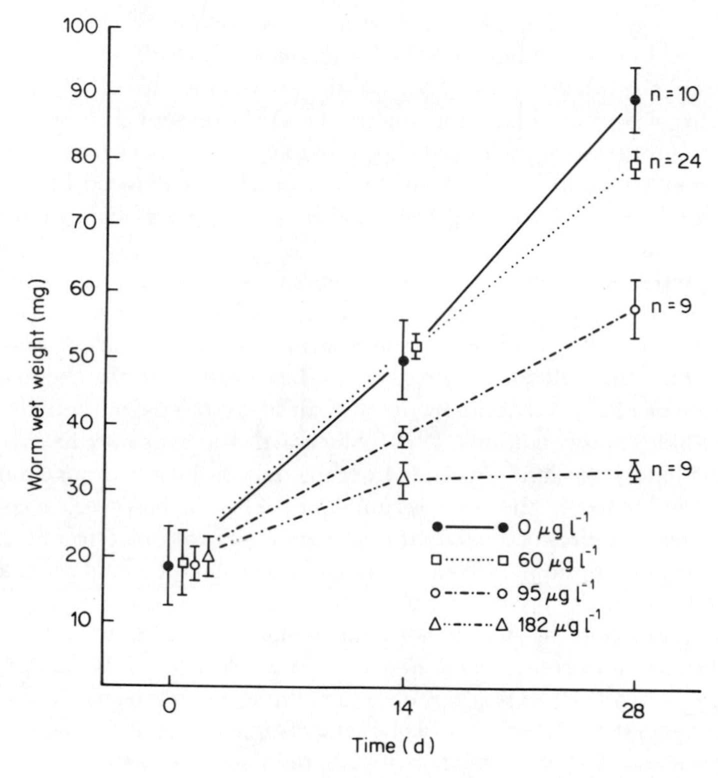

Fig. 4-19: *Neanthes arenaceodentata*. Effects of water-soluble fractions of No. 2 Fuel Oil on growth. Concentrations of total naphthalenes in the exposure water ($\mu g\ l^{-1}$) are listed; n: number of bowls in each group. (After ANDERSON, 1977; modified; reproduced by permission of the American Petroleum Institute)

Growth

This overall response measure has been studied for many animals under a large variety of experimental conditions by numerous authors (e.g. Volumes I to IV, *Marine Ecology*). ANDERSON (1977) provides an example of the growth response of the polychaete *Neanthes arenaceodentata* (Fig. 4-19).

Growth can also be measured in the sea by the use of captive or marked animals, tentatively in relation to the monitoring of oil water discharges at oil terminals and to determine the after-effects of single exposures of marked gastropod molluscs to oil dispersants (PERKINS, 1970; GRIBBON, 1973). In many cases the laboratory (or cage) environments in themselves represent a major stress influence.

Biochemical systems

The reaction of the mixed function oxidase (MFO) system to single hydrocarbons and oils has been extensively explored. Knowledge of dose–response relationships is increasing, and a good deal is known about which animals possess this system and which are responsive to the introduction of contaminating hydrocarbons. However, in addition to hydrocarbons, many other molecules—including contaminants such as some of the halogenated biphenyl isomers—induce MFO reactions.

If it can be established that hydrocarbons are the dominant cause of MFO induction for sites remote from land influence, the amount of MFO present in target organisms might be useful as a measure of prolonged exposure to increased background concentrations of oil. More information is needed on the way in which enhanced MFO modifies normal metabolic efficiency, scope for growth and perhaps the organism's vulnerability to carcinogens and pathogens.

In relation to present considerations, blood is less of a specific enzyme substrate than a complex medium. It is capable of revealing a great deal about the reproductive state and condition of the animal—its responses to physical and chemical stress, to pathogens and parasites, and to other influences. Stress or condition can affect the chemistry of the blood, the balance of blood cell components and an array of enzyme functions in the blood (e.g. THURBERG and co-authors, 1978). Since hydrocarbons may be presumed to affect more immediately the lipid-associated organs and pathways in an animal, they may reveal their influence in the total serum lipid. This is, however, a speculative suggestion and it seems more reasonable at this stage to regard blood as one of the less easy and definitive study options. Virtually any pollutant could be conceived as having an effect on blood.

Similarly, our knowledge of steroid hormone metabolism in marine organisms is largely confined to fish and considerable new ground would need to be explored before enough knowledge is available so that response to pollutants could be usefully identified. Some pollutants suggest themselves as likely subjects causing interference; the role of pesticides has been established in relation to salmonid steroid metabolism, and some petroleum hydrocarbons have structural similarities both with pesticides and with natural steroids and their metabolites. There may, therefore, be grounds for some research in this area. The methods of testing are relatively sophisticated and presumably the state of reproductive development in the animal may have a strong influence on any stress response.

Animal activity

The motive for studying behaviour in the present application is to gain insight into sublethal responses. However, animal activities are subject to numerous complicating factors. In what ways could the properties of oil be anticipated to interact with activities of a sessile, a residential, and a migratory animal? It has already been mentioned that crude oils and some oil products stimulate or interfere with chemoreceptors at low concentrations. There are many other responses affecting, e.g. motility, filtering capacity, orientation, sexual behaviour and release of gametes, and burrowing. Many oil dispersants also disrupt normal animal activity (BNCOR, 1980).

It is conventional to explore and quantify individual behavioural responses first under controlled conditions in the laboratory which has the advantage of 'excluding' most of the environmental variables as well as reducing the array of intra- and interspecific responses.

Orientation in space and time is a most important manifestation of animal activity. It has received detailed attention in Volume II: General introduction: SCHÖNE (1975); Invertebrates: CREUTZBERG (1975); Fishes: TESCH (1975); and Mammals: KINNE (1975). Animal activity and behaviour are sensitive and useful parameters for assessing potential effects of oil pollution.

(8) Oil Pollution as a Dynamic Process

A dynamic budget is offered in Table 4-30 A,B,C in order to see which oil items make the greatest impact. The data sources are well known and the rates of the various oil processes are probably of the right magnitude for North Sea waters. Nothing is special or critical about the selection of the values used and it is very easy to substitute other magnitudes and other rates to suit other warmer or colder climes.

Various gross assumptions have been made about the North Sea circulation, the fate of the oil and the important parts of the oil involved in movement between surface film, water column, and sediment.

What the standing stock assessment (Table 4-30 A) emphasizes is the large impact of volatile hydrocarbons which are probably in active flux with regard to mass balance between sediment ⇌ water column ⇌ surface film ⇌ atmosphere and also with regard to in-sediment production and aquatic biodegradation. Volatile hydrocarbons are dominated by biogenic production and are excluded on this basis.

The apparent agreement between the n-alkane-based and naphthalenes-based (GRAHL-NIELSEN and co-authors, 1979) estimates is fortuitous and no guarantee whatever that either is correct. Oil in sediments is the second largest component of the standing stock.

Most of the basic numbers relating to degradation rates (Table 4-30 B) are derived from laboratory experiments. If these rates are remotely realistic for the open sea they give rise to corresponding values for the potential degradative capacity of the sea which are very large in relation to the estimated standing stock. Only the degradative capacity of the water column shows a short fall. On this basis it appears that there is abundant natural capacity (Table 4-30 C) in the surface layer and at the seabed to cope with a very large part of the present total and foreseeable inputs, including those brought in by circulation. For the standing stock distribution to be held steady a very large part of input must travel rapidly to the sea floor with only minor change during transit. This may be interpreted as yet another argument for not using oil dispersants which place the oil in the least capable compartment.

(9) Oil Spills—Remedial Action and the Environment

(a) The Oil Spill Scene

The bald figures in Table 4-1, although impressive in magnitude, give little insight into the factors contributing to day-to-day oil spills and fail to show the close relationships between several of the contributing categories. It is difficult for authorities to collect accurate and representative routine oil spill data. For example, under-reporting must occur for spills that disperse quickly or escape observation, over-reporting of drifting oil slicks can occur and repeated reporting when they give rise to sibling slicks.

The Oil Spill Intelligence Report strives to compile a weekly complete summary of all major oil spills world-wide whether originating from land-based facilities or from tankers and other shipping at sea, in port or in inland waterways. Their First International Annual Summary and Review (OSIR, 1979), lists all incidents involving more than 20,000 US gallons of crude oil or oil products compiled from correspondents of the Center for Short-Lived Phenomena incorporating verification and additional information from the US Coast Guard, Lloyds of London, and the Canadian Protection Service.

Table 4-30

Dynamic budget for oil in the North Sea (area: $0{\cdot}575 \times 10^{12}$ m^2; volume: 54×10^{12} m^3). (A) standing stock of hydrocarbons; (B) rates of biodegradation affecting 50% of the total oil and corresponding potential; (C) summary oil account (Original)

(A) **Standing stock of hydrocarbons**

Surface film = 90 μg m^{-2} n-alkanes equivalent to $3 \times 90 = 270$ μg m^{-2} total oil;
hence oil in surface film
$= 270 \times 10^{-6} \times 0{\cdot}575 \times 10^{12} \times 10^{-6}$ t
$= 155$ t

Tar balls = 10 μg m^{-2};
hence surface tar balls
$= 10 \times 10^{-6} \times 0{\cdot}575 \times 10^{12} \times 10^{-6}$ t
$= 6$ t

Water volume (i) C_1–C_4 volatile n-alkanes dissolved in water column = 70 μg l^{-1};
hence total volatile alkanes
$= 70 \times 10^{-6} \times 54 \times 10^{12} \times 10^{3} \times 10^{-6}$ t
$= 3{,}780{,}000$ t

(ii) C_{15}–C_{33} n-alkanes = 1 μg l^{-1} equivalent to oil 3 μg l^{-1};
hence total oil
$= 3 \times 10^{-6} \times 54 \times 10^{12} \times 10^{-6}$ t
$= 162{,}000$ t*

(iii) Naphthalenes, phenanthrenes, and dibenzothiophenes equivalent to 2% oil = 60 ng l^{-1};
hence total oil

$$= \frac{100}{2} \times 60 \times 10^{-9} \times 54 \times 10^{12} \times 10^{3} \times 10^{-6} \text{ t}$$

$\sim 162{,}000$ t

Sediment All oil in dynamic exchange is in top 1 cm
1 cm layer × 1 m^2
$= 0{\cdot}02$ t (SG = 2)
total sediment
$= 0{\cdot}02 \times 0{\cdot}575 \times 10^{12}$ t
$= 1{\cdot}150 \times 10^{10}$ t
C_{15}–C_{33} n-alkanes
= 2 μg l^{-1} or 2 g t^{-1}
= 69,000 t

(B) **Rates of biodegradation affecting 50% of the total oil† and corresponding potential**

Sea surface
= 0·05 g m^{-2} d^{-1}

Water column
= 0·6 mg m^3 yr^{-1}

Seashore and on seabed
= 0·01 g m^{-2} d^{-1}

Hence potential amount of oil degraded annually:

at the sea surface
$= 0{\cdot}05 \times 365 \times 0{\cdot}575 \times 10^{12} \times 10^{-6}$ t
$= 10{\cdot}5 \times 10^{6}$ t

in water column
$= 0{\cdot}6 \times 10^{-3} \times 54 \times 10^{12} \times 10^{-6}$ t
$= 32{,}000$ t

at seashore (20 m band × 1500 km = 3×10^{7} m^2)
$= 0{\cdot}01 \times 365 \times 3 \times 10^{7} \times 10^{-6} = 110$ t

on seabed
$= 0{\cdot}01 \times 365 \times 0{\cdot}575 \times 10^{12} \times 10^{6} = 2{\cdot}1 \times 10^{6}$ t

Table 4-30—*contd.*

(C) **Summary oil account**

Standing stock of oil (t)			*Annual turnover (tentative estimates)*		*Note*
Surface film	155		186,000	(155 × 600 × 2)	1
Tar balls	6		25	(6 × 2 × 2)	2
Water column (non-volatile hydrocarbons)	162,000		324,000	(162,000 × 2)	3
Phyto + zooplankton (min)	900		82,030	(season)	4
(max)	9000				
Benthos	20		500	(20 × 25)	5
Fish	32		800	(32 × 25)	6
Sediments (1 cm depth)	69,000		829,000	(unknown turn-over rate) say monthly 12 × 69,000	7
Environment (E)	85,060	ΣE	1,378,325		8
Biota (B)	952 to 9052	ΣB	83,330		9
		ΣE + ΣB =	1,461,455		10
		Input	1,448,340	(from Note 1)	

1. The oil input figure has been derived from the global oil usage (3×10^9 tons yr^{-1}) which is associated with $4{\cdot}951 \times 10^6$ tons yr^{-1} oil spillage to the sea, correspondingly northern European oil usage (10^9 tons yr^{-1}) by proportion gives an input of 1,448,340 tons yr^{-1} to the North Sea. The surface film has to be renewed 600 times per year to match the atmospheric input. The North Sea water mass is renewed twice per year.
2. For the dominant micro-tar balls the half-life has been taken as 6 months and the North Sea turnover is twice (this is a trivial oil component).
3. North Sea water mass renewed twice, only 1 volume considered and the volatile hydrocarbons excluded
4 & 5. Based on average biomass figures, turnover about fortnightly.
6. Mean retention time 14 d.
7. This item must take into account atmospheric and fresh water inputs, etc. Turnover rate taken as monthly.
8. Subtotal for all environmental inputs: 1 + 2 + 3 + 8 + ΣE.
9. Subtotal for oil in biota: 4 + 5 + 6 = ΣB.
10. $\frac{\Sigma B}{\Sigma E + \Sigma B} = 5{\cdot}7\%$. Rate of oil sedimentation is 4 mg m^{-2} d^{-1} of which 0·635 mg m^{-2} d^{-1} is from the atmosphere.

* Compare GRAHL-NIELSEN and co-authors (1979).
† Based on JOHNSTON (1977).

An abbreviated and simplified version of the OSIR summary of the 1978 spills is given in Table 4-31. Reservations regarding the entries are fully explained in the original version. It is important to note the high credibility of the sources of original data but one cannot but question their comprehensiveness as a full and equal report of all major spills world-wide. On the basis of the limited geographical cover these surely must be incomplete.

Inevitably, observance of a lower limit of 20,000 US gallons in these records allows a very large number of minor spills through the net. There are only 2 'unknown' category entries; yet in the UK statistics, for oil slicks observed in the UK sector of the North Sea and around other coasts, the proportion of unattributable spills is much higher. Many spills become apparent only when oiled seabirds are found on the shore. Indeed it is recognized that accurate recording of oil slicks is quite a complex problem, with considerable disparity between the numbers recorded by the central authority (coastguards, aerial observation, shipping reports, etc.) and those collected by responsible non-governmental conservation organizations. At the scale of dimension and frequency of these oil slicks it is extremely difficult to ensure that each slick or series of slicks represents one incident, and that incidents are recorded only once even if not by successive observers. Despite vigilance there will always be some undetected and unlisted slicks in isolated areas and some double counting. Table 4-31 witnesses how diverse are the origins, types, and causes of oil spills. The high frequency of land-based spills adds credence to the high position in the inputs (Table 4-1) for fresh water sources. Even allowing for success in oil salvage and oil recovery from land-based facilities, a considerable proportion must find its way into waterways, municipal and industrial waste waters, and into the atmosphere, on its way to the sea.

Although the vast majority of spills are associated with errors of technical operation or negligence or equipment failure not spotted beforehand in routine maintenance, significant numbers involve accidental damage or deliberate sabotage by third parties. Quite a few are associated with natural causes such as earthquake and lightning.

The quantities in each spill cover several orders of magnitude. This explains why global estimates of the averages in each spill category (Table 4-1) are so uncertain and why oil spill contingency planning must be flexible and be ready on call to cope with 20,000 or 20,000,000 gallons of thick or thin oil.

In assessing the ecological impact of oil pollution it is relevant to recollect the effects on the fresh water environment, although our main interest here is in estuaries, seas, and oceans. Oil spills causing spectacular environmental damage are not restricted to very large crude-oil carriers; quite small spills of heavy fuel oil from ordinary shipping have caused equal damage.

(b) Oil Spills at Sea

A catalogue of the outstanding recent oil spills at sea is given in Table 4-32. The list outlines the damage that resulted, the method used in clean-up, and some of the costs and claims associated with each spill. Such lists have value in determining managerial policy in the reduction of marine oil pollution.

On a more detailed scale the classified statistics of spills at an oil terminal are a useful guideline in directing policing effort. The port of Milford Haven, Pembroke, UK is used for crude-oil imports and also handles the export by sea of some of the resulting refinery products. Table 4-33 and Fig. 4-20 show how the port has performed with respect to the handling of crude-oil imports over the past 16 yr. This study of terminal operations suggests that there is now no single outstanding cause of oil spills. By continuously developing control policies the various operational steps are demonstrably uniformly supervised and there are no easy pronouncements regarding where better controls are needed.

There is a long history of international effort to control marine pollution from operational discharges of oil and oily wastes by shipping (MOORE, 1976). The first draft convention was outlined in Washington in 1926 but never adopted. Since then pro-

Table 4-31

Outline of international oil spills during 1978. SF: shore-based facility; DR: drilling rig; NGS: onshore natural gas plant; 'Ramming': violent contact with dock, terminal or object; 'Collision': violent contact with one or more vessels (Original compilation; based on OSIR, 1979: annual summary, 1978; reproduced by permission of The Center for Short-Lived Phenomena, Cahners Publishing Company)

Date		Type	Location	U.S. gallons	Oil	Cause/Casualties
Jan	1	SF	Quebec, Canada	48,000	Gasoline	Leak, explosion
	5	SF	Montreal, Canada	195,000	Bunker C	Pipe rupture
	6	SF	Texas City, USA	Undetermined		Refinery fire
	6	SF	Saskatchewan, Canada	76,000	Heating	Pipe rupture
	7	NGS	Sonatrach, Algeria	Undetermined		Blow-out
	8	SF	NW Territory, Canada	24,000	Mar. diesel	Tank overfill
	9	Barge	New York, USA	210,000	Heating	Open hatch, aground
	10	SF	N Carolina, USA	30,000	Heating	Explosion, fire
	10	Tanker	Sao Paulo, Brazil	Est. 3,660,000	Crude	Ramming
	15	SF	Massachusetts, USA	57,000	Diesel	Pipe rupture
	23	SF	Montreal, Canada	50,000	Heating	Mechanical failure
	23	Barge	Galveston Bay, USA	50,000	No 6 Fuel	Collision
	26	SF	Ohio, USA	23,000	Gasoline & heating	Pipe rupture
	27	SF	Tennessee, USA	25,000	Asphalt	Valve failure
	31	SF	New Jersey, USA	3,000,000	No 4 fuel	Tank rupture
	31	Barge	Louisiana, USA	252,000	No 6 fuel	Collision
Feb	1	SF	Louisiana, USA	Undetermined		Refinery fire
	4	SF	Pennsylvania, USA	40,000	Diesel	Tank overflow
	5	SF	Badgranau, F.R. Germany	Undetermined		Refinery explosion
	6	Tanker	Massachussetts, USA	40,000	Lubricating & fuel	Grounding
	6	Tanker	Maine, USA	23,000	Diesel	Grounding
	8	SF	Massachussetts, USA	1,500,000	Gasoline	Tank rupture
	8	SF	Illinois, USA	21,000	Crude	Pipe rupture
	9	Tanker	New Jersey, USA	Est. 60,000	Gasoline	Grounding
	9	SF	Pensylvania, USA	25,000	Gasoline	Pipe rupture
	10	Barge	Texas, USA	38,000	Jet fuel	Grounding
	11	Barge	Houston, USA	20,000	Jet fuel	Grounding
	13	SF	Saskatchewan, USA	50,000	Crude	Pipe rupture
	15	SF	Alaska, USA	64,000	Crude	Sabotage
	19	SF	Maine, USA	30,000	Diesel	Valve failure
	21	SF	California, USA	3,350,000	Gasoline	Explosion
	25	SF	Quebec, Canada	24,000	Diesel	Pipe rupture
	25	Tanker	Penang, Malaysia	Undetermined		Fire, 5 dead
	25	SF	California, USA	100,000	Crude	Fire
	27	Barge	Chesapeake Bay, USA	25,000	No 6 fuel	Sinking, deliberate grounding
	28	SF	Florida, USA	Undetermined		Depot fire
Mar	1	Barge	Duisberg, F.R. Germany	Undetermined	Gasoline	Collision
	2	SF	Quebec, Canada	30,000	Diesel	Tank overfill
	16	*Amoco Cadiz*	Brittany, France	Est. 68,658,000	Crude	Grounding
	16	Barge	Black Is. Sound, USA	682,000	Gasoline	Grounding
	17	SF	Montana, USA	21,000	Crude	Pipe rupture
	17	Ship	Cape Town, S Africa	Undetermined	Fuel	Grounding
	20	Barge	Delaware, USA	630,000	Jet fuel	Explosion, 2 dead

Table 4-31—*contd.*

Date		Type	Location	U.S. gallons	Oil	Cause/casualties
	21	SF*	Quebec, Canada	58,000	Fuel	Railway
	22	Ship	off Sumatra	Est. 155,000	Fuel	Grounding
	25	Barge	California, USA	23,000	Gasoline	Hull failure
	27	SF	Massachusetts, USA	65,000	No 6 fuel	Tank rupture
	28	SF	Syria–Israel border	Undetermined	Crude	Pipeline fire
	28	SF	Saskatchewan, USA	84,000	Asphalt	Tank rupture
	28	SF	Arkansas, USA	Undetermined	Gasoline	Railway fire
	30	Barge	Louisiana, USA	42,000	No 6 fuel	Ramming
April	2	Tanker	Strait of Malacca	1,008,000	Special fuel	Ramming
	10	SF	Oklahoma, USA	29,000	Refined products	Mechanical failure
	15	SF	Abqaiq, Saudi Arabia	Undetermined	Crude	Explosion & fire
	16	Tanker	Bangkok, Thailand	Undetermined	Bunker fuel	Explosion & fire Sinking
	16	SF	Massachusetts, USA	34,000	Undetermined	Pipeline rupture
	21	SF	Oklahoma, USA	21,000	Crude	Pipeline rupture
	22	Barge	Missouri, USA	210,000	Diesel	Collision
	24	SF	St Johns, Canada	63,000	Diesel	Valve failure
	27	SF	Ras Shukheir, Egypt	2,520,000	Crude	Explosion & fire
May	2	Barge	Minnesota, USA	124,000	Fuel	Grounding
	4	Barge	Louisiana, USA	21,000	Gasoline	Collision
	6	*Eleni V*	Gt Yarmouth, UK	Up to 1,408,000	Heavy fuel	Collision
	8	SF	Badak Field, Indonesia	976,000	Crude	Explosion & fire
	9	Ship	St Johns, Canada	78,000	Marine diesel	Collision
	17	SF	Wyoming, USA	105,000	Crude	Pipeline rupture
	22	SF	Al Magyagish, Kuwait	Up to 2,100,000	Crude	Fire
	25	SF	Ahvazin, Iran	Up to 28,000,000	Crude	Fire
	26	SF	Alabama, USA	85,000	Fuel & crude	Ramming, hose rupture
	30	SF	Texas City, USA	Up to 110,000	Refined products	Refinery fire & explosion
June	1	SF	California, USA	22,000	Crude	Oil well mech. failure
	1	SF	Ontario, Canada	Undetermined	Refined products	Refinery explosion & fire
	4	SF	Quebec, Canada	78,000	Gasoline & diesel	Railway
	12	SF	Sendai, Japan	15,000,000	Undetermined	Earthquake
	13	SF	NW Territory, Canada	35,000	Jet fuel	Pipe rupture
	14	SF	Ontario, Canada	715,000	Crude	Pipeline rupture
	14	SF	Tennessee, USA	65,000	Diesel	Tank overfill
	14	SF	Missouri, USA	Undetermined	Undetermined	Explosion & fire
	17	Ship	Oregon, USA	Up to 60,000	Diesel	Sinking
	19	SF	California, USA	Undetermined	Crude	Refinery explosion & fire
	19	Unknown	Florida, USA	62,000	Marine diesel	Unknown
	23	SF	Burgan Field, Kuwait	Undetermined	Crude	Oil well blowout
	24	Barge	Rotterdam, Netherlands	20,000	Gasoline	Unloading mishap
	27	Ship	Oregon, USA	More than 55,000	Diesel	Hull rupture
	29	SF	Alabama, USA	Up to 34,000	Lubrication	Lighting
July	7	*Cabo Tamar*	Talcahuano, Chile	Est. 2,135,000	Crude	Grounding
	25	SF	Texas, USA	Undetermined	Refined products	Lightning
	31	SF	Teheran, Iran	Undetermined	Crude	Blowout & fire
	31	Ship	New York, USA	38,000	Fuels	Sinking

Table 4-31—*contd.*

Date		Type	Location	U.S. gallons	Oil	Cause/casualties
Aug	2	SF	New Brunswick	126,000	Bunker C	Tank rupture
	9	SF	Port Dickson, Malaysia	Undetermined	Undetermined	Explosion
	12	SF	Delaware, USA	100,000	Jet fuel	Refinery, lightning
	19	SF	New York, USA	30,000	Gasoline	Handling mishap
	19	SF	Brit. Columbia, Canada	105,000	Crude	Pipeline rupture
	21	Tanker	San Juan, Puerto Rico	Up to 60,000	Undetermined	Collision
		Ship		Est. 122,000	Undetermined	
	23	Tanker	C. d'Antifer, France	Est. 45,000	Crude	Unloading mishap
Sep	2	SF	New York, USA	Up to 1,000,000	Gasoline & No 6 fuel	Underground leak
	4	SF	Pernis, Netherlands	Undetermined		Refinery fire alleged intentional
	10	Tanker	off Britanny, France	Large slick	Undetermined	Discharge
	10	DR	Gulf of Mexico	Undetermined		Blowout & fire
	12	SF	New Jersey	20,000	No 2 fuel	Pipeline rupture
	14	Barge	Canal in Netherlands	Est. 37,000	Gasoline	Collision
	18	Tanker	Galle Hbr, Sri Lanka	Undetermined	Fuel	Grounding, sinking
	21	SF	Louisiana, USA	2,835,000	Crude	Tank fire
	27	SF	Pajares, Spain	Up to 220,000	Gasoline & diesel	Train explosion & fire
Oct	2	SF	Ontario, Canada	Undetermined	Fuel	Refinery fire
	3	SF	Colorado, USA	Undetermined		Refinery explosion
	4	Tanker	Anglesey, UK	Up to 73,000	Crude	Unloading mishap
	5	Ship	Florida, USA	Up to 40,000	Diesel & bunker C	Ballasting
	11	*Cristos Bitas*	St Georges Channel, UK	Part cargo	Crude	Grounding
	11	SF	Skaelskor, Denmark	Undetermined		Refinery explosion & fire
	12		off Milford Haven	Est. 915,000	Crude	Explosion & fire
	12	SF	Louisiana, USA	25,000	Crude	Pipeline puncture
	12	*Spyros*	Singapore	Undetermined		Explosion & fire 61 dead
	18	Tanker	off Ahus, Sweden	Est. 46,000	Diesel	Grounding
	19	SF	Mardin, Turkey	10,700,000	Crude	Explosion & fire Suspected sabotage
	22	SF	California, USA	Undetermined	Gasoline	Terminal fire, 9 dead
	30	SF	Pitesti, Rumania	Undetermined	Petrochemical	Explosion & fire
	31	SF	Abadan, Iran	Undetermined	Crude	Pipeline rupture
	31	Unknown	off NW Trinidad	Large slick	Undetermined	Intentional discharge
	31	*Cristos Bitas*	W of Ireland	305,000 not transhipped	Crude & fuel	Controlled sinking
Nov	8	SF	Utah, USA	107,000	Crude	Pipeline rupture
	8	Tanker	Manila Bay, Philippines	Undetermined	Crude	Explosion, 31 dead
	15	Ship	off NW Territory	73,000	Diesel	Ramming
	16	SF	Illinois, USA	84,000	Crude	Pipeline rupture
	28	Barge	New York, USA	More than 47,000	No 2 fuel	Grounding
Dec	8	Barge	New York, USA	50,000	Gasoline	Collision
	11	SF	Harare, Zimbabwe	20,000,000	Gasoline & diesel	Sabotage, explosion & fire
	11	SF	Texas City, USA	21,000	Crude	Pipeline rupture
	13	Tanker	Maryland, USA	Up to 20,000	No 6 fuel	Unloading mishap

Table 4-31—*contd.*

Date		Type	Location	U.S. gallons	Oil	Cause/casualties
Dec	14	SF	Benuelan, Puerto Rico	10,500,000	No 6 fuel	Tank rupture
	16	SF	Illinois, USA	107,000	Crude	Pipeline rupture
	16	SF	Virginia, USA	174,000	Fuel & Gasoline	Pipeline puncture
	19	Barge	C. San Juan, Puerto Rico	462,000	Bunker C	Grounding
	21	SF	Quebec, Canada	35,000	Diesel	Tank overfill
	24	SF	Louisiana, USA	3,400,000	Crude	Tank fire
	27	SF	NW Territory	102,000	Diesel	Valve failure
	30	Tanker	Shetland Is, UK	Est. 310,000	Bunker C	Ramming
	31	*Andrea Patria*	off C. Villano, Spain	Est. 14,600,000	Crude	Hull fracture, fire, explosion, 30 dead

tracted series of international conventions have been drawn up broadening the scope of marine pollution control and seeking ever tighter restrictions. Very slowly these conventions have been signed for adoption by contributing countries, and this assent has eventually emerged in national legislation designed to implement the terms of these conventions. A number of the major nations with heavy involvement with oil handling have not as yet signed or implemented the most recent IMCO Marine Pollution 1978 Convention.

Dealing with accidental spills is much more complex. International bodies have used a 4-fold approach:

(i) The Intergovernmental Maritime Consultative Organization (IMCO) has worked for a long time to prevent accidents at sea and has established conventions regarding safety of navigation, obligatory traffic separations, and codes for safe navigational practices and signals.

(ii) Considerable efforts have also gone into minimization of pollution resulting from an accident at sea. Conventions exist for construction standards, limitation of cargo-tank sizes and distribution of cargo tanks. Other constructional aspects anticipate the reduced stability which may follow hull damage. Pollution may also be reduced if early intervention to salvage, repair, modify or remove a stricken vessel can be achieved. A convention for intervention on the high seas has been mooted.

(iii) Cooperation in cleaning up spilled oil is primarily a matter for regional or local action. However, there are provisions for the reporting of incidents involving the spilling and the detection of spills of oil and other harmful substances.

(iv) Much attention has been given to determining liability for pollution damage and to setting up funds to meet such eventualities. These are discussed under 'Costs of oil pollution' (pp. 1553 to 1558).

Many of these matters relating to the design and operation of ships, written into the 1978 and earlier conventions, have IMCO approval as do standards of training and watch-keeping which would lead to a uniform minimum qualification necessary for all mariners. It requires meticulous implementation by every nation to make these IMCO agreements work.

Accidents can only be reduced if the oil tanker industry world-wide adopts a firm policy to employ only the best personnel and to provide and maintain the best vessels with the best equipment. If necessary, older vessels should be reconstructed and re-equipped to meet top standards. Similarly, even the best personnel can benefit from refresher courses and training in the use of new, safer and cleaner operational practices, navigation aids, fire-fighting, and so on. It is incumbent on the oil tanker industry, and also on governments which determine policy regarding ports, terminals, traffic systems and the broader issues, to set high standards and rigorously to enforce them (COLE, 1979).

(c) Options for Remedial Action

Three broad courses of action may be considered in relation to marine oil spills and the mitigation of damage to marine life and resources: containment and recovery of oil; exclusion of oil from the most vulnerable areas; dispersion of oil in areas where the resulting damage, if any, is much less than in more vulnerable areas about to be affected. In the strict interpretation of clean-up, only the first course of action truly qualifies, the others are palliatives.

It is scarcely relevant here to pursue the theme of pollution prevention, including oil, or the considerations relating to tanker and general ship construction and safety and ease of cargo recovery. At the same time the best way to keep oil from a stricken vessel out of the sea is prompt remedial action whether it be repair to damaged tanks, salvage of the vessel or salvage of the oil. Recoveries approaching 100% have been achieved under favourable circumstances. Ship's design is clearly implicated here also as suitable engineering can facilitate oil removal and introduce accessory cargo salvage lines. Similar principles can be developed for other major oil risks related to oil exploration, exploitation, transport and storage (e.g. VAN DEN HOEK, 1977).

It is encouraging to note that in the fields of design, preparedness for spill emergencies, speedy salvage with back-up equipment and second-line defences against such disasters, the contribution that engineering advances can make is well recognized. The appropriate features are being incorporated into new ships, rigs and platforms as they are built and into older units as they undergo major overhaul. Success has been achieved in a growing number of incidents at sea and on shore where a major oil spill would have earlier been inevitable.

Containment and recovery of oil when it is on water may be simple or impossible depending on the nature of the boundary 'wall' and the water flow. In situations with a plain hard wall and little flow of water it is feasible to retrieve the oil directly or by collecting the oil into a small area for recovery relatively speedily and easily. Where the boundary is complex, soft and absorbing, as in salt marshes and mangrove swamps, retrieval can be efficient only if the oil can be prevented from penetrating the soft boundary. Apart from practical issues such as capability of oil recovery equipment and access to spill sites the dominant environmental limiting factors for oil retrieval are the dynamics relating the oil and the water. If oil and water are in rapid and virtually linear flow the oil can be skimmed from the water using some kind of weir. This is moderately efficient at differential velocities up to a threshold at which the interposition of the weir causes an unacceptable degree of turbulence or vertical mixing. This is the criterion that determines the effectiveness of booms in a river or of oil recovery by devices based on oil flotation. Wave motion generated by wind or tide also creates vertical mixing; consequently, above a modest threshold all kinds of segregation or recovery systems relying on

Table 4-32

Causes, spillage, damage, clean-up and costs reported for some recent oil spills (Original compilation; based on OSIR, 1978, 1979)

Ship	Location	Cause	Spillage	Damage	Clean-up, etc.	Costs
Tsesis 26.10.77	Klarringklabben, Sweden	Grounded while under pilotage	1100 t No. 5 fuel oil cargo and some bunker oil	Severe coastal impact due to sheltered site and cold. Marine life including herring spawn	Remainder cargo transshipped. Manual and mechanical recovery of mousse	10 million Swedish Kronor claim for clean-up against Latvian Shipping Co. 6 to 7 million Swedish Kronor claim against Swedish Maritime Administration for damage and salvage
Amoco Cadiz 16.03.78	Portsall, France	Steering lost, ran aground and sank	21,600 t Arabian crude and ship's fuel	Massive coastal impact. Tourism, amenity, shellfish culture, fishing, marine life, sea birds	Dispersants (offshore), booms, absorbents, skimmers, vacuum trucks, oil mop, pumps, sumps, manual methods	$1500 million total claims filed
Rollnes (coal carrier) 26.05.78	Mobile Harbor, Alabama, USA	Hit shore installation, severed loading gear	275 t No. 2 and No 6 Fuel Oils	Harbour amenity, temporary fire hazard alarm	Oil mop, absorbent, vacuum trucks	$180,000 clean-up costs
World Horizon 26.05.78	Off Durban, South Africa	Heavy swell caused hull damage	600–700 t Middle East crude	None reported	Remainder cargo transhipped. Air bubble screen, floating barriers	R 250,000 surety imposed against oil clearance
Unknown (*Howard Star?*)	Port Sutton, Tampa, Florida, USA	Unknown; suspect operator error	103–160 t Bunker Canal Diesel fuel	Mangrove swamps, marine life; threat to power station	Absorbents, vacuum trucks, manual methods	$700,000 oil clearance costs

Cristos Bitas 12.10.78	Pembroke Coast, SW Wales, UK	Grounded on rocks, navigational error?	3000 t Iranian crude	Many seabirds, some marine life	1000 t scuttled with ship, remainder transhipped. Natural forces, dispersants	\$1·5 to 1·7 million suit for damages
Peck Slip 19.12.78	Cape San Juan, Porto Rico	Struck bottom in shallow water in heavy seas	1750 t Bunker C Fuel Oil	Tourism, mangrove swamps, a few birds, marine life	Booms, skimmers, vacuum trucks, absorbents, manual and mechanical clearance	\$1 million suit against Sun Oil Co. \$50,000 costs for clean-up
Esso Bernicia 30.12.78	Sullom Voe oil terminal, Shetland Isles, Scotland	Fire on tug, berthing tanker struck dolphin and holed	1100 t heavy fuel oil	Many seabirds, otters, sheep	Bulldozers, manual clearance, boom failed	
Andros Patria 31.12.78	Cape Villano, Spain	Hull failure in heavy seas, explosion and fire	40,000 t Iranian crude	Amenities, shellfish and marine life. 29 seamen killed	Remainder cargo transhipped. Dispersants, booms, vacuum trucks, pumps	\$1·5 million costs filed
Betelguese 8.1.79	Bantry Bay, Eire	Explosion during discharge of cargo and fire	20,000 t Saudi Arabian crude	Minor environmental damage. Problems in dealing with burned oil residues, 51 crew and shore staff killed	Bow section 20,000 t oil off-loaded. Wreck boomed, skimmers, absorbents, vacuum pumps. Heavy equipment, aerial spraying dispersants on shore	
Thuntank 3 30.1.79	Hatter Reef, Samsø, Denmark	Grounded	400 t heavy fuel oil	Many seabirds	Clearance hampered by sea ice and bad weather. Dispersants, absorbents	

Table 4-33

Oil spills at Milford Haven between 1963 and 1978 classified by frequency in each category (After DAVIES and co-authors, 1981b; reproduced by permission of the Milford Haven Conservancy Board)

		% of total spills
Tankers in passage and at moorings		3·0
Tankers	loading	17·4
	discharging	12·1
	ballasting	4·2
	de-ballasting	5·3
	hull defects (1972–1978)	15·2
	bunkering	9·5
	miscellaneous	12·9
Jetty spills		20·5

gravitational separation become progressively less efficient. Where the environs of an oil spill permit oil skimming with or without oil collection, the operation is relatively cheap and simple. The equipment involved is also simple and often readily available. Unfortunately, only very few marine spills occur under such protected conditions.

From such considerations most engineering developments in recent years have abandoned techniques for recovery based solely on gravitational separation. Furthermore, it is now well recognized by clean-up experts that booms have a specific role to play within limits dictated by wind, wave, and current.

Incidentally, if ice replaces water at the surface quite different criteria apply. Oil on top of a firm ice base is easy to recover; oil in quantity beneath solid or extensive ice cover can be manoeuvred into pools and channels for collection from equipment on the

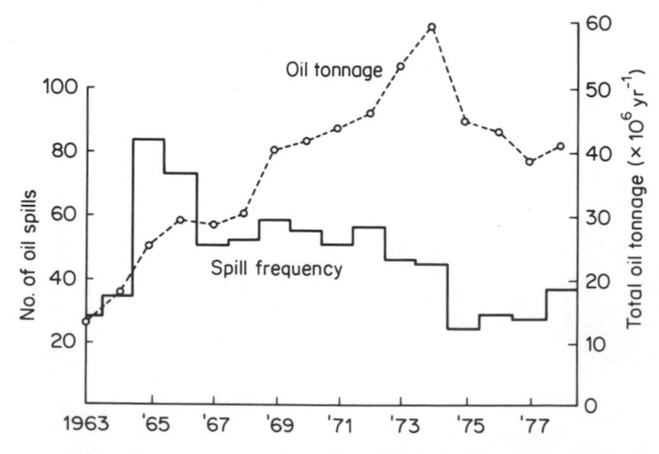

Fig. 4-20: Annual frequency of oil spills (1963–1978) in relation to tonnage of oil handled. (After DAVIES and co-authors, 1981b; reproduced by permission of the Milford Haven Conservancy Board, UK)

ice or from vessels. If ice and water are moving, or if the oil is in between heavy ice flows, clean-up may be impossible (ALLEN, 1979 and related papers in the same report).

When the dynamic relationships between oil and water are near or beyond the mixing threshold retrieval of oil from water must engage other oil characteristics to promote separation. Oil has a great propensity to cling to solid surfaces of virtually any element or composition, for example, as a lubricating layer. This characteristic can be exploited by a wide variety of mechanical contrivances usually aimed at presenting a large area for oil contact and shaped to enable the oiled surface to be wiped continuously and the oil or mousse collected.

Oleophilic materials have an enhanced capacity for collecting oil and various types of equipment have been devised to expose continuously such material in the form of rope or cloth to the oil layer which adheres and is transported on board a vessel for the enriched oily water to be squeezed out into a tank and the oil subsequently separated for reclamation.

Many substances absorb oil, particularly if unwetted. Straw and dried peat are suitable if the oil slick is continuous and not too thin. Synthetic foam materials can be oleophilic as well as absorbent and are superior. When such materials are applied loose, their subsequent recovery can cause problems and the disposal of oiled straw, peat or plastic sodden with sea water is difficult. On a small-scale absorbent, oleophilic materials held in long floating fabric or netting cylinders have proved successful. Sometimes these can be cleansed for reuse or refilled. Experiments are in progress to devise a floating rig functioning simultaneously as a sweeping boom and a collection net, to retrieve semi-solid waxy (paraffinic) oil, as for example is yielded by the Beatrice Field, Moray Firth, North Sea. A delicate balance has to be met between towing the rig as fast as possible to cover the widest area and maintaining the upper gathering edge just above the sea surface. Similarly, the speed through the water must not force the soft wax through the meshes.

The recovery of oil from beaches and shores generally has the highest priority by the public. For rocky and stony shores between low- and high-tide marks (and in bad weather often considerably beyond the high-tide mark) the arrival of oil is the start of a protracted process. Deposition on stone can be permanent (up to 8 yr have been recorded), especially if the stone becomes warm and dry after exposure to oil, enabling the oil to replace the film of moisture, evaporate, and adhere firmly. Stones covered in slimy seaweeds or microalgae may reject a coating of oil.

For sandy and pebbly beaches similar considerations hold. If the slope of the beach maintains mainly wet sand then oil penetration is slight, unless it is assisted by a considerable amount of wave energy. The coarser the beach, itself a sign of high wave energy, the greater the danger of oil penetrating it. Because the oil may subsequently associate with the sand grains it may not easily be dislodged. It is the oil that is thrown above the high-tide mark on any kind of shore that has the greatest chance of remaining there in gross and accumulating amount.

Muddy shores are invariably flat and protected. They are prone to casual invasion by vestigial oil slicks and if they are swamped by a massive slick of crude oil. They are particularly vulnerable if the mud can dry out even slightly. Soft muddy environments are probably too wet to become badly affected.

Public acclaim motivates the cleaning of beaches, not only holiday bathing and sun-bathing beaches, but compulsively also the more remote beaches visited regularly by no more than a few walkers, birdwatchers, and beachcombers. Not only sandy beaches are popular but frequently also pebble and stony beaches, and not too difficult rocky shores. To clean a beach to a standard above public criticism is a long, difficult, and costly operation. Not to clean up a holiday beach can also involve severe losses to the commercial holiday industry. Table 4-34 summarizes some of the more usual options and the incentives or reasons promoted as grounds for action. It is not too difficult to imagine how laborious and costly 'chemical restoration', as CANEVARI (1979) calls it, must be. To illustrate the point for a wider framework of oil spills Table 4-32 shows clean-up plus legal costs, actual or claimed. Clearly, money can be earned in oil spill clean-up. Records show that in relation to serious and major oil spills a staggering amount of oil-contaminated stones, pebbles, sand, soil and other beach materials (reckonable in hundreds of thousands or a million tons yr^{-1}) are moved for burial. Some shallow stony beaches have entirely disappeared, and other beach profiles have been measurably altered in the interests of good housekeeping. BELLIER and MASSART (1979, p. 141) concerning the *Amoco Cadiz* spill:

> 'The spill involved some 223,000 tons of oil spilled along 400 km of coast. Almost 10,000 persons worked on the project during the busiest period. As a result almost 200,000 tons of oil and debris were pumped and gathered. However, less than 20,000 tons of oil was finally retrieved after separation from the total mass of material obtained from the coastal zone.'

In theory, oil might be recoverable from some of these materials but most modern refineries have no access points to receive them. One could hope at best to extract only the most available portion for use as plant fuel. Methods are under development at the UK Department of Industry to cleanse beach materials with the double objectives of

Table 4-34

Some examples of shore oil pollution, clean-up options, and incentives for action (Original)

Impact of oil	Clean-up options	Incentives
Creation of oil-saturated sand and soil	Heavy equipment such as bulldozers for major incidents, manual clearance for minor incidents	Amenity reasons in tourist areas, prevent contamination of adjacent shores and property
Heavy oil on rocks and stony beaches	High power water hoses and steam lances	Remove threat of long-term contamination to adjacent amenity of sensitive areas
Oil contaminated salt marsh	Mechanical or manual cutting equipment	Reduce permanent damage to shore structure and threat to bird life
Fresh oil drifting onshore	Spray low toxicity dispersant at least 2 km offshore	Reduce shore impact, threats to seabirds, inshore fisheries and industrial plant
Sensitive shores/areas with no special features	Do nothing	Avoid damaging action/allow natural recovery

restoring the cleaned beach material and washing out, separating, and collecting the waste oil for recycling or incineration. The handling of separated but wet waste oil is much more akin to normal refinery processing than handling of low-oil solids.

(d) Impact of Clean-Up

Occasionally, the clamour of press and public for 'someone to do something' when an oil spill happens provokes otherwise responsible organizations to act hastily with less than due heed to the environmental consequences in an effort to appease the emotional outburst (e.g. COWELL and co-authors, 1979). The general public can see for themselves the damage to seabirds, the direct toxicity of oil to shore creatures, the physical coating of boats and piers, damage to sand, rock, rock pools, and so on. Since among these are the places in which familiar seaweeds and animals live, people can readily understand the disruption of food chains and the contamination of organisms, including those used as seafood. But it must be emphasized that this accounts for only part of the damage and part of the oil spilled, probably a minor part. The total harm is much more difficult for the ecologist to assess and for the non-expert to grasp.

Cycles of life in the sea and the process of oil pollution are both dynamic. Time is of paramount significance. Even if oil is physically removed, the soluble and dispersed parts continue to impose on or become incorporated into the food web. If oil is not removed but 'simply disappears' according to common parlance its influence is similar but more extensive. How much of the volatile fraction that goes into the air comes down again into the sea is not known. Use of dispersants accelerates solubilization and dispersal.

The time element in biodegradation in water is related to temperature and nutrient supply and, to a minor extent, the initial size of the population of oil degraders. In sediments it is dominated by oxygen and nutrient availability.

There is also a time element in the biota and in its latent rhythm. Sheltered communities which develop large plants such as the bigger seaweeds, shrubby tidal marshes, mangrove swamps, coral atolls, for example, have a long lifespan, high biomass and, if subjected to severe pollution, exhibit high vulnerability and slow recovery.

In less narrow and specialized niches, a mixed community will contain relatively long-lived species and others comparatively short-lived. Some of these species will be less able competitors. The effect of oil can then be to wipe out rare and endangered species. Elsewhere, oil can exert a selective action on the species most vulnerable because of feeding type or metabolic characteristics or reproductive pattern, and permit others more tolerant of oil to fill the gap. The selective action may be some extremely subtle behavioural change or impairment invoked by oil.

In the open sea, the microalgae and herbivorous plankton are usually replaceable rapidly by direct mixing of new stock or from regeneration or reproduction in less affected areas. Carnivores are more vulnerable as they are slower growing and may undergo a series of morphological transformations which can be very sensitive to oil. Highly mobile species may well avoid direct contact with oil and seek uncontaminated food species.

Hence the hidden effects of oil are much more diverse, complex, and differential than has ever been explained to the general public. When an oil spill strikes it may take an hour or two of careful thought to decide not only which of the available clean-up methods should be used in the various habitats but also what are the consequent damages from the material and equipment employed and the disturbance thus created.

It is possible to define and recognize in advance the main aspects of the varying shore habitats in the contingency planning (Table 4-35) but it nevertheless remains worth while to consult an expert with local knowledge of the communities at risk, not only in these easily recognizable places but also in the superficially homogeneous inshore and offshore waters. Because of their action, dispersants increase the water volume and seabed area over which the oil is spread. This contrasts with sinking agents which give rise to high local concentration on the sea floor.

Taking a long-term view the effects of optimum physical removal of oil from the sea can be expected to reduce the ultimate total impact of the spill by as much as one-third at most. Dispersants probably alter the ultimate distribution of oil between pelagic tar, biota, sediment, benthos, and water column, thus varying the total impact to which one must add their own impact in the water column. Sinking agents greatly concentrate oil on the bottom and this can either assist or inhibit decay, depending on the form of the aggregates. Inevitably the impact is different but not necessarily greater or less than no treatment. If no action is taken the spilled oil will gradually decay with the most resistant fractions perpetuating for 1 or a few years in the biota and on the seabed. If containment or diversion booms are used these have essentially localizing or neutral effect.

The firm conclusion, therefore, is that after the initial clamour has died down there are few sound environmental reasons for clean-up treatment at great expense. Of the motives for treatment, conservation of seabirds and preventing the further fouling of holiday beaches are the commonest. Any beach with seagulls within a 2-h journey of a major centre such as London is politically sacrosanct. Effective remedial action for really valuable wildlife sanctuaries, marine fish and shellfish farms, mangrove swamps and salt marshes almost has to be inbuilt to be timely, adequate and, just as important, to avoid serious damage due to the operations.

An analogy with a Fire Service has often been drawn. To be accurate, this analogy should also envisage the provision of fireguards, fire extinguishers, and fire escapes—inbuilt oil spill remedial measures for valuable sites—and techniques for fighting prairie and forest fires should have their marine equivalents.

(e) New Ideas on Oil Spill Monitoring

The *Argo Merchant* spill report represents the first breakthrough from established programmes based on the often inconclusive biology-oriented study of population change in plankton and benthos to a much more interdisciplinary approach. Progress of the slick of No. 6 Fuel Oil was mapped frequently from Day 2 to Day 13 and the residual fragmented slick was marked by a satellite-trackable buoy 4 d later. The buoy track was plotted for the subsequent 25 d by which time surface drift had moved it 200 miles west of the Azores Islands (MATTSON, 1978). Progress was also made towards determining subsurface oil drift. Petroleum hydrocarbon concentrations marked a high peak of 340 $\mu g\ l^{-1}$ in late December (*ca.* 7 to 9 d after initial impact) falling to less than 20 $\mu g\ l^{-1}$ in January (*ca.* 20 d) and February (*ca.* 50 d). In more remote areas, however, concentrations ranging from 10 to 100 $\mu g\ l^{-1}$ were found in February falling to 1 to 50 $\mu g\ l^{-1}$ in May and 1 to 20 $\mu g\ l^{-1}$ in August.

Table 4-35

Shoreline habitats, oil persistence and notes on clean-up; based on *Amoco Cadiz* spill experience (After GUNDLACH and co-authors, 1981; modified: incorporating the shoreline index terms of HAYES and co-authors, 1980; reproduced by permission of the American Petroleum Institute)

Sensitivity index value and shoreline type	Comments (duration of pollution)	Observed clean-up
1 Exposed rocky headlands	Composed of bedrock with high impinging wave activity; wave reflection kept most of the oil offshore; no clean-up was needed (d or wk).	Difficult access; natural processes sufficient
2 Eroding wave-cut platforms	No good example of oil interaction.	Usually difficult access.
3 Fine-grained sand beaches	Exposed to moderate-to-high wave energy; little penetration into the beach because of compact sand; thin buried layers commonly persisted in depositional areas (months to 1 yr).	Easy access; can be cleaned mechanically; buried layers difficult to remove.
4 Coarse-grained sand beaches	Common in semi-sheltered area in Brittany; greater penetration of oil due to coarser substrate; buried oil common (1 to 2 yr).	Easy access; sand removal may cause beach erosion; difficult to use mechanical means.
5 Mixed sand and gravel beaches	Found within some sheltered areas of Brittany; an asphalt pavement formed in some low energy areas of oil deposition (1 to 2 yr; more in sheltered areas).	Easy access; generally hard surface permitted some clean-up of surface oil; high-pressure hosing without sediment removal recommended.
6 Gravel beaches	Showed rapid and deep penetration of oil (1 to 2 yr).	Generally easy access; removal of sediment not recommended; high-pressure spraying with mechanical re-working of sediment into surf zone proved most effective.
7 Exposed, compacted tidal flats (moderate to high biomass)	Oil moved rapidly over flat surface and deposited along swashline; varied biological impact: in productive areas, impact was severe (months to 1 yr, oil as sheen evident after 2 yr).	Easy access; compact flats cleaned easily mechanically; trenches as part of clean-up may have caused increased oiling of interstitial water (visible after 2 yr).
8 Sheltered rocky shores	Oil sticks to rocky surfaces; pools of oil between rocks eventually turned to asphalt (up to 5 yr, but most obvious oil effects gone after 2 yr.	Access varies, but is often difficult: high-pressure spraying removed algae and organisms as well as the oil; low-pressure washing as oil comes onshore may be less damaging biologically.
9 Sheltered tidal flats	In areas of low wave energy, oil persisted on surface as mixed oil and sediment patches; contamination of interstitial water persisted even if surface was cleaned (more than 5 yr).	Access difficult on soft flats; clean-up very difficult and usually not effective; heavy machinery mixed oil into sediment.
10 Marshes	Oil pooled on surface of marsh, killing most flora and fauna. Oil still obvious 2 yr after spill (5 to 10 or possibly more yr).	Access varies; heavy equipment destroyed vegetation and natural drainage patterns; manual clean-up not very effective, but necessary in heavily oiled areas.

From December 1976 to February 1977 the shallow turbulent waters over the Nantucket shoals is well mixed and the prevailing petrogenic hydrocarbon concentration (particulate and dissolved) is about 20 $\mu g\ l^{-1}$ on the basis of Table 11 of BOEHM and co-authors (1979).

THURBERG and co-authors (1978) describe their studies on molluscs and fish following the spill of the *Argo Merchant* cargo of No. 6 Fuel Oil. Gill-tissue oxygen consumption rate of scallops from the spill area was 594 $\mu l\ h^{-1}\ g^{-1}$ for samples collected between Days 21 and 27. Scallops from an adjacent unimpacted area gave 675 $\mu l\ h^{-1}\ g^{-1}$; in February (*ca*. Day 50) scallops from previous contaminated and control areas were alike at 651 $\mu g\ h^{-1}\ g^{-1}$ and 654 $\mu g\ h^{-1}\ g^{-1}$. January samples of only 2 mussels from the affected area gave 451 $\mu g\ h^{-1}\ g^{-1}$ compared to 596 $\mu g\ h^{-1}\ g^{-1}$ for 6 clean mussels. Serum samples from scallops collected in February showed sodium and calcium ion concentrations to be elevated, the potassium osmolality remaining unaffected.

Malate dehydrogenase activity was significantly depressed in scallop muscle, lactate oxidation was lower, pyruvate reduction activity unchanged and malic enzyme activity depressed; all relating to sets of animals deep-frozen at sea. Scallops brought back live (*ca*. 3 d) and held in clean flowing sea water at least partially assumed clean condition, though lactate oxidation rate remained 20-fold higher than in controls, and pyruvate activity was depressed.

Biochemical function in scallops, flatfish (*Limanda ferruginea, Pseudopleuronectes americanus*), clupeids (*Pomolobus pseudoharengus* and *P. aestivalis*), and among the gadoids, cod (*Gadus morhua*) and haddock (*Melanogrammus aeglefinus*) from the spill area (February) was compared to individuals of the same species from outside the area (see original for sampling details).

Table 4-36

Serum measurements on blood samples from pelagic and demersal fish from clean areas and areas impacted by the *Argo Merchant* oil spill (Based on THURBERG and co-authors, 1978 (Tables 4, 5, and 6))

Fish serum	C/O	n	Osmolality (m Osm kg^{-1})	Na	K	Ca
				(meq l^{-1})		
Herring	C	9	530	245	7·44	6·51
Clupea harengus	O	7	555	241	9·45	6·66
	P		NS	NS	NS	NS
Alewife	C	6	527	230	9·24	ND
Pomolobus pseudoharengus	O	9	497	228	7·97	ND
	P		NS	NS	NS	—
Haddock	C	6	453 ± 31	200 ± 15	8·31 ± 1·20	3·92 ± 0·92
Melanogrammus aeglefinus	O	14	385 ± 5	179 ± 3	7·36 ± 0·38	3·83 ± 0·50
	P		<0·05	<0·05	<0·05	NS
Yellowtail flounder	C	7	473 ± 34	208 ± 12	7·26 ± 1·10	6·30 ± 0·38
Limanda ferruginea	O	17	425 ± 11	193 ± 4	7·07 ± 0·73	5·97 ± 0·30
	P		<0·05	<0·05	NS	NS
Winter flounder	C	5	ND	193 ± 7	7·08 ± 0·41	5·74 ± 0·40
Pseudopleuronectes americanus	O	5	ND	205 ± 2	3·97 ± 0·13	5·00 ± 0·38
	P		—	NS	<0·001	NS

C, Clean area; O, Oil-impacted area; n, Number of specimens; P, Probability;
NS, Not significant; ND, Not determined.

Table 4-37

Aryl hydrocarbon hydroxylase (AHH) activities in samples of cod, whiting, and haddock from trawls made in close proximity to the North Sea oil fields; FRV *Scotia* cruise—29.10.80 to 7.11.80 (After DAVIES and co-authors, 1981b; reproduced by permission of the authors)

Species	Trawl 1 Cormorant field	Trawl 2 Brent field	Trawl 3 Beryl field	Trawl 4 Forties field	Trawl 5 Baseline position 20 miles west of Forties
Cod	173·7	127·8	65·9	80·0	56·9
		109·4	119·6	51·2	52·4
		125·6	92·8	72·8	57·6
		81·7	81·2	53·5	44·1
		95·0	54·3	75·3	44·9
Mean AHH value	173·7	107·9	82·8	66·5	51·2
Whiting	98·8	129·8	83·8	113·6	5·9
	114·9	265·2	87·8	83·6	56·4
		123·0	233·8	241·0	102·0
			142·7	109·1	37·6
			85·9	125·0	70·7
Mean AHH value	106·8	172·7	126·8	134·4	54·5
Haddock	528·2	357·0	205·2	181·6	255·5
	194·2	134·0	294·7	287·1	76·1
	331·1		440·2	275·7	247·7
	225·0			303·1	172·9
	378·9			317·7	108·9
Mean AHH value	331·5	245·5	313·4	273·0	172·2

Serum measurements showed no significant differences for the clupeids. For the other groups any change of variables measured was a consistent (but not significant) trend to lower osmolality levels (Table 4-36). THURBERG and co-authors (1978) regard the scarcity of sample materials available as a severe limitation to establishing the statistical validity of their unique measurements. A much wider data base is needed to determine the range of spatial and seasonal variability in uncontaminated animals.

The potential value of biochemical characteristics signalling sublethal responses to very low levels of oil in the open ocean is suggested by this work. It has been positively demonstrated in the enhanced levels of aryl hydrocarbon hydroxylase (Table 4-37) in 3 commerical fish species (cod *Gadus morhua;* whiting *Merlangius merlangus;* haddock *Melanogrammus aeglefinus*) taken by trawl in close proximity to oily water discharges from the Cormorant, Brent, Beryl, and Forties oilfields in the North Sea (DAVIES and co-authors, 1981b).

(10) Hydrocarbons and Human Health

There are 2 good reasons for considering here hydrocarbons in relation to human health. Firstly, oil pollution is a global threat to which man is possibly more exposed and more vulnerable than marine life in the sea. Secondly, the response of the human body to hydrocarbon exposures may help our understanding of the response of perhaps seabirds and marine mammals, if not marine organisms more generally.

Considering man and marine animals, hydrocarbon intake by ingestion is essentially comparable but it is the lung function in man that matches gill function in most of the larger marine animals, e.g. fish.

(a) Background Exposures

Petroleum is a characteristic global contaminant of the past century with dramatically increased inputs of hydrocarbons and combustion products. A few aspects are well known, especially the hazards of the internal combustion engine, but broader historical appraisal of the enhanced background and its effects on health are generally not known.

Few people have ever thought about the hydrocarbons they encounter in everyday life. Hydrocarbons and their combustion products are everywhere. Vapours and combustion products are in the air we breathe, and hydrocarbons are integral components of crops, animals raised for food or captured; this food may be further contaminated by storage, transport, processing, marketing, cooking, sources of heat, additives, wrappings and so on. Hydrocarbons are in rainwater supplies and in drinking water made from them; hence they are in the diet, on the skin and on our clothes. Additional sources are in cosmetics, polishes, printing inks, solvents, paints, fabric finishes, and many do-it-yourself sport, motoring, and domestic items. It must be about the most difficult thing in the world to avoid petrogenic hydrocarbons, and totally impossible to avoid natural hydrocarbons. It is fortunate, therefore, that allergy to background hydrocarbons is a very rare disorder.

Medical response to background hydrocarbons has been highly selective, almost entirely concerned with a few special cases. Regarding air pollution, traffic fumes and smog have commanded attention because of their occasional or local acute effects. Traffic fumes can contain high lead levels, carbon monoxide, and residual hydrocarbons including benz [α] pyrene (B[α]P) which give rise to a complex of health hazards most evident in busy urban centres. Smog is a chemically modified form of vehicular pollution combined with other emissions which is familiar in some southwestern parts of USA and in Mexico. Smog has a range of effects on man, from eye irritation and breathing difficulties experienced by a healthy person to severe or even fatal respiratory problems for the asthmatic, sick or aged. Forests and food crops can also be affected.

Not all countries recognize a threat to health from hydrocarbons in water supplies. It is difficult to recognize a threat from epidemiological studies on drinking water since oil is transient and complex. Recently, the European Community (EEC) has identified a need for standards to be set on raw water supplies and on drinking water. Both total hydrocarbons and polycyclic aromatic hydrocarbons (PCAH) are considered. The limit for hydrocarbon taint on drinking water is not greater than 10 $\mu g\ l^{-1}$. Further numerical values from the Directives are listed inn Table 4-38. Unfortunately the chemical specification of the hydrocarbons to be measured betrays a half-hearted approach to the enterprise. The recommended chemical methodology for dissolved or emulsified hydrocarbons is non-specific regarding the contribution of groups such as alkanes, aromatics or PCAH and probably bears little relation to taint and no relationship to health hazard. The methodology for PCAH is somewhat more specific but it does not venture to get to grips with the real health problem by avoiding strict identification and quantifi-

Table 4-38

Standards for hydrocarbons*; (a) surface water intended for the abstraction of drinking water; (b) drinking water. All units are $\mu g\ l^{-1}$; GL: guideline level; MAC: maximum admissible concentration (Based on EEC Directives 75/440/EEC, 16 June, 1975 and 80/778/EEC, 15 July, 1980)

	A 1	A 2	A 3
(a) Water for abstraction			
Dissolved or emulsified hydrocarbons (after extraction by petroleum ether)	50	200	1000
Polycyclic aromatic hydrocarbons	0·2	0·2	1·0
(b) Drinking water	GL		MAC
Dissolved or emulsified hydrocarbons (after extraction by petroleum ether)	Not specified		10
Polycyclic aromatic hydrocarbons	Not specified		0·2

* Suggested treatment options (for details consult original)

A 1: to be treated by rapid filtration and disinfection;

A 2: to be treated by pre-chlorination, coagulation, flocculation, decantation, filtration, and disinfection;

A 3: to be treated by chlorination to breakpoint, coagulation, flocculation, decantation, filtration, adsorption on activated carbon, disinfection.

cation of recognized carcinogens. No attempt is made to categorize in any way the petrogenic and biogenic components.

In medical health terms, the monitoring exercise is purely political and of no positive value. It is seldom publicized that 'food grade mineral hydrocarbons' are customary additives (adulterants) to a range of foodstuffs commonly and widely used, such as oranges and other citrus fruits, dried fruits, confectionery, chewing gum (60%), rind of processed cheese, and preserved eggs.

Sometimes these hydrocarbons are rejected in preparing the food (on skins, shells, etc.) but for other items the hydrocarbons are consumed. One of the major sources of B[α]P is medicinal paraffin if used habitually as a laxative. There are precise standards for PCAHs in food grade mineral hydrocarbon but nevertheless these additives are potentially capable of exceeding all other normal dietary PCAH intakes in prepared natural foodstuffs. (Based on UK Statutory Instrument (1966) No. 1073 Food and Drugs, Composition. The Mineral Hydrocarbons in Food Regulation, 1966.)

Few countries have legislation relating to hydrocarbon contamination of food and foodstuffs. An often quoted guideline of uncertain authority is that food that is not tainted with oil to the extent that it is unacceptable is not likely to be harmful. Tainting of food has threshold limits for perception and for rejection as unacceptable which vary with the individual and can be developed with training. Oily taint varies in intensity with the nature of the hydrocarbon and it is held that certain sulphur- and nitrogen-containing hydrocarbons have the highest flavour or smell. For crude oils the lower limit for a detectable taint is in the region of 10 to 50 mg kg^{-1} fresh weight fish: the upper limit of acceptability has not been studied but is probably below 1 g kg^{-1}. Most reports of unacceptable taint in marine fishes relate to the mullet which is characteristically a harbour and inshore species living near the coastline, and sprats and herring which have a particularly high lipid content and turnover. All these species feed near the surface. In addition to taint acquired through food organisms and water, fish of all species can also become tainted in the event of an oil spill from contact with an oil slick or oiled fishing gear, or on the deck of the vessel.

(b) Carcinogens from Marine Sources

The influence on man of the hydrocarbons in sea water and in seafoods has conventionally been focused entirely on carcinogens, usually B[α]P. We do not know if this is an adequate basis of assessment. CHANDLER (1973) estimated that some 7500 kg of B[α]P was discharged into the sea annually from petrogenic sources but WILSON (1974) put this figure nearer 750 tons. Neither figure includes the potential carcinogenic input from marine sources and the processes of hydrocarbon decay and rearrangement, nor do they take into account the breakdown and eventual mineralization of carcinogens which must also take place. Polycyclic aromatic hydrocarbons have been identified in various shellfish and can be expected to occur more widely; the concentrations found or expected are little different from those in foodstuffs from land or fresh water (KING 1977). There are indications that PCAH may be substantially enhanced in shellfish from areas chronically exposed to oil pollution by discharges and dumping. Apart from known areas of oil pollution there is considerable uncertainty about validity of many of the published analyses for B[α]P and other PCAH (BLUMER, 1971). The uncertainties stem both from lack of specificity and inadequate quantification in many analyses using older techniques and impure reference materials. As a result much is unclear not only about B[α]P and PCAH but also the role of long-lived hydrocarbons and residues generally. This uncertainty cannot be resolved without dedicated research using the very best analytical, biochemical, and physiological techniques to determine not only occurrence and risk to man but also turnover and the influence of pulses of high hydrocarbon input from oil spills.

Apart from rather vague associations between neoplasms, carcinomas, and more general pathobiological symptoms in fish and shellfish and the demonstrable presence of domestic and industrial pollution, there is little firm evidence on specific causal effects attributable to hydrocarbons in the sea or the occasion of crude-oil spills. Studies on areas of oily water input, natural oil seeps, and refinery effluents have not convincingly established enhanced occurrence of neoplastic diseases.

Most authorities agree that one should avoid all contacts with enhanced levels of benzpyrene and other carcinogens and that responsible authorities should take steps to control, and wherever possible reduce, entry of polycyclic aromatic hydrocarbons into the environment including the marine environment.

(c) High-Level Exposure

Hydrocarbons in wide variety are encountered in work situations such as in the oil industry, engineering, vehicular transport, painting and paint manufacture, industrial cleaning, etc., and as lubricants and heating fuels in many workplaces.

Effects on man depend on the frequency, degree, and manner of exposure to vapours and liquids. Symptoms range from mild temporary effects to permanent disablement. Fatalities can arise in severe cases from both acute and chronic exposures. There is ample experience on the toxicology of the common products kerosene, mineral oils, diesel oil, naphtha, petroleum ether and paint thinners. Such liquid hydrocarbons are fat solvents and alter the function of nerves to produce depression, coma, and sometimes convulsions. Direct aspiration into the lungs during ingestion causes pulmonary irritation as does aspiration during vomiting. Acute poisoning gives rise to pulmonary oedema, bronchial pneumonia, gastrointestinal irritation and other conditions.

Prolonged inhalation of high concentrations causes degenerative changes in the liver and kidneys and hyperplasia of the bone marrow.

It was reported in the local press and by witnesses that people living in the near vicinity of the *Amoco Cadiz* oil spill and those involved in clean-up operations experienced dizziness, headache, and nausea during the early stages of the spill.

Benzene, toluene, and xylene are common ingredients of adhesives and plastic cements, and toluene is the usual solvent used for glue sniffing. The toxicity of benzene for human beings is around $0{\cdot}2$ g kg^{-1} body weight and for toluene and xylene $0{\cdot}5$ to 1 g kg^{-1}. In large amounts, aromatic hydrocarbons depress the central nervous system and on repeated exposures depress bone marrow functions.

The effects of single acute doses from glue sniffing, inhalation or ingestion of hydrocarbons by accident or design, range from dizziness, euphoria, nausea and staggering as a result of mild exposure to tremors, paralysis, violent excitement or delirium and unconsciousness in severe cases and kidney damage may result.

Chronic exposure usually in a work situation has similar initial effects leading in time to anaemia, petechiae, and abnormal bleeding. Chronic exposure to benzene can lead to anaemia and complete ataxia of the bone marrow.

The fatal dose of naphthalene is 2 g kg^{-1}; less for young children. Naphthalene can cause haemolysis and kidney damage for severe exposures; the earlier symptoms include nausea, vomiting, oliguria, coma, and convulsions (BREISBACH, 1980).

Thus the response of man to high levels is not unlike the responses of seabirds, mammals and other animals of the sea.

The pathways of hydrocarbons, including mode of entry and transport, between organs and tissues have many similarities and the underlying biochemistry has much in common. Thus in looking at hydrocarbons and human health it has been possible only to confirm that our understanding of the effects of background exposure to man and to the marine biota in their separate domains leave much unexplored; but with regard to higher levels of acute or chronic exposure the findings are mutually reinforcing to a reassuring degree.

(11) The Costs of Oil Pollution

(a) Damage

There are the unknown true environmental costs of an oil spill and the legal assessments of damage which can be categorized and translated into financial terms in countless different ways; with fingers crossed they might sometimes be equal. More often than not these assessments are based on philosophical decisions such as : 'Who owns that oiled seabird?' and value decisions such as 'What is its market value?' Damage valuation varies greatly depending on standpoint and self-interest, and sometimes, quite blatantly, 'How much can I make out of it?'

In the limited realm of marine oil pollution the following list picked out of one publication (API, 1979) cannot claim to be exhaustive. Attributable damages in relation to oil spills have been considered in relation to legal expenses, research expenditure, fish and shellfish farming, commercial fisheries, human health, agriculture, seaweeds, all kinds of marine organisms from worms to mangrove trees, commercial and pleasure craft, docks, wharves, public and private properties, sand dunes, beach sand, driftwood, tourist industry, sport fishing, birdwatching, and clean-up costs. Someone forgot the cost of the tanker and its cargo. In addition there are the intangible costs of, say, the drop in the market value of fish from a certain area because the catch may be tainted or 'unhealthy' due to an oil spill.

Not surprisingly, all sorts of legislation have been concocted to attempt to formulate rules for assessing damage, and some of these have led to quite farcical implementation. Some impression of the various costs to oil polluters, couched in a variety of terminology are shown in Table 4-32 already presented. How these costs were arrived at would fill a book. The schemes suggested range from tacit agreement, various forms of licence or assurance, 'dollar-per gallon-rough justice' and the meticulous counting of oily corpses of marine animals all double checked for identity and priced individually male and female, juvenile and adult with and without eggs.

The latter fully elaborated system is at present in someone's pending tray but the State of California has succeeded in getting close to this system. According to DUBIEL (1979, who may be joking) the State legislation allows for punitive damages for the wilful destruction of animals and damages to the personal property of the State. Because of the amount of work in assessing each individual oil spill in its own unique habitat, complete lists of animals in the inshore environment have been developed. The prices are kept current and reflect the present state of inflation. For one oil spill this price list was a book 2·5 cm thick. The price list has evolved from the dealer catalogues for biological specimens (live or dead, not stated). When a spill occurs dozens of biologists comb the area to count the dead and dying organisms and on the basis of these samples and baseline populations each species is costed individually according to the estimate of the number killed. To this amount is added the numbers removed or killed in carrying away oiled sand or debris or other clean-up side effects and also the psychic value of those losses not immediately quantifiable. Sand (in 1978) was charged at $11 per cubic yard, and driftwood (priced as firewood) at $35 per cord. Damage to birds was found to be difficult to assess as stuffed specimens only had been marketed.

Add to this an injunctive action for replacement of all damaged animals so that if the polluter contests the cost of an animal he can elect to replace it. Presumably the State of California will use the damages to purchase specimens and replace them in their habitat. It has been found difficult to replace barnacles and limpets.

In the case of the *Zoe Colocotroni* oil spill, settled in Puerto Rico in 1978, on this basis, the cost for damage to animals amounted to $5,526,583. In the Santa Barbara oil spill, sand and driftwood were actually costed with extra charges for transportation and replacement. Counts were made of many animals removed in the oily beach material including sand crabs from which totals were estimated and charged accordingly. In all, the documents submitted as prosecution evidence totalled 10,000 for this case.

The putative reasoning behind this crazy method of assessing damages and even crazier idea of restoration is that only a method established on real terms is acceptable. The US legal tradition is that damages, including environmental injury, be gauged in dollars and cents whether or not this makes sense.

Belatedly top officials are having second thoughts on these principles. Nevertheless DU BEY and FIDELL (1979), in seeking out alternatives, persistently maintain that baseline data can be established in advance of spills and transformed into an account that can be discussed by a forum and shown to the public. A reliable set of baseline data is regarded as paramount to estimating real losses of each species to which values can be attached. Many others outline systems of socio-economic and environmental damages assessment relying on baselines.

It is somewhat discouraging that in nearly 100 yr of dedicated research into estimating commercial fish stocks and forecasting year-to-year recruitment into the fisheries, European scientists have often failed to predict correctly whether stocks would increase or decrease in the North Sea and related waters, and are reluctant to give an estimate of the error even by the best of the current methods of assessment (see also Volume V: GULLAND, 1983). The error range in terms of quantity of fish would undoubtedly be vastly greater than the damage done by all the oil spills ever recorded for European waters.

The relative errors for stocks of crabs, lobsters, and all kinds of molluscs would be at least as large. Lawyers and administrators cannot believe that plankton nets are highly variable in catch efficiency and at best are less than 20% efficient; trawls vary from less than 1% to perhaps 30% efficient according to species encountered, grabs and corers require 5 or 10 replicates to get even a vague estimate of the number of species present and their abundance.

On this basis it is fair to dismiss as fantasy all assessment methods based on alleged baseline surveys. If further discussion is needed, consider what is known about the behaviour of oil, the reaction of animals to oil, the dynamics of oil decay, and the natural perpetuation of marine species. If an oil spill is to be evaluated as a static event, when exactly is the moment when assessment must be made? If it is a dynamic event to be integrated to infinity, how do you account for the dispersal of oil, the sum of its effects over its life time relative to the movement of populations, seasonal effects, natural mortalities, predation, and other factors?

Whether or not the lawyers are satisfied with the Alaska 'Dollar-per-Gallon Act' it has the attributes of ease and speed. It also may be associated with 2 of the main findings in oil pollution: 'oil hurts most where it is least dispersed' and 'the more oil the more damage'. The scale of damages ranges from $1 per gallon for unconfined salt water, to $2·50 for estuaries and intertidal environments and $10 for fresh water with identifiable resources; these are maximum rates. It would not be difficult to elaborate a little on this skeleton system by including separate charges for habitat types such as salt marsh, sand, rock, etc., and to add a little more flexibility. The Alaska system avoids the situation where the costs of litigation and 'research' amount to more than the damages caused by the spill.

REED and SPAULDING (1979) have attempted to construct a computer model for the interactions between an oil spill, its predicted behaviour and fate and the cod fishery (especially egg survival) on Georges Bank. In the model only direct effects of oil on egg development and larval mortality are envisaged. It is well recognized that the success of brood renewal up to this stage is only a small factor in determining survival of recruitment to the commercial fishery.

A major part of the model relates to simulating the behaviour and fate of oil and simultaneously a model of the fish stock, spawning and the advection and diffusion of the larvae. The results include predicted effects of the cloud of oil when it is over the spawning area. The extent of interaction depends on the relative siting of oil spill and spawning and the effect of wind in promoting differential advection and vertical mixing. The introduction of oil dispersant interaction could not be adequately resolved with this model.

Again the real world distribution of oil and eggs/larvae would not be as smooth as in the model and the interaction of patches of oil and patches of larvae could not be assessed. Hence if the model (of field sampling) had been used as a basis for assessing damages a considerable number of large errors could not have been eliminated.

The Organization for Economic Cooperation and Development (OECD) has attempted to (i) appraise the cost of making structural modifications to part of the existing world tanker fleet in order to eliminate oil discharge from tank washing; (ii) estimate the resulting reduction in oily water discharges; (iii) assess the likely environmental benefit in relation to amenity and the short-term and long-term effects on marine life, including commercial demersal fish species (JUHASZ, 1979). JUHASZ found that oil pollution due to oil tankers in 3 selected areas (New York Bight, southwest coast of Wales, northeast coast of Scotland) had no effect on shore amenity; in consequence, any reduction could have no detectable effect.

Using average demersal fish landings by UK vessels for an area of the northern North Sea and a model of North Sea oil tanker traffic growth, with and without the introduction of segregated ballast tankers, computer simulations indicated no direct hazard to fish or via plankton and the food chain. JUHASZ qualified this conclusion by emphasizing the inadequacy of knowledge of short-term hydrocarbon effects on aquatic organisms combined with uncertainty in the predicted oil concentrations. Considerable doubt also attached to the adaptability of organisms to sustained enhanced exposure. There were great errors in predicting the severity of beach pollution. The overall conclusion was that the costs of modification to the tankers would not be offset by benefits to the environment. However, from the magnitudes of the predicted reductions in oil loss, the study did suggest that major oil spills might product effects of significant economic importance.

(b) Costs to Fishermen and Fisheries

Oil spills vary greatly in frequency and size. This was taken into account by JOHNSTON (1977) in his model of the average annual cost of North Sea oil spills to North Sea fisheries. The costings (for 1976) are based on an annual input of 15,000 tons. Short-term interaction is regarded under 3 sub-heads: (i) toxicity due to fresh crude oil; (ii) inhibition of phytoplankton photosynthesis by volatile fractions (model: pentane); (iii) inhibition of phytoplankton photosynthesis by less volatile fractions (model: octane); (iv) mortalities of fish eggs and larvae.

The model substances pentane and octane were chosen on the basis of known n-alkane occurrence and toxicity; more toxic models such as benzene and methyl naphthalenes, now established as more important pollutants, would lead to higher costs to the environment. Zooplankton findings based on enclosed ecosystem experiments would also increase this item in the estimate.

Long-term interactions are many: smothering, attachment to oil globules, lethal contamination, damage to growth and reproduction are all regarded as slow processes in which oil is carried to the bottom. An arbitrary low rate was adopted for this collection of processes.

Table 4-39

Cost of model North Sea oil spills to fisheries and fishermen. F: number of spills per year (After JOHNSTON*, 1977; reproduced by permission of the author)

		Loss as equivalent fish (tons yr^{-1})				
Spill size (tons)	F	(i)	(ii)	(iii)	(iv)	(v)
2·5	100	0·001	0·001	3·1	0·03	4
50	10	0·002	0·11	6·2	0·40	8
1000	1	0·004	1·7	12·5	1·91	17
10,000	0·2	0·009	4·0	24·9	9·3	33
100,000	0·04	0·018	8	50	25·8	66
400,000	0·002	0·035	16	100	59·7	131
Mean 15,000	111·242	0·069	29·8	195·7	97	259

* Consult this paper for full details.
(i) Equivalent fish loss due to mortalities caused by fresh oil.
(ii) Equivalent fish loss due to short-term productivity loss.
(iii) Equivalent fish loss due to protracted productivity loss.
(iv) Equivalent fish loss due to losses of fish eggs and larvae.
(v) Equivalent fish loss due to multiple long-term effects.
Annual mean total fish equivalent = 581·5 tons
1976 value = £71,757
Hence average environmental cost is £5 per ton of oil spilled.
In disaster year when 400,000 tons + mean is spilled, the weight of equivalent fish lost is 15,328·5 tons.
1976 value = £1·9 million.

The model is simple and utilizes values available at that time for the toxicity of oil and the inhibition of $^{14}CO_2$ uptake by natural phytoplankton by pentane and octane. Losses of phytoplankton and zooplankton were converted into losses of fish using conventional factors and costed as average commercial species.

The calculations have been reorganized from the orginal in Table 4-39 and have been independently checked. The equivalent fish values for 1981 would be at least double those quoted.

The overall conclusion is that direct mortalities due to oil and effects caused by volatile fractions are negligible. The more persistent fractions and any effects on zooplankton appear as an appreciable loss of equivalent fish. It is important to stress that losses of fish eggs and larvae do not give rise to numerically equivalent losses of 1-year-old recruits to the subsequent fish stock. No good relationship has been established between spawning numbers for a given species and the recruits. This model relies mainly on food-web relationships which may be more representative because they are less seasonally and locally variable.

In the average year, 15,000 tons of oil would be spilled causing the loss of 581·5 tons of equivalent fish valued at £71,757 or approximately £5 per ton of oil spilled. In the year in which a 400,000-ton oil spill was additional to the usual spills, the loss of equivalent fish would be 15,300 tons valued at £1·9 million also approximately £5 per ton of oil spilled. These 'rough justice' rates work out at about $2 per barrel, compared with the Alaska rates of $1 a gallon. Considered relative to the total value of all North Sea fisheries the average 'damages' would be negligible and 'disaster damages' only a fraction of 1%. If fishing were forbidden in areas of oil slicks for the period they persisted, it is possible to calculate the approximate average loss of earnings which turns out to be £28,000 y^{-1} (1976).

None of these estimates allows for mitigating processes which would greatly reduce these calculated losses. JOHNSTON (1977) goes on to evaluate biodegradation (summarized in Table 4-30C) but does not attach financial values. He concludes with an appraisal of other costs of North Sea oil operations to North Sea fishermen.

There are endless possibilities of modelling the costs of marine oil pollution. There are also endless ways of using the concept of damages to punish the offender, but any hopeful schemes to tax oil operators to recoup unattributable damages would be of negligible benefit to the multinational fishermen affected and would inevitably and needlessly increase the costs of oil fuels to the general public. None of the money could effectively go to replacing the losses of marine life.

(12) Science and Management of Oil Pollution—A Concluding Discussion

An interpretive review should not only assemble the best of the information but also get the best out of it and identify trends, pressures, new openings, and future scope. To use scientific evidence as the basis for prevention, control, and alleviation of oil pollution fulfils this aim with a managerial slant.

Oil pollution control in every country has been adopted as a subject of central government policy. Consequently, decision based on political dictat, national managerial strategy, by default or whatever, is the ultimate arbiter of attitude and response. Today's ministers are often credited with the blame for all related shortcomings no matter how historical. It is more realistic for industrial and environmental managers and scientists to accept this blame since, in general, they already have the continuing responsibility and the capacity to rectify shortcomings by their own efforts and actions within existing remits. Often, effort rather than more legislation and more money is needed to bring about the very many improvements needed.

In oil pollution, responsibilities, control measures, data acquisition, and understanding are all complex. Progress will need close cooperation and mutual understanding between managers and scientists dealing with the many relevant aspects.

(a) Background on Global Oil Pollution

One might conclude from people's widespread lack of concern about low-level global oil pollution, continuously replenished, that there are no effects and no hazards. It is a conclusion that has never been critically put to test.

Some of the arguments in support might be:

(i) Many and varied natural hydrocarbons and their breakdown products and residues exist everywhere and life has evolved in their presence over countless years.

(ii) Environmental hydrocarbon concentrations are much lower than those known to have demonstrable biological effect.

(iii) Man's health over more than 50 yr of this petroleum epoch has suffered no attributable reduction due to oil pollution except in the restricted realms of occupational exposure and very local effects such as smog.

Against these arguments may be set:

(i) Petroleum hydrocarbons and their associated products of manufacture, use, and oxidation are not identical with ambient natural hydrocarbons.

(ii) Where petroleum and biogenic hydrocarbons are alike, the introduced material enhances the background level locally and perhaps globally.

(iii) Hydrocarbons, especially their used and combustion products, increase the pool of polycyclic aromatic compounds some of which are carcinogenic and, according to some experts, there is no safe level for carcinogens.

Clearly, there is scope for debate and so far no competent group has been assigned the task of critically examining the evidence and devising tests to settle the issues.

Some of the responsibility for background hydrocarbon pollution as it affects everyone is by popular choice and some is personal decision. Consider the many modes of exposure. For many people, the dose acquired in voluntary exposure—for exposure in cosmetics, smoking, driving a car—far exceeds a normal day's involuntary exposure through air, water, and food. To control or eliminate the countless voluntary and involuntary exposure routes for man amounts virtually to complete re-thinking of the twentieth-century lifestyle firmly established around petroleum with its great benefits as well as its environmental costs.

Responsibility for some of the other contributions to the generality of global oil pollution can be more specifically identified. Among these there are some where improvement is possible with benefit to the wider public and the environment.

It is history and taxation policy that have dictated the evolution of the types of petroleum for power generation, domestic heating, and the internal combustion engine. It may now be much too late in this epoch to reconsider the rational ergonomic and economic choices that would be needed to optimize the use of petroleum. Without doubt great savings over the period could have evolved if rational decisions rather than financial and political manipulation had determined the course of oil use development. In the face of the eclipse of oil, pressures in this direction must grow, led by scientists concerned with energy conservation. Short of such a revolution, very significant reduction in environmental oil pollution would follow wholehearted efforts to optimize oil recovery and recycling.

The oil in industrial waste waters, domestic sewage, and surface runoff has the characteristics of lubricating oils and greases, much of which is derived from the dumping of used sump oils and from the unattended overflows of oil interceptors and separators in trade premises. Similarly, used lubricants from shipping are a major source of direct input of PCAH to the sea. Not only would reuse of this oil contribute to sparing the scarce resource but it would reduce the problems and costs of sewage treatment, grease disposal by incineration or dumping, and the incidence of heavy oil slicks. If the sale of new lubricant were conditional on the return of a high percentage of the used oil, this problem would be solved. Environmental managers in the oil industry have it in their power to initiate such a logical move. Strategic taxation could make it feasible.

Among many examples of the possible better utilization of petroleum as an energy source, the outstanding example might be the wasteful protracted flaring of natural gas supplies at oil production centres, refineries, and petrochemical plants. To a very small extent these may be an operational necessity. In many cases the disposal of 'unsaleable' product or heat is a convenient financial or political option.

Based on one current political dogma, burning off gas is beneficial. It creates a belt of warm air round the earth. The water formed irrigates the Sahara better. The carbon dioxide is opening up transarctic shipping routes. Burning off is to the benefit of rich and poor alike. To harvest marginal gas supplies, waste heat and small volume resources of useful gas, needs effort and shrewd economic exploitation. Very often it is ill-directed, out-moded taxation that impedes such exploitation but this will change.

There are innumerable luxury uses of energy and petroleum that could be harmlessly expunged or curtailed with no intolerable loss to the general public. For example, plastic wrappings and single-use containers, paper unnecessarily used and not recycled, countless applications relating to aircraft, shipping, public transport, and the movement of goods. A great deal of unnecessary waste and energy consumption results in a waste of oil and a source of added pollution. Energy and materials conservation is a major contribution to reducing oil pollution.

There is an inkling of awareness of the importance of global background hydrocarbons in the decisions of the World Health Organization and the EEC to issue standards for drinking water and surface water for abstraction. But the terms of these standards on closer examination are demonstrably lax and the associated chemical methods and underlying chemistry woefully half-hearted. The influence of biogenic hydrocarbons is totally ignored. These measures in the long-run are of trivial value for public health evaluation. The same may be said of other standards for food and air. There is also a real need for all petroleum hydrocarbon standards relating to discharges to the air, land or aquatic realm to have meaning as regards threat to human and environmental health, to be harmonious and to be available in useful form to those striving to understand and control global oil pollution.

There are many underlying practical problems but the present status is so abyssmal that almost any change towards purposeful control and evaluation would be a major advance and a start to getting something for the money at present spent most ineffectually. Change can only start with managers and scientists currently engaged in these realms.

Most of these examples of global oil pollution relate to the contamination by man. This is a deliberate choice since land-based effects are of much more immediate consequence to man than any that may originate from the sea. Also the responsibilities for this pollution are very generally attributable to life today.

It is difficult to integrate all these contributions to global oil pollution. This unseen global presence of oil as it permeates the sea, land, and air is virtually unquantified, virtually unknown chemically, and lacks any presently definable threat to man or marine creature. Yet this aura of oil represents millions of tons of a major pollutant. Millions of tons of new oil pollution is added yearly to what remains from previous years. At the very least, enough dedicated research should be done to reassure the enquirer that this kind of oil pollution is harmless or at least is no more harmful than the environment contaminated only with geological and biogenic hydrocarbons. This is a managerial task suitable for public health institutions, universities, and specialized environmental science establishments throughout the world perhaps under the aegis of a United Nations' organization.

(b) Small Spills

Between the continuous yet unseen presence of diffusely distributed oil and the infrequent catastrophic lake of oil on the sea occur the multitudinous spills and deliberate discharges by the puff or spoonful or hundreds of tons. These occur for example, as emissions to the air, in garages, homes, workshops, fuel depots, oil terminals and everywhere oil is stored, moved or used on land or on water and also from boats and ships of all sizes and designations in harbours, coastal waters, and open seas. A typical cross-section virtually for one country was shown in Table 4-31 for spills greater than 20,000 US gallons (*ca.* 80,000 l); countless more spills of this size and smaller occur in all countries.

Evaporation, absorption by the ground, spreading, and dilution in fresh water and at sea gradually diminish the impact of these spills, but a proportion remains intact long enough to impart for some creature a perceptible exposure, for another damage, and for yet another a lethal dose.

The magnitude of this kind of oil pollution of the sea is difficult to measure. On a small scale it is seldom declared since it is probably regarded as insignificant by those responsible for it. On a larger scale even if it is notified, the costs of impact assessment could not be justified. This is the kind of oil pollution that creates the uncomfortably high number of random peaks in surveys of hydrocarbons in estuaries, seas, and oceans. One or two high peaks may outweigh all the oil indicated in the rest of the survey.

Responsibility for small oil spills may be usefully categorized into the work and the non-work sectors. In the work situation a useful management strategy is to make a carefully documented record of all spills over a period to identify spill size, frequency, and location. These data enable the manager to put in hand remedial measures which might be associated with fuel loading, storage and transfer, for example. Control, and incidentally safety, in these operations can be assisted by installing valves with interlocking arrangements and alarm systems for incorrect use and overflow. Equipment of this type can be obtained for a wide range of applications on land, on ships, and on oil platforms resulting in greater safety and savings in time, money, and materials and reduced contamination of the workplace and the environment. Accidents such as slipping and falling due to oil spilled in work areas may rank higher as health risks than the effects of oil on body function. This applies on ships as much as in factories.

Among the frequent small spills in the non-work situation are spills or leaks of domestic heating oil, spills from the home-servicing of motor cars, car accidents, and so on. Because it is the public that is responsible and the opportunities for spills are diverse, there is no single remedy. For unused oil products, financial loss is a strong disincentive. Some provision might be made by local authorities for the reception of oil-contaminated wastes from unused and used oil spills and disposals, since it is not always safe or reasonable to rely on the sewerage system.

Small and not so small spills and emissions during routine operations of oil tankers have become conspicuous targets for criticism. The discharge of oil-laden ballast, tank washings, and tank bottoms is wasteful and environmentally unacceptable. It is also unnecessary, since much less wasteful and messy methods such as load-on-top and crude-oil-washing can readily and cheaply be implemented in any conventional crude-oil carrier. Compared to oil handling at oil ports and terminals the operations on board certain tankers are greatly lacking in discipline and management. At ports and terminals, spills are readily seen and attributable and the port manager can impose strict routines for each component operation. Tankers with a record of oil spills can be banned from ports with strong pollution control policies. At sea, there is less chance of being seen and found guilty. Often, illicit tank cleaning is only apparent when oil or tar arrives on the beaches and oiled seabirds are found. Identification of the culprit after the even may be difficult and costly.

A number of strategies have been found successful in combating rogue tankers. Surveillance by aircraft aids prosecutions and acts as a deterrent and for more general monitoring of oil slicks space satellites can be used with a range of sensors. By insistence on the use of oily ballast reception facilities, where provided, it becomes uneconomic for tanker operators to adopt cost-reducing short cuts. Another disincentive is the mandatory keeping of each vessel's oil and oil ballast record books. In all these steps to reduce marine oil pollution IMCO has taken a leading part and merits full support. IMCO participation is also evident in the improvement of ships' design in relation to safety, minimization of damage in the event of collision or stranding, improved salvage arrangements, and so on. Modern tankers should have duplicate radar and navigational equipment, and electronic aids to simplify safe oil handling during tank cleaning, loading, and unloading.

Ultimately the responsibility lies with the tanker captains and with the operating companies if they should wish to assert it. It can be very expensive to lose a tanker and its cargo by a careless or thoughtless mistake and many ports inflict heavy fines even for operational oil spills.

As CAHILL and co-authors (1979) have carefully explained, public clamour for penalties on those responsible for oil spills does nothing for the curtailment of spillages but simply adds to the cost to the consumer. A more effective approach is to demand better management and where appropriate pursue for damages. Ports and terminals have encouraged discipline by refusing to service tankers with a poor record for operational care. If national standards for tanker operations, officer and crew proficiency, and ships specification and maintenance are below those recommended by IMCO and trade associations a ban on the whole fleet might be necessary.

(c) Large Spills

There is little that need to be added here about the responsibility for major oil spillages from tankers, sunken ships, and blowouts on oil rigs and production platforms. Those responsible have been meticulously sought out by legal examination and every conceivable managerial strategy has been tried.

It seems that the *Torrey Canyon* disaster and the dramatic attempts to salvage and latterly blow up the vessel to combat oil pollution at sea and on the beaches, touched a sensitive spot with the public in many countries and initiated a major effort to prevent further events of its kind.

The national representatives in IMCO sought solutions to some of the immediate problems of damages and liability. The 1969 IMCO International Convention on Civil Liability covers damage to the coast and territorial seas of signatory nations and imposes insurable liability on the tanker owner. A second IMCO convention established an International Fund for Oil Pollution Damage to provide a supplemented fund through assessments against the owner of the oil cargo. Unfortunately, these conventions were not ratified until 1975 and 1978 respectively. In the interim it was the tanker owners who first established a voluntary undertaking to assure availability of compensation for oil pollution damages in TOVALOP and to establish a fund (CRISTAL) maintained by contributions from member oil companies. These industrial agreements relate to more than 90% of the world tanker tonnage and the corresponding crude oil and fuel oils moved by water. Most member nations that have not ratified the IMCO conventions have at least some patchwork legislation establishing liability for damages caused by oil spills and a tax pool levied on petroleum set aside to provide for clean-up funds and stocks of equipment.

With a view to reducing oil spills, IMCO has established the 1978 Protocol relating to Tanker Safety and Pollution Prevention which develops the earlier 1973 Marine Pollution Convention and the 1974 Safety of Life and Sea Convention both of which were a tremendous improvement over the earlier regimes in relation to measures for accident prevention and reducing the likelihood of major oil spills. Parallel to these the 1978 Convention on Standards of Training, Certification and Watchkeeping for Seafarers developed under the auspices of IMCO tackled one of the key problems relating to the training and competency of tanker personnel.

The IMCO 1973 Marine Pollution Convention made a start to the reduction and, in some cases, the elimination of deliberate operational pollution of the seas from vessels by oil and other noxious substances. This was superseded by wider recommendations in the 1978 Marine Pollution Convention.

Thus, on the international front there is strong pressure encouraging nations to adopt and, in particular, implement national and local measures or legislation to overhaul radically the oil tanker industry. Just as the evolution of these conventions was a slow and laborious process, so also the implementation of these agreements by national decree and action is equally slow, laborious, and fraught with cumbersome patchwork measures and legislation which may already incorporate some ideas and terms of the new conventions. Meantime, while these stately matters grind on, major tanker accidents continue.

(d) Contingency Planning

Our preparedness and ability to cope with small spills, particularly in tranquil waters in the absence of complicating factors, has greatly improved and some countries at least can slip clean-up measures quietly into gear without alarm and chaos.

For the large crude carrier that runs at full speed fully laden on to shoal ground or rocks there is very little the master and crew on board can do to stem the flood of escaping oil from ruptured tanks. It is at this point that the most recent advances come into effect. Backed by the 1973 Protocol to the Intervention on the High Seas Convention, nations have prepared contingency plans including intervention. Arrangements for rapid action by salvage or by removal of oil cargo using self-powered pumps exist in each country, with international support if called for. In a few areas vessels specially designed and equipped to provide rescue, salvage, cargo removal, and storage/transfer for stricken vessels or damaged rigs or production platforms are maintained by major oil companies. It is only on this scale that meaningful action is possible in relation to major oil spills. Inevitably such operations are hazardous and expensive.

It is much easier and much less expensive to be in readiness for and operate small-scale inshore and beach clean-up. The costs of 1 special oil salvage service vessel and its maintenance on station would more than equip most coastal nations for small spills and pay for much of the labour costs. The principles and practicalities of oil spill contingency planning and of the various practical measures to alleviate the effects of oil spills in inshore waters and beaches have been exhaustively discussed elsewhere (see literature appendix) and reviewed here earlier.

Management for oil spill clean-up presents problems. For major spills, central government (or state) departments have the resources necessary, and the public expects them to work efficiently and harmoniously. For minor spills detailed organization in different countries varies widely, ranging from relatively simple to highly complex structures for command, materials, equipment, labour, waste disposal, and finance. Whatever the structure, it must be maintained in effective operation by paper and simulation exercises and training programmes. Occasionally, cooperative mock disaster exercises on a larger scale should be introduced and should be sampled, documented, and costed as real in order to iron out the important problems involving legal issues, finance, and damage to the environment. The final record should be reviewed in relation to time of response, efficiency, and probable success of the (mock) remedial measures.

(e) Treated and Controlled Oily Water Discharges

Discharges of treated or processed oily waters have specified volumes and qualities, set and enforced by appropriate environmental managers in cooperation with the industry. What is invariably lacking at present are standards that are a direct index of environmental damage or hazard which conventional routine analyses do not provide. Similarly, conventional biological monitoring is often a slow and insensitive tool and may not warn of damage until it is serious or irremediable. None of the environmental studies for discharges associated with treatment plants, refineries or petrochemical plants relates the parameters of the discharge to corresponding hydrocarbon contamination of the water and sediment and of the biota, not even on the grossest basis of routine oil measurement. As a result none of the great benefits of real site budgets and impact assessments pass into the pool of environmental knowledge. The great expense of countless routine analyses and recordings and of chemical and biological monitoring is virtually thrown away with no useful outcome. Even if the daily cost were 100 times greater, it would be much more valuable to have detailed and meaningful oil analysis with effluent flow records just for 1 week or month with corresponding field surveys in place of decades of worthless incomplete data and incomprehensible monitoring.

Expertise and instrumentation for these tasks may not exist at sites or with the local environmental unit but these could be organized on a wider basis since they are already available in the oil industry, its trade organizations, and centres of environmental expertise. It is generally untrue that industry is neglectful of the environment, and it is easier to eliminate or treat a pollutant that is known than all conceivable pollutant streams.

A revolution is overdue in the management of oil pollution. Major participation is required from chemists and biologists as well as environmental managers within industry and control authorities.

(f) Direction of Scientific Effort

Chemistry

The role of the analytical chemist is particularly challenging in relation to oil but this does not account for the organization of chemical effort having fallen far behind the requirements of the present time. Leaving aside the pursuit of specific research pro grammes, the routine measurement of so-called 'oil and grease' or 'total oil' is based largely on 3 methods which do not satisfy the primary criteria for environmental quality assessment yet these are widely used to this end. Within the oil industry there may be good reason for using the gravimetric method for oily waters in contact with heavy fractions; the IMCO infra-red method was evolved for calibrating continuous recording oil-in-water meters and the IGOSS ultra-violet method could have practical value in routine testing of oily waters in contact with known aromatic fractions. All of these lack the refinement needed for water quality determination which demands resolution, sensitivity, and speed of identification. Not all samples need to undergo fully sophisticated chemical analysis, but enough must be analysed in detail to establish the mean composition and its range of variation. Thereafter, a cheap and easy method which is quantitatively satisfactory may be adequate. These 3 methods fail to meet this requirement.

Biology

Before meaningful chemical analytical techniques can be evolved for routine work it is necessary to devise a corresponding meaningful bioassay. There are 2 broad domains to be covered: oil as it occurs in the water column and its effects; and the different effects of oil resident on superficial sediments. The bioassay organisms selected should be relevant and representative for the area to be monitored, the organisms should be sensitive and a useful response or amount of uptake should be discernible within and throughout a convenient period of, say, 5 to 10 d. The form of response should be applicable to a wide range of organisms in the community.

For the water column, some form of fish or invertebrate eggs would be ideal but these cannot be secured readily all year round. Pure cultures of marine microalgae might be considered (HUTCHINSON and co-authors, 1979) or a hydroid (STEBBING, 1980). For the sea floor, much is known about *Mytilus edulis* (see section on 'Mussel Watch' p. 1518) and some polychaetes (e.g. ÅKESSON, 1980) offer also the possibility of studying genetical effects.

Monitoring Outlook

Research results do not exclude significant effects from the n-alkanes and the heavier residues, but most toxicity has been attributed to the mono-, di-, and tricyclic aromatics. It may be a useful starting point to concentrate analytical effort on accurately determining these aromatic fractions which on current evidence are absent or rare in biogenic sources. Similarly, some idea of environmental hazard may be derived from measuring the uptake and the biochemical response of selected organisms to these hydrocarbons in mixtures appropriate to average or selected oily water discharges.

This is suggested as an interim approach pending a more thorough appraisal by experts at national and international level. The present haphazard approach to quantitative water quality assessments is getting nowhere; some uniformity in target and reference compounds is urgently needed.

Background Hydrocarbons

There is also a particular need, quite apart from monitoring of areas subject to oily water discharges, to begin to look at the chemistry and long-term toxicology of soluble and insoluble weathered (residual) oil components and their effects. At present, studies have been almost exclusively devoted to the more soluble, more toxic, more biodegradeable component groups of oil to the exclusion of those fractions less obviously toxic, less amenable to chemical isolation and biological testing, and more likely to survive and accumulate year to year in the sea and at the sea floor.

The biodegradeable fractions have a role possibly in the depression of primary production, in the survival and reproductive capacity of herbivores, in altered behaviour patterns, biodegradation of detritus, and so on. Quantification of these sublethal effects is basic for the fuller understanding of the impact of marine oil pollution and indeed global oil pollution. The study of persistent fractions may also be relevant to the understanding of cancer and disease in man and marine organisms.

In chemical terms, these components are at present reliably known in part for a few components clouded by a legacy of imprecise and inaccurate data. There is profound confusion about PCAH and the capacity of various animals especially fish and shellfish of commerce, to take up and accumulate or metabolize them on a time scale compatible with the pulsing pattern of oil exposure and especially in the aftermath of oil spills. There are questions about oil and neoplasms in marine animals, and to what extent these neoplasms are transmissible to other animals. Possibly, one must take a wider view of the causes of neoplasms and their transmissibility since it appears that oil may not be the sole causative substance. Interactions of a number of common pollutants may be important for mutagenic effects where the single compounds are demonstrably weak or not active.

Dose–response relationships for chemical carcinogens are unclear for man, and the detailed chemical and biochemical stimuli and transformation are complex. It follows, therefore, that it is difficult at present to establish guidelines for acceptable levels in water and foodstuffs including seafoods. Equally, if cancer has viral causes much remains unkown about the processes of transmission. Some evidence exists for the transfer of cancerous disease in bivalves which may open up new pathways for research into the disease.

In the general context of human health, exposures and encounters with hydrocarbons of all types in ordinary life on land enormously exceed exposures from oceanic or even inshore or intertidal environments. Whether it is reasonable to divert medical health resources from more immediate health risks to man to this peripheral aspect of global oil pollution is a managerial decision for the controlling bodies. Equivalent expertise for similar studies on marine animals is concentrated in North America and, unless marine scientiests in other countries carry equal conviction of the importance of such work, progress will be confined along the narrow lines which lie within the scope of these few laboratories.

The subtle organic chemistry of the sea is on the threshold of yielding much new knowledge about the hydrophobic components using the computer gc–ms techniques recently evolved for hydrocarbons. One can only speculate that hydrocarbons may dominate this class of compounds and what we learn about hydrocarbons may lead to deeper understanding of how organic matter influences cell wall function in the aquatic world.

Benthic Animals, Including Shellfish

The fate of oil as it is becoming better known directs attention more and more towards the bottom sediments where oil residues suffering various degrees of weathering become deposited. This sediment layer offers the opportunity to study the ageing of oil (e.g. by studying a range of diastereoisomers) and for evaluating the roles of biogenic and petrogenic inputs. From a simplistic viewpoint all incoming hydrocarbons merely add carbon to fuel the degradative system, but the highly modified and characteristic profile of hydrocarbons from bottom sediments suggests that the diagenic processes are much more subtle. Utilization of carbon sources at the sea floor is an important process in the cycling of oxygen, carbon dioxide, and nutrients. It is evident that hydrocarbons impose an exceptionally heavy oxygen burden and, in addition, have selective effects on bacteria. How then does oil pollution modify or interrupt normal mineralization processes in the upper sediments?

The benthos is a very important food component for many commercial species of fish and shellfish. Virtually nothing is known quantitatively about the effects of sedimented oil on benthos and fish after an oil spill, or about hydrocarbons transferred to benthos and food species. Results for the persistent of heavy oils in beach sediments are probably a poor guide to persistence at the sediment surface in the open sea.

An understanding of the rate of turnover of background and additional oil from discharges and spills is important in appraising the impact on benthos and the related issues of food contamination, taint, and human health.

Of all the vast literature on oil spills least effort has gone into impact assessment outside the intertidal zone. Of particular interest are the biological transformation of hydrocarbons, the formation and transfer of PCAH, and the environmental hazard created, at least temporarily, by oil spills and maintained locally by oily water discharges.

Costs of Research

A point has been made frequently throughout this chapter that it was difficult or impossible to relate a great deal of the experimental studies to a real-world situation. Frequently, the physical mode of hydrocarbon presentation has not been adequately evaluated; the chemical and quantitative distinctions between the oil introduced, the hydrocarbons presented to the organism, and the hydrocarbons absorbed are scarcely ever recognized along with changes with time; important considerations relating to test organisms are not published (PERKINS, 1976). Perhaps scrutineers of papers being submitted to scientific journals could insist on these vital matters being satisfied before accepting the work.

Oil pollution has attracted numerous short-term research projects of very limited scale, especially relating to toxicology. Many have proved useful, and together they have yielded a comprehensive picture of various oils and products on many aspects of embryology, development, metamorphosis and, of course, survival for a whole circus of animals. A good proportion of this work has been generously sponsored by oil companies and associations. Scope for individual research on oil pollution continues to exist in other directions but it is clear that more fruitful work can be achieved by organizing researchers into multidisciplinary teams, provided the objectives are clearly pertinent to real situations (preferably not transient oil spills).

Whether any oil spill situation is now worth the effort and cost of study is debatable: the intertidal scene has been described *ad nauseam*. The oil spill situation is like an ill-conducted experiment, nothing is adequately controlled or reproducible nor can it even be regarded as novel.

Future Outlook

Surely, the future of hydrocarbon research lies in the direction of consolidating existing fragmentary knowledge. For chemistry, this means widening the coverage of oil components to embrace all the main fractions (Fig. 4.14). On this wider basis a cooperative target would be the development of a dynamic model for the fate of oil in the sea, biota, surface film, and sediments. Oil spill research could then enlarge this model to include the perturbation that is caused.

For biology, existing studies have revealed a number of useful as well as many intriguing problems. Marine oil pollution has achieved importance because of its potential threat to the welfare of marine biota, therefore, this must be the objective of biological studies. The key problem is the assessment of day-to-day oil pollution world-wide. This cannot be done by reference to controlled zero hydrocarbon sites, so one must accept extrapolation from measured hydrocarbon gradients. Inevitably, this leads back to relating oil input to biological effect which is essentially the monitoring situation evaluated meaningfully. Therefore, future progress in understanding the impact of oil pollution must start with very much better understanding of a representative range of monitoring sites preferably where the focus of oil pollution merges progressively into shelf waters and open sea. Some ideas on these programmes have been outlined. Oil discharges to enclosed sites might be fruitfully compared with regions of oil seeps which themselves provide natural field sites of great research potential. All of these possible projects must have full chemical support.

New biochemical approaches and 'scope for growth' need to be developed more constructively with adequate replication in association with more conventional programmes, and some additional effort is needed to complete the coverage with respect to fish diseases, public health, and quality of fish and shellfish as food.

Clearly much of this is not new. These suggestions are a plea for better integration and perhaps for the establishment of sound chemical information for a few sites and their surroundings to provide a meaningful framework for a varied pattern of biological assessement.

Where resources cannot meet such a demanding programme, team use of large plastic enclosures or natural ponds is much preferred to test-tube experiments by the lone research worker.

Finally, the background picture is the one that is most normal, least studied and, moreover, most difficult to study. Options for study may be chemical, biochemical or biological.

An integrated study of environmental hydrocarbon assemblages in uttermost detail such as the simultaneous evaluation of carbon isotope ratios together with characterization using known biogenic components and ratios of steroid diastereomers (for a start) has not been attempted. This, along with geological evidence, would throw new light on hydrocarbon long-term decay or maturation in sediment cores and could build up a broad picture from representative sites. Can the impact of the present hydrocarbon epoch be identified and its distribution defined? It may be also possible to develop a parallel theme for hydrocarbons as an element of the organic matter in oceanic circulation.

Surely this is not just fanciful research but a real target for measuring the health of the oceans in the truest sense. If research into toxicological aspects of background hydrocarbons proves no significant hazard for the sea or man, the work will nevertheless greatly advance knowledge of this component of organic matter in the sea.

(13) Conclusion—A Summary for the Non-Expert

Conclusions are only valid as the facts on which they are based and are appropriate and reliable. Public opinion, based mainly on oil seen on the sea, in harbours, on seabirds and on beaches, is that oil pollution is a matter for serious concern.

Scientific grounds, which are intolerably inaccurate not because of lack of skill in assembling data but because of old and unreliable chemical methodology, nevertheless establish oil as a widespread, essentially global, pollutant of important magnitude.

The impact of oil spills or small discharges is highly demonstrable on seabirds and, on the basis of much experimental evidence, reduces the capacity of the open sea to produce phytoplankton and zooplankton which are the original food sources for the fish and shellfish we eat. In the shore areas, oil spills have potential to cause damage to all forms of plant and animal life to an extent depending on the kind and amount of oil, the type of shore, and the way the oil behaves. The seriousness of the damage in both cases depends on how readily natural processes restore the organisms killed or damaged.

Oil kills or damages life in many different ways, ranging from complete encasement in oil to the merest trace of oil revealed by the disturbance of ordinary enzyme balance within a single cell. As a result, the evidence of damage from an oil spill—or more generally for any discharge of oil, oily substance or oily water—only starts with the evident casualties.

It becomes very difficult to trace how far this train of damages, which extends down to subtle sublethal responses and effects on behaviour, affects the health of the oceans because of the great variety of life forms and their responses and because oil is chemically very complex and is continually changing and decaying from the moment it enters the water. Oil, which is a matured and modified natural substance, probably has marginal effects on a global scale and measurable effects only within a mile or less of major industrial sites.

This is an epoch in which oil from the subterranean rock strata has been liberated for many years into the air, water, soil, and sea. Every human being encounters oil in one shape or form every day. Although people are known to be killed or seriously affected by exposure to large amounts of oil and oil products, understanding of lesser and lesser doses, like the understanding of pollution of the sea away from an oil spill, becomes less and less clear.

Carcinogens like those in tobacco smoke are associated with all forms (air, land, sea) of oil pollution. The sea is the final sump for many pollutants. Hence there is some concern that these carcinogens may contaminate food from the sea. Such reliable evidence as exists (and there is much that is of doubtful quality) indicates that seafoods as normally prepared and consumed are no higher in these substances than food from land,

and all are much less of a hazard than smoking, bearing in mind that knowledge about cancer is incomplete.

The indications are that marine carcinogens enter the sea mainly from the air and from waste, used lubricants reaching rivers, sewage, and ships' bilges. It would take minimal community effort to remove this pollution and to recycle the oil.

Other very minor but important components of oil and oil residues are the sulphur and nitrogen compounds which have highly pronounced smell and taste and may cause tainting. They are probably also biologically active. Not much is known about them.

Research on oil pollution has tended to be fragmentary, much of it related to oil spills or experimental situations involving high levels of oil. This is useful for the occasional incident but it tells nothing about the normal situation. Studies on sites where treated oily waters are discharged have not always contributed usefully to establishing broad cause-and-effect associations usually because the chemical data on the oil were inadequate, and sometimes because the biology did not pursue well-directed lines of approach which are quantifiable.

This chapter presents a review of the science behind oil pollution and attempts to show how the sources and control of oil pollution can be better managed. Managing oil spill clean-up is complex and costly. Relative to the totality of oil pollution, is it really worth while? Perhaps the money could be better spent in preventive measures.

The chapter also discusses the scientific information at hand and identifies needs. These needs include much better data and guidance for environmental managers. Some areas for new study are indicated. Probably future advance in understanding oil pollution requires the sacrifice of individual short-term projects in order to sustain more disciplined and orderly team research.

Our modern civilization uses oil for power, heat, transport, mechanical movement, plastics, drugs, cosmetics, and many other things. Oil is a finite fossil resource which cannot be renewed. Within 50 yr it will be scarce and, depending on the control of these remaining stocks, will sooner or later dwindle to nothing.

No government wants to plan 50 yr ahead, and it will be on the deathbed of oil that realization will strike that it takes half a century to find, develop, and build alternative power sources on a scale to replace oil and to provide substitutes for its applications.

Long before that stage, oil will have attained a huge value. The thought of spilling, wasting or not reclaiming oil will be beyond belief. Hi-jacking oil will replace diamond thefts and bank raids. There will be no fears of oil spills. Oil decay processes will quite soon put an end to oil pollution and at the end of the oil era if not before, the sea will show not a trace.

Literature Cited (Chapter 4)

AALUND, L. R. (1976). Wide variety of crudes gives refiners range of charge stocks. Oil and Gas Journal **74** (13), 87–122; **74** (15), 72–78; **74** (17), 112–126; **74** (19), 85–94; **74** (21), 80–87; **74** (23), 139–148; **74** (25), 137–152; **74** (27), 98–198.

ÅKESON, B (1980). The use of certain polychaetes in bio-assay studies. *Rapp. P. -v. Réun. Cons. int. Explor. Mer*, **179**, 315–321.

ALLEN, A. A. (1979). Containment and recovery techniques for cold weather inland oil spills. In J. O. Ludwigson (Ed.), *Proceedings of the Oil Spill Conference, 1979* (Los Angeles, California). American Petroleum Institute, Washington, D. C. pp. 345–353.

ALLEN, H. (1971). Effects of petroleum fractions on the early development of a sea urchin. *Mar. Pollut. Bull., N. S.*, **2**, 138–140.

AMERICAN PETROLEUM INSTITUTE (API) (1979). Socio-economic–legal aspects. In J. O. Ludwigson (Ed.), *Proceedings of the Oil Spill Conference, 1979*. American Petroleum Institute, Washington, D. C. pp. 3–137 (Contributions by various authors).

AMERICAN PETROLEUM INSTITUTE (API) (1979). Oil pollution by tankers—issues and progress. In J. O. Ludwigson (Ed.), *Proceedings of the Oil Spill Conference, 1979*. American Petroleum Institute, Washington, D. C. pp. 3–137 (Contributions by various authors).

ANDERSON, J. W. (1977). Effects of petroleum hydrocarbons on the growth of marine organisms. *Rapp. P. -v. Reun. Cons. int. Explor. Mer*, **171**, 157–165.

ANDERSON, J. W., KIESSER, S. L., and BLAYCOCK, J. W. (1979). Comparative uptake of naphthalenes from water and oiled sediment by benthic aniphipods. In J. O. Ludwigson (Ed.), *Proceedings of the Oil Spill Conference, 1979*. American Petroleum Institute, Washington, D. C. pp. 579–584.

BARBIER, M. D., JOLY, A. S., and TOURRES, D. (1973). Hydrocarbons from sea water. *Deep Sea Res.*, **20**, 305–314.

BARSDATE, R. (1972). Natural oil seeps at Cape Simpson, Alaska: aquatic effects. *Sci. Alaska Pro. Alaskan Sci. Conf.*, **23** (52), 91–95.

BELLIER, P. and MASSART, G. (1979). The *Amoco Cadiz* oil spill clean-up operations. In J. O. Ludwigson (Ed.), *Proceedings of the Oil Spill Conference, 1979*. American Petroleum Institute, Washington, D.C. pp. 141–146.

BENTZ, A. P. and SMITH, S. L. (1979). The legal aspects of oil spill finger printing. In J. O. Ludwigson (Ed.), *Proceedings of the Oil Spill Conference, 1979*. American Petroleum Institute, Washington, D.C. pp. 3–6.

BERDUGO, V., HARRIS, R. P., and O'HARA, S. C. (1977). The effect of petroleum hydrocarbons on reproduction of an estuarine planktonic copepod in laboratory cultures. *Mar. Pollut. Bull., N.S.*, **8**, (6), 138–143.

BERGHOFF, W. (1968). Erdölverarbeitung und Petrochemie. Ein Wissensspeicher. VEB Deutscher Verl. für Grunstoffindustrie, Leipzig. pp. 44–64.

BERRY, W. O. and BRAMMER, J. D. (1977). Toxicity of water soluble gasoline fractions to fourth-instar larvae of the mosquito *Acedes aegypti*, L.—*Environ. Pollut.*, **13**, 229–234.

BLACKMAN, R. A. A. and LAW, R. J. (1981). The *Eleni V* oil spill; return to normal conditions. *Mar. Pollut. Bull., N.S.*, **12** (4), 126–130.

BLUMER, M. (1971). Scientific aspects of the oil spill problem. *Environmental Affairs*, **1**, 54–73.

BNCOR (THE BRITISH NATIONAL COMMITTEE ON OCEANIC RESEARCH) (1980). *The Effects of Oil Pollution. Some Research Needs* (a memorandum prepared by the Marine Pollution Subcommittee of the BNCOR), The Royal Society, London.

BOEHM, P. D. (1977). *The Transport and Fate of Hydrocarbons in Benthic Environments*, Ph. D. Dissertation, University of Rhode Island, Kingston.

BOEHM, P. D., PERRY, G. H., and FIEST, D. L. (1978). Hydrocarbon chemistry of the water column of Georges Bank and Nantucket shoals (February to November 1977). In *Proceedings of the Symposium 'In the Wake of the Agro Merchant', 1978*. Centre for Ocean Management Studies, University of Rhode Island, Kingston. pp. 58–64.

BOEHM, P. D., STEINHAUER, W. G., FIEST, D. L., MOSESMAN, N., BARAK, J. E., and PERRY, G. H. (1979). A chemical assessment of the present levels of hydrocarbon pollutants in the Georges Bank region. In J. O. Ludwigson (Ed.), *Proceedings of the Oil Spill Conference, 1979*. American Petroleum Institute, Washington, D.C. pp. 333–341.

BOLAND, W., MARNER, F., and JAENICKE, L. (1981). *Dictyota dichotoma* (Phaeophyceae): identification of the sperm attractant. *Science, N.Y.*, **212**, 1040–1041.

BOURNE, W. R. P. (1976). Seabirds and pollution. In R. Johnston (Ed.), *Marine Pollution*. Academic Press, London. pp. 403–502.

BOURNE, W. R. P. (1980). Scottish developments. *Mar. Pollut. Bull., N.S.*, **11**, 147.

BOURQUIN, A. W. and PRITCHARD, P. H. (Eds) (1979). Microbial degradation of pollutants in marine environments. In *Proceedings of the Workshop at Pensacola Beach, Florida, 1978*. EPA, Gulf Breeze, Florida. pp. 1–551 (609/91–79–062).

BREISBACH, R. H. (1980). *Handbook of Poisoning* (10th ed.), Lange, London.

BROWN, R. A.; SEARL, T. D., ELLIOTT, J. J., PHILLIPPS, B. G., BRANDON, D. E., and MONAGHAN, P. H. (1973). Distribution of heavy hydrocarbons in some Atlantic Ocean waters. In *Proceedings of the Oil Spill Conference 1973*. American Petroleum Institute, Washington, D.C. pp. 505–519.

BROWN, R. A., SEARL, T. D., and PRESTRIDGE, E. G. (1975). Measurement of volatile and non-volatile hydrocarbons in selected areas of the Atlantic Ocean. Final Report Prepared for the U.S. Department of Commerce, Maritime Administration, Washington, D.C.

BUNT, J. S. (1975). Primary-productivity of marine ecosystems. In H. Lieth and R. H. Whittaker (Eds), *Primary Productivity of the Biosphere*. Springer-Verlag, Berlin. pp. 169–183.

BURNS, K. A. and SMITH, J. L. (1977). Distribution of petroleum hydrocarbons in the Westernport Bay (Australia). In D. A. Wolfe (Ed.), *Fate and Effects of Petroleum Hydrocarbons in Marine Organisms and Ecosystems*. Pergamon Press, New York. pp. 442–453.

BURNS, K. A. and SMITH, J. L. (1978). *Biological Monitoring of Ambient Water Quality*, Marine Chemistry Unit, Ministry of Conservation, Australia (Unpublished manuscript quoted in NRC, Canada, 1980).

CABIOCH, L., DAUVIN, J. C., MORA BERMUDES, J., and RODERIGUES BABID, C. (1980). Effects de la marée noise de l' *Amoco Cadiz* sur le benthos sublittoral du nord de la Bretagne. *Helgoländer Meeresunters.*, **33**, 193–208.

CAHILL, E. J., SMITH, L. R., and HALEY, G. P. (1979). Penalties on oil spills with unquantified damage—the hidden tax and economic deterrent concept. In J. O. Ludwigson (Ed.), *Proceedings of the Oil Spill Conference, 19789*. American Petroleum Institute, Washington, D.C. pp. 95–97.

CALDER, J. A., LAKE, J., and LASETER, J. (1978). Chemical composition of selected environmental and petroleum samples from the *Amoco Cadiz* oil spill. A preliminary scientific report. In *The Amoco Cadiz Oil Spill, NOAA—EPA* Special Report. National Oceanic and Atmospheric Administration, Washington, D.C. pp. 21–85.

CANEVARI, G. P., 1979. The restoration of oiled shorelines by the proper use of chemical dispersants. In J. O. Ludwigson (Ed.), *Proceedings of the Oil Spill Conference, 1979*. American Petroleum Institute, Washington, D.C. pp. 443–446.

CENTRE OCEANOLOGIQUE DE BRETAGNE (1981). *Amoco Cadiz*—fates and effects of the oil spill. In *Proceedings of an International Symposium, 1979.* Centre Oceanologique de Bretagne, Brest, Paris, pp. 1–881.

CHANDLER, G. (1973). The oil industry. In *Proceedings of the Fuel and Environmental Conference, 1973*, Vol. I. Institute of Fuel, London. pp. 23–40.

CLARK, R. C. and BROWN, D. W. (1979). Petroleum properties and analyses in biotic and abiotic systems. In D. C. Malins (Ed.), *Effects of Petroleum in Arctic and Subarctic Marine Environments and Organisms*, Vol. I. Academic Press, New York. pp. 1–69.

COLE, J. A. (1979). Oil pollution by tankers—industrial view. In J. O. Ludwigson (Ed.), *Proceedings of the Oil Spill Conference, 1979*. American Petroleum Institute, Washington, D.C. pp. 116–117.

COLWELL, R. R. (1974). Microorganisms of the outer continental shelf. In *Proceedings of the Estuarine Research Federation*, Outer Continental Shelf Conference and Workshop. Estuarine Research Foundation, pp. 259–264.

CONCAWE CONSERVATION OF CLEAN AIR AND WATER—EUROPE (1980). The environmental impact of refinery effluents. *Report No. 1—80*. Assessment prepared by CONCAWE. Water Pollution Management Group. CONCAWE, Den Haag. pp.1–33.

CORMACK, D. and NICHOLS, J. A. (1977). The concentrations of oil in sea water resulting from naturally and chemically induced dispersion of oil slicks. In J. O. Ludwigson (Ed.), *Proceedings of the Oil Spill Conference, 1977*. American Petroleum Institute, Washington, D.C. pp. 381–385.

CORNER, E. D. S., HARRIS, R. P., KILVINGTON, C. C., and O'HARA, S. C. M. (1976a). Petroleum compounds in the marine food web: Short-term experiments on the fate of naphthalene in *Calanus*. *J. mar. biol. Ass. U.K.*, **56**, 121–133.

CORNER, E. D. S., HARRIS, R. P., WHITTLE, K. J., and MACKIE, P. R. (1976b). Hydrocarbons in marine zoo plankton and fish. In A. P. M. Lockwood (Ed.), *Effects of Pollutants in Aquatic Organisms*, Society for Experimental Biology (Seminar Series), Vol. 2. Cambridge United Press, London. pp. 28–35.

CORNER, E. D. S., KILVINGTON, C. C., and O'HARA, S. C. M. (1973). Qualitative studies on the metabolism of naphthalene in *Maia squinodo* (spider crab). *J. mar. biol. Ass. U.K.*, **53**, 819–832.

COWELL, E. B. (1976). Oil pullution of the sea. In R. Johnston (Ed.), *Marine Pollution*. Academic Press, London. pp. 353–401.

COWELL, E. B. (1978). Evidence prepared for a European Parliamentary Hearing, Paris, 4 July 1978 (Unpubl.).

COWELL, E. B., COX, G. V., and DUNNET, G. M. (1979). Applications of ecosystem analysis to oil spill impact evaluation. In J. O. Ludwigson (Ed.), *Proceedings of the Oil Spill Conference, 1979*. American Petroleum Institute, Washington, D.C. pp. 517–519.

CREUTZBERG, F. (1975). Orientation in space: animals. Invertebrates. In O. Kinne (Ed.), *Marine Ecology*, Vol. II, Physiological Mechanisms, Part 2. Wiley, London. pp. 555–655.

CROSS, F. A., DAVIS, W. P., HOSS, D. E., and WOLFE, D. A. (1978). Biological observations, Part 5. In W. N. Hess (Ed.), *The Amoco Cadiz Oil Spill*. NOAA/EPA Special Report, US Department of Commerce and US Environmental Protection Agency, Washington, D.C. pp. 197–215.

CROW, S. A., BOWMAN, P. I., and AHEARN, D. G. (1977). Isolation of a typical *Candida albicans* from the North Sea. *Appl. environ. Microbiol.*, **33** (3), 738–739.

DASTILLUNG, M. and ALBRECHT, P. (1976). Molecular test for oil pollution in surface sediments. *Mar. Pollut. Bull., N.S.*, **7**, 13–15.

DAVIES, J. M. and GAMBLE, J. C. (1979). Experiments with large enclosed ecosystems. *Phil. Trans. R. Soc. Lond. (Series B)*, **286**, 523–544.

DAVIES, J. M., HARDY, R., and MCINTYRE, A. D. (1981a). Environmental effects of North Sea Oil Operations. *Mar. Pollut. Bull., N.S.*, **12**, 412–414.

DAVIES, J. M., JOHNSTON, R., WHITTLE, K. J., and MACKIE, P. R. (1981b). Origin and fate of hydrocarbons in Sullom Voe. *Proc. R. Soc. Edinb.*, **80** (B), 135–154.

DESHIMARU, O. (1971). Studies on the pollution of fish meal by mineral oils. I. Deposition of crude oil in fish meat and its detection. *Bull. Jap. Soc. scient. Fish.*, **37**, (4), 297–301.

DIXIT, D. and ANDERSON, J. W. (1977). Distribution of naphthalenes within exposed *Fundulus simulus* and correlations with stress behaviour. In J. O. Ludwigson (Ed.), *Proceedings of the Oil Spill Conference, 1977*. American Petroleum Institute, Washington, D.C. pp. 633–636.

DOW, R. L. (1978). Size-selective mortalities of clams in an oil spill site. *Mar. Pollut. Bull., N.S.*, **9** (2), 45–48.

DUBEY, R. A. and FIDELL, E. R. (1979). Damage assessment rulemaking: issues and alternatives. In J. O. Ludwigson (Ed.), *Proceedings of the Oil Spill Conference, 1979*. American Petroleum Institute, Washington, D.C. pp. 79–89.

DUBIEL, E. J. (1979). The practical aspects of litigating an oil (plaintiff's viewpoint). In J. O. Ludwigson (Ed.), *Proceedings of the Oil Spill Conference, 1979*. American Petroleum Institute, Washington, D.C. pp. 91–94.

DUCE, R. A., QUINN, J. C., OLNEY, C. E., PITROWICZ, S. R., RAY, B. J., and WADE, T. L. (1972). Enrichment of heavy metals and organic compounds in the surface microlayer of Narragansett Bay, Rhode Island. *Science, N.Y.*, **176**, 161–163.

DUNN, B. P. and STICH, H. F. (1975). The use of mussels in estimating B [α] P contamination of the marine environment. *Proc. Soc. exp. Biol. Med.*, **150**, 49–51.

DUNN, B. P. and STICH, H. F. (1976). Release of the carcinogen benzo (α) pyrene from environmentally contaminated mussels. *Bull. environ. Contam. Toxicol.*, **15**, 398–401.

EATON, P. and ZITKO, V. (1979). Polycyclic aromatic hydrocarbons in marine sediments and shellfish near coastal wharf structures in eastern Canada. International Council for the Exploration of the Sea. I.C. Paper (minieo; unpubl.).

FISCHER, P. J. (1978). Natural gas and oil seeps, Santa Barbara Basin, California. In *California Offshore Gas Oil and Tar Seeps*. California State Lands Commission Report, Sacramento. pp. 1–62.

FUCIK, K. W., ARMSTRONG, H. W., and NEFF, J. M. (1977). Uptake of naphthalenes by the clam *Rangia cuneata* in the vicinity of an oil separation platform in Trinity Bay, Texas. In J. O. Ludwigson (Ed.), *Proceedings of the Oil Spill Conference, 1977*. American Petroleum Institute, Washington, D.C. pp. 627–632.

GASSMANN, G. (1978). Verfahrer zum beschleunigten mikrobiologischen Abkau von offenen und latenten Verölungen. *Patentschrift*, **25 33 755**, Bundesrepublic Deutschland.

GASSMANN, G. (1981). Chromatographic separation of diastereomeric isoprenoids for the identification of fossil oil contamination. *Mar. Pollut. Bull., N.S.*, **12**, 78–84.

GESAMP (GROUP OF EXPERTS ON THE SCIENTIFIC ASPECTS OF MARINE POLLUTION) (1977). Impact of oil on marine environment. *Reports and Studies, FAO January 1977*, **6**.

GETTLESTON, D. A. (1980). Effects of oil and gas drilling operations on the marine environment. In R. A. Geyer (Ed.), *Marine Environmental Pollution*, Vol. 1, Hydrocarbons. Elsevier, Amsterdam. pp. 371–412.

GEYER, R. A. and GIAMMONA, C. P. (1980). Naturally occurring hydrocarbons in the Gulf of Mexico and Caribbean Sea. In R. A. Geyer (Ed.), *Marine Environmental Pollution*, Vol. I, Hydrocarbons. Elsevier, Amsterdam. pp. 37–106.

GOLDBERG, E. D., BOWEN, V. T., FARRINGTON, J. W., HARVEY, G. R., MARTIN, J. H., PARKER, P. L., RISEBROUGH, R. W., ROBERTSON, W. SCHNEIDER, E., and GAMBLE, E. (1978). The mussel watch. *Environ. Conserv.*, **5**, 101–125.

GORDON, D. C. and KEIZER, P. D. (1974). Estimation of petroleum hydrocarbons in sea water by fluorescence spectroscopy: improved sampling and analytical methods. Environment Canada, Fisheries and Marine Service, Technical Report No. 481, 1–28.

GRAHL-NIELSON, O., SUNDBY, S., WESTREIM, K., and WILHELMSEN, S. (1980). Petroleum hydrocarbons in sediments resulting from drilling discharges from a production platform in the North Sea. In *Proceedings of the Symposium Research on Environmental Fate and Effects of Drilling Fluids and Cuttings, 1980, Lake Buena Vista, Florida.* Washington, D.C.

GRAHL-NIELSEN, O., WESTRHEIM, K., and WILHELMSEN, S. (1979). Petroleum hydrocarbons in the North Sea. In J. O. Ludwigson (Ed.), *Proceedings of the Oil Spill Conference, 1979.* American Petroleum Institute, Washington, D.C. pp. 629–632.

GRIBBON, E. (1973). *Toxicological Study of a Series of Oil Emulsifiers with Respect to Their Short and Long-term Effects on Selected Marine Animals, M.Sc. Thesis, University of Strathclyde, Glasgow, Scotland.*

GULLAND, J. A. (1983). World resources of fisheries and their management. In O. Kinne (Ed.), *Marine Ecology*, Vol. V, Ocean Management, Part 2. Wiley, Chichester. pp. 839–1061.

GUNDLACH, E. R., BERNE, S., D'OZOUVILLE, L., and TOPINKA, J. A. (1981). Shoreline oil two years after *Amoco Cadiz*. In R. B. Parrette (Ed.), *Proceedings of the Oil Spill Conference, 1981.* American Petroleum Institute, Washington, D.C. pp. 525–534.

GUNKEL, W. (1973). Distribution and abundance of oil-oxidizing bacteria in the North Sea. In D. G. Ahearn and S. P. Meyers (Eds), *The Microbial Degradation of Oil Pollutants*. Louisiana State University, Baton Rouge, La. pp. 127–139.

GUNKEL, W. and GASSMANN, G. (1977). Kohlenwasserstoffe und Mikro-organismen in der Nordsee. *Jahresbesicht 1976*, Biologischen Anstalt Helgoland, Hamburg.

GUNKEL, W. and GASSMANN, G. (1980). Oil, oil dispersants and related substances in the marine environment. *Helgoländer Meeresunters.*, **33**, 164–181.

GUNKEL, W., GASSMANN, G., OPPENHEIMER, C. H., and DUNDAS, I. (1980). Preliminary results of baseline studies of hydrocarbons and bacteria in the North Sea, 1975, 1976, 1977. In *Sixth National Congress of Microbiology, 1977*. Santiago de Compostela. pp. 223–247.

HANN, J. W. (1977). Fate of oil from supertanker *Metula*. In J. O. Ludwigson (Ed.), *Proceedings of the Oil Spill Conference, 1977*. American Petroleum Institute, Washington, D.C. pp. 465–468.

HARDY, R., MACKIE, P. R., and WHITTLE, K. J. (1977a). Hydrocarbons and petroleum in the marine ecosystem: a review. *Rapp. P.-v. Réun. Cons. int. Explor. Mer*, **171**, 17–26 and 53–54.

HARDY, R., MACKIE, P. R., WHITTLE, K. J., MCINTYREE, A. D., and BLACKMAN, R. A. A. (1977b). Occurrence of hydrocarbons in the surface film, sub-surface water and sediment around the United Kingdom. *Rapp. P.-v. Réun. Cons. int. Explor. Mer*, **171**, 61–65.

HARRIS, R. P., BERDUGO, V., CORNER, E. D. S., KILVINGTON, C. G., and O'HARA, S. C. M. (1977). Factors affecting the retention of a petroleum hydrocarbon by marine copepods. In D. A. Wolfe (Ed.), *Fate and Effects of Petroleum Hydrocarbons in Marine Ecosystems and Organisms*. Pergamon, New York. pp. 286–304.

HARVEY, G. R., REQUEJO, A. G., MCGILLIVARY, P. A., and TOKAR, J. M. (1979). Conservation of a subsurface oil-rich layer in the open ocean. *Science, N.Y.*, **205** (4410), 999–1001.

HASE, A. and HITES, R. A. (1976). On the origin of polycyclic aromatic hydrocarbons in recent sediments: biosynthesis by anaerobic bacteria. *Geochim. Cosmochim. Acta*, **40**, 1141–1143.

HAYES, M. O., GUNDLACH, E. R., and GETTER, C. D. (1980). Sensitivity ranking of energy port shore line. In *Proceedings of Ports, 1980*. American Society of Civil Engineers, Norfolk, Virginia.

HEBER, R. and POULET, S. A. (1980). Effect of modification of particle size of emulsions of Venezuelan crude oil on feeding, survival and growth of marine zooplankton. *Mar. environ. Res.*, **4**, 121–134.

VAN DEN HOEK, A. W. J. (1977). Drilling well control. *United Nations Environment Programme Industry Sector Seminars*, Petroleum Industry Meeting, Paris.

HOWGATE, P., MACKIE, P. R., WHITTLE, K. J., FARMER, J., MCINTYRE, A. D., and ELEFTHERIOU, A. (1977). Petroleum tainting in fish. *Rapp. P.-v. Réun. Cons. int. Explor. Mer*, **171**, 143–146.

HUTCHINSON, T. C., HELLEBUST, J. A., MACKAY, D., TAM, D., and KNAUSS, P. (1979). Relationship of hydrocarbon solubility to toxicity in algae and cellular membrane effects. In J. O. Ludwigson (Ed.), *Proceedings of the Oil Spill Conference, 1979*. American Petroleum Institute, Washington, D.C. pp. 541–547.

ILIFFE, T. M. and CALDER, J. A. (1974). Dissolved hydrocarbons in the eastern Gulf of Mexico Loop Current and the Caribbean Sea. *Deep Sea Res.*, **20**, 305–314.

IMCO (INTERGOVERNMENTAL MARITIME CONSULTATIVE ORGANIZATION) (1978). Test specification for oily water separating equipment and oil content meters. *Intergovernmental Maritime Consultative Organization Document*, **78. 02 E**

JAMES, M. C. (1925). Oil Pollution of Navigable Waters. A preliminary investigation on the effect of oil pollution on marine pelagic eggs (Report of the U.S. Bureau Fish Interdepartmental Committees, Washington). *App.*, **6**, 85–92.

JENSEN, K. (1981). Levels of hydrocarbon in mussel, *Mytilus edulis*, and surface sediments from Danish coastal areas. *Bull. environ. Contam. Toxicol.*, **26**, 202–206.

JOHANNES, R. E. (1972). Coral reefs and pollution. In M. Ruivo (Ed.), *Marine Pollution and Sea Life*. Fishing News Books, London. pp. 364–375.

JOHNSTON, C. S. (1980). Sources of hydrocarbons in the marine environment. In C. S. Johnston and R. J. Morris (Eds), *Oily Water Discharges*. Applied Science, London. pp. 41–62.

JOHNSTON, C. S. and MORRIS, R. J. (Eds) (1980). *Oily Water Discharges, Applied Science,* London.

JOHNSTON, R. (1977). What North Sea oil might cost fisheries. *Rapp. P.-v. Réun. Cons. Int. Explor. Mer*, **171**, 212–223.

JOLLEY, R. L. (Ed.) (1978). *Water Chlorination—Impact and Health Effects*, Vol. 1, Ann Arbor Science, Ann Arbor, Michigan.

JOLLEY, R. L., GORCHEV, H., and HAMILTON, D. H., Jr (Eds) (1978). *Water Chlorination—Impact and Health Effects*, Vol. 2, Ann Arbor Science, Ann Arbor, Michigan.

JONES, H. R. (1973). *Pollution Control in the Petroleum Industry (Noyes Data Corporation)*, Park Ridge, New Jersey.

JUHASZ, F. (1979). Economic evaluation of the environmental effects of oil pollution: a practical lesson using three case histories. In J. O. Ludwigson (ED.), *Proceedings of the Oil Spill Conference, 1979*. American Petroleum Institute, Washington, D.C. pp. 53–58.

KEIZER, P. D. and GORDON, D. C. (1973). Detection of trace amounts of oil in sea water by fluorescence spectroscopy. *J. Fish. Res. Bd Can.*, **30**, 1039–1046.

KING, P. J. (1977). An assessment of the potential carcinogenie hazard of petroleum hydrocarbons in the marine environment. *Rapp. P.-v. Réun. Cons. int. Explor. Mer*, **171**, 202–211.

KINNE, O. (1975). Orientation in space: animals. Mammals. In O. Kinne (Ed.), *Marine Ecology*, Vol. II, Physiological Mechanisms, Part 2. Wiley, London. pp. 709–916.

KINNE, O. (1977). Cultivation: research cultivation. In O. Kinne (Ed.), *Marine Ecology*, Vol. III, Cultivation, Part 2. Wiley, Chichester. pp. 579–1293.

KINSEY, D. W. (1973). Small scale experiments to determine the effects of crude oil films on gas exchange over the coral back-reef at Heron Island *Environ. Pollut.*, **4**, 167–82.

KITTREDGE, J. S. (1973). The effects of crude oil in the behaviour of marine invertebrates. *Nat. Tech. Inform. Serv.* (US Dept of Commerce), AD–762047, 1–12.

KITTREDGE, J. S., TAKAHASHI, F. T., and SARINANA, F. O. (1974). Bioassays indicative of some sublethal effects of oil pollution. Proceedings of the Marine Technology Society, Washington, D.C. 23–25 Sept. 1974. pp. 891–897.

KOONS, C. B. and THOMAS, J. P. (1979). Hydrocarbons in the sediments of New York Bight. In J. O. Ludwigson (Ed.), *Proceedings of the Oil Spill Conference, 1979*. American Petroleum Institute, Washington, D.C. pp. 625–628.

KORTE, F. (1977). Potential impact of petroleum on the environment. *United Nations Energy Programme Industry Sector Seminars*, Petroleum Industry Meeting, Paris.

KREBS, C. T. and BURNS, K. A. (1978). Long-term effects of an oil spill on populations of the salt-marsh crab *Uca pugnax*. *J. Fish. Res. Bd Can.*, **35**, 648–649.

KÜHNHOLD, W. W. (1969a). Effect of WSF of crude oil on eggs and larvae of cod and herring. *C.M.—ICES*/**E: 17**, 1–15.

KÜHNHOLD, W. W. (1966b). Der Einfluss wasserlöslicher Bestandteile von Rohölen und Rohölfrakionen auf die Entwicklung der Heringsbrut. *Ber. dt. wiss. Kommn Meeresforsch.*, **20** (H2), 165–171.

KÜHNHOLD, W. W. (1973). The influence of crude oils on fish fry. In M. Ruivo (Ed.), *Marine Pollution and Sea Life*. Fishing News (Books) Ltd, London. pp. 315–317.

KÜHNHOLD, W. W. (1977). The effect of mineral oils on the development of eggs and larvae of marine species: A review. *Rapp. P.-v. Réun. Cons. int. Explor. Mer*, **171**, 175–183.

LACAZE, J. C., (1974). Ecotoxicology of crude oils and the use of experimental marine ecosystems. *Mar. Pollut. Bull., N.S.*, **5**, 153–156.

LAKE, J. L. and HERSHNER, C. (1977). Petroleum sulphur-containing compounds and aromatic hydrocarbons in the marine molluscs *Modiolus demissus* and *Crassostrea virginica*. In J. O. Ludwigson (Ed.), *Proceedings of the Oil Spill Conference, 1977*. American Petroleum Institute, Washington, D.C. pp. 627–632.

LAUGHLIN, R. B. and NEFF, J. M. (1979). The interactive effects of temperature, salinity and sub-lethal exposure to phenanthrene on the respiration rate of the juvenile mud crab *Rhithropanopeus harrisii*. In J. O. Ludwigson (Ed.), *Proceedings of the Oil Spill Conference, 1979*. American Petroleum Institute, Washington, D.C. pp. 585–590.

LAW, R. J. (1980). Analytical methods and their problems in the analysis of oil in water. In C. S. Johnston and R. J. Morris (Eds), *Oily Water Discharges*. Applied Science, London. pp. 167–176.

LEE, R. F. (1975). Fate of petroleum hydrocarbons in marine zooplankton. In *Proceedings of the Oil Spill Conference, 1975*. American Petroleum Institute, Washington, D.C. pp. 549–553.

LEE, R. F. (1977). Fate of petroleum components in estuarine waters of the southeastern United States. In J. O. Ludwigson (Ed.), *Proceedings of the Oil Spill Conference, 1977*. American Petroleum Institute, Washington, D.C. pp. 611–616.

LEE, R. F., HINGA, K., and ALMQUIST, G. (1982). Fate of radiolabelled polycyclic aromatic hydrocarbons and pentachlorophenol in enclosed marine ecosystems. In G. D. Grice and M. R. Reeve (Eds), *Marine Mesocosms*. Biological and Chemical Research in Experimental Ecosystems. Springer-Verlag, New York. pp. 123–135.

LEE, R. F. and RYAN, C. (1976). Biodegradation of petroleum hydrocarbons by marine microbes. In J. M. Sharpley and A. M. Kaplan (Eds), *Proceedings of the Third International Biodegradation Symposium*. Applied Science, London. pp. 119–125.

LEE, R. F., SAUERHEBER, R., and DODDS, G. H. (1972). Uptake, metabolism and discharge of PCAH by marine fish. *Mar. Biol.*, **17**, 201–208.

LEE, R. F. and TAKAHASHI, M. (1977). The fate and effect of petroleum in controlled ecosystem enclosures. *Rapp. P.-v. Reun. Cons. int. Explor., Mer*, **171**, 150–156.

LEVY, E. M. and WALTON, A. (1973). Dispersed and particulate petroleum residues in the Gulf of St Lawrence. *J. Fish. Res. Bd Can.*, **30**, 261–267.

LIGHT, M. and LANIER, J. J. (1978). Biological effects of oil pollution. A comprehensive bibliography with abstracts. *Rep. No. CG-D-75-78*, U.S. Department of Transportation, U.S. Coast Guard, National Technical Information Service.

LINDEN, O. (1975). Acute effects of oil and oil dispersant mixture on larvae of Baltic herring. *Ambio*, **4** (3), 130–133.

LINDEN, O., LAUGHLIN, R. B., SHARP, J. R., and NEFF, J. M. (1980). The combined effect of T, S and oil on the growth pattern of embryos of the killifish, *Fundulus heteroclitus*. *Mar. environ. Res.*, **3**, 129–144.

LÖNNING, S. (1977). The sea urchin egg as a test object in oil pollution studies. *Rapp. P.-v. Réun. Cons. int. Explor. Mer*, **171**, 186–188.

LOPEZ, S. M. (1978). Ecological significance of petroleum spillage in Puerto Rico. In *Proceedings of the Conference on Assessment of Ecological Impacts of Oil Spills*. Keystone, Col. pp. 1–17.

LOYA, Y. and RINKEVICH, B. (1979). Abortion effect in corals induced by oil pollution. *Mar. Ecol. Prog. Ser.*, **1**, 77–80.

LUND, H. F. (Ed.))1971). *Industrial Pollution Control Handbook, McGraw-Hill, New York.*

LYSYJ, I. PERKINS, G., FARROW, J. S., and LAMOREAUX, W. (1981). *Effectiveness of offshore produced water treatment*. In R. B. Parrette (Ed.), *Proceedings of the Oil Spill Conference*, Atlanta, Georgia, American Petroleum Institute, Washington, D.C. pp. 63–68.

MACGREGOR, C. and MCLEAN, A. Y. (1977). Fate of crude oil spilled in a simulated arctic environment. In J. O. Ludwigson (Ed.), *Proceedings of the Oil Spill Conference*. American Petroleum Institute, Washington, D.C. pp. 461–463.

MCINTYRE, A. D. and WHITTLE, K. J. (Eds) (1977). Petroleum hydrocarbons in the marine environment. *Rapp. P.-v. Réun. Cons. int. Explor. Mer*, **171**, 1–230.

MACKIE, P. R., HARDY, R., WHITTLE, K. J., BRUCE, C., and MCGILL, A. S. (1980). The tissue hydrocarbon burden of mussels from various sites around the Scottish coast. In A. Bjørseth and A. J. Dennis (Eds), *Polynuclear Aromatic Hydrocarbons, Chemistry and Effects*. Batelle Press, Columbus, Ohio. pp. 379–393.

MACKIE, P. R., WHITTLE, K. J., and HARDY, R. (1974). Hydrocarbons in the marine environment. I.n-Alkanes in the Firth of Clyde. *Estuar. coast. mar. Sci.*, **2**, 359–374.

MASSIE, L. C., WARD, A. P., BELL, J. S., MACKIE, P.R., and SALTZMANN, H. A. (1981). The levels of hydrocarbons in water and sediments in selected areas of the North Sea and the assessment of their biological effect. C.M.–ICES/**E: 44**,

MATTSON, J. S. (1978). Chronology of events and oil slicks from *Argo Merchant*. In V. Desjardins (Ed.), *In the wake of the Argo Merchant*, Center for Ocean Management Studies, University of Rhode Island, Kingston, USA. pp. 15–18.

MERTENS, E. W. (1976). The impact of oil on marine life: a summary of field studies. In *Sources, Effects and Sinks of Hydrocarbons in the Aquatic Environment*. The American Institute of Biological Sciences, Washington, D.C., pp. 507–514.

MIKHAYLOV, V. L. (1979). Results of determination of hydrocarbons and pesticides in the surface microlayer of the Mediterranean Sea. *Oceanology*, **19** (5), 541–543.

MIRONOV, O. G. (1967). Wirkung niedriger Konzentrationen von Öl und Rohölprodukte ouf die Entwicklung der Eier des Schwarzmeer Steinbutts. *Vop. Ichtiol.*, **7**, 577–580.

MIRONOV, O. G. (1969). The development of some Black Sea fishes in sea water polluted by petroleum products (Russ.). *Voprosý ikhtio logii*, **9**(6), 108–110.

MIRONOV, O. G. (1973). *Oil Pollution and Life in the Sea*, Naukova Dumka, Kiev.

MIX, M. C., RILEY, R.T., KING, K. I., TRENHOLM, S. R., and SCHAFFER, R. L. (1977). Chemical carcinogens in the marine environment. In D. A. Wolfe (Ed.), *Fate and Effects of Petroleum Hydrocarbons in Marine Organisms and Ecosystems*. Pergamon Press, New York. pp. 421–431.

MONAGHAN, P. H., MCAULIFFE, C. D., and WEISS, F. T. (1980). Environmental aspects of drilling muds and cuttings from oil and gas operations in offshore and coastal waters. In R. A. Geyer (Ed.), *Marine Environmental Pollution*, Vol. I, Hydrocarbons. Elsevier, Amsterdam. pp. 413–432.

MOORE, G. (1976). Legal aspects of marine pollution control. In R. Johnston (Ed.), *Marine Pollution*. Academic Press, London. pp. 589–697.

MOORE, S. F., CHIRLIN, G. R., PUGGIA, G. J., and SCHRADER, B. P. (1974). Potential biological effects of hypothetical oil discharges on the Atlantic coast and Gulf of Alaska. *Sea Grant Rep* (*Massachusetts Institute of Technology*)., **74–19**, 1–121.

MOORE, S. F. and DWYER, R. L. (1974). Effects of oil on marine organisms: a critical assessment of published data. *Wat. Res.*, **8**, 819–827.

MOORE, S. F., DWYER, R. L., and KATZ, A. M. (1973). *A Preliminary Assessment of Environmental Vulnerability*, R. M. Parsons Laboratory, Massachusetts Institute of Technology, Cambridge, Massachusetts (Technical Report 162).

MYERS, E. P. and GUNNERSON, C. G. (1976). *Hydrocarbons in the Ocean*, U.S. Department of Commerce, Boulder, Colarado (Marine Ecosystems Analysis Program; Special Report).

NAS (NATIONAL ACADEMY OF SCIENCES) (1975). *Petroleum in the Marine Environment*, National Academy of Sciences, Washington, D.C.

NRC (NATIONAL RESEARCH COUNCIL) (1980). The international mussel watch. *Report of Workshop*, National Academy of Sciences, Washington, D.C. pp. 1–248.

OPPENHEIMER, C. H., GONKEL, W., and GASSMANN, G. (1977). Micro-organisms and hydrocarbons in the North Sea during July–August, 1975. In *Proceedings of the Oil Spill Conference, 1977*. American Petroleum Institute, Washington, D.C. pp. 593–609.

OSIR (OIL SPILL INTELLIGENCE REPORTS) (1978). *Oil Spill Intelligence Reports*, Vols. I and II, Inter-weekly Newsletter, Centre for Short-lived Phenomena, Cahners Publishing Company, Boston, Massachusetts.

OSIR (OIL SPILL INTELLIGENCE REPORTS) (1979). Annual summary and review, 1978. *Oil Spill Intelligence Reports*, **2** (12).

OTTWAY, S. (1971). The comparative toxicities of crude oils. In E. B. Cossel (Ed.), *The Ecological Effects of Oil Pollution on Littoral Communities*. Institute of Petroleum, London. pp. 171–180.

PANDIAN, T. J. (1975). Mechanisms of heterotrophy. In O. Kinne (Ed), *Marine Ecology* Part 2. pp. 61–249.

PARKER, P. L., WINTERS, J. K., and MORGAN, J. (1972). A baseline study of petroleum in the Gulf of Mexico. In *Baseline Studies of Pollutants in the Marine Environment*. National Science Foundation, 1 DOE, Washington, D.C. pp. 555–581.

PERKINS, E. J. (1970). Some effects of detergents in the marine environment. *Chemy Ind.*, **1970**, 14–22.

PERKINS, E. J. (1976). The evaluation of biological response by toxicity and water quality assessments. In R. Johnston (Ed.), *Marine Pollution*. Academic Press, London. pp. 505–585.

PLATT, H. M. and MACKIE, P. R. (1980). Distribution and fate of aromatic hydrocarbons in Antarctic fauna and enviroment. *Helgoländer Meeresunters.*, **33**, 236–245.

POWERS, K. D. and RUMAGE, W. T. (1978). Effect of *Argo Merchant* oil spill on bird populations off the New England coast. In *Proceedings of Symposium 'In the Wake of the Argo Merchant', 1978*. Center for Ocean Management Studies, University of Rhode Island, Kingston. pp. 142–147.

QUIRK, M. M., PATIENCE, R. L., MAXWELL, J. R., and WHEATLEY, R. E. (1980). Recognition of the sources of isoprenoid alkanes in recent environments. In J. Albaiges (Ed.), *Analytical Techniques in Environmental Chemistry*. Pergamon Press, New York. pp. 13–31.

RAY, J. P. (1981). The effect of petroleum hydrocarbons on corals. Petromar '80, Shell Oil Company, Environmental Affairs, Houston, Texas. *European Rep.*, **3**, 6.

READ, A. D. (1980). Treatment of oily water at North Sea oil installations. In C. S. Johnston and R. J. Morris (Eds), *Oily Water Discharges*. Applied Science, London. pp. 127–136.

REED, W. E. and SPAULDING, M. L. (1979). A fishery/oil interaction model. In J. O. Ludwigson (Ed.), *Proceedings of the Oil Spill Conference, 1979*. American Petroleum Institute, Washington, D.C. pp. 63–73.

RENZONI, A. (1973). Influence of crude oil, derivatives and dispersants on larvae. *Mar. Pollut. Bull., N.S.*, **4**, 9–13.

RESCORLA, A. R. (1977). *Environmental Conservation in the Petroleum Industry*, United Nations Energy Programme Seminars, Petroleum Industry Meeting, Paris.

RISEBROUGH, R. W. and co-authors (1979). Pattern of hydrocarbon contamination in Californian coastal waters. In J. Albaiges (Ed.), *Analytical Techniques in Environmental Chemistry*, Proceedings of the International Congress. Pergamon Press, New York. pp. 33–40.

ROSSI, S. S. (1977). Bioavailability of petroleum hydrocarbons from water, sediments and detritus to the marine annelid *Neanthes arenaceodentata*. In J. O. Ludwigson (Ed.), *Proceedings of the Oil Spill Conference, 1977*. American Petroleum Institute, Washington, D.C. pp. 621–625.

ROUBAL, W. T., COLLIER, I. K., and MALINS, D. C. (1977). Accumulation and metabolism of ^{14}C-labelled benzene, naphthalene and anthracene by young coho salmon. *Archs environ. Contam. Toxicol.*, **5**, 513–529.

RUMPF, K. K. (1969). Mineralöle und verwandte Produkte, 2nd ed., Springer, Berlin. pp. 306–348.

SCHONE, H. (1975). Orientation in Space: animals. In O. Kinne (Ed.), *Marine Ecology*, Vol. II, Physiological Mechanisms, Part 2. Wiley, London. pp. 499–553.

SCHREINER, D. (1980). Discharge of oil bearing waste water from the production of petroleum on the Norwegian Continental Shelf. In C. S. Johnson and R. J. Morris (Eds), *Oily Water Discharges*. Applied Science, London. pp. 137–153.

SHAW, D. G., PAUL, A. J., and SMITH, E. R. (1977). Responses of the clam *Macoma balthica* to Prudhoe Bay Crude Oil. In J. O. Ludwigson (Ed.), *Proceedings of the Oil Spill Conference, 1977*. American Petroleum Institute, Washington D.C. pp. 493–494.

SIDHU, G. S., VALE, G. L., SHIPTON, J., and MURRAY, K. E. (1972). A kerosene-like taint in the mullet *(Mugil cephalus)*. In M. Ruivo (Ed.), *Marine Pollution and Sea Life*. Fishing News (Books) Ltd, London. p. 546–550.

SIMONOV, A. I., ZUBAKINA, A. N., and MIKHALEVA, I. M. (1980). Hydrocarbons isolated from sea water and ice. *Oceanology*, **20** (2), 161–163.

SLEETER, T. D., BUTLER, J. N., and BARBASH, J. E. (1978). Hydrocarbons in the sediments of the Bermuda region. In *Symposium on Analytical Chemistry of Petroleum Hydrocarbons in the Marine/Aquatic Environment, Miami, 1978*. Advances in Chemistry Series, American Chemical Society.

SLEETER, T. D., BUTLER, J. N., and BARBASH, J. E. (1979). Hydrocarbons in sediments from the edge of the Bermuda Seamount. In J. O. Ludwigson (Ed.), *Proceedings of the Oil Spill Conference, 1979*. American Petroleum Institute, Washington, D.C. pp. 615–623.

SMITH, C. A. and CARPENTER, J. H. (1977). Reactions in chlorinated sea water. In R. L. Jolley, H. Gorchev, and D. H. Hamilton (Eds), *Water Chlorination—Environmental Impact and Human Health*, Vol. 2. Ann Arbor Science, Ann Arbor, Michigan. pp. 195–207.

SMITH, J. E. (1970). *'Torry Canyon.'Pollution and Marine Life*, United Press, Cambridge.

SPIES, R. B., DAVIES, P. H., and STUERMER, D. H. (1980). Ecology of a submarine petroleum seep off the Californian coast. In R. A. Geyer (Ed.), *Marine Environmental Pollution*, Vol. I, Hydrocarbons. Elsevier, Amsterdam. pp. 229–264.

SPOONER, M. (1969). Some ecological effects of marine oil pollution. In *Proceedings of the Joint Conference on Prevention and Control of Oil Spills*. American Petroleum Institute, Washington, D.C. pp. 313–316.

SPOONER, M. and CORKETT, C. J. (1974). A method for testing the toxicity of suspended oil droplets on planktonic copepods used at Plymouth. In L. R. Beynon and E. B. Cowell (Eds), *Ecological Aspects of Toxicity, Testing Oils and Dispersants*. Wiley, New York. pp. 69–74.

STEBBING, A. R. D. (1980). The biological measurement of water quality. *Rapp. P.-v. Réun. Cons. int. Explor. Mer*, **179**, 310–314.

STEGEMAN, J. J. and TEAL, J. M. (1973). Accumulation, release and retention of petroleum hydrocarbons by the oyster *Crassostrea virginica*. *Mar. Biol.*, **22**, 37–44.

STRAUGHAN, D. (Ed.) (1971). Biological and Oceanographical Survey of the Santa Barbara Channel Oil Spill, 1969–70, Vol. 1, Biology and Bacteriology. Allan Hancock Foundation, University of Southern California, Los Angeles.

STRAUGHAN, D. (1976). Sublethal effects of natural chronic exposure to petroleum in the marine environment. *Publication No: 4280*, American Petroleum Institute, Washington, D.C. pp. 1–119.

STRAUGHAN, D. (1977). The sublethal effects of natural chronic exposure to petroleum on marine invertebrates. In J. O. Ludwigson (Ed.), *Proceedings of the Oil Spill Conference, 1977*. American Petroleum Institute, Washington, D.C. pp. 563–568.

STRAUGHAN, D. (undated). *Analysis of Mussel Communities in Areas Chronically Exposed to Natural Oil Seepage*, American Petroleum Institute,Washington, D.C. (Publication No. 4319).

TAKAHASHI, F. T. and KITTREDGE, J. S. (1973). Sublethal effects of the water soluble components of oil *vis à vis* chemical communication in the marine environment. In D. G. Ahearn and S. P. Meyers (Eds), *The Microbial Degradation of Oil Pollutants*. Center for Wetland Resources, Louisiana State University, Baton Rouge. pp. 259–264.

TAKAHASHI, M., THOMAS, W. H., SEIBERT, D. L. R., BEERS, J., KOELLER, P., and PARSONS, T. R. (1975). The replication of biological events in enclosed water columns. *Arch. Hydrobiol.*, **76**, 5–23.

TELANG, S. A., HODGSON, G. W., and BAKER, J. L. (1980). Hydrocarbons of aquatic and terrestrial origin in mountain streams of the Marmat Basin (Rocky Mountains, Alberta, Canada). *J. environ. Qual.*, **10** (1), 103–107.

TESCH, F.-W. (1975). Orientation in space. Animals: fishes. In O. Kinne (Ed.), *Marine Ecology*. Vol. II, Physiological Mechanisms, Part 2. Wiley, London. pp. 657–707.

THURBERG, F. P., GOULD, E., and DAWSON, M. A. (1978). Some physiological effects of the *Argo Merchant* Oil spill on several marine teleosts and bivalve molluscs. In *Proceedings of the Symposium 'In the Wake of the Argo Merchant, 1978*. University Rhode Island, Kingston. pp. 103–108.

UMWELTPROBLEME DER NORDSEE (1980). *Sondergutachten der Rat von Sachverständigen für Umweltfragen*, Kohlhammer GmbH, Stuttgart.

VARANASI, U. and MALINS, D. C. (1977). Metabolism of petroleum hydrocarbons; accumulation and biotransformation in marine orgnisms. In D. C. Malins (Ed.), Effects of petroleum on Arctic and Subarctic marine environments and organisms. Vol. II, *Biological Effects*. pp. 175–270.

VAUK, G. (1973). Beobachtungen am Seehund (*Phoca vitulina*) auf Helgoland. *Z. Jagdwiss.*, **19**, (3), 117–121.

WADE, T. L. and QUINN, J. G. (1975). Hydrocarbons in the Sargasso Sea surface microlayer. *Mar. Pollut. Bull., N.S.*, **6**, 54–57.

WADE, T. L. and QUINN, J. G. (1979). Geochemical distribution of hydrocarbons in sediments from mid-Narragansett Bay and Rhode Island Sound. *Org. Geochem.*, 157–167.

WADE, T. L. and QUINN, J. G. (1980). Incorporation, distribution and fate of saturated hydrocarbons in sediments from a controlled marine ecosystem. *Mar. environ. Res.*, **3**, 15–33.

WENT, F. W. (1960). Organic matter in the atmosphere and its possible relation to petroleum formation. *Proc. natn. Acad. Sci. U.S.A.*., **46**, 212–221.

WHITTLE, K. J. (1978). Tainting in marine fish and shellfish with reference to the Mediterranean Sea. In *Data Profiles for Chemicals for the Evaluation of their Hazards to the Environment of the Mediterranean Sea*, Vol. III, United Nations Energy Programme, Geneva. pp. 89–108.

WHITTLE, K. J., MACKIE, P. R., HARDY, R., and MCINTYRE, A. D. (1974). The fate of n-alkanes in marine organisms. *International Council for the Exploration of the Sea*, ICES CM 1974/E: 33 (mimeo).

WHITTLE, K. J., MURRAY, J., MACKIE, P. R., HARDY, R., and FARMER, J. (1977). Fate of hydrocarbons in fish *Rapp. P.-v. Réun. Cons. int. Explor. Mer*, **171**, 139–142.

WILSON, K. W. (1972). The toxicity of oil spill dispersants to the embryos and larvae of some marine fish. In M. Ruivo (Ed.), *Marine Pollution and Sea Life*. Fishing News Books, London. pp. 218–322.

WILSON, K. W. (1974). Toxicity testing for ranking oils and oil dispersants. In L. R. Beynon and E. B. Cowell (Eds), *Ecological Aspects of Toxicity Testing of Oils and Dispersants*. Wiley, New York. pp. 11–22.

WILSON, K. W. (1976). Effect of oil dispersants on the developing embryos of marine fish. *Mar. Biol.*, **36**, 259–268.

WINTERS, K. and PARKER, P. L. (1977). Water soluble components of crude oils, fuel oils and used crankcase oils. In J. O. Ludwigson (Ed.), *Proceedings of the Oil Spill Conference, 1977*. American Petroleum Institute, Washington, D.C. pp. 579–581.

ZOBELL, C. E. (1964). Occurrence, effects and fate of oil polluting the sea. *Adv. Wat. Pollut. Res.*, **3**, 85–118.

ZSOLNAY, A. (1972). Preliminary study of the dissolved hydrocarbons and hydrocarbons on particulate materials in the Gotland Deep of the Baltic. *Kieler Meeres for sch.*, **27**, 129–134.

Literature Appendix (Chapter 4)
(Reviews and Major Sources Complementing Text References)

General

Impact of oil on the marine environment. *GESAMP Report and Study*, **6**, 1–250 (IMCO, FAO, UNESCO, WHO, IAEA, UN; published by FAO, Rome, 1977).

Meetings

Petroleum hydrocarbons in the marine environment. In A. D. McIntyre and K. J. Whittle (Eds), *Proceedings of the ICES Workshop, Aberdeen, 9 to 12 September 1975. Rapp. P.-v. Réun. Cons. int. Explor. Mer*, **171**.

Sources, Effects and Sinks of Hydrocarbons in the Aquatic Environment, American Institute of Biological Sciences, Washington, D.C. (Proceedings of a Symposium, American University, Washington, D.C. 9 to 11 August 1976).

Recovery potential of oiled marine northern environments. *J. Fish. Res. Bd Can.* (special issue), **35** (5), 1978.

Inputs

Petroleum in the Marine Environment, National Academy of Sciences, Washington, D.C., 1975 (Workshop on Inputs, Fates and Effects of Petroleum in the Marine Environment, Airlie, Virginia, 21 to 23 May, 1973).

JOHNSTON, C. S. and MORRIS, R. J. (Eds) (1980). *Oily Water Discharges. Regulatory, Technical and Scientific Considerations*, Applied Science, London.

Oil Industry

Series of Trade Research Reports on the Oil Industry in Europe, 1972 Onwards, CONCAWE, Van Hogenhoucklaan 60, 2596 TE, Den Haag, Netherlands.

Bacterial Degradation, etc.

ZOBELL, C. E. (1964). Occurrence, effects and fate of oil polluting the sea. *Adv. Wat. Pollut. Res.*, **3**, 85–118.

Microbial degradation of pollutants in the marine environment. In A. W. Bourquin and P. H. Pritchard (Eds), *Proceedings of a Workshop, Pensacola, Florida, 9 to 14 April 1978*. EPA 600/9—79–012 April, 1979.

WATKINSON, R. J. (Ed.) (1979). *Developments in the Biodegradation of Hydrocarbons*, Applied Science, London.

Effects

BREISBACH, R. H. (1980). *Handbook of Poisoning* (2nd ed.), Lange, London.

HUNTER, D. (1978). *The Diseases of Occupation* (6th ed.), Hodder and Stoughton, London.

JOHNSTON, R. (1976). Mechanisms in marine pollution. In R. Johnston (Ed.), *Marine Pollution*. Academic Press, London.

COWELL, E. B. (1976). Oil pollution of the sea. In R. Johnston (Ed.), *Marine Pollution*. Academic Press, London.

BOURNE, W. R. P. (1976). Seabirds and pollution. In R. Johnston (Ed.), *Marine Pollution*. Academic Press, London.

Fate and effects of petroleum hydrocarbons in marine ecosystems and organisms. In D. A. Wolfe (Ed.), *Proceedings of a Symposium, Seattle, Washington 10 to 12 November 1976*. Pergamon Press, Oxford, New York.

BAKER, J. M. (Ed.) (1976). *Marine Ecology and Oil Pollution*, Applied Science, London.

MALINS, D. C. (Ed.) (1977). *Effects of Petroleum on Arctic and Subarctic Marine Environments and Organisms*, Academic Press, New York (2 vols.).

SMITH, J. E. (Ed.) (1968). *'Torrey Canyon'—Pollution and Marine Life*, Cambridge University Press, Cambridge.

In the Wake of the 'Argo Merchant', 1978, Center for Ocean Management Studies, University of Rhode Island, Kingston.

CENTRE OCEANOLOGIQUE DE BRETAGNE (1981). *Amoco Cadiz*—fates and effects of the oil spill. In *Proceedings of an International Symposium, Brest, 19 to 22 November 1979*. Centre Océanologique de Bretagne, Brest, Paris. pp. 1–881 (CNEXO).

Clean-up

WARDLEY-SMITH, J. (Ed.) (1979). *The Prevention of Oil Pollution*, Graham and Trotman, London.

Legal

MOORE, G. (1976). Legal aspects of marine pollution. In R. Johnston (Ed.), *Marine Pollution*. Academic Press, London.

TROMP, D. (1980). The Paris Convention. In C. S. Johnston and R. J. Morris (Eds), *Oily Water Discharges*. Applied Science, London.

KING, N. (1980). Oil and the Paris Convention. In C. S. Johnston and R. J. Morris (Eds), *Oily Water Discharges*. Applied Science, London.

Bibliographies

FILION-MYKLEBUST, C. H. and JOHANNESSEN, K. (1982). *Biological Effects of Oil Pollution in the Marine Environment*, FOH Laboratory for Ecotoxicology, University of Oslo (see Norwegian Marine Pollution Research and Environmental Programmes; FOH).

NTIS Oil Spill and Oil Pollution Reports, U.S. National Technical Information Service, Springfield, Virginia: PB-272 689 June 1977; PB-276 691 April 1978 (continuing).

Oil Spillage, U.S. National Technical Information Service, Springfield, Virgina: PB-22 108 May 1973 (2 vols.).

Biodegradation of Oil Spills, U.S. National Technical Information Service, Springfield, Virginia: PS-78/0043 January 1978.

LIGHT, M. and LANIER, J. J. (1975). *Biological Effects of Oil Pollution*, U.S. Coastguard Research and Development Center, Groton, Connecticut (NTIS AD-A064).

SAMSON, A. L., VANDERMEULEN, J. H., WELLS, P. G., and MOYSE, C. (1980). *A Selected Bibliography on the Fate and Effect of Oil Pollution Relevant to the Canadian Marine Environment*)2nd ed.), Environment Canada, Environmental Protection Service, Ottawa.

MOULDER, D. S. and co-authors (1971, 1975, 1977 onwards). *A Bibliography on Marine and Estuarine Oil Pollution*, The Marine Biological Association, Plymouth, U.K.

AUTHOR INDEX

Numbers in italics refer to those pages on which the author's work is stated in full.

TAXONOMIC INDEX

SUBJECT INDEX